# Fungal Biology

***Series Editors:***

Vijai Kumar Gupta, PhD
Molecular Glycobiotechnology Group, Department of Biochemistry,
School of Natural Sciences, National University of Ireland Galway,
Galway, Ireland

Maria G. Tuohy, PhD
Molecular Glycobiotechnology Group, Department of Biochemistry,
School of Natural Sciences, National University of Ireland Galway,
Galway, Ireland

More information about this series at http://www.springer.com/series/11224

De-Wei Li
Editor

# Biology of Microfungi

Springer

*Editor*
De-Wei Li
The Connecticut Agricultural Experiment
Station Valley Laboratory
Windsor, CT, USA

Co-Innovation Center for Sustainable
Forestry in Southern China
Nanjing Forestry University
Nanjing, Jiangsu, China

ISSN 2198-7777 ISSN 2198-7785 (electronic)
Fungal Biology
ISBN 978-3-319-29135-2 ISBN 978-3-319-29137-6 (eBook)
DOI 10.1007/978-3-319-29137-6

Library of Congress Control Number: 2016933679

Springer Cham Heidelberg New York Dordrecht London


Printed on acid-free paper

Springer International Publishing AG Switzerland is part of Springer Science+Business Media (www.springer.com)

# Preface

To me, mycology is not only a branch of biological science but also an acquired interest and passion. It has led me to cross the Pacific Ocean from China to Canada 27 years ago and later to the USA to pursue my professional dream. It was a long trip, full of challenges and sometimes doubts. When I was teaching plant pathology at Olds College in Alberta, I was very grateful to Olds College for offering me my first job in North America after my graduate studies and postdoctoral work. At the same time, I missed mycological research and assumed that mycology might have parted from me. During the pinnacle of public health concerns to indoor molds along with a mounting number of indoor mold-related litigations in the USA in the early 2000s, an industry job provided me an opportunity to return to mycology by working on indoor molds, most of which are microfungi. I have been working on microfungi and aeromycology ever since. This is one of the major reasons why the book was titled "Biology of Microfungi."

The focus of this book is principally on covering the latest development of research on microfungi from both systematic and practical aspects. In a broad sense, Mesofungi were also covered. It is not an overstatement that microfungi are in our daily life, but we normally do not realize the presence of these microfungi. It is almost impossible and impractical for us to live a microfungi-free environment, whether we recognize it or not. Fungi are ubiquitous. Spores of microfungi are present in the air. We breathe them in and out all the time. Microfungi are both our friends and foes. Without microfungi, we would not be able to enjoy our bread, mantou (steamed bun), other baked food, fermented food, preserved food, alcoholic beverages (beer, wine, liquor, tequila, rice wine, etc.), and access to some modern medication, such as penicillin and cyclosporin, etc., which are secondary metabolites of microfungi. Microfungi have been directly used as medicinal herbs in Chinese medicine to treat various diseases for over 1000 years. A number of microfungi have been used as biocontrol agents to manage plant insects and diseases, such as *Trichoderma* spp., *Clonostachys rosea* (Link) Schroers et al. (≡ *Gliocladium rosea* Bainier) for plant disease control, *Beauveria bassiana* (Bals.-Criv.) Vuill., and *Metarhizium anisopliae* (Metchnikoff) Sorokin for controlling insects. Without microfungi as decomposers, our planet would be buried by numerous mountains of plant litter and nutrient flow/recycle in our ecosystem would be interrupted or even stopped. Without the microfungi of Glomeromycota to form arbuscular mycorrhizae, the host plants' adaptability to adverse environments would be significantly reduced and some may not even be able to survive. For these aspects, microfungi are our friends. On the negative aspects, some microfungi are pathogenic to humans, animals, or plants. Some microfungi, such as *Stachybotrys chartarum*, *Fusarium* spp., *Aspergillus flavus*, are able to produce mycotoxins, which are detrimental to the health of humans and animals. Microfungi, especially airborne fungal spores, can be allergens to some individuals. Thus, these microfungi are our foes. Some microfungi can be both beneficial and detrimental to human beings. For example, *Tilletia hordei*, *Ustilago crameri*, *Ustilago maydis* (DC.) Corda, and *Ustilago nuda* (C.N. Jensen) Rostrup cause smuts on a number of cereal crops and grasses, but they are used as medicinal remedies in Chinese medicine.

Have we fully explored the resources of microfungi? The answer is definitely NO.

This book is a collective effort of a team of mycologists who either contributed chapter(s) to this book or reviewed manuscripts to make this book possible. I am so fortunate that such a wonderful group of mycologists accepted my invitation and committed themselves and their time to contribute to this book. At the time of completing this book, I am very grateful to have had this opportunity to edit it. I have learnt so much from what the chapter authors have covered. However, any errors in the book are mine.

The editor would like to express his gratitude to Dr. Bryce Kendrick for his mentorship and friendship. Without his courage and support, I would not have been able to take on this editorial task. The editor is very appreciative to the authors for their excellent contributions. Without their collaboration, it would be impossible to see this book in print. I am very grateful to the former and current directors of The Connecticut Agricultural Experiment (CAES), Dr. John Anderson, the late Dr. Louis A. Magnarelli, and Dr. Theodore G. Andreadis and Chief Scientist of Valley Laboratory, Dr. James LaMondia, for their support. I am very privileged to work at CAES as a mycologist and enjoy my mycological research. I am very much indebted to my wife, Jin Zhang (张瑾) for her unreserved love and persistent support.

Windsor, CT
Nanjing, Jiangsu, China

De-Wei Li
李德伟

# Contents

# Contributors

**Mohd. Aamir, Ph.D.** Department of Botany, Banaras Hindu University, Varanasi, India

**Kessy Abarenkov, Ph.D.** Natural History Museum, University of Tartu, Tartu, Estonia

**Mercedes Amado, M.D.** Pediatric Allergy & Immunology, Allergy Asthma and Immunology, Children's Mercy Hospitals & Clinics, Kansas City, MO, USA

University of Missouri-Kansas City School of Medicine, Kansas City, MO, USA

**Harriet M. Ammann, Ph.D.** Department of Environmental & Occupational Health Sciences, School of Public Health and Community Medicine, University of Washington, Seattle, WA, USA

**Charles Barnes, Ph.D.** University of Missouri-Kansas City School of Medicine, Kansas City, MO, USA

Division of Allergy/Immunology Laboratory, Children's Mercy Hospitals and Clinics, Kansas City, MO, USA

**Christiane Baschien, Ph.D.** Department of Microbial Ecology and Diversity, Leibniz Institute DSMZ-German Collection of Microorganisms and Cell Cultures, Braunschweig, Germany

**Gerald L. Benny, Ph.D.** Department of Plant Pathology, University of Florida, Gainesville, FL, USA

**Rafael F. Castañeda-Ruiz, Ph.D.** Instituto de Investigaciones Fundamentales en Agricultura, Tropical 'Alejandro de Humboldt' (INIFAT), Académico Titular de la Academia de Ciencias de Cuba, Santiago de Las Vegas, C. Habana, Cuba

**Yucheng Dai, Ph.D.** Institute of Microbiology, Beijing Forestry University, Beijing, China

**Carla C.C.R. de Carvalho, Ph.D.** iBB-Institute for Bioengineering and Biosciences, Department of Bioengineering, Instituto Superior Técnico, Universidade de Lisboa, Av. Rovisco Pais, Lisbon, Portugal

**Wade H. Elmer, Ph.D.** Department of Plant Pathology and Ecology, The Connecticut Agricultural Experiment Station, New Haven, CT, USA

**Walter Gams, Ph.D.** Formerly The CBS-KNAW Fungal Biodiversity, The Netherlands

Molenweg, CK Baarn, The Netherlands

**Teik-Khiang Goh, Ph.D.** School of Biological Sciences, Faculty of Integrative Sciences & Technology, Quest International University Perak, Perak Darul Ridzuan, Malaysia

**Vijai Kumar Gupta, Ph.D.** Molecular Glyco-biotechnology Group, Discipline of Biochemistry, School of Natural Sciences, National University of Ireland Galway, Galway, Ireland

**Luis F.P. Gusmão, Ph.D.** Laboratorio de Micologia, Departamento de Ciências Biológicas, Universidade Estadual, Feira de Santana, Brazil

**Gabriela Heredia, Ph.D.** Red de Biodiversidad y Sistemática, Instituto de Ecología, Xalapa, Veracruz, Mexico

**Richard A. Humber, Ph.D.** USDA-ARS Emerging Pests and Pathogens Research Unit, Robert W. Holley Center for Agriculture and Health, Ithaca, NY, USA

**Timothy Y. James, Ph.D.** Department of Ecology and Evolutionary Biology, University of Michigan, Ann Arbor, MI, USA

**E.B. Gareth Jones, Ph.D.** Botany and Microbiology Department, College of Science, King Saud University, Riyadh, Saudi Arabia

**Bryce Kendrick, Ph.D., D.Sc., F.R.S.C.** 8727 Lochside Drive, Sidney-by-the-Sea, BC, Canada

Department of Biology, University of Waterloo, Waterloo, ON, Canada

**Paul M. Kirk, Ph.D.** Mycology Section, Kew Gardens, Royal Botanic Gardens, Kew, Richmond, Surrey, UK

**John Klironomos, Ph.D., F.R.S.C.** Department of Biology, University of British Columbia, Okanagan campus, Kelowna, BC, Canada

I.K. Barber School of Arts and Sciences, The University of British Columbia, Okanagan Campus, Kelowna, BC, Canada

**Urmas Kõljalg, Ph.D.** Natural History Museum, University of Tartu, Tartu, Estonia

**James LaMondia, Ph.D.** The Connecticut Agricultural Experiment Station Valley Laboratory, Windsor, CT, USA

**De-Wei Li, Ph.D.** The Connecticut Agricultural Experiment Station Valley Laboratory, Windsor, CT, USA

Co-Innovation Center for Sustainable Forestry in Southern China, Nanjing Forestry University, Nanjing, Jiangsu, China

**Donát Magyar, Ph.D.** Department of Air Hygiene and Aerobiology, National Public Health Center, Budapest, Hungary

**J. David Miller, Ph.D.** Department of Chemistry, Carleton University, Ottawa, ON, Canada

**R. Henrik Nilsson, Ph.D.** Department of Biological and Environmental Sciences, University of Gothenburg, Göteborg, Sweden

**Sepideh Pakpour** Department of Biology, University of British Columbia, Okanagan campus, Kelowna, BC, Canada

I.K. Barber School of Arts and Sciences, The University of British Columbia, Okanagan Campus, Kelowna, BC, Canada

**Ka-Lai Pang, Ph.D.** Institute of Marine Biology and Center of Excellence for the Oceans, National Taiwan Ocean University, Keelung, Taiwan (R.O.C.)

**Richa Raghuwanshi, Ph.D.** Department of Botany, Mahila Mahavidyalaya, Banaras Hindu University, Varanasi, India

**Amrita Saxena, Ph.D.** Department of Botany, Banaras Hindu University, Varanasi, India

**James A. Scott, Ph.D.** Division of Occupational and Environmental Health, Dalla Lana School of Public Health, University of Toronto, Toronto, ON, Canada

Sporometrics Inc., Toronto, ON, Canada

**Shalini Singh, Ph.D.** Department of Botany, Banaras Hindu University, Varanasi, India

**Matthew E. Smith, Ph.D.** Department of Plant Pathology, University of Florida, Gainesville, FL, USA

**Richard C. Summerbell, Ph.D.** Sporometrics Inc., Toronto, ON, Canada

Dalla Lana School of Public Health, Gage Occupational and Environmental Health Unit, University of Toronto, Toronto, ON, Canada

**Deanna A. Sutton, Ph.D.** Fungus Testing Laboratory, Department of Pathology, University of Texas Health Science Center at San Antonio, San Antonio, TX, USA

**Patricia Timper, Ph.D.** USDA ARS, Tifton, GA, USA

**Merje Toome-Heller, Ph.D.** Plant Health and Environment Laboratory, Ministry for Primary Industries, Auckland, New Zealand

**Jeffrey P. Townsend, Ph.D.** Department of Biostatistics, Yale School of Public Health, Yale University, New Haven, CT, USA

Department of Ecology and Evolutionary Biology, Yale University, New Haven, CT, USA

Program in Computational Biology and Bioinformatics, Yale University, New Haven, CT, USA

**Eric D. Tretter, M.S.** Department of Biological Sciences, Boise State University, Boise, ID, USA

**Clement K.M. Tsui, Ph.D.** Department of Pathology and Laboratory Medicine, University of British Columbia, Vancouver, BC, Canada

**R.S. Upadhyay, Ph.D.** Department of Botany, Banaras Hindu University, Varanasi, India

**Máté Vass** Department of Ecology and Genetics, Limnology, Uppsala University, Uppsala, Sweden

**Zheng Wang, Ph.D.** Department of Biostatistics, Yale School of Public Health, Yale University, New Haven, CT, USA

**Merlin M. White, Ph.D.** Department of Biological Sciences, Boise State University, Boise, ID, USA

**Nathan P. Wiederhold, Ph.D.** Fungus Testing Laboratory, Department of Pathology, University of Texas Health Science Center at San Antonio, San Antonio, TX, USA

**Chin Yang, Ph.D.** Prestige EnviroMicrobiology, Inc., Voorhees, NJ, USA

# Chapter 1
# Introduction: Advances and Predicament

**De-Wei Li**

## What Are Microfungi?

Microfungi are fungi that develop small (microscopic) fruiting bodies (Kirk et al. 2008). These fungi are also referred to as *Micromycetes* or microscopic fungi. They need to be observed and morphologically studied under a microscope. Fungi and mycology were at one point considered a branch of botany. In 1969, Robert Whittaker segregated fungi from Kingdom Plantae and erected a new fifth kingdom: Kingdom Fungi, or *Eumycota*, to the biological classification when phylogenetic studies indicated that fungi (true fungi) are closer to animal than to plants (Whittaker 1969). This is the reason why Dr. Bryce Kendrick named his popular mycology textbook as *The Fifth Kingdom* (Kendrick 1985). Some fungi traditionally studied by mycologists belong to two other Kingdoms (Protozoa and Chromista). However, this book will mainly cover the microfungi in the Kingdom Fungi.

At present, fungi are still covered in textbooks of botany or plant sciences. The nomenclature of fungi remains governed under the latest code: The International Code of Nomenclature for algae, fungi, and plants (Melbourne Code) adopted by the Eighteenth International Botanical Congress Melbourne, Australia, July 2011 (formerly The International Code of Botanical Nomenclature) (McNeill et al. 2012). The change to Article 59 to terminate the two name system for sexual and asexual states of fungi in the Melbourne Code and that "one fungus equals to one name" was adopted. The significance and subsequent implications of the change are discussed in the book.

---

D.-W. Li, Ph.D. (✉)
The Connecticut Agricultural Experiment Station Valley Laboratory,
153 Cook Hill Road, Windsor, CT 06095, USA

Co-Innovation Center for Sustainable Forestry in Southern China, Nanjing Forestry University, Nanjing, Jiangsu 210037, China
e-mail: dewei.li@ct.gov



D.-W. Li (ed.), *Biology of Microfungi*, Fungal Biology,
DOI 10.1007/978-3-319-29137-6_1

The history of mycology has been covered in detail by several authors (Lütjeharms 1936; Ramsbottom 1941; Braun 1965; Ainsworth 1976; Rippon 1982; Bridge 2002), but early evolutionary history remains less studied due to the lack of fungal fossils. However, new molecular technology has made molecular clock dating possible and provides a much better tool to study the evolution of fungi. No doubt, DNA sequencing, next-generation sequencing (high-throughput sequencing technologies), and genomic studies of fungi have led to major developments in the systematics, diversity, biochemistry, and ecology of fungi in the last 20 years. Fungal fossils are scarce in comparison with organisms in the Kingdoms of plants and animals for studies of fungal evolution. New technology allows us to overcome the difficulties in insufficient fossil records to understand fungal evolution. At the same time, more and more new data from research on fossils and archeological evidence allow us to better understand the relationship between fungi and human beings, especially the role fungi played in pre- and early historic periods of human beings.

Studies in the last two decades have led to discoveries of several major fungal groups previously unknown to science such as *Cryptomycota* (Jones et al. 2011) or substantial changes to certain taxonomic groups, such as former *Zygomycota* (zygomycetous fungi) from which new phyla *Glomeromycota* (Schüßler et al. 2001) and *Entomophthoromycota* and subphyla *Kickxellomycotina*, *Mortierellomycotina*, *Mucoromycotina*, and *Zoopagomycotina* were segregated (Benny et al. 2014). Several new phyla have been erected. Several chapters will cover the latest development of fungal systematics and the subsequent changes to the major taxonomic groups related to microfungi. *Microsporidia* (ca. 1500 species described) is a phylum newly transferred to Kingdom Fungi from Kingdom Protista based on phylogenetic studies. However, this new development has an unwanted implication. It would render ca. 1000 species invalid, since these species were named following zoological nomenclature based on the view of *Microsporidia* belonging to Protista, because fungi are subject to the rules of botanical nomenclature (Keeling 2009). Such destabilization of these names will ultimately lead to huge chaos to these taxa of *Microsporidia*. To avoid an unnecessary large number of name changes and a contentious chaotic situation for *Microsporidia*, the phylum *Microsporidia* is excluded from governance by the Melbourne Code and continue to be covered by the International Code of Zoological Nomenclature, despite their phylogenetic position (McNeill et al. 2012). This is a necessary solution, but an imperfect one. It has brought back the old question should all nomenclature codes for life [International Code of Nomenclature for algae, fungi, and plants (ICN), International Code of Zoological Nomenclature (ICZN), International Code of Nomenclature of Bacteria (ICNB), International Committee on Taxonomy of Viruses (ICTV)] be unified? Such unification will break down some barriers or boundaries and expedite cross-field collaborations. No doubt this is a huge undertaking.

A recent fungal biogeographical study by Tedersoo et al. (2014) on patterns of biodiversity of soil fungi worldwide using a mixture of six forward primers analogous to ITS3 and a degenerate reverse primer analogous to ITS4 demonstrated an exponential decline in fungal species richness and the plant-to-fungus richness ratio with distance from the equator, and at the same time the diversity of a number of fungal groups depended more on the abundance of host plants than host diversity or geography. This study indicated an enormous gap between described species and the actual numbers of distinct fungi in the soils at the global scale.

Discussions on this study indicated that the fungal diversity of certain fungal groups may not be accurately estimated due to limitations of the molecular method used. Tedersoo et al. (2014) acknowledged that *Tulasnellaceae* and *Microsporidia* would not be amplified with the primer set they used, and ca. 30 % *Tulasnella* have approximately two mismatches and *Archaeorhizomycetes*. Thus, Schadt and Rosling (2015) pointed out that there is a likely bias caused by mismatched polymerase chain reaction (PCR) primer for *Archaeorhizomycetes* in the soils. It was previously shown that the ITS4 reverse primer has at least two mismatches to all known species in the *Archaeorhizomycetes* (Rosling et al. 2011). These mismatches resulted in at least a tenfold underrepresentation of *Archeaorhizomycetes*. Thus, Schadt and Rosling (2015) opined that *Archaeorhizomycetes* are not low-diversity, low-abundance soil fungi as reported by Tedersoo et al. (2014), but the opposite.

Fungal diversity can be overestimated also. Cloning or massive parallel sequencing may significantly overestimate fungal diversity if there are sufficient differences among these sequences within a fungal individual or taxon due to the fact that these sequences were derived from single alleles (Lindner and Banik 2011; Lindner and Banik 2011; Lücking et al. 2014; Větrovský et al. 2015). Větrovský et al. (2015) indicated that operational taxonomic unit (OTU) counts and relative abundance of the *Basidiomycota* were highly overestimated by ITS. The variable length of the ITS region represents another important problem as there is a strong PCR bias against species with longer amplicons (Ihrmark et al. 2012). Thus, the methods to study fungal diversity should be improved to better reflect fungal diversity. At present this is the area where more studies should be conducted on validation or calibration of certain methods.

It is our intention that all authors can use this book as a platform to discuss not only the development in their areas, but also to express their opinions on weaknesses or gaps in microfungi research and identify the future directions and areas which remain overlooked.

There are several different estimates of the number of fungi (Blackwell 2011). A widely accepted estimate for fungal diversity is that there are 1.5 million species of fungi on our planet (Hawksworth 1991). At present, approximately 100,000 species (<7 % of total fungal taxa) have been described. Among the undescribed fungal taxa, majority are microfungi. There is no doubt that mycologists have a long way to go and a large number of fungi are waiting for us to find and describe. Tropics and subtropics are very rich in mycota with a higher fungal diversity. Microfungal Diversity of Central and South America covered in Chap. 9 will show you not only the beauty of microfungi and their diversity and habitats, but also the methodology to collect and study them. Among these fungi, many were found from the areas as new taxa in the last three decades. Unfortunately, the tropic and subtropical areas in Central and South America, Asia, and Africa remain less studied at present. Not only do we have to describe a huge number of fungi which remain unknown to science, but also to rescue the endangered fungal species from extinction due to global warming and loss of habitats from human activities such as population growth, overexploitation, and overdevelopment. Deforestation in the Amazon River area and South Asia is depleting the habitats of not only fungi, but all life in the areas and is reducing biodiversity (Foley et al. 2007; Davidson et al. 2012; Zhai et al. 2014). Amazonian forests have a significant impact on regional and global climates and deforestation in the area can be a driver of climate change (Malhi et al. 2008). Reduced habitats will result in decline of fungal diversity and cause some fungal taxa to disappear prior to even having a chance to describe them.

There are about two dozen fungi listed as rare, threatened, and endangered fungi in the webpage of Mushroom Observer. All these species on the list are macrofungi (Mushroom Observer 2015). The International Union for Conservation of Nature (IUCN) red list of endangered species listed four lichens: *Anzia centrifuga*, *Cladonia perforata* (Florida perforate reindeer lichen), *Erioderma pedicellatum* (boreal felt lichen), *Gymnoderma insulare*, and *Pleurotus nebrodensis* (white Ferula mushroom) as endangered fungal species (IUCN 2014). At present, 77 species are suggested for a Global Red List Assessment (GFRLI 2015). Among those species, they are all either macrofungi or lichens while not a single microfungus is present. The main embarrassment is that we do not have enough data on microfungi to assess and propose to the Global Red List. One simple assessment method is that for each plant species we lose, at least six fungal species would be lost. Among the lost species, about 89 % would be microfungi. This conservative estimate is based on the number of described fungi and the ratio of macrofungi and microfungi described (Kirk et al. 2008). This is an area which is severely overlooked. New initiatives for research on endangered microfungi should be promoted. The key issue is how to balance nature conservation and economic development.

The biggest challenge to mycology is the continuous decline of the mycological profession. It is rather depressing to see that many mycologist positions have disappeared in the academic and teaching institutions in the past two decades in North America. The membership of Mycological Society of America (MSA) has dropped 29 % from 1342 to 953 members between 2002 and 2015 (Cantrell 2014, 2015). It is shocking to watch MSA lose so many members in a decade. Crisis may not be an overstatement. Since so many advances have been made in mycology in the past decade, why have so many mycologists left the fields of mycology and why is there not enough new members to fill in? No doubt it is an alarming and embarrassing trend to all mycologists and society. Without fungi, the biosphere and biodiversity would be incomplete. What are the key reasons or leading factors behind this inclination? Lack of research funding and shrinking job market for mycologists may be considered the major contributing factors by a majority of mycologists, but what else? Can we reverse it? How? Microfungi, like macrofungi, are a fascinating group of organisms to study from morphology, ecology, and phylogenetics. It is our obligation to inspire more young talented individuals' interests in mycology, and these promising future mycologists can be trained to explore and study not only the biological significance, but also the beauty of microfungi from both morphological and molecular perspectives.

This book comprises of 24 chapters. Several chapters update the latest development of fungal systematics, especially the phyla related to microfungi, and fungal diversity. The perspectives and challenges we are facing in the molecular age, such as data curating and quality control, are also discussed. At the same time, this book focuses on the ecological and functional aspects of microfungi. This part covers a broad spectrum of areas from indoor to freshwater fungi, nematode fungi, and allergenic fungi to mycotoxins. However, there are several areas that were less covered in the past, such as marine fungi, whisky fungus, and spartina fungi. If you are wondering what is whisky fungus, please read Chap. 16 to find out how it was discovered and described and its relationships with people and whisky making. It is a fascinating story about an alcoholphilic fungus. Can a microfungus become alcoholic? This chapter will answer this question.

It is our intention that information on recent advances in mycology presented in this book will be useful to identify the needs in mycological research and to determine the direction or niches for future research. If the book can strike some sparks for novel ideas and in-depth discussion, its objective has been reached.

## References


Ainsworth GC (1976) Introduction to the history of mycology. Cambridge University Press, Cambridge

Benny G, Humber R, Voigt K (2014) 8 Zygomycetous fungi: Phylum Entomophthoromycota and subphyla Kickxellomycotina, Mortierellomycotina, Mucoromycotina, and Zoopagomycotina. In: McLaughlin DJ, Spatafora JW (eds) Systematics and evolution, vol 7A, The Mycota. Springer, Berlin, pp 209–250. doi:10.1007/978-3-642-55318-9_8

Blackwell M (2011) The fungi: 1, 2, 3 … 5.1 million species? Am J Bot 98(3):426–438. doi:10.3732/ajb.1000298

Braun H (1965) Geschichte der phytomedizin. P. Parey, Berlin

Bridge P (2002) The history and application of molecular mycology. Mycologist 16(3):90–99. doi:10.1017/S0269-915X(02)00201-X

Cantrell SA (2014) Annual report of the treasurer. Inoculum 65(6):13–15

Cantrell SA (2015) Annual report of the treasurer. Inoculum (Newsletter of Mycological Society of America) 66(5):13–15

Davidson EA, de Araujo AC, Artaxo P, Balch JK, Brown IF, Bustamante MM, Coe MT, DeFries RS, Keller M, Longo M, Munger JW, Schroeder W, Soares-Filho BS, Souza CM, Wofsy SC (2012) The Amazon basin in transition. Nature 481(7381):321–328. doi:10.1038/nature10717

Foley JA, Asner GP, Costa MH, Coe MT, DeFries R, Gibbs HK, Howard EA, Olson S, Patz J, Ramankutty N, Snyder P (2007) Amazonia revealed: forest degradation and loss of ecosystem goods and services in the Amazon Basin. Front Ecol Environ 5(1):25–32. doi:10.1890/1540-9295(2007)5[25:ARFDAL]2.0.CO;2

GFRLI (2015) The global fungal red list initiative. http://iucn.ekoo.se/en/iucn/welcome

Hawksworth DL (1991) The fungal dimension of biodiversity: magnitude, significance, and conservation. Mycol Res 95(6):641–655. doi:10.1016/s0953-7562(09)80810-1

Ihrmark K, Bödeker IT, Cruz-Martinez K, Friberg H, Kubartova A, Schenck J, Strid Y, Stenlid J, Brandström-Durling M, Clemmensen KE (2012) New primers to amplify the fungal ITS2 region–evaluation by 454-sequencing of artificial and natural communities. FEMS Microbiol Ecol 82(3):666–677

IUCN (2014) The IUCN red list of threatened species. Version 2014.3. International Union for Conservation of Nature. http://www.iucnredlist.org/. Accessed 26 Jan 2015

Jones MDM, Forn I, Gadelha C, Egan MJ, Bass D, Massana R, Richards TA (2011) Discovery of novel intermediate forms redefines the fungal tree of life. Nature 474(7350):200–203. doi:10.1038/nature09984

Keeling P (2009) Five questions about microsporidia. PLoS Pathog 5(9):e1000489. doi:10.1371/journal.ppat.1000489

Kendrick B (1985) The Fifth Kingdom. Mycologue Publications, Waterloo, Ontario, Canada

Kirk PM, Cannon PF, Minter D, Stalpers JA (2008) Dictionary of the fungi, 10th edn. CABI, Wallingford

Lindner DL, Banik MT (2011) Intragenomic variation in the ITS rDNA region obscures phylogenetic relationships and inflates estimates of operational taxonomic units in genus *Laetiporus*. Mycologia 103(4):731–740

Lücking R, Lawrey JD, Gillevet PM, Sikaroodi M, Dal-Forno M, Berger SA (2014) Multiple ITS haplotypes in the genome of the lichenized basidiomycete *Cora inversa* (Hygrophoraceae): fact or artifact? J Mol Evol 78(2):148–162

Lütjeharms WJ (1936) Zur Geschichte der Mykologie: das 18. Jahrhundert. Koch & Knuttel

Malhi Y, Roberts JT, Betts RA, Killeen TJ, Li W, Nobre CA (2008) Climate change, deforestation, and the fate of the Amazon. Science 319(5860):169–172

McNeill J, Barrie F, Buck W, Demoulin V, Greuter W, Hawksworth D, Herendeen P, Knapp S, Marhold K, Prado J (2012) International code of nomenclature for algae, fungi, and plants (Melbourne Code). Koeltz Scientific Books, Koenigstein

Mushroom Observer (2015) http://mushroomobserver.org/species_list/show_species_list/265

Ramsbottom J (1941) The expanding knowledge of mycology since Linnaeus: Presidential address delivered at the anniversary meeting of the Linnean Society of London on the 24th of May, 1939. Taylor & Francis, London

Rippon JW (1982) Medical mycology; the pathogenic fungi and the pathogenic actinomycetes. WB Saunders, Eastbourne

Rosling A, Cox F, Cruz-Martinez K, Ihrmark K, Grelet G-A, Lindahl BD, Menkis A, James TY (2011) Archaeorhizomycetes: unearthing an ancient class of ubiquitous soil fungi. Science 333(6044):876–879

Schadt CW, Rosling A (2015) Comment on "Global diversity and geography of soil fungi". Science 348(6242):1438. doi:10.1126/science.aaa4269

Schüßler A, Schwarzott D, Walker C (2001) A new fungal phylum, the Glomeromycota: phylogeny and evolution. Mycol Res 105(12):1413–1421

Tedersoo L, Bahram M, Põlme S, Kõljalg U, Yorou NS, Wijesundera R, Ruiz LV, Vasco-Palacios AM, Thu PQ, Suija A, Smith ME, Sharp C, Saluveer E, Saitta A, Rosas M, Riit T, Ratkowsky D, Pritsch K, Põldmaa K, Piepenbring M, Phosri C, Peterson M, Parts K, Pärtel K, Otsing E, Nouhra E, Njouonkou AL, Nilsson RH, Morgado LN, Mayor J, May TW, Majuakim L, Lodge DJ, Lee SS, Larsson K-H, Kohout P, Hosaka K, Hiiesalu I, Henkel TW, Harend H, Guo L-d, Greslebin A, Grelet G, Geml J, Gates G, Dunstan W, Dunk C, Drenkhan R, Dearnaley J, De Kesel A, Dang T, Chen X, Buegger F, Brearley FQ, Bonito G, Anslan S, Abell S, Abarenkov K (2014) Global diversity and geography of soil fungi. Science 346(6213):1256688. doi:10.1126/science.1256688

Větrovský T, Kolařík M, Žifčáková L, Zelenka T, Baldrian P (2015) The rpb2 gene represents a viable alternative molecular marker for the analysis of environmental fungal communities. Mol Ecol Resour. 16(2):388–401. doi:10.1111/1755-0998.12456

Whittaker RH (1969) New concepts of kingdoms of organisms. Science 163(3863):150–160

Zhai D-L, Cannon C, Dai Z-C, Zhang C-P, Xu J-C (2014) Deforestation and fragmentation of natural forests in the upper Changhua watershed, Hainan, China: implications for biodiversity conservation. Environ Monit Assess 187(1):1–12. doi:10.1007/s10661-014-4137-3

# Chapter 2
# Recent Changes in Fungal Nomenclature and Their Impact on Naming of Microfungi

**Walter Gams**

## The Previous Situation

As documented in the *International Code of Botanical Nomenclature* (the ICBN, Article 59, latest edition by McNeill et al. 2006), fungal nomenclature had an unusual situation for several decades: the often very different forms of sporulation (morphs) of a fungus could have different valid and legitimate names in different genera, which were either teleomorph or anamorph typified. The name attached to the teleomorph had precedence over older anamorph-based names. The basic idea behind this rule was that only with the knowledge of the teleomorph could a fungus be inserted in a natural taxonomic system while that of anamorphs was deemed to remain more or less artificial (e.g., Gams 1995). The most complete compilation of anamorph–teleomorph connections is presented for hyphomycete anamorph genera by Seifert et al. (2011). Another comparable compilation was provided by Wijayawardene et al. (2012).

This situation of dual nomenclature was in conflict with the time-honored principle IV of the *Code*: one organism–one name. It was also abnormal because a quality was demanded for the type: presence of a (sexual) teleomorph in the diagnosis and specimen to make it acceptable as a teleomorph name; otherwise it was anamorphic. In the era of molecular work, this abnormality appeared inappropriate, although the dual system had many advantages for the (morphological) identification of a fungus. A change toward unification was postulated mainly by molecular phylogeneticists, while other taxonomists were afraid of a flood of name changes inevitably following such a change.

W. Gams, Ph.D. (✉)
Formerly The CBS-KNAW Fungal Biodiversity, 3584 CT Utrecht, The Netherlands

Molenweg 15, 3743 CK Baarn, The Netherlands
e-mail: walter.gams@online.nl

 
D.-W. Li (ed.), *Biology of Microfungi*, Fungal Biology,
DOI 10.1007/978-3-319-29137-6_2

## Preparations for a Unification

In *Studies in Mycology* 45 (Seifert et al. 2000), several authors explored how molecular studies affected the classification of *Ascomycetes* and showed how far the discrepancy between anamorph and teleomorph taxonomy still persisted. This incongruence of anamorph-defined and teleomorph-defined genera was in fact one of the major obstacles toward a unification. Grégoire Hennebert, in a philosophical text published temporarily on the CBS website, outlined several scenarios for a possible change, according to which the nomenclatural changes might be reduced to a bearable minimum. Keith Seifert and Paul Cannon convened a symposium at the Oslo IMC (Seifert et al. 2003), during which two teams, defending either unification or the conservative side, debated heavily. The audience then was requested to vote, resulting in a majority of 121:84 for retaining the dual system.

Since 2000 the molecular-based classification of fungi has made dramatic progress. The Genealogical Concordance Phylogenetic Species Recognition (GCPSR) concept introduced by Taylor et al. (2000) has become broadly recognized and allows an almost final phylogenetic reconstruction, only inferior to an entire genome-based classification. In contrast to plant species, some 95 % of fungal species are estimated to be undescribed (Hawksworth 2004b). Less than 20 % of the known fungal species are represented with sequences in GenBank (Hawksworth 2004b). Only ca. 15.5 % of the known (<3 % of the estimated) species have known ITS sequences which are considered as universal barcodes (Schoch et al. 2012), although they lack resolution in some species-rich groups such as *Colletotrichum*, *Fusarium*, or *Trichoderma*. The largely varying quality of available DNA data is not sufficiently considered in molecular-based classification. Up to 20 % of fungal ITS sequences in GenBank are labeled with wrong species names (Nilsson et al. 2006). The different technical quality of sequencing (reading errors, ambiguous positions, simple reads vs. double reads, chimeric sequences) also leads to ambiguity of data interpretation. Sequences available in GenBank are often used without checking whether they have undergone a review process or are unpublished. Sequences derived from vouchers identified by other (primary) means than comparisons with DNA data are obviously superior to sequences only identified (secondarily) by comparison with other already available sequences. Schoch et al. (2014) are making laudable efforts to select and re-annotate a set of marker reference sequences that represent each currently accepted order of fungi.

At the International Botanical Congress at Vienna in 2005, just one issue was resolved toward unification: A proposal (Hawksworth 2004a) was passed that allowed epitypification of a previously anamorphic taxon with teleomorphic material in order to avoid unnecessary name changes. For this procedure Redhead (2010b) introduced the term teleotypification. Gams et al. (2010a, b) discussed some problems associated with the procedure and wanted to restrict the mechanism to cases where no appropriate teleomorph genus was yet available.

At the Vienna Congress, a Special Committee on the Nomenclature of Fungi with a Pleomorphic Life History, headed by Scott Redhead, was established and given the mandate of finding solutions for the problems surrounding Article 59.

The members of this committee were rather evenly distributed from both opposing camps and did not reach a conclusion (Redhead 2010a), in contrast to the permanent Nomenclature Committee for Fungi (NCF), which in a ballot reached a majority favoring unification. Redhead (2010b) then prepared a set of alternative proposals to be presented and voted on at the forthcoming International Botanical Congress (IBC18) in Melbourne.

Pedro Crous, for a long time one of the strongest defenders of the dual system, became converted to the unifying camp under the influence of David Hawksworth, John Taylor, Keith Seifert, and others. Rossman and Seifert (2011) edited a further volume 68 of *Studies in Mycology* with contributions honoring the retirement of Gary J. Samuels. The contributors experimented with various modes of moving toward a unified nomenclature, and some of them introduced genera for holomorphs typified by anamorphic species. In April 2011, Crous convened a symposium in Amsterdam "*One fungus–one name*." At this occasion the *Amsterdam Declaration* (Hawksworth et al. 2011) was signed by some 80, mainly practically oriented mycologists as a strong plea for immediate unification. The opponents, who regarded the time for unification as not yet ripe, were no less active and published cogent opposed arguments (Gams et al. 2011) in a paper signed by some 70 mycologists, mainly those involved in morphological taxonomy. All this was done to prepare for the crucial decisions at the IBC in Melbourne in July 2011.

## The Crucial Sessions at Melbourne

In the week preceding the main part of the Melbourne Congress, the nomenclature delegates convened and discussed 338 proposals to modify the *Code*. Hawksworth's proposal (Hawksworth et al. 2009) and lively discussions intended to do justice to mycology in the title of the *Code* resulted in the new name: *International Code of Nomenclature for algae, fungi, and plants* (the ICN, McNeill et al. 2012), in which the fungi and algae (in the broadest sense) appear for the first time in the title. A further important decision concerned a compulsory registration of all nomenclatural novelties for their validity (Hawksworth et al. 2010). Any fungal specimen can now serve as type material (including permanently preserved, inactivated fungal cultures). But environmental samples (in spite of Hibbett et al. 2011) and DNA material alone (in spite of Reynolds and Taylor 1992) remain unacceptable as types of formal fungal names. Norvell (2011), McNeill and Turland (2011), Gams (2013), and for phytopathogenic fungi Zhang et al. (2013) have surveyed the major results of this congress.

To introduce the most contentious item, Article 59, Scott Redhead, chair of the Special Committee dealing with Article 59, had prepared a set of proposals to be presented in succession. He started with the most drastic one, which implied the complete abolition of dual nomenclature and precedence of teleomorph-based names over those for anamorphs. A short presentation of pros and cons was permitted before the vote. Expecting that this drastic step would fail, some less drastic procedures would then be presented subsequently. To the great surprise of all participants, the first vote ended immediately in an overwhelming yes result. The botanical majority of voters did not seem sufficiently aware of the intricacies of the situation and simply voted for a simple and drastic solution. This caused great consternation among the many mycologists who had opposed the move and regarded this step as premature, while too few fungi were still sufficiently characterized in the phylogenetic system. Others (Hawksworth 2011; Wingfield et al. 2012) did afterwards not hide their triumph.

The new situation briefly implies the following: Names introduced independently for different morphs of a pleomorphic fungus remain legitimate but a choice must be made for one of them.

> *59.1.* A name published prior to 1 January 2013 for a taxon of non-lichen-forming *Ascomycota* and *Basidiomycota*, with the intent or implied intent of applying to or being typified by one particular morph (e.g. anamorph or teleomorph), may be legitimate even if it otherwise would be illegitimate under Art. 52 on account of the protologue including a type (as defined in Art. 52.2) referable to a different morph. If the name is otherwise legitimate, it competes for priority (Arts. 11.3 and 11.4; see also Art. 57.2).

On or after 1 Jan 2013, the introduction of a name for a morph different from that previously named for the same species is illegitimate, at least if the later-named morph includes in its protologue the type (or name) of the earlier-named morph. This ruling does therefore not preclude the possibility that two morphs of one fungus inadvertently receive two different legitimate names also after 2012 (Braun 2012). When two names for different morphs pertaining to the same new fungus are simultaneously introduced, neither is validly published (Art. 36.2). Quite generally the principle of priority prevails, no matter whether the original type of a taxon was teleomorphic or anamorphic. The present *Code* still contains the clause:

> *57.2.* In pleomorphic fungi (including lichenicolous fungi, but excluding lichen-forming fungi and those fungi traditionally associated with them taxonomically, e.g. *Mycocaliciaceae*), in cases where, prior to 1 January 2013, both teleomorph-typified and anamorph-typified names were widely used for a taxon, an anamorph-typified name that has priority is not to displace the teleomorph name(s) unless and until a proposal to reject the former under Art. 56.1 or 56.3 or to deal with the latter under Art. 14.1 or 14.13 has been submitted and rejected.

This is a remnant of the previous time-honored rule of precedence for teleomorph-based names; it now finds little appreciation and may soon be abolished (Hawksworth 2014).

Whatever the implementation is, the new ruling has started to bring about drastic nomenclatural changes. To avoid complete chaos, two new additions were made to Articles 14 and 56:

> *14.13.* In the interest of nomenclatural stability, for organisms treated as fungi (including lichenicolous fungi, but excluding lichen-forming fungi and those fungi traditionally associated with them taxonomically, e.g. *Mycocaliciaceae*), lists of names may be submitted to the General Committee, which will refer them to the Nomenclature Committee for Fungi (see Div. III) for examination by subcommittees established by that Committee in consultation with the General Committee and appropriate international bodies. Accepted names on these lists, which become Appendices of the *Code* once reviewed and approved by the Nomenclature Committee for Fungi and the General Committee, are to be listed with their types together with those competing synonyms (including sanctioned names) against which they are treated as conserved (see also Art. 56.3).

> *56.3.* In the interest of nomenclatural stability, for organisms treated as fungi (including lichenicolous fungi, but excluding lichen-forming fungi and those fungi traditionally associated with them taxonomically, e.g. *Mycocaliciaceae*), lists of names to be rejected may be submitted to the General Committee, which will refer them to the Nomenclature Committee for Fungi (see Div. III) for examination by subcommittees established by that Committee in consultation with the General Committee and appropriate international bodies. Names on these lists, which become Appendices of the *Code* once reviewed and approved by the Nomenclature Committee for Fungi and the General Committee, are to be treated as rejected under Art. 56.1 and may become eligible for use only by conservation under Art. 14 (see also Art. 14.13).

These paragraphs imply that the previous, entirely rule-dominated nomenclature is now changed to a committee- and list-based classification and nomenclature (Hawksworth 2012b). Suggestions to minimize the effects were presented by Gams et al. (2012) and Braun (2012).

## The Committee-dominated Era

For several major groups of pleomorphic fungi, ad hoc working groups of experts began to establish themselves or were convened at the 2012 and 2013 CBS spring symposia or commissioned by the ICTF. Hawksworth (2012a) proposed a time schedule for this work in order to promote the activity of various committees in view of the International Mycological Congress in Bangkok in 2014 and the International Botanical Congress in China in 2017, when the outcome of this work should be vetted.

The International Commission on the Taxonomy of Fungi (ICTF) is strongly involved in coordinating these efforts (Seifert and Miller 2012, 2013), and most of the discussion papers listed below are also located on the ICTF site www.fungaltaxonomy.org/lists, where they will be most easily found. The lists resulting from these efforts will subsequently remain freely accessible through the Internet.

The committees work with the terms *accepted* names (which are *protected*) and *suppressed* names. There is an important difference from the other system of conserved vs. rejected names, which are irreversible and universally binding. The new lists will remain open for additions and possibly changes. This system is also different from the previously proposed and then defeated system of "*Names in current use*" (Hawksworth and Greuter 1998), where only current use would determine a rather arbitrary selection of names to be retained and protected.

Several groups of mycologists have already done their work. The guiding principles for the choice of genera are well-supported monophyletic clades which comprise morphologically and ecologically rather homogeneous taxa. Normally the priority of names, no matter whether teleomorph- or anamorph-based, decides the choice. But deviations are proposed in some cases in the spirit of nomenclatural parsimony (Seifert and Miller 2012; Rossman 2014). The frequency of usage of a name may also be taken into account, but a more important criterion is the number of necessary name changes when a particular name is given preference.

A problem is whether in cases where the oldest epithet for a species was given in the suppressed genus, this epithet must be recombined, replacing a well-known binomial in the accepted genus. Thus *Trichoderma reesei* would have to be replaced by a newly combined *T. jecorinum* and *T. citrinoviride* by a new *T. schweinitzii*. This was not done by Jaklitsch and Voglmayr (2014), and Gams et al. (2012) also advised against it. Subsequently Samuels (2014) prepared formal proposals for conservation of these younger names and a few similar cases and all recognized *Trichoderma* names were listed by Bissett et al. (2015). For preferential genera of the *Leotiomycetes* Johnston et al. (2014), quite generally made the new combinations for the oldest available epithet.

## Discussion Papers

*Discussion papers* (published or unpublished) with lists of preferred generic names are now available for the following groups of ascomycetes (*Pezizomycotina*) dealing mainly with *genera*, while the phylogenetic delimitation of *species* would deserve priority (Braun 2012).

### *Orbiliomycetes*

Debates are ongoing concerning the *Orbiliales* (Baral et al. 2016). A generic segregation of the nematode trapping, so far mainly anamorph-based species from the large genus *Orbilia*, would be possible in the narrowest of the generic concepts discussed.

### *Pezizomycetes*

Tedersoo et al. (2013) and for the largest family, the *Pyronemataceae*, Hansen et al. (2013) provide phylogenetic insights, and in spite of the occurrence of anamorphs, teleomorph classification clearly dominates.

### *Dothideomycetes*

An account of preferential names for pleomorphic genera of the class was compiled by Rossman et al. (2015b).

#### *Dothideales*

Thambugala et al. (2014) delimit the *Dothideaceae* (including the *Dothioraceae*) against the newly coined *Aureobasidiaceae*, a family now comprising seven genera.

#### *Pleosporales*

In the review by Hyde et al. (2013) and complete table by Wijayawardene et al. (2014), some genera are still controversial, e.g., the speciose and still heterogeneous *Pleospora* vs. the well-delimited *Stemphylium*. The genus *Alternaria* will have to comprise species of so far separate genera like *Ulocladium*, *Embellisia*, *Nimbya*, etc. (Woudenberg et al. 2013), when only phylogeny counts.

#### *Capnodiales*

Crous et al. (2009a, b), Crous (2010). The family name *Cladosporiaceae* is resurrected and *Cladosporium* obviously deserves preference over the associated teleomorph genus *Davidiella*. Species of *Mycosphaerella* s. str. are placed in *Ramularia*, whereas a bulk of species still remains in the teleomorph genus.

#### *Botryosphaeriales*

Phillips et al. (2013) place the *Phyllostictaceae* in this order and give preference to *Phyllosticta* over *Guignardia*. In the *Botryosphaeriaceae* Slippers et al. (2013) recognize 17 genera on a phylogenetic basis, among which *Diplodia*, *Neodeightonia*, *Lasiodiplodia*, *Sphaeropsis*, *Macrophomina*, *Neoscytalidium*, and *Neofusicoccum* are all sufficiently distinct and keyed out based on anamorph features. Teleomorph features are insufficient to distinguish phylogenetically significant genera morphologically.

### *Eurotiomycetes*

#### *Chaetothyriomycetidae*, *Chaetothyriales*

Réblová and Untereiner (2013) introduce the new anamorph-based family *Cyphellophoraceae* for the expanded genus *Cyphellophora*, which now comprises species with septate and nonseptate conidia. Two new genera are introduced for some former *Cyphellophora* species. Gueidan et al. (2014) include four orders, *Celotheliales* ad int., *Chaetothyriales*, *Pyrenulales*, and *Verrucariales*, and ten families (*Adelococcaceae*, *Celotheliaceae*, *Chaetothyriaceae*, *Cyphellophoraceae*, *Epibryaceae* fam. nov., *Herpotrichiellaceae*, *Pyrenulaceae*, *Requienellaceae*, *Trichomeriaceae*, and *Verrucariaceae*) to the subclass. To resolve the very difficult complex of *Capronia–Exophiala–Rhinocladiella–Phialophora* in the *Herpotrichiellaceae*, no solution is yet offered.

### *Eurotiales*

The *Aspergillaceae* are now distinguished from the *Trichocomaceae* (Samson et al. 2011). *Penicillium* (including the teleomorph genus *Eupenicillium* and anamorphs previously classified in *Eladia*, *Torulomyces*, and *Thysanophora*) is clearly separated from *Talaromyces*, which now also incorporates anamorphic taxa (formerly *Penicillium* subgen. *Biverticillium*) (Samson et al. 2011). Debates are going on between a majority of members of the *Penicillium–Aspergillus* working group who prefer recognizing one large genus *Aspergillus* (so far linked to over ten teleomorph genera) and other mycologists who prefer several, mostly teleomorph-linked and ecologically very distinct genera, among which *Aspergillus* s. str. would have to retain a conserved type that represents the former section *Circumdati* and not the original genus in the sense of *Eurotium* (Pitt and Taylor 2014). Another debatable case is the choice between *Byssochlamys* and *Paecilomyces*.

## *Leotiomycetes*

In a voluminous survey Johnston et al. (2014) propose several cases of preferential teleomorph names that are younger than the associated anamorph names, such as *Ascocoryne* over *Coryne*, *Dematioscypha* over *Haplographium*, *Dermea* over *Sphaeronaema*, *Diplocarpon* over *Entomosporium*, *Gremmeniella* over *Brunchorstia*, *Monilinia* over *Monilia*, *Neofabraea* over *Phlyctema*, and *Pyrenopeziza* over *Cylindrosporium*, but they retain the older *Hyphodiscus* over *Catenulifera*, *Pezicula* over *Cryptosporiopsis*, *Phacidium* over *Ceuthospora*, *Phialocephala* over *Phaeomollisia*, *Pilidium* over (*Disco-*)*Hainesia*, *Rhytisma* over *Melasmia*, and *Vibrissea* over *Anavirga*.

### *Erysiphales*

Braun (2013) proposes conservation of the teleomorph-based name *Blumeria* over the older anamorph-based *Oidium* and several more teleomorph-based epithets over correlated older anamorph names. Thus the names of powdery mildew genera are now consistently teleomorph-based, although anamorph features also contribute to genus delimitation.

### Phacidiales

Crous et al. (2014) raise the family *Phacidiaceae* to ordinal level, segregated from the formerly paraphyletic *Helotiales* and synonymize the younger anamorph genus *Ceuthospora* with *Phacidium*.

## *Sordariomycetes*

### *Xylariomycetidae, Xylariales*

In the *Xylariales* the teleomorph-based taxonomy is quite clearly the guiding rule of generic classification negating that of anamorphs (Stadler et al. 2013). The authors are retaining teleomorph-generic names throughout the order (exception *Virgaria* preferred over *Ascovirgaria*), while certain anamorph features also correlate with generic delimitation. Debatable cases include *Arthrinium* vs. *Apiospora*, *Monographella* vs. *Microdochium*, *Seiridium* vs. *Eutypa*, and a few others.

### *Hypocreomycetidae*

#### *Hypocreales*

Rossman et al. (2013) list several genera of *Hypocreales* as candidates for protection while avoiding many hot irons in this group. *Nectria* clearly deserves preference over *Tubercularia*. The so far vaguely defined genus *Cylindrocarpon* (although a nomen conservandum) is sacrificed in favor of several associated teleomorph-based genera. *Gliocladium* s.str. is replaced by *Sphaerostilbella*, but the morphologically similar *Clonostachys* will outlive the associated older teleomorph name *Bionectria*. The very important genus *Fusarium* cannot be sacrificed for its teleomorph *Gibberella* (Geiser et al. 2013), but how many clades will remain in this genus is still uncertain, while some of them were already excluded and transferred to other genera by Gräfenhan et al. (2011). The controversy over the taxonomic identity of the speciose *F. solani* clade and the appropriate name for such a clade, if it were to be recognized as distinct from *Fusarium*, remain to be solved. Lombard et al. (2015) propose the name *Neocosmospora* for this clade and follow a narrow generic concept.

In *Hypocrea*, according to the former rule of teleomorph precedence, several species have been described recently, some of which lacked an anamorph altogether or had deviating anamorphs, but all of them are now transferred to the broadly preferred anamorph-based genus *Trichoderma* (Jaklitsch and Voglmayr 2014) as listed for the whole genus by Bissett et al. (2015). The introduction of a teleomorph genus for *Volutella* by Luo and Zhuang (2012) obviously becomes redundant after the work by Gräfenhan et al. (2011), but its publication in 2012 does not render the name *Volutellonectria* illegitimate. For the difficult complex of *Acremonium–Emericellopsis–Stilbella–Gliomastix* (Summerbell et al. 2011), no workable solution is yet in sight.

The *Clavicipitaceae* s. l. are treated in several papers: In the *Clavicipitaceae* s. str., the teleomorph-based genus *Epichloë* is older than *Neotyphodium* for associated anamorphs, and these taxa can easily be merged (Leuchtmann et al. 2014). Kepler et al. (2013) broaden the concept of *Polycephalomyces* and Kepler et al. (2014) create a large genus *Metarhizium* including former *Nomuraea* and, although strongly deviating, some paecilomyces-like species. Dealing with the *Ophiocordycipitaceae*, Quandt et al. (2014) give preference to the younger teleomorph name *Ophiocordyceps* over several older anamorph-based generic names. *Tolypocladium* is now conceived in a wider frame that required numerous new combinations, including some from a teleomorph genus.

The genera to be distinguished in the *Cordycipitaceae* are the most controversial. Cladistically minded mycologists wish to recognize no more than 11 rather widely defined monophyletic genera, while the morphology-trained and ecologically oriented mycologists demand a much finer distinction of genera.

### *Microascales*

For the medically relevant genera around *Pseudallescheria*, Lackner et al. (2014) recognize the generic names *Parascedosporium*, *Lomentospora*, *Petriella*, *Petriellopsis*, and *Scedosporium* (displacing *Pseudallescheria*, but still debated). For the mainly phytopathogenic taxa around *Ceratocystis*, de Beer et al. (2014) distinguish several genera, *Ceratocystis sensu stricto*, *Chalaropsis*, *Endoconidiophora*, *Thielaviopsis*, and *Ambrosiella*, and the new genera, *Davidsoniella* and *Huntiella*, most of which have chalara-like anamorphs.

### *Glomerellales*

The older genus name *Colletotrichum* is to be protected against the teleomorph genus *Glomerella* (Cannon et al. 2012).

## ***Sordariomycetidae***

### *Diaporthales*

In the *Magnaporthaceae* Luo et al. (2014) distinguish three major lines: (1) *Ophioceras*, (2) *Pyricularia* (suppressing *Magnaporthe*, but still debated), (3) *Gaeumannomyces* (*Harpophora*), *Magnaporthiopsis*, and distinct anamorph-based genera *Nakataea* and *Pseudophialophora*. A survey of preferential names in pleomorphic genera of the order was compiled by Rossman et al. (2015a).

*Ophiostomatales*

De Beer et al. (2013, see Conclusions below) give a nomenclature of all presently recognized ophiostomatoid taxa.

Some of these lists were briefly presented at the 10th International Mycological Congress in Bangkok during three nomenclature sessions (Redhead et al. 2014), which were too short for a detailed discussion. The lists still have to be scrutinized by the Nomenclature Committee for Fungi before being published in their final form in the Internet and presented to and sanctioned by the next Botanical Congress in China (Hawksworth 2012b). Thus the present years can only be regarded as a transitional period (Zhang et al. 2013), but it is likely that the examples listed here will be fixed as described.

## Conclusions

The unification of fungal nomenclature has been pushed through in order to provide for the *Fungi* a natural system just like for plants and animals. Sooner or later this move had to come, ideally at a time when a majority of fungal species is known to science. The most urgent task of mycology—discovery and careful description of new species—is now placed into second place by raising the issue of unified nomenclature to the top. At this moment phylogenetic knowledge is not sufficiently developed throughout the fungal system to provide clear-cut solutions for problematic cases. The present hectic activity at least enforces a useful stocktaking of what is so far known.

Mycologists are making enormous efforts to minimize the chaos ensuing from the somewhat prematurely introduced unification by generating meaningful lists of names. Lists of protected names are being produced and will become established. They are not the last word in fungal taxonomy, and mycologists cannot be forced to adopt a particular taxonomy when they do not agree with it. It is presently impossible to effectively squeeze all known species into recognized, available, and strictly monophyletic genera. Braun (2012) rightly emphasizes the permanent legitimacy of "suppressed" generic names as long as for many species evidence for their affinity with a list-accepted genus is missing. In addition I wish to emphasize that recognition of paraphyletic genera as being a natural phenomenon will do much more justice to a classification based on morphological and ecological criteria.

Unification was expected to facilitate the study of fungal systematics by students. However, this is not the case, as the knowledge of both sexual and asexual morphs of a fungus (and associated names) remains indispensable, even when only one generic name is recognized and the alternative one retains the role of a morphological descriptor and often also is the basis for the names of higher taxonomic ranks. As suggested by Seifert et al. (2000), some of these presently suppressed names will continue to be used as descriptive adjectives or nouns. We will just have phialophora-like and acremonium-like, but not phaeoacremonium-like or simplicillium-like because the latter two are recognized genera. Thus a certain duality of names for one genus is bound to persist after this move.

A complete move to unified nomenclature will require the recombination of all included species into a single recognized genus for a particular group. This has so far only been achieved consistently in *Trichoderma* (Jaklitsch and Voglmayr 2014; Bissett et al. 2015) and in *Tolypocladium* (Quandt et al. 2014). It has also been done for *Penicillium* and *Talaromyces* (Samson et al. 2014), but for *Aspergillus* remains controversial (Pitt and Taylor 2014). An interesting solution is chosen by de Beer et al. (2013) for ophiostomatoid fungi, where the authors list the species of monophyletic clades but leave them with their original binomial; e.g., in *Ophiostoma* they retain species of *Sporothrix*, *Raffaelea*, and even "*Leptographium*," although that genus phylogenetically belongs to a different clade. This procedure has much to recommend it, especially when not all fungi in question have yet been revised phylogenetically. This is preferable to imposing countless new generic combinations for controversial cases which at present can hardly be satisfactorily resolved.

A debatable proposal concerns morph pairs with identical epithets, for which Hawksworth et al. (2013) suggest a mechanism by which newly discovered alternate morphs are generally to be declared as new combinations based on the type of the older name. This would replace the hitherto compulsory introduction of a new species in the appropriate genus for the newly discovered morph, which created heterotypic names for contiguous morphs. However, this heterotypic situation has the advantage that the connection can be viewed critically in each case and need not always be recognized (Braun 2012); a global adoption of the proposed mechanisms has little chance of stabilizing names.

Taxonomic decisions can never be made by rules of nomenclature. If a name is placed on a list for suppression, this does not mean that it cannot be used when required. This is the important difference of the other system of officially conserving vs. rejecting names that is universally binding and irreversible (also Seifert and Miller 2012). These subtle differences are, however, not easily understood by authors, editors, and reviewers, not to say students, and may cause unnecessary conflicts about how to proceed. Presently, lists of names to be protected are prepared, and parallel official conservation proposals are published often by members of the same team (e.g., Samuels 2014), which further confuses these two different ways of formalizing names.

Publication of new names is now facilitated by eliminating the hurdles of Latin diagnoses and print publication so that the number of validly but qualitatively poorly published names is increasing. Some knowledge of Latin does remain indispensable for understanding the old literature and correctly coining new names for new taxa.

Taxonomic knowledge of numerous fungal groups is still quite inadequate and often does not yet allow decisions about the delimitation of natural taxa. The tendency by many mycologists to ignore morphological characters when introducing fungal taxa is not helpful when striving for a natural classification. The morphological knowledge gathered by older workers and that to be gathered for recent material remains an indispensable basis for establishing a natural system for fungi that guarantees stability and allows predictions of properties of related taxa. A careful morphological analysis and permanent preservation of the material studied are indispensable prerequisites to assure that a sequence obtained really applies to the fungus in question. Do not forget that genotype and phenotype are two sides of the same coin. Much more material needs to be collected and thoroughly studied to enhance mycological knowledge. Are the taxonomists of the future prepared to meet this challenge in all of its dimensions? The fungal world remains alive in its native environment, awaiting discovery. Conditions must urgently be created to enable mycologists to get out of the boardrooms and back into the field.

**Acknowledgments** For critical reading of the manuscript, I am indebted to Keith A. Seifert, Martina Réblová, Roland Kirschner, Marc Stadler, Uwe Braun, and for a final correction to Nick Turland.

## References

Baral H-O, Weber E, Stadler M, Gams W (2016) Recommendations about genera to be protected or suppressed in the *Orbiliaceae* (*Orbiliomycetes*) (in prep)

Bissett J, Gams W, Jaklitsch WM, Samuels GJ (2015) Accepted *Trichoderma* names in the year 2015. IMA Fungus 6:263–295

Braun U (2012) The impacts of the discontinuation of dual nomenclature of pleomorphic fungi: the trivial facts, problems, and strategies. IMA Fungus 3(1):81–86

Braun U (2013) (2210–2232) Proposals to conserve the teleomorph-typified name *Blumeria* against the anamorph-typified name *Oidium* and twenty-two teleomorph-typified powdery mildew species names against competing anamorph-typified names (*Ascomycota*: *Erysiphaceae*). Taxon 62:1328–1331

Cannon PF, Damm U, Johnston PR, Weir BS (2012) *Colletotrichum*: current status and future directions. Stud Mycol 73:181–213

Crous PW (2010) Taxonomy and phylogeny of the genus *Mycosphaerella* and its anamorphs. Fungal Divers 38:1–24

Crous PW, Hyde KD, Wood AR, Gueidan C, de Hoog GS, Groenewald JZ (2009a) Phylogenetic lineages in the *Capnodiales*. Stud Mycol 54:17–47

Crous PW, Summerell BA, Carnegie AJ, Wingfield MJ, Hunter GC, Burgess TI, Andjic V, Barber PA, Groenewald JZ (2009b) Unravelling *Mycosphaerella*: do you believe in genera? Persoonia 23:99–118

Crous PW, Quaedvlieg W, Hansen K, Hawksworth DL, Groenewald JZ (2014) *Phacidium* and *Ceuthospora* (*Phacidiaceae*) are congeneric: taxonomic and nomenclatural implications. IMA Fungus 5:173–193

De Beer ZW, Seifert KA, Wingfield MW (2013) A nomenclator for ophiostomatoid genera and species in the *Ophiostomatales* and *Microascales*. In: Seifert KA, De Beer ZW, Wingfield MW (eds) Ophiostomatoid fungi, expanding frontiers, CBS Biodiversity Series No. 12, pp 245–322

De Beer ZW, Duong TA, Barnes I, Wingfield BD, Wingfield MJ (2014) Redefining *Ceratocystis* and allied genera. Stud Mycol 79:187–219. doi:10.1016/j.simyco.2014.10.001

Gams W (1995) How natural should anamorph genera be? Can J Bot 73(Suppl 1):S747–S753

Gams W (2013) A new nomenclature for fungi. Mycol Iran 1:5–8

Gams W, Jaklitsch WM, Kirschner R, Réblová M (2010a) Teleotypification of fungal names and its limitations. Taxon 59:1197–1200

Gams W, Jaklitsch WM, Kirschner R, Réblová M (2010b) (172–174) Three proposals to amend Article 59 of the *Code* concerning teleotypification of fungal names. Taxon 59:1297

Gams W, Jaklitsch WM et al (2011) Fungal nomenclature 3. A critical response to the 'Amsterdam Declaration'. Mycotaxon 116:501–512

Gams W, Humber RA, Jaklitsch WM, Kirschner R, Stadler M (2012) Minimizing the chaos following the loss of Article 59: suggestions for a discussion. Mycotaxon 119:495–507

Geiser DM, Aoki T, Bacon CW, Baker SE, Bhattacharyya MK, Brandt ME et al (2013) One fungus, one name: defining the genus *Fusarium* in a scientifically robust way that preserves longstanding use. Phytopathology 103:400–408

Gräfenhan T, Schroers H-J, Nirenberg HI, Seifert KA (2011) An overview of the taxonomy, phylogeny, and typification of nectriaceous fungi in *Cosmospora*, *Acremonium*, *Fusarium*, *Stilbella*, and *Volutella*. Stud Mycol 68:79–113

Gueidan C, Aptroot A, da Silva Cáceres ME, Badali H, Stenroos S (2014) A reappraisal of orders and families within the subclass *Chaetothyriomycetidae* (*Eurotiomycetes*, *Ascomycota*). Mycol Prog 13:1027–1039

Hansen K, Perry BA, Dranginis AW, Pfister DH (2013) A phylogeny of the highly diverse cup-fungus family *Pyronemataceae* (*Pezizomycetes*, *Ascomycota*) clarifies relationships and evolution of selected life history traits. Mol Phylogenet Evol 67:311–335

Hawksworth DL (2004a) (183–187) Limitation of dual nomenclature for pleomorphic fungi. Taxon 53:596–598

Hawksworth DL (2004b) Fungal diversity and its implications for genetic resource collections. Stud Mycol 50:9–18

Hawksworth DL (2011) A new dawn for the naming of fungi: impacts of decisions made in Melbourne in July 2011 on the future publication and regulation of fungal names. MycoKeys 1:7–20

Hawksworth DL (2012a) Managing and coping with names of pleomorphic fungi in a period of transition. IMA Fungus 3(1):15–24

Hawksworth DL (2012b) Organizing mycology. IMA Fungus 3(2):(43)

Hawksworth DL (2014) Possible house-keeping and other draft proposals to clarify or enhance the naming of fungi within the International Code of Nomenclature for algae, fungi, and plants (ICN). IMA Fungus 5:31–37

Hawksworth DL, Greuter W (1998) Report of the first meeting of a working group on lists of names in current use. Taxon 38:142–148

Hawksworth DL, Crous PW, Dianese JC, Gryzenhout M, Norvell LL, Seifert KA (2009) (016–020) Proposals to amend the Code to make clear that it covers the nomenclature of fungi, and to modify its governance with respect to names of organisms treated as fungi. Taxon 58:658–659

Hawksworth DL, Cooper JA, Crous PW, Hyde KD, Iturriaga T, Kirk PM, Lumbsch HT, May TW, Minter DW, Misra JK, Norvell LL, Redhead SA, Rossman AY, Seifert KA, Stalpers JA, Taylor JW, Wingfield MJ (2010) Proposals to make the pre-publication deposit of key nomenclatural information in a recognized repository a requirement for valid publication of organisms treated as fungi under the *Code*. Taxon 59:660–662

Hawksworth DL, Crous PW, Redhead SA, Reynolds DR, Samson RA, Seifert KA, Taylor JW, Wingfield MJ et al (2011) The Amsterdam Declaration on fungal nomenclature. Mycotaxon 116:491–500

Hawksworth DL, McNeill J, de Beer ZW, Wingfield MJ (2013) Names of fungal species with the same epithet applied to different morphs: how to treat them. IMA Fungus 4:53–56

Hibbett DS, Ohman A, Glotzer D, Nuhn M, Kirk P, Nilsson RH (2011) Progress in molecular and morphological taxon discovery in fungi and options for formal classification of environmental sequences. Fungal Biol Rev 25:38–47

Hyde KD, Jones EBG, Liu J-L, Ariyawansa H, Boehm E, Boonmee S, Braun U, Chomnunti P, Crous PW et al (2013) Families of *Dothideomycetes*. Fungal Divers 63:1–313

Jaklitsch WM, Voglmayr H (2014) New combinations in *Trichoderma* (*Hypocreaceae*, *Hypocreales*). Mycotaxon 126:143–156

Johnston P, Seifert K, Stone J, Rossman A, Marvanová L (2014) Recommendations for generic names competing for use in *Leotiomycetes* (*Ascomycota*). IMA Fungus 5:91–120

Kepler R, Ban S, Nakagiri A, Bischoff J, Hywel-Jones NL, Owensby CA, Spatafora JW (2013) The phylogenetic placement of hypocrealean insect pathogens in the genus *Polycephalomyces*: an application of one fungus one name. Fungal Biol 117:613–622

Kepler RM, Humber RA, Bischoff J, Rehner S (2014) Clarification of generic and species boundaries for Metarhizium and related fungi through multigene phylogenetics. Mycologia 106:811–829

Lackner M, de Hoog GS, Yang L, Moreno LF, Ahmed SA, Andreas F, Kaltseis J, Nagl M, Lass-Flörl C, Risslegger B, Rambach G, Speth C, Robert V et al (2014) Proposed nomenclature for *Pseudallescheria*, *Scedosporium* and related genera. Fungal Divers 67:1–10

Leuchtmann A, Bacon CW, Schardl CL, White JF Jr, Tadych M (2014) Nomenclatural realignment of *Neotyphodium* species with genus *Epichloë*. Mycologia 106:202–215

Lombard L, van der Merwe NA, Groenewald JZ, Crous PW (2015) Generic concepts in Nectriaceae. Stud Mycol 80:189–245

Luo J, Zhuang W-Y (2012) *Volutellonectria* (*Ascomycota*, Fungi), a new genus with *Volutella* anamorphs. Phytotaxa 44:1–10

Luo J, Walsh E, Zhang N (2014) Four new species in *Magnaporthaceae* from grass roots in New Jersey Pine Barrens. Mycologia 106:580–588

McNeill J, Turland NJ (2011) Major changes to the Code of Nomenclature – Melbourne, July 2011. Taxon 60:1495–1497

McNeill J, Barrie FR, Burdet HM, Demoulin V, Hawksworth DL, Marhold K, Nicolson DH, Prado J, Silva PC, Skog JE, Wiersema JH, Turland NJ (2006) International Code of Botanical Nomenclature (Vienna Code). Regnum Vegetabile 146. A.R.G. Gantner Verlag, Ruggell, Liechtenstein, 568 pp

McNeill J, Barrie FR, Buck WR, Demoulin V, Greuter W, Hawksworth DL, Herendeen PS, Knapp S, Marhold K, Prado J, Prud'homme van Reine WF, Smith GF, Wiersema JH, Turland NJ (2012) International Code of Nomenclature for algae, fungi, and plants (Melbourne Code), adopted by the Eighteenth International Botanical Congress Melbourne, Australia, July 2011. Koeltz, Königstein

Nilsson RH, Ryberg M, Kristiansson E, Abarenkov K, Larsson K-H, Kõljalg U (2006) Taxonomic reliability of DNA sequences in public sequence databases: a fungal perspective. PLoS One 1:e59

Norvell LL (2011) Melbourne approves a new *Code*. Mycotaxon 116:481–490

Phillips AJL, Slippers B, Groenewald JZ, Crous PW (eds) (2013) Plant pathogenic and endophytic *Botryosphaeriales* known from culture. Stud Mycol 76, 167 pp

Pitt JI, Taylor JW (2014) *Aspergillus*, its sexual states and the new International Code of Nomenclature. Mycologia 106:1051–1062

Quandt A, Kepler R, Gams W, Araújo JPM, Ban S, Evans HC, Hughes D, Humber R, Hywel-Jones N, Li Z, Luangsa-ard JJ, Rehner SA, Sanjuan T, Sato H, Shrestha B, Sung G-H, Yao Y-J, Zare R, Spatafora JW (2014) Phylogenetic-based nomenclatural proposals for *Ophiocordycipitaceae* (*Hypocreales*) with new combinations in *Tolypocladium*. IMA Fungus 5(1):121–134

Réblová M, Untereiner WA (2013) Novel evolutionary lineages revealed in the *Chaetothyriales* (Fungi) based on multigene phylogenetic analyses and comparison of ITS secondary structure. Plos One 8(5):1–28

Redhead SA (2010a) Report on the special committee on the nomenclature of fungi with a pleomorphic life cycle. Taxon 59:1863–1866

Redhead SA (2010b) (294–306) Proposals to define the new term 'teleotype', to rename Chapter VI, and to modify Article 59 to limit dual nomenclature and to remove conflicting examples and recommendations. Taxon 59:1907–1933

Redhead SA, Demoulin V, Hawksworth DL, Seifert KA, Turland NJ (2014) Fungal nomenclature at IMC10: Report of the nomenclature sessions. IMA Fungus 5(1):449–462

Reynolds DR, Taylor JW (1992) Article 59: Reinterpretation or revision? Taxon 41:91–98

Rossman AY (2014) Lessons learned from moving to one scientific name for fungi. IMA Fungus 5:81–89

Rossman AY, Seifert KA (eds) (2011) Phylogenetic revision of taxonomic concepts in the *Hypocreales* and other *Ascomycota* – a tribute to Gary J. Samuels. Stud Mycol 68:1–256
Rossman AY, Seifert KA, Samuels GJ, Minnis AM, Schroers H-J, Lombard L, Crous PW, Põldmaa K, Cannon PF, Summerbell RC, Geiser DM, Zhuang W-Y, Hirooka Y, Herrera C, Salgado-Salazar C, Chaverri P (2013) Genera in *Bionectriaceae*, *Hypocreaceae*, and *Nectriaceae* (*Hypocreales*) proposed for acceptance or rejection. IMA Fungus 4:41–51
Rossman AY, Adams GC, Cannon PF, Castlebury LA, Crous PW, et al. (2015a) Recommendations of generic names in Diaporthales competing for protection or use. IMA Fungus 6:145–154
Rossman AY, Crous PW, Hyde KD, Hawksworth DL, et al. (2015b) Recommended names for pleomorphic genera in Dothideomycetes. IMA Fungus 6:507–523
Samson RA, Yilmaz N, Houbraken J, Spierenburg H, Seifert KA, Peterson SW, Varga J, Frisvad JC (2011) Phylogeny and nomenclature of the genus *Talaromyces* and taxa accommodated in *Penicillium* subgenus *Biverticillium*. Stud Mycol 70:159–183
Samson RA, Visagie CM, Houbraken J (eds) (2014) Species diversity in *Aspergillus*, *Penicillium* and *Talaromyces*. Stud Mycol 78:1–451
Samuels GJ (2014) (2305–2309) Proposals to conserve the names *Trichoderma catoptron* against *Hypocrea catoptron*, *H. sulfurella*, and *H. flavovirens*; *T. citrinoviride* against *Sphaeria schweinitzii* (*H. schweinitzii*), *S. contorta*, *H. repanda*, and *H. minima*; *H. lutea* against *Gliocladium deliquescens* (*T. deliquescens*) with a recommendation to reject the proposal (cf. Art. 57.2); *H. pezizoides* (*T. pezizoides*) against *T. pezizoideum*, and *T. reesei* against *H. jecorina* (*Ascomycota*: *Pezizomycotina*: *Sordariomycetes*: *Hypocreales*: *Hypocreaceae*). Taxon 63:936–938
Schoch CL, Seifert KA, Huhndorf S, Robert V, Spouge JL, Levesque CA, Chen W, Fungal Barcoding Consortium (2012) Nuclear ribosomal internal transcribed spacer (ITS) region as a universal DNA barcode marker for fungi. Proc Natl Acad Sci USA 109:6241–6246
Schoch CL, Robbertse B, Robert V, Vu D, Cardinali G, Irinyi L, Meyer W et al (2014) Finding needles in haystacks: linking scientific names, reference specimens and molecular data for fungi. Science & Mathematics Database. doi:10.1093/database/bau061
Seifert KA, Miller AN (2012) One fungus – which name? IMA Fungus 3(1):(10)–(14)
Seifert KA, Miller AN (2013) International Commission on the taxonomy of fungi: preparing for IMC10. IMA Fungus 4(1):(7)
Seifert KA, Gams W, Crous PW, Samuels GJ (eds) (2000) Molecules, morphology and classification: Towards monophyletic genera in the Ascomycetes. Stud Mycol 45:1–230
Seifert KA, Gams W, Korf RP, Pitt JI, Hawksworth DL, Berbee ML, Kirk PM (2003) Has dual nomenclature for fungi run its course? The Article 59 debate. Mycotaxon 88:493–508
Seifert KA, Morgan-Jones G, Gams W, Kendrick B (2011) The genera of hyphomycetes. CBS-KNAW Fungal Biodiversity Centre, Utrecht, 997 pp
Slippers B, Boissin E, Phillips AJL, Groenewald JZ, Lombard L, Wingfield MJ, Postma A, Burgess T, Crous PW (2013) Phylogenetic lineages in the *Botryosphaeriales*: a systematic and evolutionary framework. Stud Mycol 76:31–49
Stadler M, Kuhnert E, Persoh D, Fournier J (2013) The *Xylariaceae* as model example for a unified nomenclature following the "One Fungus–One Name" (1F1N) concept. Mycology 4:5–21
Summerbell RC, Gueidan C, Schroers H-J, de Hoog GS, Starink M, Arocha Rosete Y, Guarro J, Scott JA (2011) *Acremonium* phylogenetic overview and revision of *Gliomastix*, *Sarocladium*, and *Trichothecium*. Stud Mycol 68:139–162
Taylor JW, Jacobsen DJ, Kroken S, Kasuga T, Geiser DM, Hibbett DS, Fisher MC (2000) Phylogenetic species recognition and species concepts in fungi. Fungal Genet Biol 31:21–32
Tedersoo L, Arnold AE, Hansen K (2013) Novel aspects in the life cycle and biotrophic interactions in *Pezizomycetes* (Ascomycota, Fungi). Mol Ecol 22:1488–1493
Thambugala KM, Ariyawansa HA, Li Y-M, Boonmee S, Hongsanan S, Tian Q, Singtripop C, Bhat DJ, Camporesi E, Jayawardena R, Liu Z-Y, Xu J-C, Chukeatirote E, Hyde KD (2014) Dothideales. Fungal Divers 68:105–158
Wijayawardene DNN, McKenzie EHC, Hyde KD (2012) Towards incorporating anamorphic fungi in a natural classification – checklist and notes for 2011. Mycosphere 3:157–228. doi:10.5943/mycosphere/3/2/5/157

Wijayawardene NN, Crous PW, Kirk PM, Hawksworth DL Dai D, Boehm E, Boonmee S, Braun U, et al (2014) Naming and outline of *Dothideomycetes* – 2014. http://www.fungaltaxonomy.org/files/6813/9241/1345 Naming_and_Outline_of_*Dothideomycetes*_2014.pdf

Wingfield MJ, de Beer ZW, Slippers B, Wingfield BD, Groenewald JZ, Lombard L, Crous PW (2012) One fungus, one name promotes progressive plant pathology. Mol Plant Pathol 13:604–613

Woudenberg JHC, Groenewald JZ, Binder M, Crous PW (2013) *Alternaria* redefined. Stud Mycol 75:171–212

Zhang N, Rossman AY, Seifert K, Bennett JW, Cai G, Cai L, Hillman B, Hyde KD, Luo J, Manamgoda D, Meyer W, Molnar T, Schoch C, Tadych M, White JF Jr (2013) Impacts of the International Code of Nomenclature for algae, fungi and plants (Melbourne Code) on the scientific names of plant pathogenic fungi (Online) (APSnet Feature). American Phytopathological Society, St. Paul, MN

# Chapter 3
# Future Perspectives and Challenges of Fungal Systematics in the Age of Big Data

**Zheng Wang, R. Henrik Nilsson, Timothy Y. James, Yucheng Dai, and Jeffrey P. Townsend**

## Fungal Diversity and Systematics

*The beginning of wisdom is to call things by their proper name* (Chinese proverb: 名正言顺)

Z. Wang, Ph.D. (✉)
Department of Biostatistics, Yale School of Public Health, Yale University,
135 College Street, New Haven, CT 06510, USA
e-mail: wangzhengff@gmail.com; wang.zheng@yale.edu

R.H. Nilsson, Ph.D.
Department of Biological and Environmental Sciences, University of Gothenburg,
Box 461, Göteborg 405 30, Sweden
e-mail: henrik.nilsson@bioenv.gu.se

T.Y. James, Ph.D.
Department of Ecology and Evolutionary Biology, University of Michigan,
1147 Kraus Natural Science Building, 830 North University, Ann Arbor,
MI 48109-1048, USA
e-mail: tyjames@umich.edu

Y. Dai, Ph.D.
Institute of Microbiology, Beijing Forestry University, Beijing 100083, China
e-mail: yuchengd@yahoo.com

J.P. Townsend, Ph.D. (✉)
Department of Biostatistics, Yale School of Public Health, Yale University,
135 College Street, New Haven, CT 06510, USA

Department of Ecology and Evolutionary Biology, Yale University,
165 Prospect Street, New Haven, CT 06520, USA

Program in Computational Biology and Bioinformatics, Yale University,
165 Prospect Street, New Haven, CT 06520, USA
e-mail: jeffrey.townsend@yale.edu



D.-W. Li (ed.), *Biology of Microfungi*, Fungal Biology,
DOI 10.1007/978-3-319-29137-6_3

The mission and agenda of fungal systematists are to discover, describe, and inventory the global species diversity of one of the most diverse groups on earth. The circumscription of the fungi has evolved over time. Fungi are most closely related to animals and share a more recent common ancestor with them than with all other major groups of eukaryotes. The majority of the fungal kingdom is composed of heterotrophic, non-photosynthetic eukaryotes with cell walls containing chitin and β-glucans and, when present, a single flagellum (Stajich et al. 2009). Fungi can occur both as single-celled and multicelled organisms and can reach sizes typically associated with plants and animals. For example, the largest single fungal fruiting body on record was found to be nearly 500 kg in weight (Dai and Cui 2011), and the oldest and largest mycelium described covers 15 ha of area and is over 1500 years old (Smith et al. 1992). Life cycles of many fungi include a vegetative growth phase that spreads throughout its environment by extension of hyphae and/or release of a large number of asexual spores from simple structures and by a more complex, transient sexual phase producing smaller numbers of resistant sexual spores from well-developed fruiting bodies. Fungal diversity is estimated to comprise 1.5–7.1 million species. An increasing number of new taxa continue to be reported worldwide (Blackwell 2011; Bass and Richards 2011), and fungi have been isolated from almost all kinds of ecosystems on Earth (Stajich et al. 2009). This fungal diversity is described by systematics, which is the science not only of naming fungi but also of positioning the species among other existing names to represent their evolutionary relationships. To properly describe the substantial diversity of the Kingdom Fungi, mycologists have been updating its classification and systematics, based on accumulated knowledge of fungal biology interpreted within new concepts and approaches that are emerging from evolutionary biology.

In contrast to large aboveground organisms that can be easily spotted and counted, fungi are major components of underground diversity. Their study is often made difficult by their microscopic structures and shortage of discriminatory morphological characters. Traditional biological information used for classifying fungi into major groups includes morphology, ultrastructure, physiology, tissue biochemistry, and ecological traits. Early synthesis of this information yielded major fungal groups that have remained comparatively stable over a very long time period in the twentieth century. Some morphological and ecological traits, such as the structure of the cell wall and hyphal septa, sexual reproduction and meiotic spores, nutritional modes, as well as geographic distribution, have proven to be relatively conserved and informative, especially for high-level classification. However, phenotypic plasticity of traits and fast-evolving traits have caused considerable uncertainty regarding lower-level phylogenies based on morphology and ecology (Lutzoni et al. 2004). Starting in the 1970s, but gaining momentum in the late 1990s, the use of DNA sequence data to infer phylogenetic relationships among fungal lineages brought about a revolution in terms of taxonomic resolution and scientific reproducibility (de Bertoldi et al. 1973; Bruns et al. 1991; Bridge et al. 2005; Blackwell et al. 2006). Initial molecular studies, typically based on a single gene region, were followed by a wave of multilocus phylogenetic studies including all major fungal groups. The new phylogenies facilitated several major taxonomic revisions, including new lineages at the phylum and class level (James et al. 2006a; Hibbett et al. 2007; Kirk et al. 2008; Schoch et al. 2009a, b; Rosling et al. 2011; James and Berbee 2012; Matheny et al. 2007). More changes and many new taxa were added to lower-level fungal groups. In addition, much novel diversity was revealed in sequence data collected from environmental samples and identified as operational taxonomic units (OTUs) (Blaxter et al. 2005). The quantity of novel OTUs in most environmental samples hints at a massive, inconspicuous, undescribed, and thriving fungal diversity (Hibbett et al. 2011). Classifying and naming this huge fungal diversity is a necessary step toward understanding the functions of these fungi in the ecosystems. Thus, finding ways to take full advantage of the power afforded by next-generation sequencing approaches to integrate environmental DNA sequences has become one of the major challenges for fungal systematics. Simultaneously, a complementary aspect of the future of the fungal systematics is the integration of systematics, the evolution of complex traits, and functional genomics to understand the comparative biology of fungi and to create a holistic view of the fungus and how it evolves.

Currently, efficient communication regarding fungal species rests upon on the use of scientific names constructed based on a system of hierarchical ranks. Within this system, one of the major purposes of fungal taxonomy and systematics is to create and position nomenclatural units unique for each fungal species. While a stable name as a symbol for communication is always appreciated by researchers—especially for the widely used industrial, medical, plant pathogenic, and model species—systematics must also continue to refine and revise the application of names to reflect continual gains in knowledge about the evolutionary histories of all taxa. We make no attempt here to cite all papers on development of fungal taxonomy and systematics nor to summarize recent systematic progress within and among the major fungal phyla. Instead, we have chosen to highlight recent research that enables us to illustrate specific points about perspectives and challenges of fungal systematics in the age of big data.

## Integrative Taxonomy and Current Fungal Systematics

Traditionally, morphological and sometimes ecological traits have been used to classify fungi into hierarchical ranks and groups. However, evolutionary relationships derived from these traits, whose ontology is often inferred from a phylogenetic hypothesis, can be problematic, especially for lower-level phylogeny, where diverse fungal groups can have plesiomorphic or convergent morphologies. One problem in reconstructing fungal evolutionary history is a lack of paleontological information due to the scarcity of well-preserved fungal fossils (Bidartondo et al. 2011). This scarcity makes it extremely difficult to evaluate the evolutionary history of morphological traits for fungal systematics, especially for morphologically simple groups. The meager fossil record also makes it difficult to precisely calibrate molecular phylogenies. Nevertheless, information on molecular evolutionary events, such as mutation and gain or loss of nucleotide characters, has been well preserved in gene sequences. Molecular phylogenies using single genetic markers or multilocus data have led to dramatic advances in the systematics of a range of taxonomic levels within the fungal kingdom over the past three decades. However, systematic hypotheses based on molecular phylogenetic data alone can be questioned, especially when in conflict with morphological evidence. Ideally, evidence from different lines, such as morphology, ecology, and molecular data, can be evaluated jointly to robustly define taxa at all ranks. This approach has been called integrative taxonomy and has been advocated by Will et al. (2005) and Pante et al. (2015).

A major driver of new advances in the molecular phylogeny of fungi was the Assembling the Fungal Tree of Life (AFTOL) project, funded by the National Science Foundation (NSF) of the United States and organized by mycologists at several leading laboratories. This project sprung out of an NSF-funded research coordination network known as Deep Hypha and culminated in significant gains in the study of evolution and molecular phylogenetics of fungi (Lutzoni et al. 2004; Blackwell et al. 2006; James et al. 2006a; Hibbett et al. 2007; Schoch et al. 2009a). Among the very first multilocus phylogenies targeting the majority of major fungal lineages, Lutzoni et al. (2004) highlighted two major challenges in fungal systematics in the molecular age. One is achieving a balanced sampling of taxa and genetic markers. The other is identifying and interpreting inconsistency between the evolution of morphology and molecular phylogeny. When standard PCR using degenerate primers and Sanger sequencing were the major tools for recovering DNA sequences from fungal tissue, loci such as nuclear and mitochondrial rRNAs and several widely used protein-coding genes, including subunits of elongation factors and RNA polymerases, were selected by the AFTOL project. A six-gene phylogeny using these markers, including data from 52 sequenced fungal genomes, was generated to assess early evolution of fungi, and ecological characters were mapped onto the tree.

Groups recognized in the six-gene phylogeny were generally consistent with traditional views of fungal systematics prior to the molecular systematic age, but only for the fungi in *Dikarya* (James et al. 2006a). Non-monophyly of two of the six recognized phyla led to the abandonment of one (*Zygomycota*) and the description of two new phyla *Blastocladiomycota*, by James et al. (2006b), and *Neocallimastigomycota*, by Hibbett et al. (2007). Similar sequence datasets, which were often incomplete with missing sequences, were generated for a more inclusive taxon sampling within each major fungal group at class level, and the resulting phylogenetic classifications were collected in the special Deep Hypha issue of Mycologia in 2006 (Blackwell et al. 2006). A comprehensive phylogenetic classification of the fungi kingdom was later proposed by Hibbett et al. (2007), featuring 16 new taxa above the level of order. This classification was adopted by the latest version of the Dictionary of the Fungi (Kirk et al. 2008). A more taxonomically complete six-gene dataset for 420 ascomycetes was subsequently assembled and analyzed. Key morphological and ecological characters were evaluated for usefulness in ascomycete systematics, and a new class was differentiated for two earthtongue genera: *Geoglossum* and *Trichoglossum* (Schoch et al. 2009a, b). This dataset made it possible to quantify phylogenetic informativeness (Townsend et al. 2008; Townsend and Lopez-Giraldez 2010; Lopez-Giraldez and Townsend 2011) for several widely used genetic markers (Schoch et al. 2009a, b). With the release of more fungal genome sequences and the ever-growing availability of data from additional genetic markers, several multilocus phylogenies inferred using partially or solely from genomic data (phylogenomics) have been published (Ebersberger et al. 2012; Binder et al. 2013; Ortiz-Santana et al. 2013; Dutilh et al. 2007). Updated classifications for major fungal groups were collected in Mycota VII—Systematics and Evolution (McLaughlin et al. 2014).

The vast majority of molecular systematic studies of fungi have been based on annotated (voucher) specimens of primarily sexual (teleomorphic) but also asexual (anamorphic) collections. The accuracy of voucher specimens is particularly important now, because in many modern studies, only molecular data are shared and examined: fungal herbaria thus play important roles in keeping records for well-annotated specimens (Bidartondo 2008; Schoch et al. 2014). Best-practice guidelines on how to appropriately use molecular data in mycology are readily available (Lindahl et al. 2013; Hyde et al. 2013; Nilsson et al. 2012). Nevertheless, these guidelines are not sufficiently frequently adhered to fungal molecular phylogeneticists. Well-preserved and annotated collections are now mandatorily required by journals for newly published morphological and molecular data (Seifert and Rossman 2010). However, there has been no guarantee of accurate identification of fungal collections, especially for microfungi, partially due to the problematic outcomes of applying species concepts in fungi.

Morphological, biological, or phylogenetic species concepts all have limitations when they are applied to fungal species (Taylor et al. 2000, 2006). In particular, different mycologists often have different quantitative or qualitative interpretations of data used to define species boundaries. For example, using several genetic markers, multiple species were identified within the single morphological and biological species commonly known as the "turkey tail" fungus *Trametes versicolor* (Carlson et al. 2014), and two species were recognized for North American *Heterobasidion annosum*, which has been considered one of the most important forest pathogens in the world (Otrosina and Garbelotto 2010). Another extraordinary and exciting example would be that of the morel fungi, for which tens, if not hundreds, of new species have been recognized within several original common names (Du et al. 2012; Richard et al. 2014). An increasing number of low-level classifications are based on integrative approaches using both morphological and molecular data. These approaches have been applied to solve identification issues of several commercially important fungi (Cao et al. 2012; Wu et al. 2014; Zhang et al. 2005).

In many cases, the reference molecular data are directly downloaded from various databases, assuming accurate identification without checking the resource specimens. Cryptic species complexes are particularly likely for many species of microfungi, in which case, dense samples from accurately annotated specimens will be especially critical for proper species taxonomy. However, phylogenetic recognition of fungal species has proved to be reliable, reproducible, and increasingly widely applicable, facilitating convenient naming of species or strains, especially for microfungi. The huge undisclosed fungal diversity and the difficulty of reconciling species concepts in fungi can make the application of the International Code of Nomenclature (McNeill and Turland 2012) very challenging—to the extent that it can ironically slow down, rather than speed up, mycological progress. Recently, for instance, instead of following the code to use the teleomorph genus name for monophyletic groups, mycologists advocated recognizing the genus *Fusarium* as the sole name for groups that have been studied under that name but are not monophyletic (Geiser et al. 2013). Such challenges will become more significant as more invisible diversity is discovered within diverse environmental samples. These challenges should aid the community in pushing for the development of standards for sequence-based classification (Hibbett and Taylor 2013). A recent review of the impacts of the nomenclatural code on the scientific names that have been adopted is available for plant pathogenic fungi (Zhang et al. 2013).

## Systematics and Classification for Invisible Diversity

Fungi are widely distributed in all terrestrial and aquatic ecosystems. About 100,000 fungal species have been discovered and documented. They play critical roles in inorganic and organic nutrition, nutrient cycling, and especially in the decay of carbon compounds that were fixed and integrated into complex compounds by plants. Furthermore, fungi are frequently intimate partners in coevolving biotic and trophic relationships with other organisms, notably through mycorrhizal associations with plants; almost all land plants form symbiotic associations with mycorrhizal fungi (Stajich et al. 2009; van der Heijden et al. 2015). However, only a small portion of the total fungal diversity has been documented based on specimens/strains deposited in herbaria, culture collections, or in personal collections all over the world. Indeed, a modest ~1000 new species are described per year (Hibbett et al. 2011), which would require 5000 years of cataloging at this rate, should the 5.1 million estimate of species diversity hold.

The challenges to description of this undescribed fungal diversity are threefold. First, there are few mycological researchers and little research to study this undescribed diversity. Second, many of these undescribed species whose morphology can be characterized are actually cryptic species hidden within species previously described on the basis of morphological characters; morphological characters might not separate the genetic species, as discussed for *Trametes versicolor* and *Morchella* spp. above. Third, the majority of the extant fungal diversity produces no distinguishing morphological structures that are visible or describable, e.g., these fungi carry out their lives mostly or entirely as unculturable and morphologically indistinguishable yeasts or vegetative hyphae that cannot be described formally. If these fungi are unculturable as well as morphologically and biochemically indistinguishable, only can molecular identification be used as a tool to classify this potentially huge diversity. This kind of molecular-only identification leads to the absurd situation where next-generation sequencing efforts of environmental substrates reveal the existence of thousands upon thousands of new species of very high relevance to phylogenetic and ecological characterization of the fungal kingdom—and yet this huge diversity of species cannot be described. This inability to describe these species effectively excludes them from further scientific scrutiny. Such sequences are typically submitted to sequence databases labeled as "uncultured fungus," making unambiguous reference to those species across datasets and studies problematic at best. This lack of linkage across studies in turn makes it difficult to assemble data for these species; what countries, hosts, and substrates these individual, unnamed species are known from cannot easily be compiled. In turn, this lack of synthetic inferential power further complicates the eventual formal description of these species.

The UNITE database for molecular identification of fungi recently presented a solution to this problem (Kõljalg et al. 2013). All fungal ITS sequences are clustered to approximately the species level based on sequence similarity, and each such OTU—called a *species hypothesis*—is assigned a unique, stable name of the accession number. Thus, regardless of whether the OTU has a formal Latin name or not, unambiguous reference across publications—as well as data assembly for individual species—is possible and even automated. A recent study, based on 365 soil samples collected from across the globe, identified 80,486 fungal OTUs and used the UNITE species hypothesis system to analyze them. Although a very modest 4353 of the OTUs could be linked to highly similar reference sequences from herbarium specimens or described culture collections, the underlying sequences of the full results of the study are now integrated in UNITE for standardized reference (Tedersoo et al. 2014; Wardle and Lindahl 2014). At the time of this writing, GenBank has a collection of more than 600,000 fungal sequences from environmental samples, chiefly the nuclear ribosomal internal transcribed spacer (ITS) region. Among these, there are about 200,000 that have been identified as stemming from an "uncultured fungus," without an affiliation to any existing ranks.

It is hard to estimate how inclusion of this huge invisible diversity would affect the fungal systematics that so far encompassed only just over 100,000 accepted fungal species. Despite the challenges, it is clear that not including these extant but unnamed species in molecular studies of fungi and fungal communities is detrimental to mycology. Nilsson et al. (2011) examined the topological effects of including such environmental sequences in phylogenetic analyses that featured only sequences from vouchered fruiting bodies and cultures. Their inclusion made a significant difference to the inferred topology and to the support of internodes. Similarly, the relatively recent realization that aquatic ecosystems abound with uncharted fungal diversity, particularly in the *Chytridiomycota* and *Cryptomycota*, could provide taxonomic sampling that might provide resolution of this part of the fungal tree of life, which has been plagued by low resolution and poor branch support (Wurzbacher and Grossart 2012; Ishii et al. 2015). Recently a whole new class, *Archaeorhizomycetes*, comprising hundreds of cryptically reproducing culturable filamentous fungi of poorly understood ecology, has been discovered from soil samples (Rosling et al. 2011). Using multilocus analyses, they have been phylogenetically placed into the species-poor group *Taphrinomycotina* of the *Ascomycota*. The recognition of the *Archaeorhizomycetes* represents a major step forward in our understanding of soil fungi, as these fungi seem to be common in soil samples throughout the world (Porter et al. 2008; Rosling et al. 2013). At an even higher rank, the new fungal phylum *Cryptomycota*, rich particularly in aquatic environments, is also known almost exclusively from environmental DNAs (James and Berbee 2012; Jones et al. 2011). The systematics of the *Archaeorhizomycetes* and *Cryptomycota* will remain hindered by the absence of complete genome sequences, which will be challenging to obtain from these minute fungi. On the other hand, recent advances in obtaining near-complete genome sequences from single cells hold promise for both placing uncultured fungal lineages on the tree of life and for inferring their ecological roles (Rinke et al. 2013).

For the majority of fungal lineages, ITS sequences provide a powerful and efficient means of identification. Therefore, the ITS has been proposed and accepted as a universal DNA barcode marker for fungi (Schoch et al. 2012). A DNA barcode, however, is nothing more than a sequence that can be unambiguously linked to a taxonomic label for a species. DNA barcodes do not promise a solution for nomenclatural classification of diversity. Such a solution might arise from digital codes such as PhyloCode (de Queiroz and Gauthier 1994). However, this concept still lacks a standardized real-life implementation (de Queiroz and Gauthier 1994; De Oliveira Martins et al. 2014; Money 2013). While ITS is generally considered as only informative for species recognition and low-lever phylogenetic analysis, classification of the environmental diversity typically relies on observations of high sequence similarity to reference sequences from annotated specimens (Schoch et al. 2014). However, with the use of new tools to address some serious alignment issues regarding the ITS region (Liu et al. 2009, 2012), ITS alignments have shown promise in use for intermediate-level phylogeny (Koetschan et al. 2010), providing comparable classification accuracy to some other frequently used gene markers, such as the large subunit of rRNA sequence (Wang et al. 2011). Including proper reference sequences would provide insights into evolutionary history and ecology for these so-called invisible fungi (Wang et al. 2011; Porras-Alfaro et al. 2014; Del Olmo-Ruiz and Arnold 2014). Automatic phylogenetic approaches, such as those implemented in MOR (Hibbett et al. 2005) and WASABI (Kauff et al. 2007) would be able to efficiently filter and classify environmental sequence data. Still, there might be many environmental species that have no comparable characterized lineages, such that they cannot be morphologically defined or easily systematically positioned. Moreover, the absence of barcodes of the ITS region associated with this phylum is also an impediment, as many barcodes that cannot be assigned to a phylum may belong to these poorly sampled basal lineages, which exist in databases primarily as 18S rDNA sequences. To incorporate these taxa into fungal systematics requires developing methods for gathering informative sequence data that link barcodes to darker regions of the fungal phylogeny and performing efficient phylogenetic analysis on large datasets.

Given the deep divergence of the major fungal lineages, plodding through taxa using PCR with degenerate primers to fish for loci is a challenging, if not impossible, approach toward recovering an effective diversity of protein-coding genes that will prove informative for deep phylogeny. Moreover, establishing linkages among multiple independent genes that derive from the same OTU defined from environmental DNA is nearly impossible at present. Thus, with the development of single-cell genome sequencing, phylogenomic approaches might provide an alternative and more powerful means to reconstruct a systematics of both the visible and the invisible fungal diversity.

## Fungal Genomes, Phylogenomics, and Phylotranscriptomics

The very first sequenced fungal genome was also the first sequenced eukaryotic genome: that of the wine yeast *Saccharomyces cerevisiae*, an important genetic model and an industrial workhorse. This comparatively small genome was published in 1996 (Goffeau et al. 1996). Since then, following the technical progress in genome sequencing, fungal genomes have been released at an ever-accelerating rate. The number of available fungal genome sequences has increased by another order of magnitude (Galagan et al. 2005). In GenBank (http://www.ncbi.nlm.nih.gov) alone, there are currently fungal genomes representing 451 species. The recently launched 1000 Fungal Genomes (1KFG) project (http://1000.fungalgenomes.org) plans to sequence representatives from more than 650 recognized families of fungi (Kirk et al. 2008; Hibbett et al. 2013). The released genomes facilitate assembly of closely related genomes against the reference genomes even in small laboratories, and the sampled genomes of closely related organisms are designed to enable comparative studies. Comparative genomics of closely related organisms can provide a powerful approach to ascertain the genetic basis of diverse phenotypes, such as fungi-host associations, secondary metabolic pathways, morphological development, and fungal responses to environmental signals (Galagan et al. 2005; Hibbett et al. 2013; Sikhakolli et al. 2012; Andersen et al. 2011; Lehr et al. 2014; Nishant et al. 2010; Rodriguez-Romero et al. 2010; Heitman 2007). Many comparative genomic studies focus on the biology and evolution of model fungi to make inferences about basic biological processes in all eukaryotes. Studies that analyze genomes in the context of their phylogenetic and evolutionary relationships are accelerating research into the fundamental aspects of eukaryotic biology. As stated in Delsuc et al. (2005) "…nothing in genomics makes sense except in the light of evolution."—large numbers of genomes alone do not provide much insight into organismal biology, however. Many features of genomes need to be related to organismal knowledge and understood in the context of their evolutionary history.

How can these fungal genomes empower fungal systematic research? The genome itself comprises all informative genetic markers that could be sampled for any individual. Access to this scale of genomic data for phylogenetic purposes could potentially alleviate previous and present problems of phylogenetics that arise from insufficient or biased sampling of genetic markers. With this massive increase of potentially useful characters, the focus of phylogenetic inference must shift toward development of new methodologies that can efficiently, accurately, and reliably handle big data and toward approaches that facilitate a powerful sampling of taxa (Philippe et al. 2011). Basic approaches and future challenges in phylogenomics toward reconstruction of the larger tree of life were addressed 10 years ago (Delsuc et al. 2005), and phylogenomic approaches and tree reconstruction methods have been tested using different sets of fungal genomic data (Ebersberger et al. 2012; Dutilh et al. 2007; Medina et al. 2011). Development of phylogenomic approaches for fungal phylogenetic inference has been addressed recently (Hibbett et al. 2013; Taylor and Berbee 2014) and is beyond the scope of this review. Current genome projects have sampled representative taxa in major lineages across fungal kingdom, providing extensive datasets for resolving relationships between major lineages of higher fungi. The current genomic projects might provide sufficient taxon sampling to resolve some of the unsolved polytomies within *Basidiomycota* and *Ascomycota*, as summarized in Hibbett et al. (2007). However, to resolve the phylogeny of the earliest fungal lineages, it is already clear that densely sampled genomes and the development of novel culture-independent methods will be critical. Recent phylogenomic analyses support the supergroup *Opisthosporidia* (*Micro sporidia* + *Cryptomycota* + *Aphelida*) as the basal branch of all sequenced fungi (Capella-Gutierrez et al. 2012; Haag et al. 2014; James et al. 2013; Karpov et al. 2014). This group is known to be highly diverse on the basis of environmental DNA studies (Jones et al. 2011; Karpov et al. 2014) and also completely unculturable in the absence of a host. Sufficient sampling of genomes is also important for understanding divergence and recent adaptation among very closely related species, especially to reveal cryptic species and enable genome-wide population studies (Ellison et al. 2011; Park et al. 2011; Padamsee et al. 2012; Neafsey et al. 2010). Taking advantage of next-generation sequencing techniques, genome-wide expressed mRNA sequences can be easily generated without previous knowledge of genome sequence or of specific gene regions. Phylotranscriptomics, the use sequences of expressed messenger RNA sequences to infer phylogeny, has been shown to be a promising approach to infer phylogenies in several non-fungal groups (Breinholt and Kawahara 2013; Wickett et al. 2014). Similar applications in the fungal kingdom are certainly looming on the horizon.

Despite increasing sequencing capacity, it remains the case that for the majority of fungal species, genome-scale sequence is unlikely to be available soon. In most of these cases, a multilocus phylogeny is now realistically affordable and is expected to be informative enough for most systematic questions about these taxa. However, previously used genetic markers for phylogenetic analysis were originally identified by a trial and error process based on very limited data and often subsequently sequenced in other taxa solely motivated by the desire for completion of particular datasets. Thus, the phylogenetic usefulness of some genetic markers can be far from optimal (Robert et al. 2011). Sequenced genomes make it possible to assess the potential phylogenetic utility of many genetic markers as well as to enable more successful primer design and PCR efficiency (Ye et al. 2012). Knowledge regarding gene ontology and substitution rates is also critical for selecting proper markers for resolution of divergences occurring on diverse time scales during disparate epochs.

Approaches for selecting robust sets of phylogenetic markers based on sequenced genomes are starting to emerge and are urgently needed. For example, ranking genes for their usefulness in phylogenetic inferences showed promise as a means of solving phylogenies for some problematic fungal groups (Schoch et al. 2009a; Binder et al. 2013; Robert et al. 2011; Hyde et al. 2014; Capella-Gutierrez et al. 2014).

## Experimental Design and Analysis for Systematics Using Genome Data

Phylogenetic inference can be improved either by use of better models or by obtaining better data. For phylogenetic problems corresponding to short, deep internodes, quality of data is often the limitation to successful resolution (Townsend et al. 2008; Philippe et al. 2011; Su et al. 2014). Early fungal phylogenetic research expanded the repertoire of genetic markers beyond the common rRNA markers by testing and developing gene markers that had been found useful in other organisms. The first AFTOL project selected six markers to sample from major fungal groups after attempting to widely amplify more than 10 markers (Lutzoni et al. 2004; James et al. 2006a; Hibbett et al. 2007; Matheny et al. 2007; Liu and Hall 2004). Testing a small number of genetic markers on a small number of taxa using degenerate PCR amplification is laborious but feasible; however, its use for evaluating a genome-scale pool of genes for diverse taxonomic sampling would be infeasible. Identifying the most informative candidate loci across the genome in advance can provide a prioritized list for identification by degenerate PCR of novel promising markers or for use in deciding on reference gene sets for genome-scale targeted capture methodologies (e.g., Li et al. 2013). By adopting relaxed assumptions regarding the model of molecular evolution and deriving theory based on asymptotic interest in resolving short deep internodes of four taxon trees, a method for profiling phylogenetic informativeness over time of diverse gene markers was developed (Townsend 2007) and applied to the task of identifying better markers during the second AFTOL project (Schoch et al. 2009a; Townsend et al. 2008).

This theory was generalized to resolve nodes based on rates of evolution of individual characters or sets of characters onto the molecular evolutionary or chronological time scale of interest, weighing the accumulation of signal with internode length versus the loss of signal on subtending branches of the phylogenetic tree (Taylor and Berbee 2014; Su et al. 2014; Townsend et al. 2012; Feau et al. 2011; Miadlikowska et al. 2014; Walker et al. 2012). Binder et al. (2013) performed a thorough analysis of candidate loci to identify optimal experimental design for resolution of phylogenetic hypotheses. In this comprehensive study, among 356 single-copy genes, 25 markers ranked at the top for phylogenetic informativeness and probability to resolve key epochs were selected to resolve the problematic phylogeny of wood-decay fungi. As demonstrated in that study, gene markers selected from sequenced genomes should be evaluated for their site rate distributions, phylogenetic informativeness, and predicted signal and noise. Markers then can be quantified for predicted utility compared to the worst possible performance or random sampling of taxa and genes. For a given phylogenetic hypothesis, the process should rank additional taxa whose genome sequences would provide the most power for resolving these nodes and then predict which taxon-gene elements of a presumed data matrix would provide the most power for resolving these nodes. The result minimizes the effort for resolving the given nodes (and simultaneously minimizes the probability of error) by assessing phylogenetic performance for top taxon-gene combination until a robust phylogeny is reached.

The advent of big data in phylogenomics has brought renewed attention not only to issues of phylogenetic signal but also to issues of phylogenetic noise and bias (Townsend and Lopez-Giraldez 2010; Lopez-Giraldez and Townsend 2011; Lopez-Giraldez et al. 2013). In data-limiting analyses, it was always possible to quiet concerns about the relative efficacy of some data over other data with a plaintive call for more data. In the genomic era, with the availability of big data, due to known issues such as inconsistency of substitution rates, horizontal gene transfer, and unclear gene ontology, it has become clear that big data results bulwarked by the traditional hallmarks of strong support are sometimes in conflict with each other (Salichos and Rokas 2013). The resolution of this conflict requires rigorous thought about the sources of noise and consequently the relative power of data to address phylogenetic hypotheses. At the same time, the growing resource of publicly available sequenced genes and genomes should in principle provide some guidance as to how to optimally design a phylogenetic sequencing study. For example, genes can be chosen from sequenced genomes of known phylogeny and then ranked for their performance in accurately inferring phylogenetic relationships—this approach is an extension of the practice of traditional marker selection facilitated by automatic computer programs (Capella-Gutierrez et al. 2014). Performance of these analyses is facilitated by the web application PhyDesign (Fig. 3.1) (Lopez-Giraldez and Townsend 2011). PhyDesign evaluates gene performance based on sequence alignment and a chronogram to predict signal and noise and the best-possible performance, where the metric of interest is the amount of support provided for the given nodes. Providing a means for prioritizing gene sequencing and taxon sampling and for sorting the "wheat from the chaff" in large phylogenomic studies, this application of the theory for phylogenetic study design would robustly improve the scope of data collection and analysis, the overall cost-effectiveness, and the probability of correct inference of a phylogenetic study. In addition, phylogenetic inferences are increasingly required to be robust to differential gene divergence under the multispecies coalescent, necessitating informed choices not only on what genetic markers to employ but also on what analysis approaches to take (Hyde et al. 2013).

Theoretical tools are still needed to address long-standing controversies in experimental design that have occasionally engendered contentious academic debate, including (1) the power of different genetic markers, (2) the relative utility of taxon sampling versus gene sampling, (3) the differentiation between soft and hard polytomies, and (4) the design of taxonomically dense phylogenetic studies

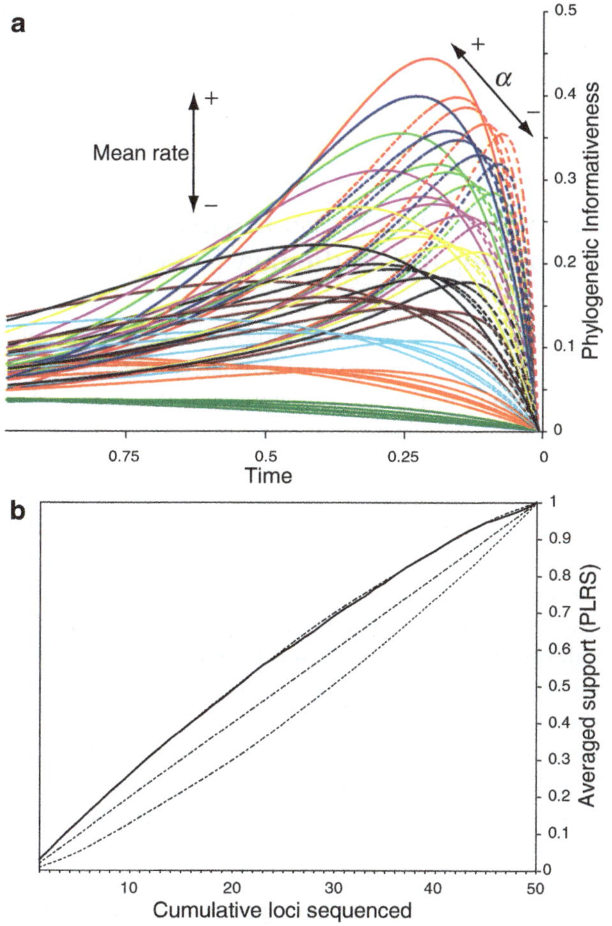


**Fig. 3.1** Utility to phylogenetics of extraction of informative genes from genome sequence data. (**a**) Phylogenetic informativeness can be estimated and compared among different genes for given epochs (modified from López-Giráldez et al. 2013). Phylogenetic informativeness profiles for simulated sequence alignments on a single molecular evolutionary unit. Each of the ten different colors represents a different mean rate, from 0.0001 (slowest, bottom) to 0.001 (fastest, top) substitutions per site per time unit. *Dashed lines* are profiles from alignments simulated with gamma rate heterogeneity ($\alpha$=0.5, 1, 2, and 3). (**b**) Cumulative proportionate likelihood-ratio support (PLRS), averaged across nodes for simulated amino acid datasets. Genes are ranked by differential phylogenetic informativeness encompassing all branches in the tree. The *upper dashed line* and the *lower dashed line* separately represent cumulative PLRS when loci are prioritized, post hoc, from highest to lowest PLRS values and from lowest to highest PLRS values. The *intermediate dashed line* is the hypothetical average one would achieve sampling at random from loci available. The *solid line* is the performance when genes are selected by their phylogenetic informativeness based on inferred rates of sites only

optimized by taxonomically sparse genome-scale data (Lopez-Giraldez et al. 2013; Moeller and Townsend 2013). A robust fungal phylogeny would provide a solid framework for fungal systematics that would, in turn, be of increasing significance in modern mycological research.

## Fungal Systematics in the Future: Integration of Fungal Systematics and Fungal Evolutionary Biology

Systematics is fundamental to organismal biology and is the discipline that synthesizes achievements from all of biology and ultimately underlies all research in evolutionary biology. Arising in part from systematics, the theory of evolution is the basis of modern biology. A robust phylogeny and reliable classification is the first step for the development of fungal systematics, and systematists should not be satisfied only with describing the evolutionary history of fungal lineages (Hibbett and Taylor 2013). More importantly it is our responsibility to qualitatively and quantitatively explain how this history led to the diversity we observe today, a question that brings us to the integration of systematics and evolutionary biology. In fact, from taxonomy, diversity, molecular phylogeny, to the tree of life, the study of systematics of all organism groups has itself been evolving, and new contents from evolutionary biology have been continually if controversially incorporated into modern systematics (Losos et al. 2013).

Fungal systematics is critical for understanding the evolution of genes and their functions in fungal genetics, and multigene analysis provides an opportunity to avoid the pitfalls associated with assuming a single-gene phylogeny represents a true species phylogeny. Genetics has long focused on gene behavior and function within species, especially for model organisms, until recently the availability of sequenced genomes and robust fungal phylogeny made data available to trace gene ontology among different lineages within a long evolutionary history. Like many other eukaryotic organisms, horizontal gene transfer and gene/genome duplication are main contributors for new genes and gene functions in many fungal species (Bruto et al. 2014; Cohen-Gihon et al. 2011; Fitzpatrick 2012; Wapinski et al. 2007), and horizontal transfer of toxic gene clusters among fungal species was discovered based on sequenced fungal genomes across lineages of fungal tree of life (Slot and Rokas 2011; Wisecaver et al. 2014). For many fungi, the dominant form of their life history is haploid, and mitotic and meiotic recombination can happen via parasexual and sexual reproductions in fungal species (Schoustra et al. 2007; van de Vondervoort et al. 2007). Thus, the reconciliation of gene phylogeny and species phylogeny in low-level species taxonomy in fungi could provide insights into the modeling of speciation events (Taylor et al. 2000).

Fungal systematics and genome-enabled mycology are linked through evolutionary biology. Sequenced genomes provide a huge amount of data that can be brought to bear on all branches of fungal research. Recent progress has been especially interesting in efficiently addressing the genetic basis of various phenotypes (Hibbett et al. 2013; Taylor and Berbee 2014). Genomic research based on fungal models, such as *S. cerevisiae*, *Neurospora*, and *Aspergillus* species, has been focused on fundamental biology with implications that extend toward many non-fungal branches of the tree of life, including meiosis, cell cycle, and internal oscillation (Galagan et al. 2005). In contrast, with an increase of released fungal genomes, genome-enabled mycology has emerged: early studies have focused either on specific ecology or on metabolic pathways or functional gene families and their evolution (Spanu et al. 2010; Vogel and Moran 2013; O'Connell et al. 2012; Stajich et al. 2010; Pel et al. 2007; Martin et al. 2010; Morin et al. 2012). In most of these early studies, fungal systematics generally serves not only as a guide for what taxa to sample and study independently but also as a reference for tracking gene history. With the expected robust phylogeny and well-sampled genomes that could come as an outcome from the 1000 Fungal Genome project, a reliable gene ontology should be inferred that would facilitate inference of how fungal morphologies and ecologies have evolved, knowledge of which is one of the overarching goals of fungal systematics. For example, one-celled (yeast) stages are distributed throughout the fungal kingdom, and comparative genomics has revealed that yeast forms arose early and independently in multiple fungal clades via parallel diversification of a fungal-specific transcription factor family involved in regulating yeast-filamentous switches (Nagy et al. 2014). Reliable gene ontology is critical to the reconstruction of gene networks and the assessment of gene functions, especially for systems biology investigation that attempt to answer how complexity can be developed while essential functions are maintained.

The importance of robust phylogenies to infer the evolution of fungal ecology is clear, but fungal systematics is also an essential component of any complete understanding of fungal ecology. Inorganic and organic components of the environment impose significant selection on fungal phenotypes (Tedersoo et al. 2014). Ecological factors, such as host types, nutrient resources, or geographic distribution, have long been considered characters that are important for fungal classification. With well-resolved molecular phylogenies, we could evaluate the role of ecology in fungal evolution, reconcile the ontology of specific gene function groups, and infer the genetic basis of ecological success. Recent discoveries on the evolution of wood decay among polypore species and mushroom-forming fungi have demonstrated how this strategy can work (Binder et al. 2013; Floudas et al. 2012; Eastwood et al. 2011). Applying principles of systematics to metagenomics makes it possible to monitor the dynamics of biological processes involving diverse fungal species on both spatial and temporal scales to understand the contributions of those fungal species to the process of interest. For instance, a study on global soil sampled by (Tedersoo et al. 2014) demonstrates direct and indirect effects of climatic and edaphic variables on plant and fungal richness. The National Science Foundation has launched a program called Dimensions of Biodiversity, which "takes a broad view of biodiversity and focuses on the intersection of genetic, phylogenetic, and functional dimensions of biodiversity."

Further extension of the broad impacts of fungal systematic research requires experienced mycologists with broad training in traditional fungal classification, molecular systematics, and bioinformatics/genomics. Mycologists have long been considered as naturalists. Training of fungal systematics has been provided in many institutes, especially in colleges or departments for plant pathology. Fungal classification and taxonomy training usually via monographic work require a lot of time in both field and laboratory, while molecular systematic training requires a decent facility for sequencing and/or computation. Significant computational needs are especially required for phylogenomics. Funding resources are heavily biased toward molecular research, leading to a scarcity of high-quality training in traditional fungal systematics, especially at the graduate level (Pearson et al. 2011). In the long run, the lack of well-trained mycological systematists would be a problem not only holding back the development of fungal systematics but also impeding many other research fields that rely on knowledge of fungal biodiversity and evolutionary biology. Well-trained mycologists are also critical for helping the public to understand the gaps between the quickly developing "omics" sciences and the long-developed traditional senses of fungi and fungal biology.

***The greatest challenge for fungal systematics has always been to be able to take disparate pieces of knowledge from diverse kinds of studies of fungi to make synthetic biological inference, and only in this way will fungal systematics be of maximum benefit to the whole community conducting research on fungi and the scientific community at large.***

## References

Andersen MR, Salazar MP, Schaap PJ, van de Vondervoort PJ, Culley D et al (2011) Comparative genomics of citric-acid-producing *Aspergillus niger* ATCC 1015 versus enzyme-producing CBS 513.88. Genome Res 21:885–897

Bass D, Richards TA (2011) Three reasons to re-evaluate fungal diversity 'on Earth in the ocean'. Fungal Biol Rev 25:159–164

Bidartondo MI (2008) Preserving accuracy in GenBank. Science 319:1616

Bidartondo MI, Read DJ, Trappe JM, Merckx V, Ligrone R et al (2011) The dawn of symbiosis between plants and fungi. Biol Lett 7:574–577

Binder M, Justo A, Riley R, Salamov A, Lopez-Giraldez F et al (2013) Phylogenetic and phylogenomic overview of the Polyporales. Mycologia 105:1350–1373

Blackwell M (2011) The fungi: 1, 2, 3 … 5.1 million species? Am J Bot 98:426–438

Blackwell M, Hibbett DS, Taylor JW, Spatafora JW (2006) Research coordination networks: a phylogeny for kingdom fungi (Deep Hypha). Mycologia 98:829–837

Blaxter M, Mann J, Chapman T, Thomas F, Whitton C et al (2005) Defining operational taxonomic units using DNA barcode data. Philos Trans R Soc Lond B Biol Sci 360:1935–1943

Breinholt JW, Kawahara AY (2013) Phylotranscriptomics: saturated third codon positions radically influence the estimation of trees based on next-gen data. Genome Biol Evol 5:2082–2092

Bridge P, Spooner B, Roberts P (2005) The impact of molecular data in fungal systematics. Adv Bot Res 42:33–67

Bruns T, White T, Taylor J (1991) Fungal molecular systematics. Annu Rev Ecol Syst 22:525–564

Bruto M, Prigent-Combaret C, Luis P, Moenne-Loccoz Y, Muller D (2014) Frequent, independent transfers of a catabolic gene from bacteria to contrasted filamentous eukaryotes. Proc Biol Sci 281:20140848

Cao Y, Wu SH, Dai YC (2012) Species clarification of the prize medicinal *Ganoderma* mushroom "Lingzhi". Fungal Divers 56:49–62

Capella-Gutierrez S, Marcet-Houben M, Gabaldon T (2012) Phylogenomics supports microsporidia as the earliest diverging clade of sequenced fungi. BMC Biol 10:47

Capella-Gutierrez S, Kauff F, Gabaldon T (2014) A phylogenomics approach for selecting robust sets of phylogenetic markers. Nucleic Acids Res 42, e54

Carlson A, Justo A, Hibbett DS (2014) Species delimitation in *Trametes*: a comparison of ITS, RPB1, RPB2 and TEF1 gene phylogenies. Mycologia 106:735–745

Cohen-Gihon I, Sharan R, Nussinov R (2011) Processes of fungal proteome evolution and gain of function: gene duplication and domain rearrangement. Phys Biol 8:035009

Dai YC, Cui BK (2011) *Fomitiporia ellipsoidea* has the largest fruiting body among the fungi. Fungal Biol 115:813–814

de Bertoldi M, Lepidi AA, Nuti MP (1973) Significance of DNA base composition in classification of *Humicola* and related genera. Trans Br Mycol Soc 60:77–85

De Oliveira ML, Mallo D, Posada D (2014) A Bayesian supertree model for genome-wide species tree reconstruction. Syst Biol. doi:10.1093/sysbio/syu082

de Queiroz K, Gauthier J (1994) Toward a phylogenetic system of biological nomenclature. Trends Ecol Evol 9:27–31

Del Olmo-Ruiz M, Arnold AE (2014) Interannual variation and host affiliations of endophytic fungi associated with ferns at La Selva, Costa Rica. Mycologia 106:8–21

Delsuc F, Brinkmann H, Philippe H (2005) Phylogenomics and the reconstruction of the tree of life. Nat Rev Genet 6:361–375

Du XH, Zhao Q, O'Donnell K, Rooney AP, Yang ZL (2012) Multigene molecular phylogenetics reveals true morels (*Morchella*) are especially species-rich in China. Fungal Genet Biol 49:455–469

Dutilh BE, van Noort V, van der Heijden RT, Boekhout T, Snel B et al (2007) Assessment of phylogenomic and orthology approaches for phylogenetic inference. Bioinformatics 23:815–824

Eastwood DC, Floudas D, Binder M, Majcherczyk A, Schneider P et al (2011) The plant cell wall-decomposing machinery underlies the functional diversity of forest fungi. Science 333:762–765

Ebersberger I, de Matos SR, Kupczok A, Gube M, Kothe E et al (2012) A consistent phylogenetic backbone for the fungi. Mol Biol Evol 29:1319–1334

Ellison CE, Hall C, Kowbel D, Welch J, Brem RB et al (2011) Population genomics and local adaptation in wild isolates of a model microbial eukaryote. Proc Natl Acad Sci USA 108:2831–2836

Feau N, Decourcelle T, Husson C, Desprez-Loustau ML, Dutech C (2011) Finding single copy genes out of sequenced genomes for multilocus phylogenetics in non-model fungi. PLoS One 6, e18803

Fitzpatrick DA (2012) Horizontal gene transfer in fungi. FEMS Microbiol Lett 329:1–8

Floudas D, Binder M, Riley R, Barry K, Blanchette RA et al (2012) The Paleozoic origin of enzymatic lignin decomposition reconstructed from 31 fungal genomes. Science 336:1715–1719

Galagan JE, Henn MR, Ma LJ, Cuomo CA, Birren B (2005) Genomics of the fungal kingdom: insights into eukaryotic biology. Genome Res 15:1620–1631

Geiser DM, Aoki T, Bacon CW, Baker SE, Bhattacharyya MK et al (2013) One fungus, one name: defining the genus *Fusarium* in a scientifically robust way that preserves longstanding use. Phytopathology 103:400–408

Goffeau A, Barrell BG, Bussey H, Davis RW, Dujon B et al (1996) Life with 6000 genes. Science 274(546):563–567

Haag KL, James TY, Pombert JF, Larsson R, Schaer TM et al (2014) Evolution of a morphological novelty occurred before genome compaction in a lineage of extreme parasites. Proc Natl Acad Sci USA 111:15480–15485

Heitman J (2007) Sex in fungi: molecular determination and evolutionary implication. In: Heitman J, Kronstad JW, Taylor JW, Casselton LA (eds). ASM Press, Washington, DC
Hibbett DS, Taylor JW (2013) Fungal systematics: is a new age of enlightenment at hand? Nat Rev Microbiol 11:129–133
Hibbett DS, Nilsson RH, Snyder M, Fonseca M, Costanzo J et al (2005) Automated phylogenetic taxonomy: an example in the homobasidiomycetes (mushroom-forming fungi). Syst Biol 54:660–668
Hibbett DS, Binder M, Bischoff JF, Blackwell M, Cannon PF et al (2007) A higher-level phylogenetic classification of the fungi. Mycol Res 111:509–547
Hibbett DS, Ohman A, Glotzer D, Nuhn M, Kirk PM et al (2011) Progress in molecular and morphological taxon discovery in Fungi and options for formal classification of environmental sequences. Fungal Biol Rev 25:38–47
Hibbett DS, Stajich JE, Spatafora JW (2013) Toward genome-enabled mycology. Mycologia 105:1339–1349
Hyde KD, Udayanga D, Manamgoda D, Tedersoo L, Larsson E et al (2013) Incorporating molecular data in fungal systematics: a guide for aspiring researchers. Currt Res Environ Appl Mycol 3:1–32
Hyde KD, Nilsson RH, Alias SA, Ariyawansa HA, Blair JE et al (2014) One stop shop: backbones trees for important phytopathogenic genera: I. Fungal Divers 67:21–125
Ishii N, Ishida S, Kagami M (2015) PCR primers for assessing community structure of aquatic fungi including Chytridiomycota and Cryptomycota. Fungal Ecol 13:33–43
James TY, Berbee ML (2012) No jacket required—new fungal lineage defies dress code: recently described zoosporic fungi lack a cell wall during trophic phase. Bioessays 34:94–102
James TY, Kauff F, Schoch CL, Matheny PB, Hofstetter V et al (2006a) Reconstructing the early evolution of fungi using a six-gene phylogeny. Nature 443:818–822
James TY, Letcher PM, Longcore JE, Mozley-Standridge SE, Porter D et al (2006b) A molecular phylogeny of the flagellated fungi (Chytridiomycota) and description of a new phylum (Blastocladiomycota). Mycologia 98:860–871
James TY, Pelin A, Bonen L, Ahrendt S, Sain D et al (2013) Shared signatures of parasitism and phylogenomics unite Cryptomycota and *microsporidia*. Curr Biol 23:1548–1553
Jones MD, Forn I, Gadelha C, Egan MJ, Bass D et al (2011) Discovery of novel intermediate forms redefines the fungal tree of life. Nature 474:200–203
Karpov SA, Mamkaeva MA, Aleoshin VV, Nassonova E, Lilje O et al (2014) Morphology, phylogeny, and ecology of the aphelids (Aphelidea, Opisthokonta) and proposal for the new superphylum Opisthosporidia. Front Microbiol 5:112
Kauff F, Cox CJ, Lutzoni F (2007) WASABI: an automated sequence processing system for multigene phylogenies. Syst Biol 56:523–531
Kirk PM, Cannon PF, Minter DW, Stalpers JA (eds) (2008) Ainsworth and Bisby's dictionary of the fungi. CAB International, Wallingford
Koetschan C, Forster F, Keller A, Schleicher T, Ruderisch B et al (2010) The ITS2 Database III—sequences and structures for phylogeny. Nucleic Acids Res 38:D275–D279
Kõljalg U, Nilsson RH, Abarenkov K, Tedersoo L, Taylor AFS, Bahram M et al (2013) Towards a unified paradigm for sequence-based identification of fungi. Mol Ecol 22:5271–5277
Lehr NA, Wang Z, Li N, Hewitt DA, Lopez-Giraldez F et al (2014) Gene expression differences among three *Neurospora* species reveal genes required for sexual reproduction in *Neurospora crassa*. PLoS One 9, e110398
Li C, Hofreiter M, Straube N, Corrigan S, Naylor GJ (2013) Capturing protein-coding genes across highly divergent species. Biotechniques 54:321–326
Lindahl BD, Nilsson RH, Tedersoo L, Abarenkov K, Carlsen T et al (2013) Fungal community analysis by high-throughput sequencing of amplified markers—a user's guide. New Phytol 199:288–299
Liu YJ, Hall BD (2004) Body plan evolution of ascomycetes, as inferred from an RNA polymerase II phylogeny. Proc Natl Acad Sci USA 101:4507–4512

Liu K, Raghavan S, Nelesen S, Linder CR, Warnow T (2009) Rapid and accurate large-scale coestimation of sequence alignments and phylogenetic trees. Science 324:1561–1564
Liu K, Warnow TJ, Holder MT, Nelesen SM, Yu J et al (2012) SATe-II: very fast and accurate simultaneous estimation of multiple sequence alignments and phylogenetic trees. Syst Biol 61:90–106
Lopez-Giraldez F, Townsend JP (2011) PhyDesign: an online application for profiling phylogenetic informativeness. BMC Evol Biol 11:152
Lopez-Giraldez F, Moeller AH, Townsend JP (2013) Evaluating phylogenetic informativeness as a predictor of phylogenetic signal for metazoan, fungal, and mammalian phylogenomic data sets. Biomed Res Int 2013:621604
Losos JB, Arnold SJ, Bejerano G, Brodie ED 3rd, Hibbett D et al (2013) Evolutionary biology for the 21st century. PLoS Biol 11, e1001466
Lutzoni F, Kauff F, Cox CJ, McLaughlin D, Celio G et al (2004) Assembling the fungal tree of life: progress, classification, and evolution of subcellular traits. Am J Bot 91:1446–1480
Martin F, Kohler A, Murat C, Balestrini R, Coutinho PM et al (2010) Perigord black truffle genome uncovers evolutionary origins and mechanisms of symbiosis. Nature 464:1033–1038
Matheny PB, Wang Z, Binder M, Curtis JM, Lim YW et al (2007) Contributions of *rpb2* and *tef1* to the phylogeny of mushrooms and allies (Basidiomycota, Fungi). Mol Phylogenet Evol 43:430–451
McLaughlin DJ, Spatafora JW, Esser K (eds) (2014) The Mycota. A comprehensive treatise on fungi as experimental systems for basic and applied research. Springer, Heidelberg
McNeill J, Turland NJ (2012) International code of nomenclature for algae, fungi, and plants (Melbourne Code). PREFACE. 154:Ix–Xxii
Medina EM, Jones GW, Fitzpatrick DA (2011) Reconstructing the fungal tree of life using phylogenomics and a preliminary investigation of the distribution of yeast prion-like proteins in the fungal kingdom. J Mol Evol 73:116–133
Miadlikowska J, Kauff F, Hognabba F, Oliver JC, Molnar K et al (2014) A multigene phylogenetic synthesis for the class Lecanoromycetes (Ascomycota): 1307 fungi representing 1139 infrageneric taxa, 317 genera and 66 families. Mol Phylogenet Evol 79:132–168
Moeller AH, Townsend JP (2013) Response to: the relative utility of sequence divergence and phylogenetic informativeness profiling in phylogenetic study design. Mol Phylogenet Evol 66:436
Money NP (2013) Against the naming of fungi. Fungal Biol 117:463–465
Morin E, Kohler A, Baker AR, Foulongne-Oriol M, Lombard V et al (2012) Genome sequence of the button mushroom *Agaricus bisporus* reveals mechanisms governing adaptation to a humic-rich ecological niche. Proc Natl Acad Sci USA 109:17501–17506
Nagy LG, Ohm RA, Kovacs GM, Floudas D, Riley R et al (2014) Latent homology and convergent regulatory evolution underlies the repeated emergence of yeasts. Nat Commun 5:4471
Neafsey DE, Barker BM, Sharpton TJ, Stajich JE, Park DJ et al (2010) Population genomic sequencing of *Coccidioides* fungi reveals recent hybridization and transposon control. Genome Res 20:938–946
Nilsson RH, Ryberg M, Sjokvist E, Abarenkov K (2011) Rethinking taxon sampling in the light of environmental sequencing. Cladistics 27:197–203
Nilsson RH, Tedersoo L, Abarenkov K, Ryberg M, Kristiansson E et al (2012) Five simple guidelines for establishing basic authenticity and reliability of newly generated fungal ITS sequences. MycoKeys 4:37–63
Nishant KT, Wei W, Mancera E, Argueso JL, Schlattl A et al (2010) The baker's yeast diploid genome is remarkably stable in vegetative growth and meiosis. PLoS Genet 6, e1001109
O'Connell RJ, Thon MR, Hacquard S, Amyotte SG, Kleemann J et al (2012) Lifestyle transitions in plant pathogenic *Colletotrichum* fungi deciphered by genome and transcriptome analyses. Nat Genet 44:1060–1065
Ortiz-Santana B, Lindner DL, Miettinen O, Justo A, Hibbett DS (2013) A phylogenetic overview of the antrodia clade (Basidiomycota, Polyporales). Mycologia 105:1391–1411

Otrosina WJ, Garbelotto M (2010) *Heterobasidion occidentale* sp. nov. and Heterobasidion irregulare nom. nov.: a disposition of North American Heterobasidion biological species. Fungal Biol 114:16–25

Padamsee M, Kumar TK, Riley R, Binder M, Boyd A et al (2012) The genome of the xerotolerant mold *Wallemia sebi* reveals adaptations to osmotic stress and suggests cryptic sexual reproduction. Fungal Genet Biol 49:217–226

Pante E, Schoelinck C, Puillandre N (2015) From integrative taxonomy to species description: one step beyond. Syst Biol 64:152–160

Park B, Park J, Cheong KC, Choi J, Jung K et al (2011) Cyber infrastructure for *Fusarium*: three integrated platforms supporting strain identification, phylogenetics, comparative genomics and knowledge sharing. Nucleic Acids Res 39:D640–D646

Pearson DL, Hamilton AL, Erwin TL (2011) Recovery plan for the endangered taxonomy profession. Bioscience 61:58–63

Pel HJ, de Winde JH, Archer DB, Dyer PS, Hofmann G et al (2007) Genome sequencing and analysis of the versatile cell factory *Aspergillus niger* CBS 513.88. Nat Biotechnol 25:221–231

Philippe H, Brinkmann H, Lavrov DV, Littlewood DT, Manuel M et al (2011) Resolving difficult phylogenetic questions: why more sequences are not enough. PLoS Biol 9, e1000602

Porras-Alfaro A, Liu KL, Kuske CR, Xie G (2014) From genus to phylum: large-subunit and internal transcribed spacer rRNA operon regions show similar classification accuracies influenced by database composition. Appl Environ Microbiol 80:829–840

Porter TM, Schadt CW, Rizvi L, Martin AP, Schmidt SK et al (2008) Widespread occurrence and phylogenetic placement of a soil clone group adds a prominent new branch to the fungal tree of life. Mol Phylogenet Evol 46:635–644

Richard F, Bellanger JM, Clowez P, Courtecuisse R, Hansen K et al (2014) True morels (*Morchella*, Pezizales) of Europe and North America: evolutionary relationships inferred from multilocus data and a unified taxonomy. Mycologia 107:359–382.

Rinke C, Schwientek P, Sczyrba A, Ivanova NN, Anderson IJ et al (2013) Insights into the phylogeny and coding potential of microbial dark matter. Nature 499:431–437

Robert V, Szoke S, Eberhardt U, Cardinali G, Meyer W et al (2011) The quest for a general and reliable fungal DNA barcode. Open Appl Inf J 5(Suppl 1-M6):45–61

Rodriguez-Romero J, Hedtke M, Kastner C, Muller S, Fischer R (2010) Fungi, hidden in soil or up in the air: light makes a difference. Annu Rev Microbiol 64:585–610

Rosling A, Cox F, Cruz-Martinez K, Ihrmark K, Grelet GA et al (2011) Archaeorhizomycetes: unearthing an ancient class of ubiquitous soil fungi. Science 333:876–879

Rosling A, Timling I, Taylor DL (2013) Distribution and abundance of Archaeorhizomycetes. In: Horwitz A, Mukherjee PK, Mukherijee M, Kubicek CP (eds) Genomics of soil- and plant-associated fungi. Springer, Heidelberg, pp 333–349

Salichos L, Rokas A (2013) Inferring ancient divergences requires genes with strong phylogenetic signals. Nature 497:327–331

Schoch CL, Sung GH, Lopez-Giraldez F, Townsend JP, Miadlikowska J et al (2009a) The Ascomycota tree of life: a phylum-wide phylogeny clarifies the origin and evolution of fundamental reproductive and ecological traits. Syst Biol 58:224–239

Schoch CL, Wang Z, Townsend JP, Spatafora JW (2009b) Geoglossomycetes cl. nov., Geoglossales ord. nov. and taxa above class rank in the Ascomycota tree of life. Persoonia 22:129–138

Schoch CL, Seifert KA, Huhndorf S, Robert V, Spouge JL et al (2012) Nuclear ribosomal internal transcribed spacer (ITS) region as a universal DNA barcode marker for Fungi. Proc Natl Acad Sci USA 109:6241–6246

Schoch CL, Robbertse B, Robert V, Vu D, Cardinali G et al (2014) Finding needles in haystacks: linking scientific names, reference specimens and molecular data for fungi. Database (Oxford). doi:10.1093/database/bau061

Schoustra SE, Debets AJ, Slakhorst M, Hoekstra RF (2007) Mitotic recombination accelerates adaptation in the fungus *Aspergillus nidulans*. PLoS Genet 3, e68

Seifert KA, Rossman AY (2010) How to describe a new fungal species. IMA Fungus 1:109–116

Sikhakolli UR, Lopez-Giraldez F, Li N, Common R, Townsend JP et al (2012) Transcriptome analyses during fruiting body formation in *Fusarium graminearum* and *Fusarium verticillioides* reflect species life history and ecology. Fungal Genet Biol 49:663–673

Slot JC, Rokas A (2011) Horizontal transfer of a large and highly toxic secondary metabolic gene cluster between fungi. Curr Biol 21:134–139

Smith ML, Bruhn JN, Anderson JB (1992) The fungus *Armillaria bulbosa* is among the largest and oldest living organisms. Nature 356:428–431

Spanu PD, Abbott JC, Amselem J, Burgis TA, Soanes DM et al (2010) Genome expansion and gene loss in powdery mildew fungi reveal tradeoffs in extreme parasitism. Science 330:1543–1546

Stajich J, Berbee M, Blackwell M, Hibbett D, James T et al (2009) The fungi. Curr Biol 19:R840–R845

Stajich JE, Wilke SK, Ahren D, Au CH, Birren BW et al (2010) Insights into evolution of multicellular fungi from the assembled chromosomes of the mushroom *Coprinopsis cinerea* (*Coprinus cinereus*). Proc Natl Acad Sci USA 107:11889–11894

Su Z, Wang Z, Lopez-Giraldez F, Townsend JP (2014) The impact of incorporating molecular evolutionary model into predictions of phylogenetic signal and noise. Front Ecol Evol 2:1–11

Taylor JW, Berbee ML (2014) Fungi from PCR to Genomics: the spreading revolution in evolutionary biology. In: Esser K, McLaughlin D, Spatafora J (eds) Mycota VII. A comprehensive treatise on fungi as experimental systems for basic and applied research: systematics and evolution, 2nd edn. Springer, Berlin, pp 1–18

Taylor JW, Jacobson DJ, Kroken S, Kasuga T, Geiser DM et al (2000) Phylogenetic species recognition and species concepts in fungi. Fungal Genet Biol 31:21–32

Taylor JW, Turner E, Townsend JP, Dettman JR, Jacobson D (2006) Eukaryotic microbes, species recognition and the geographic limits of species: examples from the kingdom Fungi. Philos Trans R Soc Lond B Biol Sci 361:1947–1963

Tedersoo L, Bahram M, Põlme S, Kõljalg U, Yorou NS et al (2014) Global diversity and geography of soil fungi. Science 346(6213):1256688

Townsend JP (2007) Profiling phylogenetic informativeness. Syst Biol 56:222–231

Townsend JP, Lopez-Giraldez F (2010) Optimal selection of gene and ingroup taxon sampling for resolving phylogenetic relationships. Syst Biol 59:446–457

Townsend JP, Lopez-Giraldez F, Friedman R (2008) The phylogenetic informativeness of nucleotide and amino acid sequences for reconstructing the vertebrate tree. J Mol Evol 67:437–447

Townsend JP, Su Z, Tekle YI (2012) Phylogenetic signal and noise: predicting the power of a data set to resolve phylogeny. Syst Biol 61:835–849

van de Vondervoort PJ, Langeveld SM, Visser J, van Peij NN, Pel HJ et al (2007) Identification of a mitotic recombination hotspot on chromosome III of the asexual fungus *Aspergillus niger* and its possible correlation with [corrected] elevated basal transcription. Curr Genet 52:107–114

van der Heijden MG, Martin FM, Selosse MA, Sanders IR (2015) Mycorrhizal ecology and evolution: the past, the present, and the future. New Phytol 205:1406–1423

Vogel KJ, Moran NA (2013) Functional and evolutionary analysis of the genome of an obligate fungal symbiont. Genome Biol Evol 5:891–904

Walker DM, Castlebury LA, Rossman AY, White JF Jr (2012) New molecular markers for fungal phylogenetics: two genes for species-level systematics in the Sordariomycetes (Ascomycota). Mol Phylogenet Evol 64:500–512

Wang Z, Nilsson RH, Lopez-Giraldez F, Zhuang WY, Dai YC et al (2011) Tasting soil fungal diversity with earth tongues: phylogenetic test of SATe alignments for environmental ITS data. PLoS One 6, e19039

Wapinski I, Pfeffer A, Friedman N, Regev A (2007) Natural history and evolutionary principles of gene duplication in fungi. Nature 449:54–61

Wardle DA, Lindahl BD (2014) Disentangling global soil fungal diversity. Science 346:1052–1053

Wickett NJ, Mirarab S, Nguyen N, Warnow T, Carpenter E et al (2014) Phylotranscriptomic analysis of the origin and early diversification of land plants. Proc Natl Acad Sci USA 111:E4859–E4868

Will KW, Mishler BD, Wheeler QD (2005) The perils of DNA barcoding and the need for integrative taxonomy. Syst Biol 54:844–851

Wisecaver JH, Slot JC, Rokas A (2014) The evolution of fungal metabolic pathways. PLoS Genet 10, e1004816

Wu F, Yuan Y, Malysheva VF, Du P, Dai YC (2014) Species clarification of the most important and cultivated *Auricularia* mushroom "Heimuer": evidence from morphological and molecular data. Phytotaxa 186:241–253

Wurzbacher C, Grossart HP (2012) Improved detection and identification of aquatic fungi and chitin in aquatic environments. Mycologia 104:1267–1271

Ye J, Coulouris G, Zaretskaya I, Cutcutache I, Rozen S et al (2012) Primer-BLAST: a tool to design target-specific primers for polymerase chain reaction. BMC Bioinf 13:134

Zhang LF, Yang ZL, Song DS (2005) A phylogenetic study of commercial Chinese truffles and their allies: taxonomic implications. FEMS Microbiol Lett 245:85–92

Zhang N, Rossman AY, Seifert K, Bennett JW, Cai G et al (2013) Impacts of the international code of nomenclature for algae, fungi, and plants (Melbourne Code) on the scientific names of plant pathogenic fungi. Online, APSnet Feature: American Phytopathoglogical Society, St. Paul, MN

# Chapter 4
# Molecular Techniques in Mycological Studies and Sequence Data Curating: Quality Control and Challenges

**R. Henrik Nilsson, Kessy Abarenkov, and Urmas Kõljalg**

## Introduction

Mycology is tasked with characterizing and cataloging the fungal kingdom. Molecular (DNA-based) studies of substrates, such as soil and wood, have shown this task to be enormous; each new study typically recovers a hundred or more species of fungi that do not seem to be known from before and that do not produce close matches to any fully identified reference sequence (Tedersoo et al. 2014). Some 100,000 species are recognized formally, which is less than a tenth of the estimated 1.5 million extant species of fungi (Hawksworth 2001). Given that only about 1000 new fungi are described each year, it is not surprising that much of the fungal diversity being unearthed through DNA sequencing is not easily fitted into the framework of already described species (Hibbett et al. 2011). Incorporating also these poorly known lineages into mycology and the fungal tree of life—often in total absence of morphological structures and any data on ecology, trophic mode, or other species traits—is perhaps the greatest challenge for the early twenty-first-century mycology.

The nature of fungal life puts many obstacles in the way of expedient mycological progress. Many species are subterranean or substrate dwelling and rarely—if ever—manifest themselves through morphological structures such as fruiting bodies (Stajich et al. 2009; Rosling et al. 2011). Morphological characters in fungi can be deceptively similar or dissimilar, such that even expert observers may be easily

R.H. Nilsson, Ph.D. (✉)
Department of Biological and Environmental Sciences, University of Gothenburg, Box 461, 405 30 Göteborg, Sweden
e-mail: henrik.nilsson@bioenv.gu.se

K. Abarenkov, Ph.D. • U. Kõljalg, Ph.D.
Natural History Museum, University of Tartu, Vanemuise 46, Tartu 51014, Estonia
e-mail: kessy.abarenkov@ut.ee; urmas.koljalg@ut.ee



D.-W. Li (ed.), *Biology of Microfungi*, Fungal Biology,
DOI 10.1007/978-3-319-29137-6_4

lead astray (Yang 2011). Many fungi cannot readily be cultured in the laboratory, which excludes them from the rich array of examination techniques available for fungal cultures (Hawksworth 1991). These and other challenges have hampered mycology for far too long, and it is critical that we adopt a different view of what criteria must be fulfilled before a fungal lineage merits scientific attention. Observable life stages and cultivability should not be among those criteria. Fortunately, access to molecular data has done a lot to facilitate this transition in thinking.

Molecular data gradually gained ground as a mycological research implement in the 1990s; the first applications included inference of phylogenetic relationships and taxonomic identification of mycorrhiza (Förster et al. 1990; Gardes and Bruns 1993). The low-throughput Sanger DNA sequencing approaches of those days soon saw substantial improvement in speed, quality, and degree of automation, and they were recently complemented by next-generation sequencing (NGS) methods with the capacity to generate millions—even billions—of (typically short) sequence reads over the course of a few hours to days (Liu et al. 2012). Today, molecular data underpin much of our understanding of fungi and fungal diversity, and a molecular thinking permeates most mycological research efforts. It is, however, a mistake to believe that molecular data per se vouch for increased resolution in the results and for objective answers to the research questions being pursued. Much like other sources of data, DNA sequences will not give rise to appropriate, correct results unless they are generated and analyzed in an appropriate, correct way.

The most popular genetic marker for research at the fungal species/genus level is the nuclear ribosomal internal transcribed spacer (ITS) region. In addition to its prominent role in phylogenetic inference, the ITS region is the formal fungal barcode and the primary target for countless molecular ecology and environmental sequencing efforts of fungal communities (Schoch et al. 2012; Lindahl et al. 2013). To date, some 500,000 (Sanger sequencing-derived) fungal ITS sequences have been generated by the scientific community and are available for reference in the International Nucleotide Sequence Database Collaboration (INSDC: GenBank, ENA, and DDBJ; Nakamura et al. 2013). The corpus of public ITS sequences is, however, not devoid of complications. Only about half of the sequences are identified to species level, the rest being given names of various degrees of informativeness, ranging from, e.g., *Cryptococcus* sp. to "uncultured fungus." Second, estimates show that more than 10 % of the sequences annotated to species level may in fact carry incorrect names (Bidartondo et al. 2008). Third, a nontrivial portion of the sequences suffers from technical problems: some have reduced read quality/incorrect base-calling (Hyde et al. 2013), some are chimeric (Nilsson et al. 2010), some represent other genes and markers than the ITS region (Bengtsson-Palme et al. 2013), and some are reverse complementary (given backward and with purines and pyrimidines transposed; Hartmann et al. 2011). This makes uncritical use of public DNA sequences for, e.g., molecular species identification prone to errors and suboptimal results. Many users of those data may not be in a position to spot such errors, which increases the pressure on the mycological community to provide ITS sequences of high integrity and usefulness for the scientific community at large.

The INSDC fills an invaluable function in providing long-term storage and global access to DNA and protein sequences (and other data). The INSDC takes many measures to highlight their data to researchers and to accommodate the wishes of an increasingly molecular research community (Schoch et al. 2014). The INSDC is to some degree interactive and gives users the opportunity to comment on certain parts of the data (http://www.ncbi.nlm.nih.gov/pubmedcommons/). The balance between long-term stability and immediate flexibility is however a hard one to navigate. Users cannot change INSDC data directly even if they come across incorrect or substandard information; such changes will have to go through the database curators. Certain changes—such as taxonomic re-annotation of incorrectly identified sequences—require the explicit permission of the original sequence authors, whose present contact information may or may not be available to the curators. Whereas we have found the INSDC curators very helpful and knowledgeable, there is a limit to the changes they can implement, particularly in the short-term perspective. The field of molecular ecology, in contrast, sees continual bursts of technical progress and new taxonomic and ecological reference data. This brings about the need for frequent additions and changes in the reference data—changes that, ideally, should take effect immediately.

## The UNITE Database

The UNITE database (http://unite.ut.ee; Kõljalg et al. 2013) was devised to facilitate molecular identification of fungi. It mirrors the INDSC to offer all public fungal ITS sequences as a reference corpus. This makes UNITE unusual; other reference databases typically focus on select sequences known to be of particularly high technical quality, metadata richness, and taxonomic reliability. UNITE takes the position that such a limited-taxon approach would not serve mycology well. The majority of fungal operational taxonomic units (OTUs; Blaxter et al. 2005) known from ITS sequences do not carry full species names, yet they are every bit as real as those who do. Excluding the majority of the known species from molecular identification procedures would lead to massive losses of precision and scientific explanatory power. Instead, UNITE provides all public sequences and offers several ways to highlight particularly reliable entries as well as to remove substandard ones. The user is presented with several ways to interact with the sequences and modify them in various regards. The inclusive taxonomic scope of the database means that a limited number of database developers and curators cannot be expected to possess the expertise to revise all taxonomic lineages equally well in the database. Thus, the database comes with a web-based sequence management environment such that all registered users can participate in data curation through a regular web browser.

Two times a year, all fungal ITS sequences are clustered to approximately the genus level. A second round of clustering inside each such cluster seeks to produce OTUs at approximately the species level. These OTUs are called species hypotheses (SHs). By default, a 98.5 % similarity threshold (1.5 % distance to the closest

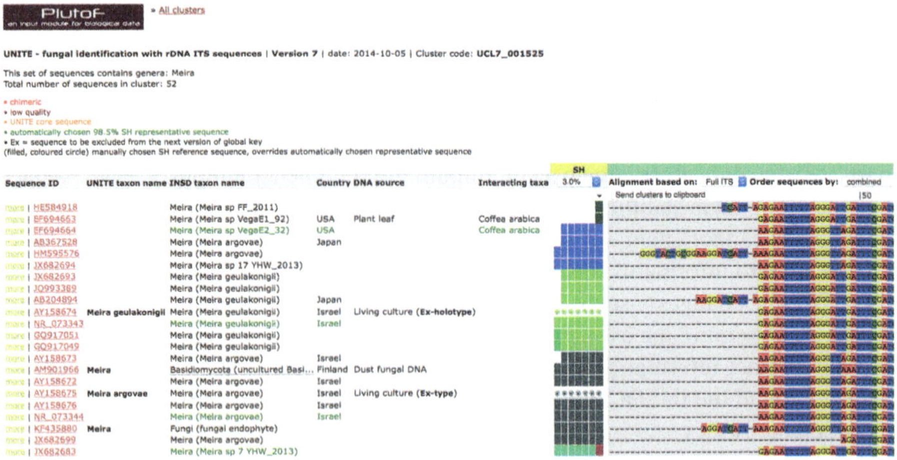

Fig. 4.1 A genus-level alignment of the basidiomycete genus *Meira*. The sequences are displayed together with available metadata on country and host of collection. Sequences from type material are used to fix the application of species names through reference sequences. Two examples of taxonomic re-annotation are shown; for instance, sequence KF435880, originally deposited in the INSDC as "fungal endophyte," is now annotated as *Meira*. The *filled*, *colored boxes* indicate the SHs and their scope under different similarity thresholds. The SHs point to the existence of hitherto unrecognized diversity in the genus

distinct SH) is used as proxy for the species level, but a user can reset the threshold (in the interval 90 %, 95 %, and 97–100 % similarity in 0.5 % steps) to better reflect the species level for individual SHs. A genus-level cluster of parts of the genus *Meira* is shown in Fig. 4.1 together with the constituent sequences, SHs, and their metadata on ecology and geography. The genus-level alignments form perhaps the most straightforward arena for the user to navigate, explore, and interact with the sequence data in UNITE. Tools such as BLAST (Altschul et al. 1997) allow users to query newly generated sequences against the database. UNITE also serves as primary or secondary data provider for a range of external applications and resources, including the next-generation sequencing analysis pipelines QIIME (Caporaso et al. 2010), mothur (Schloss et al. 2009), CREST (Lanzén et al. 2012), UTAX (http://drive5.com/usearch/manual8/cmd_utax.html), and SCATA (http://scata.mykopat.slu.se/) as well as the EUBOLD and ISHAM databases (http://www.eubold.org/ and http://its.mycologylab.org/, respectively). This heterogeneous user base highlights the need for high reliability and richness of the data in UNITE. As a consequence, UNITE is the subject of intense sequence curation in various regards. These curation and quality control processes are described below and are given together with recommendations and ideas on how users can exercise quality control also on personal, not-yet-public sequences.

## Names and Taxonomic Re-annotation

Names of organisms and organism groups are central devices for communication in biology. They allow precise and unambiguous reference and knowledge building across studies and over time. Incorrectly applied names in sequence databases counteract these processes and invite further misidentifications as users adopt and incorporate erroneous names from sequence similarity searches. Any user with a newly generated set of DNA sequences should take steps to ensure that any hypothesized taxonomic affiliations of the sequences check out—a BLAST search in UNITE or the INSDC is usually enough to rule out contamination or severe cases of misidentification. BLAST searches are often informative also on the reference sequences; most searches will reveal at least one reference sequence that either seems incorrectly identified (such as cf. *Puccinia* sp. for an ascomycete) or that lack meaning taxonomic annotation altogether (such as the common "uncultured fungus"). A friendly email to the original sequence authors or the database curators may resolve such problems, at least if the user provides sufficient data and detail to make the validity of the claim easy to verify.

The original sequence authors are however not always known or available, which makes it difficult for, e.g., the INDSC to seek permission to effectuate name changes. This slows down the renaming process significantly. UNITE offers one-click taxonomic re-annotation of public sequences using the names and classification of Index Fungorum (http://www.indexfungorum.org/; Fig. 4.2). The changes take effect immediately, and the user will be credited with the name change. Incremental name changes are supported and encouraged; if a sequence cannot be assigned to species level, a genus or order name will still be helpful. The same—or another—user can then provide the full species or genus name if and when additional, explanatory data emerge. All taxonomic re-annotations done in UNITE are made available to the INSDC, and at least a subset of them are implemented there too.

Many species hypotheses cannot be given full species names for lack of reference data and/or taxonomic progress in those fungal lineages. Assigning them as far as possible will be of great value to many users: *Puccinia* sp. is a great deal more informative than "uncultured fungus." Even so, these and similar progressional names may be applied to hundreds or even thousands of other, non-conspecific sequences. This makes precise reference to those species across publications and datasets problematic and forms an obstacle to knowledge building and data assembly for individual species. As a remedy, UNITE assigns names of the accession number type to all species hypotheses. Whereas the shortest form of those names is simply an accession number—such as SH203822.06FU—the longer form, *Hymenoscyphus pseudoalbidus* | SH203822.06FU | 98.5 | GU586904, comprises any name (Latin or otherwise) given to the sequence, the SH accession number, the similarity threshold at which the SH was designated, and the INSDC accession number of the sequence chosen to represent the SH in situations when only one sequence per SH is used. (The latter includes the nonredundant BLAST databases

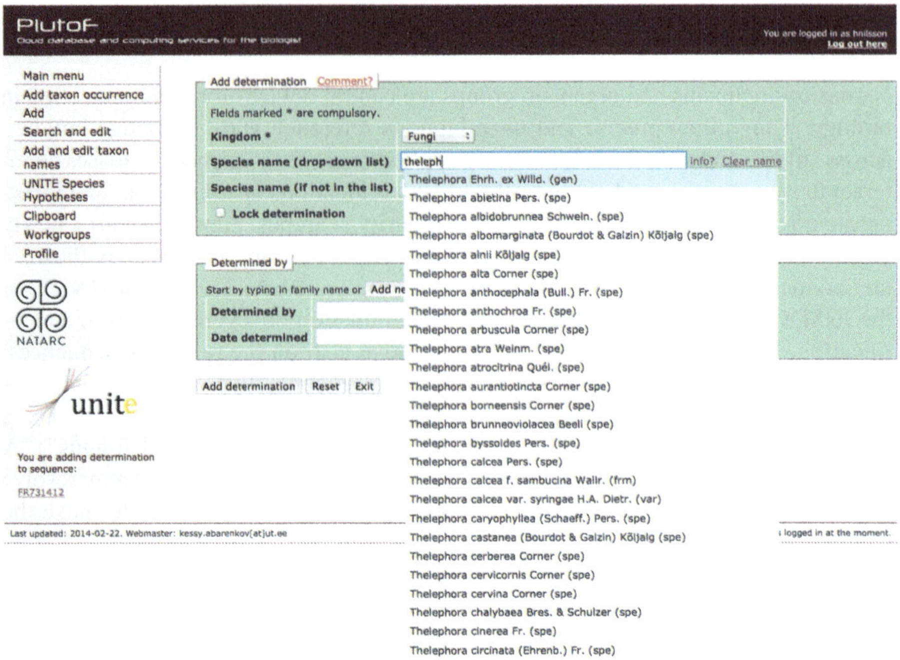


**Fig. 4.2** Taxonomic re-annotation of INSDC entry FR731412, originally deposited under the low-resolution name "uncultured ectomycorrhizal fungus." Sequence analysis shows that it belongs to the *Thelephoraceae*, and the according taxonomic annotation is a one-click process. Upon starting to type the new name, the user is presented with a drop-down list of accepted names and ranks to choose from

of many sequence analysis pipelines.) These SH names are implemented throughout UNITE and are forwarded to the downstream resources and applications that use UNITE. We hope that users will consider using these names for OTUs that otherwise would lack an informative name (see, e.g., Morgado et al. 2014); this will allow them and other users to trace those OTUs across publications and datasets, much to the purpose of reproducibility and data harvesting.

## Chimeras

Chimeras are artificial sequences composed of two or more sequence fragments that do not belong together but that were joined in the PCR or sequence assembly steps (Qiu et al. 2001). The typical chimera is produced when the PCR enzyme switches template from one species to the other in mixed-template PCR reactions; any presence of a conserved sequence segment in the target marker—such as the 5.8S gene in ITS sequences—increases the risk of chimeric unions (Fonseca et al. 2012).

Chimeras lack a meaningful biological interpretation and introduce noise and bias to studies featuring them; molecular identification, richness estimations, multiple sequence alignment, and phylogenetic inference are examples of processes that are adversely affected by chimeric sequences. Steps can be taken to reduce the chances of chimera formations in the PCR phase (Lindahl et al. 2013). Equally important is to screen for chimeras in newly generated datasets; this should be done on a routine basis for all new fungal ITS datasets.

Screening small-to-midsize, relatively homogeneous datasets for at least severe cases of chimeras is relatively easy and amounts to a multiple sequence alignment of the query sequences. Any sequence found to align well in the first half of the alignment but to align poorly in the second half (or the other way around) merits further scrutiny (Fig. 4.3; Nilsson et al. 2012). A BLAST search followed by manual inspection of the results is another approach with a similar potential for chimera discovery (Fig. 4.4). Separate BLAST searches of the two sequence halves are then

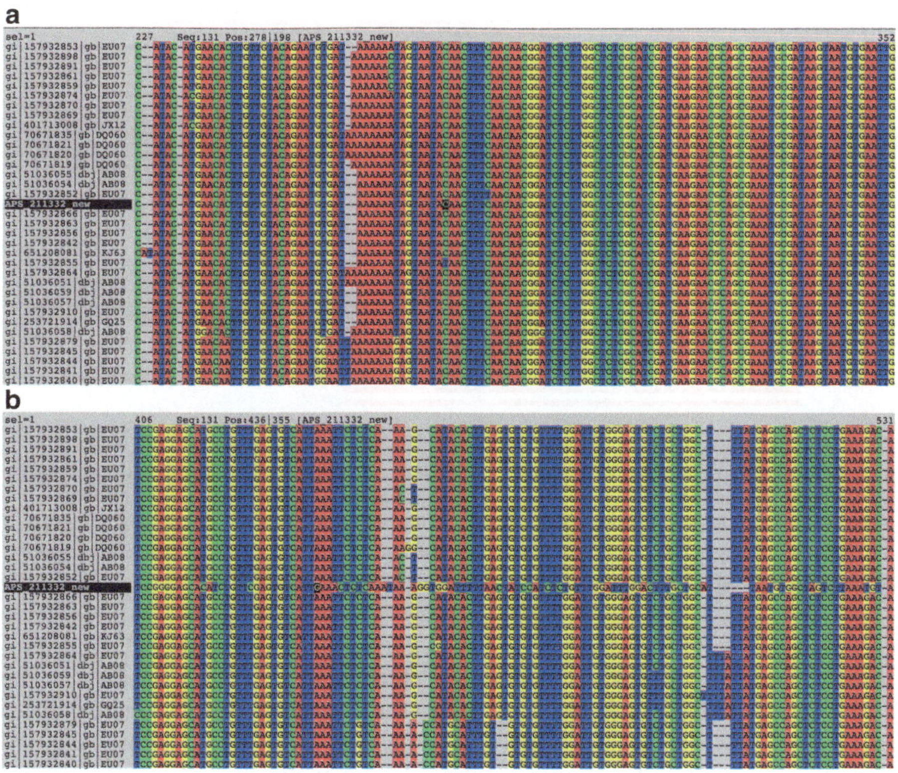

**Fig. 4.3** A chimeric sequence showing (**a**) perfect alignment in the first half of the multiple sequence alignment (the cursor indicates the start of the 5.8S gene) but (**b**) poor alignment in the second half (the cursor marks the end of the 5.8S). Particularly conserved sequence elements—the 5.8S gene in this case—often serves as chimeric breakpoints. SeaView (Gouy et al. 2010) was used to display the multiple sequence alignment

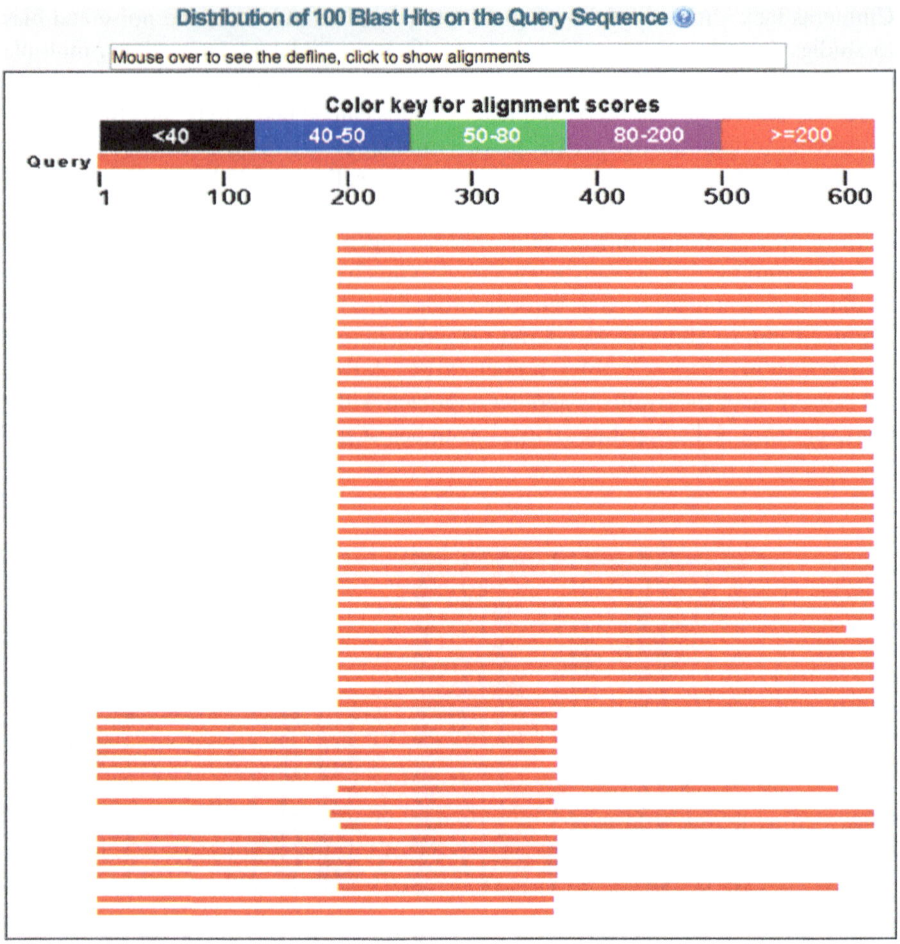


**Fig. 4.4** A chimeric sequence often produces odd-looking BLAST results, such as the one shown here. Either the last part of the sequence, or the first part of the sequence, but never both at the same time, account for the alignment to the sequences in the reference database. Manual examination of the query sequence and the BLAST results are needed in cases like this. Chimeras between closely related species is unlikely to give rise to BLAST results as clear-cut as this one

usually enough to convince oneself of the chimeric—or authentic—nature of the sequence. For users with larger datasets and/or some experience of the computer command line, we recommend UCHIME (Edgar et al. 2011), which is a fast and accurate software tool for chimera control. An automatically updated and reasonably chimera-free reference file for fungal ITS sequences is available from the UNITE download page (http://unite.ut.ee/repository.php). (UCHIME also supports de novo, reference-free chimera calling for NGS datasets.) Chimera detection is nevertheless not a trivial undertaking, and elusive chimeras are certain to exist in both the INSDC and UNITE. Users finding chimeras in UNITE are asked to mark these as chimeric through a simple click, which will prevent their ulterior use as representative sequences and in sequence identification efforts. The INSDC staff is similarly very quick to remove chimeras once alerted to their presence. UNITE and the INDSC exchange data on chimeras regularly.

## Technical Quality of Sequences

Once sequences are deposited in public databases, quality control becomes hard. The sequences are typically deposited without chromatograms and other auxiliary data, leaving other researchers struggling to tell sequences of low technical quality from sequences that differ from other sequences for the reason that they represent biological novelties. DNA ambiguity codes, such as N, R, and S (Cornish-Bowden 1985), scattered across a sequence are a sure sign that the sequence should be dropped from most research efforts. Lengthy homopolymer regions (such as AAAAAAAAAA), particularly in the distal parts of a sequence, similarly hint at a substandard entry. Distal ends are generally of lower read quality compared to interior parts of sequences (owing to the nature of the Sanger sequencing method), and researchers should make it a habit to trim such noisy parts prior to sequence submission. A regular BLAST search coupled with manual inspection of the results is usually enough to detect all the above problems (Fig. 4.5). It is often more or less impossible to *prove* a sequence to be compromised, but for most purposes, it is enough to establish that a compromised nature seems very likely. Such entries can then be left out from analyses. To routinely exclude all sequences that differ from known sequences is however not a good way to expand our understanding of the world around us, and the user is probably best off excluding only sequences for which a compromised nature is deemed beyond reasonable doubt.

Sequences can be of very high read quality and still suffer from technical problems. Sanger sequences are typically assembled from two or more primer reads. Although (semi)automated, this process requires some understanding on part of the user of the relative order and read direction of the primers. Failure to inspect—or understand—the assembly results regularly leads to submission of chimera-like sequences to the public databases. Sometimes the first half of the sequence is given in the correct orientation, whereas the second half is given in the reverse complementary orientation. At other times, sequence fragments were assembled in the wrong order or with a fragment missing. Insertion of fragments that do not belong in the sequence to begin with has also been reported (Nilsson et al. 2012). These sequences typically produce odd-looking results in BLAST. The alignment may be divided into sections or cover only a part of the query sequence (Fig. 4.6); the "strand" flag of the BLAST output is helpful in detecting reverse complementary insertions. The technically inclined user will find that the software tools ITSx (Bengtsson-Palme et al. 2013) and UCHIME (Edgar et al. 2011), although not explicitly designed to find sequences of these types, work surprisingly well for the

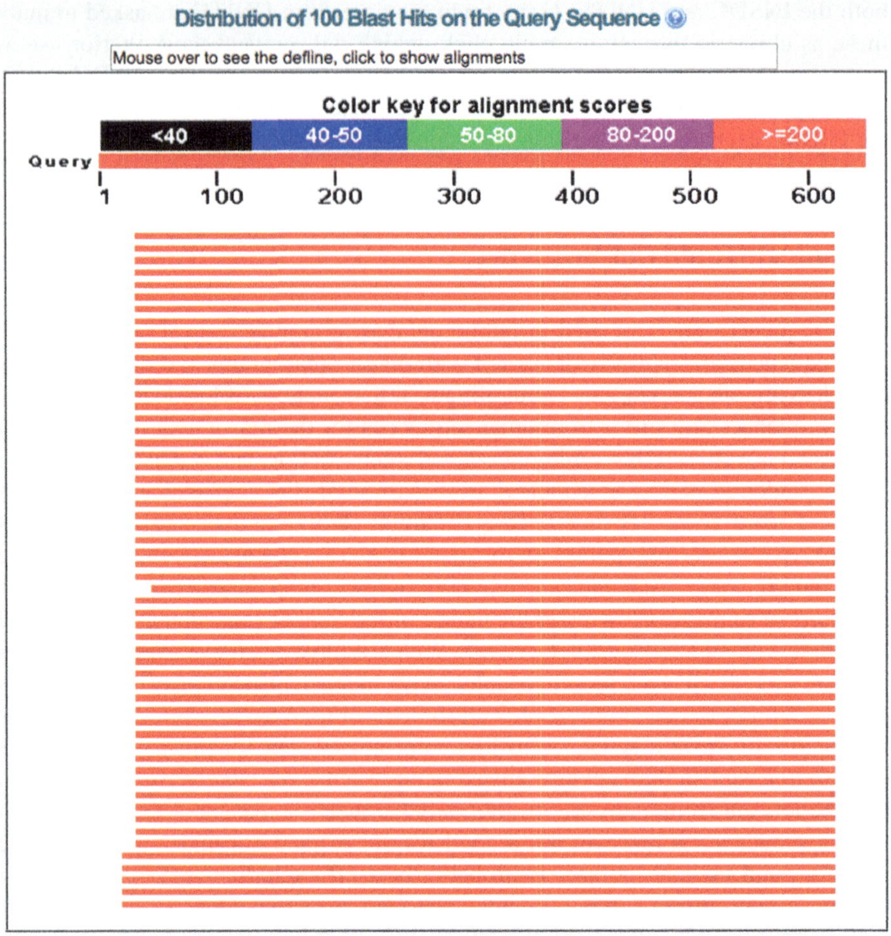


**Fig. 4.5** Failure to trim noisy ends of a sequence gives rise to BLAST results looking like this; note how the distal ends of the sequence do not find matches in the counterpart sequences in the reference database. The reader should keep in mind that a large proportion of public fungal ITS sequences were submitted with noisy ends, such that it is not always straightforward to tell if it is the query or the reference sequences that do not meet quality expectations

task. The INSDC is quick to take action when alerted to such substandard entries; one of the actions available to their staff is to mark such sequences "UNVERIFIED." Users of UNITE similarly have the option to mark sequences as being of low technical quality. The INSDC and UNITE exchange data on substandard entries.

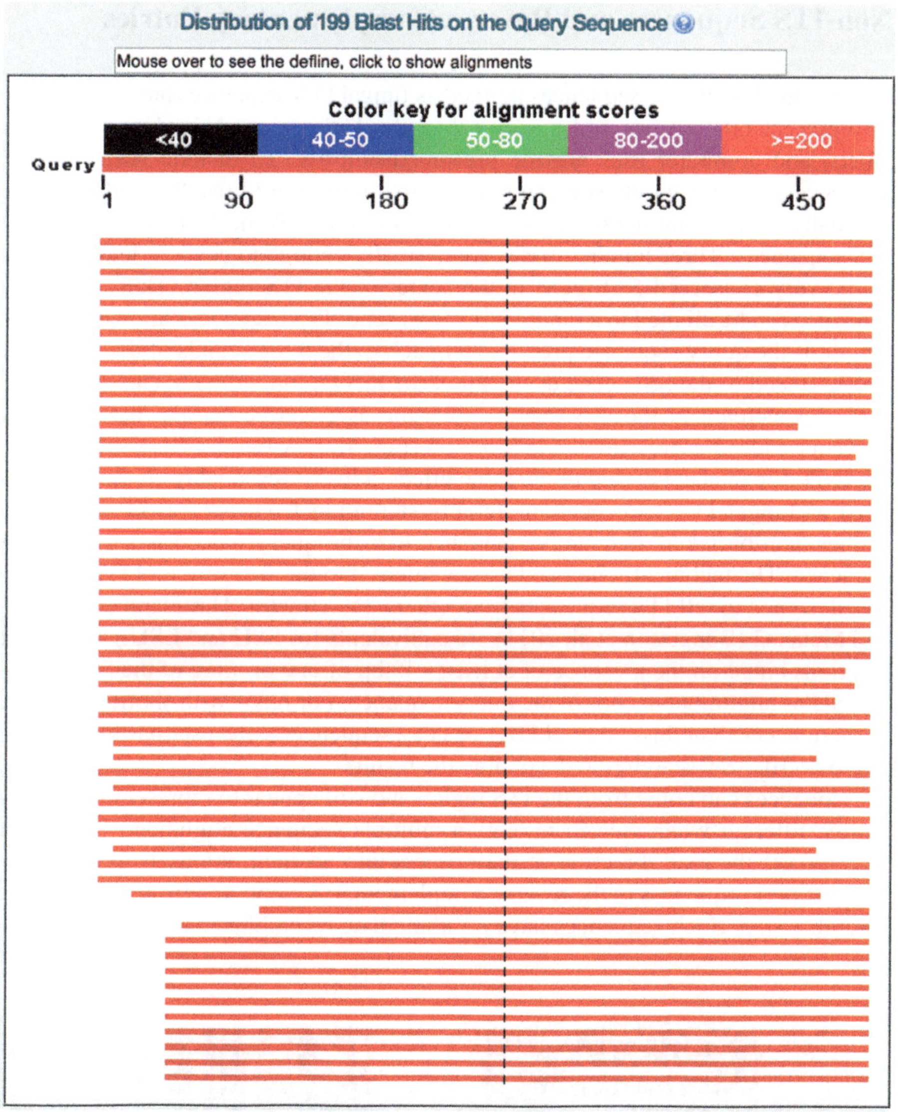


**Fig. 4.6** The *black vertical line* in the BLAST results means that BLAST was unable to produce a straightforward alignment involving the query sequences and the topmost reference sequences. A segment was missing (or was extraneous) in the alignment; alternatively, the query sequence was assembled incorrectly. Manual examination of the sequences involved is needed in cases like this

## Non-ITS Sequences and Reverse Complementary Entries

Several hundred public sequences marked as fungal ITS sequences have been found to represent other genes and markers than the ITS region. Mixed-up test tubes, labels, and computer files are the presumed culprits. To include such non-ITS sequences in ITS datasets is certain to lead to compromised results. Fortunately, it is usually easy to find out whether sequences of at least some 300 bases indeed are ITS sequences. Three hundred bases are normally enough to cover at least one end of the very conserved 5.8S gene in the center of the ITS region. Thus, the query sequence can be aligned to a random (diverse) set of ITS sequences known to be full length. If the query sequence produces a match to the 5′ or 3′ ends of the 5.8S gene (or the 3′ end of the SSU gene upstream of the ITS region, or the 5′ end of the LSU gene downstream of the ITS region), then the user can be certain that the query sequence indeed represents the ITS region (Fig. 4.7). A regular BLAST search in the INDSC can also be used to the same effect: a list of reasonably close hits annotated as fungal ITS sequences, preferably stemming from two or more different studies, can be taken as tentative evidence that the query indeed is a fungal ITS sequence. The technically inclined user is referred to the software tool ITSx, which was designed to tell ITS sequences from others. (V-Xtractor (Hartmann et al. 2010) and Metaxa2 (Bengtsson et al. 2011) are equivalents for SSU and LSU.) The above solutions hinge on the presence of at least ~45 bp of one or more of the SSU, 5.8S, or LSU as alignment anchor in the query sequence. For reads with shorter coverage of these genes than this—partial ITS1 or ITS2 sequences—the user has little choice but to do BLAST searches and examine the results.

Sequences can come from the ITS region and still appear to represent something totally different when viewed in, e.g., a multiple sequence alignment. If the user fails to take the read direction of the primers into account, the user may inadvertently export sequences in the reverse complementary orientation from the sequence assembly step. Such sequences have no immediate resemblance to their true coun-

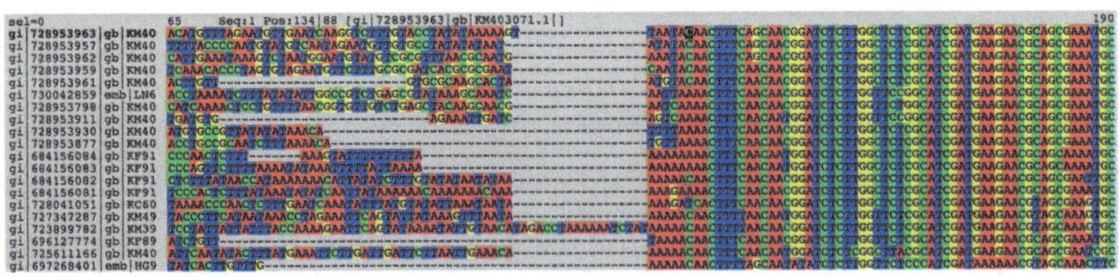

**Fig. 4.7** Whether or not a sequence represents the ITS region can be established in several ways. One of them is to focus on the very conserved 5.8S gene, particularly its 5′ end. Shown is the 5′ end (starting at alignment position 134, *highlighted*) of five random sequences from each of the *Basidiomycota*, *Ascomycota*, *Glomeromycota*, and 5 the former *Zygomycota*. A sequence that produces a satisfactory alignment to the first part of such a 5.8S reference alignment is certain to represent the ITS region. Strictly speaking, it does not necessarily indicate that the sequence is of fungal origin though

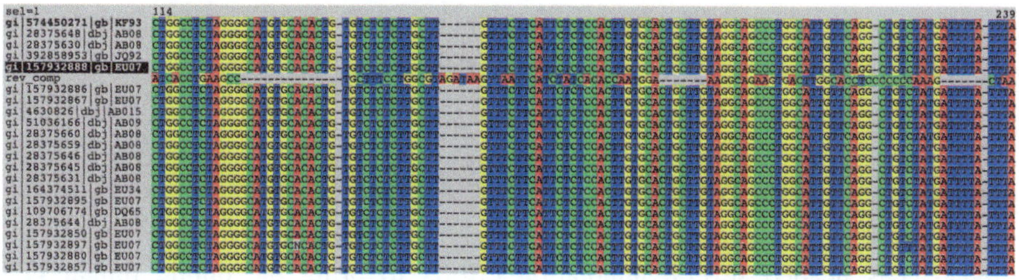

**Fig. 4.8** A reverse complementary sequence was computed from the entry *highlighted in black*. It is shown immediately below that entry. Even though the two entries are identical in terms of information content, there is little resemblance in the multiple sequence alignment. With the realization that such odd-looking sequences may in fact be reverse complementary, their proper integration into the alignment is easily overseen in modern alignment viewers

terpart (Fig. 4.8). Fortunately, BLAST supports reverse complementary sequences by default and accounts for their nature in the alignment stage. A "strand" flag of "plus/minus" or "minus/plus" in the BLAST output indicates opposing read directions of the query and the reference sequence. The user can reorient reverse complementary sequences in alignment editors such as AliView (Larsson 2014) or in any of a number of online tools (e.g., http://www.bioinformatics.org/sms/rev_comp.html). The tool ITSx can be used to track down reverse complementary entries in ITS datasets of any size.

## Metadata and Species Traits

Many sequences are submitted to sequence databases with little more extra data than the name and address of the sequence authors. Tedersoo et al. (2011) showed, for instance, that country of origin was provided for a modest 43 % of the public fungal ITS sequences. Along the same line, Ryberg et al. (2009) found that less than a quarter of the public ITS sequences were annotated with host of collection. This is most unfortunate since data on geography and ecology are often needed to make informed decisions in molecular identification and systematics. To be able to provide more metadata than those present in the original sequence entries, UNITE offers the user the possibility to enter such data for entries lacking them. While the two most important data items are probably country and host of collection, a long range of parameters are available for specification. These include nutritional mode, lifestyle, substrate type, mycorrhizal lineage, primers used, and voucher specimen/culture. DarwinCore (http://rs.tdwg.org/dwc/) specifications are followed as applicable. Associating individual sequences with voucher specimens and/or cultures is straightforward and particularly valuable in the case of type material. The mycological community has not always been consistent in the way specimen/culture data are specified during the sequence submission process (and whether or not they are provided at all), which makes subsequent searches difficult. Recent developments in the INSDC show significant promise in this regard (Schoch et al. 2014).

## Results and Data Sharing

UNITE has headed several annotation efforts to bring the public fungal ITS sequences up to standards. Two efforts for mycorrhizal (Tedersoo et al. 2011) and plant-pathogenic fungi (Nilsson et al. 2014) focused on improving taxonomic names and affiliations given to sequences as well as providing ecological and geographical metadata that were missing. An annotation effort headed by GenBank (Schoch et al. 2014) fixed the application or more than 3000 species names throughout the fungal kingdom by associating sequences with type specimens/ex-type cultures. Efforts focusing on specific technical problems of public ITS sequences include chimera detection (Nilsson et al. 2014), reverse complementary entries (Nilsson et al. 2011), and non-ITS sequences (Bengtsson-Palme et al. 2013). These efforts, together with all changes contributed by UNITE users, translate into a full 31,593 taxonomic re-annotations, which corresponds to 7.4 % of the number of public fungal ITS sequences as of December 2014. Data on ecology and geography have been provided for 52,807 and 58,414 sequences (25,974 of these corresponding to sequences for which both metadata on ecology and geography were added). Specimen/culture associations have been established or clarified for 12,183 sequences, of which 3657 are related to type material. A full 7141 sequences have been identified as being of low read/technical quality, and 2554 cases of chimeras have been found and marked as such. Some 3034 sequences were originally deposited in the reverse complementary orientation; these now appear in their correct orientation.

These additions and improvements benefit anyone who uses UNITE for molecular identification or other analyses of fungal ITS sequences. UNITE furthermore serves as data provider of fungal ITS sequences for a range of other databases and sequence analysis tools; these resources, too, benefit from any improvements done to the data in UNITE. The sharing of the species hypotheses and their unique, non-ambiguous names with the INSDC and the largest NGS analysis pipelines furthermore solves a long-standing problem in fungal molecular ecology: how to refer to specific, unidentified species in a standardized way across studies, datasets, and software pipelines. This brings promise of an end to the large proportion of "unidentified/fungi sp." species recovered in NGS-based environmental sequencing efforts. Any user deciding to dig a bit deeper on the taxonomic affiliation of such unidentified species is likely to uncover additional resolution; nearly all SHs we have examined, for example, can be assigned at least to the phylum—and often also order—level. Once such a re-annotation is implemented in UNITE, the corresponding change in the data export files of UNITE will reach the users of the NGS analysis pipelines. UNITE also provides downloadable FASTA files of all sequences (http://unite.ut.ee/repository.php) for the use in, e.g., local BLAST searches and sequence analysis efforts. Various FASTA versions are available—it is our intention to suit all needs, and we would be happy to consider requests for specific formats or data items to support for regular exports. UNITE finally exchanges data—bidirectionally—with the INSDC. This maintains some level of data synchronization among the datasets, although UNITE typically lags behind in terms of the latest sequences and their integration into BLAST searches. The species hypothesis platform is recomputed twice a year, meaning that some six months may elapse before a sequence is assigned to an SH.

## Discussion

Fungi play significant ecological roles—notably as nutrient cyclers—and form key players in most ecosystems. This makes fungi attractive research objects in a range of scientific fields in addition to mycology, including forestry, soil biology, and agriculture. Disturbingly, when researchers in these fields sequence their substrates for fungal diversity, they often get results that are low in resolution ("uncultured fungus") or that are directly misleading. In our opinion, this makes mycology look bad. There are, after all, several thousand mycologists worldwide, and their joint knowledge and expertise are remarkable. And yet mycologists do not seem to have found a way to communicate this knowledge to the scientific community at large—at least not in such a way that it could be put to direct use. Although you could argue that every environmental sequencing effort targeting fungal communities should have at least one mycologist onboard, this will not always be the case. Thus, the mycological community should see to it that analysis and interpretation of DNA sequencing efforts of fungal communities are as straightforward and easy as possible also for non-mycologists. There are numerous ways in which mycologists can contribute toward this goal. For instance, we feel that mycologists should provide a DNA sequence (preferably from the ITS region) along with every new species they describe. When mycologists describe species and deposit fungal sequence data in public databases, they should make an effort to provide rich metadata and to establish the basic integrity of the sequence data (see guidelines in Seifert and Rossman 2010; Nilsson et al. 2012; Hyde et al. 2013; Schoch et al. 2014). Finally, as this chapter has shown, a lot would be gained if individual mycologists made sure that public sequence data from their particular fungal lineage of expertise were as correct and richly annotated as possible in the public sequence databases.

Another reason to ensure that your fungi of expertise are tidy and properly annotated in the sequence databases is that your research is likely to benefit from it. When other researchers recover those species in their samples, those species will then be named and annotated correctly owing to your expert curation. This will facilitate knowledge building for individual species and will prevent publication of results attached to spurious names. The user is, furthermore, likely to find that browsing the species hypothesis alignments in UNITE is a rewarding undertaking in terms of data visualization and exploration. We often hear from users who have found hitherto unrecognized patterns in the data as a result of the simultaneous exploration of species hypothesis boundaries, sequence data, and information on country and host of collection. These patterns are typically not apparent in regular multiple sequence alignments, let alone in the lists of sequences in INSDC that some researchers monitor to stay updated on developments in their fungal lineage of interest. With this chapter, we welcome all researchers to examine the species hypothesis alignments of their fungal groups of interest.

Most researchers would probably agree that you should try to review approximately as many manuscripts as you submit for review on an annual basis. That way, some sort of balance is attained in that you both add to, and draw from, the collective expertise of the scientific community. The same line of thinking should be applied to sequence data; if you plan to undertake a series of NGS studies of fungal communities and thus draw from the pool of available reference sequences, then it would be good if you could allot a small amount of money to sequence, say, a handful of previously unsequenced fungal type specimens/ex-type cultures in your local herbarium/culture collection and make those sequences publicly available. Similarly, when you find a public fungal sequence that is severely compromised, it would be very helpful if you could take action. To do nothing, which sadly is all too common, is an indefensible position in the long run. It makes mycology look bad and invites further mistakes and suboptimal results based on that particular compromised entry. Mycology, we feel, is not a field whose name should conjure up visions of suboptimal results and failure to resolve research hypotheses in the mind of the general scientific audience. Indeed, mycology struggles for funding in competition with fields that are often deemed larger or more fashionable, and we simply cannot afford public fungal DNA sequences to remain in a suboptimal state. Fortunately, at least part of the remedy comes at no cost and is no further away than the nearest Internet connection.

## References

Altschul SF, Madden TL, Schäffer AA, Zhang J, Zhang Z, Miller W, Lipman DJ (1997) Gapped BLAST and PSI-BLAST: a new generation of protein database search programs. Nucleic Acids Res 25:3389–3402

Bengtsson J, Eriksson KM, Hartmann M et al (2011) Metaxa: a software tool for automated detection and discrimination among ribosomal small subunit (12S/16S/18S) sequences of archaea, bacteria, eukaryotes, mitochondria, and chloroplasts in metagenomes and environmental sequencing datasets. Antonie Van Leeuwenhoek 100:471–475

Bengtsson-Palme J, Ryberg M, Hartmann M et al (2013) Improved software detection and extraction of ITS1 and ITS2 from ribosomal ITS sequences of fungi and other eukaryotes for analysis of environmental sequencing data. Methods Ecol Evol 4:914–919

Bidartondo M, Bruns TD, Blackwell M et al (2008) Preserving accuracy in GenBank. Science 319:1616

Blaxter M, Mann J, Chapman T, Thomas F, Whitton C, Floyd R, Abebe E (2005) Defining operational taxonomic units using DNA barcode data. Philos Trans R Soc Lond B Biol Sci 360:1935–1943

Caporaso JG, Kuczynski J, Stombaugh J et al (2010) QIIME allows analysis of high-throughput community sequencing data. Nat Methods 7:335–336

Cornish-Bowden A (1985) Nomenclature for incompletely specified bases in nucleic acid sequences: recommendations 1984. Nucleic Acids Res 13:3021–3030

Edgar RC, Haas BJ, Clemente JC, Quince C, Knight R (2011) UCHIME improves sensitivity and speed of chimera detection. Bioinformatics 27:2194–2200

Fonseca VG, Nichols B, Lallias D, Quince C, Carvalho GR, Power DM, Creer S (2012) Sample richness and genetic diversity as drivers of chimera formation in nSSU metagenetic analyses. Nucleic Acids Res 40:e66

Förster H, Coffey MD, Elwood H, Sogin ML (1990) Sequence analysis of the small subunit ribosomal RNAs of three zoosporic fungi and implications for fungal evolution. Mycologia 82:306–312

Gardes M, Bruns TD (1993) ITS primers with enhanced specificity for basidiomycetes - application to the identification of mycorrhizae and rusts. Mol Ecol 2:113–118

Gouy M, Guindon S, Gascuel O (2010) SeaView Version 4: a multiplatform graphical user interface for sequence alignment and phylogenetic tree building. Mol Biol Evol 27:221–224

Hartmann M, Howes CG, Abarenkov K, Mohn WW, Nilsson RH (2010) V-Xtractor: an open-source, high-throughput software tool to identify and extract hypervariable regions of small subunit (16S/18S) ribosomal RNA gene sequences. J Microbiol Methods 83:250–253

Hartmann M, Howes CG, Veldre V et al (2011) V-RevComp: automated high-throughput detection of reverse complementary 16S ribosomal RNA gene sequences in large environmental and taxonomic datasets. FEMS Microbiol Lett 319:140–145

Hawksworth DL (1991) The fungal dimension of biodiversity: magnitude, significance, and conservation. Mycol Res 95:641–655

Hawksworth DL (2001) The magnitude of fungal diversity: the 1.5 million species estimate revisited. Mycol Res 105:1422–1432

Hibbett DS, Ohman A, Glotzer D, Nuhn M, Kirk P, Nilsson RH (2011) Progress in molecular and morphological taxon discovery in Fungi and options for formal classification of environmental sequences. Fungal Biol Rev 25:38–47

Hyde KD, Udayanga D, Manamgoda DS et al (2013) Incorporating molecular data in fungal systematics: a guide for aspiring researchers. Curr Res Environ Appl Mycol 3:1–32

Kõljalg U, Nilsson RH, Abarenkov K et al (2013) Towards a unified paradigm for sequence-based identification of fungi. Mol Ecol 22:5271–5277

Lanzén A, Jørgensen SL, Huson DH, Gorfer M, Grindhaug SH, Jonassen I, Øvreås L, Urich T (2012) CREST – classification resources for environmental sequence tags. PLoS One 7:e49334

Larsson A (2014) AliView: a fast and lightweight alignment viewer and editor for large datasets. Bioinformatics 30:3276–3278

Lindahl BD, Nilsson RH, Tedersoo L, Abarenkov K, Carlsen T, Kjöller R, Koljalg U, Pennanen T, Rosendahl S, Stenlid J, Kauserud H (2013) Fungal community analysis by high-throughput sequencing of amplified markers - a user's guide. New Phytol 199:288–299

Liu L, Li Y, Li S, Hu N, He Y, Pong R, Lin D, Lu L, Law M (2012) Comparison of next-generation sequencing systems. J Biomed Biotechnol 2012:1–11

Morgado LN, Semenova TA, Welker JM, Walker MD, Smets E, Geml J (2014) Summer temperature increase has distinct effects on the ectomycorrhizal fungal communities of moist tussock and dry tundra in Arctic Alaska. Glob Chang Biol 21(2):959–972. doi:10.1111/gcb.12716

Nakamura Y, Cochrane G, Karsch-Mizrachi I (2013) The International Nucleotide Sequence Database Collaboration. Nucleic Acids Res 41(D1):D21–D24

Nilsson RH, Abarenkov K, Veldre V, Nylinder S, De Wit P, Brosche S, Alfredsson JF, Ryberg M, Kristiansson E (2010) An open source chimera checker for the fungal ITS region. Mol Ecol Resour 10:1076–1081

Nilsson RH, Veldre V, Wang Z et al (2011) A note on the incidence of reverse complementary fungal ITS sequences in the public sequence databases and a software tool for their detection and reorientation. Mycoscience 52:278–282

Nilsson RH, Tedersoo L, Abarenkov K et al (2012) Five simple guidelines for establishing basic authenticity and reliability of newly generated fungal ITS sequences. MycoKeys 4:37–63
Nilsson RH, Hyde KD, Pawlowska J et al (2014) Improving ITS sequence data for identification of plant pathogenic fungi. Fungal Divers 67:11–19
Qiu X, Wu L, Huang H, McDonel PE, Palumbo AV, Tiedje JM, Zhou J (2001) Evaluation of PCR-generated chimeras, mutations, and heteroduplexes with 16S rRNA gene-based cloning. Appl Environ Microbiol 67:880–887
Rosling A, Cox F, Cruz-Martinez K, Ihrmark K, Grelet GA, Lindahl BD, Menkis A, James TY (2011) Archaeorhizomycetes: unearthing an ancient class of ubiquitous soil fungi. Science 333:876–879
Ryberg M, Kristiansson E, Sjökvist E, Nilsson RH (2009) An outlook on the fungal internal transcribed spacer sequences in GenBank and the introduction of a web-based tool for the exploration of fungal diversity. New Phytol 181:471–477
Schloss PD, Westcott SL, Ryabin T et al (2009) Introducing mothur: open-source, platform-independent, community-supported software for describing and comparing microbial communities. Appl Environ Microbiol 75:7537–7541
Schoch CL, Seifert KA, Huhndorf S et al (2012) Nuclear ribosomal internal transcribed spacer (ITS) region as a universal DNA barcode marker for Fungi. Proc Natl Acad Sci USA 109:6241–6246
Schoch CL, Robbertse B, Robert V et al (2014) Finding needles in haystacks: linking scientific names, reference specimens and molecular data for Fungi. Database (Oxford). doi:10.1093/database/bau061
Seifert KA, Rossman AY (2010) How to describe a new fungal species. IMA Fungus 1:109–116
Stajich JE, Berbee ML, Blackwell M, Hibbett DS, James TY, Spatafora JW, Taylor JW (2009) The fungi. Curr Biol 19:R840–R855
Tedersoo L, Abarenkov K, Nilsson RH et al (2011) Tidying up International Nucleotide Sequence Databases: ecological, geographical, and sequence quality annotation of ITS sequences of mycorrhizal fungi. PLoS One 6:e24940
Tedersoo L, Bahram M, Põlme S et al (2014) Global diversity and geography of soil fungi. Science 346:1256688
Yang ZL (2011) Molecular techniques revolutionize knowledge of basidiomycete evolution. Fungal Divers 50:47–58

# Chapter 5
# Challenges and Future Perspectives in the Systematics of *Kickxellomycotina*, *Mortierellomycotina*, *Mucoromycotina*, and *Zoopagomycotina*

**Gerald L. Benny, Matthew E. Smith, Paul M. Kirk, Eric D. Tretter, and Merlin M. White**

## Introduction

The basal lineages or early-diverging fungi contain the aquatic (zoosporic) chytrids (*Chytridiomycota* (including *Neocallimastigomycetes*), *Blastocladiomycota*) and the terrestrial (aplanosporic) zygomycetes (*Entomophthoromycota*, *Glomeromycota*, *Kickxellomycotina*, *Mortierellomycotina*, *Mucoromycotina*, *Zoopagomycotina*) (reviewed by Shelest and Voigt 2014). The members of the final four subphyla, *Kickxellomycotina*, *Mortierellomycotina*, *Mucoromycotina*, and *Zoopagomycotina*, which constitute four of the six predominantly terrestrial clades of the early-diverging fungi, have been discussed elsewhere (Benjamin 1979; Benny 2001, 2012; Benny et al. 2001, 2014; Voigt 2012; Voigt and Kirk 2014). The fifth member of the terrestrial early-diverging fungi, the phylum *Entomophthoromycota*, is discussed by (Gryganskyi et al. 2012, 2013; Humber 2016) and in the following chapter (Humber 2016). These fungi often are presented only in ordinal discussions or more recently as subphyla. The *Mortierellales* were previously combined with *Mucoromycotina* before the description of the *Mortierellomycotina* (Hoffmann et al. 2011). Information on the Internet for the *Asellariales* and *Harpellales* can be found at this URL that

G.L. Benny, Ph.D. (✉) • M.E. Smith Ph.D.
Department of Plant Pathology, University of Florida, PO Box 110680, Gainesville, FL 32611-0680, USA
e-mail: gbenny@ufl.edu

P.M. Kirk, Ph.D.
Mycology Section, Kew Gardens, Royal Botanic Gardens, Kew, Richmond, Surrey TW9 3AB, UK
e-mail: p.kirk@kew.org

E.D. Tretter, M.S. • M.M. White, Ph.D.
Department of Biological Sciences, Boise State University, Boise, ID 83725-1515, USA
e-mail: merlinwhite@boisestate.edu


D.-W. Li (ed.), *Biology of Microfungi*, Fungal Biology,
DOI 10.1007/978-3-319-29137-6_5

contains Lucid keys (http://keys.lucidcentral.org/key-server/data/0b08020c-0f0c-4908-8807-030c020a0002/media/Html/home.htm/), and the remaining zygomycetes exclusive of the *Asellariales*, *Harpellales*, and *Entomophthoromycota* can be located at the following URL (http://www.zygomycetes.org/).

A few taxa of the *Mortierellomycotina* and *Mucoromycotina* are economically important as the causative agents of plant diseases, including *Choanephora* blight on cucurbits and a few other vegetables [*Choanephora cucurbitarum*], postharvest diseases on a variety of fruits (*Gilbertella persicaria*, *Mucor piriformis*, *Rhizopus stolonifer*), and mucormycosis on humans and other mammals (Michailides and Spotts 1990; Ribes et al. 2000; Gomes et al. 2011; Benny et al. 2014)]. Other taxa are used in industry for biotechnology, biodegradation, biosorption, bioremediation, and biotransformation. Some *Mucorales* are used in the production of fermented Asian foods and for laboratory studies on photobiology and physiology, including zygospore formation, the production of polyunsaturated fatty acids and β-carotene, as well as subsequent derivative products such as lycopene and various sterols (Benny 2012; Benny et al. 2014).

Many species of the *Kickxellomycotina*, *Mortierellomycotina*, *Mucoromycotina*, and *Zoopagomycotina* are deposited in culture collections around the world, although their ability to grow and persist in culture is highly variable among taxa. The most complete representations are in CBS (Centraalbureau voor Schimmelcultures, Utrecht, The Netherlands), IMI (CABI Europe UK, Bakeham Lane, Egham, Surrey, United Kingdom), and NRRL (USDA, ARS, NCAUR, Peoria, Illinois, USA). Other taxa or isolates may be obtained from soil, dung, and other organic material or host organisms using the methods described below. Some methods are widely useful for many taxa, whereas others are specialized for specific species or lineages within these subphyla.

The phylogenetic relationships of these fungi have been elucidated using molecular techniques, initially, using sequences from one or a few genes for: (1) *Mucorales* (O'Donnell et al. 2001; Voigt and Wöstemeyer 2001), (2) *Harpellales* and *Kickxellales* (O'Donnell et al. 1998), (3) *Dimargaritales* and *Zoopagales* (Tanabe et al. 2000), (4) *Harpellales* (White 2006), and (5) *Zygomycota sensu lato* (White et al. 2006a).

James et al. (2006) included six genes from approximately 200 taxa analyzed by Bayesian analysis to reveal the early evolution of fungi. The phylogeny of James et al. (2006) and other multilocus analyses of the fungi have presented a relatively stable classification at the ordinal level and above (Hibbett et al. 2007).

Karpov et al. (2014) and James et al. (2013) analyzed phylogenetic and phylogenomic data to determine that members of the *Cryptomycota* (synonym = *Rozellomycota*; Corsaro et al. 2014) such as *Amoeboaphelidium protococcarum* and *Rozella allomyces* belong to a clade that also includes the *Microsporidia*. Their results also suggest that this lineage is the sister clade of all other fungi. The branching order of the zoosporic chytrid lineages (*Chytridiomycota*, *Neocallimastigomycetes*, *Blastocladiomycota*) has not yet been confidently determined. In this chapter, we focus on isolation, culture preservation, detection in the environment, and evolution of those zygomycotan early-diverging fungi in the *Kickxellomycotina*, *Mortierellomycotina*, *Mucoromycotina*, and *Zoopagomycotina*.

## Subphylum *Kickxellomycotina* (Hibbett et al. 2007; Tretter et al. 2013, 2014) (Table 5.1)

Fungi produce hyphae that are regularly septate; each septum possesses a single, more or less central lenticular cavity containing a plug. Asexual reproduction by the formation of arthrospores, one- or two-spored merosporangia bearing

**Table 5.1** Classification of the four orders and four clades of the *Kickxellomycotina*

| |
|---|
| *Kickxellomycotina* Benny |
| *Asellariales* Manier ex Manier & Lichtw. |
| *Asellariaceae* Manier ex Manier & Lichtw. |
| *Asellaria* R.A. Poiss. |
| *Orchesellaria* Manier ex Manier & Lichtw. |
| Genus of unknown affinity |
| *Baltomyces* Cafaro emend. Oman and M.M. White |
| *Dimargaritales* R.K. Benj. |
| *Dimargaritaceae* R.K. Benj. |
| *Dimargaris* Tiegh. |
| *Dispira* Tiegh. |
| *Tieghemiomyces* R.K. Benj. |
| Genus of unknown affinity |
| *Spinalia* Vuil. |
| *Harpellales* Lichtw. & Manier |
| *Harpellaceae* L. Léger & Duboscq ex P.M. Kirk & P.F. Cannon |
| *Carouxella* Manier, J.-A. Rioux & Whisler ex Manier, J.-A. Roux & Lichtw. |
| *Harpella* L. Léger & O. Duboscq |
| *Harpellomyces* Lichtw. & S.T. Moss emend. Lichtw., M.M. White and Colbo |
| *Stachylina* L. Léger & M. Gauthier |
| *Stachylinoides* Lichtw. & López-Lastra |
| *Legeriomycetaceae* Pouzar |
| = *Genestellaceae* L. Léger & M. Gauthier |
| *Allantomyces* M.C. Williams & Lichtw. |
| *Austrosmittium* Lichtw. & M.C. Williams |
| *Bactromyces* R.T. William & Strongman |
| *Baetimyces* L.G. Valle & Santam. |
| *Barbatospora* M.M. White, Siri & Lichtw. |
| *Bojamyces* Longcore emend. L.G. Valle & Santam. |
| *Capniomyces* S.W. Peterson & Lichtw. |
| *Caudomyces* Lichtw., Kobayasi & Inhoh |
| *Coleopteromyces* Ferrington, Lichtw. & López-Lastra |
| *Dacryodiomyces* Lichtw. |
| *Ejectosporus* S.W. Peterson, Lichtw. & M.C. Williams emend. Strongman |

(continued)

**Table 5.1** (continued)

| |
|---|
| *Ephemerellomyces* M.M. White & Lichtw. |
| *Furculomyces* Lichtw. & M.C. Williams |
| *Gauthieromyces* Lichtw. |
| *Genistelloides* S.W. Peterson, Lichtw. & B.W. Horn |
| *Genistellospora* Lichtw. |
| *Glotzia* M. Gauthier ex Manier & Lichtw. |
| *Graminella* L. Léger & M. Gauthier ex Manier |
| *Graminelloides* Lichtw. |
| *Klastostachys* Lichtw., M.C. Williams & M.M. White |
| *Laculus* R.T. William & Strongman |
| *Lancisporomyces* Santam. |
| *Legeriodes* M.M. White |
| *Legerioides* M.M. White |
| *Legeriomyces* Pouzar |
| = *Genistella* L. Léger & M. Gauthier |
| *Legeriosimilis* M.C. Williams, Lichtw., M.M. White & J.K. Misra |
| *Orphella* L. Léger & M. Gauthier emend. Santam. & Girbal |
| *Pennella* Manier ex Manier |
| *Plecopteromyces* Lichtw., Ferrington & López-Lastra |
| *Pseudoharpella* Ferrington, M.M. White & Lichtw. |
| *Pteromaktron* Whisler |
| *Simuliomyces* Lichtw. |
| *Sinotrichium* Juan Wang, S.Q. Xu & Strongman |
| *Smittium* R.A. Poiss. |
| *Spartiella* Tuzet & Manier ex Manier |
| *Stipella* L. Léger & M. Gauthier |
| *Tectimyces* L.G. Valle & Santam. |
| *Trichozygospora* Lichtw. |
| *Trifoliellum* Strongman & M.M. White |
| *Zancudomyces* Yan Wang, Tretter, Lichtw. & M.M. White |
| *Zygopolaris* S.T. Moss, Lichtw. & Manier |
| *Kickxellales* Kreisel ex R.K. Benj. |
| *Kickxellaceae* Linder |
| *Coemansia* Tiegh. & G. Le Monn. |
| *Dipsacomyces* R.K. Benj. |
| *Kickxella* Coem. |
| *Linderina* Raper & Fennell |
| *Martensella* Coem. |
| *Martensiomyces* Meyer |
| *Mycoëmilia* Kurihara, Degawa & Tokum. |
| *Myconymphaea* Kurihara, Degawa & Tokum. |
| *Pinnaticoemansia* Kurihara & Degawa |
| *Ramicandelaber* Y. Ogawa, S. Hayashi, Degawa & Yaguchi |
| *Spirodactylon* R.K. Benj. |

(continued)

**Table 5.1** (continued)

| |
|---|
| *Spiromyces* R.K. Benj. |
| From the *Harpellales* |
| Clade 1. *Barbatospora* M.M. White, Siri & Lichtw. |
| Clade 2. *Orphella* L. Léger & M. Gauthier emend. Santam. & Girbal |
| From the *Kickxellales* |
| Clade 3. *Mycoëmilia* Kurihara, Degawa & Tokum. and *Spiromyces* R.K. Benj. |
| Clade 4. *Ramicandelaber* Y. Ogawa, S. Hayashi, Degawa & Yaguchi |
| Possible members of the *Kickxellomycotina* |
| *Ballocephala* Drechsler |
| *Zygnemomyces* Miura |

sporangiospores, or trichospores. Sexual reproduction by zygospores that are more or less globose, fusiform lanceolate, or long cylindrical. Suspensors either: (1) somewhat differentiated and with only one produced or (2) hyphoid, with two or three produced in the substratum. Mostly saprobes with the exception of a putative parasite, *Martensella* (*Kickxellales*), haustorial parasites of fungi (*Dimargaritales*), or symbiotic and attached to the gut lining of an arthropod by a cellular or noncellular holdfast (*Asellariales*, *Harpellales*).

Four orders, *Asellariales*, *Dimargaritales*, *Harpellales*, and *Kickxellales*, and an additional four distinct clades may deserve formal recognition in the future. Two other genera may belong in this subphylum based on septal ultrastructure.

The *Kickxellomycotina* contained four orders when the subphylum was proposed: *Asellariales*, *Dimargaritales*, *Harpellales*, and *Kickxellales* (Hibbett et al. 2007). The cardinal feature of the subphylum is the formation of regularly septate hyphae that have a lenticular cavity containing a variously shaped plug, depending on the species. Previous studies have suggested that three genera of the *Kickxellales* s.l. likely deserve to be recognized as separate orders: (1) *Mycoëmilia* + *Spiromyces* and (2) *Ramicandelaber* (Kurihara 2002; Kurihara et al. 2005; Ogawa et al. 2005). A phylogenetic construction by Tretter et al. (2013, 2014) recovered the *Mycoëmilia* + *Spiromyces* clade and the *Ramicandelaber* clade as well as two additional groups (the *Barbatospora* clade and the *Orphella* clade). Two genera, *Ballocephala* and *Zygnemomyces*, were transferred to the *Kickxellomycotina* from the *Meristacraceae* (*Entomophthoromycota*). Humber (2016) made this transfer because one member of each genus produces septal plugs in lenticular cavities (Saikawa 1989; Saikawa et al. 1997b).

## *Order* Asellariales *(Lichtwardt 1986; Benny 2001; Valle and Cafaro 2008; Benny et al. 2014)*

Mycelium branched, composed of regularly formed, plugged septa. Asexual reproduction by arthrospores. Globose, intercalary, or terminal chlamydospores formed by some species. Sexual reproduction by formation of conjugation tubes that result in the formation of more or less globose, thick-walled zygospores. Thalli are attached to the hindgut by a cellular holdfast or cells secreting a holdfast. Symbionts in the gut of either Collembola or Isopoda.

One family: *Asellariaceae* with two or three genera.

Members of the *Asellariales* are attached to the hindgut of their hosts by a cellular or acellular holdfast and species identification for some species hinges on holdfast morphology (Valle 2006; Valle and Cafaro 2008). Taxa of *Asellaria* are associated only with Isopoda (Arthropoda), whereas species of *Orchesellaria* (Valle 2006) are specific to Collembola (Insecta). *Baltomyces styrax*, a putative member of the order, also is associated with isopods (Oman and White 2012). These fungi, along with the *Harpellales*, are often referred to collectively as the trichomycete fungi.

Morphological evidence for including *Asellariales* in the *Kickxellomycotina* includes the presence of septa containing plugs in a lenticular cavity (Manier 1973; Saikawa et al. 1997a, b) and the discovery of zygospores in *Asellaria jatibonicua* (Valle and Cafaro 2008). Chlamydospores bearing some resemblance to the zygospores of *Asellaria* (Valle and Cafaro 2008) have been reported in *Orchesellaria* (Lichtwardt and Moss 1984; Degawa 2009). Sequence data also support the inclusion of the *Asellariales* in the *Kickxellomycotina* (Tretter et al. 2014).

Entomological forceps have been recommended for collecting isopods under moist organic substrata and rocks for terrestrial species (Valle 2006). Common suburban "roly-polies" or pill bugs and woodlice from gardens or compost piles are hosts that are easy to collect. Other than handpicking with forceps, some freshwater hosts may otherwise be dislodged and swept up with an aquatic net using a kick sampling technique (Oman and White 2012). Marine hosts may be collected in tide pools or near the shoreline using a net. These hosts are typically placed in a jar containing ocean water and native vegetation and then supplemented with an oxygen tablet. "Essentially terrestrial" marine hosts, such as rock lice, might be found near the high tide zone on rocks or among shoreline debris. All samples should be placed in a cooler for transport to the laboratory.

Trichomycetologists are increasingly paying attention to fungi outside of actual animal guts. Because the digestive tract serves as conduit for the attachment and establishment of the thallus, spores will be dispersing and cycling from inside the gut to the environment and back again. In many cases, critical life history stages of the fungus might be found in the molted exoskeleton. This is especially true for hindgut-dwelling fungi, as the gut lining is ectodermal in origin and is shed with the exuvium at each stage of host development. This might include individual parts of the thallus or spores, including sexual spores, which can be key to some species identifications. Additionally, fungal development can parallel that of the host, and therefore the most mature specimens are often found with mature hosts or those nearing the end of an instar stage.

Some isopods are large enough to dissect by eye, but typically a microscope and fine tools are needed for further manipulation and dissection. Most hyaline fungi are best spotted with a dissecting scope equipped with an indirect backlighting. As soon as possible after collecting, candidate hosts are dissected to remove the digestive tract. Different regions of the gut might be specifically targeted before opening to look for endobionts attached along the inside. The gut is opened by slicing along the length or tearing smaller portions of it open. Depending on content, the gut may need to be rinsed with distilled water to flush it and subsequently reveal the fungi. It is common to dissect larger hosts in a Petri dish before moving specific sections of the digestive tract to glass slides as wet mounts. Dissecting needles aid final opening and teasing of gut sections of interest as well as orienting the microfungi (with practice these can be seen at 40–60x). A coverslip is then added to scan the specimen for trichomycete fungi using a compound light microscope. Thalli are preserved by replacing the water with lactophenol cotton blue, which is infiltrated from under the edge of the coverslip that has minimal distilled water (i.e., after excess has evaporated or been wicked away gently with a tissue). Infiltration can be in as little as 20 min or slides can be left 24–48 h before double sealing with clear fingernail polish. Wet-mounted specimens may be imaged before (living) or after fixation and staining (with lactophenol cotton blue). Many such "semipermanent" vouchers have been revisited and imaged decades later. Some slides will dry over time if not sealed well, so care and diligence with sealing coverslips is important. Drawings and measurements can be performed later (Valle 2006; Valle and Cafaro 2008), although some gut fungi suffer from distortion due to storage on slides (Kandel and White 2012).

Using similar methods, fungi may be obtained from collections of Collembola. Springtails offer a challenge due to their small size but they can occur in large numbers and varied habitats. Many could be obtained with modifications of a Baermann funnel approach, but most collections have been obtained during larger surveys from aquatic habitats where they can be present as clusters on the surface of water near the edges of creeks, streams, or pools. The use of a small aspirator can aid their collection as well, but in many cases, springtails can occur in large enough numbers to be swept by hand into a net, bag, or jar. Except for their occasional mention in survey-based reports, perhaps no group of trichomycete hosts has been more overlooked in recent years. Degawa (2009) extended our understanding of their developmental potential, which may play into possible future attempts to culture *Orchesellaria* species and understand their ecology. The phylogenetic placement of this understudied group is clearly in need of further research (Tretter et al. 2014).

*Baltomyces* is one of the smallest trichomycetes endobionts, but has been observed across multiple states in the USA and over extended sampling occasions in Idaho (Oman and White 2012). Spores have been observed with appendage-like filaments. They are also present in freshwater isopods (Oman and White 2012).

There is no report of any member of the *Asellariales* being successfully isolated in axenic culture, nor is there any known commercial value for species in this group. However, it seems likely that some species could be cultivated under the proper conditions.

## *Order* Dimargaritales *(Benjamin 1979)*

Vegetative hyphae septate, simple, or branched, giving rise to simple or branched merosporangiophores. Septa with a lenticular cavity containing a biconvex plug bearing polar protuberances; plugs dissolving in dilute KOH or NaOH. Asexual reproduction by two-spored merosporangia borne directly on a fertile vesicle or on fertile branchlets arising from vesiculate or avesiculate merosporangiophore branches. Sexual reproduction by more or less globose zygospores borne on sexual hyphae. In nature obligate, haustorial mycoparasites. Haustorial parasites of species of *Chaetomium* (*Ascomycota*), *Mortierella* (*Mortierellales*, *Mortierellomycotina*), and *Mucorales* (*Mucoromycotina*).

One family: *Dimargaritaceae* with three or four genera.

Members of the *Dimargaritales* are typically isolated from dung and rarely from soil. These fungi are regularly septate and produce a septal plug bearing a protuberance that extends from each surface into the cytoplasm of the cells on each side of the septum (Benjamin 1959; Benny 1972; Brain et al. 1982). The merosporangia are two-spored and the spores are either dry or wet at maturity, depending on the species. The zygospores are formed directly in the substratum and are more or less globose with a hyaline wall and with typical hyphoid suspensors.

Members of the *Dimargaritales* are not common but appear to be worldwide in distribution. The *Dimargaritales* are all haustorial mycoparasites of *Chaetomium* (*Dispira implicata*, *D. simplex*; Benjamin 1959, 1961; Misra and Lata 1979), *Mortierellales* (Mehrotra and Baijal 1963), or *Mucorales* (Benjamin 1959, 1961, 1963, 1965; Misra and Lata 1979). All species of *Dimargaris* and *Tieghemiomyces* and both of the other species of *Dispira* (*D. cornuta*, *D. parvispora*) parasitize species of *Mortierella* or the *Mucorales* in nature. In the laboratory, Benjamin (1959, 1961, 1963, 1965) cultivated *Dispira simplex* on cultures of *Chaetomium botrychodes*. Normal development of both host and parasite occurred when grown on MEYE agar, PDA, and YpSs agar, but the parasite could not be grown without the host (Benjamin 1961). The remaining species were grown on YpSs agar using *Cokeromyces recurvatus* as the host (Shanor et al. 1950; Benny and Benjamin 1976). These fungi grow and sporulate well from 21 °C to room temperature (23–24 °C), but *Dimargaris cristalligena* grows optimally at 18 °C.

Culture of *Dimargaritales* has been reported several times in the literature. Ayers (1933) grew *D. cornuta* in culture without a host on high protein content culture medium (beef, egg, swordfish), but the egg medium proved to be the best for *D. cornuta*. Benjamin (1959, p. 384) was able to grow *D. cornuta* in pure culture through four transfers on MEYE agar. As with *D. simplex*, *Dispira parvispora* cannot grow or sporulate in culture without a host (Benjamin 1961, 1963). *Dimargaris cristalligena*, *D. bacillispora*, and *D. verticillata* grow saprobically on PAB-DEX agar, PDA, and YpSs agar, but optimal growth occurs on MEYE agar (Benjamin 1959, p. 368).

Benjamin reported that species of *Tieghemiomyces* can be grown on MEYE agar without host fungi (Benjamin 1959, p 392; Benjamin 1961, p. 11). Some members of the *Dimargaritales* can be grown on a culture medium (YGCH agar) emended with vitamins (biotin, pyridoxine, thiamine) using glycerol as the carbon source (O'Donnell et al. 1998; see also Barnett 1970). Axenic cultures of these taxa are extremely slow growing, and only five of the taxa (*Dimargaris verticillata* [RSA 527], an undescribed species of *Dimargaris* [RSA 2174], *Dispira cornuta* [RSA 632], *Tieghemiomyces californicus* [RSA 1194], *T. parasiticus* [RSA 2429B]) sporulate under these conditions. Methods of isolation and cultivation can be found in Benjamin (1959) and Benny (2008). The fastest recovery of members of the *Dimargaritales* from lyophils occurs on V8 juice agar and possibly also on clarified V8 (cV8) juice agar.

The best dishes to observe maximum sporophore height of *Dimargaris cristalligena* with *Cokeromyces recurvatus* are 7 cm high and contain YpSs agar incubated at 21 °C. When *D. cristalligena* is grown in test tubes containing unslanted YpSs agar, the sporophores are shorter (5–8 mm long), but zygospores are formed abundantly.

*Spinalia radians* (Vuillemin 1904; Benjamin 1959; Wrzosek and Gajowniczek 1998) may also be a member of the *Dimargaritales*, but its final placement will depend on obtaining an axenic culture and sequencing informative DNA regions. *Spinalia radians* was reported from France growing on *Mucor fragilis* (Vuillemin 1904) and Poland as a parasite of *Mucor hiemalis* (Wrzosek and Gajowniczek 1998).

There is no known economic use for species of *Dimargaritales* although as mycoparasites they have potential uses as biocontrol agents of spoilage fungi.

## *Order* Harpellales *(Lichtwardt 1986; Benny 2001; Lichtwardt et al. 2001)*

Vegetative hyphae simple or branched, septate. Each septum with a lenticular cavity containing a biconvex plug. Asexual reproduction by the formation, in basipetal succession, of unispored trichospores with one or more long, thin appendages or appendages lacking. Sexual reproduction by variously attached biconical or fusiform zygospores, lanceolate zygospores attached basally, or spores that are elongate-cylindrical or somewhat coiled at one end. Symbionts typically attached with an acellular holdfast to the digestive tract lining of larval aquatic insects or, more rarely, isopods.

Two families: *Harpellaceae* with six genera and *Legeriomycetaceae* with 44 genera.

*Harpellales* are found in guts of their hosts using the methods described by Lichtwardt (1986). Members of several genera of the *Legeriomycetaceae* have been isolated and cultured on 10 % BHIv overlaid with a thin layer of sterile, distilled water (Benny 2001; Benjamin et al. 2004; Benny et al. 2014; Lichtwardt 1986; Lichtwardt et al. 2001). A second culture medium (TGv) also has been used for many years (Benjamin et al. 2004; Lichtwardt 1986). More recently, success has been found by mixing these two media equally (BHIGTv) as an agar layer under a distilled water overlay. The key with this method, regardless of the agar medium, is to include an antibiotic rinse at the time of dissection and/or with the addition of the water overlay, especially during the initial or earliest attempts to move the gut fungi (and invariably minimal gut content or other host tissue) to the Petri dish. Petri dishes can be stored with or without parafilm around the lid, but care should be taken to not splash the water overlay near the lid.

Most of these fungi grow well at room temperature (20–22 °C), but certain taxa do better at specific and/or lower temperatures (e.g., many *Genistelloides* isolates do best at 18 °C). It is likely that our understanding of optimal culture conditions and temperatures will continue to evolve. For example, should we not be attempting isolations of fungi from winter-emerging insects in streams where the hosts and their fungi are at 1–5 °C in dishes that are held in a 4 °C fridge? Or with reduced oxygen? The latter do not represent typical approaches or modifications but are suggestions worthy of further consideration.

Daily monitoring of fungal growth is extremely important. Once growth of the fungus is noted, it should be monitored for thallus colony enlargement and spore production. As soon as possible, the colony should be partitioned into other dishes, eventually without antibiotic. Once growth is established axenically, some fungi (e.g., many *Smittium* species) will produce trichospores that extrude their sporangiospores in vitro, with each one potentially serving to produce a branching mass of typically asexually fertile thalli. Depending on the genus or species, the degree of vegetative growth versus sporulation will vary by culture medium employed. Current trends suggest that 10 % BHIv offers more spore production per colony compared to enhanced vegetative growth with TGv. Thus, once cultures are established, some attention to the degree of sporulation, ease of thallus fragmentation (either naturally with plate or slant agitation or by actively breaking up "tougher" thalli), and whether the spores extrude in vitro versus those that must be vegetatively propagated via thallus fragmentation are important details to consider. Some cultures of *Smittium* were first isolated in the early 1960s and maintained (at the laboratory of Robert Lichtwardt, University of Kansas) for years at 4 °C in living stock culture collections (as test tube slants with a distilled water overlay) and/or under liquid nitrogen. These collections are invaluable resources, deserving of the time, diligence, and efforts to maintain them.

Several new genera have been described in the *Legeriomycetaceae* since the recent list in Benny (2012): *Bactromyces*, *Dacrodiomyces*, *Laculus*, *Sinotrichum*, *Trifoliellum*, and *Zancudomyces* (Wang et al. 2010, 2013; Lichtwardt 2011; Strongman and White 2011; William and Strongman 2012). One new genus, *Klastostachys* (Lichtwardt et al. 2011), was recently described in the *Harpellaceae*. It is noteworthy that the disproportionate success and resultant bias in our attempts to culture gut fungi have been among members of the branched family, the *Legeriomycetaceae*, from the hindguts of aquatic insects. Historically, fungi were isolated first from larval Diptera (mosquitoes, midges, black flies, etc.) and then from stoneflies. Lesser known successes have been from mayflies and caddis worms (White, unpublished). The number of possible host groups worthy of consideration as candidates that could host putative culturable taxa of gut fungi will undoubtedly broaden with future attempts. All of these have been dissected from hindguts; the pH of the media used for these successful isolation attempts has always been slightly acidic and therefore approximating the conditions of the hindgut in this broad range of hosts.

Why has the *Harpellaceae* not been isolated under similar conditions? Is it just that fewer attempts have been made, in the pursuit of the "more likely" success among the hindgut-dwelling taxa, or could it be that just adjusting the in vitro conditions to better match the midgut (where unbranched members of the *Harpellaceae* reside) would improve success? Undoubtedly the key is triggering the release of the sporangiospores from the sporangium or trichospore (or the zygospore for that matter), and these fungi have adapted to specific gut triggers such as pH and specific ions. Once in culture, they exhibit growth patterns and rates that rival fungi in the *Dikarya*. Future efforts to culture members of both families of these fungi are worthwhile.

Zygospores are not observed in culture, but have been described for several of the genera. Four zygospore types are currently recognized (Moss et al. 1975; Lichtwardt 1986), but there are several genera where zygospore formation has not been observed. The zygospores of species of *Orphella* (Valle and Santamaria 2005) are unique in that they are long, cylindrical, and coiled rather than fusiform or lanceolate as is characteristic of the other taxa where sexual reproduction is known (Lichtwardt 1986).

An undescribed stage (ovarian cysts) in the *Harpellales* life cycle was first revealed by Moss and Descals (1986), and the life cycle of *Harpellales* in infected blackfly larvae with ovary cysts was published based on material from New York, USA (Labeyrie et al. 1996). Overnight incubations of cysts as wet mounts on slides can promote germ tube development and even spore production. Clearly there is potential for consideration of these stages as sources of material for culture isolation as well—most studies of gut fungi in insects focus first on immature stages, but the adult "dispersive" stage should not be overlooked. For example, infected adult blackflies swarm and can be field collected with healthy blackflies, but in the lab and to the trained eye, slight discoloration and distention of their abdomen can be used as a clue to the thousands of fungal cysts inside infected individuals (White unpublished and see White et al. 2006b).

White et al. (2006b) used molecular techniques to identify the *Harpellales* that are the causative agents of ovary cysts of blackfly larvae, including some of the same specimens from eastern North America. It is tempting to consider if future methods might be devised to induce ovarian cysts of *Harpellales* in blackfly larvae as a means of biocontrol.

Lipid content has been studied in *Smittium culisetae* (Patrick et al. 1973). There is also a wealth of possible comparative physiological studies that could be undertaken, especially as more cultures become available.

## *Order* Kickxellales *(Benjamin 1979; Benny 2012; Benny et al. 2014)*

Vegetative hyphae septate, simple, or branched, giving rise to simple or branched merosporangiophores. Septa with a lenticular cavity containing a biconvex plug; plugs do not dissolve in dilute KOH or NaOH. Asexual reproduction by unispored merosporangia arising from pseudophialides borne on uni- or multicelled sporocladia or directly from sporocladia and pseudophialides not formed. Sexual reproduction by more or less globose zygospores borne on sexual hyphae that is often hyphoid. Most species saprobes but one genus (*Martensella*) may be parasitic on *Corticium radiosum*, a resupinate member of the *Basidiomycota*.

One family: *Kickxellaceae* with 12 genera.

Four new monotypic genera (*Mycoëmilia* [*M. scoparia*], *Myconymphaea* [*M. yatsukahoi*], *Pinnaticoemansia* [*P. coronantispora*], *Ramicandelaber* [*R. longisporus*]) have been added to the *Kickxellales* since 2001 (Ogawa et al. 2001; Kurihara et al. 2001, 2004; Kurihara and Degawa 2006). *Myconymphaea yatsukahoi* was isolated from earwig dung in Japan (Kurihara and Degawa 2006; Kasuhiro and Degawa 2013). Kasuhiro and Degawa (2013) described another unnamed member of the *Kickxellales* with an unusual spore, a stalked sporulating head with several long sterile spines, and hyphae that appear to enclose the fertile head. This is likely an undescribed species but no other information is given, and the illustrations are extremely small and not diagnostic (Kasuhiro and Degawa 2013).

*Ramicandelaber* is currently the second largest genus in the order with four species (Ogawa et al. 2001; Kurihara et al. 2004; Chuang et al. 2013).

Three new species have been added to *Coemansia* in the last 15 years (*C. asiatica*, *C. furcata*, *C. javanensis*) (Kurihara et al. 2000, 2008). Kwaśna et al. (2002) renamed three species of *Coemansia* with spiral sporangiophores.

Both species of *Linderina* were reported from Indonesia (Kurihara et al. 2008) and Taiwan (Ho et al. 2007; Chuang and Ho 2009). These fungi (*L. macrospora*, *L. pennispora*) can be found in soil from the tropics to temperate regions with high humidity.

The majority of the taxa in the *Kickxellales* release spores in a droplet of fluid at maturity, including all of the species in the newly described genera *Mycoëmilia*, *Myconymphaea*, *Pinnaticoemansia*, and *Ramicandelaber* (Kurihara et al. 2001, 2004; Kurihara and Degawa 2006). Wet-spored species that have been studied by electron microscopy have spines embedded in the wall (Benny and Aldrich 1975; Young 1968; Zain et al. 2012). Taxa with spores that are dry at maturity are species of *Spirodacylon* (*S. aureum*) and *Spiromyces* (*S. aspiralis*, *S. minutus*) (Benjamin 1963; O'Donnell et al. 1998). These fungi produce spore walls that are covered with spines or warts (O'Donnell et al. 1998; Young 1968), but spines do not appear to be embedded in the spore wall.

The asexual apparatus of the *Asellariales*, *Harpellales*, and *Kickxellales* was discussed by Moss and Young (1978). Young (1999) reviewed the radiation of the asexual apparatus in *Kickxellales*, sporocladia and sporangiophores, and the morphology of the merosporangiospores of the eight genera treated by Benjamin (1958, 1959, 1961, 1963, 1965, 1966).

Konova et al. (2002) found that the fatty acid cis-9-hexadecenoic acid constituted 37 % of the total fatty acids produced by *Linderina pennispora*, and Konova et al. (2005) found that this compound was also present in *Kickxella alabastrina*. This and other fatty acids can be used as precursors for biofuel production. The sterol most frequently isolated from *Kickxellales*, *Dimargaritales*, and *Harpellales* is 22-dihydroergosterol (Weete et al. 2010).

## *Isolation and Culture of* Kickxellales

The most commonly encountered genus of the *Kickxellales* is *Coemansia* with 21 described taxa. The species of *Coemansia* can be isolated from both dung and soil. *Coemansia* cannot be found in all dung or soil collections, but sometimes one or two species can be isolated from a single soil sample. *Ramicandelaber* occasionally can be isolated from soil. The other taxa are rarely encountered or are only known from the original descriptions.

*Kickxellales* also can be isolated from soil that has been placed in Petri dishes (60 or 100 mm; glass may be better than plastic). The soil is moistened using a water mist. Other soil can be enriched with a sterile nutrient-poor broth (5 % yeast extract or 2.5 % peptone/2.5 % soytone). These soil plates can be baited with dried and broken edible shrimp, dried mealworm larvae, dried krill (food for birds or pet amphibians), dead aphids, or *Drosophila*. Soil plates are typically kept for approximately 30 days and examined for growth of *Kickxellales* or *Mortierella* using a dissecting microscope. Spores are then transferred to a nutrient-rich solid medium for isolation such as MEYE agar.

Members of the *Kickxellales* can be isolated from dung, soil, and other organic substrata. Isolation and culture of these fungi can be conducted using the methods listed in Kurihara and Degawa (2006), Kurihara et al. (2001, 2004), Krug (2004), Krug et al. (2004), and Benny (2008). Many taxa described in these publications were isolated from soil using crustacean baiting or enrichment (Kurihara et al. 2008). Benjamin (1959) recommended the use of MEYE or YpSs agars for growth and sporulation of many *Kickxellales*. Benjamin (1959) used a variety of media formulations to obtain isolates of *Coemansia*, depending on the species and isolate, including: carrot, CM, CM-S, 2 % ME, MED, ME-P, MEYE, OMS, PDA, PG, PYED, PYEDS, SDY, STA, Wg, Wg-DD, Wg-S, V8, WSH-DD, YGCH, and YpSs (Benny et al. 2008 and see Supplement 1).

Wg10 agar supplemented with antibiotics and benomyl is also excellent for isolating *Coemansia* (G. Benny, unpublished). The use of benomyl is critical to retard the growth of *Trichoderma*, *Penicillium*, and other rapidly sporulating *Ascomycota*.

## Subphylum *Mortierellomycotina* (Hoffmann et al. 2011) (Table 5.2)

Fungi that produce coenocytic hyphae, septa are formed to wall-off reproductive structures or old or damaged hyphae. Sporangiophores not cylindrical, base usually wider than apex, somewhat constricted basally. Asexual reproduction by chlamydospores or unicelled sporangiospores produced in uni-, few-, or mutispored sporangia. Columella not well developed, convex, or septoid. Sexual reproduction, where known, by zygospores formed in the substratum on apposed, heterogamous, or more or less isogamous suspensors. Saprobes, colony may appear to have irregular

**Table 5.2** Classification of the taxa of the *Mortierellomycotina*

| |
|---|
| *Mortierellomycotina* Kerst. Hoffm., K. Voigt & P.M. Kirk |
| *Mortierellales* Caval.-Sm. |
| *Mortierellaceae* A. Fischer |
| *Aquamortierella* Embree & Indoh |
| *Dissophora* Thaxt. |
| *Echinochlamydosporium* X.Z. Jiang, X.Y. Liu, Xing Z. Liu |
| *Gamsiella* (R.K. Benj.) Benny & M. Blackw. |
| *Lobosporangium* M. Blackw. & Benny |
| = *Echinosporangium* Malloch, non *Echinosporangium* Kylin |
| *Modicella* Kanouse |
| *Mortierella* Coem. |
| = *Haplosporangium* Thaxt. |
| = *Azygozygum* Chesters |
| = *Actinomortierella* Chalab. |
| Genus of unknown affinity |
| *Nothadelphia* Degawa & W. Gams |

zones and undulate margins on nutrient-rich media and may produce a garlic- or onion-like odor.

One order, *Mortierellales*, and one family *Mortierellaceae* (Benjamin 1978, 1979; Benny 2012; Benny et al. 2014) with eight or nine genera.

Members of the genus *Mortierella* are among the most common zygomycete fungi encountered in soil. These fungi also can be found on dung and other organic substrata. Procedures for the isolation and culture of *Mortierella*, and some of the other members of *Mortierellales*, can be found in Kuhlman (1972), Benny and Blackwell (2004), and Benny (2008). Gams (1976) recommended growing species of *Mortierella* at 18–22 °C for at least 7 days. Colony growth rate and macroscopic characters are determined by culture on MEA (2 % malt extract agar), and the microscopic structures (hyphae, sporophore morphology and branching, sporangiola, sporangiospores) are observed after growth on PCA (potato-carrot agar, CBS formulation) (http://www.cbs.knaw.nl/index.php/food-mycology/101-mycological-media-for-food-and-indoor-fungi) or SEA (soil extract agar). Some species of *Mortierella* do not readily sporulate in culture. However, these isolates can sometimes be induced to sporulate by transferring an agar plug from MEA containing the *Mortierella* colony to 2 % water agar. Often the *Mortierella* will then sporulate on the edges of the MEA plug.

Degawa and Tokumasu (1997) recommended a method for isolating species of *Mortierella*. They suggest collecting soil where the isopod *Armadillidium vulgare* was plentiful, placing it in 101 × 44 mm plastic sample cups, keeping it moist, and then placing a piece of sterile, dry shrimp on the soil surface. Spores of *Mortierella capitata* that formed on the shrimp were then transferred to Miura agar (LcA) for isolation (Sugiyama et al. 2003). Two disks of agar (5 mm diam.) that contain

*Mortierella* hyphae from different colonies were placed 10 mm apart on 0.3 % shrimp agar (ShA, Supplement 1) plates, incubated at various temperatures. When cultures were found to represent the opposite mating types, plugs of each isolate were placed 20 mm apart on Czapek's agar (CZA, Supplement 1) and a dead, sterile isopod placed in the center. The temperature optimum for *M. capitata* zygospore formation is 15–20 °C on Czapek's agar amended with an isopod (approximately 1 week) or ShA (approximately 2 weeks). These methods, with slight modification, were used to isolate and induce zygospore formation in *Mortierella cogitans*, *M. microzygospora*, and *M. umbellata* (Degawa and Tokumasu 1998a, b). Degawa and Gams (2004), however, placed pieces of bat dung on the surface of ShA and found that *Mortierella hypsicladia* was a good host for a new monotypic genus of unknown zygomycete affinity, *Nothadelphia mortierellicola. Mortierella indohii* also was a good host for *N. mortierellicola*, but other species of *Mortierella* that were tested did not function as hosts.

The other genera are known from only a single isolate or a few reports in the literature, including *Gamsiella multidivaricata*, *Lobosporangium transversale*, and *Modicella malleola* (Smith et al. 2013). Many of these taxa may be more common than currently thought because of exacting requirements for sporulation. *Lobosporangium transversale* will only sporulate on selected culture media, including CMA, CZA, HAS, LA, MEA, PAB, ShA, and TSM. In contrast, little or no sporulation is induced on AM, MAM, MEYE, PDA, PGA, V8, Wg, and YpSs when incubated at 25 °C, either in continuous light or 12 h light/12 h dark cycle (Benny and Blackwell 2004). *Modicella malleola* was isolated in pure culture on water, soil extract, or other media (Walker 1923). The spores readily germinate in water but cornmeal agar was used for cultural studies (Walker 1923).

*Echinochlamydosporium variable* (Jiang et al. 2011) was originally described as a member of the *Mortierellales*, but it is the sister species of a taxon in the *Mucoromycotina* (Hirose et al. 2014).

The culture of *Dissophora decumbens* (Thaxter 1914; Benny 1995a) was lost when Thaxter was away from his lab on an extended collecting trip. No culture was available when the description of *D. decumbens* was prepared. This fungus is a psychrophile (Carreiro and Koske 1992), and it was isolated from leaf litter using MYP-ps agar incubated at 0 °C when grown for 1–4 weeks. Sporulation occurred on PCA (CBS formulation; http://www.cbs.knaw.nl/index.php/food-mycology/101-mycological-media-for-food-and-indoor-fungi) incubated at 15–20 °C. *Dissophora decumbens* also was found in cloned sequences from beetle guts (Zhang et al. 2003); it is now in culture (Gams and Carreiro 1989). *Aquamortierella* (Embree and Indoh 1967) has not knowingly been cultured and is rarely collected or reported.

In a phylogenetic study published by Petkovits et al. (2011), the *Mortierellaceae* are distributed among 12 clades. Species of *Dissophora*, *Gamsiella*, and *Lobosporangium* are dispersed among species of *Mortierella*, making the latter genus paraphyletic. The sectional classification of *Mortierella* (Gams 1977), which is based on morphology, was not supported. Additional data that included ITS and LSU sequences from 400 cultures was added to the analysis of the *Mortierellomycotina* published by Wagner et al. (2013), and only seven clades were revealed. This study also demonstrated that the *Mortierella* sections were not maintained and that morphology of these fungi was variable depending on culture criteria.

*Mortierella wolfii* is the only species in the *Mortierellomycotina* that is the causative agent of zygomycosis and the only taxon that can grow at 37 °C (48 °C maximum). This species usually is known only as an animal pathogen, especially of cattle (Papp et al. 2011), but recently it was reported as the causative agent of mycosis in a human (Layios et al. 2014).

Crude glycerol, a major biodiesel production by-product, can be used as a carbon source instead of glucose, to produce arachidonic acid (AA) by *Mortierella alpina*. Several other species of *Mortierella* can use bioglycerol for the production of both AA and dihomo-γ-linolenic acid (Hou 2008). Münchberg et al. (2013) found that the hyphal oil composition is variable in the first 600 μm of the hyphae in *Mortierella alpina* and *M. elongata*.

Weete et al. (2010) reported that ergosterol is the major sterol of *Mucorales*. One of the following or combinations of 24-methyl-cholesterol, 24,25-methylene-cholesterol, and desmosterol are formed by *Mortierellales* (Weete et al. 2010). Sterol composition of members of the *Mortierellales* and *Mucorales* further supports the separation of these orders into different subphyla.

*Mortierellomycotina* (*Mortierellales*) and *Mucoromycotina* (*Mucorales*) grow readily, and at least one isolate of most genera in these subphyla is currently in culture collections. These fungi can be isolated from many substrata including dung (using a moist chamber) and soil sprinkled on agar (using nutrient-poor culture media emended with antibiotics and benomyl). Other procedures can be used to isolate these fungi from soil. The largest number of taxa of the *Mortierellales* is present in the Centraalbureau voor Schimmelcultures (CBS) filamentous fungi collection although other culture collections (ATCC, IMI, NRRL) also contain isolates of *Mortierella* and other members of the *Mortierellales*.

## Subphylum *Mucoromycotina* (Hibbett et al. 2007) (Table 5.3)

Fungi that produce coenocytic hyphae, septa are formed to wall-off reproductive structures or old or injured hyphae. Asexual reproduction by unicelled conidia arising from conidiogenous cells or unicelled sporangiospores produced in sporangia, sporangiola, merosporangia, or spores lacking or less commonly by arthrospores, chlamydospores, or yeast cells. Columella usually well defined and easily observed using the light microscope, hemispherical or obovoid to obpyriform, difficult to observe, or lacking in some unispored taxa. Sexual reproduction unknown or by the formation of zygospores produced in the aerial hyphae on opposed suspensors or in the substratum or in sporocarps on apposed suspensors. Saprobes, facultative gall-forming parasites, or ectomycorrhizal.

Two orders: *Endogonales* and *Mucorales* (as well as two additional unnamed clades).

**Table 5.3** Synopsis of the classification of the two orders and two clades of *Mucoromycotina* to family and subfamily

| |
|---|
| *Mucoromycotina* Benny |
| *Endogonales* Moreau ex R.K. Benj. emend. Morton & Benny |
| *Endogonaceae* Paoletti emend. J.B. Morton & Benny |
| *Endogone* Link |
| *Peridiospora* C.G. Wu & J. Lin |
| *Sclerogone* Warcup |
| *Youngiomyces* Y.J. Yao |
| Genus of unknown affinity |
| *Densospora* McGee |
| *Mucorales* Fr. |
| *Backusellaceae* K. Voigt & P.M. Kirk |
| *Backusella* Hesselt. & J.J. Ellis |
| *Choanephoraceae* J. Schröt. |
| = *Gilbertellaceae* Benny |
| *Choanephoroideae* K. Voigt & P.M. Kirk |
| *Blakeslea* Thaxt. |
| *Choanephora* Curr. |
| *Poitrasia* P.M. Kirk |
| = *Abradeosporangium* Subrahm. & Swathi Sri |
| *Gilbertelloideae* K. Voigt & P.M. Kirk |
| *Gilbertella* Hesselt. |
| *Cunninghamellaceae* R.K. Benj. emend. Benny, R.K. Benj. & P.M. Kirk |
| ≡ *Absidiaceae* Arx |
| *Cunninghamelloideae* K. Voigt & P.M. Kirk |
| *Cunninghamella* Matr. |
| *Absidioideae* K. Voigt & P.M. Kirk |
| *Absidia* Tiegh. s.s. |
| = *Tieghemella* Berl. & De Toni |
| = *Proabsidia* Vuill. |
| *Chlamydoabsidia* Hesselt. & J.J. Ellis |
| *Gongronella* Ribaldi |
| *Halteromyces* Shipton & Schipper |
| *Hesseltinella* H.P. Upadhyay |
| *Lentamycetaceae* K. Voigt & P.M. Kirk |
| *Lentamyces* Kerst. Hoffm. & K. Voigt |
| ?*Siepmannia* Kwaśna & Nirenberg ex Nirenberg & Kwaśna |
| *Lichtheimiaceae* Kerst. Hoffm., G. Walter & K. Voigt |
| *Dichotomocladioideae* K. Voigt & P.M. Kirk |
| *Dichotomocladium* Benny & R.K. Benj. |
| *Lichtheimioideae* K. Voigt & P.M. Kirk |
| *Lichtheimia* Vuill. |
| *Rhizomucoroideae* K. Voigt & P.M. Kirk |
| *Rhizomucor* Lucet & Costanin |

(continued)

**Table 5.3** (continued)

| |
|---|
| *Thermomucor* Subrahm., B.S. Mehrotra & Thirum. |
| *Mucoraceae* Dumort. |
| = *Chaetocladiaceae* A. Fisch. |
| = *Thamnidiaceae* Fitzp. |
| = *Dicranophoraceae* J.H. Mirza |
| *Dicranophoroideae* K. Voigt & P.M. Kirk |
| *Dicranophora* J. Schröt. |
| *Chaetocladioideae* K. Voigt & P.M. Kirk |
| *Chaetocladium* Fresen. |
| *Mucoroideae* K. Voigt & P.M. Kirk |
| *Actinomucor* Schostak. |
| = *Glomerula* Bainier |
| *Ambomucor* R.Y. Zheng & X.Y. Liu |
| *Circinomucor* Arx |
| = *Mucor* Fresen. |
| *Hyphomucor* Schipper & Lunn |
| ?*Isomucor* J.I. Souza, Pires-Zottar. & Harakava |
| *Kirkiana* L.S. Loh, Kuthub. & Nawawi |
| *Mucor* Fresen. |
| = ?*Zygorhynchus* Vuill. |
| *Nawawiella* L.S. Loh & Kuthub. |
| *Parasitella* Bainier |
| *Pilaira* Tiegh. |
| *Tortumyces* L.S. Loh |
| ?*Zygorhynchus* Vuill |
| *Zygambella* S. & A. Subrahm. |
| *Thamnidioideae* K. Voigt & P.M. Kirk |
| *Ellisomyces* Benny & R.K. Benj. |
| *Helicostylum* Corda emend. Benny |
| *Pirella* Bainier |
| *Thamnidium* Link, non *Thamnidium* Tuck. ex Schwend |
| *Mycocladaceae* Kerst. Hoffmann, S. Discher & K. Voigt |
| *Mycocladus* Beauverie |
| *Mycotyphaceae* Benny & R.K. Benj. |
| *Cokeromycetoideae* K. Voigt & P.M. Kirk |
| *Benjaminiella* Arx |
| *Cokeromyces* Shanor |
| *Kirkomycetoideae* K. Voigt |
| *Kirkomyces* Benny |
| = *Kirkia* Benny, non *Kirkia* Oliv. |
| *Mycotyphoideae* K. Voigt & P.M. Kirk |
| *Mycotypha* Fenner |
| *Phycomycetaceae* Arx |
| *Phycomyces* Kunze |
| *Spinellus* Tiegh. |

(continued)

**Table 5.3** (continued)

| |
|---|
| *Pilobolaceae* Corda |
| *Pilobolus* Tode |
| *Utharomyces* Boedijn |
| *Radiomycetaceae* Hesseltine & J.J. Ellis |
| *Radiomyces* Embree |
| *Rhizopodaceae* K. Schum. |
| *Amylomyces* Calmette |
| *Rhizopus* Ehrenb. |
| *Sporodiniella* Boedijn |
| *Syzygites* Ehrenb. |
| *Saksenaeaceae* Hesselt. & J.J Ellis |
| *Apophysomyces* P.C. Misra |
| *Saksenaea* S.B. Saksena |
| *Syncephalastraceae* Naumov ex R.K. Benj. |
| *Circinella* Tiegh. & G. Le Monn. |
| *Fennellomyces* Benny & R.K. Benj. |
| *Phascolomyces* Boedijn |
| *Protomycocladus* Schipper & Samson |
| *Syncephalastrum* J. Schröt. |
| *Thamnostylum* Arx & H.P. Upadhyay |
| *Zychaea* Benny & R.K. Benj. |
| *Umbelopsidaceae* W Gams & W. Meyer |
| *Umbelopsis* R.E. Amos & H.L. Barnett |
| = *Micromucor* (W. Gams) Arx |
| Clade 1. *Calcarisporiella* de Hoog |
| Clade 2. *Sphaerocreas* Sacc. & Ellis |
| Genera of unknown family affiliations |
| ?*Isomucor* J.I. Souza, Pires-Zottar. & Harakava |
| *Rhizopodopsis* Boedijn |
| ?*Siepmannia* Kwaśna & Nirenberg ex Nirenberg & Kwaśna |

## *Unnamed Clade 1*

Hyphae fragile, relatively wide and more or less undulating, aerial hyphae forming one or two conidiogenous cells that are undifferentiated. Conidiogenous cell swollen slightly above the base, two to several conidia-bearing denticles formed apically. Each denticle more or less cylindrical giving rise to conidia sympodially. Conidia ellipsoid to ovoid with a basal hilum and hyaline and walls smooth and thin. Sexual reproduction unknown.

A family, not described, contains only a monotypic genus, *Calcarisporiella thermophila*, although several unnamed taxa have been differentiated based on molecular data (Hirose et al. 2012).

The growth temperature range of *C. thermophila* is 25–40 °C. Isolation is possible when the plates are incubated at 35 °C. Sporulation was good at 25–30 °C. The temperature range for the unnamed species of *Calcarisporiella* (NBRC 105922) is 15–35 °C with the best growth occurring between 20 °C and 30 °C (Hirose et al. 2012). These isolates of *Calcarisporiella* grow on several ordinary fungal culture media when incubated in the optimal temperature range. This monotypic genus, *C. thermophila*, was originally thought to be an anamorphic member of the *Pezizomycotina* [Evans 1971; de Hoog 1974; Index Fungorum (http://www.indexfungorum.org/Names/Names.asp) accessed 11 October 2014]. A phylogenetic analysis of 18S rDNA sequences, however, demonstrated that *C. thermophila* is separate from the *Endogonales* and *Mucorales* clades. Another study of the relationship of *C. thermophila* (Hirose et al. 2014) revealed an unnamed *Calcarisporiella* that is the sister species to *Echinochlamydosporium variable* (Jiang et al. 2011; described as a member of the *Mortierellales*). Petkovits et al. (2011) and Wagner et al. (2013) did not have a culture or sequences of *E. variable* to include in their analyses of the *Mortierellales*.

## *Unnamed Clade 2*

Sporocarps, 0.2–2 mm in diameter, formed on dead plant material (branches, leaves, twigs), covered with a light-colored hyphal tomentum 60–100 µm long, borne in tightly attached bundles, 12–30 µm wide basally, and tapering to 2 µm apically. Sporocarps contain branched refractive hyphae ca. 2 µm in diameter, which give rise to numerous, randomly arranged chlamydospores, broadly ellipsoid to subspherical, 18 × 15 to 25 × 22 µm, walls yellowish, 1.5–2.5 µm thick, contents hyaline, with many granules and one or more oil droplets. Sporocarps hard when dry (Thaxter 1922).

A family, not named, contains only a monotypic genus: *Sphaerocreas pubescens.*

*Sphaerocreas pubescens* has been included in several genera by previous authors, including *Endogone*, *Glomus*, and *Sclerocystis*. Various sources believe that the current name should be *Glomus pubescens* (or *Sclerocystis pubescens* (*Glomeraceae* [Thaxter 1922; Index Fungorum (http://www.indexfungorum.org/Names/Names.asp) accessed 11 October 2014; Schüßler and Walker 2010; Hirose et al. 2014]).

*Sphaerocreas pubescens* may be a saprobe but the formation of sporocarps being produced in culture was not mentioned. The species was cultured on MNC (Modified Norkrans' C medium). The fungus was grown in the dark on MNC agar with added thiamine (100 mg) according to Hirose et al. (2014) before DNA extraction.

A phylogenetic analysis (Hirose et al. 2014) revealed that *S. pubescens* is the sister taxon to those fungi in *Mucoromycotina* that are symbionts of hornworts and liverworts but are known only from environmental DNA sequences (Bidartono et al. 2011). These latter sister taxa form a clade that is separate from the members of the *Endogonales* and *Mucorales* and a third clade composed of *Echinochlamydosporium* and *Sphaerocreas* (Hirose et al. 2014); *Calcarisporiella thermophila* is the fourth clade in this subphylum.

## *Order* Endogonales *(Benjamin 1979)*

Hyphae coenocytic, with few septa. Asexual reproduction unknown. Sexual reproduction usually by many zygospores with apposed suspensors that are formed in a sporocarp. Saprobes or ectomycorrhizal.

One family: *Endogonaceae* with four genera.

The majority of the *Endogonales* must be collected as sporocarps that are found on or in the substrate. One species, *Endogone pisiformis* Link, has been grown (*Endogone* agar) and forms sporocarps in culture (Berch and Fortin 1983; Berch and Castellano 1986; Dalpé 1990). A culture of *Endogone pisiformis* produces hyphal swellings and chlamydospore-like structures; mature zygospores were not produced. Optimal growth was attained when thiamine was added to the growth medium (Dalpé 1990). Sporocarps of *Endogone pisiformis* were produced in axenic cultures in vitro (Berch and Castellano 1986).

*Endogonales* all form sporocarps that contain only zygospores. The sporocarps can be collected intact in the field on leaves or other organic substrata or in soil. Some sporocarps are eaten and dispersed by rodents. These zygospores may be found in dung of these rodents. Most species are apparently saprobes but a few taxa are ectomycorrhizal.

Descriptions and illustrations for the four genera in the *Endogonales*—*Endogone*, *Pteridiospora*, *Sclerogone*, and *Youngiomyces*—can be found in Gerdemann and Trappe (1974), Yao et al. (1996), and Wu and Lin (1997). A species of *Endogone*, *E. maritima*, was described later (Błaszkowski et al. 1998).

Bidartono et al. (2011) discussed the possibility that a member of the *Mucoromycotina*, and not the *Glomeromycota* (Schüssler et al. 2001), was the first organism to form a symbiotic association with hornworts and liverworts. Desirò et al. (2013a), after conducting a worldwide survey, discovered that fungi from the *Glomeromycota* and *Mucoromycotina* both formed symbioses with hornworts. *Mollicutes*-related endobacteria (Mre) were found in almost 45 % (13 of 29) of the *Endogone* isolates examined by Desirò et al. (2014a); they also are known in *Glomeromycota* (Desirò et al. 2013b). Several isolates of *Gigaspora margarita* from soil collected in Cameroon (Desirò et al. 2014b) contained both Mre and *Candidatus Glomeribacter gigasporarum* (CaGg).

A fossil was recently described from the Middle Triassic of Antarctica that resembles a member of the *Endogonaceae* (*Jimwhitea circumtecta*). Another fossil species of *Endogonaceae*, *Mycocarpon asterinum*, also was described from Triassic sediments in Antarctica (Krings et al. 2012, 2013).

Jabaji-Hare et al. (1988) examined the fatty acid profile in *Endogone pisiformis*. The composition varied with the time in culture. Several fatty acids were produced with oleic acid being the most abundant, and both isomers (ω3, ω6) of linolenic acid were also formed, among others. Although it is possible to use *E. pisiformis* for fatty acid production, the isolates of this species are difficult to maintain in culture, and obtaining new isolates can be challenging because fresh sporocarps are difficult to find.

## *Order* Mucorales *(Benjamin 1979)*

Fungi that produce coenocytic hyphae, septa are formed to wall-off reproductive structures or old or damaged hyphae. Asexual reproduction by unicelled sporangiospores produced in sporangia, sporangiola, merosporangia, or spores lacking or less commonly by arthrospores, chlamydospores, or yeast cells. Columella usually well defined and easily observed using the light microscope, hemispherical of obovoid to obpyriform, difficult to observe, or lacking in some unispored taxa. Sexual reproduction by the formation of zygospores produced in the aerial hyphae on opposed suspensors or in the substratum on apposed suspensors. Saprobes or facultative parasites that form galls on the host.

Fifteen families, *Backusellaceae*, *Choanephoraceae*, *Cunninghamellaceae*, *Lentamycetaceae*, *Lichtheimiaceae*, *Mucoraceae*, *Mycocladiaceae*, *Mycotyphaceae*, *Phycomycetaceae*, *Pilobolaceae*, *Radiomycetaceae*, *Rhizopodaceae*, *Saksenaeaceae*, *Syncephalastraceae*, and *Umbelopsidaceae*, and at least 57 genera.

The families of the *Mucorales* are listed because some taxa require special conditions for culture. The abbreviations for the media are listed below and can be found in the Appendix.

### Families

*Backusellaceae* The sporangiophore is initially curved immediately below the sporangium but is erect at maturity. Sporangiola may be produced. One genus is included and the species are isolated from soil and other organic debris. Several species were transferred to *Backusella* (Walther et al. 2013), but three—*Backusella circina*, *B. lamprospora*, and *B. recurva*—were included by Hoffmann et al. (2013). Asexual growth and sporulation and zygospore formation of all species of *Backusella* occur at 26 °C on 2 % ME, MEYE, MSMA, and YpSs (Benny and Benjamin 1975). MSMA and other nutrient-poor media are optimal for observing branching patterns. One species, *B. ctenidia* (syn. = *Mucor ctenidia*), grows and sporulates asexually on most media, but zygospores form only on LYE agar.

*Choanephoraceae* As currently accepted, the *Choanephoraceae* is composed of two subfamilies (Voigt and Kirk 2012), *Choanephoroideae* (*Blakeslea*, *Choanephora*, *Poitrasia*; Kirk 1984) and *Gilbertelloideae* (*Gilbertella*; Benny 1991). All members form sporangiospores bearing several hyaline appendages at each end, and the sporangia have a dark, persistent, spinose wall that separates into two or more parts at maturity (Kirk 1984; Benny 1991). *Choanephoroideae* form zygospores in the substratum with apposed (parallel), coiled suspensors, whereas the *Gilbertelloideae* produce zygospores in the aerial hyphae with opposed (grow toward one another) suspensors. Nutrient-poor culture media, such as CHA (*Choanephora* agar), PCA, and Wg10, can be used to both isolate and culture members of the *Choanephoroideae*. *Gilbertella persicaria* grows and sporulates on any ordinary culture media used in the laboratory, including PDA, but observations of reproductive structures are best on nutrient-poor media.

*Cunninghamellaceae* The genera in this family are dispersed in two subfamilies (Voigt and Kirk 2012): *Cunninghamelloideae* (*Cunninghamella*) and *Absidioideae* (*Absidia* s.s., *Chlamydoabsidia*, *Gongronella*, *Halteromyces*, *Hesseltinella*). There are no morphological characters that are cardinal features of the genera in the family. *Cunninghamella* (Zheng and Chen 2001) produces pedicellate, unispored sporangiola on the surface of fertile vesicles. The *Absidia*-like taxa (*Absidia* s.s., *Chlamydoabsidia*, *Halteromyces*, *Gongronella*—Hesseltine and Ellis 1961, 1964, 1966; Ellis and Hesseltine 1965) produce apophysate, columellate sporangia on stolons but not opposite a rhizoid; a septum is formed in the sporangiophore. The zygospores, where known, have appendaged (*Absidia* s.s.) or non-appendaged suspensors (*Gongronella*). *Hesseltinella* (Benny and Benjamin 1991) is similar morphologically to *Gongronella* (the subsporangial vesicle of *Hesseltinella* resembles the constricted apophysis of *Gongronella*).

These taxa all grow and sporulate on most culture media, but V8 and cV8 agars are excellent for studying branching, spore formation, and other morphological characteristics. Zygospore formation can be induced on both nutrient-rich and nutrient-poor media. Zygospore ontogeny can be observed optimally on nutrient-poor agar media. A few species of *Absidia* s.s. are homothallic, but *Gongronella butleri* and other taxa of *Absidia* s.s. are heterothallic.

*Lentamycetaceae* Members of *Lentamycetaceae* (*Lentamyces*) are *Absidia*-like and homothallic. Zygospores are produced that have opposed suspensors that lack appendages (Hoffmann and Voigt 2009). The upper limit for growth temperature is 30 °C for both species (*Lentamyces parricida*, *L. zychae*). These taxa were discussed by Ellis and Hesseltine (1966) and Hoffmann and Voigt (2009). *Siepmannia* (Kwaśna and Nirenberg 2008a, b) may be transferred eventually to the *Lentamycetaceae* if supported by molecular data. Hoffmann and Voigt (2009), however, transferred two of the four species of *Siepmannia* to *Lentamyces*. Species of *Lentamyces* were crossed on SUP medium at 20 °C (Hoffmann and Voigt 2009).

*Lichtheimiaceae* The *Lichtheimiaceae* (Kirk 2012) is divided into three subfamilies (Voigt and Kirk 2012), *Dichotomocladioideae* (*Dichotomocladium*; Benny and Benjamin 1975, 1993), *Lichtheimioideae* (*Lichtheimia*; Alastruey-Izquierdo et al. 2010), and *Rhizomucoroideae* (*Rhizomucor*, *Thermomucor*; Schipper 1978b, 1979).

*Dichotomocladium* initially was treated as a member of the *Thamnidiaceae*. The former genus resembles *Chaetocladium* because members of both genera form sterile spines and unispored sporangiola. *Dichotomocladium* was includedin the *Chaetocladiaceae* by Benny and Benjamin (1993); both of the aforementioned families are now considered synonyms of *Mucoraceae* (Hoffmann et al. 2013). All known species of *Dichotomocladium*, except *D. hesseltinei*, have been isolated from dung (Benny and Benjamin 1975, 1993). The descriptions and drawings of all species of *Dichotomocladium* were made on MSMA. However, good fertile head formation occurs on LYE, TPO, and WSH but not on MEYE. Sexual reproduction is optimal on TPO, WSH, and whey. All species of *Dichotomocladium* are mesophiles (optimal growth at 26 °C). Zygospores of *D. hesseltinei* were observed when cultures were crossed on TPO and grown at 26 °C in the dark (Benny and Benjamin 1993).

*Lichtheimia* (Hoffmann et al. 2009; Santiago et al. 2014) is an *Absidia*-like genus in which five of six species are thermotolerant whereas *Absidia* s.s. is mesophilic (Hoffmann et al. 2007; Hoffmann 2010). There are six species in *Lichtheimia* and three (*L. corymbifera*, *L. ornata*, *L. ramosa*) are the causative agents of mucormycosis (Alastruey-Izquierdo et al. 2010; Schwartz et al. 2014).

*Rhizomucor* and *Thermomucor* (Schipper 1978b, 1979) are thermophilic *Mucorales* that comprise the remaining genera of the *Lichtheimiaceae*. *Rhizomucor*, *R. pusillus*, and *Thermomucor indicae-seudaticae* can cause mucormycosis. *Thermomucor*, however, is rarely reported as the causative agent of mycosis.

Species of *Lichtheimia*, *Rhizomucor*, and *Thermomucor* can be grown and will sporulate on almost any culture medium discussed above under *Dichotomocladium*.

*Mucoraceae* The *Mucoraceae* is, by a number of genera and species, the largest family of the *Mucorales*. Currently, three families (*Chaetocladiaceae*, *Dicranophoraceae*, *Thamnidiaceae*) are treated as synonyms of the *Mucoraceae* (Hoffmann et al. 2013). The *Mucoraceae* is divided into three subfamilies (Voigt and Kirk 2012): *Chaetocladioideae* (*Chaetocladium*), *Dicranophoroideae* (*Dicranophora*), and *Mucoroideae* (*Actinomucor*, *Ellisomyces*, *Helicostylum*, *Hyphomucor*, *Mucor*, *Parasitella*, *Pilaira*, *Pirella*, *Thamnidium*, *Zygorhynchus*, and possibly *Ambomucor*, *Isomucor*, *Kirkiana*, *Nawawiella*, *Tortumyces*, and *Zygambella*) (Benjamin and Hesseltine 1957; Benny and Benjamin 1975, 1976; Schipper 1975, 1978a, b, 1986; Benny and Schipper 1992; Benny 1992, 1995c; Volgmayr and Krisai-Greilhuber 1996; Loh et al. 2001; Nagalakhsmi et al. 2008; Zheng and Liu 2009, 2014; de Souza et al. 2012). A subfamily for the thamnidiaceous *Mucoraceae*, *Ellisomyces*, *Helicostylum*, *Pirella*, and *Thamnidium* currently in the *Mucoroideae*, may be justified.

The type species for the family and order is *Mucor mucedo* (Benny and Benjamin 1975; Benny and Schipper 1992; Benny 1992, 1995c). De Bary (1865) treated *Mucor mucedo* and *Thamnidium elegans* as the same species. Some taxa in the *Mucoraceae* are isolated from soil (*Ambomucor*, *Isomucor*, *Hyphomucor*, *Zygorhynchus*). In contrast, the thamnidiaceous *Mucoraceae* are best isolated from rodent dung: *Chaetocladium* and *Ellisomyces* (asexual reproduction on MSMA, zygospore formation on MSMA and YpSs), *Helicostylum* (asexual reproduction is optimal on MSMA at 18 °C, zygospore formation on whey in the dark at 7 °C for 60 days), *Pilaira* (asexual reproduction on MEYE and YpSs, zygospore formation on PDA at 25 °C), *Pirella* (asexual reproduction on MSMA at 22 °C, optimal zygospore formation on TPO at 15 °C), and *Thamnidium* (optimal asexual reproduction on MSMA at 18–22 °C, zygospore formation on MEYE, YPD, or YpSs when incubated at 7–10 °C for 3–5 weeks). Three genera are mycoparasites in nature (*Dicranophora* on mushrooms, *Chaetocladium* and *Parasitella* on *Mucorales*).

There are two genera (*Actinomucor, Mucor*) that can be isolated from dung, soil, and other substrates. *Dicranophora fulva*, a rare mushroom parasite, must be grown at 19 °C or less and optimal sporulation occurs on V8. Three other species (*Chaetocladium jonesii, Helicostylum elegans, H. pulchrum*) are psychrotolerant. *Chaetocladium brefeldii* and *C. jonesii* are gall-forming facultative parasites; both grow and sporulate well on MEYE. Zygospore formation of *C. brefeldii* occurs on MEYE and YpSs. Several taxa are most commonly encountered when it is cool (more northern latitude, higher elevation, or during winter in warm regions) including *Ambomucor, Chaetocladium, Dicranophora, Helicostylum, Pirella, Thamnidium*, and some species of *Mucor* and *Zygorhynchus*. Except for the taxa where optimal conditions are mentioned for asexual and sexual reproduction, the majority of the other species will grow and sporulate on almost any culture medium at 25 °C.

*Pilaira* has been included with several classical genera of thamnidiaceous fungi including *Helicostylum, Pirella*, and *Thamnidium* based on the analysis of sequence data (Hoffmann et al. 2013; Walther et al. 2013). *Pilaira* lacks a subsporangial vesicle and trophocyst, but it has been classically treated in the *Pilobolaceae* based on other morphological criteria (dark, persistent sporangium wall and a circumscissile zone of dehiscence between the sporangial wall and the columella; Grove 1934; Zheng and Liu 2009). All species of *Pilaira* (Benny and O'Donnell 1978; Zheng and Liu 2009), however, are morphologically distinct from *Helicostylum, Pirella*, and *Thamnidium* (Benny 1992, 1995b; Benny and Schipper 1992).

*Mycocladiaceae* The family was described by Hoffmann et al. (2007) and later discussed by Hoffmann et al. 2009). The type and only species in the *Mycocladiaceae* is *Mycocladus verticellatus* (Beauverie 1900). This species was probably a mixture of two fungi, *Lentamyces parricida* (*Lentamycetaceae*) and a host fungus, a species of *Absidia* (Hoffmann et al. 2009). A name based on a mixed type (http://www.iapt-taxon.org/nomen/main.php?page=art9#9.11) is not invalid. The fungus is not known in culture and, therefore, no recommendations for media, temperature, etc. can be given.

*Mycotyphaceae* *Mycotyphaceae* has four genera (*Benjaminiella, Cokeromyces, Kirkomyces, Mycotypha*) in three subfamilies: (1) *Cokeromycetoideae* (*Benjaminiella, Cokeromyces*), (2) *Kirkomycetoideae* (*Kirkomyces*), and (3) *Mycotyphoideae* (*Mycotypha*). Members of *Benjaminiella, Cokeromyces*, and *Mycotypha* grow and sporulate readily on MEYE and YpSs agars although other media will suffice (LYE, MSMA, PDA, TPO, V8, whey, YpD). *Benjaminiella, Cokeromyces*, and *Mycotypha* are mesophilic or thermotolerant. In nature these three taxa are usually found on dung but they are not commonly encountered. Members of the *Mycotyphaceae*, except *Kirkomyces*, form a culture with relatively short sporangiophores that is easily overlooked and probably readily overgrown by other fungi. They can all form yeast cells on the surface of MEYE agar (Benny and Benjamin 1976; Benny et al. 1985) at 26 °C. *Cokeromyces recurvatus* can cause mucormycosis in humans and other animals but it is rarely reported. *Kirkomyces cordense* was isolated from soil collected in India (Benny 1995b, c). The only known culture of *K. cordense* is a mesophile that will readily grow and sporulate on all of media listed above, but the morphological features are best observed on MSMA or another nutrient-poor culture medium.

*Phycomycetaceae* *Phycomycetaceae* has no subfamilies and two genera: *Phycomyces* and *Spinellus*. These fungi are found in nature on organic substrata (*Phycomyces*, dung, pressed seedcakes for birds, mushrooms, etc.; *Spinellus*, on fungi, especially *Collybia* or *Mycena*). *Phycomyces* is usually found when the weather is cool (winter in Florida) although it can grow at room temperature; zygospores form best at 15–20 °C on PDA, TPO agar, or YpSs agar (Benjamin and Hesseltine 1959). *Phycomyces blakesleeanus* is the most common of the three known species in nature and often can be found on dung. *Spinellus* contains five species (Zycha et al. 1969) but only *S. fusiger* is in culture collections. *Spinellus fusiger* is a psychrophile and must be grown on MEYE agar at 19 °C or below.

*Pilobolaceae* *Pilobolaceae* has no subfamilies and two genera. This family has traditionally contained two genera, *Pilaria* and *Pilobolus*, with *Utharomyces* added later. Molecular evidence has resulted in the transfer of *Pilaira* to the *Mucoraceae* (Hoffmann et al. 2013; Benny et al. 2014). The remaining genera, *Pilobolus* and *Utharomyces* (Grove 1934; Kirk and Benny 1980; Hu et al. 1989), are characterized by the production of a trophocyst that forms a simple sporophore bearing an apical or subapical vesicle with an apical columellate, multispored sporangium having a persistent black wall. *Pilaira* does not produce a trophocyst or a subsporangial vesicle. These genera, *Pilobolus* and *Utharomyces*, are coprophilous. *Pilobolus* is an obligate coprophile that forcibly disperses the sporangium. Culture of *Utharomyces* is best on a nutrient-rich medium such as YpSs agar or 2 % malt agar (Kirk and Benny 1980); dung extract is not required. Most species of *Pilobolus*, however, requires a culture medium that contains dung extract or hemin (SHM, Levetin and Caroselli 1976). Transfer the entire sporangium of *Pilobolus* with sterile forceps, preferably a pair of watchmaker's forceps, before being forcibly dispersed. This sporangium is then placed on the surface of an agar plate emended with dung extract or hemin (SHM). Place the inoculated plate in a 37 °C incubator overnight. Germination of at least some spores will occur as a result of the overnight 37 °C incubation. Optimal sporulation is probably best observed if the culture is transferred to sterile dung, but observation of the trophocyst and zygospore formation, if crosses are made, will be impaired.

*Radiomycetaceae* *Radiomycetaceae* has no subfamilies and one genus. Initially the *Radiomycetaceae* included two genera, *Hesseltinella* and *Radiomyces* (Ellis and Hesseltine 1974). This pair of genera also was the subject of three papers (Benny and Benjamin 1991; Benny and Khan 1988; Benny and Samson 1989). Molecular analysis reveals, however, that these taxa belong in different mucoralean families (O'Donnell et al. 2001; White et al. 2006a). Based on current molecular studies, *Hesseltinella* belongs in the *Cunninghamellaceae*, whereas *Radiomyces* is the only genus in the *Radiomycetaceae*.

The three known species of *Radiomyces* usually are found on dung (Benny and Benjamin 1991) but Ranzoni (1968) made two isolations of *R. embreei* from soil collected in Arizona and California (USA). None of the species of *Radiomyces* are common in nature. Sporulating head and sporangiolar formation are promoted by growth on nutrient-poor culture media (V8, Wg), whereas zygospores are formed optimally on MEYE agar. YpSs agar produces variable results for both asexual and sexual reproduction depending on the species and probably the isolate (Benny and Benjamin 1991).

*Rhizopodaceae* *Rhizopodaceae* has no subfamilies. The *Rhizopodaceae* (Schumacher 1894) includes four genera: *Amylomyces*, *Rhizopus*, *Sporodiniella*, and *Syzygites* (Ellis et al. 1976; Ekpo and Young 1979; Schipper 1984; Schipper and Stalpers 1984; Gbaja and Young 1985; Zheng et al. 2007). These fungi, except *Amylomyces*, produce ornamented spores, the sporangia are apophysate, and the sporangial wall is fugacious or deliquescent; sporangia and spores are not produced by *Amylomyces rouxii* (Ellis et al. 1976). *Sporodiniella umbellata* is only found on insect larvae, mature insects, and spiders in nature; *Syzygites megalocarpus* is a facultative parasite of mushrooms in nature. *Amylomyces* and some species of *Rhizopus* are used in the production of Asian foods or indigenous fermented products in Guyana (South America) and can be isolated from these substrates (Hesseltine 1965, 1985; Henkel 2005). Many other species of *Rhizopus* can be isolated from dung, soil, and other organic materials. *Rhizopus stolonifer* can cause postharvest disease of strawberries, sweet potatoes, and other crops. A few species of *Rhizopus* are the causative agents of mucormycosis. Observation of *Rhizopus* and possibly *Amylomyces* branching patterns is best made on nutrient-poor media.

*Syzygites megalocarpus* can be cultured on CM+ (Kovacs and Sundberg 1999), GYP (Kaplan and Goos 1982a), MEA (Baker 1931), MEYE [the same as yeast-malt agar, Difco] (Davis 1967), and PDA (Hesseltine 1957; Benny and O'Donnell 1978). The formation of sporangia versus zygospores can be manipulated depending on the temperature, media, and light. Idnurm (2011) found that some isolates of *S. megalocarpus* do not produce zygospores, whereas others will reproduce sexually. Zygospores may or may not germinate to produce germspores. However, one culture (CBS 108947) produces germ sporangia containing germspores that do germinate.

*Sporodiniella umbellata* is a monotypic genus that is found in nature on insects, larvae of Homoptera and Lepidoptera, and spiders (Evans and Samson 1977). Growth and sporulation can be induced on CA (Chien and Hwang 1997) and CMA (after plates solidify place a sterile mealworm larvae on the agar surface; Chien and Hwang 1997) with isolation conducted in the dark at 20 °C and the artificial infection at 24 °C. Evans and Samson (1977) used MEA and MWA (Samson 1974) in order to isolate, and for growth and sporulation of *S. umbellata* at 25 °C in diffuse light or a light/dark period, but sporulation was not observed on any of the aforementioned culture media. *Sporodiniella umbellata* is heterothallic (Degawa 2006).

*Saksenaeaceae* *Saksenaeaceae* has no subfamilies and two genera. When originally described, *Saksenaeaceae* (Ellis and Hesseltine 1974) contained two unusual but unrelated monotypic genera, *Saksenaea* and *Echinosporangium* (now

*Lobosporangium* (Benny and Blackwell 2004), *Mortierellales*). Currently, *Saksenaeaceae* contains *Apophysomyces* and *Saksenaea*. Additional species were described in both genera (Alvarez et al. 2010a, b; Bonifaz et al. 2014). The pathogenic species are found in both genera (Guarro et al. 2011; Hospenthal et al. 2011; Austin et al. 2013).

Members of both *Apophysomyces* and *Saksenaea* do not sporulate readily when grown using ordinary culture methods. Padhyay and Ajello (1988) reported a simple method that can be used to induce sporulation. This method involved growing the fungus on a nutrient-rich culture medium, such as Sabouraud dextrose agar (SDA), and then transferring a piece of the culture to 20 ml of sterile distilled water supplemented with 0.2 ml of filter-sterilized 10 % yeast extract solution. These latter cultures were then incubated for 7–10 days at 37 °C; sporulation is abundant and reliable.

*Syncephalastraceae* There are seven genera (*Circinella*, *Fennellomyces*, *Phascolomyces*, *Protomycocladus*, *Syncephalastrum*, *Thamnostylum*, *Zychaea*: Hesseltine and Fennell 1955; Hesseltine and Ellis 1961; Benny and Benjamin 1975, 1976; Schipper and Stalpers 1983; Zheng et al. 1988; Schipper and Samson 1994) in the family. *Syncephalastraceae* can be divided into three clades: (1) *Protomycocladus* (2 % MEA for both asexual and sexual reproduction); (2) *Circinella* (SMA (asexual reproduction), steep agar (zygospore formation)), *Fennellomyces* (MSMA (asexual reproduction, no zygospores produced)), *Phascolomyces* (MSMA, MEYE, YpSs (asexual reproduction varies among isolates, zygospores unknown)), *Thamnostylum* (MSMA [Wort + 3.5] for *T. repens* sporangial formation, whereas zygospores are produced on MEYE, TPO, YPD, and YpSs), and *Zychaea* reproduces asexually on MSMA but zygospores are unknown; and (3) *Syncephalastrum* produces merosporangia on PDA and YpS, but sexual reproduction is optimal on YpSs when compatible cultures are grown at 25 °C. According to Hoffmann et al. (2013), these genera were all maintained in the latter family because of low branch support. The final distribution of the aforementioned members of the *Syncephalastraceae* will depend on additional sequencing and phylogenetic analysis.

Members of the genera of *Syncephalastraceae* can all be isolated from dung, soil, and other organic substrata. Initial isolation probably is best done on a clear, nutrient-rich culture medium such as MEYE in order to observe spore germination and also to check for contamination. Sporulating cultures can then be transferred to any one of several nutrient-poor culture media in order to observe branching, sporulating structures, etc.

*Umbelopsidaceae* This family (Meyer and Gams 2003) contains only the type of genus, *Umbelopsis*, with 14 species (Meyer and Gams 2003; Sugiyama et al. 2003; Mahoney et al. 2004; Wang et al. 2014). The majority of species of *Umbelopsis* are isolated from leaf litter and soil (Meyer and Gams 2003). Culture can be done on 2 % MEA, CMA, CZA, Miura agar (LcA, same as MA), PCA, and PDA (Sugiyama et al. 2003; Mahoney et al. 2004; Wang et al. 2014). Zygospores have never been reported.

A synopsis of the isolation and culture of species of the *Mucorales* from dung, soil, and any other organic substratum is presented in the preceding section. Additional methods for isolation and culture of *Mucorales* are detailed in Krug et al. (2004) and Benny (2008).

## Subphylum *Zoopagomycotina* (Hibbett et al. 2007) (Table 5.4)

Fungi produce hyphae that are coenocytic or septate. Asexual reproduction by arthrospores, chlamydospores, conidia, or sporangiospores in multispored merosporangia. Sexual reproduction, where known, by zygospores borne on apposed suspensors. Obligate haustorial, endo- or ectoparasites of amoebae, fungi, nematodes, and rotifers.

One order, *Zoopagales*, and five families with 19 or 20 genera.

**Table 5.4** Synopsis of the classification of the *Zoopagomycotina*

| |
|---|
| *Zoopagomycotina* Benny |
| *Zoopagales* Bessey ex R.K. Benj. |
| *Cochlonemataceae* Dudd. |
| *Amoebophilus* P.A. Dang. |
| *Aplectosoma* Drechsler |
| *Bdellospora* Drechsler |
| *Cochlonema* Drechsler |
| *Endocochlus* Drechsler |
| *Euryancale* Drechsler |
| Possible member of the family |
| *Aenigmatomyces* R.F. Castañeda & W.B. Kendr. |
| *Helicocephalidaceae* Boedijn |
| *Brachymyces* G.L. Barron |
| *Helicocephalum* Thaxt. |
| *Rhopalomyces* Corda |
| *Verrucocephalum* Degawa |
| *Piptocephalidaceae* J Schröt. |
| *Kuzuhaea* R.K. Benj. |
| *Piptocephalis* de Bary |
| *Syncephalis* Tiegh. & G. Le Monn. |
| *Sigmoideomycetaceae* Benny, R.K. Benj. & P.M. Kirk |
| *Reticulocephalis* Benny, R.K. Benj. & P.M. Kirk |
| *Sigmoideomyces* Thaxt. |
| *Thamnocephalis* Blakeslee |
| *Zoopagaceae* Drechsler emend. Duddington |
| *Acaulopage* Drechsler |
| *Cystopage* Drechsler |
| *Stylopage* Drechsler |
| *Zoopage* Drechsler |
| *Zoophagus* Sommerst. emend. M.W. Dick |
| Possible member of the subphylum |
| *Basidiolum* Cienk |

## *Order* Zoopagales *(Benjamin 1979)*

With the description of the subphylum.

Five families: *Cochlonemataceae*, *Helicocephalidaceae*, *Piptocephalidaceae*, *Sigmoideomycetaceae*, and *Zoopagaceae*.

Members of the *Zoopagomycotina* can be isolated from any organic substrata that support the hosts because all taxa are obligate parasites of small animals or other fungi in nature. Only the members of the *Piptocephalidaceae*, especially *Piptocephalis* and *Syncephalis*, are usually found in nature and grow readily in culture (Benjamin 1959, 1961, 1963, 1966) as mycoparasites, usually on a member of the *Mortierellales* or *Mucorales*.

Two of these families, *Cochlonemataceae* and *Zoopagaceae*, will be presented together because the methods of isolation and culture are similar. Members of the remaining families may be encountered when looking for the aforementioned taxa.

In the culture of species of the *Cochlonemataceae* and *Zoopagaceae*, members of these two families parasitize small animals (amoebae, nematodes). Drechsler (1936) recommended making media (CMA?) with 1.5–2.0 % agar to keep the hosts on the surface of the substratum. The agar must be kept moist by adding a small amount of sterile water to each plate and then placing the culture in containers that prevent water loss to promote both bacterial growth and host migration. When the hosts have reached a critical population, then the zoopagalean parasites also will be more readily observed (Drechsler 1936). They first appear after 2 weeks but must be kept and observed for a few months.

Conditions must be promoted to maximize the growth of these host animals on agar in a Petri dish of 2 % water agar or NN-agar emended with Page's amoeba saline solution. *Enterobacter cloacae* has been used to feed amoebae (Michel et al. 2014) although it is possible that other bacteria could be used instead. Incubation of plates is probably best at 18–20 °C. Higher or lower temperature may stop growth or induce changes in the cytoplasm of either the host or parasite (Drechsler 1946). *Stylopage cephalote* (Drechsler 1938) can appear at higher temperatures. Conidia are produced when the host dies on the agar surface, but zygospores form when the dead host is in the agar (Drechsler 1935a).

The presence of many members of the *Cochlonemataceae* and *Zoopagaceae* initially can be detected by shinning a light laterally over its surface. Taxa with many spore chains, such as *Cochlonema agamum* (Drechsler 1946), *C. eryblastum* (Drechsler 1942), and *C. symplocum* (Drechsler 1941), will be noted because of the white, scattered conidial tufts. The conidial tufts of other species may appear thin and scattered (*Zoopage pachyblasta*, Drechsler 1947) or restricted to a 15 mm circle (*Z. mitospora*; Drechsler 1938). *Cystopage lateralis* (Drechsler 1941) sporulates on or below the agar surface but not aerially. Single spores formed by some species (*Acaulopage rhicnospora* (Drechsler 1935b), *Endocochlus asteroides* (Drechsler 1935b)) can be located in the plate with a low or medium power light microscope objective. Many of Drechsler's zoopagalean fungi are probably easier to find after the first one or two have been observed (see references in Lumsden 1987).

Drechsler (1929) inoculated cornmeal agar (CMA) with a species of *Pythium* and then added decaying vegetable material to the plate (Drechsler 1936). His later cultures often were inoculated a second time with compost, leaf mulch, moss, etc. The details of Drechsler's techniques of both isolation and observation are discussed in Benny (2008). Members of the *Cochlonemataceae* and *Zoopagaceae* are not in culture collections. Drechsler also reported the appearance of species of *Helicocephalum* and *Rhopalomyces* (*Helicocephalidaceae*; see below) using these same methods.

*Cochlonemataceae* *Cochlonemataceae* (Duddington 1973) was described for the endo- and ectoparasitic members of the order. These microfungi are obligate parasites of amoebae or nematodes. *Cochlonemataceae* contains six genera: two that are ectoparasites (*Amoebophilus*, *Bdellospora*) and four endoparasites (*Aplectosoma*, *Cochlonema*, *Endocochlus*, *Euryancale*). The thallus can arise from the infecting conidium that produces a haustorium in the host and chains of conidia or zygospores (*Amoebophilus*, *Bdellospora*) and is internal and coiled (*Cochlonema*), cushion-shaped (*Aplectosoma*), or fertile hyphae forming single conidia per hypha (*Endocochlus*, *Euryancale*). Asexual reproduction is by conidia formed singly or in chains of spores. Sexual reproduction via zygospores formed on apposed suspensors.

*Amoebophilus simplex* (Barron 1983) has been reported several times recently, both on the Internet and in the literature (see Benny et al. 2014). Siemensma (2012) believes that *Mayorella penardi* may be the only host for *A. simplex*. Leidy (1879) may have illustrated *A. simplex* as *Ouramoeba botulicauda*.

Saikawa and Kadowaki (2002) illustrated amoebae capture by two species of *Acaulopage*, *A. dichotoma* and *A. tetraceros* (*Cochlonemataceae*). These two species of *Acaulopage* were isolated from debris collected from a fire reservoir. Some of the debris was placed in a Petri dish containing 10 ml of water, the *Acaulopage* sporulated on the water surface, was transferred to the surface of 2 % water agar with amoebae that was cultured on SA agar (formula in Benny 2008). *Cochlonema euryblastum* (Drechsler 1942) was cultured on a host, *Thecamoeba quadrilineata*, and 18S rRNA was sequenced by Koehsler et al. (2007). The phylogenetic analysis demonstrated that *C. euryblastum* is the sister species of, but basal to, *Kuzuhaea moniliformis* (Benjamin 1985) and *Piptocephalis corymbifera* (*Piptocephalidaceae*, *Zoopagales*).

*Zoopagaceae* The *Zoopagaceae* as defined by Duddington (1973) was reserved for the taxa that are predaceous. Hyphae are coenocytic and formed externally. Conidia are not produced (*Cystopage*), formed singly (*Acaulopage*, *Stylopage*, *Zoophagus*) or in chains (*Zoopage*). *Zoopagaceae* are haustorial parasites of amoebae, nematodes, and rotifers. Sexual reproduction is by zygospore formation.

*Zoophagus* is a genus formerly thought to be a zoosporic fungus (Dick 1990). Zygospores were reported in *Acaulopage pectospora*, and it was transferred to the zygomycetes as *Zoophagus pectosporus* (Dick 1990). This transfer was verified by the phylogenetic study, based on an 18S rDNA data set, by Tanabe et al. (2000). The host for *Zoophagus* spp., in water, is usually a rotifer but nematodes also can be hosts (Karling 1936; Saikawa et al. 1988; Glocking 1997; Liu et al. 1998). Cultures of a species of *Zoophagus* may be in one or more culture collections. Shimada and Saikawa (2006) observed nematode capture and chlamydospore germination in *Cystopage cladospora*.

A review of the literature on the ultrastructure of members of *Cochlonemataceae* and *Zoopagaceae* has been published (Saikawa 2011b). Other publications (Saikawa 2011a, 2012) reviewed the morphology and presented keys to taxa in the latter two families. The first color illustrations are included for several taxa (Saikawa 2012). The techniques described by Drechsler below for finding *Cochlonemataceae* and *Zoopagaceae* also were efficacious for isolating some members of the *Entomophthoromycota* (*Meristacrum*, Drechsler 1940), *Kickxellomycotina* (*Ballocephala*, Drechsler 1951), and other *Zoopagomycotina* (*Helicocephalum*, *Rhopalomyces*, *Syncephalis*; Drechsler 1934, 1943, 1961).

*Cochlonemataceae* and *Zoopagaceae*—information that applies to both families.

Hirotane-Akane and Sakawa (2010) studied zygospore morphology and germination in *Cochlonema cerasphorum*, *C. megalosomum* (*Cochlonemataceae*), and *Acaulopage lophospora* (*Zoopagaceae*). One year later, Saikawa (2011) reviewed his ultrastructural studies of selected taxa in the *Cochlonemataceae* (*Acaulopage dichotoma*, *A. tetraceros*, *Stylopage cephalote*, *Zoophagus insidians*, *Z. tenticulum*) and *Zoopagaceae* (*Cochlonema odontosperma*, *Endocochlus gigas*).

*Helicocephalidaceae* This family is reserved for four genera (*Brachymyces*, *Helicocephalum*, *Rhopalomyces*, *Verrucocephalum*; Thaxter 1891a, b; Barron 1975, 1980a, b; Drechsler 1943; Ellis 1963; Cano et al. 1989; Roux 1996; Degawa 2014) that are parasites of nematodes or their eggs. The sporangiophores are erect, have basal rhizoids, and either form spores blastosporically on pedicels directly from the sporophores or pedicellate sporangiola on fertile vesicles or arthrospores resulting from disarticulation of the sporophore apex. The spores are relatively large and pigmented. Sexual reproduction has not been reported. The members of the *Helicocephalidaceae* are very rarely reported in the literature. *Rhopalomyces elegans* (Corda 1839, 1840; Ellis 1963) probably is the most common member of the family.

Ellis (1963) reported that the cultures of *R. elegans* were initially isolated on hay extract agar (HEA) or 2 % water agar plates, inoculated with soil-plant debris, and incubated at 25 °C for 5 days to 6 weeks. Ellis (1963) found three varieties of *R. elegans* (*apiculatus*, *crassus*, *elegans*). He described the first two varieties as new; they were cultured on his *Rhopalomyces* medium (RM) that consisted of two media: (1) TKY and (2) LFK.

His studies of *R. elegans* var. *apiculatus* revealed that spore germination was 10 % when the medium (TKY—see under RM) alone was used. *Bacillus cereus* "wild B" NRRL B509 was inoculated on the TKY agar in a glass plate and incubated at 25 °C for 20 h. These plates were autoclaved for 20 min at 15 psi. When cooled and solidified, the plate containing the sterilized *Bacillus* culture was inoculated with the *R. elegans* var. *apiculatus* spores and placed in a 25 °C incubator. Another culture medium (LFK) needs to be prepared, also in a glass Petri dish (100×15 mm).

This culture medium contains 2 % agar with the addition of $K_2HPO_4$ 0.4 % before being autoclaved. The glass dish contained a 5 mm³ piece of baby beef liver, washed in three changes of distilled water, and placed in the middle of the dish and the dish autoclaved. The hot agar (ca. 25 ml) is added to the dish with liver, two drops of 10 % KOH are added to one side of the liver, and lamb fat is heated until liquefied, and two small drops are added using a hot Pasteur pipet, and then gently mix to distribute the lamb fat droplets. Let the LFK (see RM in Supplement) agar solidify and cool and then inoculate one side of the liver with the *R. elegans* spores that germinated on the TKY agar; incubate at 20–25 °C. The *R. elegans* hyphae will fill the dish in 4 days, and then the sporophores form near the fat droplets in 5–6 days (Ellis and Hesseltine 1962; Ellis 1963).

*Piptocephalidaceae* Aerial hyphae not produced or relatively thin, giving rise to appressoria and haustoria. Sporophores are unbranched or if branched, then branching is usually dichotomous. Merosporangia uni- or multispored. Spores remaining dry or released in a droplet of fluid at maturity. Sexual reproduction by zygospores usually formed in the substratum on apposed suspensors. Three genera are currently in the *Piptocephalidaceae* (*Kuzuhaea*, *Piptocephalis*, *Syncephalis*; Benjamin 1959, 1985a, b; Kirk 1978; Jeffries 1979; Ho and Kirk 2009; Santiago et al. 2011).

These fungi are all haustorial mycoparasites in nature. Two species are parasites of the *Ascomycota* [*Piptocephalis xenophila* on *Penicillium waksmanii*, *Syncephalis wynneae* on *Wynnea macrotis* (Dobbs and English 1954; Thaxter 1897)], but the remainder grow on hosts in *Mortierellales* (*Mortierellomycotina*) and *Mucorales* (*Mucoromycotina*). Many species are found in nature on dung, humus, soil, and other organic substrates. Both the host and parasite usually grow well in culture and these taxa are relatively large and readily recognized. The *Piptocephalidaceae* are the most common members of the *Zoopagales* recognized in nature. A phylogenetic study, based only SSU rDNA, indicates that *Syncephalis* may not be closely related to *Piptocephalis* and *Kuzuhaea moniliformis* may be a species of *Piptocephalis* (Hou 2011; White et al. 2006a).

Isolation of *Piptocephalidaceae* differs for *Piptocephalis* and *Syncephalis* so they are discussed separately: (1) *Syncephalis* can be isolated on sterile Wg10 that was then cooled to 55 °C and emended with sterile chlortetracycline hydrochloride (50 ppm), streptomycin sulfate (100 ppm), and benomyl (10 ppm). After the plates are solidified, soil was sprinkled on each plate and then incubated at room temperature (ca. 21–22 °C) on a laboratory bench near a south-facing window. The benomyl inhibits some species of *Mortierella* (Strauss et al. 2000) but also soil-inhabiting members of *Penicillium* and *Trichoderma* that could overgrow the plate before the parasites (and species *Coemansia*) normally appear (6–14 days). The *Syncephalis* is transferred to the host that it initially parasitized, a species of *Absidia*, *Cunninghamella*, *Mortierella*, *Mucor*, *Zygorhynchus*, etc. Some species of *Syncephalis* also can be transferred to *Cokeromyces recurvatus* or another slow-growing host. Growth on *C. recurvatus* or on Ellis' medium (Ellis 1966) without a host could be used to collect *Syncephalis* hyphae and sporangiophores for molecular studies. *Syncephalis* can be transferred, with the appropriate host, to Wg10 with or without benomyl, and then zygospores also can be studied, if they are present. When more than one culture of a species is available, then they can be crossed on Wg10 or YpSs/5. As many as four species of *Syncephalis* can occur in each soil sample, based on examination of over 500 soil samples (Benny et al. 2016). (2) *Piptocephalis* can be isolated from dung, soil, or other organic debris. After isolation cultures should be incubated at 18 °C because some species of *Piptocephalis* may not sporulate above this temperature (R.K. Benjamin, pers. comm.). Cultures can be transferred to *Cokeromyces recurvatus*, if available, or to the original or another suitable host. The majority of known species of *Piptocephalis* will grow and sporulate on *C. recurvatus*, but *P. minuta* (Kuzuha 1976) only parasitizes a species of *Mortierella*, *P. cruciata* on *Mycotypha microspora*, and an undescribed species (literature name *P. digitata*) on *Umbelopsis ramanniana* (Gräfenhan 1998). Isolation of *Piptocephalis* from soil can be performed on 2 % WA emended with the antibiotics mentioned above. These plates then are inoculated with *C. recurvatus*, either pieces of colony culture from the agar, sporangiola scraped from the colony surface, or using the technique and broth (PG) recommended to produce Y-phase yeast cells by Jeffries and Kirk (1976). When the isolates of *Piptocephalis* are in culture, then different media can be used to study the anamorph and teleomorph. Gräfenhan (1998) made species descriptions on MEA for *Piptocephalis* parasitizing *C. recurvatus* and *M. microspora*, PCA for *Mortierella* (*P. minuta*), YpSs for *Umbelopsis*, and PDA for *Penicillium waksmanii* (*P. xenophila*). Benjamin (1959) used MEYE and YpSs for routine culture of many parasites including *Piptocephalis*. Appearance of the parasite is relatively slow on MEYE or YpSs (2–3 weeks), versus on cV8 and 50 % PCA (7–12 days). The appearance of zygospores, when produced, often is obscured by the luxuriant growth of the host on MEA, MEYE, and YpSs. Zygospores, if formed, were readily observed when the *Piptocephalis* culture was grown on cV8 and PCA. One isolate of *Piptocephalis tieghemiana* (NRRL A-19043) produced zygospores on 2 % WA.

The third genus, *Kuzuhaea*, is known only in culture from the initial isolation made from soil in Japan (Benjamin 1985a). *Kuzuhaea moniliformis* was cultured on *Umbelopsis ramanniana* on YpSs agar at 25 °C. *Kuzuhaea moniliformis* will grow over the host culture and then sporulate on the surface of the colony. Zygospore formation was studied on YpSs/5 agar. *Kuzuhaea moniliformis* (11.2 % of the fungal community) has been reported from Alaska permafrost soil using 454 pyrosequencing (Penton et al. 2013).

*Sigmoideomycetaceae* The family contains three genera (*Reticulocephalis*, *Sigmoideomyces*, *Thamnocephalis*; Benny et al. 1992). The fertile hyphae are dichotomously branched and coenocytic when young but multiseptate in age. Pairs of stalked fertile vesicles are produced from the cell at the branching point of the fertile hyphae. Fertile vesicles are more or less globose and covered with unispored sporangiola borne on small denticles. Sexual reproduction has never been observed.

The hosts for these fungi are generally unknown because they have only been studied as herbarium specimens (Benny et al. 1992), whereas those in culture are mycoparasites (hosts *Basidiobolus ranarum*, *Cokeromyces recurvatus*; Benny and

Benjamin 1992; Chien 2000). *Basidiobolus ranarum* was grown on 2 % cornmeal agar and then transferred to MEA (2 % ME agar?) and then inoculated with the spores from a growing colony of *Thamnocephalis quadrupedata* (Blakeslee 1905). Infection of *B. ranarum* was observed within 48 h (Chien 2000).

*Thamnocephalis sphaerospora* was isolated from frog dung; *Basidiobolus ranarum* might have been the original host. This *Thamnocephalis* then was transferred to *Cokeromyces recurvatus*. It also grew on a culture of *Microascus singularis* (Malloch and Cain 1971) with lunate ascospores that did not produce conidia. The *T. sphaerospora* cultures grown on *C. recurvatus* were placed in a lighted incubator set at 26 °C with a light/dark (12 h/12 h) cycle. Those cultures grown on YpSs agar produced fruiting heads and very little aerial mycelium, whereas when on MEYE or 2 % ME + 0.5 % YE agars, the aerial hyphae were abundant and yellow orange, reached the Petri dish lid, and contained many fertile heads. Recently, one of us (GLB) grew the *T. sphaerospora/C. recurvatus* coculture on V8 juice agar at 24 °C on the laboratory bench near a south-facing window, and the formation of both fertile heads and aerial hyphae was better than that observed on the media containing malt extract.

Suyama and Degawa (2013) reported that there were two members of the *Sigmoideomycetaceae* found in Japan *Thamnocephalis sphaerospora* and *Sphondylocephalum verticillatum* (Stalpers 1974). The latter species was previously an anamorphic member of the *Pezizomycotina*. It was described originally as *Oedocephalum verticillatum* (Thaxter 1891a).

There is no known commercial value for the *Zoopagales*. In the future, however, species of *Syncephalis* may be utilized as biocontrol agents for those members of the *Mucorales* that are plant pathogens (*Choanephora cucurbitarum*) or that cause storage rots (*Gilbertella persicaria*, *Mucor piriformis*, *Rhizopus stolonifer*) in the USA. Species of *Syncephalis* and not *Piptocephalis* should be used because some taxa of the former species can completely overgrow or inhibit sporulation of the host or both. Selected species of *Syncephalis* can reduce asexual reproduction and, as a result, could delay or prevent the plant disease or postharvest diseases caused by *C. cucurbitarum*, *G. persicaria*, *M. piriformis*, and *R. stolonifer.*

Other *Zoopagales* can be isolated from bat dung (*Verrucocephalum*), frog dung (species of *Thamnocephalis*), or nematode eggs (*Rhopalomyces elegans*); these fungi are not commonly encountered in nature. One possible member of the Zoopagomycotina is *Basidiolum fimbriatum* (Cienkowski 1863; White 2003) but its affinities are unknown at this time.

## Discussion

The early-diverging fungi include all of the fungi with the exception of the *Dikarya* (Grigoriev et al. 2011). These early-diverging fungi can be either aquatic (*Blastocladiomycota*, *Chytridiomycota*, *Monoblepharidiomycota*, *Neocallimastigomycetes*, *Olpidium*) or terrestrial (*Glomeromycota*, *Entomophthoromycota* and

*Kickxellomycotina*, *Mortierellomycotina*, *Mucoromycotina*, *Zoopagomycotina*) (Benny et al. 2014; Corsaro et al. 2014; Didier et al. 2014; James and Berbee 2011; James et al. 2014; Powell and Letcher 2014; Redecker et al. 2013; Sekimoto et al. 2011).

## *DNA Barcoding and Environmental Sampling*

DNA barcoding using the internal transcribed spacer ribosomal DNA (the ITS1-5.8S-ITS2 region, known most frequently as simply ITS) has become an invaluable approach to facilitate comparisons among fungal species and to enhance identification of fungi from environmental samples (Schoch et al. 2012; Nilsson et al. 2011; Tsui et al. 2011). Barcodes are available for most members of the *Mortierellales* (*Mortierellomycotina*) and *Mucorales* (*Mucoromycotina*) that are currently in culture (Wagner et al. 2013; Walther et al. 2013). Unfortunately, the ITS region is often too divergent to be aligned across zygomycete orders or sometimes even within genera and, therefore, is not phylogenetically informative (Smith and Benny, unpublished). Practically, this means that even for taxa for which other informative loci such as RPB1, RPB2, 18S and 28S rDNA, and EF-1α are available, the ITS sequences are quite often missing from public DNA sequence databases. For example, two major papers on the *Kickxellomycotina* (*Asellariales*, *Dimargaritales*, *Harpellales*, *Kickxellales*) were recently published, but ITS was not included in these analyses (Tretter et al. 2013, 2014).

Samples should be collected using the techniques presented by Barron (2004), Krug et al. (2004), and Benny (2008). Dung and other larger samples can be incubated in moist chambers (large samples, Krug 2004; small samples, Benny 2008). Many specialized techniques have been devised to find and culture fungi of the *Kickxellomycotina*, *Mortierellomycotina*, *Mucoromycotina*, and *Zoopagomycotina* (Benny 2008).

A paper with complete coverage of the techniques used to study *Zoopagales* (*Zoopagomycotina*) has not been published because many of the organisms are parasites of small animals (amoebae, rotifers, nematodes, water bears) and the hosts require specialized isolation and culture techniques. This includes fungi in the *Cochlonemataceae*, *Helicocephalidaceae*, *Sigmoideomycetaceae*, and *Zoopagaceae*. One species of *Helicocephalidaceae* (*Rhopalomyces elegans*) and one species of *Sigmoideomycetaceae* (*Thamnocephalis sphaerospora*) have been successfully cultured. The apparent scarcity of these species may be due more to lack of optimal modes of collecting and culturing rather their true rarity in nature. Members of the *Piptocephalidaceae* (*Kuzuhaea moniliformis* and many species of both *Piptocephalis* and *Syncephalis*) are common in culture collections and can be isolated from nature. Members of both genera can be readily isolated and grown in culture on appropriate host fungi, but only a handful of sequences are available for these taxa due to the difficulty of obtaining pure cultures and/or pure DNA (e.g., without contamination from the host fungi). In contrast, *Kuzuhaea moniliformis* is known only from the original isolate from soil in Japan (Benjamin 1985a). However, 28S rDNA sequences similar to that of the type strain of *Kuzuhaea* were found in Alaska permafrost soil, suggesting that this genus may be more widespread than previously thought (Penton et al. 2013). A few species of *Zoopagales* do have ITS sequences deposited in GenBank, but many more DNA barcodes are needed to enhance our understanding of their ecology based on environmental sequencing.

The lack of DNA barcode information for the zygomycete fungi is a serious limitation for studies that seek to use environmental DNA sequences to examine fungal communities. Molecular fungal community studies are only as accurate as the curated and correctly identified sequences in GenBank that are used as a library for identification. In addition to an incomplete database of fungal sequences, preliminary data from our studies of the genus *Syncephalis* (*Zoopagomycotina*) suggest that "fungal-specific" primers such as ITS1F may also be problematic (Smith and Benny, unpublished). PCR primers such as ITS1F were often designed based on a small number of DNA sequences from a biased pool of species (e.g., Gardes and Bruns 1993) and therefore may be inappropriate to use in targeting all fungi. Our data suggest that the ITS1F primer site is altered in some species of zygomycetes and therefore that PCR with mixed fungal templates is likely to preferentially amplify sequences of *Dikarya* and not those of zygomycete targets. This problem is further enhanced for some zygomycete taxa that may have long ITS rDNA sequences (Lazarus et al., unpublished). Future molecular approaches that use specific primers to focus on zygomycetes and/or account for the longer sequence length in some zygomycetes are likely to find a much greater diversity of these fungi than studies that attempt to document all fungi from complex substrata. We suspect that a more focused approach is likely to be effective when combined with next-generation DNA sequencing (e.g., 454, MiSeq, PacBio).

## *Genome Sequencing*

Resolving evolutionary relationships among the early-diverging fungi will be dependent on genome sequencing and phylogenetic comparison of available representatives as discussed for the 1000 Fungal Genomes project in the MycoCosm portal (Grigoriev et al. 2013; http://genome.jgi-psf.org/programs/fungi/index.jsf). Kuo et al. (2014) discussed the procedures used to sequence and annotate fungal genomes. These procedures are based on the platforms used for next-generation sequencing (Quail et al. 2012). A long list of fungal genomes has been sequenced or nominated for sequencing (http://genome.jgi.doe.gov/pages/fungi-1000-projects.jsf/), including species in *Kickxellomycotina*, *Mortierellomycotina*, *Mucoromycotina*, and *Zoopagomycotina* (Table 5.5). Hopefully, more species will be sequenced before the Fungi-1000 project is complete.

Phylogenetically informative genes will be extracted from these genomes and the sequences aligned and then analyzed using phylogenetic methods. A data set composed of the genes from the terrestrial early-diverging fungi can also be

**Table 5.5** Taxa of zygomycota with genomes sequenced or to be sequenced as part of the 1000 Fungal Genomes project (F1000)

| |
|---|
| *Kickxellomycotina* |
| *Asellariales* |
| None nominated |
| *Dimargaritales* |
| *Dimargaritaceae* |
| *Dimargaris cristalligena* Tiegh. |
| *Dispira cornuta* Tiegh. |
| *Tieghemiomyces californicus* R.K. Benj. |
| *Tieghemiomyces parasiticus* R.K. Benj. |
| *Harpellales* |
| *Harpellaceae* |
| None nominated |
| *Legeriomycetaceae* |
| *Capniomyces stellatus* S.W. Peterson & Lichtw. |
| *Furculomyces boomerangus* (M.C. Williams & Lichtw.) M.C. Williams & Lichtw. |
| *Smittium angustum* M.C. Williams & Lichtw. |
| *Smittium annulatum* Licht. |
| *Smittium culicis* Tuzet & Manier ex Kobayasi |
| *Smittium culicioides* Licht. |
| *Smittium fecundum* Lichtw. & M.C. Williams |
| *Smittium megazygosporum* Manier & F. Coste |
| *Smittium simulii* Licht. |
| *Smittium simulatum* Licht. & Arenas |
| *Zancuomyces culisetae* (Lichtw.) Yan Wang, Tretter, Lichtw. & M.M. White |
| *Kickxellales* |
| *Kickxellaceae* |
| *Coemansia mojavensis* R.K. Benj. |
| *Coemansia pectinata* (Coem.) Bainier |
| *Coemansia reversa* Tiegh. & G. Le Monn. |
| *Coemansia spiralis* Eidam |
| *Dipsacomyces acuminosporus* R.K. Benj. |
| *Kickxella alabastrina* Coem. |
| *Linderina pennispora* Raper & Fennell |
| *Martensiomyces pterosporus* J.A. Mey. |
| *Myconymphaea yatsukahoi* Kurihara, Degawa & Tokum. |
| *Pinnaticoemansia coronantispora* Kurihara & Degawa |
| *Spirodactylon aureum* R.K. Benj. |
| Clade 1 (*Barbatospora* sp.) |
| Clade 2 (*Orphella* spp.) |
| Clade 3 (*Mycoemilia scoparia*, *Spiromyces* spp.) |
| *Spiromyces aspiralis* Benny & R.K. Benj. |
| Clade 4 (*Ramicandelaber* spp.) |

(continued)

**Table 5.5** (continued)

| |
|---|
| *Ramicandelaber brevisporus* Kurihara, Degawa & Tokum. |
| *Mortierellomycotina* |
| *Mortierellales* |
| *Mortierellaceae* |
| *Gamsiella multidivaricata* (R.K. Benj.) Benny & M. Blackw. |
| *Lobosporangium transversale* (Malloch) M. Blackw. & Benny |
| *Mortierella alpina* Peyronel |
| *Mortierella elongata* Linnem. |
| *Mortierella verticillata* Linnem. |
| *Nothadelphia mortierellicola* Degawa & W. Gams |
| *Mucoromycotina* |
| Clade 1 |
| *Calcarisporiella thermophile* (H.C. Evans) de Hoog |
| Clade 2 |
| None nominated |
| *Endogonales* |
| *Endogonaceae* |
| *Endogone flammicorona* Trappe & Gerd. |
| *Endogone lactiflua* Berk. |
| *Endogone pisiformis* Link |
| *Endogone* sp. |
| *Mucorales* |
| *Backusellaceae* |
| *Backusella circina* J.J. Ellis & Hesselt. |
| *Choanephoraceae* |
| C *hoanephoroideae* |
| *Choanephora cucurbitarum* (Berk. & Ravenel) Thaxt. |
| *Blakeslea trispora* Thaxt. |
| G *ilbertelloideae* |
| *Gilbertella hainanensis* J.Y. Cheng & F.M. Hu |
| *Gilbertella persicaria* var. *persicaria* (E.D. Eddy) Hesselt. |
| *Cunninghamellaceae* |
| A *bsidioideae* |
| *Absidia repens* Tiegh. |
| *Chamydoabsidia padenii* Hesselt. & J.J. Ellis |
| *Gongronella butleri* (Lendn.) Peyronel & Dal Vesco |
| *Halteromyces radiatus* Shipton & Schipper |
| *Hesseltinella vesiculosa* H.P. Upadhyay |
| C *unninghamelloideae* |
| *Cunninghamella bertholletiae* Stadel |
| *Cunninghamella echinulata* (Thaxt.) Thaxt. |
| *Cunninghamella elegans* Lendn. |
| *Lentamycetaceae* |

(continued)

**Table 5.5** (continued)

| |
|---|
| *Lentamyces parricida* (Renner & Muskat ex Hesselt. & J.J. Ellis) Kerst. Hoffm. & K. Voigt |
| *Lichtheimiaceae* |
| D *ichotomomycetoideae* |
| *Dichotomocladium elegans* Benny & R.K. Benj. |
| *Dichotomocladium robustum* Benny & R.K. Benj. |
| L *ichtheimioideae* |
| *Lichtheimia corymbifera* (Cohn) Vuill. |
| *Lichtheimia hyalospora* (Saito) Kerst. Hoffm., Walther & K. Voigt |
| *Lichtheimia ramosa* (Zopf) Vuill. |
| *Rhizomycetoideae* |
| *Rhitzomucor miehei* (Cooney & R. Emers.) Schipper |
| *Rhizomucor pusillus* (Lindt) Schipper |
| *Rhizomucor variabilis* R.Y. Zheng & G.Q. Chen (see *Mucor irregularis*) |
| *Mucoraceae* |
| C *haetocladioideae* |
| *Chaetocladium jonesii* (Berk. & Broome) Fresen. |
| D *icranophoroideae* |
| *Dicranophora fulva* J. Schröt. |
| M *ucoroideae* |
| *Mucor circinelloides* |
| *Mucor indicus* Lendn. |
| *Mucor irregularis* Stchigel, Cano, Guarro & Ed. Álvarez |
| *Mucor racemosus* Fresen. |
| *Mucor ramosissimus* Samouts. |
| *Mucor* sp. |
| *Mucor velutinosus* Ed. Álvarez, Stchigel, Cano, Deanna A. Sutton & Guarro |
| *Parasitella parasitica* (Bainier) Syd. |
| *Zygorhynchus heterogamous* (Vuill.) Vuill. |
| T *hamnidioideae* |
| *Ellisomyces anomalus* (Hesselt. & P. Anderson) Benny & R.K. Benj. |
| *Helicostylum pulchrum* (Preuss) Pidopl. & Milko |
| *Pilaira anomala* (Ces.) J. Schröt. |
| *Pirella circinans* Bainier |
| *Thamnidium elegans* Link |
| *Mycotyphaceae* |
| C *okeromycetoideae* |
| *Benjaminiella poitrasii* (R.K. Benj.) Arx |
| *Cokeromyces recurvatus* Poitras |
| K *irkomycetoideae* |
| *Kirkomyces cordense* (B.S. Mehrotra & B.R. Mehrotra) Benny |
| M *ycotyphoideae* |
| *Mycotypha africana* R.O. Novak & Backus |
| *Phycomycetaceae* |

(continued)

**Table 5.5** (continued)

| |
|---|
| *Phycomyces blakesleeanus* Burgeff |
| *Spinellus fusiger* (Link) Tiegh. |
| *Pilobolaceae* |
| *Pilobolus umbonatus* Buller |
| *Utharomyces epallocaulus* Boedijn |
| *Radiomycetaceae* |
| *Radiomyces spectabilis* Embree |
| *Rhizopodaceae* |
| *Rhizopus delemar* Boidin ex Wehmer & Hanzawa |
| *Rhizopus microsporus* var. *chinensis* (Saito) Schipper & Stalpers |
| *Rhizopus microsporus* var. *microsporus* Tiegh. |
| *Rhizopus microsporus* var. *oligosporus* (Saito) Schipper & Stalpers |
| *Rhizopus microsporus* var. *rhizopodiformis* (Cohn) Schipper & Stalpers |
| *Rhizopus oryzae* Went & Prins. Geerl. |
| *Rhizopus stolonifer* (Ehrenb.) Vuill. |
| *Syzygites megalocarpus* Ehrenb. |
| *Saksenaeaceae* |
| *Apophysomyces elegans* P.C. Misra, K.J. Srivast. & Lata |
| *Apophysomyces trapeziformis* E. Álvarez, Stchigel, Cano, Deanna A. Sutton & Guarro |
| *Saksenaea oblongispora* E. Álvarez, Stchigel, Cano & Guarro |
| *Saksenaea vasiformis* S.B. Saksena |
| *Syncephalastraceae* |
| Clade 1 |
| *Syncephalastrum monosporum* R.Y. Zheng, G.Q. Chen & F.M. Hu |
| *Syncephalastrum racemosum* Cohn ex J. Schröt. |
| Clade 2 |
| *Protomycocladus faisalabadensis* (J.H. Mirza, S.M. Khan, S. Begum & Shagufta) Schipper & Samson |
| Clade 3 |
| *Circinella umbellate* Tiegh. & G. Le Monn. |
| *Fennellomyces linderi* (Hesselt. & Fennell) Benny & R.K. Benj. |
| *Phascolomyces articulosus* Boedijn |
| *Thamnostylum lucknowense* (J.N. Rai, J.P. Tewari & Mukerji) Arx & H.P. Upadhyay |
| *Zychaea mexicana* Benny & R.K. Benj. |
| *Umbelopsidaceae* |
| *Umbelopsis isabellina* (Oudem.) W. Gams |
| *Umbelopsis ramanniana* (Möller) W. Gams |
| *Zoopagomycotina* |
| *Zoopagales* |
| *Cochlonemataceae* |
| None nominated |
| *Helicocephalidaceae* |
| *Rhopalomyces elegans* Corda |

(continued)

**Table 5.5** (continued)

| |
|---|
| *Piptocephalidaceae* |
| *Piptocephalis corymbifera* Vuill. |
| *Syncephalis fuscata* Indoh |
| *Syncephalis plumigaleata* Embree |
| *Syncephalis pseudoplumigaleata* Benny & H.M. Ho |
| *Sigmoideomycetaceae* |
| *Thamnocephalis sphaerospora* R.K. Benj. & Benny |
| *Zoopagaceae* |
| None nominated |

supplemented with data from zoosporic early-diverging fungi and *Dikarya*, aligned and analyzed to provide a phylogeny of the fungi. As aptly pointed out by Stajich (2015), phylogenomics will undoubtedly revolutionize our view of the fungal tree of life. We suspect that the most exciting discoveries in the fungal tree of life during the coming decade will be centered on the zygomycete subphyla *Kickxellomycotina*, *Mortierellomycotina*, *Mucoromycotina*, and *Zoopagomycotina*.

## Supplement 1: Culture Media Other sources for media can be found in Bills and Foster (2004) and Benny (2008)

2 % ME (same as 2 % MEA (2 % malt extract agar): malt extract, 20 g; agar, 20 g; distilled water, 1 l (Benny and Benjamin 1975).

2 % ME + 0.5 % YE—malt-yeast agar: malt extract, 20 g; yeast extract, 5 g; agar, 15 g; distilled water, 1 l (Benny et al. 1992; Benny and Schipper 1992).

2 % WA—2 % water agar: agar, 20 g; distilled water, 1 l.

BHIv/10—10 % brain-heart infusion agar + vitamins: brain-heart infusion agar (Difco), 3.7 g; thiamine-HCl, 200 μg; biotin, 50 μg; 50 agar, 15 g; glass-distilled water, 1 l (Lichtwardt 1986).

CA—carrot agar: carrots, 200 g, blended (boil 20 min in 500 ml distilled water, filter, make volume to 1 l); agar, 16 g (Chien and Hwang 1997).

CHA—*Choanephora* agar: dextrose, 3 g; casamino acids [Difco], 2 g; $KH_2PO4$, 1 g; $MgSO_4 \cdot 7H_2O$, 0.5 g; thiamine-HCl, 25 mg; agar, 20 g; distilled water, 1 l; pH 6.0 (Benny 2008—as CH).

CM or CMA—cornmeal agar: yellow cornmeal, 20 g (boil 10 min in 700 ml distilled water, filter, and add distilled water to make 1 l); dextrose, 10 g; agar, 15 g; adjust pH to 6.0 (Benjamin 1958, 1959).

CM+—cornmeal agar+: cornmeal agar (Difco), 17 g; yeast extract, 1.0 g; malt extract, 1.0 g; distilled water, 1 l (Kovacs and Sundberg 1999).

CM-S—cornmeal-steep agar: CM + corn steep liquor (Sigma), 5 ml [adjust pH to 6.0 with 1N NaOH] (Benny et al. 1992).

cV8—clarified V8 juice agar—filter v8 juice [regular] through Miracloth® to remove pulp; add $CaCO_3$ [powdered], 3 g/l; mix, decant off supernatant, make volume to original amount with distilled water, add ca. 45–50 ml plastic tubes, and freeze; when media is needed, thaw and dilute one 45 ml tube to 1 l, agar 20 g.

CDY—Czapek-Dox agar [CZA, Difco] + yeast extract: CZA + yeast extract, 6.5 g; pH 5.0; adjust with HCl (Gams and Williams 1963).

CZA—Czapek's solution agar: $NaNO_3$, 3.0 g; $K_2HPO_4$, 1.0 g; $MgSO_4 \cdot 7H_2O$, 0.5 g; KCl, 0.5 g; $FeSO_4 \cdot 7H_2O$, 0.1 g; sucrose, 30 g; agar, 15 g; distilled water, 1 l.

DD—dung decoction: horse dung [moist], 125 g; tap water, 1 l (soak 2–3 h, filter, make volume to 1 l); pH 6.0 or above 7.0 for *Pilobolus* (use in place of water for media needing DD) (Benny and Schipper 1992).

GYP—glucose-yeast-peptone agar: glucose, 10 g; yeast extract, 1 g; peptone, 2 g; agar, 15 g; distilled water, 1 l (Kaplan and Goos 1982b).

LA—Leonian's agar: malt extract, 6.25 g; peptone, 0.625 g; maltose, 6.25 g; $KH_2PO_4$, 1.25 g; $MgSO_4 \cdot 7H_2O$, 1.25 g; agar, 15 g; distilled water, 1 l (Leonian 1924; Benny and Blackwell 2004).

LcA—Miura agar (see "MA" below).

LYE—Leonian's agar + yeast extract: malt extract, 6 g; yeast extract, 1 g; peptone, 0.6 g; maltose, 6 g; $KH_2PO_4$, 1.2 g; $MgSO_4 \cdot 7H_2O$, 1.2 g; agar, 15 g; distilled water, 1 l (Malloch and Cain 1971; Benny and Benjamin 1975).

MED—2 % MEA + 5 g dextrose per liter.

ME-P—malt extract-peptone agar: malt extract, 20 g; peptone, 5 g; agar, 15 g; distilled water, 1 l (Richard K. Benjamin—pers. comm. 1988).

MEYE—malt extract-yeast extract agar: malt extract, 3 g; yeast extract, 3 g; peptone, 5 g; dextrose, 10 g; agar, 15 or 20 g; distilled water, 1 l (Benjamin 1958, 1959; Benny and Benjamin 1975).

MA—Miura agar: glucose, 1 g; $KH_2PO_4$, 1 g; $MgSO_4 \cdot 7H_2O$, 0.2 g; KCl, 0.2 g; $NaNO_3$, 2 g; yeast extract, 0.2 g; agar, 15 g; distilled water 1 l (Sugiyama et al. 2003—no $KH_2PO_4$; yeast extract, 2 g; agar, 13 g).

MWA—mealworm agar: dried mealworms (ground), 200 g (boil in tap water 3 h, filter, add tap water to make supernatant to 1 l); sucrose, 20 g; agar, 20 g (Samson 1974).

MEYE/2—malt extract-yeast extract agar one-half strength: malt extract, 1.5 g; yeast extract, 1.5 g; peptone, 2.5 g; dextrose, 5 g; agar, 15 g; distilled water, 1 l (Kurihara et al. 2001).

MNC+T—modified Norkrans' C medium + thiamine: $K_2HPO_4$, 1.0 g; $MgSO_4 \cdot 7H_2O$, 0.5 g; $ZnSO_4$, 0.5 ml 2 % sol; $NH_4$-tartrate, 0.5 g; Fe-tartrate, 0.5 ml 1.0 % sol; thiamine, 0.5 ml of 100 ppm stock sol; casein hydrolysate, 0.23 g; yeast extract, 0.5 g; dextrose, 10 g; agar, 15 g; distilled water, 1 l; streptomycin, 100 ppm; tetracycline, 50 ppm + thiamine 100 mg (Hirose et al. 2014).

MSMA—modified synthetic *Mucor* agar: dextrose, 10 g; $NaNO_3$, 4 g; $K_2HPO_4$, 0.5 g; $MgSO_4 \cdot 7H_2O$, 0.5 g; thiamine-HCl, 0.5 mg; agar, 15 g; distilled water, 1 l (Benny and Benjamin 1975).

MRN + thiamine—*Endogone* agar + thiamine: saccharose, 5 g [replace dextrose, 10.7 g]; $NH_4(SO4)_2$, 1.15 g; $CaCl_2.4H_2O$, 55.5 mg; $MgSO_4 \cdot 7H_2O$, 0.5 g; KCl,

0.67 g; ferric citrate, 5 mg; $ZnSO_4 \cdot 4H_2O$, 4.4 mg; $MnSO_4$, 5 mg; $CaCl_2 \cdot 4H_2O$, 55.5 mg; ferric or aluminum phytate, 200 mg; sodium inositol hexaphosphate, 550 mg; thiamine, 100 mg; distilled water, 1 l; agar, 15 g (Dalpé 1990; Mitchell and Read 1981).

MWA—mealworm agar: dried mealworms (ground), 200 g (boil in tap water 3 h, filter, add tap water to make supernatant to 1 l); sucrose, 20 g; agar, 20 g (Samson 1974).

MYP-ps—malt extract-yeast extract-peptone agar with antibiotics: malt extract, 7 g; yeast extract, 0.5 g; peptone, 1.0 g; penicillin G, 0.5 g; streptomycin sulfate, 0.5 g; agar, 15 g; distilled water, 1 l; autoclave medium, cool to 43 °C, and add antibiotics aseptically using a Millipore syringe filter (0.22 μm)—used for both dilution and direct inoculation on chilled plates [Carreiro and Koske 1992])).

NN-agar + PAS—non-nutrient agar + Page's amoeba saline (PAS) solution: agar (Oxoid L11), 15 g; plus PAS (stock 1 (NaCl, 12 g; $MgSO_4 \cdot 7H_2O$, 0.40 g; $CaCl_2 \cdot 6H_2O$, 0.60 g; distilled water, 500 ml) and stock 2 ($Na_2HPO_4$, 14.20 g; $KH_2PO_4$, 13.60 g; distilled water, 500 ml)); combine stocks 1 and 2, 5.0 ml each; distilled water, 990 ml. Mix agar, 15 g; with PAS 1 l, slowly bring to a boil; autoclave 15 min (from Culture Collection of Algae and Protozoa—accessed 6 January 2015, http://www.ccap.ac.uk/pdfrecipes.htm/).

OMA—oatmeal agar: rolled oats, 30 g; distilled water, 1 l (heat to boiling and simmer 2 h, filter, and bring volume to 1 l); agar, 15 g (Gams et al. 1975).

PAB—pablum agar: pablum, 50 g (boil in 700 ml distilled water, filter, adjust final volume to 1 l); agar, 15 g (Benjamin 1959).

PAB-DEX—pablum agar: pablum, 50 g (boil in 700 ml distilled water, filter, adjust final volume to 1 l); dextrose, 10 g; agar, 15 g (Benjamin 1959).

PCA—potato-carrot agar: potato (peeled and diced), 20 g; carrots (peeled and diced), 20 g (boil carrots and potatoes in 300 ml of tap water, filter, and add water to adjust volume to 1 l); agar, 20 g (Bawcutt 1983).

PDA—potato dextrose agar: potatoes (peeled and cut), 200 g (boil extract 10 min in 700 ml distilled water, filter, adjust final volume to 1 l); dextrose, 20 g; agar, 15 g (Schipper 1969—pH 6.6; Benjamin 1958, 1959—pH not mentioned).

PG—peptone-glucose agar: peptone [Difco], 10 g; dextrose, 20 g; agar, 20 g; distilled water, 1 l (Gauger 1961).

PYED—peptone-yeast extract-dextrose agar: peptone, 1 g; yeast extract, 1 g; dextrose, 0.5 g; agar, 15 g; distilled water, 1 l; adjust pH to 6.5 (Benjamin 1978).

PYEDS—peptone-yeast extract-dextrose-corn steep agar: PYED + corn steep liquor, 5 ml; adjust pH to 6.0 with 1N NaOH.

RM—*Rhopalomyces* medium:

(1) TKY (tryptone, 0.5 %; yeast extract, 0.5 %; $K_2HPO_4 \cdot H_2O$, 0.4 %; agar, 2 %; pH 9 with KOH). *Bacillus cereus* Frankland & Frankland "wild B" NRRL B509 was inoculated on the TKY agar in a glass plate and incubated at 25 °C for 20 h. These plates were autoclaved for 20 min at 15 psi. When cooled and solidified, the plate containing the sterilized *Bacillus* culture was inoculated with the *R. elegans* var. *apiculatus* spores and placed in a 25 °C incubator.

(2) LFK needs to be prepared, also in a glass Petri dish (100×15 mm). This culture medium contains 2 % agar with the addition of $K_2HPO_4$ (0.4 %) before being autoclaved. The glass dish contained a 5 $mm^3$ piece of baby beef liver, washed in three changes of distilled water, and placed in the middle of the dish and the dish autoclaved. The hot agar (ca. 25 ml) is added to the dish with liver, two drops of 10 % KOH are added to one side of the liver, and lamb fat is heated until liquefied, and two small drops are added using a hot Pasteur pipet, and then gently mix to distribute the lamb fat droplets. Let the LFK agar solidify and cool and then inoculate one side of the liver with the *R. elegans* spores that germinated on the TKY agar; incubate at 20–25 °C. The *R. elegans* hyphae will fill the dish in 4–6 days, and then the sporophores form near the fat droplets in 5–6 days (Ellis and Hesseltine 1962; Ellis 1963).

SA—Sato and Aoki's agar: $KNO_3$, 0.2 g; $MgSO_4•7H_2O$, 0.02 g; $KH_2PO_4$, 0.1 g; $KHPO_4$, 0.3 g; $NaHCO_3$, 0.02 g; $Na_2SiO_3$, 0.02 g; agar, 20 g; distilled water, make volume to 1 l; unadjusted pH 7.8 (Saikawa and Kadowaki 2002).

SDA—Sabouraud dextrose agar (BBL™ or Difco™): peptic digest of animal tissue, 5 g; pancreatic digest of casein, 5 g; dextrose, 40 g; agar, 15 g; distilled water, 1 l.

SDA+YE—Sabouraud dextrose agar + yeast extract: 20 ml sterile distilled water, cool and add 0.2 ml filter-sterilized, 10 % yeast extract solution; transfer a small piece of growing culture of a species of *Apophysomyces* or *Saksenaea* to the latter solution and incubate at 37 °C for 7–10 days, and the fungus will reliably sporulate (Padhyay and Ajello 1988).

SDY—soil-dextrose-yeast extract agar: loam soil extract, 100 g (boil 10 min in 800 ml distilled water, filter, make volume to 1 l); dextrose, 2 g; yeast extract, 1 g; $K_2HPO_4$, 1 g; $MgSO_4•7H_2O$, 0.5 g; agar, 15 g.

SEA—soil extract agar: garden soil and distilled water, equal weights; autoclave 30 min, let soil settle to the bottom of flask, and filter; agar, 15 g; soil extract, 1 l (Gams et al. 1975).

ShA—shrimp agar: dried ground edible shrimp, 3 g; agar, 15 g; distilled water, 1 l (Degawa and Tokumasu 1997, 1998a, b).

SHM—simplified hemin medium for *Pilobolus*: hemin, 10 mg (dissolved in 37.5 ml of 0.1N NaOH); sodium acetate ($CH_3COONa•3H_2O$), 10 g; thiamine-HCl, 10 mg; $(NH_4)_2SO_4$, 0.66 g; $K_2HPO_4$, 1.0 g; $MgS0_4•7H_2O$, 0.5 g; deionized distilled water to bring the volume to one liter; agar, 15 g (Levetin and Caroselli 1976).

SMA—synthetic *Mucor* agar: dextrose, 40 g; asparagine, 2 g; $KH_2PO_4$, 0.5 g; $MgSO_4•7H_2O$, 0.25 g; thiamine HCl, 0.5 mg; agar, 15 g; distilled water, 1 L (Benny 2008).

STA—steep agar (Czapek's solution agar [Difco™, CZA, or CZB broth Difco™] + corn steep liquor): CZA, 49 g, or CZ broth, 35 g; corn steep liquor [Sigma© C4648-500G], 10 ml; adjust pH to 7.0 with NaOH before autoclaving; agar, 15 g [if needed]; distilled water, 1 l (Raper and Thom 1949).

SUP—SUP medium: glucose, 50 mg; potassium dihydrogen phosphate, 30 mg; ammonium chloride, 20 mg; dipotassium hydrogen phosphate, 5 mg; magnesium sulfate, 1 mg; calcium chloride, 100 mg (Hoffmann and Voigt 2009), or dextrose 10 g; yeast extract, 5 g; $KH_2PO_4$, 4 g; $K_2HPO_4$, 0.9 g; $NH_4Cl$, 1.0 g; $MgSO_4•7H_2O$, 0.25 g; agar, 15 g; distilled water, 1 l (Wöstemeyer 1985).

TGv—tryptone-glucose medium + salts and vitamins: tryptone (Difco), 20 g; glucose, 5 g; $KH_2PO_4$, 0.28 g; $K_2HPO_4$, 0.35 g; $(NH_4)_2SO_4$, 0.26 g; $MgCl_2•6H_2O$, 0.10 g; $CaCl_2•2H_2O$, 0.07 g; thiamine-HCl, 200 µg; biotin, 50 µg; 50 g agar, 15 g; glass-distilled water, 1 l (Lichtwardt 1986).

TPO—tomato paste-oatmeal agar: tomato paste, 20 g; instant baby oatmeal, 20 g; agar, 15 g; distilled water, 1 l (Hesseltine 1960; Benny and Benjamin 1975).

TSM—Thorton's standardized medium: $K_2HPO_4$, 1 g; $MgSO_4•7H_2O$, 0.2 g; $CaCl_2$, 0.1 g; NaCl, 0.1 g; $FeCl_3$, 0.002 g; $KNO_3$, 0.5 g; asparagine, 0.5 g; mannitol, 1 g; agar, 15 g; distilled water, 1 l (Thorton 1922; Benny and Blackwell 2004).

V8—V8 juice agar [modified]: V8® juice [original], 5.5 fluid oz (163 ml) can; adjust volume to 1 l with distilled water; $CaCO_3$ [powdered], 3 g; agar, 15 g (Benny and Benjamin 1991).

Wg—wheat germ agar: wheat germ, 15 g (boil 10 min in 700 ml distilled water, filter, and adjust volume to 1 l); dextrose, 5 g; agar, 15 g (Benny 1972).

Wg10—one-tenth strength wheat germ agar: wheat germ, 1.5 g; heat in a microwave oven for 3 min in 300 ml of distilled water, and then filter through Miracloth® (Calbiochem 475855) or cheese cloth, take the supernatant, and add distilled water to adjust volume to 1 l; dextrose, 0.5 g; agar, 15 g.

WgDD—wheat germ-dung decoction agar: Wg broth, 500 ml, + DD made with distilled water, 500 ml.

Wg-S—wheat germ-steep agar: Wg + corn steep liquor, 5 ml; adjust pH to 6.0 with 1N NaOH.

Whey—whey agar: powdered whey, 20 g; dextrose, 10 g; agar, 15 g; distilled water, 1 l (Schipper 1969; Benny et al. 1985).

WSH—Weitzman and Silva-Hutner medium: Alpha-Cel [powdered cellulose], 20 g; $MgSO_4•7H_2O$, 1 g; $KH_2PO_4$, 1.5 g; $NaNO_3$, 1 g; tomato paste, 10 g; baby oatmeal, 10 g; agar, 18 g; distilled water, 1 l; pH 5.6 (Weitzman and Silva-Hutner 1967).

WSH-DD—Weitzman and Silva-Hutner medium (WSH) in distilled water 500 ml + dung decoction (DD) 500 ml.

YpSs—Emerson's yeast-phosphate-soluble starch agar: soluble starch, 15 g; yeast extract, 4 g; $K_2HPO_4$, 1.0 g; $MgSO_4•7H_2O$, 0.5 g; agar, 20 g [15 g used later; Benny and Benjamin 1975]; distilled water, 1 l (Benjamin 1959).

YpSs/5—one-fifth strength YpSs agar: soluble starch, 3 g; yeast extract, 0.8 g; $MgSO_4•7H_2O$, 0.5 g; $KH_2PO_4$, 0.5 g; $K_2HPO_4$, 0.5 g; agar, 15 g; distilled water, 1 l (Benjamin 1985a, b).

YGCH—yeast extract-glycerol-casein hydrolysate agar: yeast extract, 10 g; glycerol, 15 ml; casein hydrolysate, 15 g; $K_2HPO_4$, 1.0 g; $MgSO_4•7H_2O$, 0.5 g; agar, 15 g; distilled water, 1 l (O'Donnell et al. 1998).

## Supplement 2: Isolation and Transfer

Members of the four subphyla discussed here all have asexual reproductive structures that are either dry- or wet-spored when mature (Ingold and Zoberi 1963). Dry-spored taxa can be isolated using watchmakers forceps, whereas wet-spored species can be collected using stainless steel, minuten insect pins (0.20 mm in diameter; Carolina #654370) in one of several pin-vise handles or with forceps (Benny 2008). Transfer the spores to a clear, nutrient-rich culture medium such as MEYE and observe spore germination. For best results be sure that the bottom of the plate is pre-marked with several small circles, and additionally be sure to note which are inoculated first. The zygomycotan fungi, especially *Mucorales*, will germinate rapidly, and the mycelium will be very robust; members of the *Mortierellales* are much small in diameter. Contaminated cultures will grow relatively slow and sporulation will be indicated by the color of the colony. A species of *Penicillium* is one of the most common contaminants, but other imperfect fungi such as *Aspergillus* and *Trichoderma* also can appear instead.

## Supplement 3: Microscope Slides

Benjamin (1959) recommended that the zygomycotan fungi be mounted in KOH-phloxine or 2 % KOH for the observation of vegetative and both asexual and sexual reproductive structures. An aqueous solution of phloxine is used if the fungi are hyaline and they are examined using a microscope not equipped with interference-phase contrast. Young portions of an actively growing colony are used to make the slide mounts for taking photographs, making drawings and observation of the characteristic structures of each species. The process can be accomplished using the following steps:

1. Take a small amount of the fungus from the colony.
2. Transfer the material to the surface of a clean microscope slide.
3. Add a drop of 95 % ethyl alcohol to wet the specimen and minimize air bubbles.
4. Tilt the slide on a paper towel or filter paper to blot away excess alcohol.
5. Add a drop of 2.0 % KOH to the slide before the alcohol dries and a drop of distilled water or phloxine to dilute the KOH to ca. 1.0 %.
6. The specimen can be manipulated at this stage to reveal the best material for documentation.
7. Carefully add an 18 $mm^2$ coverslip by placing it on one side at a 30–45° angle and slowly lowering it over the specimen without migrating it to the edge if excess mounting agent has been used.
8. Blot away excess mounting agent using bibulous or filter paper.
9. Seal the slide with paraffin (I use canning paraffin melted in a baby food jar) using a large paper clamp opened to expose one straight side.
10. Heat the paper clamp above warm enough to melt the paraffin, touch it to the paraffin, and transfer it to one side of the slide, while still warm, so that paraffin covers both the edge of the coverslip and adjacent slide.
11. Repeat until all four edges of the coverslip are sealed with paraffin; reheat the paper clamp, if necessary, to seal any gaps in the paraffin coating.
12. Seal the slide again using clear fingernail polish so that a slight band of polish extends from the coverslip, completely covers the paraffin, and also reaches to the subtending slide.

13. This slide should last 3–7 days if no gaps are present in the fingernail polish coating.
14. Any photographs should be taken as soon as possible to avoid recording anomalies in the specimen due to preservation or heating. This is especially true for the cytoplasm, which could be altered during preservation or storage.

## References


Alastruey-Izquierdo A, Hoffmann K, de Hoog SG, Rodriguez-Tudela JL, Voigt K, Bibashi E, Walther G (2010) Species recognition and clinical relevance of the zygomycetous genus *Lichtheimia* (syn. *Absidia* pro parte, *Mycocladus*). J Clin Microbiol 48:2154–2170. doi:10.1128/JCM.01744-09

Alvarez E, Garcia-Hermoso D, Sutton DA, Cano JF, Stchigal AM, Hoinard D, Fothergill AW, Rinaldi MG, Dromer F, Guarro J (2010a) Molecular phylogeny and proposal of two new species of the emerging pathogenic fungus *Saksenaea*. J Clin Microbiol 48:4410–4416. doi:10.1128/JCM.01646-10

Alvarez E, Stchigel AM, Cano J, Sutton DA, Fothergill AW, Chander J, Salas V, Rinaldi MG, Guarro J (2010b) Molecular phylogenetic diversity of the emerging mucoralean fungus *Apophysomyces*: proposal of three new species. Rev Iberoam Micol 27:80–89. doi:10.1016/j.riam.2010.01.006

Austin CL, Finley PJ, Mikkelson DR, Tibbs B (2013) Mucormycosis: a rare fungal infection in tornado victims. J Burn Care Res 35:e164–e171. doi:10.1097/BCR.0b013e318299d4bb

Ayers TT (1933) Growth of *Dispira cornuta* in artificial culture. Mycologia 25:333–341. doi:10.2307/3754008

Baker RED (1931) Observations on the conditions for spore formation in *Sporodinia grandis* Link. New Phytol 30:303–316. doi:10.1111/j.1469-8137.1931.tb07425.x

Barnett HL (1970) Nutritional requirements for the axenic growth of some haustorial mycoparasites. Mycologia 62:750–760. doi:10.2307/3757664

Barron GL (1975) Nematophagus fungi: *Helicocephalum*. Trans Br Mycol Soc 65:309–310. doi:10.1016/S0007-1536(75)80015-5

Barron GL (1980a) A new genus of Zygomycetes. Can J Bot 58:2450–2453. doi:10.1139/b80-284

Barron GL (1980b) The biological role of *Rhopalomyces magnus*. Mycologia 72:427–430. doi:10.2307/3759270

Barron GL (1983) A new *Amoebophilus* (Zygomycetes) ectoparasitic on amoebae. Can J Bot 61:3091–3094. doi:10.1139/b83-347

Barron GL (2004) Fungal parasites and predators of rotifers, nematodes and other invertebrates. In: Mueller GM, Bills GF, Foster MS (eds) Biodiversity of fungi. Inventory and monitoring methods. Elsevier, Burlington, MA, pp 435–450

Bawcutt RA (1983) Biology of *Syncephalis leadbeateri* sp. nov. (Mucorales) and observations on *Syncephalis*. Trans Br Mycol Soc 80:219–230. doi:10.1016/S0007-1536(83)80004-7

Beauverie J (1900) *Mycocladus verticillatus* (gen. nov. sp. nov.). Annales de l'université de Lyon, Séries 2, Sciences, Médicine 3:162–180

Benjamin RK (1958) Sexuality in the Kickxellaceae. Aliso 4:149–169

Benjamin RK (1959) The merosporangiferous Mucorales. Aliso 4:321–433

Benjamin RK (1961) Addenda to "The merosporangiferous Mucorales". Aliso 5:11–19

Benjamin RK (1963) Addenda to "The merosporangiferous Mucorales" II. Aliso 5:273–288

Benjamin RK (1965) Addenda to "The merosporangiferous Mucorales" III. Dimargaris. Aliso 6:1–10

Benjamin RK (1966) The merosporangium. Mycologia 58:1–42. doi:10.2307/3756986

Benjamin RK (1978) *Gamsiella*, a new subgenus of *Mortierella* (Mucorales: Mortierellaceae). Aliso 9:157–170

Benjamin RK (1979) Zygomycetes and their spores. In: Kendrick B (ed) The whole fungus. National Museum of Natural Sciences and the Kanaskis Foundation, Ottawa, pp 573–616

Benjamin RK (1985a) A new genus of the Piptocephalidaceae (Zoopagales) from Japan. Bot J Linn Soc 91:117–133. doi:10.1111/j.1095-8339.1985.tb01139.x

Benjamin RK (1985b) A novel new species of *Syncephalis* (Zoopagales: Piptocephalidaceae) from California that forms hypogenous merosporangia. Aliso 11:1–15

Benjamin CR, Hesseltine CW (1957) The genus *Actinomucor*. Mycologia 49:240–249, http://jstor.org/stable/3793166

Benjamin CR, Hesseltine CW (1959) Studies on the genus *Phycomyces*. Mycologia 51:751–771. doi:10.2307/3755828

Benjamin RK, Blackwell M, Chapala IH, Humber RA, Jones KG, Klepzig KD, Lichtwardt RW, Malloch D, Noda H, Roeper RA, Spadaflora JW, Weir A (2004) Insect- and other arthropod-associated fungi. In: Mueller GM, Bills GF, Foster MS (eds) Biodiversity of fungi. Inventory and monitoring methods. Elsevier, Burlington, MA, pp 395–433

Benny GL (1972) Histochemistry of the cell wall and septum of vegetative cells of selected species of Dimargaritaceae and Kickxellaceae. Mycologia 64:854–862. doi:10.2307/3757940

Benny GL (1991) Gilbertellaceae, a new family of the Mucorales (Zygomycetes). Mycologia 83:150–157. doi:10.2307/3759930

Benny GL (1992) Observations on Thamnidiaceae (Mucorales). V. Thamnidium. Mycologia 84:834–842. doi:10.2307/3760280

Benny GL (1995a) Classical morphology in zygomycete taxonomy. Can J Bot 73(Suppl. 1):S725–S730. doi:10.1139/b95-315

Benny GL (1995b) *Kirkomyces*, a new name for *Kirkia* Benny. Mycologia 87:922. doi:10.2307/3760870

Benny GL (1995c) Observations on Thamnidiaceae. VII. *Helicostylum* and a new genus *Kirkia*. Mycologia 87:253–264. doi:10.2307/376091

Benny GL (2001) Zygomycota: Trichomycetes. In: McLaughlin DJ, McLaughlin EG, Lemke PA (eds) The Mycota, vol VIIA. Springer, Berlin, pp 147–160

Benny GL (2008) Methods used by Dr. R. K. Benjamin, and other mycologists, to isolate Zygomycetes. Aliso 26:37–61 (Iss. 1, Article 8). http://scholarship.claremont.edu/aliso/vol26/iss1/8

Benny GL (2012) Current systematics of the zygomycotan fungi with a brief discussion of their biology. In: Misra JK, Tewari JP, Deshmukh SK (eds) Systematics and evolution of fungi. Progress in mycological research. Science Publishers, Enfield, NH, pp 55–105

Benny GL, Aldrich HC (1975) Ultrastructural observations on septal and merosporangial ontogeny in *Linderina pennispora* (Kickxellales; Zygomycetes). Can J Bot 53:2325–2335. doi:10.1139/b75-258

Benny GL, Benjamin RK (1975) Observations on Thamnidiaceae (Mucorales). New taxa, new combinations, and notes on selected species. Aliso 8:301–351

Benny GL, Benjamin RK (1976) Observations on Thamnidiaceae (Mucorales). II. *Chaetocladium*, *Cokeromyces*, *Mycotypha*, and *Phascolomyces*. Aliso 8:391–424

Benny GL, Benjamin RK (1991) The Radiomycetaceae (Mucorales: Zygomycetes. III. A new species of *Radiomyces*, and cladistic analysis and taxonomy of the family: with a discussion of evolutionary ordinal relationships on Zygomycotina. Mycologia 83:713–735. doi:10.2307/3760429

Benny GL, Benjamin RK (1993) Observations on Thamnidiaceae (Mucorales). VI. Two new species of *Dichotomocladium* and the zygospores of *D. hesseltinei* (Chaetocladiaceae). Mycologia 85:660–671. doi:10.2307/3760511

Benny GL, Khan SR (1988) The Radiomycetaceae (Mucorales; Zygomycetes). Calcium oxalate crystals on the sporangiolar wall and aerial hyphae. Scann Microsc 2(2):1199–1206

Benny GL, O'Donnell KL (1978) *Pilaira anomala*. In: Fuller MS (ed) Lower fungi in the laboratory. Department of Botany, University of Georgia, Athens, GA, pp 129–130

Benny GL, Samson RA (1989) The Radiomycetaceae (Mucorales, Zygomycetes). II. Ultrastructural morphology of the mature zygospore and comparative morphology of the sporangiola and sporangiolar appendages. Mem NY Bot Gard 49:11–19

Benny GL, Schipper MAA (1992) Observations on Thamnidiaceae (Mucorales). IV. *Pirella*. Mycologia 84:52–63. doi:10.2307/3760401

Benny GL, Blackwell M (2004) *Lobosporangium*, a new name for *Echinosporangium* Malloch, and *Gamsiella*, a new genus for *Mortierella multidivaricata*. Mycologia 96:143–149. doi:10.2307/3761995

Benny GL, Kirk PM, Samson RA (1985) Observations on Thamnidiaceae (Mucorales). III. Mycotyphaceae fam. nov. and a re-evaluation of *Mycotypha* sensu Benny & Benjamin illustrated by two new species. Mycotaxon 22:119–148

Benny GL, Benjamin RK, Kirk PM (1992) A reevaluation of Cunninghamellaceae (Mucorales). Sigmoideomycetaceae fam. nov. and *Reticulocephalis* gen. nov.; Cladistic analysis and description of two new species. Mycologia 84:615–641. doi:10.2307/3760372

Benny GL, Humber RA, Morton JB (2001) Zygomycota: Zygomycetes. In: McLaughlin DJ, McLaughlin EG, Lemke PA (eds) The Mycota, vol VIIA. Springer, Berlin, pp 113–146

Benny GL, Humber RA, Voigt K (2014) Zygomycetous fungi: Phylum Entomophthoromycota and subphyla Kickxellomycotina, Mortierellomycotina, Mucoromycotina, and Zoopagomycotina. In: Mclaughlin DJ, Spatafora JW (eds) The Mycota, vol VII. Part A, Systematics and evolution, 2nd edn. Springer, Berlin, pp 208–250. doi:10.1007/978-3-642-55318-9_8

Benny GL, Ho H-M, Lazarus KL, Smith ME (2015) Five new species of the obligate mycoparasite *Syncephalis* (Zoopagales, Zoopagomycotina) from soil. Mycologia 15-073 (in review)

Berch SM, Castellano MA (1986) Sporulation of *Endogone pisiformis* in axenic and monoxenic culture. Mycologia 78:292–295. doi:10.2307/3793176

Berch SM, Fortin JA (1983) Germination of zygospores of *Endogone pisiformis*. Mycologia 75:328–332. doi:10.2307/3792820

Bidartono MI, Read DJ, Trappe JM, Merckx V, Ligrone R, Duckett JG (2011) The dawn of symbiosis between plants and fungi. Biol Lett 7:574–577. doi:10.1098/rsbl.2010.1203

Bills GF, Foster MS (2004) Formulae for selected materials used to isolate and study fungi and fungal allies. In: Mueller GM, Bills GF, Foster MS (eds) Biodiversity of fungi. Inventory and monitoring methods. Elsevier, Burlington, MA, pp 595–618

Blakeslee AF (1905) Two conidia-bearing fungi *Cunninghamella* and *Thamnocephalis*, n. gen. Bot Gaz 40:161–170, http://www.jstor.org/stable/3466377

Błaszkowski J, Tadych M, Madej T (1998) *Endogone maritima*, a new species in the Endogonales from Poland. Mycol Res 102:1096–1100. doi:10.1017/S0953756298006170

Bonifaz A, Stchigel AM, Guarro J, Guevra E, Pintos L, Sanchis M, Cano-Lira JF (2014) Primary cutaneous mucormycosis produced by the new species *Apophysomyces mexicanus*. J Clin Microbiol 52:4428–4431. doi:10.1128/JCM.02138-14

Brain APR, Jeffries J, Young TWK (1982) Ultrastructure of septa in *Tieghemiomyces californicus*. Mycologia 74:173–181. doi:10.2307/3792883

Cano J, Fort F, Guarro J (1989) Zygomycetes from Spain I. Nova Hedw 427–434

Carreiro MM, Koske RE (1992) Room temperature isolations can bias against selection of low temperature microfungi in temperate forest soils. Mycologia 84:886–900. doi:10.2307/3760287

Chien C-Y (2000) *Thamnocephalis quadrupedata* (Mucorales) as a mycoparasite of the entomophthoraceous fungus *Basidiobolus ranarum*. Cytobios 103:71–78

Chien C-Y, Hwang B-C (1997) First record of *Sporodiniella umbellata* (Mucorales) in Taiwan. Mycoscience 38:343–346. doi:10.1007/BF02464094

Chuang S-C, Ho H-M (2009) Notes on Zygomycetes of Taiwan (VII): Two kickxellalean species, *Linderina macrospora* and *Ramicandelaber brevisporus* new to Taiwan. Fungal Sci 24:23–28

Chuang S-C, Ho H-M, Benny GL, Lee C-F (2013) Two new *Ramicandelaber* species from Taiwan. Mycologia 105:320–334. doi:10.3852/11-219

Cienkowski L (1863) Ueber parasitische Schäuche auf Crustaceen und eigen Insektenlarven (*Amoebidium parasticum* m.). Bot Zeit 19:169-174 + Taf. VII

Corda ACJ (1839) Pracht-Flora europaeischer Schimmelbildungen Gerhardt Fleischer, Leipsiz, pp 1–55 + Tefel XXV

Corda ACJ (1840) Flore illustrée de Mucédinées d'Europe. Gérard Fleischer, Leipzig, pp 1–55 + Planche XXV

Corsaro D, Walochnik J, Venditti D, Steinmann J, Müller K-D, Michel R (2014) Microsporidia-like parasites of amoebae belong to the early fungal lineage Rozellomycota. Parasitol Res 113:1909–1918. doi:10.1007/s00436-014-3838-4

Dalpé Y (1990) *Endogone pisiformis*: croissance, morphologie et conservation des cultures. Can J Bot 68:910–915. doi:10.1139/b90-116

Davis EE (1967) Zygospore formation in *Syzygites megalocarpus*. Can J Bot 45:531–532. doi:10.1139/b67-055

de Bary A (1865) Zur Kenntniss der Mucorinen. Abhandl Senkenberg Naturf Ges 5:345–366, 373–375 + Tafel LXIII–LXV

de Hoog GS (1974) The genera *Blastobotrys*, *Sporothrix*, *Calcarisporium* and *Calcarisporiella* gen. nov. Stud Mycol 7:1–84

de Souza JI, Pires-Zottarelli CLA, dos Santos JF, Harakava R (2012) *Isomucor* (Mucoromycotina): a new genus from a Cerrado reserve in state of São Paulo, Brazil. Mycologia 104:232–241. doi:10.3852/11-133

Degawa Y (2006) Heterothallism in insecticolous mucoralean fungus *Sporodiniella umbellata* new to Japan. In: Abstracts of papers presented at the 50th anniversary annual meeting of the Mycological Society of Japan, Chiba, Japan. Abstract 54B (In Japanese)

Degawa Y (2009) Secondary spore formation in *Orchesellaria mauguioi* (Asellariales, Trichomycetes) and its taxonomic and ecological implications. Mycoscience 50:247–252. doi:10.1007/s10267-008-0477-6

Degawa Y (2014) *Verrucocephalum*, a new nematophagous genus in the Helicocephalidaceae (Zoopagales). Mycoscience 55:144–148. doi:10.1016/j.myc.2013.07.002

Degawa Y, Gams W (2004) A new species of *Mortierella*, and an associated sporangiferous mycoparasite in a new genus, *Nothadelphia*. Stud Mycol 50:567–572

Degawa Y, Tokumasu S (1997) Zygospore formation in *Mortierella capitata*. Mycoscience 38:387–397. doi:10.1007/BF02461677

Degawa Y, Tokumasu S (1998a) Two new homothallic species of *Mortierella*, *M. cogitans*, and *M. microzygospora*, and their zygospore formation. Mycologia 90:1040–1046. doi:10.2307/3761278

Degawa Y, Tokumasu S (1998b) Zygospore formation in *Mortierella umbellata*. Mycol Res 102:593–598. doi:10.1017/S095375629700508X

Desirò A, Duckett JG, Pressel S, Villarreal JC, Bidartondo MI (2013a) Fungal symbiosis with hornworts: a chequered history. Proc Roy Soc B Biol Sci 280:20130207. doi:10.1098/rspb.2013.0207

Desirò A, Neumann M, Epis S, Novero M, Bandi C, Genre A, Bonfante P (2013b) Mollicutes-related endobacteria thrive inside liverwort-associated arbuscular mycorrhizal fungi. Environ Microbiol 15:822–836. doi:10.1111/j.1462-2920.2012.02833.x

Desirò A, Faccio A, Karch A, Bidartono MI, Bonfante P (2014a) [2015] *Endogone*, one of the oldest plant-associated fungi, host unique Mollicutes-related endobacteria. New Phytol 205:1464–1472. doi:10.1111/nph.13136

Desirò A, Salvioli A, Ngonkeu EL, Mondo SJ, Epis S, Faccio A, Kaech A, Pawlowska TE, Bonbfante P (2014b) Detection of a novel intracellular microbiome hosted in arbuscular mycorrhizal fungi. ISME J 8:257–270. doi:10.1038/ismej.2013.151

Dick MW (1990) The systematic position of *Zoophagus insidans*. Mycol Res 94:347–354. doi:10.1016/S0953-7562(09)80360-2

Didier ES, Becnel JJ, Kent ML, Sanders JL, Weiss LM (2014) Microsporidia. In: McLaughlin DJ, Spadafora JW (eds) The Mycota, vol VII. Part A, Systematics and evolution, 2nd edn. Springer, Berlin, pp 115–140

Dobbs CG, English MP (1954) *Piptocephalis xenophila* sp. nov. parasitic on non-mucorine hosts. Trans Br Mycol Soc 37:375–389. doi:10.1016/S0007-1536(54)80021-0

Drechsler C (1929) The beet water mold and several related root parasites. J Agric Res 38:309–361

Drechsler C (1934) A new species of *Helicocephalum*. Mycologia 26:33–37. doi:10.2307/3754084

Drechsler C (1935a) Some conidial Phycomycetes destructive to terricolous amoebae. Mycologia 27:6–40. doi:10.2307/3754001
Drechsler C (1935b) Some non-catenulate conidial Phycomycetes preying on terricolous amoebae. Mycologia 27:176–205. doi:10.2307/3754052
Drechsler C (1936) New conidial Phycomycetes destructive to terricolous amoebae. Mycologia 28:363–389. doi:10.2307/3754001
Drechsler C (1938) New Zoopagaceae capturing and consuming soil amoebae. Mycologia 30:137–157. doi:10.2307/3754553
Drechsler C (1940) Three fungi destructive to free-living terricolous nematodes. J Wash Acad Sci 30:240–254
Drechsler C (1941) Four phycomyces destructive to nematodes and rhizoids. Mycologia 33:248–269. doi:10.2307/3754760
Drechsler C (1942) New species of *Acaulopage* and *Cochlonema* destructive to soil amoebae. Mycologia 34:274–297. doi:10.2307/3754641
Drechsler C (1943) A new non-helicoid bisporous *Helicocephalum* parasitizing nematode eggs. Mycologia 35:134–141. doi:10.2307/3754705
Drechsler C (1946) Three new Zoöpagaceae subsisting on soil amoebae. Mycologia 38:120–143. doi:10.2307/3755057
Drechsler C (1947) Three zoopagaceous fungi that capture and consume soil-inhabiting rhizopods. Mycologia 39:253–281. doi:10.2307/3755204
Drechsler C (1951) Various zoöpagaceous fungi subsisting on protozoans and eelworms. Mycologia 43:161–185. doi:10.2307/3755526
Drechsler C (1961) Production of large solitary sporangiospores in some species of the Mucorales. Mycologia 53:439–442. doi:10.2307/375690
Duddington CL (1973) Zoopagales. In: Ainsworth GC, Sparrow FK, Sussman (eds) The fungi, vol 4B. Academic Press, New York, pp 231–234
Ekpo EJA, Young TWK (1979) Fine-structure of the dormant and germinating sporangiospore of *Syzygites megalocarpus* (Mucorales) with notes on *Sporodiniella umbellata*. Microbios Lett 10:63–68
Ellis JJ (1963) A study of *Rhopalomyces elegans* in pure culture. Mycologia 55:183–198. doi:10.2307/3756289
Ellis JJ (1966) On growing *Syncephalis* in pure culture. Mycologia 58:465–469. doi:10.2307/3756922
Ellis JJ, Hesseltine CW (1962) *Rhopalomyces* and *Spinellus* in pure culture and the parasitism of *Rhopalomyces* on nematode eggs. Nature 193:699–700. doi:10.1038/193699a0
Ellis JJ, Hesseltine CW (1966) Species of *Absidia* with ovoid sporangiospores. Sabouraudia 5:59–77. doi:10.1080/00362176785190111
Ellis JJ, Hesseltine CW (1974) Two new families of Mucorales. Mycologia 66:87–95. doi:10.2307/3758456
Ellis JJ, Rhodes LJ, Hesseltine CW (1976) The genus *Amylomyces*. Mycologia 68:131–143. doi:10.2307/3758903
Embree RW, Indoh H (1967) *Aquamortierella*, a new genus in the Mucorales. Bull Torrey Bot Club 94:464–467. doi:10.2307/2483563
Evans HC (1971) Thermophilous fungi from coal spoil tips. I. Taxonomy. Trans Br Mycol Soc 57:241–254. doi:10.1016/S0007-1536(71)80006-2
Evans HC, Samson RA (1977) *Sporodiniella umbellata*, an entomogenous fungus of the Mucorales from cocoa farms in Ecuador. Can J Bot 55:2981–2984. doi:10.1139/b77-334
Gams W, Williams ST (1963) Heterothallism in *Mortierella parvispora* Linnemann. Nova Hedwigia 5:347–355
Gams W (1976) Some new or noteworthy species of *Mortierella*. Persoonia 9:111–140
Gams W (1977) A key to the species of *Mortierella*. Persoonia 9:381–391
Gams W, Carreiro MM (1989) Two new species of *Mortierella* and rediscovery of Thaxter's *Dissophora decumbens*. Stud Mycol 31:85–91

Gardes M, Bruns TD (1993) ITS primers with enhanced specificity for basidiomycetes – application to the identification of mycorrhizae and rusts. Mol Ecol 2:113–118. doi:10.1111/j.1365-294X.1993.tb00005.x

Gauger WL (1961) The germination of zygospores of *Rhizopus stolonifer*. Am J Bot 48:427–429. doi:10.2307/2439336

Gbaja IS, Young TWK (1985) Ultrastructure of the anamorph of *Sporodiniella umbellata* (Mucorales). Microbios 42:263–272

Gerdemann JH, Trappe JM (1974) The Endogonaceae in the Pacific Northwest. Mycol Mem 5:1–76

Glocking SL (1997) *Zoophagus cornus*: a new species from Japan. Mycol Res 101:1179–1182. doi:10.1017/S095375629700378X

Gomes MZR, Lewis RE, Kontoyiannis (2011) Mucormycosis caused by Mucormycetes, non-*Rhizopus*, *-Mucor*, and *Lichtheimia* isolates. Clin Microbiol Rev 24:411–445. doi:10.1128/CMR.00056-10

Gräfenhan T (1998) Taxonomic revision of the genus *Piptocephalis* (Fungi). Master's thesis. Mathematisch Naturwissenschaftlichen Fakultät der ernst-Moritz-Arndt-Universität, Greifswald, Germany

Grigoriev IV, Cullen D, Goodwin SB, Hibbett D, Jeffries TW, Kubicek CP, Kuske C, Magnuson JK, Martin F, Spatafora JW, Tsang A, Baker SE (2011) Fueling the future with fungal genomics. Mycology 2:192–209. doi:10.1080/21501203.2011.584577

Grigoriev IV, Nikitin R, Haridas S, Kuo A, Ohm R, Otillar R, Riley R, Salamov A, Zhao X, Korzeniewski F, Smirnova T, Nordberg H, Dubchak I, Shabalov I (2013) MycoCosm portal: gearing up for 1000 fungal genomes. Nucleic Acids Res 42(Database Issue):D699–D704. doi:10.1093/nar/gkt1183

Grove WB (1934) A systematic account and arrangement of the Pilobolidae. In: Buller AHR (ed) Researches on fungi, vol VI. Longmans, Green and Company, London, pp 190–224

Gryganskyi AP, Humber RA, Smith ME, Miadlikovska J, Wu S, Voigt K, Walther G, Anishchenko IM, Vilgalys R (2012) Molecular phylogeny of the Entomophthoromycota. Mol Phylogenet Evol 65:682–694. doi:10.1016/j.ympev.2012.07.026

Gryganskyi AP, Humber RA, Smith ME, Hodge K, Huang B, Voigt K, Vilgalys R (2013) Phylogenetic lineages in Entomophthoromycota. Persoonia 30:94–105. doi:10.3767/003158513X666330

Guarro J, Chander J, Alvarez E, Stchigel AM, Robin K, Dalal U, Rani H, Punia RS, Cano JF (2011) *Apophysomyces variabilis* infections in humans. Emerg Infect Dis 17:134–135. doi:10.3201/eid1701.101139

Henkel TW (2005) Parakari, an indigenous fermented beverage using amylolytic *Rhizopus* in Guyana. Mycologia 97:1–11. doi:10.3852/mycologia.97.1.1

Hesseltine CW (1957) The genus *Syzygites* (Mucoraceae). Lloydia 20:228–237

Hesseltine CW (1965) A millennium of fungi, food, and fermentation. Mycologia 57:149–197. doi:10.2307/3756821

Hesseltine CW (1985) Fungi, people, and soybeans. Mycologia 77:505–525. doi:10.2307/3793349

Hesseltine CW, Ellis JJ (1961) Notes on Mucorales, especially *Absidia*. Mycologia 53:406–426. doi:10.2307/3756584

Hesseltine CW, Ellis JJ (1964) The genus *Absidia*: *Gongronella* and cylindrical-spored species of *Absidia*. Mycologia 56(4):568–601. doi:10.2307/3756362

Hesseltine CW, Ellis JJ (1966) Species of *Absidia* with ovoid sporangiospores. I. Mycologia 58:761–785. doi:10.2307/3756851

Hesseltine CW, Fennell DI (1955) The genus *Circinella*. Mycologia 47:193–212. doi:10.2307/3755410

Hibbett DS, Binder M, Bischoff JF, Blackwell M, Cannon PF, Eriksson OE, Huhndorf S, James T, Kirk PM, Lücking R, Lumbsch H, Lutzoni F, Matheny PB, McLaughlin DJ, Powell MJ, Redhead S, Schoch CL, Spatafora JW, Stalpers JA, Vilgalys R, Aime MC, Aptroot A, Bauer R, Begerow D, Benny GL, Castlebury LA, Crous PW, Dai YC, Gams W, Geiser DM, Griffith GW, Gueidan C, Hawksworth DL, Hestmark G, Hosaka K, Humber RA, Hyde KD, Ironside

JE, Kõljalg U, Kurtzman CP, Larsson KH, Lichtwardt R, Longcore J, Miądikowska J, Miller A, Moncalvo JM, Mozley-Standridge S, Oberwinkler F, Parmasto E, Reeb V, Rogers JD, Roux C, Ryvarden L, Sampaio JP, Schüssler A, Sugiyama J, Thorn RG, Tibell L, Untereiner WA, Walker C, Wang Z, Weir A, Weiss M, White MM, Winka K, Yao YJ, Zhang N (2007) A higher-level phylogenetic classification of the Fungi. Mycol Res 111:509–547. doi:10.1016/j.mycres.2007.03.004

Hirose D, Degawa Y, Inaba S, Tokumasu S (2012) The anamorphic genus *Calcarisporiella* is a new member of the Mucoromycotina. Mycoscience 53:256–260. doi:10.1007/s10267-011-0160-1

Hirose D, Degawa Y, Yamamoto K, Yamada A (2014) *Sphaerocreas pubescens* is a member of the Mucoromycotina closely related to fungi associated with liverworts and hornworts. Mycoscience 55:221–226. doi:10.1016/j.myc.2013.09.002

Hirotani-Akabane E, Saikawa M (2010) Germination and morphology of zygospores in two *Cochlonema* and one *Acaulopage* species. Mycologia 102:39–43. doi:10.3852/08-116

Ho H-M, Kirk PM (2009) *Piptocephalis formosana*, a new species from Taiwan. Bot Stud 50:69–72

Ho H-M, Chien C-Y, Chuang S-C (2007) Notes on Zygomycetes of Taiwan (V): *Linderina pennispora* new to Taiwan. Fung Sci 22:35–38

Ho Y-H, Ho H-M, Saikawa M (2014) Two new species of *Cystopage*, *C. ovispora* and *C. irregularispora*, obtained in Taiwan. Mycoscience 56:396–401. doi:10.1016/j.myc.2014.12.003

Hoffmann K (2010) Identification of the genus *Absidia* (Mucorales, Zygomycetes): a comprehensive taxonomic revision. In: Gherbawy Y, Voigt K (eds) Molecular identification of fungi. Springer, Berlin, pp 439–460. doi:10.1007/978-3-642-05042-8_19

Hoffmann K, Voigt K (2009) *Absidia parricida* plays a dominant role in biotrophic fusion parasitism among mucoralean fungi (Zygomycetes): *Lentamyces*, a new genus for *A. parricida* and *A. zychae*. Plant Biol 11:537–554. doi:10.1111/j.1438-8677.2008.00145.x

Hoffmann K, Discher S, Voigt K (2007) Revision of the genus *Absidia* (Mucorales, Zygomycetes) based on physiological, phylogenetic, and morphological characters; thermotolerant *Absidia* spp. form a coherent group, Mycocladiaceae fam. nov. Mycol Res 111:1169–1183. doi:10.1016/j.mycres.2007.07.002

Hoffmann K, Walther G, Voigt K (2009) *Mycocladus* vs. *Lichtheimia*: a correction (Lichtheimiaceae fam. nov., Mucorales, Mucoromycotina). Mycol Res 113:277–278. doi:10.1016/j.mycres.2009.02.001

Hoffmann K, Voigt K, Kirk PM (2011) Mortierellomycotina subphyl. nov., based on multi-gene genealogies. Mycotaxon 115:353–363. doi:10.5258/115.353

Hoffmann K, Pawłowska J, Walther G, Wrzosek M, de Hoog GS, Benny GL, Kirk PM, Voigt K (2013) The family structure of the Mucorales: a synoptic revision based on comprehensive multigene-genealogies. Persoonia 30:57–76. doi:10.3767/003158513X666259

Hospenthal DR, Chung KK, Lairet K, Thompson EH, Guarro J, Renz EM, Sutton DA (2011) *Saksenaea erythrospora* infection following combat trauma. J Clin Microbiol 49:3707–3709. doi:10.1128/JCM.05095-11

Hou CT (2008) Production of arachidonic acid and dihomo-γ-linolenic acid from glycerol by oil-producing filamentous fungi, *Mortierella* in the ARS culture collection. J Ind Microbiol Biotechnol 35:501–506. doi:10.1007/s10295-008-0308-y

Hou Y-H (2011) Taxonomic and phylogenetic study of *Piptocephalis* in Taiwan. Master's Thesis, College of Science, National Taipei University of Education, Taipei, Taiwan [In Chinese]

Hu F-m, Zheng R-Y, Chen G-Q (1989) A redelimitation of the species of *Pilobolus*. Mycosystema 2:111–133

Humber RA (2012) Entomophthoromycota: a new phylum and reclassification for entomophthoroid fungi. Mycotaxon 120:477–492. doi:10.5248/120.477

Humber RA (2016) Entomophthoromycota, a new phylum. In: Li DW (ed) Biology of microfungi. Springer, Berlin

Idnurm A (2011) Sex determination in the first-described sexual fungus. Eukaryot Cell 10:14851491. doi:10.1128/EC.05149-11

Ingold CT, Zoberi MH (1963) The asexual apparatus of Mucorales in relation to spore liberation. Trans Br Mycol Soc 46:115–134. doi:10.1016/S0007-1536(63)80014-5

Jabaji-Hare S, Tremblay MF, Parent S (1988) Changes in the concentrations of total fatty acids, soluble proteins, nitrogen and phosphatase activities in mycelia of *Endogone pisiformis* (Endogonaceae) during growth. Mycologia 80:54–62. doi:10.2307/3807493

James TY, Berbee ML (2011) No jacket required – new fungal lineage defies dress code. Bioessays 34:94–102. doi:10.1002/bies.201100110

James TY, Kauff F, Schoch CL, Matheny PB, Hofstetter V, Cox C, Celio G, Gueidan C, Fraker E, Miądikowska J, Lumbsch HT, Rauhut A, Reeb V, Arnold EA, Amtoft A, Stajich JE, Hosaka K, Sung G-H, Johnson D, O'Rourke B, Crockett M, Binder M, Curtis JM, Slot JC, Wang Z, Wilson AW, Schüßler A, Longcore JE, O'Donnell K, Mozley-Standridge S, Porter D, Letcher PM, Powell MJ, Taylor JW, White MM, Griffith GW, Davies DR, Humber RA, Morton J, Sugiyama J, Rossman AY, Rogers JD, Pfister DH, Hewitt D, Hansen K, Hambleton S, Shoemaker RA, Kohlmeyer J, Volkmann-Kohlmeyer B, Spotts RA, Serdani M, Crous PW, Hughes KW, Matsuura K, Langer E, Langer G, Untereiner WA, Lacking R, Badel B, Geiser DM, Aptroot A, Diederich P, Schmitt I, Schultz M, Yahr R, Hibbett DS, Lutzoni F, McLaughlin D, Spatafora J, Vilgalys R (2006) Reconstructing the early evolution of the fungi using a six gene phylogeny. Nature 443:818–822. doi:10.1038/nature05110

James TY, Pelin A, Bonen L, Ahrendt S, Sain D, Corradi N, Stajich JE (2013) Shared signatures of parasitism and phylogenomic unite Cryptomycota and Microsporidia. Curr Biol 23:1548–1553. doi:10.1016/j.cub.2013.06.057

James TY, Porter TM, Martin WW (2014) Blastocladiomycota. In: McLaughlin DJ, Spadafora JW (eds) The Mycota, Vol VIIA Systematics and evolution, 2nd edn. Springer, Heidelberg, pp 177–207. doi:10.1007/978-3-642-55318-9_7

Jeffries P (1979) Observations on *Piptocephalis pseudocephala*. Mycologia 71:209–213. doi:10.2307/3759235

Jeffries P, Kirk PM (1976) New technique for the isolation of mycoparasitic Mucorales. Trans Br Mycol Soc 66:541–543. doi:10.1016/S0007-1536(76)80234-3

Jiang X, Yu H, Xiangf M, Liu X, Liu X (2011) *Echinochlamydosporium variabile*, a new genus and species of Zygomycota from soil nematodes. Fungal Divers 46:43–51. doi:10.1007/s13225-010-0076-7

Kandel P, White MM (2012) A new species of *Ephemerellomyces* from North America highlights its morphological plasticity and possible intergeneric similarities with other Harpellales. Fungal Biol 116:171–184. doi:10.1016/j.funbio.2011.10.003

Kaplan JD, Goos RD (1982a) Effect of water potential on zygospore formation in *Syzygites megalocarpus*. Mycologia 74:684–686. doi:10.2307/3792762

Kaplan JD, Goos RD (1982b) The effect of water potential on zygospore formation in *Syzygites megalocarpus*. Mycologia 74:684–686. doi:10.2307/3792762

Karling JS (1936) A new predaceous fungus. Mycologia 28:307–320. doi:10.2307/3753995

Karpov SA, Mikhailov KV, Mirzaeva GV, Mirabdellaev KA, Titova NN, Aleoshin VV (2014) Obligately phagotrophic aphelids turned out to branch with the earliest-diverging fungi. Protist 164:195–205. doi:10.1016/j.protis.2012.08.001

Kasuhiro O, Degawa Y (2013) Kickxellales from the gut of earwigs (Dermaptera). In: Anon. Abstracts of the 2013 annual meeting of the Japanese Alps inter-university cooperative project, 21–23 November 2013, Sugadaira, Japan, p. 79 (ES-B2_2013.pdf). http://jalps.suiri.tsukuba.ac.jp/files/6813/9478/3916/2013_meeting_Jalps_program_v2.2.pdf. Accessed 15 Oct 2014

Kirk PM (1978) A new and unusual species of *Piptocephalis* (Mucorales). Trans Br Mycol Soc 70:335–340. doi:10.1016/S0007-1536(78)80131-4

Kirk PM (1984) A monograph of the Choanephoraceae. Mycol Pap 152:1–61

Kirk PM (2012) Index fungorum no. 11, 1 p

Kirk PM, Benny GL (1980) The genus *Utharomyces* Boedijn (Pilobolaceae: Zygomycetes). Trans Br Mycol Soc 75:123–131. doi:10.1016/S0007-1536(80)80202-6

Koehsler M, Michael R, Wylezich C, Lugauer J, Walochnik J (2007) Molecular identification and classification of *Cochlonema euryblastum*, a zoopagalean parasite of *Thecamoeba quadrilineata*. Mycologia 99:215–221. doi:10.3852/mycologia.99.2.215

Konova IV, Galanina LA, Kochkina GA, Pan'kina OI (2002) Fatty acids in species of several zygomycete taxa. Microbiology (New York) 71:550–557. doi:10.1023/A:1020506702892

Konova IV, Kockhina GA, Galanona LA (2005) The Prevalence of cis-9-hexadecenoic acid is a specific feature of the fatty acid profile of Zygomycetes from the order Kickxellales. Microbiology (New York) 74:86–90. doi:10.1007/s11021-005-0033-8

Kovacs RL, Sundberg WJ (1999) *Syzygites megalocarpus* (Mucorales, Zygomycetes) in Illinois. Trans IL State Acad Sci 92:181–190

Krings M, Taylor TN, Dotzler N, Persichini G (2012) Fossil fungi with suggested affinities to the Endogonaceae from the Middle Triassic on Antarctica. Mycologia 104:835–844. doi:10.3852/11-384

Krings M, Taylor TN, Dotzler N (2013) Fossil evidence of the zygomycetous fungi. Persoonia 30:1–10. doi:10.3767/003158513X664819

Krug JC (2004) Moist chambers for the development of fungi. In: Mueller GM, Bills GF, Foster MS (eds) Biodiversity of fungi. Inventory and monitoring methods. Elsevier, Burlington, MA, pp 589–593

Krug JC, Benny GL, Keller HW (2004) Coprophilous fungi. In: Mueller GM, Bills GF, Foster MS (eds) Biodiversity of fungi. Inventory and monitoring methods. Elsevier, Burlington, MA, pp 467–499

Kuhlman EG (1972) Variation in zygospore formation among species of *Mortierella*. Mycologia 64:325–341. doi:10.2307/3757836

Kuo A, Bushnell, Grigoriev IV (2014) In: Martin F (ed) Fungal genomics: sequencing and annotation. Adv Bot Res 70:1-52. doi:10.1016/B978-0-12-397940-7.00001-X

Kurihara Y (2002) Taxonomic study on the order Kickxellales (Zygomycetes, Zygomycota). Doctoral Dissertation. University of Tsukuba, Tsukuba, Japan. http://hdl.handle.net/2241/6878. Accessed 16 Oct 2014

Kurihara Y, Degawa Y (2006) *Pinnaticoemansia*, a new genus of Kickxellales, with a revised key to the genera of Kickxellales. Mycoscience 47:205–211. doi:10.1007/s10267-006-0294-8

Kurihara Y, Tokumasu S, Chien C-Y (2000) *Coemansia furcata* sp. nov. and its distribution in Japan and Taiwan. Mycoscience 41:579–583. doi:10.1007/BF02460924

Kurihara Y, Degawa Y, Tokumasu S (2001) A new genus, *Myconymphaea* (Kickxellales) with peculiar septal plugs. Mycol Res 105:1397–1402. doi:10.1017/S0953756201005032

Kurihara Y, Degawa Y, Tokumasu S (2004) Two novel kickxellalean fungi, *Mycoëmilia scoparia* gen. sp. nov. and *Ramicandelaber brevisporus* sp. nov. Mycol Res 108:1143–1152. doi:10.1017/S0953756204000930

Kurihara Y, Ogawa Y, Degawa Y, Tokumasu S (2005) A proposal for the division of the order Kickxellales based on the comparison with its related orders. Inoculum 56(4):34

Kurihara Y, Sukarno N, Ilyas M, Yuniarti E, Mangunwardoyo W, Park J-Y, Saraswati R, Widyastuti Y, Ando K (2008) Indonesian Kickxellales: two species of *Coemansia* and Linderina. Mycoscience 49:250–257. doi:10.1007/s10267-008-0417-5

Kuzuha S (1976) A new acephalis species of *Piptocephalis* (Mucorales). J Jpn Bot 51:123–127

Kwaśna H, Richardson MJ, Bateman GL (2002) Morphological variation in the *Coemansia spiralis* complex. Mycol Res 106:252–256. doi:10.1017/SS0953756201005421

Kwaśna H, Nirenberg HI (2008a) *Siepmannia*, a new genus in the Mucoraceae. Mycologia 100:259–274. doi:10.3852/mycologia.100.2.259

Kwaśna H, Nirenberg HI (2008b) Validation of the genus *Siepmannia* (Mucoraceae) and it four species. Polish Bot J 53:187–188

Labeyrie ES, Molloy DP, Lichtwardt RW (1996) An investigation of Harpellales (Trichomycetes) in New York state blackflies (Diptera: Simuliidae). J Invertebr Pathol 68:293–298. doi:10.1006/jipa.1996.0099

Layios N, Canivet J-L, Baron F, Moutschen M, Hayette M-P (2014) *Mortierella wolfii*-associated invasive disease. Emerg Infect Dis 20:1591–1592. doi:10.3201/eid2009.140469

Leidy J (1879) Freshwater rhizopods of North America. In: U.S. geological survey of the territories, vol XII. U.S. Government Printing Office, Washington, DC

Leonian LH (1924) A study of factors promoting pycnidium formation in some Sphaeropsidales. Am J Bot 11:19–50. doi:10.2307/2446570

Levetin E, Caroselli NE (1976) A simplified medium for growth and sporulation of Pilobolus species. Mycologia 68:1254–1258. doi:10.2307/3758760

Lichtwardt RW (1986) The Trichomycetes. Fungal associates of arthropods. Springer, New York

Lichtwardt RW (2011) *Dacryodiomyces*, a new genus of Harpellales in Chironomidae larvae. Mycologia 103:912–914. doi:10.3852/10-395

Lichtwardt RW, Moss ST (1984) New Asellariales (Trichomycetes from the hindguts of aquatic isolapods and springtails. Mycotaxon 20:259–274

Lichtwardt RW, Cafaro MJ, White MM (2001) The Trichomycetes. Fungal associates of arthropods, Revised edition. http://www.nhm.ku.edu/fungi/monograph/mono.htm/. Accessed 1 June 2014

Lichtwardt RW, Williams MC, White MM (2011) *Klastostachys*, a new genus of Harpellales in Chironomidae larvae. Mycologia 103:915–917. doi:10.3852/10-414

Liu X-s, Miao Z-q, Gao R-h, Zhang K-q (1998) Nematode-capturing Zygomycetes: *Zoophagus tylopagus* sp. nov. and *Z. pectosporus* from China. Mycosystema 17:105–108

Loh LS, Nawawi A, Kuthubutheen AJ (2001) Mucoraceous fungi from Malaysia. Institute of Biological Sciences, University of Malaya, Kuala Lumpur, Malaysia. ISSN: 983208525X, 9789832085256 (original not seen)

Lumsden RD (1987) Charles Drechsler, 1892-1986. Mycologia 79:345–352. doi:10.2307/3807454

Mahoney D, Gams W, Meyer W, Starink-Willemse M (2004) *Umbelopsis dimorpha* sp. nov., a link between *U. vinacea* and *U. versiformis*. Mycol Res 108:107–111. doi:10.1017/S0953756203 008876

Malloch D, Cain RF (1971) The genus Kernia. Can J Bot 49:855–867. doi:10.1139/b71-126

Manier J-F (1973) Quelques aspects ultrastructuraux du Trichomycite Asellariale, Asellaria ligiae Tuzet et Manier, 1950 ex Manier, 1968. C R Acad Sci Paris Ser D 276:3429–3431

Mehrotra BS, Baijal U (1963) A new species of *Dimargaris* from India. Can J Bot 41:499–502. doi:10.1139/b63-043

Meyer W, Gams W (2003) Delimitation of *Umbelopsis* (Mucorales, Umbelopsidaceae fam. nov.) based on ITS sequence and RFLP data. Mycol Res 197:339–350. doi:10.1017/ S0953756203007226

Michailides TJ, Spotts RA (1990) Postharvest diseases of pome and stone fruits caused by *Mucor piriformis* in the Pacific Northwest and California. Plant Dis 74:537–543. doi:10.1094/ PD-74-0537

Michel R, Walochnik J, Scheid P (2014) Article for the "Free-living amoebae Special Issue": Isolation and characterisation of various amoebophagous fungi and evaluation of their prey spectrum. Exp Parasitol 145(Suppl):S131–S136. doi:10.1016/j.exppara.2014.10.005

Misra PC, Lata K (1979) Studies on Dimargaritaceae (Mucorales) II. A new Dispira parasitic on ascomycetous hosts. Mycotaxon 8:372–376

Mitchell DT, Read DJ (1981) Utilization of inorganic and organic phosphates by the mycorrhizal endophytes of *Vaccinium macrocarpon* and *Rhododendron ponticum*. Trans Br Mycol Soc 76:255–260. doi:10.1016/S0007-1536(81)80147-7

Moss ST, Young TWK (1978) Phyletic considerations of the Harpellales and Asellariales (Trichomycetes, Zygomycotina) and the Kickxellales (Zygomycetes, Zygomycotina). Mycologia 70:944–963. doi:10.2307/3759130

Moss ST, Descals E (1986) A previously undescribed state in the life cycle of Harpellales (Trichomycetes). Mycologia 78:213–222. doi:10.2307/3793166

Moss ST, Lichtwardt RW, Manier J-F (1975) *Zygopolaris*, a new genus of Trichomycetes producing zygospores with polar attachment. Mycologia 67:120–127. doi:10.2307/3758234

Münchberg U, Wagner L, Spielberg ET, Voigt K, Rösch P, Popp J (2013) Spatially resolved investigation of the oil composition in single intact hyphae of *Mortierella* spp. with micro-Raman spectroscopy. Biochim Biophys Acta 1831:341–349. doi:10.1016/j.bbalip.2012.09.015

Nagalakhsmi S, Jothi RV, Subrahmanyam A (2008) *Zygambella phosphophila* gen. & sp. nov. Curr Trends Biotechnol Pharm 2:198–200

Nilsson RH, Abarenkov K, Larsson K-H, Kõljalg U (2011) Molecular identification of fungi: rationale, philosophical concerns, and the UNITE database. Open Inform J 5(Suppl1-M9):81–86.

O'Donnell K, Cigelnik E, Benny GL (1998) Phylogenetic relationships among the Harpellales and Kickxellales. Mycologia 90:624–639. doi:10.2307/3761222

O'Donnell K, Lutzoni FM, Ward TJ, Benny GL (2001) Evolutionary relationships among mucoralean fungi (Zygomycota): evidence for family polyphyly on a large scale. Mycologia 93:286–297. doi:10.2307/3761650

Ogawa Y, Hayashi S, Degawa Y, Yaguchi Y (2001) *Ramicandelaber*, a new genus of the Kickxellales, Zygomycetes. Mycoscience 42:193–199. doi:10.1007/BF02464137

Ogawa Y, Kurihara Y, Suda A, Kusama-Eguchi K, Watanabe K, Tokumasu S (2005) Taxonomic position of the genus *Ramicandelaber*, Kickxellales, inferred from 18S rDNA. Nippon Kingakukai Kaiho 46:13–17 [In Japanese]

Oman SJ, White MM (2012) Extended studies of *Baltomyces styrax* in Idaho and expanded distribution of this isopod gut fungus in USA. Mycologia 104:313–320. doi:10.3852/11-204

Padhyay AA, Ajello L (1988) Simple method of inducing sporulation by *Apophysomyces elegans* and *Saksenaea vasiformis*. J Clin Microbiol 26:1861–1863

Papp T, Hoffmann K, Nyilasi I, Petkovits T, Wagner L, Vágvölgyi C, Voigt K (2011) Mortierella. In: Liu D (ed) Molecular detection of human fungal pathogens. Taylor & Francis, Boca Raton, FL, pp 749–758

Patrick MA, Sangar VK, Lichtwardt RW (1973) Lipids of *Smittium culisetae*. Mycologia 65:122–127. doi:10.2307/3757792

Penton CR, StLouis D, Cole JR, Luo Y, Wu L, Schuur EAG, Zhou J, Tiedje JM (2013) Fungal diversity in permafrost and tallgrass prairie soils under experimental warming conditions. Appl Environ Microbiol 79:7063–7072. doi:10.1128/AEM.01702-13

Petkovits T, Nagy LG, Hoffmann K, Wagner L, Nyilasi I, Griebel T, Schnabelrauch D, Vogel H, Voigt K, Vágvölgyi C, Papp T (2011) Data partitions, Bayesian analysis and phylogeny of the zygomycetous fungal family Mortierellaceae, inferred from nuclear ribosomal DNA sequences. PLoS One 6(11):e27057

Powell MJ, Letcher PM (2014) Chytridiomycota, Monoblepharidomycota, and Neocallimastigomycota. In: Mclaughlin DJ, Spadafora JW (eds) The Mycota, vol. VIIA. Systematics and evolution, vol 2, Ed. Springer, Heidelberg, pp 141–175. doi:10.1007/978-3-642-55318-9_6

Quail MA, Smith M, Coupland P, Otto TD, Harris SR, Conner TR, Bertoni A, Swerdlow HP, Gu Y (2012) A tale of three next generation sequencing platforms: comparison of Ion Torrent, Pacific Biosciences and Illumina MiSeq sequencers. BMC Genomics 13:341. doi:10.1186/1471-2164-13-341

Ranzoni FV (1968) Fungi isolated in culture from soils of the Sonoran Desert. Mycologia 60:356–371. doi:10.2307/3757166

Raper KB, Thom C (1949) A manual of the penicillia. Williams & Wilkins, Baltimore, MD

Redecker D, Schüßler A, Stocklinger H, Stürmer SL, Morton JL, Walker C (2013) An evidence-based consensus for the classification of arbuscular mycorrhizal fungi (Glomeromycota). Mycorrhiza 23:515–531. doi:10.1007/s00572-013-0486-y

Ribes JA, Vanover-Sams CL, Baker DJ (2000) Zygomycetes in human disease. Clin Microbiol Rev 13:236–301. doi:10.1128/CMR.13.2.236-301.2000

Roux C (1996) A new species of *Heli*[*co*]*cephalum* (Zygomycotina) from South Africa. S Afr J Bot 62:104–107

Saikawa M, Yamaguchi K, Morikawa C (1988) Capture of rotifers by *Acaulopage pectospora*, and further evidence for its similarity to *Zoophagus insidians*. Mycologia 80:880–884. doi:10.2307/3807574

Saikawa M (1989) Ultrastructure of the septum in *Ballocephala verrucospora* (Entomophthorales, Zygomycetes). Can J Bot 67:2484–2488. doi:10.1139/b89-318

Saikawa M (2011a) On fungi in the Zoopagaceae and Cochlonemataceae. Nihon Kinoakkai Kaiho [Trans Mycol Soc Japan = Japan J Mycol] 52:19–27 (In Japanese)

Saikawa M (2011b) Ultrastructural studies on zygomycotan fungi in the Zoopagaceae and Cochlonemataceae. Mycoscience 52:83–90. doi:10.1007/s10267-010-0083-2

Saikawa M (2012) Morphological studies on fungi in the Zoopagaceae and Cochlonemataceae. Bull Tokyo Gakugei Univ Div Nat Sci 64:55–76. http://hdl.handle.net/2309/131820 (In Japanese)

Saikawa M, Kadowaki T (2002) Studies on *Acaulopage dichotoma* and *A. tetraceros* (Zoopagales, Zygomycota) capturing amoebae. Nova Hedw 74:365–371. doi:10.1127/0029-5035/2002/0074-0365

Saikawa M, Oguchi M, Castañeda Ruiz RF (1997a) Electron microscopy of two nematode destroying fungi, *Meristacrum asterospermum* and *Zygnemomyces echinulatus* (Meristacraceae, Entomophthorales). Can J Bot 75:762–768. doi:10.1139/b97-086

Saikawa M, Sugiura K, Sato H (1997b) Electron microscopy of two trichomycetous fungi attached to the hindgut lining of pill bugs. Can J Bot 75:1479–1484. doi:10.1139/b97-862

Samson RA (1974) *Paecilomyces* and some allied Hyphomycetes. Stud Mycol 6:1–119

Santiago ALCMA, Benny GL, Maia LC (2011) *Syncephalis aggregata*, a new species from semiarid region of Brazil. Mycologia 103:135–138. doi:10.3852/10-049

Santiago ALCMA, Hoffmann K, Lima DX, de Oliveira RJV, Biera HEE, Malosso E, Maia LC, da Silva GA (2014) A new species of *Lichtheimia* (Mucoromycotina, Mucorales) isolated from Brazilian soil. Mycol Prog 13:343–352. doi:10.1007/s11557-013-0920-8

Schipper MAA (1969) Zygosporic stages in heterothallic *Mucor*. Antonie van Leeuwenhoek 35:189–208. doi:10.1007/BF02219130

Schipper MAA (1975) *Mucor mucedo*, *Mucor flavus* and related species. Stud Mycol (Baarn) 10:1–33

Schipper MAA (1978a) On certain species of *Mucor* with a key to all accepted species. Stud Mycol 17:1–52

Schipper MAA (1978b) On the genera *Rhizomucor* and *Parasitella*. Stud Mycol 17:53–71

Schipper MAA (1979) *Thermomucor* (Mucorales). Antonie van Leeuwenhoek 45:275–280. doi:10.1007/BF00418590

Schipper MAA (1984) A revision of the genus *Rhizopus*. I. The *Rhizopus stolonifer*-group and *Rhizopus oryzae*. Stud Mycol 25:1–19

Schipper MAA (1986) *Hyphomucor*, a new genus in the Mucorales for *Mucor assamensis*. Mycotaxon 27:83–86

Schipper MAA, Samson RA (1994) Miscellaneous notes on Mucoraceae. Mycotaxon 50:475–491

Schipper MAA, Stalpers JA (1983) Spore ornamentation and species concept in *Syncephalastrum*. Persoonia 12:81–85

Schipper MAA, Stalpers JA (1984) A revision of the genus *Rhizopus*. II. The *Rhizopus microsporus*-group. Stud Mycol 25:20–34

Schoch CL, Seifert KA, Huhndorf S, Robert V, Sprouge JL, Levesque CA, Chen W, Fungal Barcode Consortium (2012) Nuclear ribosomal internal transcribed spacer (ITS) region as a universal DNA barcode marker for Fungi. Proc Natl Acad Sci USA 109:6241–6246. doi:10.1073/pnas.1117018109

Schumacher K (1894) Lehrbuch der Systematischen Botanik Phytopalaontologie und Phytogeographie. Verlag von Ferdinand Enke, Stuttgart, p 70

Schüßler A, Walker C (2010) The Glomeromycota. A species list with new families and new genera. Published at the libraries of the Royal Botanic Garden Edinburgh, The Royal Botanic Garden Kew, Botanische Staatssammlung Munich, and Oregon State University. http://www.lrz.de/schuessler/amphylo/. Accessed 13 Oct 2014

Schüssler A, Schwarzott D, Walker C (2001) A new fungal phylum, the Glomeromycota: phylogeny and evolution. Mycol Res 105:1413–1421. doi:10.1017/S0953756201005196

Schwartz VU, Santiago ALCMA, Jacobson ID, Voigt K (2014) The pathogenic potential of the *Lichtheimia* genus revisited: *Lichtheimia brasiliensis* is a novel, non-pathogenic species. Mycoses 57(Suppl 3):128–131. doi:10.1111/myc.12230

Sekimoto S, Rochon D'A, Long JE, Dee JM, Berbee ML (2011) A multigene phylogeny of *Olpidium* and its implications for early fungal evolution. BMC Evol Biol 11:331. doi:10.1186/1471-2148-11-331

Shanor L, Poitras WA, Benjamin RK (1950) A new genus of the Choanephoraceae. Mycologia 42:271–278. doi:10.2307/3755440

Shelest E, Voigt K (2014) Genomics to study basal lineage fungal biology: phylogenomics suggests a common origin. In: Nowrousian M (ed) The Mycota, XIII Fungal genomics, 2nd edn. Springer, Berlin, pp 31–60. doi:10.1007/978-3-642-45218-5_2

Siemensma F (2012) *Mayorella* infected by the fungus *Amoebophilus simplex*. http://www.arcella.nl/fungus-Mayorella?. Accessed 21 Oct 2012

Smith ME, Gryganskyi A, Bonito G, Nouhra E, Moreno-Arroyo B, Benny G (2013) Phylogenetic analysis of the genus *Modicella* reveals an independent evolutionary origin of sporocarp-forming fungi in the Mortierellales. Fungal Genet Biol 61:61–68. doi:10.1016/j.fgb.2013.10.001

Stajich JE (2015) Phylogenomics enabling genome-based mycology. In: Mclaughlin DJ, Spatafora JW (eds) The Mycota VII, Part B, Systematics and evolution, 2nd edn. Springer, Berlin, pp 279–294. doi:10.1007/978-3-662-46011-5-11

Stalpers JA (1974) Revision of the genus *Oedocephalum* (Fungi Imperfecti). Proc Koninkl Nederl Akademie van Wetenschappen – Amsterdam Ser C 77:383–401

Strauss T, Botha T, Greyling D, du Preez MT, Kock JLF (2000) Development and testing of selective media for mucoralean fungi. S Afr J Sci 96:597–602

Strongman DB, White MM (2011) *Trifoliellum bioblitzii*, a new genus of trichomycete from mayfly nymphs in Nova Scotia, Canada. Mycologia 103:219–225. doi:10.3852/10-198

Sugiyama M, Tokumasu S, Gams W (2003) *Umbelopsis gibberispora* sp. nov. from Japanese leaf litter and a clarification of *Micromucor ramannianus* var. *angulisporus*. Mycoscience 44:217–226. doi:10.1007/s10267-003-0105-4

Suyama Y, Degawa Y (2013) Sigmoideomycetaceae, a family new to Japan, recorded from Japanese Alps and Ryukyu Islands. In: Anon. Abstracts of the 2013 annual meeting of the Japanese Alps inter-university cooperative project, 21–23 Nov 2013, Sugadaira, Japan, p 79 (ES-A15_2013.pdf). http://jalps.suiri.tsukuba.ac.jp/files/6813/9478/3916/2013_meeting_Jalps_program_v2.2.pdf. Accessed 15 Oct 2014

Tanabe Y, O'Donnell K, Saikawa M, Sugiyama J (2000) Molecular phylogeny of parasitic Zygomycota (Dimargaritales, Zoopagales) based on nuclear small subunit ribosomal DNA sequences. Mol Phylogenet Evol 16:253–262. doi:10.1006/mpev.2000.0775

Thaxter R (1891a) On certain new or peculiar North American Hyphomycetes. I. *Oedocephalum*, *Rhopalomyces* and *Sigmoideomyces* n. g. Bot Gaz 16:14–26. http://www.jstor.org/stable/2994435

Thaxter R (1891b) On certain new or peculiar North American Hyphomycetes. II. *Helicocephalum*, *Gonatorrhodiella*, *Desmidiospora* nov. genera and *Everhartia lignatilis* n. sp. Bot Gaz 16:201–205. http://www.jstor.org/stable/i353298

Thaxter R (1897) New or peculiar Zygomycetes. 2. *Syncephalastrum* and *Syncephalis*. Bot Gaz 24:1–15. http://www.jstor.org/stable/2464604

Thaxter R (1914) New or peculiar Zygomycetes. 3: *Blakeslea*, *Dissophora*, and *Haplosporangium*, nova genera. Bot Gaz (Crawfordsville) 58:353–366. http://www.jstor.org/stable/2467998

Thaxter R (1922) A revision of the Endogoneae. Proc Am Acad Arts Sci 57:291–341, 343–344, 346, 348, 350. doi:10.2307/20025921

Tretter ED, Johnson EM, Wang Y, Kandel P, White MM (2013) Examining new phylogenetic markers to uncover the evolutionary history of early diverging fungi: comparing MCM7, TSR1, and rRNA genes for single- and multi-gene analysis of the Kickxellomycotina. Persoonia 30:106–125. doi:10.3767/003158513X666394

Tretter ED, Johnson EM, Benny GL, Lichtwardt RW, Wang Y, Kandel P, Novak SJ, Smith White MM (2014) An eight-gene molecular phylogeny of the Kickxellomycotina, including the first phylogenetic placement of Asellariales. Mycologia 106:912–935. doi:10.3852/13-253

Tsui CKM, Woodhall J, Chen W, Lévesque A, Lau A, Schoen CD, Baschien C, Najafzadeh MJ, de Hoog SB (2011) Molecular techniques for pathogen identification and fungus detection in the environment. IMA Fungus 2(2):177–189. doi:10.5598/imafungus.2011.02.02.09

Valle LG (2006) Asellariales (Trichomycetes) from the Iberian Peninsula. Fungal Divers 21:167–179

Valle LG, Cafaro MJ (2008) First report of zygospores in Asellariales and new species from the Caribbean. Mycologia 100:122–131. doi:10.3852/mycologia.100.1.122

Valle LG, Santamaria S (2005) Zygospores as evidence of sexual reproduction in the genus *Orphella*. Mycologia 97:1335–1347. doi:10.3852/mycologia.97.6.1335

Voigt K (2012) 10. Zygomycota Moreau. In: Frey W (ed) Syllabus of plant families. A. Engler's Syllabus der Pflanzenfamilien, 13th ed. Part 1/1. Blue-green algae, Myxomycetes and Myxomycete-like organisms, phytoparasitic protists, heterotrophic Heterokontobiota and fungi p.p. Borntraeger Science, Berlin, pp 130–162

Voigt K, Kirk PM (2012) Nomenclatural novelties. Index Fungorum 12:1

Voigt K, Kirk PM (2014) Fungi classification of Zygomycetes: reappraisal as coherent class based on a comparison between traditional versus molecular systematics. In: Batt CA, Tortorello ML (eds) Encyclopedia of food microbiology, 2nd edn. Elsevier B.V, London, pp 54–67. doi:10.1016/B978-0-12-384730-0.00136-1

Voigt K, Wöstemeyer J (2001) Phylogeny and origin of 82 Zygomycetes from all 54 genera of the Mucorales and Mortierellales based on combined analysis of actin and translation elongation factor EF-1a genes. Gene 270:113–120. doi:10.1016/S0378-1119(01)00464-4

Volgmayr H, Krisai-Greilhuber I (1996) *Dicranophora fulva*, a rare mucoraceous fungus growing on boletes. Mycol Res 100:583–590. doi:10.1016/S0953-7562(96)80012-8

Vuillemin P (1904) Le Spinalia radians g. et sp. nov. et la série des Dispirées. Bull Soc Mycol France 20:26–33

Wagner L, Stielow B, Hoffmann K, Petkovits T, Papp T, Vágvölgyi C, de Hoog GS, Verkley G, Voigt K (2013) A comprehensive molecular phylogeny of the Mortierellales (Mortierellomycotina) based on nuclear ribosomal DNA. Persoonia 30:77–93. doi:10.3767/003158513X666268

Walker LB (1923) Some observations on the development of *Endogone malleoli* Hark. Mycologia 15:245–257. doi:10.2307/3753173

Walther G, Pawlowska P, Alastruey-Izquierdo A, Wrzosek M, Rodriguez-Tudela JL, Dolatabadi S, Chakrabarti A, de Hoog GS (2013) DNA barcoding in Mucorales: an inventory of biodiversity. Persoonia 30:11–47. doi:10.3767/003158513X665070

Wang J, Xu S, Strongmann DB (2010) Two new Harpellales inhabiting the digestive tracts of midge larvae and other trichomycetes from Tianshan Mountains, China. Mycologia 102:135–141. doi:10.3852/08-223

Wang Y, Tretter ED, Lichtwardt RW, White MM (2013) Overview of 75 years of *Smittium* research, establishing a new genus *for Smittium culisetae*, and prospects for future revisions of the 'Smittium' clade. Mycologia 105:90–111. doi:10.3852/11-311

Wang Y-N, Liu X-Y, Zheng R-Y (2014) *Umbelopsis changbaiensis* sp. nov. from China and the typification of *Mortierella vinacea*. Mycol Prog 13:657–669. doi:10.1007/s11557-013-0948-9

Weete JD, Abril M, Blackwell M (2010) Phylogenetic distribution of fungal sterols. PLoS One 5(5):e10899. doi:10.1371/journal.pone.0010899

Weitzman I, Silva-Hutner M (1967) Non-keratinous agar media as substrates for the ascigerous state in certain members of the Gymnoascaceae pathogenic for man and animals. Sabouraudia 5:335–340. doi:10.1080/00362176785190611

White MM (2003) First report of *Basidiolum fimbriatum* since 1861, with comments on its development, occurrence, distribution and relationship to other fungi. Mycol Res 107:245–250. doi:10.1017/S0953756203007287

White MM (2006) Evolutionary implications of a rRNA-based phylogeny of Harpellales. Mycol Res 110:1011–1024. doi:10.1016/j.mycres.2006.06.006

White MM, James TY, O'Donnell K, Cafaro MJ, Tanabe Y, Sugiyama J (2006a) Phylogeny of the Zygomycota based on nuclear ribosomal sequence data. Mycologia 98:872–884. doi:10.3852/mycologia.98.6.872

White MM, Lichtwardt RW, Colbo MH (2006b) Confirmation and identification of parasitic stages of obligate endobionts (Harpellales) in blackflies (Simuliidae) by means of rRNA sequence data. Mycol Res 110:1070–1079. doi:10.1016/j.mycres.2006.06.008

White MM, Siri A, Lichtwardt RW (2006c) Trichomycete insect symbionts in Great Smoky Mountains National Park and vicinity. Mycologia 98:333–352. doi:10.3852/mycologia.98.2.333

William RT, Strongman DB (2012) Two new genera of fungal trichomycetes, *Bactromyces* and *Laculus* (Harpellales), from Nova Scotia, Canada. Botany 90:101–111. doi:10.1139/B11-086

Wöstemeyer J (1985) Strain dependent variation in ribosomal DNA arrangement in *Absidia glauca*. Eur J Biochem 146:443–448. doi:10.1111/j.1432-1033.1985.tb08671.x

Wrzosek M, Gajowniczek Z (1998) Some zygomycetous fungi new to Poland. Acta Mycol 33:265–271

Wu G-G, Lin S-J (1997) Endogonales in Taiwan: a new genus with unizygosporic sporocarps and a hyphal mantle. Mycotaxon 64:179–188

Yao Y-J, Pegler DN, Young TWK (1996) Genera of Endogonales. Royal Botanic Gardens, Kew

Young TWK (1968) Electron microscopic study of asexual spores in Kickxellaceae. New Phytol 67:823–836. doi:10.1111/j.1469-8137.1968.tb06398.x

Young TWK (1999) Taxonomy of the Kickxellaceae and radiation of the asexual apparatus. Kew Bull 54:651–661. doi:10.2307/4110861

Zain ME, Moss ST, El-Sheikh (2012) Development of merosporangia in *Linderina pennispora* (Kickxellales, Kickxellaceae). IMA Fungus 3:103–108. doi:10.5598/imafungus.2012.03.02.01

Zhang N, Suh S-O, Blackwell M (2003) Microorganisms in the gut of beetles: evidence from molecular cloning. J Invertebr Pathol 84:226–233. doi:10.1016/j.jip.2003.10.002

Zheng R-Y, Chen G-Q (2001) A monograph of *Cunninghamella*. Mycotaxon 80:1–75

Zheng R-Y, Liu X-Y (2009) Taxa of *Pilaira* (Mucorales, Zygomycota) from China. Nova Hedw 88:255–267. doi:10.1127/0029-5035/2009/0088-0255

Zheng R-Y, Liu X-Y (2014) *Ambomucor* gen. & spp. nov. from China. Mycotaxon 126:97–108. doi:10.5248/126.97

Zheng R-Y, Chen G-Q, Hu F-M (1988) Monosporus varieties of *Syncephalastrum*. Mycosystema 1:35–52

Zheng R-Y, Chen G-Q, Huang H, Liu X-Y (2007) A monograph of *Rhizopus*. Sydowia 59:273–372

Zycha H, Siepmann R, Linnemann G (1969) Mucorales eine Beschreibung aller Gattungen und Arten dieser Pilzgruppe. J. Cramer, Lehre, Lichtenstein

# Chapter 6
# *Entomophthoromycota*: A New Overview of Some of the Oldest Terrestrial Fungi

**Richard A. Humber**

## Phylogenetic Reclassifications

The All-Fungal Tree of Life (AFTOL) project (James et al. 2006) confirmed several pertinent points about the traditional classification of the *Entomophthorales*: The phylum *Zygomycota* is unsupportable and was subsequently divided by Hibbett et al. (2007) into five phylogenetically distinct subphyla not assigned to any specific phylum or phyla. Entomophthoroid fungi were shown to include *Basidiobolus* (whose controversial placement is discussed below) and to occupy a unique and special position at the bottom of the vastly diverse tree of terrestrial (nonflagellate) fungi immediately adjacent to the flagellate fungal phyla *Chytridiomycota*, *Neocallimastigomycota*, and *Blastocladiomycota* (Hibbett et al. 2007) that form the "roots" of the fungal tree. It should be expected that entomophthoroid fungi, because of their age and phylogenetic position, show some biological traits not shared with the more recent and biologically homogeneous basidiomycetes and ascomycetes in the subkingdom Dikarya in the "canopy" of this phylogenetic tree.

The most extensive phylogenetic studies of entomophthoroid fungi (Gryganskyi et al. 2012, 2013a) analyzed more genes and taxa than any previous studies, validated the place of *Basidiobolus* in a monophyletic lineage within the traditionally classified *Entomophthorales* (Humber 1989), and showed the entomophthoroid fungi to be sufficiently divergent from all other fungi to merit their placement in a new phylum (Table 6.1) (Humber 2012). While some poorly resolved taxa still require more study, the six-family classification of these fungi (Humber 1989) was

---

R.A. Humber, Ph.D. (✉)
USDA-ARS Emerging Pests and Pathogens Research Unit, Robert W. Holley Center for Agriculture and Health, 538 Tower Road, Ithaca, NY 14853-2901, USA
e-mail: richard.humber@ars.usda.gov; rah3@cornell.edu



D.-W. Li (ed.), *Biology of Microfungi*, Fungal Biology,
DOI 10.1007/978-3-319-29137-6_6

**Table 6.1** Classification of fungi in the phylum *Entomophthoromycota* (Humber 2012)

| | |
|---|---|
| Class *Basidiobolomycetes* | |
| Order *Basidiobololales* | |
| Family *Basidiobolaceae* | *Basidiobolus*, plus two undescribed genera |
| Class *Neozygitomycetes* | |
| Order N*eozygitales* | |
| Family *Neozygitaceae* | *Neozygites, Apterivorax, Thaxterosporium* |
| Class *Entomophthoromycetes* | |
| Order *Entomophthorales* | |
| Family *Ancylistaceae* | *Ancylistes, Conidiobolus*[a], *Macrobiotophthora* |
| Family *Completoriaceae* | *Completoria complens* (monotypic) |
| Family *Entomophthoraceae* | |
| Subfam. *Entomophthoroideae* | *Batkoa, Entomophaga, Entomophthora, Massospora*[b] |
| Subfam. *Erynioideae* | *Erynia, Eryniopsis, Furia, Orthomyces, Pandora, Strongwellsea, Zoophthora*[c] |
| Family *Meristacraceae*[d] | *Meristacrum, Tabanomyces* |

[a]Still requiring phylogenetic revision that will result in recognition of more than one genus
[b]Keller and Petrini (2005) described the subfamily *Massosporoideae* for this genus; Humber (2012) found insufficient bases to accept such a subfamily
[c]*Zoophthora* remains phylogenetically supported; other genera in this subfamily may or may not be retained after more complete phylogenetic analyses, but how many or which genera will be sustained remains unclear
[d]Humber (1989) included *Ballocephala* and *Zygnemomyces* in this family, but subsequent evidence supports their reassignments from *Entomophthorales* to the unclassified subphylum *Kickxellomycotina* (see Humber 2012)

confirmed, although the families were redistributed among three new classes—*Basidiobolomycetes*, *Neozygitomycetes*, and *Entomophthoromycetes*—each of which includes a single order (*Basidiobolales*, *Neozygitales*, and *Entomophthorales*).

*Basidiobolus* has phylogenetic and biological properties that have led it to be treated variously as removed from *Entomophthorales* but nebulously related to flagellate fungi (Nagahama et al. 1995) to be placed (correctly) among the entomophthoroid fungi (James et al. 2006). The genome-based retention of *Basidiobolus* in the *Entomophthoromycota* may remain controversial for some time to come since *Basidiobolus* (and its close relatives) represent a conundrum for the understanding of the correct phylogenetic relationships of the most basal taxa of terrestrial (nonflagellate) fungi. Despite any uncertainties about interpreting the affinities of *Basidiobolus* from data for a tiny handful of genes, this genus and its close relatives appear to be indisputably correctly placed in the *Entomophthoromycota* based on their more traditional developmental, cytological, and general biological characters (Humber 2012). The similarities between basidiobolomycete and all other entomophthoroid fungi seem to be far greater than can be accounted for by convergent evolution with phylogenetically unrelated (flagellate) fungi.

## *Major Characters Recognized to Have Traditional Taxonomic Value*

Fungi in the *Entomophthoromycota* are easily differentiated from all other fungi by their broad hyphae that are coenocytic to infrequently septate (except in *Basidiobolomycetes*, whose broad hyphae have distinctly uninucleate cells) and production of forcibly discharged conidia with a strong tendency to produce secondary conidia that are also forcibly discharged. These fungi are best known for their entomopathogenic habits, but these fungi show a range of habits from saprobic to obligately pathogenic for insects, other arthropods, or even nematodes, as well as two uncommon phytoparasitic genera, *Ancylistes* and *Completoria*, known from desmid algae and the gametophytes of ferns, respectively. The most phylogenetically basal class, *Basidiobolomycetes*, is primarily saprobic, although *Basidiobolus* species can cause mycoses of humans and other mammals, and a related (still undescribed) genus is known only from mycoses of snakes in Europe and the USA.

The long-accepted taxonomy for entomophthoromycotan genera has been based on a range of traditional characters (Humber 1981; Gryganskyi et al. 2012) such as the morphologies of primary and secondary conidia, the modes of conidial discharge and dispersal, nuclear cytological characters (nuclear appearance in interphase, mode of mitosis, and the number of nuclei in primary conidia), conidiophore branching, and on the presence and morphologies of rhizoids and cystidia. Resting spores (RS), whether formed as zygospores or azygospores, have comparatively limited taxonomic value (Humber 1981).

## *Mitotic/Nuclear Characters*

Variations in mitotic patterns were once hoped to be used to discover phylogenetic relationships (e.g., Heath 1978), but those hopes withered in by the end of the last millennium. The rise of DNA-based phylogenetics, however, brought a new appreciation of nuclear cytology and mitotic patterns as newer DNA-based revisions of many fungi became available. The ancient fungi in *Entomophthoromycota* do, however, underscore the value of nuclear characters that remained unrealized for the younger fungi in *Ascomycota* and *Basidiomycota*.

Among entomophthoroid fungi, the diverse mitotic patterns and types of mitotic nuclear-associated organelles (NAO) observed in *Basidiobolus* (Robinow 1963; Tanaka 1970; Sun and Bowen 1972; McKerracher and Heath 1985), *Ancylistes* (Moorman 1976), *Entomophaga* (Murrin et al. 1984), *Strongwellsea* (Humber 1975), *Pandora* (Butt and Beckett 1984), and *Neozygites* (Butt and Humber 1989) are now seen to represent distinctive cytologies that help to differentiate the classes in this phylum. The *Basidiobolomycetes* have large nuclei with huge central nucleoli, numerous tiny chromosomes that align on a central metaphase plate in nuclei whose nuclear envelopes dissociate during mitosis but whose fragments and cytoplasmic membranes isolate the nucleoplasm from the cytoplasm; these fungi have an enigmatic NAO that is unique among all nonflagellate organisms in having embedded microtubules (McKerracher and Heath 1985). The *Neozygitomycetes* have closed mitoses (inside intact nuclear envelopes) with no known NAO and with vermiform chromosomes aligning on a central metaphase plate (Butt and Humber 1989). Entomophthoromycete nuclei are more variable, but all pre-anaphase spindles seen so far are so small that they are necessarily laterally located until elongation during anaphase; chromosomes in this class vary from not obviously condensing during mitoses (in *Ancylistaceae*) to exceptionally large chromosomes in *Entomophthoraceae* (see Gryganskyi et al. 2013a) that remain partially condensed and present a granular appearance during interphase as seen by light microscopy with or even without using such nuclear stains as aceto-orcein.

## *Entomophthoromycotan Taxa Needing Further Revisionary Study*

The latest multigenic phylogenetic studies of these fungi (Gryganskyi et al. 2012, 2013a) included a much greater range of taxa than earlier such studies, but the extraction of cleanly amplifying DNA from some entomophthoroids appears to be more difficult than for most other fungi. The problem of generating good PCR data is intensified because so many of these fungi have never been cultured, and many critically important taxa of this phylum are very rare and poorly known. It may also be germane that entomophthoraceous nuclei are so large (larger than many yeast cells), contain large amounts of condensed chromatin during interphase, and have total quantities of DNA much larger than in many other fungi (Murrin et al. 1986). The total genome sizes for *Basidiobolus meristosporus* and *Conidiobolus incongruus* (listed at http://www.ncbi.nlm.nih.gov/genome/browse) are large in comparison to other fungi. Polyploidization may have occurred often among fungi in this phylum (Humber 2012) and might account for some of the large total genome sizes. However, until the whole genomic sequences of several entomophthoroids can be compared with the more intensively studied ascomycetes and basidiomycetes, we may never understand just how unique or anomalous the genomes of entomophthoromycotan fungi may actually be.

*Basidiobolomycetes* The unprecedented nature (and meaning) of the mitotic NAO with embedded singlet microtubules (McKerracher and Heath 1985) and the demonstrated similarities between some *Basidiobolus* genes and those of flagellate fungi need further evaluation. *Basidiobolus* urgently needs more phylogenetic study to determine whether the isolates that cause vertebrate mycoses are conspecific with its saprobic species. At least two undescribed genera—one from plant detritus and another from mycoses of snakes—whose overall biology, morphology, development, and nuclear cytology place them indisputably in this class need more evaluation and documentation before their formal descriptions can be published.

*Neozygitomycetes* That this group has long been difficult to place among entomophthoroid fungi is not aided by the extreme challenge of obtaining clean DNA from them. *Neozygites* species are not common and are among the most difficult to culture of all entomophthoroid fungi. To date, the only in vitro cultures from this genus have been of mites pathogens (mostly *N. floridana* and *N. tanajoae*); neither *N. fresenii* (the generic type species) nor other aphid or hemipteran pathogens are available in vitro. These technical difficulties with *Neozygites* species are more troubling since traditional taxonomic approaches suggest a probable need to split *Neozygites* into two genera for the hemipteran and acarine species, respectively. The genera *Thaxterosporium* Ben-Ze'ev and Kenneth (Ben-Ze'ev et al. 1987) and *Apterivorax* Keller (Keller and Petrini 2005) are also poorly known and remain unavailable in culture; without any genomic information, the taxonomies of all of these genera (and their species) remain unverified.

*Entomophthoromycetes*: *Ancylistaceae*: *Conidiobolus* This genus may be the most widely distributed and commonly occurring in the phylum but has needed taxonomic revision for a very long time. Its first two described species—*C. utriculosus*, the type species, and *C. minor*—have not been re-collected since their description (Brefeld 1884) and must either be neotypified or the genus must be conserved with some other extant species. Beyond these nomenclatural issues, the last taxonomic treatment of this genus (King 1976a, b, 1977) and the descriptions of three subgenera separated by secondary conidial characters (Ben-Ze'ev and Kenneth 1982a) have resolved none of the enigmas plaguing this genus. Recent molecular evidence (Gryganskyi et al. 2012, 2013a) does not support the subgenera of this genus but do confirm that this genus includes several phylogenetically distinct lineages. The systematics of *Conidiobolus* cannot be resolved until its typification is settled and a determination made about which lineage will continue to bear the name *Conidiobolus*.

*Entomophthoromycetes*: *Entomophthoraceae*: *Zoophthoroid* **genera (subfamily** *Erynioideae*) The segregation of genera in this group whose species form uninucleate bitunicate conidia on digitately branched conidiophores (except for simple conidiophores in *Strongwellsea*) began with recognition of four subgenera in *Zoophthora* (Batko 1966) that were later treated as genera (Humber 1989) and augmented with the genera *Strongwellsea* (Batko and Weiser 1965) and *Orthomyces* (Steinkraus et al. 1998). Whether all of the Batkoan subgenera should be recognized as separate genera (compare treatments of these taxa by Remaudière and Hennebert 1980; Remaudière and Keller 1980; Ben-Ze'ev and Kenneth 1982b; Humber 1982, 1989; Bałazy 1993; Keller and Petrini 2005) remains controversial: The available phylogenetic data (Gryganskyi et al. 2012, 2013a) confirm the recognition *Zoophthora* for species producing elongate secondary capilliconidia. Too few genomic data are available for *Strongwellsea* species, marginally too few for the diversity of *Pandora* species, and distinctly too few for *Furia* or *Erynia* species, and absolutely no gene-based data are available for *Orthomyces aleyrodis* to support any phylogenetically sound conclusions about genera within this subfamily. It seems that still more genes have to be analyzed before relationships among these fungi can be fully determined.

**Miscellaneous Other Poorly Known Taxa** No fully comprehensive and stabilized phylogenetic revision of the *Entomophthoromycota* will be possible without more pertinent biological and gene-based information about some of the uncultured and extremely rare taxa in this phylum including the following:

- Two new genera of *Basidiobolaceae* from plant detritus or soil and from snakes await formal description. Publications about the snake pathogen include Ippen (1980), Jessup and Seely (1981), Kaplan et al. (1983), and Dwyer et al. (2006).
- The monotypic family *Completoriaceae*, *Completoria complens* (Fig. 6.1; also see Atkinson 1895), parasitizes fern gametophytes and strongly resembles some entomophthoraceous entomopathogens in having apparently protoplastic vegetative growth and physically large nuclei that appear to have granular contents during interphase (presumably due to condensed chromosomal material, as in the *Entomophthoraceae*).
- The taxa in the *Meristacraceae* (Humber 1989) form multiple conidia on an unbranched erect conidiophore but remain little understood and rarely collected. *Meristacrum asterospermum* is a rarely seen nematode pathogen (Drechsler 1940), and *Tabanomyces milkoi* (Couch et al. 1979) was based on a single collection of tabanid fly larvae. *Ballocephala* Drechsler and *Zygnemomyces* Miura

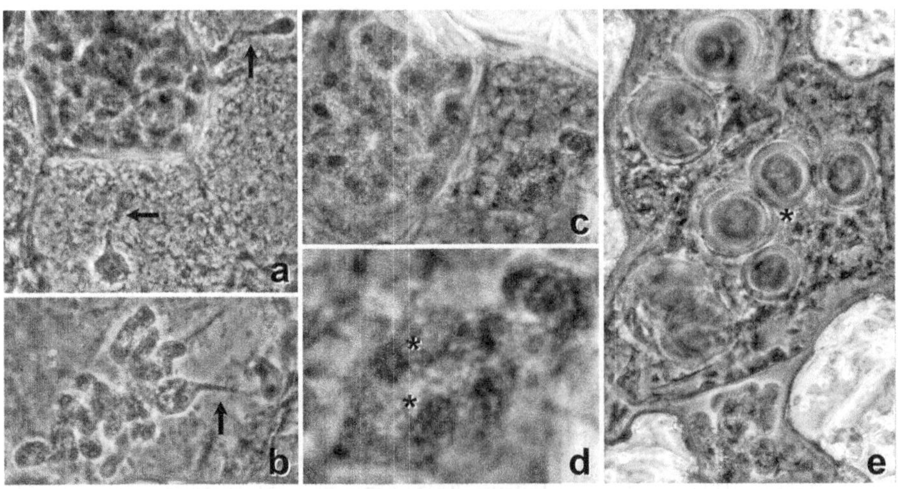


**Fig. 6.1** *Completoria complens* in cells of fern gametophyte. These first published micrographs of this fungus are of the CUP herbarium collection illustrated and discussed by Atkinson (1895). (**a, b**) Apparently wall-less extrusions (*arrows*) of hyphae from one infected cell filled with vegetative hyphal bodies through narrow holes in cell walls into adjacent cells. (**c, d**) Vegetative hyphal bodies in gametophyte cells showing nuclei; cells on right clearly showing large nuclei (at higher magnification in **1d**) with granular contents reminiscent of granular heterochromatin in nuclei of *Entomophthoraceae*. (**e**) Mature, thick-walled resting spores are binucleate (*asterisk*)

affect tardigrades and strongly resemble *Meristacrum*, but these genera have been reassigned to the *Zoopagomycotina* because of the ultrastructure of their septa. Both *Meristacrum* and *Tabanomyces* (see Humber 1989) need reevaluation and confirmation of their affinities whenever suitable collections become available.

## Nutritional Habits and Host Range

The entomophthoroid fungi exemplify the three main nutritional habits of fungi: *Saprobes* depend mostly of nonliving sources of carbon and other nutrients, although some taxa that are primarily saprobic *may* act as facultative parasites and cause mycoses of medical or veterinary importance. *Parasites* such as species in the plant-associated genera *Ancylistes* (from desmid algae) and *Completoria* (from fern gametophytes) infect living hosts but do not necessarily cause the host's death (*Completoria*) or do not seriously affect the host population (*Ancylistes*). The vertebrate mycoses caused by *Basidiobolus* and *Conidiobolus* species are also examples of parasitism. *Pathogens* infect, develop internally, and later kill their hosts. Entomopathogenicity—broadly defined as causing fatal disease for insects, mites, spiders, and, indeed, such other invertebrates as nematodes—is the most important (and best known) nutritional habit for the *Entomophthoromycota*. This habit appears to have arisen independently and repeatedly among water molds (especially in *Blastocladiomycota*), *Entomophthoromycota* (with obligate entomopathogens occurring in two of the three classes and four of the six families), and among several phylogenetically diverse classes and orders of *Ascomycota*, as well as in the zoophilic rust-like basidiomycetes in the order *Septobasidiales* (see Humber 2008). A fascinating and important more general study on fungal evolution (Moore 2013) also suggests that parasitism is a nutritional habit that has arisen easily and often among fungi.

The phylogenetic data (Gryganskyi et al. 2012, 2013a) suggest that the *Basidiobolomycetes* are basal in this phylum, followed by the *Neozygitomycetes* and *Entomophthoromycetes*, successively. With the exception of the few basidiobolomycete fungi that cause vertebrate mycoses, this class appears to be primarily saprobic in soil and plant detritus. Even *Basidiobolus ranarum*—the type species for its class, whose classic "source" is amphibian dung—is a saprobe in soil that is easily acquired (probably carried on the cuticles of insects that are also not diseased by this fungus), is ingested, and survives in the guts of amphibians or reptiles without causing disease.

The mainly saprobic species of *Basidiobolus* and *Conidiobolus* from soil or plant detritus have simple nutritional needs, enabling them to grow on almost any culture medium. Nonetheless, isolations of these fungi, despite their generalized nutritional needs, are unlikely to succeed unless using a "canopy plate" technique (Drechsler 1952) in which small amounts of detritus or soil are minced into a small amount of nutrient agar on a petri dish lid to allow the entomophthoroid fungus to develop and to discharge its conidia forcibly either upward (or, with a slightly higher chance of contamination, downward) onto the main agar surface. Some taxa of both of these genera are known to cause mycoses of humans or other vertebrates (see Humber et al. 1989). A small number of *Conidiobolus* species (especially *C. thromboides* and *C. obscurus*) are best known as entomopathogens, although *C. thromboides* was first described as a soil saprobe (Drechsler 1953). *Macrobiotophthora* affects nematodes, and all species of *Ancylistes* parasitize desmid algae.

All taxa of the *Neozygitomycetes* and *Entomophthoraceae* are obligate pathogens of arthropods. Within the genus *Neozygites*, however, there may be a taxonomically significant disjunction between species with smooth, ovoid zygospores that affect aphids and related hemipterans versus those with globose, rough-walled zygospores affecting mainly mites. There are too few cultures (all of which are from acarine hosts) or fresh collections or in vivo colonies (especially from hemipteran hosts) to allow a phylogenetic evaluation of whether these differences support separate genera.

The specificities of pathogenic entomophthoroid genera for their hosts are often generalized to the order or family of their hosts; individual fungal species may show higher levels of specificity to their host genera or species. A very common entomophthoroid entomopathogen, *Zoophthora radicans*, may represent an unresolved species complex that is known from hosts in an extremely broad range of insect orders (see Bałazy 1993). *Pandora neoaphidis*, on the other hand, is another cosmopolitan and common species on a very large diversity of aphids, but *P. neoaphidis* does not appear to be a species complex (Nielsen et al. 2001). The species of *Massospora* are completely specialized to their unique gregarious cicada hosts (Soper 1974), and *Strongwellsea* species are also apparently specific to their host fly genera (Humber 1976; Eilenberg and Michelsen 1999).

## Infection

Entomophthoroid entomopathogens work against their hosts in a manner very much like almost all other fungal entomopathogens (Humber 2009, 2011): The infection process begins with the adherence of infective spores, a primary or secondary conidium, onto the host cuticle; no entomophthoroid fungi infects any invertebrate by means of an ingested spore germinating and having the germ tube penetrate the gut wall. After landing on a potential host or any other substrate, the complex physiological interactions of the infection process begin with the spore's reaction to its substrate by germinating, forming a secondary conidium, or doing nothing. The germ tubes of these fungi may grow across the cuticle (with the cytoplasm always in the apex of a growing but empty, frequently septate germ tube), possibly growing toward higher nutrient levels that might leach through thinner areas of the cuticle or from spiracles, or it may almost immediately turn down to begin to penetrate that cuticle. Few entomopathogens from this phylum form any differentiated appressorium before growing into the cuticle. Some melanotic host reactions may be induced in the cuticle by fungal germ tubes (Humber 2009), but the most decisive fungus–host interactions occur when the invading fungus reaches the hemocoel and confronts the full force of the host's humoral and/or cellular immune reactions (Boucias and Pendland 1998; Humber 2000, 2009). No infection is truly established until the pathogenic fungus and all host responses are exhausted or evaded and fungal vegetative development has begun. The usual outcome of that growth in the host will be the death of the host and production of primary conidia or of thick-walled resting spores. The variations on this quick scenario are too numerous to be discussed in detail here.

All germ tube growth depends on nutrients and other components packaged into spores at the time of their formation. A dramatic morphological and physiological transition occurs at the physical point when an entomophthoroid cell converts from the exclusive use of internal nutrient resources (in germ tubes) to initiate vegetative growth as marked by the complete switch to using external nutrient sources (see Manners 1966).

No fungal infection has been successfully established *until* the fungus has overcome or evaded all of the host's immune responses. One great advantage for many entomophthoroid entomopathogens to grow at least initially (if not throughout all of vegetative development) as wall-free protoplasts appears to be the evasion of host immune responses (Butt et al. 1996; Humber 2009).

The infective unit for most entomophthoroid fungi is either the primary or secondary conidia. However, there seem to be separate roles for the primary and secondary conidia as dispersive and infective forms, respectively, in some taxa; a major clue to this distinction is that some primary conidia routinely begin to form secondary conidia almost immediately after discharge. *Neozygites* species almost always immediately form secondary capilliconidia with a distinctive apical mucoid droplet whose sole purpose is to attach the capilliconidium to a passing host; the primary conidia in this genus are not discharged to any great distance from the host and only very rarely form forcibly discharged secondary conidia. Species of *Zoophthora* may form forcibly discharged secondary conidia but show a very strong tendency to form secondary capilliconidia soon after the discharge of the primary conidia. *Entomophthora* species show not only an immediate production of secondary conidia from the primaries, but their conidial morphologies and modes of discharge are also distinct. The secondary conidia of *Entomophthora* are discharged by papillar eversion, but the primary conidia are affixed to the substrate by cytoplasmic contents discharged from the conidiogenous cell as part of a cannon-like discharge mechanism. However, it is necessary to acknowledge that there are conflicting interpretations of the mucoid material in which *Entomophthora* primary conidia are embedded (Humber 1981; Eilenberg et al. 1986).

Mycoses of humans or other vertebrates can be caused by some entomophthoroid fungi. These mycoses are comparatively rare, occur mostly in tropical and subtropical locations, and are generally caused by only a handful of *Conidiobolus* and *Basidiobolus* species. Humber et al. (1989) offer the most biologically sensible hypothesis about the ontogenies of the primarily nasopharyngeal conidiobolomycoses of humans, equines, and other mammals. Basidiobolomycoses are less common but have a greater tendency to become either systemic in humans or occur as disseminated subcutaneous infections not affecting the head as is the case with most conidiobolomycoses. An undescribed genus in *Basidiobolaceae* sometimes referred to (but still not formally described) as *Schizangiella* is reported only from captive snakes in Europe (Ippen 1980) and North America (Jessup and Seely 1981; Kaplan et al. 1983; Dwyer et al. 2006) and grows vegetatively with a unique pattern of repeated *internal* cleavage of globose cells into clusters of up to 16 cells by mutually perpendicular series of cell divisions that recall the yeastlike (but microaerophilic or anaerobic) darmform growth pattern of *Basidiobolus* cells. It seems highly likely that this snake pathogen is a soil- or detritus-borne saprobe that can become a facultative subcutaneous to systemic pathogen of snakes after inoculation through skin abrasions or cuts.

## Development

Established infections by entomophthoroid fungi typically grow in the hemocoel of the host as hyphal bodies (short hyphal segments) that multiply rapidly and circulate freely throughout the body cavity. Hyphal bodies are short structures whose morphologies and general behaviors will depend, among other factors, on whether they are walled or protoplastic rather than growing as long, continuous walled (or protoplastic) hyphae. Even wall-less hyphal bodies, however, can vary dramatically in their appearance from being rodlike and more or less cylindrical throughout their length to highly amoeboid shape-shifting cells typical of *Entomophaga* species affecting lepidopterans that were first reported by Tyrrell and Macleod (1972) to much more rodlike cells in many genera (Humber 1975; Latgé et al. 1988; Butt et al. 1990). Protoplastic development by entomophthoroid fungi is not universal—and is not known for any taxa in *Basidiobolomycetes* or *Ancylistaceae*—but occurs in a wide range of taxa. Some *Neozygites* species may develop in their hosts (and in culture) as protoplasts. In the *Entomophthoraceae*, protoplasts are known from *Entomophthora*, *Entomophaga*, and *Massospora* but not from *Batkoa* species within the subfamily *Entomophthoroideae*; among the zoophthoroid genera (*Erynioideae*), protoplasts are known from at least the early developmental stages (vegetative proliferation up to the point where the host is moribund or dead, and the fungus is preparing itself for reproduction) in the genera *Strongwellsea* and *Pandora* but may be absent (or uncommon) in *Zoophthora* and remain unconfirmed from the other genera of this subfamily. Curiously, vegetative development by the fern-parasitic *Completoria complens* clearly appears to be protoplastic; the fungus spreads from cell to cell injecting highly irregularly shaped cells through tiny holes in the adjacent cell walls (Fig. 6.1; also see Atkinson 1895).

Entomophthoroid entomopathogens mostly follow a common developmental pattern in their hosts involving proliferation and dispersal of cells throughout the body in the hemocoel. Few entomophthoroid entomopathogens grow primarily as longer hyphae (rather than hyphal bodies) throughout their development. Significant spatial restriction of development, usually to the abdomen, is limited mainly to *Massospora* and *Strongwellsea* species. In the case of *Strongwellsea*, whose species affect muscoid fly adults, vegetative development is as protoplastic hyphae that eventually form a hollow ball in the abdomen, become walled toward the interior of that ball so that (walled) conidiophores line the interior of the hollow ball, and discharge conidia through an expanding hole in the abdominal pleuron of living flies (Humber 1976; Eilenberg and Michelsen 1999). Despite the major development of *Strongwellsea* in the abdomen, the hyphae can interpolate themselves into the host's nervous system to proliferate in the thoracic ganglion and even throughout the brain while few if any hyphae are found in the thoracic or cephalic hemocoel (Humber 1975, 1976). *Massospora* species, all of which are known only from gregarious cicadas, produce protoplastic hyphal bodies but are spatially restricted entirely to the terminal (gonadal) segments of the abdomen that are separated from the rest of the host's body by a membrane (Speare 1921).

Entomophthoroid development in invertebrate hosts usually remains in the hemocoel until the fungal growth occludes that space and blood circulation is impeded. The host is moribund by this time, and the fungus then begins excreting enzymes that digest the host's organs; all of these events—nutrient starvation, loss of excretory clearance of now rapidly accumulating cellular breakdown products, and loss of circulation and of oxygenation of the host's body—cause the host's death. By this time, any fungal protoplasts will have acquired walls, and walled hyphae begin penetrating through the cuticle immediately prior to external conidiogenesis or, mostly inside the host body, to produce the thick-walled resting spores.

Arthropods killed by entomophthoroid fungi may produce rhizoids and/or cystidia shortly before or soon after death. These two types of specialized cells emerge before the conidiophores; their morphologies and taxonomic significance were discussed by Humber (1981). Rhizoids descend to the substrate from the ventral surfaces of the host and frequently form some type of terminal holdfast to anchor the host. Cystidia emerge ventrally and facilitate the passage of developing conidiophores by their multiple perforations of the host cuticle. Except for extending far beyond the level of the conidiogenous hymenium that will form on the surface of infected host cadavers, cystidia (much like rhizoids) may be little differentiated from vegetative hyphae in their thickness or may (e.g., in *Erynia* species) be markedly thicker and may have characteristic apical branchings (Thaxter 1888; Humber 1981).

The only entomophthoroid entomopathogens that sporulate from living hosts are the species of *Strongwellsea* and *Massospora*, as well as *Entomophthora erupta* (Hall 1959).

## Reproduction

### *Conidiogenesis and Dispersal*

Zygomycete propagules were formerly always considered to be sporangiospores—spores formed inside a cell (sporangium) whose cytoplasm is repackaged into one or more sporangiospores, each of which acquires a newly generated cell wall. However, the thin-walled (usually) forcibly discharged spores of *Entomophthoromycota* and also the asexual spores of some taxa of *Zoopagomycotina* (Saikawa 2011) are not sporangiospores: The wall layers covering these spores are identical to those of the cells producing them (Humber 1975, 1981; Benny et al. 2001); no new wall is generated inside a parental cell wall, and these spores must be treated as conidia formed in the same manner as the mitospores of innumerable ascomycete taxa. In slide preparations of uninucleate conidia from taxa in the subfamily *Erynioideae* (*Entomophthoraceae*), a wall is often seen to separate and to lift away from the spore surface; although this appearance suggests these spores might be monosporic sporangioles, transmission electron microscopy of such "bitunicate" conidia (Remaudière and Hennebert 1980) confirms that these are conidia with the same wall layers of all other entomophthoroid cells except for having an outer (electron dense) wall layer that is extensible and easily detachable from the inner (electron lucent) layer.

On arthropod hosts with comparatively thick cuticles (e.g., flies, grasshoppers, beetles, etc.), the conidiophores of entomopathogens emerge through the thinner cuticle of intersegmental membranes of the abdomen, around the eyes and mouthparts, from leg joints, or in other thin sections of cuticle to form densely packed hymenial bands of conidiogenous cells. On hosts with comparatively thin cuticles (e.g., aphids and some lepidopterans), the conidiophores often form hymenia completely covering the host cadaver. The conidiophores are unbranched (or, rarely, sparingly branched well below the conidiogenous apices) in all taxa except those in the subfamily *Erynioideae* (*Entomophthoraceae*), in which conidiophores are digitately branched just below the conidiogenous cells (although the conidiophores of *Strongwellsea* remain simple). The primary and also secondary conidiophores of many of these fungi are highly phototropic and orient both the production and discharge of conidia toward a light source (Page and Humber 1973). Positive phototropism may be very important for the dispersal of saprobes growing in soil or plant detritus (to get the spores discharged freely into the air) but may also be observed in the cultures of many entomopathogenic taxa.

Several different mechanisms for active discharge of conidia are found within the phylum, but the most common and important is that of papillar eversion (see Humber (1981) for a characterization of all of the conidial discharge mechanisms in this phylum), in which the septum between the conidiogenous cell and conidium forms by the growth of the inner layer of the cell wall and projects into the developing conidia (possibly because cytoplasm is being transferred hydraulically from the conidiogenous cell into the developing conidium while this septum is developing). Once the septum is closed, very nearly all cytoplasm has been transferred from the conidiogenous cell into the conidium. The hydrostatic pressure builds inside the conidium (possibly driven by the hydrolysis of some storage products) and is subsequently relieved by the sudden eversion of this papilla in a manner that breaks the outer cell wall layer so that the spore flies away to a considerable distance. Two other modes of primary conidial discharge are found in the genera *Basidiobolus* and *Entomophthora*; differing hypotheses about how primary conidial discharge is accomplished in *Entomophthora* species are discussed by Humber (1981) and Eilenberg et al. (1986).

## *Secondary Conidiogenesis*

Secondary conidiogenesis is a key character of *Entomophthoromycota*. Primary conidia formed by the developing vegetative fungus are forcibly discharged (except in *Massospora*), and, if they land on substrates unsuited to forming germ tubes (or are preprogrammed *not* to germinate), the primary conidium produces one (or more) of several possible types of secondary conidium that were classified by Ben-Ze'ev and Kenneth (1982a) according to their morphologies and modes of formation. Forcibly discharged secondary conidia formed singly on short conidiophores are the most typical for these fungi; the formation and phototropically oriented discharge of these conidia was illustrated and discussed in detail by Page and Humber (1973). The production and forcible discharge of multiple secondary "microconidia" from single primary conidia are seen in several species of *Conidiobolus* and one of *Basidiobolus*. The production of a single, passively dispersed capilliconidium on a long, slender capillary conidiophore is a key character of species of *Zoophthora* (Ben-Ze'ev and Kenneth 1982a, b), but a somewhat variant morphology in the companion zoophthoroid genus *Orthomyces* (Steinkraus et al. 1998) suggests that the slightly papillate secondary conidium in *Orthomyces* may, in fact, be forcibly dislodged to fall from the capillary apex, whereas all other capilliconidia produced within the phylum are passively dispersed (from much narrower capillaries on which the secondary capilliconidia have no visible basal papilla). Other forms of secondary capilliconidia with various orientations on the conidiophores and with or without terminal mucoid droplets to aid attachment to a passing host are seen in species of *Neozygites*, *Basidiobolus*, some *Conidiobolus* species, and also in the nematode-pathogenic genus *Macrobiotophthora*—in which short, flexible capillary conidiophores are caught and rapidly bent by surface tension in the fluid layer on a contacting nematode and are literally slapped onto the cuticle of the unfortunate nematode (Humber, unpublished).

A bit more about the production of passively dispersed secondary capilliconidia on long, capillary conidiophores is justified. This mode of sporogenesis does not appear to be an ancestral state for the *Entomophthoromycota*. While capilliconidiogenesis may be uncommon elsewhere, it is known for the ascomycete entomopathogen *Hirsutella aphidis* (Bałazy 1985). There is a good reason to believe that the formation of such secondary conidia in four families (*Basidiobolaceae*, *Neozygitaceae*, *Ancylistaceae*, and *Entomophthoraceae*) and all classes of the phylum suggests that this mode of sporogenesis has arisen multiple times. It probably arose easily as a modification of forcibly discharged secondary conidiogenesis with the secondary conidium forming at a greater height (to aid dispersal to possible hosts) with the consequence that elongating conidiophores necessarily became increasingly narrower, eventually too narrow to allow forcible discharge by papillar eversion. The hypothesized forcible dislodgement of minutely papillate secondary capilliconidia of *Orthomyces aleyrodis* from unusually broad capillaries (whose apices also show a correspondingly tiny eversion like that seen with discharge by papillar eversion) seems to support this supposition (Steinkraus et al. 1998).

## *Specialized Adaptations of Conidiogenesis in* Entomophthorales

Species of diverse entomophthoraceous fungi infecting aquatic larvae of nematoceran dipterans (especially Simuliidae) or their female adults (that may become waterlogged during oviposition on moss-covered rocks in splash zones and die soon thereafter while drying off on the backsides of these rocks) may produce conidia with several stout arms projecting forward from the apex of primary conidia or backward from around the papillae of secondary conidia; these spores are referred to as "coronate" or "stellate" conidia, respectively (Descals et al. 1981; Descals and Webster 1984). These branched conidia so not appear to promote spore dispersal (as for the "tetraradiate" conidia of many waterborne ascomycetes) but rather to retard or to prevent their dispersal from the site where susceptible, healthy hosts can be infected by the small, globose, forcibly discharged secondary (or tertiary) conidia discharged from these modified conidial forms.

Cryptoconidiogenesis (Humber and Ramoska 1986) is an unexpected variation on entomophthoroid fungal sporulation that confirms many basic principles about reproduction by these fungi. Populations of spur-throated grasshoppers, *Melanoplus* spp. (Acrididae: Melanoplinae), may support large epizootics of *Entomophaga calopteni*, but all infected grasshoppers appear to produce thick-walled resting spores without any customary conidial state. How such a fungus could spread disease in its host populations without producing a forcibly discharged conidia (since the resting spores are not involved in horizontal transmissions) remained unanswered until it was noted that hosts infected by this fungus often showed marked stretching of the abdominal intersegmental membranes. When these thinned and very fragile membranes rupture by the movements of other grasshoppers or by cannibalism of cadavers, the masses of hyphal bodies that had not (yet) produced resting spores are exposed to full atmospheric oxygen levels, and these hyphal bodies can produce forcibly discharged primary conidia directly without ever forming any "normal" conidiophores. These unexpected spores formed by cell types not known for such an ability once supplied with sufficient oxygen were referred to as cryptoconidia. The critical role of oxygen was proven because exposed populations of hyphal bodies from infected hosts produced no cryptoconidia when enclosed in jars with a piece of burning paper (resulting in depleted oxygen and augmented carbon dioxide) or when incubated in pure nitrogen gas (Humber and Ramoska 1986).

## *Resting Spores and Sexuality in* Entomophthoromycota

Entomophthoroid fungi challenge many mycological stereotypes, and the durable resting spore states of these ancient fungi pose some especially difficult challenges (Humber 1981, 2012). The locations and importance of sexuality in the life histories of ascomycete and basidiomycete fungi are clear. Meiosis occurs in asci or basidia; ascospores and basidiospores are haploid meiospores; and the overall life histories of these fungi are understood. The thick-walled durable spores produced by zygomycete fungi are, however, more problematic. Classic studies by Cutter (1942a, b) demonstrated several distinct patterns of sexuality among mucoroid fungi that produce thick-walled zygospores and azygospores. The thick-walled spores of zygomycetous taxa, of chytrid and blastocladian fungi, and of oomycetous straminopiles usually fit in the expected pattern of life histories, with these durable spores being sexually derived and with meiosis occurring either during the formation or the germination of these spores. Among the (formerly) zygomycete fungi, the thick-walled durable spores are regarded to be zygospores (if formed after a gametangial conjugation) or azygospores (if formed without any gametangial conjugation), and it is almost always assumed that zygospores are sexual spores while azygospores are not.

These morphological definitions of sexuality, unfortunately, ignore the issue of a diametrically opposed definition of sexuality based on genetic events in a life history: The genetic definition of sexuality requires the alternation of meiosis and karyogamy in a life history. In almost every well-known and widely studied group of organisms, there is no separation of the morphological and genetic definitions of sexuality, and most biologists have fallen into the conceptual trap of expecting the distinction between these two definitions to be trivial or nonexistent. The fungi in the *Entomophthoromycota*—especially the numerous and highly diverse fungi in the *Entomophthoraceae* (see Table 6.1)—both set and spring this trap on the unwary. These fungi shine a strong light onto the need to be aware of these disparate definitions of sexuality and on the role of sexual reproduction in phylogenetic radiation:

Unlike nearly all other fungi, homothallism is the observed rule in *Entomophthoromycota.* Heterothallism, with its obligatorily outcrossed pairings of disparate mating types, has never been definitively demonstrated in *Entomophthoromycota.* Only very indirect, poorly supported suggestions (e.g., Gryganskyi et al. 2013b) exist that outcrossing *might* occur among these fungi.

McCabe et al. (1984) showed that there is a more or less free mixture of the two important morphological and developmental hallmarks of sexuality within *Entomophthorales*—of zygosporogenesis versus azygosporogenesis and of whether nuclear numbers reduce to two or not resting spore maturation—with all four possible combinations of these two processes being demonstrable. The assumed pattern that when mature resting spores are binucleate that the nuclei do fuse is reinforced by the demonstration (McCabe et al. 1984) of a single, notably larger nucleus in the germinating resting spore of *Conidiobolus thromboides* that was interpreted as a possible zygotic nucleus that would undergo meiosis as germination proceeded; the nuclei of all other cells of all entomophthoroid fungi are haploid.

If (a)zygosporogenesis in the *Entomophthoromycota* is entirely homothallic and the involvement of gametangial conjugations immaterial, then other questions can be raised: Do these fungi contain genetic suggestions of mating type genes or other similar markers of outcrossing organisms? Are the fungi in *Entomophthoromycota* the only fungi showing evidence for an ancient abandonment of (traditional) sexuality like that demonstrated for bdelloid rotifers (Normark et al. 2003) in the animal kingdom?

An alternative route to allow gene flow among organisms with known importance for many ascomycetes, for example, is parasexuality, in which fusions of genetically different cells result in heterokaryons in which genetically dissimilar nuclei may undergo fusions and some type of genetic exchange (whether by "mitotic crossing-over" or meiosis in "unusual" places at "unusual" times). Parasexuality is, however, a mechanism that may be wholly unavailable in the *Entomophthoromycota*: No cell-to-cell fusions have been observed at any time among these fungi *except* when forming zygospores, but even these fusions may be between genetically *identical* adjacent cells (e.g., as in *Conidiobolus* or adjacent *uninucleate* cells in *Basidiobolus*) or, in an insect, between hyphal bodies that might be derived from a single nucleus (e.g., in infections by zoophthoroid genera whose conidia are uninucleate). Curiously, even in the protoplastic vegetative cells of entomophthoroid taxa, there is no verified record of cellular fusions despite intimate or prolonged cell-to-cell contacts. Dunphy and Nolan (1977) reported fusions in vitro of spindle-shaped protoplasts of *Entomophaga aulicae* (as *Entomophthora egressa*), but this report is now known to be a misinterpretation of hypertrophic swelling of single protoplasts in nutrient-deficient liquid medium; cells that progressively become large, highly multinucleate "spheroplasts" with gigantic central vacuoles will, when transferred to fresh medium, spin off new fusoid, mobile protoplasts and disappear from the culture.

## References


Atkinson GF (1895) A potting bed fungus new to America. Bull NY Agric Exp Stat 94:252–260

Bałazy S (1985) Notes on *Hirsutella aphidis*. Trans Br Mycol Soc 85:752–756

Bałazy S (1993) Entomophthorales. Flora of Poland (Flora Polska), Fungi (Mycota) 24. Polish Acad Sci, W Szafer Inst Botany, Kraków, pp 1–356

Batko A (1966) On the subgenera of the fungus genus *Zoophthora* Batko 1964 (Entomophthoraceae). Acta Mycol 2:15–21

Batko A, Weiser J (1965) On the taxonomic position of the fungus discovered by Strong, Wells, and Apple: *Strongwellsea castrans* gen. et sp. nov. (Phycomycetes: Entomophthoraceae). J Invertebr Pathol 7:455–463

Benny GL, Humber RA, Morton JB (2001) Zygomycota: Zygomycetes. In: McLaughlin DJ, McLaughlin E, Lemke PA (eds) The Mycota, vol 7A. Springer, Heidelberg, pp 113–146

Ben-Ze'ev I, Kenneth RG (1982a) Features-criteria of taxonomic value in the Entomophthorales: I. A revision of the Batkoan classification. Mycotaxon 14:393–455

Ben-Ze'ev I, Kenneth RG (1982b) Features-criteria of taxonomic value in the Entomophthorales: II. A revision of the genus *Erynia* Nowakowski 1881 (= *Zoophthora* Batko 1964). Mycotaxon 14:456–475

Ben-Ze'ev I, Kenneth RG, Uziel A (1987) A reclassification of *Entomophthora turbinata* in *Thaxterosporium* gen. nov., Neozygitaceae fam. nov. (Zygomycetes: Entomophthorales). Mycotaxon 28:313–326

Boucias DG, Pendland JC (1998) Principles of insect pathology. Kluwer Academic, Boston

Brefeld O (1884) *Conidiobolus utriculosus* und *minor*. Untersuch Gesammtgeb Mykol 6(2):35–78

Butt TM, Beckett A (1984) Ultrastructure of mitosis in the aphid pathogenic fungus *Erynia neoaphidis*. Protoplasma 120:72–83

Butt TM, Humber RA (1989) An immunofluorescence study of mitosis in a mite-pathogen, *Neozygites* sp. (Zygomycetes: Entomophthorales). Protoplasma 151:115–123

Butt TM, Beckett A, Wilding N (1990) A histological study of the invasive and developmental processes of the aphid pathogen *Erynia neoaphidis* (Zygomycotina: Entomophthorales) in the pea aphid *Acyrthosiphon pisum*. Can J Bot 68:2153–2163

Butt TM, Hajek AE, Humber RA (1996) Gypsy moth immune defenses in response to hyphal bodies and natural protoplasts of entomophthoralean fungi. J Invertebr Pathol 68:278–285

Couch JN, Andreeva RV, Laird M, Nolan RA (1979) *Tabanomyces milkoi* (Dudka and Koval) emended, genus novum, a fungal pathogen of horseflies. Proc Nat Acad Sci USA 76:2299–2302

Cutter VM Jr (1942a) Nuclear behavior in the Mucorales. I. The *Mucor* pattern. Bull Torrey Bot Club 69:480–508

Cutter VM Jr (1942b) Nuclear behavior in the Mucorales. II. The *Rhizopus*, *Phycomyces*, and *Sporodinia* patterns. Bull Torrey Bot Club 69:592–616

Descals E, Webster J (1984) Branched aquatic conidia in *Erynia* and *Entomophthora* sensu lato. Trans Br Mycol Soc 83:669–682

Descals E, Webster J, Ladle M, Bass JAB (1981) Variations in asexual reproduction in species of *Entomophthora* on aquatic insects. Trans Br Mycol Soc 77:85–102

Drechsler C (1940) Three fungi destructive to free-living terricolous nematodes. J Wash Acad Sci 30:240–254

Drechsler C (1952) Widespread distribution of *Delacroixia coronata* and other saprophytic Entomophthoraceae in plant detritus. Science 115:575–576

Drechsler C (1953) Three new species of *Conidiobolus* isolated from leaf mold. J Wash Acad Sci 43:29–43

Dunphy GB, Nolan RA (1977) Regeneration of protoplasts of *Entomophthora egressa*, a fungal pathogen of the eastern hemlock looper. Can J Bot 55:107–113

Dwyer J, Burwell B, Humber RA, Mcleod C, Fleetwood M, Johnson T (2006) *Schizangiella serpentis* infection in a Virginia ratsnake (*Elaphe obsoleta*). Vet Pathol 43:819

Eilenberg J, Michelsen V (1999) Natural host range and prevalence of the genus *Strongwellsea* (Zygomycota: Entomophthorales) in Denmark. J Invertebr Pathol 73:189–198

Eilenberg J, Bresciani J, Latgé J-P (1986) Ultrastructural studies of primary spore formation and discharge in the genus *Entomophthora*. J Invertebr Pathol 48:318–324

Gryganskyi AP, Humber RA, Smith ME, Miadlikowska J, Wu S, Voigt K et al (2012) Molecular phylogeny of the *Entomophthoromycota*. Mol Phylogenet Evol 65:682–694

Gryganskyi AP, Humber RA, Smith ME, Hodge K, Huang B, Voigt K, Vilgalys R (2013a) Phylogenetic lineages in *Entomophthoromycota*. Persoonia 30:94–105

Gryganskyi AP, Humber RA, Stajich JE, Mullens B, Anishchenko IM, Vilgalys R (2013b) Sequential utilization of hosts from different fly families by genetically distinct, sympatric populations within the *Entomophthora muscae* species complex. PloS One 8:e71168. doi:10.1371/journal.pone.0071168

Hall IM (1959) The fungus *Entomophthora erupta* (Dustan) attacking the black grass bug, *Irbisia solani* (Heidemann) (Hemiptera, Miridae), in California. J Insect Pathol 1:48–51

Heath IB (ed) (1978) Nuclear division in the fungi. Academic Press, New York

Hibbett DS, Binder M, Bischoff JF, Blackwell M, Cannon PF, Eriksson OE et al (2007) A higher-level phylogenetic classification of the fungi. Mycol Res 111:509–547

Humber RA (1975) Aspects of the biology of an insect-parasitic fungus, *Strongwellsea magna* (Zygomycetes: Entomophthorales). PhD dissertation, University of Washington, Seattle

Humber RA (1976) The systematics of the genus *Strongwellsea* (Zygomycetes: Entomophthorales). Mycologia 68:1042–1060

Humber RA (1981) An alternative view of certain taxonomic criteria used in the *Entomophthorales* (*Zygomycetes*). Mycotaxon 13:191–240

Humber RA (1982) *Strongwellsea* vs. *Erynia*: the case for a phylogenetic classification of the Entomophthorales (Zygomycetes). Mycotaxon 15:167–184

Humber RA (1989) Synopsis of a revised classification for the *Entomophthorales* (*Zygomycotina*). Mycotaxon 34:441–460
Humber RA (2000) Fungal pathogens and parasites of insects. In: Priest F, Goodfellow M (eds) Applied microbial systematics. Kluwer Academic, Dordrecht, pp 199–227
Humber RA (2008) Evolution of entomopathogenicity in fungi. J Invertebr Pathol 98:262–266
Humber RA (2009) Fungi: Entomogenous fungi. In: Schaechter M (ed) Encyclopedia of microbiology, 3rd edn. Elsevier, San Diego, pp 443–456
Humber RA (2011) Entomogenous fungi. In: Schaechter M (ed) Eukaryotic microbes. Elsevier, Amsterdam, pp 127–141
Humber RA (2012) *Entomophthoromycota*: a new phylum and reclassification for entomophthoroid fungi. Mycotaxon 120:477–492
Humber RA, Ramoska WA (1986) Variations in entomophthoralean life cycles: practical implications. In: Samson RA, Vlak JM, Peters D (eds) Fundamental and applied aspects of invertebrate pathology. Foundation for the 4th International Colloquium of Invertebrate Pathology, Wageningen, pp 190–193
Humber RA, Brown CC, Kornegay RW (1989) Equine zygomycosis caused by *Conidiobolus lamprauges*. J Clin Microbiol 27:573–576
Ippen R (1980) Ein Beitrag zu den Mykosen der Schlangen. Milu, Berlin 5:386–396
James TY, Kauff F, Schoch C, Matheny PB, Hofstetter V, Cox CJ et al (2006) Reconstructing the early evolution of Fungi using a six-gene phylogeny. Nature 443(7113):818–822
Jessup DA, Seely JC (1981) Zygomycete fungus infection in two captive snakes: gopher snake (*Pituophis melanoleucos*); copperhead (*Agkistrondon contortrix*). J Zoo Anim Med 12:54–59
Kaplan W, Chandler FW, Padhye AA, Hamm TE Jr (1983) A zygomycotic infection in captive snakes. Sabouraudia 21:85–91
Keller S, Petrini O (2005) Keys to the identification of the arthropod pathogenic genera of the families Entomophthoraceae and Neozygitaceae (Zygomycetes), with descriptions of three new subfamilies and a new genus. Sydowia 57:23–53
King DS (1976a) Systematics of *Conidiobolus* (Entomophthorales) using numerical taxonomy. I. Biology and cluster analysis. Can J Bot 54:45–65
King DS (1976b) Systematics of *Conidiobolus* (Entomophthorales) using numerical taxonomy. II. Taxonomic considerations. Can J Bot 54:1285–1296
King DS (1977) Systematics of *Conidiobolus* (Entomophthorales) using numerical taxonomy. III. Descriptions of recognized species. Can J Bot 55:718–729
Latgé J-P, Eilenberg J, Beauvais A, Prevost MC (1988) Morphology of *Entomophthora muscae* protoplasts grown in vitro. Protoplasma 146:166–183
Manners JG (1966) Assessment of germination. In: Madelin MF (ed) The fungus spore (Colston Papers No 18). Butterworths, London, pp 165–173
McCabe DE, Humber RA, Soper RS (1984) Observation and interpretation of nuclear reductions during maturation and germination of entomophthoralean resting spores. Mycologia 76:1104–1107
McKerracher LJ, Heath IB (1985) The structure and cycle of the nucleus-associated organelle in two species of *Basidiobolus*. Mycologia 77:412–417
Moore D (2013) Fungal biology in the origin and emergence of life. Cambridge University Press, Cambridge
Moorman GW (1976) Mitosis in *Ancylistes* sp. Mycologia 68:902–909
Murrin F, Heath IM, Newcomb W (1984) The taxonomic significance of the mitotic spindle and nucleus-associated organelle of *Entomophthora aulicae*. Protoplasma 120:84–90
Murrin F, Holtby J, Nolan RA, Davidson WS (1986) The genome of *Entomophaga aulicae* (Entomophthorales, Zygomycetes): base composition and size. Exp Mycol 10:67–75
Nagahama T, Sato H, Shimazu M, Sugiyama J (1995) Phylogenetic divergence of the entomophthoralean fungi: evidence from nuclear 18S ribosomal RNA gene sequences. Mycologia 87:203–209

Nielsen C, Sommer C, Eilenberg J, Hansen KS, Humber RA (2001) Characterization of aphid pathogenic species in the genus *Pandora* by PCR techniques and digital image analysis. Mycologia 93:864–874

Normark BB, Judson OP, Moran NA (2003) Genomic signatures of ancient asexual lineages. Biol J Linn Soc 79:69–84

Page RM, Humber RA (1973) Phototropism in *Conidiobolus coronatus*. Mycologia 65:335–354

Remaudière G, Hennebert GL (1980) Révision systématique de *Entomophthora aphidis* Hoffm. in Fres. Description de deux nouveaux pathogènes d'aphides. Mycotaxon 11:269–321

Remaudière G, Keller S (1980) Révision systématique des genres d'Entomophthoraceae à potentialité entomopathogène. Mycotaxon 11:323–338

Robinow CF (1963) Observations on cell growth, mitosis and division in the fungus *Basidiobolus ranarum*. J Cell Biol 17:123–152

Saikawa M (2011) Ultrastructural studies on zygomycotan fungi in the Zoopagaceae and Cochlonemataceae. Mycoscience 52:83–90

Soper RS (1974) The genus *Massospora*, entomopathogenic for cicadas. Part I, Taxonomy of the genus. Mycotaxon 1:13–40

Speare AT (1921) *Massospora cicadina* Peck: a fungous parasite of the periodical cicada. Mycologia 13:72–82

Steinkraus DC, Oliver JB, Humber RA, Gaylor MJ (1998) Mycosis of banded winged whitefly (*Trialeurodes abutilonea*) (Homoptera: Aleyrodidae) caused by *Orthomyces aleyrodis* gen. & sp. nov. (Entomophthorales: Entomophthoraceae). J Invertebr Pathol 72:1–8

Sun NC, Bowen CC (1972) Ultrastructural studies of nuclear division in *Basidiobolus ranarum* Eidam. Caryologia 25:471–494

Tanaka K (1970) Mitosis in the fungus *Basidiobolus ranarum* as revealed by electron microscopy. Protoplasma 70:423–440

Thaxter R (1888) The Entomophthoreae of the United States. Mem Boston Soc Nat Hist 4:133–201, plates xiv–xxi

Tyrrell D, Macleod DM (1972) Spontaneous formation of protoplasts by a species of *Entomophthora*. J Invertebr Pathol 19:354–360

# Chapter 7
# Latest Developments in the Research of Rust Fungi and Their Allies (Pucciniomycotina)

**Merje Toome-Heller**

## Introduction

Pucciniomycotina is one of the three subphyla of Basidiomycota, known members of which have disc-like spindle pole bodies and simple septal pores that lack pore caps and dolipores (Bauer et al. 2006; Swann et al. 2001). Pucciniomycotina species are divided into nine classes: Agaricostilbomycetes, Atractiellomycetes, Classiculomycetes, Cryptomycocolacomycetes, Cystobasidiomycetes, Microbotryomycetes, Mixiomycetes, Pucciniomycetes, and Tritirachiomycetes (Aime et al. 2014). Nearly all these fungi are microfungi, ranging from psychrophilic or animal pathogenic yeasts to moulds, various plant pathogens, and saprobes (Aime et al. 2006, 2014). There are also a few sporocarp-forming species in Pucciniomycotina, which form either minute—up to a few millimetres high—(some species of Atractiellomycetes, Agaricostilbomycetes, and Microbotryomycetes) or resupinate fruiting bodies (members of Septobasidiales and Helicobasidiales in Pucciniomycetes; Swann et al. 2001).

The vast majority of Pucciniomycotina species (ca. 7500 of nearly 8500) belong to a single order Pucciniales (formerly Uredinales) exclusively containing the obligate parasitic fungi that cause rust diseases of plants. While all rust fungi share relatively similar life cycles, morphology, and biology, their allies, i.e. the remaining members of Pucciniomycotina, are very diverse. The next largest orders, in terms of species, are the entomophilic Septobasidiales (ca. 280 species) and the phytopathogenic anther smut fungi from Microbotryales (ca. 125 species). All other described orders of Pucciniomycotina contain few known species, ranging from 1 to 70 (Aime et al. 2014; Kirk et al. 2011) and the only other larger groups in Pucciniomycotina are the polyphyletic anamorphic yeast genera *Rhodotorula* and *Sporobolomyces*, which contain 162 and 93 known species, respectively.

M. Toome-Heller, Ph.D. (✉)
Plant Health and Environment Laboratory, Ministry for Primary Industries,
Auckland 1140, New Zealand
e-mail: merje.toome@mpi.govt.nz



D.-W. Li (ed.), *Biology of Microfungi*, Fungal Biology,
DOI 10.1007/978-3-319-29137-6_7

Plant pathological and systematic research of rust fungi has a long history, mostly due to their economic significance in agriculture and forestry, and their easily noticeable symptoms (Cummins and Hiratsuka 2003). In contrast, the remaining members of Pucciniomycotina have been placed together relatively recently and comparatively less is known about them (Aime et al. 2006, 2014; Bauer et al. 2006; Swann et al. 2001). The availability of molecular research and identification tools has greatly facilitated the research of Pucciniomycotina, especially the systematics and species discovery of the anamorphic yeasts and rust fungi. Genome sequencing has opened new opportunities to study the phylogenetic relationships within Pucciniomycotina and to use rusts and their allies for better understanding the biology of fungi in general (e.g. Ianiri et al. 2011; Raffaele and Kaumon 2012; Tavares et al. 2014). Albeit, as is true for all groups of fungi (Blackwell 2011), a great deal of research is still to be completed in both the areas of biology and biodiversity.

This chapter focuses on the latest developments in Pucciniomycotina research—which are mainly in the area of biodiversity research, systematics, taxonomy, and genomics—and on highlighting some of the challenges related to working with this group. For further general information on this subphylum, please refer to previously published overviews, which provide detailed information about the major groups of Pucciniomycotina and their general characteristics (Swann et al. 2001; Aime et al. 2006, 2014).

## Biodiversity

### *Species Discovery*

While the majority of the species now classified in Pucciniomycotina were described during the 1900s, new species discovery continues to be important for this group, with 375 (or ca. 5 % of the total) new species described since year 2000. When comparing the average new species members between the 2000s and the first half of 2010s, a trend towards a rise was detected as on average 21 new species were detected for the first period and 27 for the latter (Fig. 7.1a). Although increased species discoveries could be expected due to improved accessibility to molecular data analysis and therefore greater ability to detect new to science and cryptic species, the following years and further analyses should reveal whether the species discovery is truly a rising trend in Pucciniomycotina or whether this has been an artefact of the timing of publications.

Taxonomic distribution of the new species between Pucciniomycotina classes followed greatly the distribution of already known species. More than 70 % of new species belonged to Pucciniomycetes, the most species-rich class of Pucciniomycotina, and around 17 % belonged to Microbotryomycetes, the second-most species-rich class. Over half of the remaining species were determined to be members of Cystobasidiomycetes, and a quarter belonged to Agaricostilbomycetes. The remaining species (less than 2 % of total new species) belonged to Atractiellomycetes,

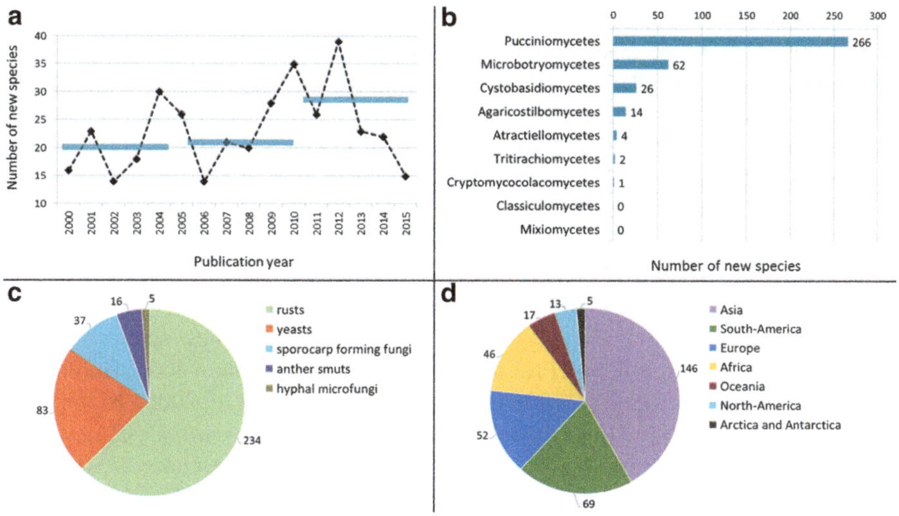


**Fig. 7.1** Information about the new Pucciniomycotina species described between 2000 and 2015. (**a**) The yearly total number of new species (*black squares*) and the 5-year average values (*blue bars*). (**b**) Distribution of the number of new species between the nine Pucciniomycotina classes. (**c**) The number of species by the type of organism. (**d**) Geographic distribution of the newly described species

Tritirachiomycetes, and Cryptomycocolacomycetes, and no new taxa were added to Classiculomycetes and Mixiomycetes (Fig. 7.1b). It remains to be seen if these trends are in accordance with the actual number of Pucciniomycotina species in nature or the smaller groups are just greatly under-sampled due to limited number of researchers working on these groups and/or their preference for under-studied environments. However, examining the rate at which new species are added to different classes, a further increase in microfungal species discovery could be expected, especially those of the saprobic yeasts. Proportionally, the greatest contribution to the total species numbers was made to Cystobasidiomycetes, of which over 50 % of known species were discovered over the last 15 years. Nearly all those new additions represent anamorphic yeast species, indicating that the discovery of the unicellular Pucciniomycotina species could be on a rise.

Although rust fungi comprise the most thoroughly studied group of Pucciniomycotina, a considerable number of new species are still described for this group every year. In fact, more than half (i.e. 234 species) of all the new Pucciniomycotina species discoveries made since 2000 were for rust fungi. These novelties were from 41 different genera, ten of which represent new genera, and therefore entirely new genetic lineages. Most of the new rust fungi species were described in the genera *Puccinia* (66 new species; Abbasi et al. 2002; Abbasi and Darvishnia 2015; Afshan and Khalid 2008, Afshan et al. 2009, 2010a, b; Aliabadi and Abbasi 2012; Bahcecioglu and Gjaerum 2003, 2004; Bahcecioglu et al. 2005, 2009; Berndt 2007, 2009, 2010, 2013a; Berndt and Freire 2004; Berndt and

Uhlmann 2006; de Carvalho and Hennen 2012; Hüseyin and Kirbag 2003; Iqbal et al. 2009; Kabaktepe 2015; Khalid and Afshan 2009; Kirbag et al. 2001, 2011; Liu and Hambleton 2012; McKenzie 2008; McKenzie and Johnston 2004; Mennicken and Oberwinkler 2004; Okane et al. 2014; Perdomo-Sanchez and Piepenbring 2008; Sotao et al. 2007; Thaung 2011; Zhuang and Wei 2001, 2011), *Uromyces* (28 new species; Agarwal 2003; Bahcecioglu 2014; Bahcecıoglu and Gjaerum 2004; Berndt 2002a, 2004, 2009, 2013b; Berndt and Baiswar 2009; Berndt and Uhlmann 2006; Berndt et al. 2007; Doungsaard et al. 2014; Hernandez et al. 2005; Mennicken and Oberwinkler 2004; Perdomo-Sanchez and Piepenbring 2014; Rezende and Dianese 2003; Thaung 2009; Walker and van der Merwe 2009; Wood and Scholler 2005; Zhuang and Wei 2003), *Uredo* (16 new species; Berndt 2002b, 2004, 2009; Berndt and Freire 2004; Berndt and Uhlmann 2006; Berndt and Wood 2012; Berndt et al. 2007; Cao et al. 2000; Hernandez et al. 2005; Mennicken and Oberwinkler 2004; Zhuang and Wei 2011, 2012), *Prospodium* (12 new species; Berndt 2002b; Berndt et al. 2007; de Carvalho and Hennen 2010), and *Phakopsora* (11 new species; Bagyanarayana et al. 2001; Beenken 2014; Berndt and Wood 2012; Berndt et al. 2008; Ferreiea et al. 2001; Maier et al. 2015; Ono 2000; Ono et al. 2012; Ritschel et al. 2007). Interestingly, species descriptions for rust genera follow the same trend as was seen for classes, i.e. the most species-rich genera (*Puccinia*, *Uredo*, and *Uromyces*) had the highest number of new species discovered.

Among the remaining genera of Pucciniomycotina, most of the new species were revealed for the insect associated hyphal fungi in *Septobasidium* (32 new species; Chen and Guo 2011a, b, c, d, 2012a, b, c; Li and Guo 2013; Li et al. 2013; *Lu and Guo* 2009a, b, c, 2010a, b, c, 2011; Lu et al. 2010), for the anamorphic red and white yeast genera *Rhodotorula* (30 new species; Bai et al. 2004; Belloch et al. 2007; Fell et al. 2011; Golubev and Scorzetti 2010; Huang et al. 2011; Laich et al. 2013; Libkind et al. 2010; Margesin et al. 2007; Nagahama et al. 2001, 2003, 2006; Pohl et al. 2011; Satoh et al. 2013; Shivaji et al. 2008; Singh et al. 2014; Thanh et al. 2004; Vishniac and Takashima 2009; Zhao et al. 2002) and *Sporobolomyces* (23 new species; Bai et al. 2002; Fell et al. 2002; Hamamoto et al. 2002; Libkind et al. 2005; Nakase et al. 2003, 2005a, b; Satoh and Makimura 2008; Takashima and Nakase 2000; Valerio et al. 2002; Wang and Bai 2004; Zhao et al. 2003), and for anther smuts in *Microbotryum* (17 new species; Chlebicki and Sukova 2005; Denchev 2007; Denchev et al. 2009; He and Guo 2008; Lutz et al. 2005, 2008; Piatek et al. 2012, 2013; Vanky 2004; Vanky and Berner 2003). These four are the remaining largest genera in Pucciniomycotina, confirming the trend that more new species are found for already known species-rich groups.

## Diverse Habitats and Geographic Distribution

When analysing the new species data by their organism type, 61 % of all new species were rust fungi, 22 % yeasts, 10 % sporocarp forming fungi, 5 % anther smuts, and the remaining 2 % hyphal microfungi (Fig. 7.1c). Most of the Pucciniomycotina species have been found to have an association with plants. While rust and anther smut fungi strictly inhabit only the above-ground parts of living plants, other groups have been isolated from plant roots, stems, as well as dead plant materials. Non-plant habitats include soil, fresh and marine water (including ice), air, animals (including insects), and other fungi. Most of the 83 new yeast species described since 2000 have been recovered from plant surfaces (47 new species; e.g. Golubev and Scorzetti 2010; Wang et al. 2003) where they are believed to be epiphytic saprobes. Other habitats of the remaining yeasts are marine environments, often associated with the sea floor animals (10 new species; e.g. Laich et al. 2013, 2014; Nagahama et al. 2003) and soil (9 new species; Vishniac and Takashima 2009). Nearly all new sporocarp-forming species were of the scale insect-associated *Septobasidium* species, which grow on various tree branches (34 new species; e.g. Chen and Guo 2012a; Li et al. 2013), three were isolated from beetle galleries (Hausner et al. 2008; Oberwinkler et al. 2006), and one from palm litter (Toome and Aime 2014). The few new hyphal microfungi species have been recovered from soil (Bauer et al. 2009; Nguyen et al. 2014), aquatic environments (Manohar et al. 2014), or in association with bark beetles (Kirschner et al. 2001).

Some Pucciniomycotina species can grow in extreme or remote environments, which are more difficult to access and study. For instance, some of the most recently described Pucciniomycotina species have been isolated from flare pit soils in Canada (Nguyen et al. 2014), anoxic costal sediments of Arabian Sea (Manohar et al. 2014), Antarctic marine sponge (Laich et al. 2014), cryoconite holes in Arctic (Singh et al. 2014), tropical rain forests in South America (Toome et al. 2013), and desert soil crusts in China (Zhang et al. 2013).

The general geographic distribution of new rusts and their allies is dominated by previously less documented areas (Fig. 7.1d). Increased mycological activity in Asia has resulted in a concomitant increase in new species discovery from these regions, totalling nearly 40 % of all new records, including all new *Septobasidium* species, 73 new rusts, and 44 new yeasts (e.g. Crane 2005a; Pohl et al. 2011; Yang et al. 2014) described since 2000. Nearly 20 % of the new species are from South America, the majority of which are rust fungi described from Brazil (e.g. Beenken et al. 2012; Rezende and Dianese 2001; Berndt and Freire 2000). The studies in Africa have also been greatly biased towards the rust fungi with 43 new rusts described mostly from South Africa (e.g. Berndt and Wood 2012; Wood and Crous 2005). In Europe, a more even distribution was seen between the new species of Pucciniomycotina with 21 yeasts, 20 rusts, six anther smuts, and four hyphal microfungi (e.g. Lutz et al. 2005; Margesin et al. 2007; Valerio et al. 2008). All species found from Arctic and Antarctic environments (2 and 3, respectively) were yeasts (e.g. Laich et al. 2013; Singh et al. 2014; Turchetti et al. 2011).

The role of many non-rust Pucciniomycotina species in nature is not yet well understood. For example, some species belonging to Atractiellomycetes were isolated from the roots of orchids and poplar trees (Bonito et al. 2010; Kottke et al. 2010). Although a mycorrhizal or a similar symbiotic relationship between those fungi and plants is suspected, additional studies are needed to clarify the exact nature of this association. The majority of the yeasts isolated from plant surfaces are believed to be saprobes, although some research also show that they could be protecting the plants from pathogens (e.g. Robiglio et al. 2011; Vero et al. 2013), and there is evidence that some may be pathogenic on cultivated mushrooms (e.g. Xu et al. 2014) or animals, including humans (Chitko-McKown et al. 2014; Tsiodras et al. 2014). In fact, more human infections by Pucciniomycotina species have been reported over the past years (Wirth and Goldani 2012). In the majority of these cases, infections are associated with a weakened immune system and central venous catheter, indicating that these fungi are opportunistic pathogens and do not affect healthy humans. One such example is the red yeast *Rhodotorula mucilaginosa* that has become a serious human pathogen over the last few years. This species appears to be present in the normal microflora of several animals (Park et al. 2012; Raggi et al. 2014) and is present in various foods, but it can also cause serious infections in hospitals because it readily adheres to plastic surfaces. The yeast can cause superficial skin or eye infections or more serious fungemia, which have led to the death of the patients in nearly half of the cases (Wirth and Goldani 2012). Pathogenicity towards humans has also been reported for a hyphal species *Tritirachium oryzae*, which can cause eye, nail, and scalp infections (Moraes et al. 2010; Naseri et al. 2013; Schell et al. 2011).

## *Challenges in the Biodiversity Research*

The availability of verified reference sequence data is critical for efficient identification of hidden biodiversity. While most new species descriptions today include the publication of sequence data for at least one or two gene regions, obtaining good sequence coverage for members of Pucciniomycotina can still be challenging. Thanks to the sequencing completed by culture collections (e.g. CBS—Centraalbureau voor Shimmelcultures, ATCC—American Type Culture Collection) and the Assembling the Fungal Tree of Life (AFTOL) project, the majority of culturable Pucciniomycotina species now have publicly available sequence data. Moreover, as a result of a yeast sequencing initiative, the LSU region (and in most cases also the ITS region) of all the known yeast species has been sequenced and made publicly available (Kurtzman et al. 2011), allowing the identification and discovery of new Pucciniomycotina yeasts based on sequence data. On the other hand, for the fungal groups that are not culturable (e.g. rust fungi) and for rarely collected sporocarp-forming and hyphal taxa that are known only from herbarium material, sequence data may not be available. Today, less than 10 % of all known rust fungal species have published sequence data (Toome and Aime 2015). Therefore, taxonomic studies and species identification in rust fungi still rely greatly on non-molecular methods. Since it can be challenging to obtain sequence data from old herbarium specimens, new collections and thorough molecular studies of all the collected material in the future (especially of the ones with few morphological characters) will allow revealing some unknown biodiversity of Pucciniomycotina with greater confidence.

There are also a few challenges related to next generation environmental sequencing projects. Several pipelines and programs for data analysis have been developed to manage the high volumes of generated data and to annotate and identify the species that are recovered (e.g. Abarenkov et al. 2010; Dannemiller et al. 2014). However, these identifications are, of course, only as good as the reference sequence databases. Therefore, the identification of inadequately sequenced lineages (e.g. rust fungi) cannot be reliable until the taxa are thoroughly sequenced. Additionally, since the anamorphic yeast genera *Sporobolomyces* and *Rhodotorula* contain species that span several orders but the types only belong to Microbotryomycetes, under- or overestimation of the representativeness of some of the taxa can occur and may not illustrate the true biodiversity of the studied communities. These issues can be resolved only via verified and improved databases.

Although new species cannot be described solely based on sequence data and without a voucher specimen, environmental community sequencing projects can provide very valuable information about the hidden biodiversity and hint, where the future studies should focus to find the hidden species of Pucciniomycotina. For example, the species *Mixia osmundae* of the monotypic Mixiomycetes was known only from fern leaves in North America and Asia, but published environmental sequencing data analysis indicates that conspecific or congeneric species are also present in Europe and could be associated with other hosts (Toome et al. 2014). Similarly, unknown members of Atractiellomycetes have been isolated from tree roots in a number of locations, indicating that in addition to the known habitats in North and South America (Bonito et al. 2010; Kottke et al. 2010), they are also present in the Seychelles (Suvi et al. 2010). Some other examples of the community studies where various Pucciniomycotina taxa have been recovered include the study of Arctic soils (Timling et al. 2014), deep-sea sediments in the Central Indian Basin (Singh et al. 2011), *Quercus macrocarpa* phyllosphere in Kansas, USA (Jumpponen and Jones 2010), acid mine drainage biofilm in California, USA (Baker et al. 2009), and ice of the Baltic Sea (Majaneva et al. 2012).

## Phylogenetic Classification

### *Higher-Level Phylogenetic Relationships*

The placement of Pucciniomycotina within Basidiomycota is not yet fully resolved. While several multigene studies have reported this subphylum to be a sister clade to Agaricomycotina and Ustilaginomycotina, i.e. the most early diverging basidiomycetes (James et al. 2006; Matheny et al. 2006; Ebersberger et al. 2012; Floudas et al. 2012), others have found that Ustilaginomycotina is the basal group and Pucciniomycotina is sister to Agaricomycotina (Medina et al. 2011). Similarly, the deeper nodes within Pucciniomycotina have not been resolved. The best resolution to date was provided in Aime et al. (2006) and in Schell et al. (2011), which both strongly support all class level lineages but do not allow clarifying the evolutionary relationships between the classes. Although the availability of fungal genome data has greatly improved our ability to perform phylogenomic studies of fungi to reach better resolution for deeper nodes (e.g. Floudas et al. 2012), this approach has not yet been applied to gain knowledge of evolutionary relationships within Pucciniomycotina. This is because the complete genomic coverage of all the Pucciniomycotina classes has not been available. Albeit, this situation is most likely to change over the next few years as new Pucciniomycotina genomes are becoming available through the 1000 Fungal Genomes Project (Grigoriev et al. 2013).

## *Taxonomic Improvements as a Result of Improved Understanding of Phylogenetic Relationships*

Over the last 16 years, 23 new genera have been described in Pucciniomycotina. Eight of those genera have been described as a result of a thorough analysis of existing species and revision of existing genera. These are the rust fungi genera *Desmosorus* (Ritschel et al. 2007), *Esalque* (Hennen et al. 2000), *Pelastoma* (Yepes and de Carvalho 2012), *Puccorchidium* (Beenken and Wood 2015), *Racospermyces* (Walker 2001), and *Sphenorchidium* (Beenken and Wood 2015); a yeast genus *Leucosporidiella* (Sampaio et al. 2003); and a mould genus *Paratritirachium* (Beguin et al. 2012). The remaining 15 new genera were described based on new species discovery. These are the sporocarp-forming genera *Basidiopycnis*, *Proceropycnis* (Oberwinkler et al. 2006), *Basidiopycnoides* (Hausner et al. 2008), and *Pycnopulvinus* (Toome and Aime 2014); the hyphal microfungi genera *Cystobasidiopsis* (Bauer et al. 2009) and *Colacosiphon* (Kirschner et al. 2001); rust fungi genera *Bibulocystis* (Walker et al. 2006), *Canasta* (de Carvalho and Hennen 2010), *Cratea* (Yepes and de Carvalho 2012), and *Diaphanopellis* (Crane 2005b); and yeast genera *Bannoa* (Hamamoto et al. 2002), *Curvibasidium* (Sampaio et al. 2004), *Glaciozyma* (Turchetti et al. 2011), *Meredithblackwellia* (Toome et al. 2013), and *Microbotryozyma* (Suh et al. 2012). These new genera have greatly improved the taxonomy of Pucciniomycotina species as several of them have enabled to re-classify members of polyphyletic and anamorphic genera. As a result, numerous new combinations have been proposed to already existing species and several species complexes have been identified to species level, greatly improving our understanding of Pucciniomycotina phylogeny (e.g. Berndt 2011, 2013b; Turchetti et al. 2011; Yurkov et al. 2015).

Above genus level classification has also seen some improvements. Since the major revision of Pucciniomycotina published by Bauer et al. (2006), a few new families and orders have been introduced to provide better higher level classification. The most significant change since the last major revision has been the description of a new class Tritirachiomycetes and lower level groups Tritirachiales and Tritirachiaceae (Schell et al. 2011). *Tritirachium* species were previously classified in Ascomycota, but a thorough molecular study revealed that these fungi are members of Pucciniomycotina, forming a separate lineage in it. The other higher level change has been within Microbotryomycetes, where Kriegeriaceae and Kriegeriales were described to provide higher level classification for a group of anamorphic yeasts and a sedge pathogen (Toome et al. 2013). More of such studies are needed as there are still numerous *incertae sedis* taxa in nearly all Pucciniomycotina classes (Aime et al. 2014).

## *Challenges in Phylogenetic Studies*

Due to their complicated life cycles including two different host plants and up to five or more different spore stages, many species of rust fungi have been given numerous names. As a result, we can find nearly 10,000 different names registered for rust fungi in MycoBank (www.mycobank.org). While new names are created, it is important to evaluate the suitability of the existing names. This is one of the challenges researchers who work with rust fungi need to face as often there are only few morphological characters on the type specimens. Moreover, since the DNA extractions from old rust herbarium specimens have a low success rate it might be nearly impossible to obtain DNA sequences for old specimens. Most of the 133 recognised genera of rust fungi have only one or less than ten known species (Cummins and Hiratsuka 2003) and examining sequence data might sometimes be the only way to determine, whether they really represent separate genera/species or they could be part of some other groups. This is likely going to be one of the greatest challenges in rust research due to the poor sequence coverage and the need for new collections.

A few studies have been already completed to address this issue. In one of them, sequence data from the two species of the rust genus *Frommeëlla* were analysed, which determined that these species do not represent a separate genus, but are actually part of a larger genus *Phragmidium* (Yun et al. 2011). A contrary situation was revealed in a recent study on mayapple rust, a fungus that was first described as *Aecidium podophylli*, then placed in genus *Allodus* and thereafter transferred to *Puccinia*, based on morphological characters. After completing phylogenetic analyses, it was determined that the species actually is different from *Puccinia* and the genus *Allodus* was resurrected (Minnis et al. 2012). In addition to genus level improvements, thorough molecular revisions are needed at the species level as well since some studies of rust fungi have uncovered a great number of cryptic species. For example, analyses of the *Melampsora epitea* complex in the North American pacific northwest determined the existence of 14 different phylotypes within this one morphospecies (Bennett et al. 2011) and studies within the *Endoraecium digitatum* (Berndt 2011) and *Dasyspora* (Beenken et al. 2012) species complexes are revealing similar patterns of cryptic speciation.

Great taxonomic changes are also needed for the yeasts that have been placed in anamorphic catch-all genera, especially *Sporobolomyces* and *Rhodotorula*. Several research groups have already started this progress by proposing new genera for anamorphic Pucciniomycotina yeasts (e.g. Bauer et al. 2009; Toome et al. 2013; Turchetti et al. 2011; Yurkov et al. 2015), while other researchers still choose to place new discovered species into the anamorphic genera until further data is available (e.g. Laich et al. 2013; Singh et al. 2014). Thanks to the complete availability of yeast sequence data, more changes are expected to be made to create meaningful taxonomy for the yeasts during the coming years.

## Genomics

As is true for mycological studies in general, the availability of draft genomes has greatly advanced the research of the biology of the members of Pucciniomycotina. For example, access to genomic data enables us to study the gene functions and the evolution of gene families, discover effector proteins, and perform detailed population studies (e.g. Duplessis et al. 2014; Hacquard et al. 2012; Persoons et al. 2014). Further availability of genomes over a range of different Pucciniomycotina lineages will also enable us to better understand the phylogeny of rust fungi and their allies and the evolutionary pathways that have led to such a diverse group of fungi. The number of sequenced genomes for Pucciniomycotina is still relatively small compared to other fungal subphyla, with published genomes being available for only 12 species to date—one of these was published in 2015, six in 2014, one in 2012, and three in 2011 (Table 7.1). Several other genomes have been completed recently as part of the 1000 Fungal Genomes project (Grigoriev et al. 2013) and other sequencing projects (see Duplessis et al. 2014) and they should become available in the next few years.

### *Genomes of Rust Fungi*

The first annotated rust genomes were published for the poplar leaf rust fungus *Melampsora laricis-populina* and the wheat stem rust fungus *Puccinia graminis* f. sp. *tritici* (Duplessis et al. 2011), followed by two genomes for the stripe rust fungus *P. striiformis* f. sp. *tritici* (Cantu et al. 2011; Zheng et al. 2013). These genomes have facilitated the research of several new aspects of rust fungi, especially the genome-based analysis of effector genes and transcriptomic studies that are elucidating interactions between rusts and their host plants (e.g. Cantu et al. 2013; Hacquard et al. 2012; Persoons et al. 2014). For a detailed overview of the genome research on these fungi, please refer to Duplessis et al. (2014). More recently, the draft genomes of four other rust species have been published. These are for the myrtle rust fungus *P. psidii* (Tan et al. 2014), the flax rust fungus *M. lini* (Nemri et al. 2014), the coffee rust fungus *Hemileia vastatrix* (Cristancho et al. 2014), and the broad bean rust fungus *Uromyces fabae* (Link et al. 2014; Table 7.1).

Several years before the rust fungi genome sequencing was completed, genome size studies by flow cytometry estimated that these fungi have large genomes, reaching hundreds of millions of base pairs (Eilam et al. 1994). The first completed rust genomes of *M. laricis-populina* and *P. graminis* f. sp. *tritici* showed that the genomes of those obligate parasites are indeed large, with over 101 and 88 Mb,

**Table 7.1** Genomic characters of the Pucciniomycotina species with published genomes

| Species | Strain | Order | Genome size (Mb) | Number of gene models | Reference |
|---|---|---|---|---|---|
| *Hemileia vastatrix* | 8 different isolates | Pucciniales | 333 | 14,445 | Cristancho et al. (2014) |
| *Melampsora laricis-populina* | 98AG31 | Pucciniales | 101 | 16,399 | Duplessis et al. (2011) |
| *Melampsora lini* | CH5 | Pucciniales | 189 | 16,271 | Nemri et al. (2014) |
| *Mixia osmundae* | IAM14324 | Mixiales | 13 | 6903 | Toome et al. (2014) |
| *Puccinia graminis* f. sp. *tritici* | CDL7S-36-700-3 | Pucciniales | 89 | 17,773 | Duplessis et al. (2011) |
| *Puccinia psidii* | PBI 115012-Mr | Pucciniales | 103–145 | 19,000 | Tan et al. (2014) |
| *Puccinia striiformis* f. sp. *tritici* | PST-130/CY32 | Pucciniales | 79/110 | 22,815/25,288 | Cantu et al. (2011)/Zheng et al. (2013) |
| *Puccinia triticina* | 1-1 (BBBD) | Pucciniales | 135 | 14,880 | Duplessis et al. (2014) |
| *Rhodosporidium toruloides* | MTCC457 | Sporidiobolales | 20 | 5993 | Kumar et al. (2012) |
| *Rhodotorula glutinis* | ATCC204091 | Sporidiobolales | 20 | 3359 | Paul et al. (2014) |
| *Sporidiobolus salmonicolor* | CBS6832 | Sporidiobolales | 20 | 5147 | Coelho et al. (2015) |
| *Uromyces fabae* | I2 | Pucciniales | 329 | 23,153 | Link et al. (2014) |

respectively (Duplessis et al. 2011). Moreover, the genome sizes of the other completed rust fungi genomes range from 80 to 333 Mb (Table 7.1). A recently published thorough flow cytometry study which examined the genome sizes of ten different genera of rust fungi provided further support that some rusts have even larger genomes (Tavares et al. 2014). Based on their data, the largest known genomes belong to *Gymnosporangium confusum* (893 Mb), *Puccinia chrysanthemi* (806 Mb), *Phakopsora pachyrhizi* (720 Mb), and *Uromyces vignae* (712 Mb). These are clearly the largest fungal genomes ever measured (Tavares et al. 2014). The availability of the genome size estimates is important for successful completion of new genome projects because this enables the researchers to calculate the sampling depth needed to cover the genome with the minimum number of gaps. Since Tavares et al. (2014) revealed the genome sizes for several rust genera that had no previous information this now facilitates further sequencing of the representatives of other rust families. Interestingly, possible connection between the rust host plant and the fungal genome size has been observed, most likely due to the close co-evolution between these plant pathogens and their hosts (Tavares et al. 2014). Future research will determine if the species with similar host range share similar genomic traits specific to the hosts, or each lineage has evolved to obtain a unique set of genes to overcome host resistance.

Although it might be assumed that these large genomes might be needed due to the macrocyclic life cycles of rust fungi, where gene sets for infecting two different host plants and producing up to five different spore stages are required, comparative analysis of protein coding genes does not seem to support this (Duplessis et al. 2014). The reasons for such large genomes of rusts are still not fully understood, but it is clear that the vast majority of their genomic content is comprised of repetitive and transposable elements (Duplessis et al. 2011, 2014). In fact, based on currently sequenced genomes, rust fungi have just between 14,000 and 25,000 protein coding genes (Table 7.1), despite their great genome sizes. Initial analyses show that many of those genes belong to unique gene families which are not shared with other basidiomycetes, including the other Pucciniomycotina species (Duplessis et al. 2014; Toome et al. 2014; Zheng et al. 2013).

## *Other Pucciniomycotina Genomes*

Only a few genomes are publicly available for non-rust Pucciniomycotina species, four of these from Microbotryomycetes. The first two, the red yeasts *Sporobolomyces roseus* and *Rhodotorula graminis*, were sequenced and assembled in 2006 and 2010, respectively (http://genome.jgi.doe.gov), but have not been officially published to date although genomic data of these species have been included in various comparative studies (e.g. Coelho et al. 2010; Horns et al. 2012). Three additional Microbotryomycetes, the yeasts *Sporidiobolus salmonicolor*, *Rhodosporidium toruloides*, and *Rhodotorula glutinis*, have been sequenced and published (Coelho et al. 2015; Kumar et al. 2012; Paul et al. 2014). Outside of Microbotryomycetes and Pucciniomycetes, genome data have been published only for the fern pathogen *Mixia osmundae*, the monotypic representative of Mixiomycetes (Toome et al. 2014).

The genome sizes of rust relatives are considerably smaller than those of the rust fungi. Both *S. roseus* and *R. graminis* have small genomes at around 21 Mb (http://genome.jgi.doe.gov) and the genomes of *S. salmonicolor*, *Rh. toruloides*, and *R. glutinis* are sized at 20 Mb (Table 7.1). While these are already fairly small genome sizes compared to other sequenced fungi (average fungal genome size is 37.7 Mb, Tavares et al. 2014), the genome of *M. osmundae* is even smaller, representing the smallest known genome among all Basidiomycota and one of the smallest genomes in fungi at only 13 Mb (Toome et al. 2014). Genomes for the representatives of five additional classes have been completed and are being analysed at the moment (Aime et al., unpublished), which should give us a better understanding of general genomic traits of other Pucciniomycotina representatives and enable to reveal unique traits of this subphylum.

**Acknowledgements** I am grateful to Dr. M. Catherine Aime for her continuous support and contribution to this chapter and for sharing her invaluable knowledge of Pucciniomycotina with me during the past years. Drs. Wellcome Ho and Gregory Heller are also acknowledged for their comments on an earlier version of this manuscript.

## References

Abarenkov K, Tedersoo L, Nilsson RH, Vellak K, Saar I, Veldre V et al (2010) PlutoF—a web based workbench for ecological and taxonomic research, with an online implementation for fungal ITS sequences. Evol Bioinform 6:189–196

Abbasi M, Darvishnia M (2015) *Puccinia fritschii* sp. nov. — a new rust species from Iran. Schlechtendalia 28:77–79

Abbasi M, Hedjaroude GA, Gjaerum HB, Scholler M (2002) *Puccinia ariorum* sp. nov. and other noteworthy graminicolous rust fungi (Uredinales) from Iran. Mycotaxon 81:435–444

Afshan NS, Khalid AN (2008) New rust fungi on noxious weeds from Pakistan. Pak J Phytopathol 20:82–87

Afshan NS, Khalid AN, Iqbal SH, Niazi AR, Sultan A (2009) *Puccinia subepidermalis* sp. nov. and new records of rust fungi from Fairy Meadows, Northern Pakistan. Mycotaxon 110:173–182

Afshan NS, Khalid AN, Iqbal SH, Niazi AR, Sultan A (2010a) *Puccinia anaphalidis-virgatae*, a new species, and a new variety of rust fungi from Fairy Meadows, Northern Pakistan. Mycotaxon 112:483–490

Afshan N-S, Khalid AN, Niazi AR (2010b) Three new species of rust fungi from Pakistan. Mycol Prog 9:485–490

Agarwal DK (2003) *Uromyces pentaceae* Agarwal sp. nov. on *Pantace burmanica* Kurz. J Mycopathol Res 41:115–116

Aime MC, Matheny PB, Henk DA, Frieders EM, Nilsson RH, Piepenbring M et al (2006) An overview of the higher level classification of Pucciniomycotina based on combined analyses of nuclear large and small subunit rDNA sequences. Mycologia 98:895–905

Aime MC, Toome M, McLaughlin DJ (2014) Pucciniomycotina. In: McLaughlin DJ, Spatafora JW (eds) The mycota, Vol 7A: Systematics and evolution, 2nd edn. Springer, Heidelberg, pp. 271–294

Aliabadi F, Abbasi M (2012) Four new rust taxa on Asteraceae from Central Alborz, Northern Iran. Mycotaxon 122:129–134
Bagyanarayana G, Ramesh P, Ono Y (2001) *Phakopsora schreberae*, sp. nov. (Uredinales) occurs on *Schrebera swietenioides* in India. Mycoscience 42:631–632
Bahcecioglu Z (2014) A new species of *Uromyces* from Turkey. Mycotaxon 129:21–23
Bahcecioglu Z, Gjaerum HB (2003) New and rare rust fungi (Uredinales) from Anatolia (Turkey). Mycotaxon 85:165–173
Bahcecıoglu Z, Gjaerum HB (2004) New and rare rust fungi (Uredinales) from Anatolia (Turkey) - 2. Mycotaxon 90:55–68
Bahcecioglu Z, Kabaktepe S, Yildiz B (2005) Three new rust species (Uredinales) from Turkey. Mycotaxon 93:327–331
Bahcecioglu Z, Brendt R, Kabaktepe S (2009) *Puccinia ardahanensis* sp. nov., a new rust fungus from Turkey. Sydowia 61:209–213
Bai FY, Zhao JH, Takashima M, Jia JH, Boekhout T, Nakase T (2002) Reclassification of the *Sporobolomyces roseus* and *Sporidiobolus pararoseus* complexes, with the description of *Sporobolomyces phaffii* sp. nov. Int J Syst Evol Microbiol 52:2309–2314
Bai FY, Cai Y, Wang QM, Ohkubo H (2004) *Rhodotorula oryzae* sp. nov., a novel basidiomycetous yeast species isolated from paddy rice. Antonie Van Leeuwenhoek 86:295–299
Baker BJ, Tyson GW, Goosherst L, Banfield JF (2009) Insights into the diversity of eukaryotes in acid mine drainage biofilm communities. Appl Environ Microbiol 75:2192–2199
Bauer R, Begerow D, Sampaio JP, Weiss M, Oberwinkler F (2006) The simple-septate basidiomycetes: a synopsis. Mycol Prog 5:41–66
Bauer R, Metzler B, Begerow D, Oberwinkler F (2009) *Cystobasidiopsis nirenbergiae*, a new agaricostilbomycete (Pucciniomycotina). Mycol Res 113:960–966
Beenken L (2014) Pucciniales on *Annona* (Annonaceae) with special focus on the genus *Phakopsora*. Mycol Prog 13:791–809
Beenken L, Wood AR (2015) *Puccorchidium* and *Sphenorchidium*, two new genera of Pucciniales on Annonaceae related to *Puccinia psidii* and the genus *Dasyspora*. Mycol Prog 14:49
Beenken L, Zoller S, Berndt R (2012) Rust fungi on Annonaceae II: the genus *Dasyspora* Berk. & M.A. Curtis. Mycologia 104:659–681
Beguin H, Pyck N, Detandt M (2012) *Tritirachium*, a hyphomycetous genus belonging to the Basidiomycota. Nova Hedwigia 94:139–152
Belloch C, Villa-Carvajal M, Alvarez-Rodriguez ML, Coque JJ (2007) *Rhodotorula subericola* sp. nov., an anamorphic basidiomycetous yeast species isolated from bark of *Quercus suber* (cork oak). Int J Syst Evol Microbiol 57:1668–1671
Bennett C, Aime MC, Newcombe G (2011) Molecular and pathogenic variation within *Melampsora* on *Salix* in western North America reveals numerous cryptic species. Mycologia 103:1004–1018
Berndt R (2002a) *Uromyces vankyorum* sp. nov. and *Uromyces atriplicis*, putative vicariant rust species of Patagonian and Australian semi-desert biomes. Mycol Prog 1:179–186
Berndt R (2002b) New species, reports and observations on rust fungi. Nova Hedwigia 75:3–4
Berndt R (2004) A checklist of Costa Rican rust fungi. In: Agerer R, Piepenbring M, Blanz P (eds) Frontiers in Basidiomycote mycology. IHW Verlag, Berlin, pp 185–236
Berndt R (2007) A global survey of *Puccinia*-rust on Cucurbitaceae. Mycol Prog 6:151–178
Berndt R (2009) New species of rust fungi (Uredinales) from South Africa and new observations on known species. Mycol Prog 8:99–114
Berndt R (2010) The *Puccinia* species of *Berkheya* (Asteraceae) with description of four new species from South Africa. Mycologia 102:1437–1449
Berndt R (2011) Taxonomic revision of *Endoraecium digitatum* (rust fungi, Uredinales) with description of four new species from Australia and Hawaii. Mycol Prog 10:497–517
Berndt R (2013a) First catalogue of the rust fungi of French Guiana, northern South America. Mycol Prog 12:193–211
Berndt R (2013b) Revision of the rust genus *Uromyces* on Cucurbitaceae. Mycologia 105:760–780

Berndt R, Baiswar P (2009) *Uromyces umiamensis*, a new rust species on *Momordica cochinchinensis* in India. Mycol Prog 8:83–86

Berndt R, Freire F (2000) *Ravenelia cenostigmatis*, a new rust fungus from Brazil. Mycoscience 41:227–232

Berndt R, Freire F (2004) New species of rust fungi from Brazil. Mycol Prog 3:257–264

Berndt R, Uhlmann E (2006) New species, reports, observations and taxonomical changes of southern African rust fungi (Uredinales). Mycol Prog 5:154–177

Berndt R, Wood AR (2012) Additions to the rust fungi of South Africa. Mycol Prog 11:483–497

Berndt R, Rössel A, Freire F (2007) New species and reports of rust fungi (Basidiomycota, Uredinales) of South America. Mycol Prog 6:27–34

Berndt R, Freire F, Piatek M, Wood AR (2008) New species of *Phakopsora* (Basidiomycota, Uredinales) from Cameroon, South Africa and Brazil. Sydowia 60:15–24

Blackwell M (2011) The fungi: 1, 2, 3 … 5.1 million species? Am J Bot 98:426–438

Bonito D, Reynolds H, Robeson MS, Nelson J, Hodkinson BP, Tuskan G et al (2010) Plant host and the soil origin influence fungal and bacterial assemblages in the roots of woody plants. Mol Ecol 23:3356–3370

Cantu D, Govindarajulu M, Kozik A, Wang M, Chen X, Kojima KK et al (2011) Next generation sequencing provides rapid access to the genome of *Puccinia striiformis* f. sp. *tritici*, the causal agent of wheat stripe rust. PLoS One 6:e24230

Cantu D, Segovia V, MacLean D, Bayles R, Chen X, Kamoun S et al (2013) Genome analyses of the wheat yellow (stripe) rust pathogen *Puccinia striiformis* f. sp. *tritici* reveal polymorphic and haustorial expressed secreted proteins as candidate effectors. BMC Genomics 14:270

Cao ZM, Li ZQ, Zhuang JY (2000) Uredinales from the Qinling mountains. Mycosystema 19:13–23

Chen SZ, Guo L (2011a) *Septobasidium sichuanense* sp. nov. (Septobasidiaceae) from China. Mycotaxon 115:481–484

Chen SZ, Guo L (2011b) *Septobasidium glycosmidis* and *S. albiziae* spp. nov. (Septobasidiaceae) from Hainan Province. Mycosystema 30:861–864

Chen SZ, Guo L (2011c) *Septobasidium atalantiae* sp. nov. (Septobasidiaceae) and *S. henningsii* new to China. Mycotaxon 117:291–296

Chen SZ, Guo L (2011d) *Septobasidium saurauiae* sp. nov. (Septobasidiaceae) and *S. pseudopedicellatum* new to China. Mycotaxon 118:283–288

Chen SZ, Guo L (2012a) Three new species and three new Chinese records of *Septobasidium* (Septobasidiaceae). Mycosystema 31:651–655

Chen SZ, Guo L (2012b) Three new species of *Septobasidium* (Septobasidiaceae) from Hainan Province in China. Mycotaxon 120:269–276

Chen SZ, Guo L (2012c) Three new species of *Septobasidium* (Septobasidiaceae) from southern and southwestern China. Mycotaxon 121:375–383

Chitko-McKown CG, Leymaster KA, Heaton MP, Griffin DD, Veatch JK, Jones SA et al (2014) *Rhodotorula minuta* fungemia in a ewe lamb. Transbound Emerg Dis 61(6):e76–e78

Chlebicki A, Sukova M (2005) Two *Microbotryum* species from the Himalayas. Mycotaxon 93:149–154

Coelho MA, Sampaio JP, Goncalves P (2010) A deviation from the bipolar-tetrapolar mating paradigm in an early diverged basidiomycete. PLoS Genet 5(8):e1001052

Coelho MA, Almeida JMGCF, Hittinger CT, Goncalvesa P (2015) Draft genome sequence of *Sporidiobolus salmonicolor* CBS 6832, a red-pigmented basidiomycetous yeast. Genome Announce 3:e00444

Crane PE (2005a) *Caeoma tsukubaense* n. sp., a rhododendron rust fungus of Japan and southern Asia, and its relationship to *Chrysomyxa rhododendri*. Mycoscience 46:143–147

Crane PE (2005b) Rust fungi on rhododendrons in Asia: *Diaphanopellis forrestii* gen. et sp. nov., new species of *Caeoma*, and expanded descriptions of *Chrysomyxa dietelii* and *C. succinea*. Mycologia 97:534–548

Cristancho MA, Botero-Rozo DO, Giraldo W, Tabima J, Riano-Pachon DM, Escobar C et al (2014) Annotation of a hybrid partial genome of the coffee rust (*Hemileia vastatrix*) contributes to the gene repertoire catalog of the Pucciniales. Front Plant Sci 5:594

Cummins GB, Hiratsuka Y (2003) Illustrated genera of rust fungi, 3rd edn. APS Press, St. Paul, MN

Dannemiller KC, Reeves D, Bibby K, Yamamoto N, Peccia J (2014) Fungal high-throughput taxonomic identification tool for use with next-generation sequencing (FHiTINGS). J Basic Microbiol 54:315–321

de Carvalho AA, Hennen JF (2010) New species and nomenclature in *Prospodium* (Uropyxidaceae, Pucciniales) and the new anamorphic genus *Canasta* in the Neotropics. Mycologia 102:1096–1113

de Carvalho AA, Hennen JF (2012) The species of *Puccinia* on *Piptocarpha* and *Vanillosmopsis* in the Neotropics. Mycologia 104:557–568

Denchev CM (2007) *Microbotryum lagerheimii* sp. nov. (Microbotryaceae). Mycol Balcan 4:61–67

Denchev CM, Giraud T, Hood ME (2009) Three new species of anthericolous smut fungi on Caryophyllaceae. Mycol Balcan 6:79–84

Doungsaard G, McTaggart AR, Geering ADW, Dalisay TU, Ray J, Shivas RG (2014) *Uromycladium falcatarium* sp. nov., the cause of gall rust on *Paraserianthes falcataria* in south-east Asia. Australas Plant Pathol 44:25–30

Duplessis S, Cuomo CA, Lin YC, Aerts A, Tissetant E, Veneault-Fourrey C et al (2011) Obligate biotrophy features unravelled by the genomic analysis of rust fungi. Proc Natl Acad Sci USA 108:9166–9171

Duplessis S, Bakkeren G, Hamelin R (2014) Advancing knowledge on biology of rust fungi through genomics. In: Francis MM (ed) Advances in botanical research, Vol 70. Elsevier, Oxford, pp 173–209

Ebersberger I, de Matos SR, Kupczok A, Gube M, Kothe E, Voigt K et al (2012) A consistent phylogenetic backbone for the fungi. Mol Biol Evol 29:1319–1334

Eilam T, Bushnell WE, Anikster Y (1994) Relative nuclear DNA content of rust fungi estimated by flow cytometry of propidium iodide-stained pycniospores. Phytopathology 94:728–735

Fell JW, Scorzetti G, Statzell-Tallman A, Pinel N, Yarrow D (2002) Recognition of the basidiomycetous yeast *Sporobolomyces ruberrimus* sp. nov. as a distinct species based on molecular and morphological analyses. FEMS Yeast Res 4:265–270

Fell JW, Statzell-Tallman A, Scorzetti G, Gutierrez MH (2011) Five new species of yeasts from fresh water and marine habitats in the Florida Everglades. Antonie Van Leeuwenhoek 99:533–549

Ferreiea FA, Gasparotto L, Lima MIPM (2001) Uma ferrugem (*Phakopsora tomentosae* sp. nov.) no oiti, em Manaus. Fitopatol Brasil 26:206–208

Floudas D, Binder M, Riley R, Barry K, Blanchette RA, Henrissat B et al (2012) The Paleozoic origin of enzymatic lignin decomposition reconstructed from 31 fungal genomes. Science 336:1715–1719

Golubev WI, Scorzetti G (2010) *Rhodotorula rosulata* sp. nov., *Rhodotorula silvestris* sp. nov. and *Rhodotorula straminea* sp. nov., novel myo-inositol-assimilating yeast species in the Microbotryomycetes. Int J Syst Evol Microbiol 60:2501–2506

Grigoriev IV, Nikitin R, Haridas S, Kuo A, Ohm R, Otillar R, Riley R, Salamof A, Zhao X, Korzeniewski F, Smirnova T, Nordberg H, Dubchak I, Shabalov I (2013) MycoCosm portal: gearing up for 1000 fungal genomes. Nucl Acids Res 42:D699–D704

Hacquard S, Joly DL, Lin YC, Tisserant E, Feau N, Delaruelle C et al (2012) A comprehensive analysis of genes encoding small secreted proteins identifies candidate effectors in *Melampsora larici-populina* (poplar leaf rust). Mol Plant Microbe Interact 25:279–293

Hamamoto M, Thanh VN, Nakase T (2002) *Bannoa hahajimensis* gen. nov., sp. nov., and three related anamorphs, *Sporobolomyces bischofiae* sp. nov., *Sporobolomyces ogasawarensis* sp. nov. and *Sporobolomyces syzygii* sp. nov., yeasts isolated from plants in Japan. Int J Syst Evol Microbiol 52:1023–1032

Hausner G, Reid J, Eyjolfsdottir GG, Iranpour M, Loewen PC (2008) *Basidiopycnides albertensis* gen. et sp. nov., a new anamorphic fungus with phylogenetic affinities in the Atractiellales (Basidiomycota). Mycotaxon 103:279–297

He S, Guo L (2008) *Microbotryum vivipari* sp. nov. and *Anthracoidea mulenkoi* new to China. Mycotaxon 104:455–458.

Hennen JF, Figueireda MB, de Carvalho Jr. AA (2000) *Esalque holwayi* gen. et comb. nov., a rust of Brazilian ironwood (*Caesalpinia* species). Mycologia 92:312–316.

Hernandez JR, Aime MC, Henkel TW (2005) The rust fungi (Uredinales) of Guyana. Sydowia 57:189–222

Horns F, Petit E, Yockteng R, Hood ME (2012) Patterns of repeat-induced point mutation in transposable elements of basidiomycete fungi. Genome Biol Evol 4:240–247

Huang C-H, Lee F-L, Tien C-J, Hsieh P-W (2011) *Rhodotorula taiwanensis* sp. nov., a novel yeast species from a plant in Taiwan. Antonie Van Leeuwenhoek 99:297–302

Hüseyin E, Kirbag S (2003) A new *Puccinia* on endemic *Phryna*. Pak J Bot 35:477–478

Ianiri G, Wright SA, Castoria R, Idnurm A (2011) Development of resources for the analysis of gene function in Pucciniomycotina red yeasts. Fungal Genet Biol 48:685–695

Iqbal SH, Afshan NS, Khalid AN, Niazi AR, Sultan A (2009) Additions to the rust fungi of Fairy Meadows, the Northern Areas of Pakistan. Mycotaxon 109:1–7

James TY, Kauff F, Schoch CL, Matheny PB, Hofstetter V, Cox CJ et al (2006) Reconstructing the early evolution of fungi using a six-gene phylogeny. Nature 443:818–822

Jumpponen A, Jones KL (2010) Seasonally dynamic fungal communities in the *Quercus macrocarpa* phyllosphere differ between urban and nonurban environments. New Phytol 186:495–513

Kabaktepe S (2015) *Puccinia yahyaliensis* (Pucciniaceae) – a new rust species on *Hypericum scabrum* L. from Aladaglar Mountains in Turkey. Nova Hedwigia 100:265–268

Khalid AN, Afshan NS (2009) Additions to the graminicolous rust fungi of Pakistan. Mycotaxon 108:175–183

Kirbag S, Civelek S, Huseyinov E (2001) A new *Puccinia* on *Centaurea* from Turkey. Mikologiya i Fitopatologiya 35:20–21

Kirbag S, Aime MC, Kursat M (2011) A new *Puccinia* on *Thymelaea* from Turkey. Mycotaxon 115:501–504

Kirk PM, Cannon PF, Minter DW, Stalpers JA (2011) Dictionary of the fungi, 10th edn. CABI, London, p 771

Kirschner R, Bauer R, Oberwinkler F (2001) *Colacosiphon*: a new genus described for a mycoparasitic fungus. Mycologia 93:634–644

Kottke I, Suarez JP, Herrera P, Cruz D, Bauer R, Haug I et al (2010) Atractiellomycetes belonging to the 'rust' lineage (Pucciniomycotina) from mycorrhizae with terrestrial and epiphytic neotropical orchids. Proc Biol Sci 22:1289–1298

Kumar S, Kushwaha H, Bachhawat AK, Raghava GP, Ganesan K (2012) Genome sequence of the oleaginous red yeast *Rhodosporidium toruloides* MTCC 457. Eukaryot Cell 11:1083–1094

Kurtzman CP, Fell JW, Boekhout T (eds) (2011) The yeasts, a taxonomic study, Vol 3, 5th edn. Elsevier, London

Laich F, Vaca I, Chavez R (2013) *Rhodotorula portillonensis* sp. nov., a basidiomycetous yeast isolated from Antarctic shallow-water marine sediment. Int J Syst Evol Microbiol 63:3884–3891

Laich F, Chavez R, Vaca I (2014) *Leucosporidium escuderoi* f.a., sp. nov., a basidiomycetous yeast associated with an Antarctic marine sponge. Antonie Van Leeuwenhoek 105:593–601

Li W, Guo L (2013) Two new species of *Septobasidium* (Septobasidiaceae) from Yunnan Province in China. Mycotaxon 125:91–96

Li W, Chen S, Guo L, Ye Y (2013) *Septobasidium hoveniae* sp. nov. and *S. rhabarbarinum* new to China. Mycotaxon 125:97–101

Libkind D, Gadanho M, van Broock M, Sampaio JP (2005) *Sporidiobolus longiusculus* sp. nov. and *Sporobolomyces patagonicus* sp. nov., novel yeasts of the Sporidiobolales isolated from aquatic environments in Patagonia, Argentina. Int J Syst Evol Microbiol 55:503–509

Libkind D, Sampaio JP, van Broock M (2010) Cystobasidiomycetes yeasts from Patagonia (Argentina): description of *Rhodotorula meli* sp. nov. from glacial meltwater. Int J Syst Evol Microbiol 60:2251–2256

Link T, Seibel C, Voegele RT (2014) Early insights into the genome sequence of *Uromyces fabae*. Front Plant Sci 5:587

Liu M, Hambleton S (2012) *Puccinia chunjii*, a close relative of the cereal stem rusts revealed by molecular phylogeny and morphological study. Mycologia 104:1056–1057

Lu C, Guo L (2009a) *Septobasidium maesae* sp. nov. (Septobasidiaceae) from China. Mycotaxon 109:103–106

Lu C, Guo L (2009b) *Septobasidium annulatum* sp. nov. (Septobasidiaceae) and *S. kameii* new to China. Mycotaxon 110:239–245

Lu C, Guo L (2009c) Two new species of *Septobasidium* (Septobasidiaceae) from China. Mycotaxon 109:477–482

Lu C, Guo L (2010a) Three new species of *Septobasidium* (Septobasidiaceae) from Gaoligong Mountains in China. Mycotaxon 112:143–151

Lu C, Guo L (2010b) Two new species of *Septobasidium* (Septobasidiaceae) and *S. pallidum* new to China. Mycotaxon 113:87–93

Lu C, Guo L (2010c) Two new species of *Septobasidium* (Septobasidiaceae) from Hainan Province in China. Mycotaxon 114:217–223

Lu C, Guo L (2011) Two new species of *Septobasidium* (Septobasidiaceae) from Gaoligong Mountains in China. Mycotaxon 116:395–400

Lu C, Guo L, Wei J, Li J (2010) Two new species of *Septobasidium* (Septobasidiaceae) from southern China. Mycotaxon 111:269–274

Lutz M, Göker M, Piatek M, Kemler M, Begerow D, Oberwinkler F (2005) Anther smuts of Caryophyllaceae: molecular characters indicate host-dependent species delimitation. Mycol Prog 4:225–238

Lutz M, Piatek M, Kemler M, Chlebicki A, Oberwinkler F (2008) Anther smuts of Caryophyllaceae: molecular analyses reveal further new species. Mycol Res 112:1280–1296

Maier W, McTaggart AR, Roux J, Wingfield MJ (2015) *Phakopsora myrtacearum* sp. nov., a newly described rust (Pucciniales) on eucalypts in eastern and southern Africa. Plant Pathol. doi:10.1111/ppa.12406

Majaneva M, Rintala J-M, Piisila M, Fewer DP, Blomster J (2012) Comparison of wintertime eukaryotic community from sea ice and open water in the Baltic Sea, based on sequencing of the 18S rRNA gene. Polar Biol 35:875–889

Manohar CS, Boekhout T, Müller WH, Stoeck T (2014) *Tritirachium candoliense* sp. nov., a novel basidiomycetous fungus isolated from the anoxic zone of the Arabian Sea. Fungal Biol 118:139–149

Margesin R, Fonteyne P-A, Schinner F, Sampaio JP (2007) *Rhodotorula psychrophila* sp. nov., *Rhodotorula psychrophenolica* sp. nov. and *Rhodotorula glacialis* sp. nov., novel psychrophilic basidiomycetous yeast species isolated from alpine environments. Int J Syst Evol Microbiol 57:2179–2184

Matheny PB, Gossmann JA, Zalar P, Kumar ATK, Hibbett DS (2006) Resolving the phylogenetic position of the Wallemiomycetes: an enigmatic major lineage of Basidiomycota. Can J Bot 84:1794–1805

McKenzie EHC (2008) Rust fungi in the subantarctic islands of New Zealand. Mycoscience 49:1–10

McKenzie EHC, Johnston PR (2004) *Puccinia embergeriae* sp. nov. on Chatham Islands sow thistle (*Embergeria grandifolia*) and a note on *Miyagia pseudosphaeria* on sow thistles (*Sonchus* spp.) in New Zealand. N Z J Bot 42:657–661

Medina EM, Jones GW, Fitzpatrick DA (2011) Reconstructing the fungal tree of life using phylogenomics and a preliminary investigation of the distribution of yeast prion-like proteins in the fungal kingdom. J Mol Evol 73:116–133

Mennicken M, Oberwinkler F (2004) A contribution to the rust flora (Uredinales) of southern Africa, with an emphasis on South Africa. Mycotaxon 90:1–28

Minnis AM, McTaggart AR, Rossman AY, Aime MC (2012) Taxonomy of mayapple rust: the genus *Allodus* resurrected. Mycologia 104:942–950

Moraes RN, Ribeiro MC, Nogueira MC, Cunha KC, Soares MM, Almeida MT (2010) First report of *Tritirachium oryzae* infection of human scalp. Mycopathologia 169:257–259

Nagahama T, Hamamoto M, Nakase T, Horikoshi K (2001) *Rhodotorula lamellibrachii* sp. nov., a new yeast species from a tubeworm collected at the deep-sea floor in Sagami Bay and its phylogenetic analysis. Antonie Van Leeuwenhoek 80:317–323

Nagahama T, Hamamoto M, Nakase T, Horikoshi K (2003) *Rhodotorula benthica* sp. nov. and *Rhodotorula calyptogenae* sp. nov., novel yeast species from animals collected from the deep-sea floor, and *Rhodotorula lysiniphila* sp. nov., which is related phylogenetically. Int J Syst Evol Microbiol 53:897–903

Nagahama T, Hamamoto M, Horikoshi K (2006) *Rhodotorula pacifica* sp. nov., a novel yeast species from sediment collected on the deep-sea floor of the north-west Pacific Ocean. Int J Syst Evol Microbiol 56:295–299

Nakase T, Tsuzuki S, Lee F-L, Sugita T, Jindamorakot S, Jan-ngam H et al (2003) *Sporobolomyces magnisporus* sp. nov., a new yeast species in the *Erythrobasidium* cluster isolated from plants in Taiwan. J Gen Appl Microbiol 49:337–344

Nakase T, Lee F-L, Am-in S, Jindamorakot S, Potacharoen W, Tanticharoen M et al (2005a) *Sporobolomyces fushanensis* sp. nov., a new species of ballistoconidium-forming yeast in the *Microbotryum* lineage isolated from a plant in Taiwan. J Gen Appl Microbiol 51:41–45

Nakase T, Tsuzuki S, Lee F-L, Jindamorakot S, Jan-ngam H, Am-in S et al (2005b) *Sporobolomyces diospyroris* sp. nov., *Sporobolomyces lophatheri* sp. nov. and *Sporobolomyces pyrrosiae* sp. nov., three new species of ballistoconidium-forming yeasts in the *Agaricostilbum* lineage isolated from plants in Taiwan. J Gen Appl Microbiol 51:277–286

Naseri A, Fata A, Najafzadeh MJ (2013) First case of *Tritirachium oryzae* as agent of onychomycosis and its susceptibility to antifungal drugs. Mycopathologia 176:119–122

Nemri A, Saunders DG, Anderson C, Upadhyaya NM, Win J, Lawrence GJ et al (2014) The genome sequence and effector complement of the flax rust pathogen *Melampsora lini*. Front Plant Sci 5:98

Nguyen HDT, Tanney JB, Chabot D, Nickerson NL, Seifert KA (2014) *Paratritirachium curvibasidium*, a new heat-resistant basidiomycete from flare pit soils in Alberta, Canada. Mycol Prog 13:575–587

Oberwinkler F, Kirschner R, Arenal F, Villarreal M, Rubio V, Begerow D et al (2006) Two new pycnidial members of the Atractiellales: *Basidiopycnis hyalina* and *Proceropycnis pinicola*. Mycologia 98:637–649

Okane I, Yamaoka Y, Kakishima M, Abe JP, Obata K (2014) *Puccinia galiiuniversa*, a new caricicolous rust fungus systemically inhabiting *Galium aparine* in its spermogonial–aecial stage. Mycoscience 55:89–97

Ono Y (2000) Taxonomy of the *Phakopsora ampelopsidis* species complex on vitaceous hosts in Asia including a new species, *P. euvitis*. Mycologia 92:154–173

Ono Y, Chatasiri S, Pota S, Yamaoka Y (2012) *Phakopsora montana*, another grapevine leaf rust pathogen in Japan. J Gen Plant Pathol 78:338–347

Park HK, Ha M-H, Park S-G, Kim MN, Kim BJ, Kin W (2012) Characterization of the fungal microbiota (mycobiome) in healthy and dandruff-afflicted human scalps. PLoS One 7(2):e32847

Paul D, Magbanua Z, Arick M II, French T, Bridges SM, Burgess SC et al (2014) Genome sequence of the oleaginous yeast *Rhodotorula glutinis* ATCC 204091. Genome Announce 2:e00046

Perdomo-Sanchez O, Piepenbring M (2008) A new species of *Puccinia* (Pucciniales, Basidiomycota) and new records of rust fungi from Panama. Mycol Prog 7:161–168

Perdomo-Sanchez O, Piepenbring M (2014) Species of *Uromyces* (Pucciniales, Basidiomycota) on Loranthaceae. Trop Plant Pathol 39:141–153

Persoons A, Morin E, Delaruelle C, Payen T, Halkett F, Frey P et al (2014) Patterns of genomic variation in the poplar rust fungus *Melampsora larici-populina* identify pathogenesis-related factors. Front Plant Sci 5:450

Piatek M, Lutz M, Ronikier A, Kemler M, Świderska-Burek U (2012) *Microbotryum heliospermae*, a new anther smut fungus parasitic on *Heliosperma pusillum* in the mountains of the European Alpine System. Fungal Biol 116:185–195

Piatek M, Lutz M, Kemler M (2013) *Microbotryum silenes-saxifragae* sp. nov. sporulating in the anthers of *Silene saxifraga* in southern European mountains. IMA Fungus 4:29–40

Pohl CH, Smit MS, Albertyn J (2011) *Rhodotorula bloemfonteinensis* sp. nov., *Rhodotorula eucalyptica* sp. nov., *Rhodotorula orientis* sp. nov. and *Rhodotorula pini* sp. nov., yeasts isolated from monoterpene-rich environments. Int J Syst Evol Microbiol 61:2320–2327

Raffaele S, Kaumon S (2012) Genome evolution in filamentous plant pathogens: why bigger can be better. Nat Rev Microbiol 10:417–430

Raggi P, Lopez P, Diaz A, Carrasco D, Silva A, Velez A et al (2014) *Debaryomyces hansenii* and *Rhodotorula mucilaginosa* comprised the yeast core gut microbiota of wild and reared carnicorous salmonids, croaker and yellowtail. Environ Microbiol 16:2791–2803

Rezende DV, Dianese JC (2001) New *Ravenelia* species on leguminous hosts from the Brazilian Cerrado. Fitopatol Brasil 26:627–634

Rezende DV, Dianese JC (2003) *Uromyces* species on leguminous hosts from the Cerrado with description of *U. galactiae* sp. nov. Fitopatol Brasil 28:495–501

Ritschel A, Berndt R, Oberwinkler F (2007) New observations of rust fungi (Uredinales) from northern Namibia. Mycol Prog 6:137–150

Robiglio A, Sosa MC, Lutz MC, Lopes CA, Sangorrin MP (2011) Yeast biocontrol of fungal spoilage of pears stored at low temperature. Int J Food Microbiol 147:211–216

Sampaio JP, Gadanho M, Bauer R, Weiß M (2003) Taxonomic studies in the Microbotryomycetidae: *Leucosporidium golubevii* sp. nov., *Leucosporidiella* gen. nov. and the new orders Leucosporidiales and Sporidiobolales. Mycol Prog 2:53–68

Sampaio JP, Golubev WI, Fell JW, Gadanho M, Golubev NW (2004) *Curvibasidium cygneicollum* gen. nov., sp. nov. and *Curvibasidium pallidicorallinum* sp. nov., novel taxa in the Microbotryomycetidae (Urediniomycetes), and their relationship with *Rhodotorula fujisanensis* and *Rhodotorula nothofagi*. Int J Syst Evol Microbiol 54:1401–1407

Satoh K, Makimura K (2008) *Sporobolomyces koalae* sp. nov., a basidiomycetous yeast isolated from nasal smears of Queensland koalas kept in a Japanese zoological park. Int J Syst Evol Microbiol 58:2983–2986

Satoh K, Maeda M, Umeda Y, Sugamata M, Makimura K (2013) *Cryptococcus lacticolor* sp. nov. and *Rhodotorula oligophaga* sp. nov., novel yeasts isolated from the nasal smear microbiota of Queensland koalas kept in Japanese zoological parks. Antonie Van Leeuwenhoek 104:83–93

Schell WA, Lee AG, Aime MC (2011) A new lineage in Pucciniomycotina: class Tritirachiomycetes, order Tritirachiales, family Tritirachiaceae. Mycologia 103:1331–1340

Shivaji S, Bhadra B, Rao RS, Pradhan S (2008) *Rhodotorula himalayensis* sp. nov., a novel psychrophilic yeast isolated from Roopkund Lake of the Himalayan mountain ranges, India. Extremophiles 12:375–381

Singh P, Raghukumar C, Verma P, Shouche Y (2011) Fungal community analysis in the deep-sea sediments of the Central Indian Basin by culture-independent approach. Microb Ecol 61:507–517

Singh P, Singh SM, Tsuji M, Prasad GS, Hoshino T (2014) *Rhodotorula svalbardensis* sp. nov., a novel yeast species isolated from cryoconite holes of Ny-Ålesund, Arctic. Cryobiology 68:122–128

Sotao HMP, Hennen JF, de Rezende DV (2007) *Puccinia caxiuanensis* sp. nov. of Uredinales on species of Burseraceae in Brazil. Hoehnea 34:493–495

Suh S-O, Maslov DA, Molestina RE, Zhou JJ (2012) *Microbotryozyma collariae* gen. nov., sp. nov., a basidiomycetous yeast isolated from a plant bug *Collaria oleosa* (Miridae). Antonie Van Leeuwenhoek 102:99–104

Suvi T, Tedersoo L, Abarenkov K, Beaver K, Gerlach J, Kõljalg U (2010) Mycorrhizal symbionts of *Pisonia grandis* and *P. sechellarum* in Seychelles: identification of mycorrhizal fungi and description of new *Tomentella* species. Mycologia 102:522–533

Swann EC, Frieders EM, McLaughlin DJ (2001) Urediniomycetes. In: McLaughlin DJ, McLaughlin EC, Lemke PA (eds) The mycota, Vol 7A: Systematics and evolution. Springer, Heidelberg, pp. 37–56

Takashima M, Nakase T (2000) Four new species of the genus *Sporobolomyces* isolated from leaves in Thailand. Mycoscience 41:357–369

Tan M-K, Collins D, Chen Z, Englezou A, Wilkins MR (2014) A brief overview of the size and composition of the myrtle rust genome and its taxonomic status. Mycology 5:52–63

Tavares S, Ramos AP, Pires AS, Azinheira HG, Caldeirinha P, Link T (2014) Genome size analyses of Pucciniales reveal the largest fungal genomes. Front Plant Sci 5(422):1–11

Thanh VN, Smit MS, Moleleki N, Fell JW (2004) *Rhodotorula cycloclastica* sp. nov., *Rhodotorula retinophila* sp. nov., and *Rhodotorula terpenoidalis* sp. nov., three limonene-utilizing yeasts isolated from soil. FEMS Yeast Res 4:857–863

Thaung MM (2009) A new species of *Uromyces* (Pucciniales) on *Trichosanthes* (Cucurbitaceae) from Burma. Australas Mycol 28:43–44

Thaung MM (2011) *Puccinia reynoldsii* sp. nov. on *Clerodendrum* (Lamiaceae) from Burma. Taiwania 56:301–304

Timling I, Walker DA, Nusbaum C, Lennon NJ, Taylor DL (2014) Rich and cold: diversity, distribution and drivers of fungal communities in patterned-ground ecosystems of the North American Arctic. Mol Ecol 23:3258–3272

Toome M, Aime MC (2014) *Pycnopulvinus aurantiacus* gen. et sp. nov., a new sporocarp-forming member of Pucciniomycotina. MycoKeys 8:43–50

Toome M, Aime MC (2015) Reassessment of rust fungi on weeping willows in the Americas and description of *Melampsora ferrinii* sp. nov. Plant Pathol 64:216–224

Toome M, Roberson RW, Aime MC (2013) *Meredithblackwellia eburnea* gen. et sp. nov., Kriegeriaceae fam. nov. and Kriegeriales ord. nov. – toward resolving higher-level classification in Microbotryomycetes. Mycologia 105:486–495

Toome M, Ohm RA, Riley RW, James TY, Lazarus KL, Henrissat B et al (2014) Genome sequencing provides insight into the reproductive biology, nutritional mode and ploidy of the fern pathogen *Mixia osmundae*. New Phytol 202:504–564

Tsiodras S, Papageorgiou S, Meletiadis J, Todas P, Pappa V, Panayiotides J et al (2014) *Rhodotorula mucilaginosa* associated meningitis: a subacute entity with high mortality. Case report and review. Med Mycol Case Rep 6:46–50

Turchetti B, Hall SRT, Connell LB, Branda E, Buzzini P, Theelen B et al (2011) Psychrophilic yeasts from Antarctica and European glaciers: description of *Glaciozyma* gen. nov., *Glaciozyma martini* sp. nov. and *Glaciozyma watsonii* sp. nov. Extremophiles 15:573–586

Valerio E, Gadanho M, Sampaio JP (2002) *Sporobolomyces odoratus* sp. nov., a new species in the *Sporidiobolus ruineniae* clade. FEMS Yeast Res 2:9–16

Valerio E, Gadanho M, Sampaio JP (2008) Reappraisal of the *Sporobolomyces roseus* species complex and description of *Sporidiobolus metaroseus* sp. nov. Int J Syst Evol Microbiol 58:736–741

Vanky K (2004) Taxonomic studies on Ustilaginomycetes - 24. Mycotaxon 89:55–118

Vanky K, Berner D (2003) *Microbotryum silybum* sp. nov. (Microbotryales). Mycotaxon 85:307–311

Vero S, Garmendia G, Gonzalez MB, Bentancur O, Wisniewski M (2013) Evaluation of yeast obtained from Antarctic soil samples as biocontrol agents for the management of postharvest diseases on apple (*Malus* x *domestica*). FEMS Yeast Res 13:189–199

Vishniac HS, Takashima M (2009) *Rhodotorula arctica* sp. nov., a basidiomycetous yeast from Arctic soil. Int J Syst Evol Microbiol 60:1215–1218

Walker J (2001) A revision of the genus *Atelocauda* (Uredinales) and description of *Racospermyces* gen. nov. for some rusts of *Acacia*. Australas Mycol 20:3–28

Walker J, van der Merwe MM (2009) Two previously undescribed rusts of *Acanthocarpus* and *Lomandra* (Lomandraceae) in Australia. Australas Plant Pathol 28:525–532

Walker J, Beilharz V, Pascoe IG, Priest MJ (2006) *Bibulocystis* gen. nov. (Uredinales) on *Daviesia* (Fabaceae) and *Albizia* (Mimosaceae) in Australasia, with comments on the genera *Cystomyces* and *Dicheirinia* and rust genera with teliospore cysts. Australas Plant Pathol 35:1–22

Wang Q-M, Bai F-Y (2004) Four new yeast species of the genus *Sporobolomyces* from plant leaves. FEMS Yeast Res 4:579–586

Wang Q-M, Bai F-Y, Zhao J-H, Jia J-H (2003) *Bensingtonia changbaiensis* sp. nov. and *Bensingtonia sorbi* sp. nov., novel ballistoconidium-forming yeast species from plant leaves. Int J Syst Evol Microbiol 53:2085–2089

Wirth F, Goldani LZ (2012) Epidemiology of *Rhodotorula*: an emerging pathogen. Interdiscipl Perspect Infect Dis 2012:465717

Wood AR, Crous PW (2005) Morphological and molecular characterization of *Endophyllum* species on perennial asteraceous plants in South Africa. Mycol Res 109:387–400

Wood AR, Scholler M (2005) *Uromyces euryopsidicola* sp. nov., a rust species that forms witches' brooms on *Euryops* (Asteraceae) in South Africa. Sydowia 57:137–143

Xu F, Wang SX, Liu Y, Ma YW, Zhang DP, Zhao S (2014) First report of *Sporobolomyces symmetricus* induced red spot disease of *Pleurotus eryngii* in China. Plant Dis 98:693

Yang T, Tian C-M, Liang Y-M, Kakishima M (2014) *Thekopsora ostryae* (Pucciniastraceae, Pucciniales), a new species from Gansu, northwestern China. Mycoscience 55:246–251

Yepes MS, de Carvalho AA (2012) *Caetea*, a new genus of Pucciniales on *Piptadenia* (Fabaceae) from Brazil. Mycologia 104:911–914

Yun HY, Minnis AM, Kim YH, Castlebury LA, Aime MC (2011) The rust genus *Frommeëlla* revisited: a later synonym of *Phragmidium* after all. Mycologia 103:1415–1463

Yurkov AM, Kachalkin AV, Daniel HM, Groenewald M, Libkind D, de Garcia V et al (2015) Two yeast species *Cystobasidium psychroaquaticum* f.a. sp. nov. and *Cystobasidium rietchieii* f.a. sp. nov. isolated from natural environments, and the transfer of *Rhodotorula minuta* clade members to the genus *Cystobasidium*. Antonie Van Leeuwenhoek 107:173–185

Zhang T, Jia RL, Zhang YQ, Liu HY, Su J, Yu LY (2013) *Kurtzmanomyces shapotouensis* sp. nov., an anamorphic, basidiomycetous yeast isolated from desert soil crusts. Int J Syst Evol Microbiol 63:3892–3895

Zhao J-H, Bai F-Y, Guo L-D, Jia J-H (2002) *Rhodotorula pinicola* sp. nov., a basidiomycetous yeast species isolated from xylem of pine twigs. FEMS Yeast Res 2:159–163

Zhao J-H, Bai F-Y, Wang Q-M, Jia J-H (2003) *Sporobolomyces bannaensis*, a novel ballistoconidium-forming yeast species in the *Sporidiobolus* lineage. Int J Syst Evol Microbiol 53:2091–2093

Zheng W, Huang L, Huang J, Wang X, Chen X, Zhao J et al (2013) High genome heterozygosity and endemic genetic recombination in the wheat stripe rust fungus. Nat Commun 4:2673

Zhuang JY, Wei S-X (2001) *Puccinia austroyunnanica* sp. nov. on *Gouania*. Mycotaxon 78:345–347

Zhuang JY, Wei S-X (2003) A new species of *Uromyces* attacking *Helleborus thibetanus*. Mycosystema 22:513–514

Zhuang JY, Wei SX (2011) Additional materials for the rust flora of Hainan Province, China. Mycosystema 30:853–860

Zhuang JY, Wei S-X (2012) Additional notes of rust fungi from southwestern China. Mycosystema 31:480–485

# Chapter 8
# Conidiogenesis: Its Evolutionary Aspects in the Context of a Philosophy of Opportunity (Lectics)

**Richard C. Summerbell and James A. Scott**

In 1682, botanist John Ray published his *Methodus Plantarum Nova* (Ray 1682), revealing a new method of classifying plants. While still holding to the traditional division of plants into trees, shrubs, and herbs, Ray devised a refined system that was based on details of plant anatomy, including the numbers and arrangement of flower parts, stem parts, and seed leaves. This method revealed the hidden, fundamental distinction between monocotyledonous and dicotyledonous angiosperms for the first time. Ray was sufficiently awed by the hidden correspondences in anatomical plant symmetries that the finding became a cornerstone of his theology. He propounded the idea that God expressed hidden wisdom throughout nature and outlined this credo in a later book *The Wisdom of God Manifested in the Works of the Creation*, 1691 (Ray 1691).

The factor that made the contrast of monocots and dicots appealing was that many different sorts of character distinctions, none of which betokened obvious functional differences, correlated to distinguish the groups. The number of cotyledons emerging from the seed differed, even though there was no obvious reason to imagine the two seed leaves of the dicot would perform significantly differently from the single seed leaf of the monocots. The numbers of parts in the flower—petals, stamens—differed, again, without clear functional significance: multiples of three for the monocots and multiples of four or five for the dicots. The leaf venation

R.C. Summerbell, Ph.D. (✉)
Sporometrics Inc., 219 Dufferin Street, Suite 20-C, Toronto, ON, Canada M6K 1Y9

Dalla Lana School of Public Health, Gage Occupational and Environmental Health Unit, University of Toronto, 223 College Street, Toronto, ON, Canada M5T 1R4
e-mail: rsummerbell@sporometrics.com

J.A. Scott, Ph.D.
Division of Occupational and Environmental Health, Dalla Lana School of Public Health, University of Toronto, 223 College Street, Toronto, ON, Canada M5T 1R4

Sporometrics Inc., 219 Dufferin Street, Suite 20-C, Toronto, ON, Canada M6K 1Y9
e-mail: james.scott@utoronto.ca



D.-W. Li (ed.), *Biology of Microfungi*, Fungal Biology,
DOI 10.1007/978-3-319-29137-6_8

tended to differ, parallel in the monocots and reticulate in the dicots, and even though this character showed a degree of correlation with elongate leaves in the monocots and more radially symmetric leaves in the dicots, the members of each angiosperm subgroup possessed a wide variety of leaf shapes. Such divergent character appositions, then, when they seemed not to indicate counterposed functional differences, could be envisioned as clues to the thought process of intelligent design. Later in biological history, after evolution was generally accepted, the same character differences could be seen as unusual factors that were so fundamental that they could be maintained through all the stresses of natural selection. They appeared to be arbitrary formats of construction that could be viable in essentially any niche inhabitable by the type of organism in question. Occasionally, one of these basic developmental formats showed variation—for example, in the yam, *Dioscorea*, reticulate leaf venation occurred in a monocot. In general, however, there was curiously little indication that natural selection was directly involved in shaping the form of the defining monocot/dicot differences.

Fungi, being arguably of lower morphological complexity than plants, showed few such characters. In the relatively complex agarics, however, a number of seemingly arbitrary construction formats were found to correlate with each other and to indicate apparently co-derived lineages. The macroscopic characters used at first included hymenial form, lamellar attachment, veil types, and basidiospore mass coloration. Emphasis on the last was one of the great contributions of Elias Fries (1825), who believed that major differences in agaric spore coloration followed a spiritual trend in nature to divide and subdivide taxa into clusters of four, with each member of a tetrad marked by fundamental distinctions in design. Later in history, as the protective functions of melanin began to be appreciated, there might have been stimulus for scientists to speculate that brown and black basidiospores would be more robust in sunlight than pale ones. To our knowledge, however, basidiospore colour, to this day, has not been shown to be a salient selective factor in any natural situation. Although most lineages once defined on basidiospore colour are now known, through DNA analysis, to include some exceptions, there remain large blocks of related species that, no matter how diverse the habitats they grow in, retain the characteristic spore colour of their lineage.

When microanatomy came to mushroom work, other quasi-arbitrary developmental formats such as lamellar tramal structures and pileipellis constructions were revealed (Fayod 1889; Singer 1986). They joined the list of characters that appeared to be lineage-revealing, selectively near-immutable evolutionary whims. Since mycologists have mostly tended to Neodarwinism, the idea that these character differences were perhaps not directly selected for by environmental exigencies has tended to receive little attention, even though biologists have generally accepted that some developmental characters appear to be constrained towards self-perpetuation.

In the microscopic conidial fungi, taxonomy for many decades remained at the level analogous to the trees, bushes, and herbs of primitive plant taxonomy. In the definitive work of Saccardo (1886), the conidial fungi were divided into form taxa called Hyphomycetes (essentially, branches), Sphaeropsideae (flasks, i.e., pycnidia), and Melanconiae (erumpent tufts, i.e., acervuli). The Hyphomycetes were then subdivided into Mucedineae (pale branches), Dematiae (dark branches),

Stilbeae (erect fascicles of branches, i.e., synnemata), and Tuberculariae (superficial tufts of short branches, i.e., sporodochia).

In 1953, Hughes (1953) revolutionized hyphomycete taxonomy by crystallizing and extending our knowledge of the microanatomy of conidiogenesis. Specifically, he discerned that conidiogenous structures provided a degree of the correlative taxonomic insight provided by developmental design characters in other organisms, such as mushrooms and plants. Additional studies in subsequent years only underscored the importance of his scheme of organization. *Aspergillus* conidiogenous states, for example, strongly featured phialides, and their associated sexual states tended to be cleistothecia producing equatorially ornamented ascospores. *Scopulariopsis* conidiogenous states featured annellides (a term originally coined by Hughes as 'annellophores') where the phialide-like conidiogenous cells extended their necks via percurrent proliferation, and any sexual states found were clearly related species of *Microascus*. Entire orders of fungi were characterized by distinct patterns of conidiogenous development, e.g. the Onygenales, with thallic conidiogenesis producing conidia dehiscing via lytic disruption of small, empty separating cells. Conidiogenesis, for the first time, gave hyphomycete experts a somewhat reliable means of predicting the phylogenetic affinity of most of the asexual fungi they studied.

Up until a few years ago, review articles dealing with conidiogenesis would have been primarily focused on reviewing the patterns of conidiogenous development and morphogenesis [an excellent example is that of Cole (1986), in tandem with a discussion of the taxonomic information these structures revealed. Now, however, conidiogenesis has been superseded as a biosystematic indicator by DNA sequences. Many misplaced or hard-to-place asexual forms have thereby been correctly assigned to higher taxa and the clades they are based on. The study of conidiogenesis, we can now see, was by no means a taxonomic panacaea (as Cole 1986) had already pointed out), but it was truly the best of the available microanatomical characters that could have been used to order the Hyphomycetes.

The eclipsing of the study of conidiogenesis by DNA-based biosystematics raises the danger that treasured information will be downgraded in importance and ultimately ignored. If our primary concerns are now divided into systematics, the discipline allowing identification; ecology, the discipline characterizing the interactions of organisms with each other and ourselves; and the -omics (genomics, proteomics, metabolomics), gauging the functioning and potential functioning of organisms as biotechnological machinery, how much attention is left to observe the development of seemingly arbitrary, industrially unexploitable characters like annellations and sympodial proliferations? In an effort, then, to show that conidiogenesis transcends being a superseded taxonomic character (still useful to microscopists in their ever-cloudier identifications of DNA-defined taxa) and a minor developmental curiosity, we would like to elaborate a succinct evolutionary classification of the characters possessed by organisms and show where the characters of conidiogenesis fit in. That is to say, we wish to accord these characters a place in the story of fungal evolution, not just in taxonomy, ecology, or development.

The classification of taxonomic characters used henceforth arises from a philosophy of the background conditions such characters address. In the broadest conception, these conditions are called 'evolutionary opportunities', and when organisms are conceived of as having settled into exploitation of a subset of these opportunities, the limited locus of recurring energy interchange is referred to as a 'niche'. In order that an 'opportunity', in the evolutionist's sense of the term, not be seen as vague hand-waving toward 'something out there', we are going to define it exactly: it is *a possibility for 'benefit', or increase, that may or may not be undertaken by a self-reconstituting entity in a modifiable system.* This definition has the following dimensions:

1. *Energy transfer and its constructive (including reproductive) use.* Biology is based upon solar energy, and the phrase 'benefit or increase' refers, ultimately, to gaining a portion of that energy and putting it to use in life and in perpetuation of the lineage.
2. *Random or chosen optionality.* Fundamental to the notion of an opportunity is the idea that the entity confronted by the opportunity is not completely constrained to undertake it. If it were to be so constrained, one would have a complex deterministic system, like air flow or water flow, not an opportunistic system like biology. The finch reaching the Galapagos Islands may evolve to specialize in eating seeds or insects, and, regardless of its level of prefigurement (e.g. arriving as a seed-eater), it is not wholly constrained a priori in its recruitment towards one or the other of these opportunities or niches. The optionality intrinsic to evolution as an opportunistic system is indicated by the 'may or may not' phrase in the definition of opportunity. Evolutionary optionality is a matter addressed by the lineage, and the arrival of the lineage at one side of the option (e.g. seed-eating) or the other (insect-eating) is a matter of chance. Terminology for the arbitration among these options is awkward and encumbered by teleology. We have, thus, borrowed a common Greek root used (sometimes via Latin) in selection, election, lectotypy, and so on and coined the word 'paralection', meaning chance-derived pseudo-choice, for this retrospectively seen process. Its antonym, 'lection' applies to any deliberate choice made by a conscious entity, if the concept of 'free will' is accepted, a matter we need not get into here. The consequences of a theory of opportunity ('lectics') for conscious beings have been adumbrated elsewhere (Rogerson 2013).
3. *Self-reconstitution.* The fundamental difference between opportunistic and deterministic systems is that the acting (discrete, causally connected) elements in the former can paralectically (or lectically) modify and regenerate the fundamental basis of their interactions. Tungsten always acts as tungsten, and water as water, but a finch lineage (in contrast to the individual finch) may change over time in any or all of its evolutionarily interacting characters and be transformed from the production of dull-coloured seed-eaters with the digestive and metabolic characters fitting that niche to the production of brightly coloured nectar-eaters with radically altered digestion, metabolism, flight, breeding habits, and so on. Insofar as entities address opportunities, they are ipso facto not static, but rather are changeably regenerative.
4. *The modifiable system.* The non-static nature of the 'paralectont' (the non-deliberating entity addressing the opportunity; in biology, the lineage) is only of

possible significance in the context of a larger, ambient system that also allows modification. In evolution, an evolving species tends to change all the species it interacts with, e.g. a nectar-feeding insect may induce a flowering species, through cyclical reinforcement of benefit, to alter its floral coloration to better attract the insect as a successful pollinator. Moreover, through shading, biochemical secretion, and so on, species also alter the conceptually abiotic elements of their environments, such as soil and water. These modifications to the abiotic substrata then redound to influence the progressive evolution of all the species interacting with those substrata. The defining 'logical' feature of these interactions is progressively altering cyclicality. If the interactions are programmed into a computerized model, they become elaborately interlocking recursive functions that are ultimately unable (extremely unlikely) to re-attain exactly any previously seen configuration. No moment in evolutionary history can ever be duplicated later. Neither can any of the interacting causal agents and causal momentum-carriers, except at the level of simplest chemistry, be duplicated. Convergence occurs but is never absolute.

In order to assess where conidiogenesis fits into the systematics of evolutionary opportunity, we need to consider the question of how the characters seen in biological organisms address or interact with the opportunities/niches from which the species obtain increase.

A consideration of the features of organisms shows that these features may be approximately divided into two conceptual groups, which we trust will soon become apparent as operationally different even though the distinction may seem abstract at first. Again, these criteria apply to opportunistic situations outside biology as well as inside it, but we will restrict the scope of our comments to biology.

1. *Characters for which the form is obligate to the performance.* Any organism evolved to propulsively fly must have wings that can be used to overcome gravity by directing air flow. Whether these wings are feathered or membranous, the character 'has functional wings' must pertain. Any fungus producing disseminating conidia must have a dehiscence mechanism, whereby disseminable cells or cell groups break away in a genetically determined manner from the fixed structures that formed them. 'Has functional wings' and 'has a conidial dehiscence mechanism' are examples of this type of character obligate to a function that is an essential element addressing the current niche. The 'character' considered here may only be an aspect of a discrete structure, e.g. the lift functionality of a wing, but it is nonetheless conceptually individuated as an adaptively necessary aspect in relation to the niche. Drawing on two etymologically related Latin verbs, both with present active form appellō and their Romance-language derivatives such as the French 'rappeler', 'to recall', we have designated this category of character 'rappellative'—the character, in its form, directly 'recalls' the form of the niche exigency to which it responds (e.g. 'has functional wings' responds to 'needs to locomote by air to survive in its niche'). The niche exigency can metaphorically be said to 'pull back' (etymology: 're'—again + 'ad'—towards + 'pello' push) the lineage if any of its members deviate from the needed form (offspring born wingless in a species that needs to fly will not survive).

(This assumes that the deviation is not of the rare serendipitous type that opens up a novel path to survival.) The metaphor of a mountain climber 'rappelling' down a cliff face and being recurrently pulled back to the rigid limit of the mountain is also apt when extended to this type of evolutionary situation. The production by any type of organism living predominantly on lactose of at least one functional lactase enzyme can be seen as an archetypical rappellative character.

2. *Characters that represent one of a number of alternate forms that can fulfil the performance.* Wings may rely on feathers, stretched skin, or chitinous membrane to support flight. Thallic conidia may dehisce by weakening of a cell wall middle lamella or by rupture of an entire dead, empty separating cell. It is not obvious how adaptive forces could selectively replace one of these characters with one of its (approximately) equally functional alternative forms, since the already functioning character will be likely to hold the advantage over any as yet undeveloped possible replacement in any case of selective pressure. As an antonym to 'rappellative', we refer to the characters juxtaposed in these contrasts as 'arbitrative', based on the Latin arbitro, 'I judge' or 'I consider'. This root word in English has long been used to allude to the concept of impartial justice and has given us the adjective 'arbitrary' to describe indifferent conscious decision processes. The indifference due to evolutionary chance is likewise alluded to in the concept of an 'arbitrative' character. Such characters could also be referred to by the tantalizing adjective 'fungible', but that word has the slight disadvantage that some might take it to imply the existence of a force or agent that could effect a switch among the different types.

   Arbitrative characters come in approximately four subtypes, listed below.

   2a. *Novel neutral alleles or mutations.* As Kimura (1983) pointed out, many evolutionary divergences from parental stock may begin as mutations in protein-coding genes that are initially synonymous or, if non-synonymous, have no marked effect on the function of the proteins involved. In some cases, non-synonymous neutral mutations may become consequential when new adaptive stresses emerge. For example, a mutation that confers fortuitously more heat tolerance than its predecessor may begin to be selected for when a habitat begins to trend as warmer over the years. Alternatively, initially neutral mutations may, through the complexities of protein folding and through reconfiguring of active domains, fortuitously lend functionality to later-arising novel mutations that occur in more evolutionarily constrained, enzymatically functional regions.

   2b. *Coordinating characters.* These are characters that are interchangeable in principle, interacting with other biological characters that are also interchangeable in principle. Such characters tend to be held in relative stasis by coordination, whether coordination within cells, or within structures, or within interacting members of the population, or within interacting species in the ecosystem. Enzymes, hormones, etc., that interact with other cellular components must in some way chemically interlock with them. The biochemical and physical forms of mating type A must be compatible with those of mating type B if mating is to occur. Where sensory perception exists, the members of species must be able to interact with one another through recognition of shape, colour, olfactory impact, and so on. Woodpecker species A has a red cap, used in visual recognition of conspecifics; its sibling-species B has a yellow cap, used for the same function. Mushroom species A, based on interactions within its own inner developmental schema, develops complex, layered lamellae widened out with interwoven hyphal growth, while species B develops equally complex lamellae widened with parallel tramal tissue. Hormone A only stimulates a signal when it meets compatible receptor A'. Hormone B in a closely related species may be co-derived genetically from a common ancestor with hormone A, and it may be identical in function and nearly so in structure; nonetheless, it only stimulates receptor B'. Such characters appear arbitrative when they are juxtaposed against one another by an observer, but in practice, they are locally near-obligatory to newly generated members of the species, unlike characters of type 2a. They can be conceived of as rappellative characters within an arbitrative framework. Based on their property of near-obligatory coordination within adaptive subsystems, they are designated 'interrappellative' characters. Interrappellative characters intergrade with rappellative characters in situations where coordination occurs as a standoff among antagonistic species, e.g. where a predator uses odour to recognize its prey, and the prey species takes steps to conceal its odour against the predator's specific methods of detection. In such cases, discoordination, such as a change of odour by the prey, may be adaptively favoured, if it occurs. Cooperative interrappellatives have evolutionary advantage in remaining synchronized, while antagonistic interrappellatives may entail advantage in developing asynchrony, especially for species being preyed upon, parasitized, or grazed. Much like inclement physical conditions, then, predators, grazers, allelopaths, and other 'hostile' species tend to promote adaptive change.

2c. *Vestigial characters*. The human vermiform appendix is a well-known example of a character that is considered to exist as a developmental relict form, arguably lacking any functional significance that would outweigh the death risk it poses. Evolution is efficient, but in complex organisms, essentially functionless developmental leftovers may remain. These, then, are purely arbitrative characters, in that closely related species may either possess them or not. We are not aware of any clear mycological examples, but see further comment below sub *Trichophyton rubrum*.

2d. *Characters without strictly specified role, elaborated by adapted processes that intrinsically generate variation*. Cells and organisms operate various processes through which the rates of adaptive change can be varied. DNA repair, for example, can adapt to be more or less rigorously accurate, lowering and raising the rate of stabilized mutation. Genes can become arranged to undergo recombination or positional reshuffling with greater or lesser facility, depending on diverse factors such as methylation, regulatory proteins, histones, repetitive regions between genes, and genomic expansion processes featuring retention and eventual redeployment of originally homologous gene copies on multiple chromosomes. Each human receives a unique face, set of fingerprints, and set of 'genetic fingerprints' that are determined by genetic characters, or by genetically enabled epigenetic processes, or by early developmental generators of quasi-random variation. The extent to which such processes operate in other species is often unclear, but can hardly be thought to be absent in most or all cases.

Some clearly arbitrative character types are remarkably evolutionarily labile, sometimes perhaps 'locking in' interrappellative relations with features of interacting organisms, and sometimes appearing functionally mysterious. A good example is the secondary metabolite spectra elaborated by all mycelial organisms that densely occupy volumes of substrate, including Hyphomycetes and Streptomycetes, among others. Each species, even in recently evolved phylogenetic clusters, generally produces a distinct array of several characteristic 'extrolites', sometimes consisting of a shuffling from a short list of metabolites found in multiple related species, and sometimes including one or more metabolites unique to the species in question. A given unusual metabolite may pop up in isolation in very distantly related species, e.g, penicillin in *Penicillium chrysogenum* and *Trichophyton rubrum*, a happenstance implying either a common exogenic origin of the character (Peñalva et al. 1990) or the existence of a randomization process in which the spate of permutations is constrained in chemical possibility and thus has some tendency towards producing coincidental results.

Though extrolites sometimes become typical examples of features locked into interrappellative relationships with competing species—for example, the case of mycotoxins adapted to deter grazing of the mycelial domain by poisoning the major grazers—many have no known interrappellative function, either within the producing species or in the ecosystem. A previous paper (Summerbell 2000) has proposed that selection for anomaly or non-recognizability, as an ecologically interactive feature in its own right, can occur, for example when an array of unusual metabolites distinguishes a mycelial colonization zone as 'unrecognizable and/or indigestible as food' to a non-specific array of grazers, even though the metabolites are neither toxic nor notably repugnant.

Heath (2000) developed and tested a sophisticated theory of host vs. non-host defence mechanisms in plants. The former ('host resistance') mechanisms become involved in a near-rappellation, based on progressive adaptation of the plant to recognize established pathogens by acquiring new means of detecting them as they, in turn, altered their chemistries to evade recognition. The latter ('non-host resistance') mechanisms consist of the generation of features generally adapted to deter a wide and possibly ever-changing group of potential, unspecialized antagonists. Some of these non-host defence features may consist of the production of chemically distinctive 'flavour' compounds arguably analogous to the extrolites of fungi—again, a group of chemicals including some strongly deterring ('antifeedent') or toxic members, and also members arguably merely anomalous: distinctive, odd chemicals, 'un-food-like' to the perceptions of many grazers, and perhaps somewhat inaccessible to many species' digestive enzymes.

Axiomatically, evolution demands the ability to adapt, that is, to change. Genomic information systems may have various mechanisms for second-order logic, controlling their own adaptability, functioning to set the rate of change. Adapted constitutive changeability has to be seen as to some degree prospective: a lineage has no way of predicting when change will become obligatory. Thus, an exigency is exerted upon evolving species to produce initially merely anomalous changes responding, in effect, to a general need for change. Species that anomalize themselves to grazers and predators may be responding to an enhanced need to generate an ever-progressing unrecognizability, like a cryptographic device using fresh encryption settings. Such changes can be viewed in models of evolution as a search of the evolutionary landscape for advantage in semi-randomized novelties. Some of these anomalies, as mentioned, may become stabilized as interrappellatives, but in anomalizing selection, some may be stabilized for a time *as anomalies*. We will make a case elsewhere that the bright colours of many mushroom species are a quintessential example of this pattern in evolution. Because these characters evolve in processes 'recalling' (responding to) the *general* evolutionary conditions favouring the constant elaboration of new diversity, they are designated 'general-rappellative' characters. This conceptually difficult character type can be conceived of as the output of adapted innovation mechanisms prefiguring the generation of other, more specific adaptations.

Character type 2a is very closely related to 2d and may be seen as a subset of it. It is not clear how much genetic mutation is truly unavoidable—no doubt some errors fail to be repaired by chance alone, but the efficiency of such repairs is itself an adaptable feature. Rapidly evolving animals such as humans, canines, and rats are notoriously cancer prone, arguably as a consequence of inadequately repaired DNA lesions. Some lineages such as sharks (Lee and Langer 1983; Luer and Luer 1982) and decapod crustacea (Vogt 2008) have at least a preliminary reputation as being less prone to carcinogenesis (In sharks, it must be noted, discussion of this idea has been clouded by tendentious argument about the extreme 'straw man' position that 'sharks never get cancer', something allegedly propounded by people touting shark cartilage as an alternative medicine remedy; see Ostrander et al. (2004). Regardless of what the true story is with the sharks, it remains very much an evolutionary possibility that our intrinsic human carcinogenic disease control, based on oncogenic resistance mechanisms involving accurate repair of damaged DNA, prompt apoptosis of affected cells, and other processes may have been partly sacrificed in the achievement of our recent evolutionary success.

An analysis of the characters seen by early botanists as showing hidden spiritual symmetries in nature, together with later-described taxonomically revelatory analogues such as mechanisms of conidiogenesis, shows that these characters are, to a large extent, developmental interrappellatives. The few purely rappellative features that are seen in the different conidiogenesis mechanisms mainly consist of features spacing newly forming conidia apart from one another in conidiogenesis adapted for airborne dispersal. Clearly, sympodulae, producing conidia well distanced from each other, are less likely to be involved in producing mucoid conidial masses favouring arthropod transmission than are phialides, which can readily be adapted to produce many conidia amassed in a small volume of space. Other features, such as the percurrent extensions seen in annellides, the cell wall splitting and separating cells seen in different subclasses of thallic conidiogenesis, and the retrogressive progress of conidiation seen in species like *Trichothecium roseum*, are relatively pure developmental patterns held in stasis by the need for conidia, as more-or-less standardized propagules, to be sequentially formed and released in an orderly manner. That these conidiogenesis mechanisms are perhaps closely ontogenetically related and, in principle, truly fungible can be seen in a few species like *Trichothecium indicum* (Summerbell et al. 2011), which produces retrogressive conidia, sympodial conidia, and phialoconidia. Another example lies in the members of the main *Exophiala-Cladophialophora-Fonsecaea-Phialophora-Rhinocladiella-Capronia* 'black yeast' clade, arguably all one genus, in which multiple species can produce two or more elements of a list including acropetally extending, branching blastoconidial chains, annelloconidia, sympodial conidia, meristematic fission cells, and phialoconidia. In the case of 'black yeasts', there may be some rappellative partitioning among the conidial types into airborne and subaqueous or arthropod-dispersed forms, with a complex ecology favouring polymorphism, but clearly, even in the relatively few organisms featuring multiple types of conidiogenesis, each type is stably reproduced with its spate of distinctive features, such as the relatively large, vase-shaped, darkened collarettes seen in many of the phialides produced by the 'black yeast' clade.

No evolution towards efficiency in the relatively stripped-down Hyphomycetes can eliminate the details of conidiogenous development, and thus most species are distinctly fixed in one of the major variants of this system. Even in yeasts and other sequentially reproducing single-celled fungal states, where conidiogenesis (whenever non-ballistic) tends to produce progeny cells immersed in highly physically comparable aqueous or oily habitats, arbitrative differences in conidiogenesis are prominent. There are splitting cells, as seen in *Schizosaccharomyces*, blastoconidia, as seen in most yeasts, phialoconidia, as seen in most *Malassezia* species and the yeasty states of *Lecythophora* species, annelloconidia as seen in the 'black yeast' clade as well as *Hortaea*, and even sympodially produced conidia, as seen in *Malassezia sympodialis, Bullera, Kurtzmannomyces*, and, most extravagantly, *Sympodiomycopsis*.

An interesting study in the fungibilities of fungal conidiogenesis is seen in the organisms in the scattered Eurotiomycete lineages that have adapted to aerially infect mammalian bodies and use them as perennating structures (Plate 8.1).

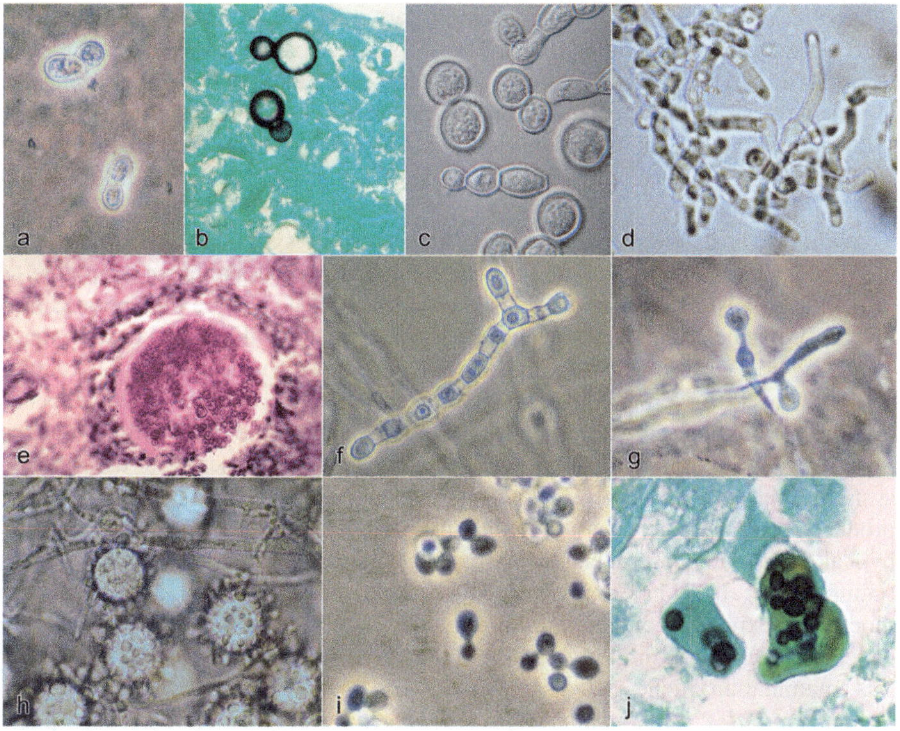

**Plate 8.1** Conserved and apomorphic conidiation in medically important Eurotiomycetes. (**a–c**) Blastoconidial particulate morph of *Blastomyces dermatitidis*. (**a**) Nanic form, in sputum. (**b**) Normal form, in tissue, Gomori methenamine silver (GMS) stain. (**c**) Conversion from filaments to particulate phase, in vitro, Nomarski. (**d**) Arthroconidial host phase of *Talaromyces marneffei*. (**e–f**) *Coccidioides posadasii*. (**e**) Spherule particulate morph in tissue, Periodic acid-Schiff stain. (**f**) Disjunctive arthoconidia. (**g**) *Blastomyces dermatitidis*. Aleurioconidium (lateral disjunctive arthroconidium) formation in filamentous form. (**h–j**) *Histoplasma capsulatum*. (**h**) Aleurioconidia. (**i**) Blastoconidial particulate morph in vitro. (**j**) Blastoconidial particulate morph in human macrophage cells, GMS stain

Although experimental confirmation of this scenario has proved elusive for these species, their inoculum appears to persist in the infected host until the animal dies, whereupon the fungus is released into nitrogenously enriched soil. The species involved make up most of the small group of human and animal invaders referred to as 'dimorphic systemic pathogens'. They are *Blastomyces* spp. (*Ajellomyces* pro parte), *Emmonsia* spp., *Paracoccidioides* spp., *Histoplasma capsulatum*, and *Coccidioides* spp. in the Onygenales, and *Talaromyces marneffei* in the Eurotiales. They have all developed forms of conidiogenesis within the host that are radically different from the conidiogenesis normally seen in vitro and, presumably, in the non-host environment.

The non-host, 'environmental form' conidia of the Onygenalean species are all variations on a theme of aleurioconidia (Plate 8.1g, h) and alternate arthroconidia

(Plate 8.1f), which are the lateral and intercalary manifestations of the same conidial release mechanism (Sigler 1989). These conidia separate from subtending cells via the lysis and breakage of an empty disjunctor cell. Single-celled conidia released in this way can readily become airborne and, in species possessing the necessary virulence characters, can infect the respiratory systems of hosts. Likewise, the typical penicillium-type catenulate phialoconidia of *T. marneffei* are specialized for airborne dissemination.

Within the host, however, the spread of the dimorphic-systemic organisms through the bloodstream and lymph system requires a planktonic assimilative form able to pass through vessels. In many species, particulate fungal cells are incorporated into host macrophages (Plate 8.1j), where any linear growth would tend to destroy these sheltering host cells. In addition, the host form, free of the need to resist desiccation and of the need to exert and integumentally restrain the turgor force of hyphal penetration, is evolutionarily free to develop a biochemically modified cellular integument that will minimize the diversity of potential antigens presented to the immune system. The yeast phase of *Histoplasma capsulatum* (Plate 8.1i, j), for example, produces proportionately less of the highly immunoreactive ß-1,3 glucan than the hyphal form and, in virulent chemotypes, tends to produce more α-1,3-glucan (Guimarães et al. 2011). Conidial reproduction in the host is thus modified to accommodate the exigencies and opportunities particular to that habitat.

The particulate assimilative phases that have evolved in rapprochement with this mammalian niche are unexpectedly diverse. *Blastomyces, Histoplasma, Paracoccidioides*, and *Emmonsia pasteuriana* have revived or re-evolved pathways for producing budding yeast structures, although the 'budding on a broad base' yeasts of *Blastomyces* (Plate 8.1a–c) and the multipolar budding yeasts of *Paracoccidioides* are highly distinctive in morphology. Only *Histoplasma* produces a yeast phase that can be confused morphologically with those seen in distant basal Ascomycete lineages in the Saccharomycetes and Taphrinomycotina. Blastoconidiation in all these cases occurs without any formation of the disjunctor cell seen in all conidiation occurring off the host.

*Coccidioides*, as seen in the host, has evolved completely differently from other Onygenales, producing structures of a type rarely seen in the fungi. Endoconidia are produced by the internal fragmentation of multicelled, spherical mother cells (Plate 8.1d). Just a very few Ascomycetous genera are known to produce endoconidia in sacs or cysts, namely *Phaeotheca*, *Phaeothecoidea*, *Hyphospora* (Seifert et al. 2011), *Endoconidioma* (Tsuneda et al. 2004), and, arguably, the obligate pulmonary pathogen *Pneumocystis*. Environmental conidia of *Coccidioides* species are alternate arthroconidia separated by disjunctors (Plate 8.1f).

Meanwhile, *T. marneffei*, in its host conidiogenesis, has evolved to recapitulate the fission-yeast form (Plate 8.1d) best known from extremely remotely related *Schizosaccharomyces*, placed by Schoch et al. (2009) in the Taphrinomycotina. *Schizosaccharomyces* is the genus most closely related to *Pneumocystis*.

*Pneumocystis* itself can be mentioned as a possible remote analogue of the host states of Eurotiomycetous dimorphics. It produces two kinds of propagules, 'intracystic bodies' and 'trophozoites'. These names hearken back to the time when this genus was thought to be protozoan. The eight intracystic bodies that form in each reproducing cyst appear to be diploid ascospores forming within an ascus (De Souza and Benchimol 2005). These ascospores, upon release from the ascus, germinate to give rise to the irregularly shaped trophozoites, which were sometimes termed 'amoeboid' in the past. These forms, however, are non-motile, mitotically dividing fission cells, with a thin integument of typical fungal cell wall polysaccharides containing mannose and N-acetyl-glucosamine residues. They produce static, filiform projections that may aid in attachment to cells, but otherwise, they are typical particulate assimilative cells reproducing by schizolytic conidiogenesis. Only shape and habitat distinguish them from the fission cells of the related *Schizosaccharomyces*. *Pneumocystis* is simply another fungus that has utilized an unusual non-filamentous form of conidiation in its particulate growth in mammalian tissue.

The 'conservatism' of the non-host conidiogenesis in the Eurotiomycetous dimorphic systemic pathogens, contrasted with the 'liberality' of the divergent conidiogenous adaptations developed in the host forms, illustrates the stabilizing effect of interrappellative developmental systems. Only when opportunities change significantly, and highly altered, quasi-rappellative host–pathogen exigencies begin to pertain, do the very stable processes of thallic-disjunctive or phialidic conidiogenesis become replaced or supplemented with completely different mechanisms. The ability to evolve radically different conidiogenous mechanisms can be said to lie latent in any of these organisms, but usually only to manifest when an environment renders the established interrappellatives non-functional or developmentally disengaged. (As an example illustrating what we mean by the latter descriptive, typical alternate arthroconidia would fail to be produced in *Coccidioides* in its growth in the host in part because formation of the filaments from which they derive is not induced.)

A relict of the hypothetical, protean transitional form through which most of these Eurotiomycete dimorphic host phases may have arisen is perhaps seen in *Emmonsia parva* and *E. crescens*. These species infect the lungs of small mammals as their relatives do, but undergo no conidiogenesis or other cellular reproduction there. (Their close relative, *Emmonsia pasteuriana,* produces budding yeast). Instead, the airborne conidia, caught in the respiratory passages, simply swell up and develop a progressively thickened cell wall. The resulting highly enlarged, tough 'adiaspores' may cause death of the host by occlusion of the lower airways. One can easily imagine that the more complex forms produced by related Onygenalean species originally arose from conidia that persisted and, while possibly shedding some outer cell wall layers in incipient germination, swelled in the lungs rather than germinating as filaments. Eventually, after this simple form of mammalian colonization became reinforced by favouring survival of the fungus in stressful habitats, new, particle-producing methods of cellular proliferation developed, increasing the success of each affected species by allowing it to disseminate through the circulatory system and colonize tissues beyond the lungs.

The fission cells produced by *T. marneffei* distinctly resemble fragmenting hyphae in culture at 37 C (Plate 8.1d) and may have arisen through a genetically relatively simple development producing cell wall schizolysis along the septal plane. Their ontogenetic similarity to *Pneumocystis* trophozoites may be purely coincidental or may carry some degree of symplesiomorphy in the retention of enzymatic programmes for lysing middle lamellae of walls in adjacent cells.

Outside the Eurotiales and Onygenales, there are other pathogen host states that illustrate the potential for conidiogenous plasticity in Hyphomycetes. Chaetothyrialean fungi of the 'black yeast' clade, e.g. *Cladophialophora carrionii* and *Phialophora verrucosa*, causing subcutaneous 'chromoblastomycosis' infections, produce schizolytic fission cells in infected tissue. Similar cells may also be produced in stems of infected plant hosts (de Hoog et al. 2007). That this is an evolutionarily highly localized form of conidiogenesis is perhaps illustrated by the failure of the nominally congeneric *Cladophialophora bantiana* to produce a particulate vegetative state in its typical cerebral infections, and by the prevalence of blastic or minutely annellidic budding yeast cells in closely related *Exophiala* species. Some typical 'black yeast' fungi, like *Exophiala phaeomuriformis*, can produce thecate 'meristematic' cell clumps under certain metabolic conditions, and these appear to bear some ontogenetic relationship to the developing fission cells of the chromoblastomycosis agents, though they also resemble pseudothecial initials. In the chromoblastomycosis agents, however, the process of schizolytic conidiogenesis producing individual cells from meristematic clumps has been regularized.

Dermatophyte fungi in the Onygenales (Arthrodermataceae), infecting the superficial layers of human and animal skin, may have life cycles where they spend part of the life history growing on shed keratinous materials such as rodent hairs in the environment, or they may be essentially entirely restricted to growth on their hosts, as far as is known. In the latter case, as seen in the most common species on humans, *Trichophyton rubrum*, their conidiogenesis is altered to a fission process involving adjacent swollen cells within the skin. This 'endoarthric' process, based on production and schizolysis of a septum derived from the inner cell wall within a friable outer cell wall (Hashimoto et al. 1984), produces conidia referred to as 'substrate arthroconidia'. In culture, however, the species produces completely distinct, dacryoid to clavate microconidia and subcylindrical macroconidia, as well as disjunctor-separated arthroconidia, typical of Onygenales. This production of classic thallic-disjunctive, aerial conidia in *T. rubrum* and other species known only from growth on the host may be a truly vestigial feature, a latent programme only elucidated by growth, after many thousands of years of diverging evolution, in a laboratory. Some human-associated dermatophytes like *Trichophyton schoenleinii* and *T. concentricum* have been completely divested of this programme and never produce aerial conidia, but can still produce the separable swollen cells in host material that serve as the ecologically functional conidia. These substrate-arthroconidia appear to produce lectin-like materials enabling them to bind efficiently to host skin (Esquenazi et al. 2004), and as small separable particles, thus have an advantage over any larger structures, such as filaments in shed skin fomites, which might also be physiologically capable of infecting a new host. Here again, then, though the typical Onygenalean conidiogenesis usually remains, it has been largely superseded in the natural ecology by a de novo form, in this case one visually suggestive of the sessile chlamydospores formed by other fungi in substrate materials.

It can take surprisingly little time for a fungal lineage to be permanently transformed by host conditions. To give an illustration: as part of studies on *Acremonium* species, we dealt early on with an unusually fast-growing species called *A. falciforme* that was known only from a slowly progressing, chronic human subcutaneous infection called mycetoma. (It grew more rapidly than other *Acremonium* species, where a slow growth rate was a genus-defining character, though it grew more slowly than any common *Fusarium* species.) It was unusual in producing some curved conidia that were occasionally one-septate. Upon reading that an inky blue to violet-purple pigment was sometimes seen in fresh isolates of this species (e.g. Gams 1971), we were seized by the idea that it was probably a host-adapted member of the *Fusarium solani* complex, members of which sometimes produce such pigments. In part, the idea that a *Fusarium* could be so morphologically transformed was influenced by years of work with aberrant host-adapted *Aspergillus fumigatus* isolates, not to mention dealing with the polymorphous skin pathogen *T. rubrum* (Summerbell 2000). It struck us also that the completely morphologically different *Cylindrocarpon lichenicola*, also occasionally making bluish pigments and involved in subcutaneous infections (Summerbell and Schroers 2002), could be an *F. solani* sensu lato. After sequencing by Hans-Josef Schroers showed both hunches to be correct, and the species were recombined into *Fusarium*, we realized that there was a likely third example on the list of medically important fungi. The species in question was named for the same blue pigment that had provided a clue in the other cases. The ex type isolate of this fungus, originally isolated from a subcutaneous infection (de Vries et al. 1984; De Bruyn et al. 1985), was revived from the CBS collection as CBS 518.82 and secondary isolation CBS 637.82. Study revealed a dense, slow-growing, off-white clump of thin-walled, colourless chlamydospores connected by short hyphal segments. Hyphae occasionally produced a relatively long, tapering phialide that yielded some unicellular, ellipsoidal conidia. No blue pigment matching that noted in the description was formed. To the essentialist eye of the medical mycologist, if not to the formalist eye of the morphotaxonomist, this enigmatic form was strikingly suggestive of *Fusarium solani*. Sequencing by Hans-Josef Schroers disclosed that indeed, this species, originally described as *Phialophora cyanescens* (de Vries et al. 1984) and later cannily transferred to *Cylindrocarpon* by Zoutman and Sigler (1991), represented a distinct haplotype in the *F. solani* complex. The *F. solani* complex has recently been segregated as the uninominate genus *Neocosmospora* by Lombard et al. (2014), and *F. falciforme* and *F. lichenicola* have become *N. falciforme* and *N. lichenicola*. *Phialophora cyanescens* is recombined here as

***Neocosmospora cyanescens*** (G.A. de Vries, de Hoog & Bruyn) Summerbell, Schroers and Scott, **comb. nov.** (MycoBank MB 813864)

Basionym: *Phialophora cyanescens* G.A. de Vries, de Hoog & Bruyn, Antonie van Leeuwenhoek 50 (2): 150 (1984) [MB 107121]

≡ *Cylindrocarpon cyanescens* G.A. de Vries, de Hoog & Bruyn) Sigler, Journal of Clinical Microbiology 29 (9): 1858 (1991) [MB 499349]

The point of describing this essentialist detective work here is to illustrate how extremely malleable Hyphomycetes may be under any selective pressure. After 33 years in human tissue (De Bruyn et al. 1985), CBS 518.82, which at the time of initial infection was probably a completely orthodox, rapidly growing *F. solani* capable of producing normal macroconidia and microconidia, had stably adapted to become a lump of aberrant chlamydospores with simplified, sparse conidiogenesis. (Evaluation of this hypothesis of rapid morphological adaptation awaits isolation of another representative of the exact same haplotype from nature.) Under immediate host selection, also, it showed hyperproduction of a pigment that may fortuitously have had some immunomodulatory effect. Fungal anthro- and naphthoquinone pigments, including some incompletely characterized metabolites from *F. solani* (You et al. 2013) as well as dermatophyte xanthomegnin (Alvi et al. 2000), may suppress the inducible nitric oxide synthase pathway of mammalian immune response. This protective pigment was soon lost or switched off in the culture collection. That hyphomycetes can be so extremely plastic under emergent selection concomitantly shows how remarkable the stability of their normal characters is.

Stability in fungal morphological characters can be achieved either through constant ecologically mediated stabilizing selection (rappellative) or through some type of developmental momentum, whether it be the functionally necessary coordination needed for morphogenesis to produce well-adapted structures, or the functionally superfluous vestigial holdovers from the production of previously functional structures (fungal equivalents of the vermiform appendix, if any). Conidiogenesis appears to be unusually rich in such arbitratives because (1) coordinated, stably reproducible development is required to produce ecologically optimized, morphologically stereotypical conidia and (2) there are relatively few rappellative exigencies that potentially come to bear on this process. Rappellative exigencies exerted on conidiogenous cells per se mainly constrain:

1. Structures capable of producing airborne conidia
2. Structures capable of producing mucoid conidia or water-borne conidia (many of these structures overlapping with items in the previous category, e.g. phialides)
3. Structures capable of producing intra-matrical and specialized host conidia
4. Structures capable of supporting very heavy conidia, unusually shaped conidia, or other conidial types exerting or subject to unusual physical stress or requiring unusual volumes of space (a theoretical possibility yielding few obvious examples; one that comes to hand, perhaps, is the production of large, rounded *Pleospora* [previously *Stemphylium*] conidia apically on well-separated annellides in a clade where most related genera, like *Bipolaris*, produce narrower conidia closely spaced together on tretic sympodulae)

This short list is in contrast to the considerably longer preliminary list of rappellative factors that can be compiled for the form of conidia themselves, including (with short mnemonic words for each category in parentheses),

1. Structures facilitating airborne dispersal, e.g. small size or small aerodynamically equivalent size, fragile attachment, surface roughening ('flying')

2. Structures and external chemistry physically facilitating arthropod dispersal ('sticking', 'hooking')
3. Structures and dimensions facilitating penetration of the physical boundary layer about a surface subject to moving air or water ('impaction', 'hooking', 'entangling')
4. Structures mediating buoyancy in water or affinity for the interface, where applicable ('floating')
5. Structures physically conferring resistance to grazing ('deterring')
6. Structures and chemistry (e.g. melanin) conferring resistance to physical factors such as drought and light ('resisting')
7. Structures facilitating germination of resistant forms ('splitting')
8. Structures facilitating detachment of mucously attached conidia from arthropod vectors ('detaching', examples being the pointed, crescent-shaped ends of many *Fusarium* macroconidia and, in general, the curvature of *Fusarium* and many *Neonectria* macroconidia)
9. Structures physically stabilizing the conidium on the conidiophore during formation ('stabilizing', e.g. the extended portion of the 'foot cells' of *Fusarium* macroconidia)
10. Structures and dimensions allowing storage of sufficient food reserves to germinate on recalcitrant substrata ('storing')

Thus, when one observes the forms of conidia, one mainly sees characters that directly reflect the physical ecology of the organism, sometimes intermixed with basic physical modifications influencing vector dispersal and anti-predation interactions. The influence of physical or ecological stressors is only minimally seen when one examines the forms of conidiogenous cells, particularly the aspects focused on in twentieth century taxonomy. This is the most likely explanation of why conidiogenous cells tend to remain true to form within lineages. The structures are probably not highly developmentally constrained: it is perhaps not a major genetic leap to interconvert sympodial and annellidic conidiogenesis, for example, and there are genera like *Pseudeurotium* where ambiguous structures resembling both of these categories have long puzzled investigators. Likewise, to diminish percurrent proliferation so that an annellide becomes a phialide may not be genetically 'difficult', if there is any selection pressure tending to impose such a transition. Because the conidiogenous cell structures, however, are probably so strongly equally functional in most habitats and niches, there are likely to be stronger interrappellative (coordination-based) genetic factors tending to maintain them than environmental (rappellative) or genetic (drift/mutational or general-rappellatively programmed-explorational) factors tending to alter them.

At this point in biosystematic history, we can look through the conidial microfungi and begin to determine how closely the taxa partly based on consistent conidiogenesis reflect the underlying phylogenetic structure. Ultimately, perhaps, a phylogenetic overview of conidiogenesis types in all ascomycetes and basidiomycetes, including lichenized forms, would be interesting, following the prototype set for ascomatal characters such as ascus dehiscence mechanisms in Schoch et al.'s

(2009) supplemental Fig. 6. Conidiogenesis structures are somewhat more variable than the ascomatal structures annotated there, and the result of such an effort would be compendious. A beautifully illustrated preliminary survey of the Hyphomycetes and their conidiogenesis has been published by Seifert et al. (2011) in the introductory overview to the Genera of Hyphomycetes. It excludes many items of interest in the context of this chapter, including yeasts, pycnidial fungi, and specialized host infection conidia.

Fungi of human medical interest make up a limited survey of species, mostly ascomycetous, that have by now been well covered by molecular biosystematics, at least at the preliminary level. Table 8.1 shows the patterns of conidiogenesis seen in a selection of these species, as well as a few of their close relatives and other fungi well known from the clinical laboratory and its impinging indoor air. Only ascomycetous fungi are included. The order followed is that seen in Schoch et al.'s Supplemental Fig. 6, and follows the biosystematic spine of the dendrogram (Schoch et al. 2009), not the cut-and-paste individual figures (6A, 6B, etc.) used to show details.

Chlamydospores that are not known to have a regular dehiscence mechanism are not included in Table 8.1.

As can be seen, the division of conidiogenesis into its airborne, moisture-borne, and intramatrical rappellatives considerably clarifies the significance of the diversity of forms. For example, Pleosporales producing the typical heavy, dark, leaf-impacting airborne conidia of the group mainly produce tretic-sympodial, cylindrical phragmoconidia or blastic or annellidic dictyoconidia. *Alternaria* (inclusive of *Ulocladium*) has an intermediate form, a tretically formed dictyoconidial type, with sympodial proliferation inconspicuous in some species. The variably disarticulating blastic sympodial chains of *Cladosporium*, producing single conidia, short chains, and longer chains, branched and unbranched, are an elegant adaptation to simultaneously make an extensive range of aerodynamically different configurations using a single mechanism. On the other hand, Pleosporales producing potentially arthropod- or splash-dispersed mucoid conidia tend to feature pycnidia with small phialides or annellides.

Phialides, an item found in members of most groups other than Onygenales and Saccharomycetes, are equally used for airborne and moisture-borne conidiation, but very rarely for intramatrical forms, the exception being found in a few organisms like *Lecythophora* with yeast-like repetitive conidiation carried out by phialidic cells. The Hypocreales are almost completely dedicated to phialidic conidiogenesis, except in a few cases where budding yeast forms are produced in infected insects (*Beauveria, Metarhizium*) or where warty, resistant aleurioconidia are produced (*Sepedonium, Mycogone*). We have classed the latter as substrate conidia even though they are produced on aerial hyphae in culture, in that their heavy form and relatively sturdy attachment suggest they may persist in the substrate (if soil) or near its location (if it was, for example, an infected agaric) rather than dispersing. Such structures may be adapted for heat resistance in fire ecosystems (Hambleton et al. 2005).

The overall impression given by Table 8.1 is that multiple class-level groups of conidial ascomycetes have reproduced, whether as symplesiomorphies or as a series

**Table 8.1** A survey of conidiogenesis types seen in a phylogenetic arrangement of medically important ascomycetous fungi, common clinical contaminants, and selected related genera

| Clade | Genus | Aerial conidia | Mucoid/aquatic conidia (including pycnidia with desiccating mucoid cirrhi) | Substrate or host conidia |
|---|---|---|---|---|
| Sordariomycetes Hypocreales | *Beauveria* | Blastic, sympodial, sometimes synnematal | | Blastic (budding yeast) |
| | *Purpureocillium* | Phialidic, chains with connectives | | |
| | *Stachybotrys* | Phialidic, chains | Phialidic, heads | |
| | *Metarhizium* | Phialidic, chains, sporodochial | | Blastic (budding yeast) |
| | *Fusarium* | Phialidic, chains | Phialidic, heads, single, and sporodochial | |
| | *Acremonium* | Phialidic, chains | Phialidic, heads | |
| | *Sarocladium* | Phialidic, chains | Phialidic, heads | |
| | *Trichothecium* | Blastic, retrogressive; Blastic, sympodial | Phialidic, heads | |
| | *Sepedonium* | | Phialidic, heads | Blastic, aleurioconidia |
| Sordariomycetes Microascales | *Scopulariopsis* | Annellidic, chains | | |
| | *Scedosporium* | | Annellidic, mucoid | |
| | *Ceratocystis* | Phialidic, chains | | Blastic, aleurioconidia |
| | *Thielaviopsis* | Phialidic, chains | | Blastic, aleurioconidia |
| Sordariomycetes Calosphaeriales | *Togninia* (*Phaeoacremonium*), *Pleurostoma* (*Pleurostomophora*) | | Phialidic, heads | |

(continued)

**Table 8.1** (continued)

| Clade | Genus | Aerial conidia | Mucoid/aquatic conidia (including pycnidia with desiccating mucoid cirrhi) | Substrate or host conidia |
|---|---|---|---|---|
| Sordariomycetes, Ophiostomatales | *Sporothrix* | Blastic, sympodial | | Blastic, aleurioconidia in substrate; Blastic budding yeast in host |
| | *Leptographium* | | Annellidic, mucoid | |
| Sordariomycetes, unnamed order | *Phialemoniopsis* | | Phialidic, heads, single, and sporodochial | |
| Sordariomycetes, Sordariales | *Chaetomium* (*Taifanglania*) | Phialidic, chains | | Blastic, aleurioconidia |
| Sordariomycetes, Coniochaetales | *Lecythophora* | | Phialidic, heads | Phialoconidiation, repetitive |
| Sordariomycetes Cephalothecaceae | *Phialemonium* | Phialidic, chains | Phialidic, heads | |
| Sordariomycetes, Xylariales | *Arthrinium* | Blastic, basauxic | | |
| | *Ascotricha* (*Dicyma*) | Blastic, sympodial | | |
| Leotiomycetes, Sclerotiniaceae | *Botrytis* | Blastic, simultaneous | | |
| Leotiomycetes, Helotiales | *Cadophora, Phialocephala* | | Phialidic | |
| Leotiomycetes, unnamed order | *Scytalidium* | Arthric-schizolytic | | |
| Leotiomycetes, Myxotrichaceae | *Oidiodendron, Geomyces* | Arthric-disjunctive | | |
| | *Amorphotheca* | Blastic, branching chains | | |
| Leotiomycetes, Pseudeurotiaceae | *Pseudeurotium* | | Blastic, obscurely sympodial | |

| | | | | |
|---|---|---|---|---|
| Eurotiomycetes, Chaetothyriales | *Exophiala/Rhinocladiella* (often synanamorphs) | Blastic, sympodial | Annellidic, blastic (budding yeast) | |
| | *Cladophialophora, Fonsecaea* | Blastic, sympodial branching chains | Phialidic | Fission cells schizolytic |
| | *Phialophora, Cyphellophora, Phaeomoniella* | | Phialidic | Fission cells schizolytic (*Phial. verrucosa*), Blastic budding cells (*Phaeom. zymoides,* etc.) |
| Eurotiomycetes, Eurotiales | *Aspergillus* | Phialidic, chains with connectives | | Blastic, aleurioconidia (*A. terreus*) |
| | *Talaromyces* | Phialidic, chains with connectives | | Fission cells schizolytic (*T. marneffei*) |
| | *Penicillium* | Phialidic, chains with connectives | | |
| | *Paecilomyces* | Phialidic, chains with connectives | | |
| | *Phialosimplex* | Phialidic, chains | Phialidic, mucoid heads | |
| | *Monascus* | Blastic, retrogressive | | Arthric –schizolytic |
| Eurotiomycetes, Onygenales | *Trichophyton, Microsporum, Epidermophyton* | Arthric-disjunctive (empty separating cell) | | Arthric –schizolytic |
| | *Histoplasma, Blastomyces, Paracoccidioides* | Arthric-disjunctive | | Blastic (budding yeast) |
| | *Coccidioides* | Arthric-disjunctive | | Thecate, endoconidial fission |
| | *Emmonsia* | Arthric-disjunctive | | Blastic (budding yeast) (*E. pasteuriana*) |

(continued)

**Table 8.1** (continued)

| Clade | Genus | Aerial conidia | Mucoid/aquatic conidia (including pycnidia with desiccating mucoid cirrhi) | Substrate or host conidia |
|---|---|---|---|---|
| Dothidiomycetes, Pleosporales | *Cochliobolus* (*Bipolaris, Curvularia*) | Tretic (blastic-cicatrized) with sympodial proliferation | | |
| | *Alternaria* | Tretic (blastic-cicatrized) with sympodial proliferation | | |
| | *Pleospora* (*Stemphylium*) | Tretic (blastic-cicatrized) with percurrent proliferation | | |
| | *Phoma* | | Phialidic, in pycnidia | |
| | *Coniothyrium* | | Annellidic, in pycnidia | |
| | *Nigrograna* | | Phialidic, in pycnidia | |
| | *Paraconiothyrium* | | Phialidic, in pycnidia | |
| | *Epicoccum* | Blastic or polyblastic aleurioconidia in sporodochia | Phialidic, in pycnidia | |
| Dothidiomycetes, Botryosphaeriales | *Lasiodiplodia* | | Blastic, in pycnidia | |
| | *Neoscytalidium* | Arthric–schizolytic | Blastic, in pycnidia | |
| Dothideomycetes, Dothideales | *Aureobasidium* | | Blastic, synchronous | |
| Dothidiomycetes, Capnodiales, Davidiellaceae | *Cladosporium* | Blastic, cicatrized, in sympodial branching chains | | |

| | | | | |
|---|---|---|---|---|
| Dothidiomycetes, Capnodiales, Teratosphaer iaceae | *Hortaea* | | Annellidic | Annellidic yeast-like particulate assimilative phase |
| | *Stenella* | Blastic, sympodial | | |
| Dothidiomycetes, Capnodiales, Mycosphaerellaceae | *Septoria* | | Blastic, sympodial, with few annellations, in pycnidia | |
| Saccharomycetes, Saccharomycetales | *Candida, Saccharomyces* | | Blastic; mainly budding yeast | Blastic; mainly budding yeast |
| Taphrinomycota, Pneumocystidiales | *Pneumocystis* | | | Fission cells, schizolytic |

of apomorphies, most of the basic possibilities of conidiogenous construction, viz., phialide, blastic, annellidic, sympodial, and the two major forms of arthric conidiogenesis, schizolytic and disjunctive [of which 'rhexolytic', in the strict sense, is a subcategory that lacks obvious autolytic processes, fide (Sigler 1989). At the same time, there is extensive stability seen, with orders, families, and large generic complexes sharing consistent patterns in conidiogenesis, sometimes partitioned to different mechanisms for airborne, liquid-borne, and intramatrical dispersal. In the future, genome sequencing and subsequent data mining studies, combined with genetic knockout and targeted mutation, may clarify how many times features such as the percurrent proliferation seen in annellides have evolved independently. This type of study will yield insight not only into the degree of convergent evolution that has occurred historically in conidiogenous morphogenesis, but also into the degree of intrinsic informational complexity involved in engendering interrappellative stability in the continuing production of such structures. If features like annellidic and sympodial proliferations turn out to be long-lasting, highly conserved symplesiomorphies that are, in effect, switched on and off within major lineages, that suggests that these processes entail complex coordination that is not easily convergently evolved. If, on the other hand, such features emerge repeatedly as apomorphies on the genetic level, then the degree of coordination involved in the morphogenetic processes involved should be found to be relatively low.

As a final note, to anyone speculating that this information is all less interesting than a dendrogram precisely locating conidial fungi into their evolutionary historical context, we would only point out that lectical considerations apply just as much to DNA sequences as to morphological forms. Pure lineal history is indicated by arbitrations, minimally constrained DNA mutations, assumed to reflect, via an assumption of near-randomness, a time clock. When aligning regions like ribosomal DNA sequences, one sees regions that are 'conserved', highly functionally constrained, either via interrappellation (selection for correct ribosomal folding and coordination with other biomolecules) or rappellation (assuming a configuration with adequate tolerance of heat, osmotic stress, and efficient speed of protein production). In coding genes, the third bases of codons may, of course, be a fraction less constrained. These areas are subject to rappellative and interrappellative stabilizing selection and, when they do evolve significantly, may do so in a punctuated way, where a change in a small number of items may entail an acceleration of coordination-optimizing changes in many others. One evolutionary question that we know next to nothing about is whether rappellative exigencies or opportunities can entail significant convergent evolution even in DNA sequences. For example, if two organisms evolve a lactase enzyme completely independently of one another, is there sufficient fungibility in the potential form of the actively catalytic protein regions involved that the sequence of enzyme A can be completely different from that of enzyme B? What about cases where the organisms involved are related at the level of the taxonomic class and share some common genetic background—will their independently evolved lactases be highly likely to share sequence motifs? Nothing, of course, *necessarily* constrains such a predisposed convergence as the organism's only solution to the opportunity presented by lactose, but the actual responses of organisms to such opportunities are subject to statistics and probability. We know relatively little, so far, about the statistics of convergent and semi-convergent (convergence from shared symplesiomorphic basis) evolution in DNA sequences coding for functional proteins. We do not know the extent to which including coding regions in phylogenies will, in effect, lead us back towards a systematics based on trees, shrubs, and herbs.

In marked contrast to the constrained sequence regions, some areas within spacer regions are so labile that they may be poorly alignable except within tightly related clades. They are subject to coincidental convergence at the level of the individual base pair. Microsatellite repeat regions likewise are predictably vulnerable to statistical coincidence, e.g. when investigators contrast lineages whose sequences show six repeat motifs to those with sequences showing seven repeats. Pure neutrality, arbitrative condition 2a, above, may be inscrutable when it is found in DNA sequences.

Fortunately, there are regions of possible low-specificity rappellative and/or interrappellative constraint, as seen in the more highly alignable regions of the ribosomal internal transcribed spacer. Mutation about these regions may be constrained by inter- and intramolecular interactions arising from functionally necessary nucleic acid folding or twisting, or from coordination (whether amino-acid-specific or at the general van der Waals level) with regulatory proteins or nucleic acids, or from a need for precise physical qualities (e.g. pH). The somewhat constrained DNA-base fungibility applying to these regions arguably provides the best available molecular clock for dating the divergence of lineages. Ultimately, understanding of the subcategories of constraint on DNA sequences may provide a mathematical means of selecting regions that truly give the best independent phylogenetic signal, optimally free (to use metaphors) of both the flywheel noise arising from excessive mutation and the brake noise arising from excessive rappellative constraint.

The bottom line, however, is that DNA is a highly complex, self-interacting construction template and, as such, is a vast repository of developmental, interrappellatively modulated taxonomic characters. It, thus, provides the same fundamental type of information that the study of conidiogenous mechanisms provided, but does so in much higher quantity. Awareness of this typological correspondence should allow us to maintain our respect for the insights Hughes and contemporaries gained from studying conidiogenesis. Even if such revelatory structures are not seen as the mycelial needlework of God, their discovery should still stimulate due awe at the ability of good scientists to unlock the deep secrets of reality.

## References

Alvi KA, Baker DD, Stienecker V, Hosken M, Nair BG (2000) Identification of inhibitors of inducible nitric oxide synthase from microbial extracts. J Antibiot (Tokyo) 53(5):496–501

Cole GT (1986) Models of cell differentiation in conidial fungi. Microbiol Rev 50:95–132

De Bruyn HP, Broekman JM, de Vries GA, Klokke AH, Greep JM (1985) Een patiënt met eumycetoma in Nederland. Ned Tijdschr Geneeskd 129:1099–1101

de Hoog GS, Nishikaku AS, Fernandez-Zeppenfeldt G, Padín-González C, Burger E, Badali H et al (2007) Molecular analysis and pathogenicity of the *Cladophialophora carrionii* complex, with the description of a novel species. Stud Mycol 58:219–234
De Souza W, Benchimol M (2005) Basic biology of *Pneumocystis carinii*—a mini review. Mem Inst Oswaldo Cruz, Rio de Janeiro 100:903–908
de Vries GA, de Hoog GS, de Bruyn HP (1984) *Phialophora cyanescens* sp. nov., with *Phaeosclera*-like synanamorph, causing white-grain mycetoma in man. Antonie Van Leeuwenhoek 50:149–153
Esquenazi D, Alviano CS, de Souza W, Rozental S (2004) The influence of surface carbohydrates during in vitro infection of mammalian cells by the dermatophyte *Trichophyton rubrum*. Res Microbiol 155:144–153
Fayod V (1889) Prodrome d'une histoire naturelle des Agaricinées. Ann Sci Nat Botan VII 9:181–411
Fries E (1825) Systema orbis vegetabilis. Primas lineas novae constructionis periclitatur Elias Fries. Pt. 1. Plantae homonemeae. Typographia Academica, Lund, Sweden
Gams W (1971) *Cephalosporium*-artige Schimmelpilze (Hyphomycetes). Gustav Fischer Verlag, Stuttgart
Guimarães AJ, de Cerqueira MD, Nosanchuk JD (2011) Surface architecture of *Histoplasma capsulatum*. Front Microbiol 2:225
Hambleton S, Nickerson NL, Seifert KA (2005) *Leohumicola*, a new genus of heat-resistant hyphomycetes. Stud Mycol 53:29–52
Hashimoto T, Emyanitoff RG, Mock RC, Pollock JH (1984) Morphogenesis of arthroconidiation in the dermatophyte *Trichophyton mentagrophytes* with special reference to wall ontogeny. Can J Microbiol 30:1415–1421
Heath MC (2000) Non-host resistance and non-specific plant defences. Curr Opin Plant Biol 3:315–319
Hughes SJ (1953) Conidiophores, conidia and classification. Can J Bot 31:577–659
Kimura M (1983) The neutral theory of molecular evolution. Cambridge University Press, Cambridge
Lee A, Langer R (1983) Shark cartilage contains inhibitors of tumor angiogenesis. Science 221:1185–1187
Lombard L, van der Merwe NA, Groenewald JZ, Crous PW (2014) Generic concepts in Nectriaceae. Stud Mycol 80:189–245
Luer CA, Luer WH (1982) Acute and chronic exposure of nurse sharks to aflatoxin B1. Fed Proc 41:925
Ostrander GK, Cheng KC, Wolf JC, Wolfe MJ (2004) Shark cartilage, cancer and the growing threat of pseudoscience. Cancer Res 64:8485–8491
Peñalva MA, Moya A, Dopazo J, Ramón D (1990) Sequences of isopenicillin N synthetase genes suggest horizontal gene transfer from prokaryotes to eukaryotes. Proc Biol Sci 241:164–169
Ray J (1682) Methodus plantarum nova. H. Faitborne & J. Kersey, London
Ray J (1691) The wisdom of God manifested in the works of the creation. Samuel Smith, London
Rogerson M (2013) This moonless sky. FriesenPress, Vancouver
Saccardo PA (1886) Sylloge fungorum 4:1–807. Pavia, Italy
Schoch CL, Sung G-H, López-Giráldez F, Townsend JP, Miadlikowska J, Hofstetter V et al (2009) The Ascomycota tree of life: a phylum wide phylogeny clarifies the origin and evolution of fundamental reproductive and ecological traits. Syst Biol 58:224–239
Seifert K, Morgan-Jones G, Gams W, Kendrick B (2011) The genera of Hyphomycetes. CBS-KNAW Fungal Biodiversity Centre, Utrecht
Sigler L (1989) Problems in application of the terms 'blastic' and 'thallic' to modes of conidiogenesis in some onygenalean fungi. Mycopathologia 106:155–161
Singer R (1986) The Agaricales in modern taxonomy, 4th edn. Koeltz Scientific Books, Koenigstein, 981 p
Summerbell RC (2000) Form and function in the evolution of dermatophytes. In: Kushwaha RKS, Guarro J (eds) Biology of dermatophytes and other keratinophilic fungi. Revista Iberoamericana de Micologia supplement, Bilbao, Spain, pp 30–43

Summerbell RC, Schroers HJ (2002) Analysis of phylogenetic relationship of *Cylindrocarpon lichenicola* and *Acremonium falciforme* to the *Fusarium solani* species complex and a review of similarities in the spectrum of opportunistic infections caused by these fungi. J Clin Microbiol 40:2866–2875

Summerbell RC, Gueidan C, Schroers H-J, de Hoog GS, Starinck M, van Iperen A, Rehner S, Arocha Rosete Y, Guarro J, Scott JA (2011) *Acremonium* phylogenetic overview and revision of *Gliomastix, Trichothecium* and *Sarocladium*. Stud Mycol 68:139–162

Tsuneda A, Hambleton S, Currah RS (2004) Morphology and phylogenetic placement of *Endoconidioma*, a new endoconidial genus from trembling aspen. Mycologia 96:1128–1135

Vogt G (2008) How to minimize formation and growth of tumours: potential benefits of decapod crustaceans for cancer research. Int J Cancer 123:2727–2734

You X, Feng S, Luo S, Cong D, Yu Z, Yang Z, Zhang J (2013) Studies on a rhein-producing endophytic fungus isolated from *Rheum palmatum* L. Fitoterapia 85:161–168

Zoutman DE, Sigler L (1991) Mycetoma of the foot caused by *Cylindrocarpon destructans*. J Clin Microbiol 29:1855–1859

# Chapter 9
# Fungal Diversity of Central and South America

**Rafael F. Castañeda-Ruiz, Gabriela Heredia, Luis F.P. Gusmão, and De-Wei Li**

## Introduction

Since the last two decades of the twentieth century, numerous novelties of hyphomycetes have been published from tropical Central and South American forests and semiarid areas. These fungi were collected from tropical forests, jungles, and semiarid vegetation as well as from decaying plant debris accumulated on the leaf litter surface or submerged in streams and rivers (Castañeda-Ruiz et al. 1998; da Cruz et al. 2008; Gusmão et al. 2008; Heredia et al. 2007; Monteiro et al. 2014, 2015; Pereira-Carvalho et al. 2009). The quantity and diversity of fungi found suggests the

R.F. Castañeda-Ruiz, Ph.D. (✉)
Instituto de Investigaciones Fundamentales en Agricultura, Tropical "Alejandro de Humboldt" (INIFAT), Académico Titular de la Academia de Ciencias de Cuba, Calle 1 Esq. 2, Santiago de Las Vegas, C. Habana C. P. 17200, Cuba
e-mail: rfcastaneda@inifat.co.cu; rfcastanedaruiz@gmail.com

G. Heredia, Ph.D.
Red de Biodiversidad y Sistemática, Instituto de Ecología, A. C. Km. 2.5 Antigua Carretera a Coatepec No. 351 Congregación El Haya, C.P. 91070 Xalapa, Veracruz, Mexico
e-mail: gabriela.heredia@inecol.mx

L.F.P. Gusmão, Ph.D.
Laboratorio de Micologia, Departamento de Ciências Biológicas, Universidade Estadual, de Feira de Santana, BR116 KM03, 44031-460 Feira de Santana, BA, Brazil
e-mail: lgusmao@uefs.com

D.-W. Li, Ph.D.
The Connecticut Agricultural Experiment Station Valley Laboratory, 153 Cook Hill Road, Windsor, CT 06095, USA

Co-Innovation Center for Sustainable Forestry in Southern China, Nanjing Forestry University, Nanjing, Jiangsu 210037, China
e-mail: Dewei.Li@ct.gov



D.-W. Li (ed.), *Biology of Microfungi*, Fungal Biology,
DOI 10.1007/978-3-319-29137-6_9

existence of an indefinite number of undiscovered fungi still remaining in the tropical and subtropical forests in these geographical regions (Mueller and Schmit 2007) according to the estimation that 1–1.5 million of fungi are present on Earth (Hawksworth 1990, 2001).

In this chapter, the authors discuss the novelties of fungal diversity of Central and South America described during 2000–2015, with a focus on hyphomycetes, and provide the detailed methodology proposed by Castañeda-Ruiz (2005) and used by the authors in the collection and incubation of samples colonized by hyphomycetes.

## Collection of Samples

The choice of location to collect study materials depends on the collection area and its surroundings. Small streams in the forests of mountainous regions sometimes produce cracks in the soil where plant residues accumulate. The plant materials are often in the areas near the course of the stream subject to constant washing. Generally, the degree of colonization in these substrates is lower than those on the sides or near the shore, farther from the current course, and receiving a soft wash, sometimes almost imperceptible. The fragments of bark, branches, pods, flowers, and other debris submerged in relatively quiet areas close to the rapids, cascades, and waterfalls of the stream where bubbles or foam accumulates are ideal for collecting hyphomycete-colonized materials (Figs. 9.1, 9.2, 9.3, 9.4 and 9.5).

The samples may consist of leaves, twigs, bark, stems, flowers, pods, or other structures of plant origins and are collected in plastic bags or plastic flasks for transportation to the laboratory.

The terrestrial samples of the hyphomycete-colonized plant materials are usually found on steep hillsides, which are clad in dense, almost impenetrable vegetation. The trees are covered in epiphytic bromeliads and orchids, and the dead leaves that yield many of the new hyphomycetes are found lying in small crevices in the rocks, or hidden underneath the branches of shrubs or creepers. Cryptic, undisturbed microhabitats appear to maintain the leaves, branches, bark, and other plant debris at a relatively high humidity during much of the year (Castañeda-Ruiz and Kendrick 1991). Numerous uncommon or new hyphomycetes can be collected under the big trunks of the fallen trees in the rainforest, and sometimes several hyphomycetes are growing on a piece of debris about 2–3 $cm^2$ (Castañeda-Ruiz et al. 2011).

### *Sample Washing*

Undesirable particles and sediments are washed off with tap water. It is important that the jet of tap water does not directly hit the plant materials because in some cases the material is weakly colonized by the assimilative hyphae. It should be enough to let a flow of water passing through the whole sample for 20–45 min to

**Fig 9.1** Collection site of tropical hyphomycete diversity

**Fig 9.2** Collection sites (a different site) of tropical hyphomycete dicersity

**Fig. 9.3** A boat view showing plant debris accumulating along river banks in a tropical area

**Fig. 9.4** Sampling materials on the soil in the riparian zone in a rainforest

**Fig. 9.5** Sampling aquatic hyphomycetes

wash off surface soil particles, protozoa, nematodes, insect, and other organisms in the samples. A soil sieve and a vegetable washing basket are used to wash different diameters and sizes of the samples for the study of anamorphic fungi, since tiny colonies in fragments or reduced size can be recovered and will not be lost (Fig. 9.6).

## *Sample Drying*

Plant debris is placed on a paper towel or a filter paper for 1–2 h to remove excess water before putting them in moist chambers or sterile water according to the origin of the sample and the ecological group of hyphomycetes studied. The tap water film on the surface should completely disappear before the incubation (Figs. 9.7, 9.8 and 9.9).

## *Incubation*

The washed plant material is divided or cut into two parts. A part of the washed material is dipped into plates or containers filled with sterile distilled water in a Petri dish and placed in an incubator at a temperature of 20–25 °C or a cooler (Fig. 9.10) at room temperature.

**Fig. 9.6** Samples washed with tap water

**Fig 9.7** Samples placed on paper towels to remove excessive water

**Fig 9.8** A close view of the samples placed on paper towels to remove excessive water

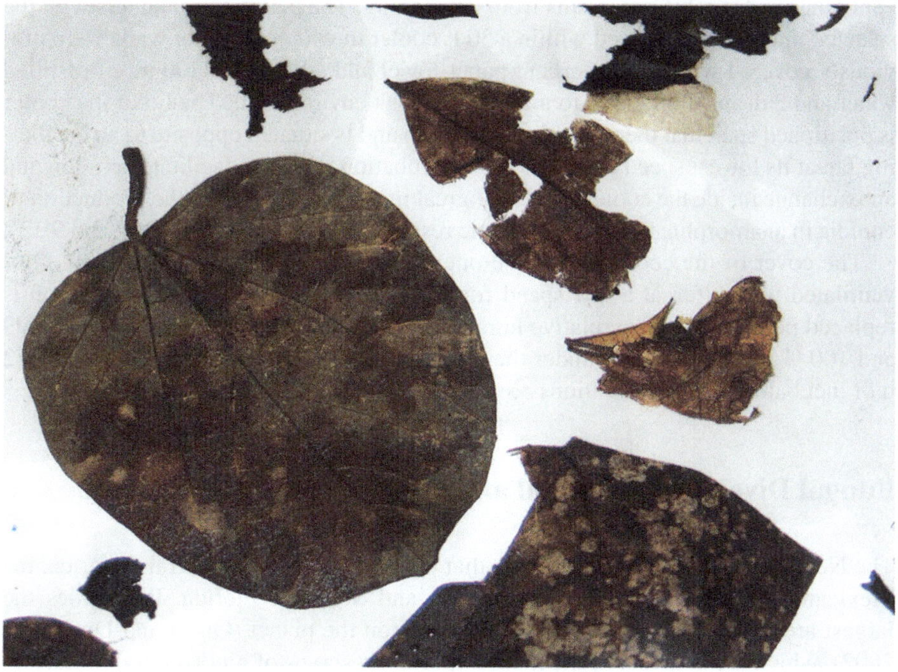

**Fig 9.9** Another close view of the samples placed on paper towels to remove excessive water

**Fig. 9.10** A cooler used for sample incubation

The other part of the sample is placed in a Petri dish or a flask with a perforated lid prepared with a filter paper and absorbent paper or cotton at the bottom. Several drops of water are added to provide the necessary moisture within the container. The surface of litter incubated should be free of water to facilitate sporulation and prevent other undesirable organisms from developing. The plates or containers with the selected materials are placed within a 50 L cooler in which the inner walls were previously covered with wet absorbent paper towel and added 100–400 ml of distilled water and 2 drops of glycerol to maintain a moist environment. The lid of the cooler is positioned such that there is a slit of 2–3 mm and is situated opposite to an oscillating fan at its lowest speed throughout the incubation period to facilitate aeration and air exchange inside the cooler. This is of great importance to ensure the production of conidia in anamorphic fungi from aquatic resources (Figs. 9.10, 9.11, 9.12 and 9.13).

The cover of the cooler should be open for 30 min after 24 h of incubation or ventilated with a fan at a low speed for 30 min. The water lost by evaporation is replaced periodically. The relative humidity in the cooler is maintained between 99 and 100 %, the observation under the stereoscopic microscope starts after 48 or 72 h of incubation, and observations are repeated for up to 30 days.

## Fungal Diversity in Central and South American Neotropic

The Neotropic is a terrestrial ecozone that includes South and Central America, the Mexican lowlands, the Caribbean islands, and Southern Florida. It includes the largest area of tropical and subtropical forests on the planet (Olson and Dinerstein 2002), which are considered the most important reserves of biodiversity on Earth.

**Fig. 9.11** Petri dishes with plant materials on the filter papers in the incubating cooler

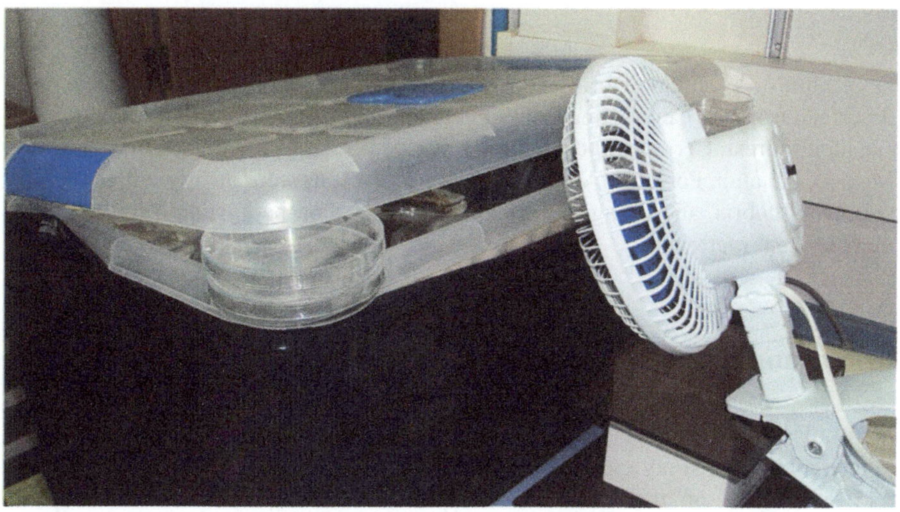

**Fig. 9.12** The partially opened incubating cooler with a fan for ventilation

Among the 17 countries considered as megadiverse, six are located in the Neotropic including Brazil, Colombia, Ecuador, Mexico, Peru, the United States, and Venezuela. The diversity of microfungi inhabiting the rainforest and the correlations with host plants and their debris have been estimated in several studies (Hawksworth 2001; Blackwell 2011), and the authors have remarked on the importance of collecting and documenting the fungal diversity of tropical regions because such regions have the greatest diversity of plants and we can therefore also expect a

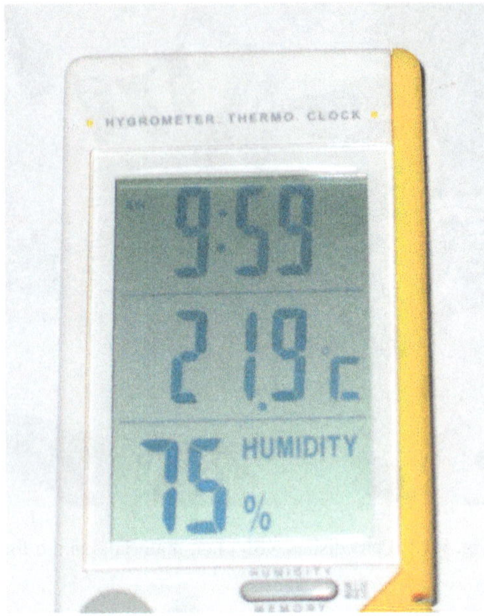


**Fig. 9.13** A hygrometer used to monitor temperature and relative humidity in the incubating cooler

large number of fungi. During the last 30 years, 3757 taxa of microfungi were described from the Central and South American Neotropic, mainly from the rainforest, cloud forest, Cerrado, and semiarid areas of Brazil, Costa Rica, Cuba, Ecuador, Mexico, Panama, Peru, and Venezuela. These novelties were mostly collected from soil, plant debris, and living plants, sometimes as plant pathogenic fungi and submerged plant materials.

## Microfungi from Living Plants

Most of the microfungi novelties were described from living plants as plant pathogenic fungi, sooty molds, and endophytic fungi or associated with trichomes; 445 new taxa of the genera *Cercospora*, *Cercostigmina*, *Distocercospora*, *Mycovellosiella*, *Passalora*, and *Pseudocercospora* and other cercosporoid fungi were described by numerous authors such as Braun and Castañeda-Ruiz (1991), Braun and Urtiaga (2012, 2013a, 2013b), Hernández-Gutiérrez and Dianese (2008), Castañeda-Ruiz and Braun (1989), Crous et al. (1997, 1999), and Pons and Sutton (1988). Other nectrioid fungi associated with living plants were discovered mainly from Brazil by Alfenas et al. (2013) and Crous et al. (1993, 1998). Sooty molds and other epiphyllous fungi were treated by Pinho et al. (2012a, 2012b, 2013, 2014), Rodríguez-Hernández (1985), Rodríguez-Hernández and Camino (1987), Rodríguez and Piepenbring (2007), and Rodríguez et al. (2015). Several interesting novel genera as *Alveariospora*, *Echinoconidiophorum*, *Globoconidiopsis*, *Globoconidium*,

*Helminthosporiomyces*, *Microtrichosphaera*, *Phaeoidiomyces*, *Phragmoconidium*, *Trichomatoclava*, *Trichomatomyces*, *Trichosporodochium*, and *Vesiculohyphomyces* occurring on trichomes of plants in diverse localities of Brazil were added by da Silva et al. (2012), Dornelo-Silva and Dianese (2004), and Pereira-Carvalho et al. (2009), and some of these genera showed a curious combination of sympodial and percurrent extensions of conidial development.

## Microfungi from Soil, Plant Debris, and Submerged Plant Materials

Along with its enormous diversity and endemism, Neotropical forests are subject to many kinds of threats, which occur in other tropical regions of the world, consisting of conversion of natural habitats to farmland and grasslands and forest degradation as a result of overexploitation of hunting and logging (Zunino and Zullini 2003).

The documentation of fungal diversity-associated plant debris, soil, and submerged plant materials in the forests of the Central and South American Neotropic has increased due to the effort of mycologists of several institutions in Brazil, Cuba, Mexico, Venezuela, and other countries. They described more than 2800 taxa which included 104 new genera (Table 9.1, Figs 9.14, 9.15, 9.16, 9.17, 9.18, 9.19, 9.20,

**Table 9.1** Genera of hyphomycetes and coelomycetes described from soil, plant debris, and submerged plant materials from Central and South America (1985–2015)

| Genera |
|---|
| *Acarocybiopsis* J. Mena et al. 1999 |
| *Amoenomyces* R.F. Castañeda et al. 1996 |
| *Amphophialis* R.F. Castañeda et al. 1998 |
| *Ampullicephala* R.F. Castañeda et al. 1998 |
| *Anabahusakala* L.T. Carmo et al. 2014 |
| *Anacraspedodidymum* C.R. Silva et al. 2014 |
| *Anaexserticlava* T.S. Santa Izabel et al. 2015 |
| *Anaselenosporella* Heredia et al. 2009 |
| *Anaseptoidium* Heredia et al. 2012 |
| *Ancorasporella* J. Mena et al. 1998 |
| *Anungitopsis* R.F. Castañeda & W.B. Kendr. 1990 |
| *Arachnospora* R.F. Castañeda et al. 2003 |
| *Arthrodochium* R.F. Castañeda & W.B. Kendr. 1990 |
| *Arthrowallemia* R.F. Castañeda et al. 1998 |
| *Atrogeniculata* J.S. Monteiro et al. 2014 |
| *Barretomyces* Klaubauf et al. 2014 |
| *Bibanasiella* R.F. Castañeda & W.B. Kendr. 1991 |
| *Bisseomyces* R.F. Castañeda 1985 |
| *Brachycephala* J.S. Monteiro et al. 2015 |
| *Brachyconidiella* R.F. Castañeda & W.B. . Kendr. 1990 |

(continued)

**Table 9.1** (continued)

| Genera |
|---|
| *Brachyconidiellopsis* Decock et al. 2004 |
| *Brevicatenospora* R.F. Castañeda et al. 2006 |
| *Briansuttonia* R.F. Castañeda et al. 2004 |
| *Brunneodinemasporium* Crous & R.F. Castañeda 2012 |
| *Bulbocatenospora* R.F. Castañeda & Iturr. 2000 |
| *Carrismyces* R.F. Castañeda & Heredia 2000 |
| *Castanedaea* W.A. Baker & Partr. 2001 |
| *Cheiromyceopsis* Mercado & J. Mena 1988 |
| *Chlamydopsis* Hol.-Jech. & R.F. Castañeda 1986 |
| *Chrysofolia* Crous & M.J. Wingf. 2015 |
| *Circinoconiopsis* A. Hern.-Gut.2014 |
| *Cubasina* R.F. Castañeda 1986 |
| *Curvicladiella* Decock & Crous 2006 |
| *Cylindrosympodium* W.B. Kendr. & R.F. Castañeda 1990 |
| *Dentocircinomyces* R.F. Castañeda & W.B. Kendr.1990 |
| *Digicatenosporium* S. M. Leão et al. 2015 |
| *Digitoramispora* R.F. Castañeda & W.B. Kendr. 1990 |
| *Distophragmia* R.F. Castañeda et al. 2015 |
| *Ellisembiopsis* T.S. Santa Izabel & Gusmão 2013 |
| *Elotespora* R.F. Castañeda & Heredia 2010 |
| *Endogenospora* R.F. Castañeda et al. 2010 |
| *Enridescalsia* R.F. Castañeda & Guarro 1998 |
| *Falcocladium* S.F. Silveira et al. 1994 |
| *Helensiella* Minter et al. 2015 |
| *Helicodochium* J.S. Monteiro et al. 2014 |
| *Herreromyces* R.F. Castañeda & W.B. Kendr. 1991 |
| *Hyalocladosporiella* Crous & Alfenas 2014 |
| *Hyalopleiochaeta* R.F. Castañeda et al. 1996 |
| *Idriellopsis* Hern.-Restr. & Crous 2015 |
| *Inesiosporium* R.F. Castañeda & W. Gams 1997 |
| *Inifatiella* R.F. Castañeda 1985 |
| *Kostermansindiopsis* R.F. Castañeda 1985 |
| *Lauriomyces* R.F. Castañeda 1990 |
| *Luxuriomyces* R.F. Castañeda 1988 |
| *Luzfridiella* R.F. Castañeda & W.B. Kendr. 1991 |
| *Menidochium* R.F. Castañeda & W.B. Kendr. 1990 |
| *Mercadomyces* J. Mena 1988 |
| *Micropustulomyces* R. W. Barreto 1995 |
| *Minimelanolocus* R.F. Castañeda & Heredia 2001 |
| *Minteriella* Heredia et al. 2012 |
| *Mirimyces* Nag Raj 1993 |
| *Mycelephas* R. F. Castañeda 2009 |

(continued)

**Table 9.1** (continued)

| Genera |
|---|
| *Nagrajia* R.F. Castañeda & W.B. Kendr. 1991 |
| *Neopenidiella* Quaedvlieg & Crous 2013 |
| *Neosporidesmium* Mercado & J. Mena 1988 |
| *Nigrolentilocus* R.F. Castañeda & Heredia 2001 |
| *Paraceratocladium* R.F. Castañeda 1987 |
| *Paraidriella* Hern.-Restr. & Crous 2015 |
| *Paraulocladium* R.F. Castañeda 1986 |
| *Perelegamyces* R.F. Castañeda & W.B. Kendr. 1990 |
| *Phaeocandelabrum* R.F. Castañeda et al. 2009 |
| *Phaeomoniella* R.F. Castañeda et al. 2007 |
| *Phaeoschizotrichum* Silva et al. 2015 |
| *Phellinocrescentia* Crous & Decock 2014 |
| *Phialosporostilbe* Mercado & J. Mena 1985 |
| *Phragmospathulella* J. Mena & Mercado 1986 |
| *Physalidiopsis* R.F. Castañeda & W.B. Kendr. 1990 |
| *Piricaudiopsis* J. Mena & Mercado 1987 |
| *Polyphialoseptoria* Quaedvlieg et al. 2013 |
| *Porobeltraniella* Gusmão 2004 |
| *Pseudocanalisporium* R.F. Castañeda & W.B. Kendr. 1991 |
| *Pycnovellomyces* R.F. Castañeda 1987 |
| *Readerielliopsis* Crous & Decock 2015 |
| *Repetoblastiella* R.F. Castañeda et al. 2010 |
| *Rogergoosiella* A. Hern.-Gut. & J. Mena 1996 |
| *Sanjuanomyces* R.F. Castañeda & W.B. Kendr. 1991 |
| *Scolecobeltrania* Iturr. et al. 2013 |
| *Selenodriella* R. F. Castañeda & W.B. Kendr. 1990 |
| *Selenosporopsis* R. F. Castañeda & W.B. Kendr. 1991 |
| *Solicorynespora* R. F. Castañeda & W.B. Kendr. 1990 |
| *Soloacrospora* W.B. Kendr. & R.F. Castañeda 1991 |
| *Stephembruneria* R. F. Castañeda 1988 |
| *Ticosynnema* R. F. Castañeda et al. 2013 |
| *Trinosporium* Crous & Decock 2012 |
| *Venustosynnema* R. F. Castañeda & W.B. Kendr. 1990 |
| *Veracruzomyces* Mercado et al. 2002 |
| *Wojnowiciella* Crous et al. 2015 |
| *Xenocylindrocladium* Decock et al. 1997 |
| *Xenopenidiella* Quaedvlieg & Crous 2014 |
| *Ypsilomyces* D.A.C. Almeida & Gusmão 2014 |
| *Zeloasperisporium* R. F. Castañeda 1996 |
| *Zelodactylaria* A.C. Cruz et al. 2012 |
| *Zelosatchmopsis* Nag Raj 1991 |
| *Zelotriadelphia* R. F. Castañeda et al. 2005 |

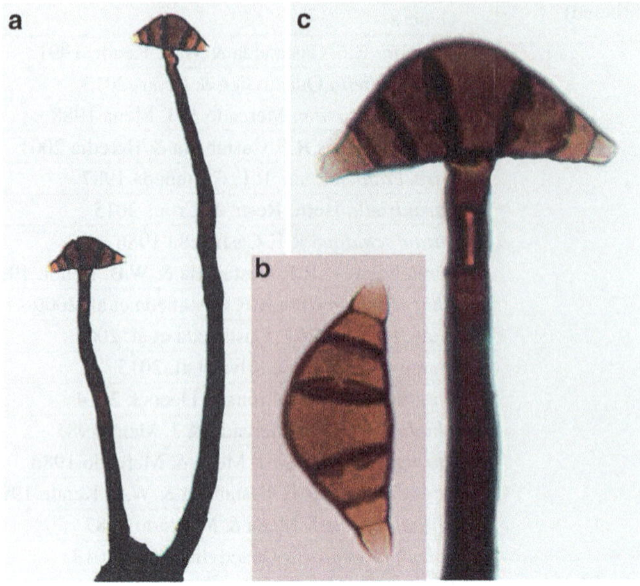


**Fig. 9.14** *Ancorasporella mexicana*. (**a**) Conidiophores and conidia. (**b**) Conidium. (**c**) Conidium and conidiogenous cell

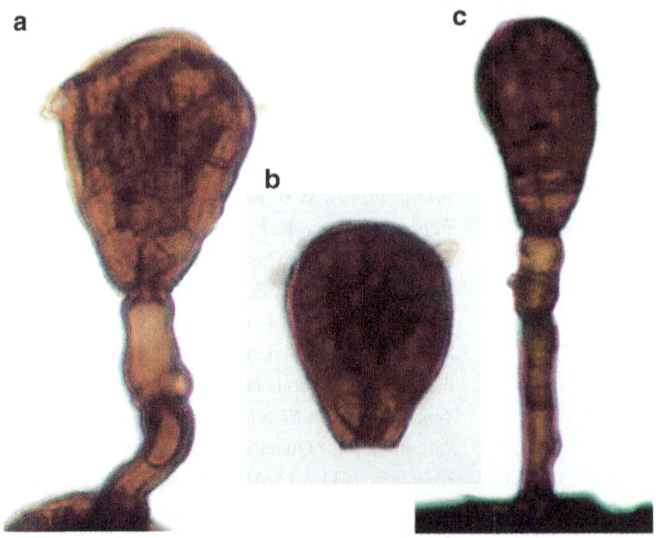


**Fig. 9.15** *Carrismyces proliferatus*. (**a**) Conidiophore and conidium. (**b**) Conidium. (**c**) Conidiophore and conidium

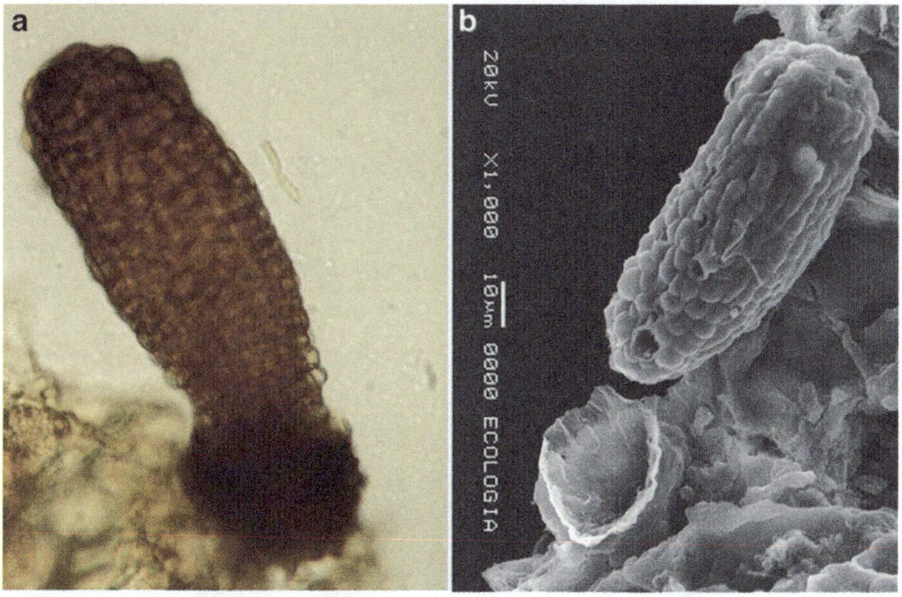


**Fig. 9.16** *Elotespora mexicana*. (**a**) Conidioma and conidium and conidium. (**b**) Conidioma "cup" and conidium

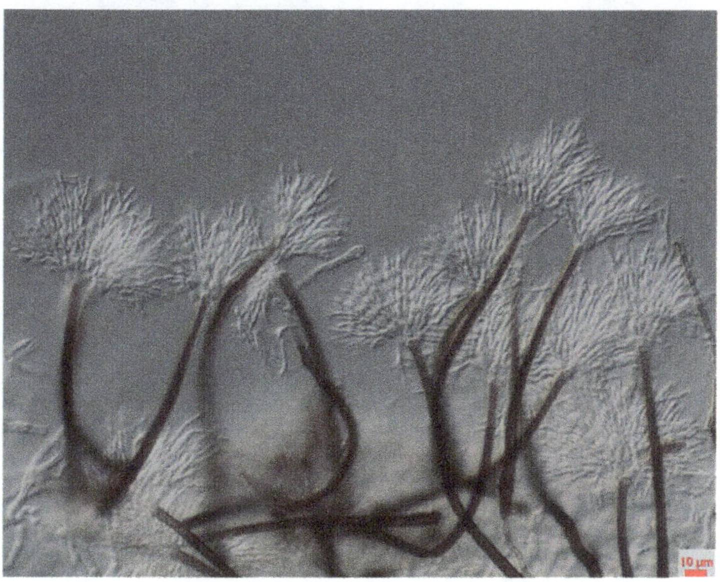


**Fig. 9.17** *Lauriomyces bellulus*. (**a**) Conidiophores, conidiogenous cells, and conidia

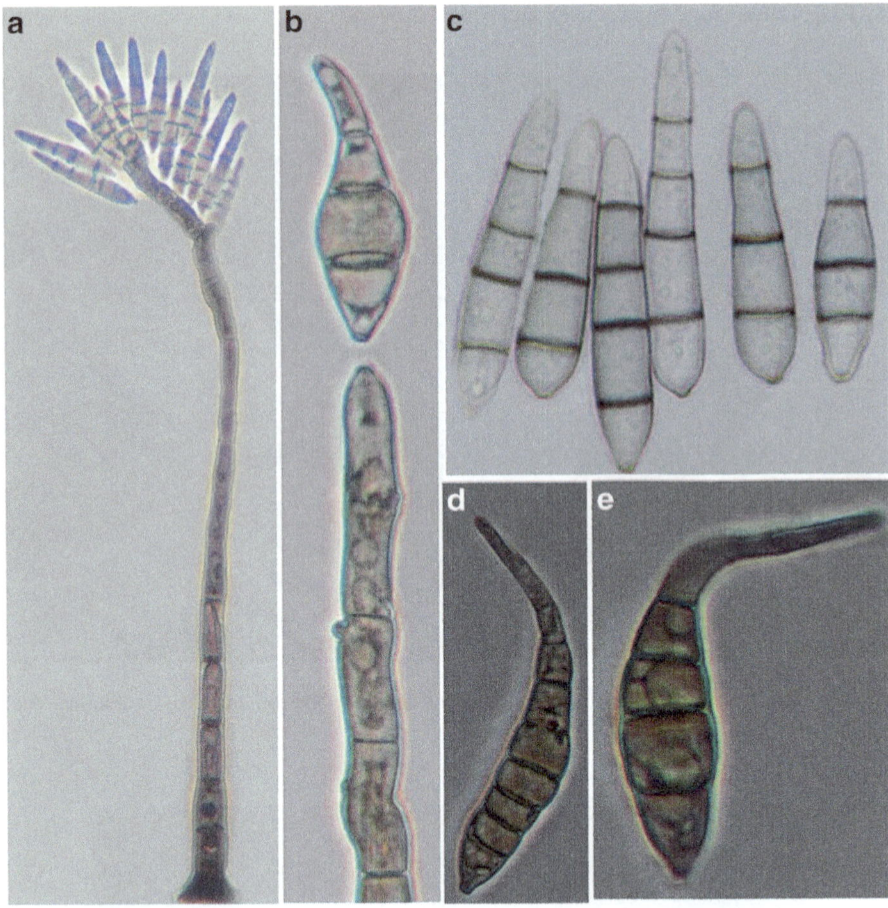


**Fig. 9.18** *Minimelanolocus limpidus*. (**a**) Conidiophore and conidia. (**c**) Conidiogenous cells and conidium. *M. curvisporus*. (**b**) Conidiogenous cells and conidium. (**d**, **e**) Conidium

**Fig. 9.19** *Minimelanolocus navicularis*, conidiophore, and conidia

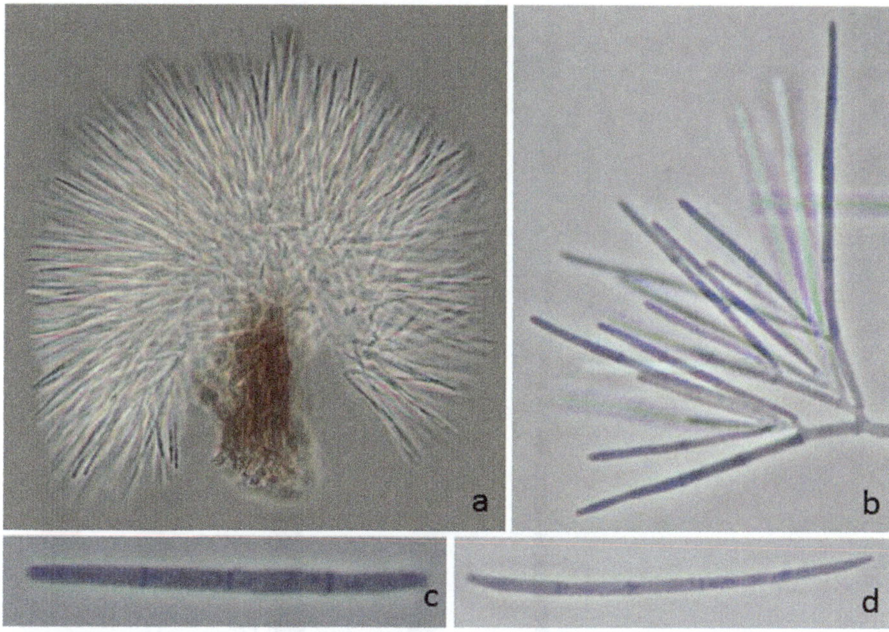


**Fig. 9.20** *Minteriella cenotigena.* (**a**) Conidioma and conidia. (**b**) Conidiophore, conidiogenous cells, and conidia. (**c**, **d**) Conidium

9.21, 9.22 and 9.23) in the past 30 years (Seifert et al. 2011; Index Fungorum 2015; MycoBank 2015), and some mycological resources on the Internet provide useful information about the regional inventories of Central and South American Neotropic (Cybertruffle 2015).

**Acknowledgments** The authors are very grateful to Drs. Josep Guarro, Marcos Fabio Marques, Margarita Hernández-Restrepo, and Xiuguo Zhang for providing photos.

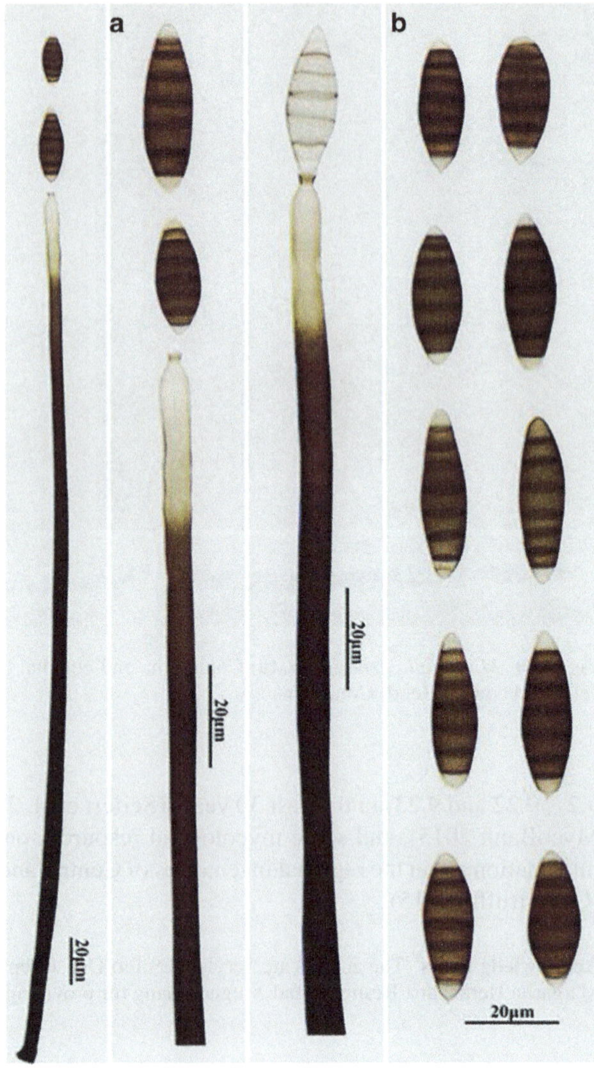


**Fig. 9.21** *Stephembruneria elegans*. (**a**) Conidiophores, conidiogenous cells, and conidia. (**b**) Conidia

**Fig. 9.22** *Phaeocandelabrum joseiturriagae*. (**a**) Conidiophore, conidiogenous cell, and conidium. (**b**) *Venustosynnema ciliatum*. Conidioma

**Fig. 9.23** *Venustosynnema ciliatum*. Conidioma, marginal setae, conidiogenous cells, and conidia

## References


Alfenas RF, Pereira OL, Jorge VL, Crous PW, Alfenas AC (2013) A new species of *Calonectria* causing leaf blight and cutting rot of three forest tree species in Brazil. Trop Plant Pathol 38(6):513–521

Blackwell M (2011) The fungi: 1,2,3 …5.1 million species? Am J Bot 98(3):426–438

Braun U, Castañeda-Ruiz RF (1991) *Cercospora* and allied genera of Cuba (II). Cryptogam Bot 2(2–3):289–297

Braun U, Urtiaga R (2012) New species and new records of cercosporoid hyphomycetes from Cuba and Venezuela (Part 1). Mycosphere 3:301–329

Braun U, Urtiaga R (2013a) New species and new records of cercosporoid hyphomycetes from Cuba and Venezuela (Part 3). Mycosphere 4:591–614

Braun U, Urtiaga R (2013b) New species and new records of cercosporoid hyphomycetes from Cuba and Venezuela (Part 2). Mycosphere 4:174–214

Castañeda-Ruiz RF (2005) Metodología en el estudio de los hongos anamorfos. In: Anais do V Congresso Latino Americano de Micología, Brasilia, pp 182–183

Castañeda-Ruiz RF, Braun U (1989) *Cercospora* and allied genera of Cuba (I). Cryptogam Bot 1(1):42–55

Castañeda-Ruiz RF, Kendrick B (1991) Ninety-nine conidial fungi from Cuba and three from Canada. University of Waterloo Biology Series 35, 132 p

Castañeda-Ruiz RF, Guarro J, Mayayo E (1998) *Enridescalsia*, a new genus of conidial fungi from submerged leaves in Cuba. Mycol Res 102(1):42–44

Castañeda-Ruiz RF, Heredia G, Arias RM, McKenzie EHC, Hyde KD, Stadler M, Saikawa M, Gené J, Guarro J, Iturriaga T, Minter DW, Crous PW (2011) New species and re-disposed taxa in *Repetophragma*. Mycosphere 2(3):273–289

Crous PW, Alfenas AC, Wingfield MJ (1993) *Calonectria scoparia* and *Calonectria morganii* sp. nov., and variation among isolates of their *Cylindrocladium* anamorphs. Mycol Res 97(6):701–708

Crous PW, Alfenas AC, Barreto RW (1997) Cercosporoid fungi from Brazil. 1. Mycotaxon 64:405–430

Crous PW, Alfenas AC, Junghans TG (1998) Variability within *Calonectria ovata* and its anamorph *Cylindrocladium ovatum* from Brazil. Sydowia 50:1–13

Crous PW, Braun U, Alfenas AC (1999) Cercosporoid fungi from Brazil. 3. Mycotaxon 72:171–193

Cybertruffle (2015) http://www.cybertruffle.org.uk/eng/index.htm. Accessed 30 Sep 2015

da Cruz ACR, Leão-Ferreira SM, Barbosa FR, Gusmão LFP (2008) Conidial fungi from semi-arid Caatinga biome of Brazil. New and interesting *Dictyochaeta* species. Mycotaxon 106:15–27

da Silva M, Castañeda-Ruiz RF, Pereira OL, Barreto RW (2012) *Alveariospora*, a new anamorphic genus from trichomes of *Dimorphandra mollis* in Brazil. Mycotaxon 119:109–116

Dornelo-Silva D, Dianese JC (2004) New hyphomycete genera on *Qualea* species from the Brazilian cerrado. Mycologia 96(4):879–884

Gusmão LFP, Leão-Ferreira SM, Oliveira Marques MF, Carneiro de Almeida DA (2008) New species and records of *Paliphora* from the Brazilian semi-arid region. Mycologia 100(2):306–309

Hawksworth DL (1990) The fungal dimension of biodiversity: magnitude, significance, and conservation. Mycol Res 95:641–655

Hawksworth DL (2001) The magnitude of fungal diversity: the 1.5 million species estimate revisited. Mycol Res 105:1422–1432

Heredia Abarca G, Castañeda-Ruiz RF, Arias RM, Saikawa M, Stadler M (2007) Anamorphic fungi from submerged plant material: *Acumispora verruculosa* sp. nov., *Pleurophragmium aquaticum* sp. nov and *Pleurophragmium miniumbonato* comb. nov. Mycotaxon 101:89–97

Hernández-Gutiérrez A, Dianese JC (2008) New cercosporoid fungi from the Brazilian Cerrado 1. Species on hosts of the families *Anacardiaceae*, *Araliaceae*, *Bombacaceae*, *Burseraceae* and *Celastraceae*. Mycotaxon 106:41–63

Hernández-Gutiérrez A, Dianese JC (2014) Cercosporoid hyphomycetes on malpighiaceous hosts from the Brazilian Cerrado: 1. New taxa on *Banisteriopsis* species. Mycol Prog 13:365–371
Index Fungorum (2015) http://www.indexfungorum.org/names/Names.asp. Accessed 30 Sep 2015
Monteiro JS, Castañeda-Ruiz RF, Gusmão LFP (2014) *Helicodochium*, a new microfungus from submerged wood from Brazil. Mycotaxon 127:5–9
Monteiro JS, Castañeda-Ruiz RF, Gusmão LFP (2015) A new species of *Matsushimiella* from submerged leaves in the Brazilian Amazon Forest. Mycotaxon 130(2):311–314
Mueller GM, Schmit JP (2007) Fungal biodiversity: what do we know? What can we predict? Biodivers Conserv 16(1):1–5. doi:10.1007/s10531-006-9117-7
MycoBank (2015) http://www.mycobank.org/. Accessed 30 Sep 2015
Olson DM, Dinerstein E (2002) The Global 200: priority ecoregions for global conservation. Ann Mo Bot Gard 89:199–224
Pereira-Carvalho RC, Sepúlveda-Chavera G, Armando EAS, Inácio CA, Dianese JC (2009) An overlooked source of fungal diversity: novel hyphomycete genera on trichomes of cerrado plants. Mycol Res 113(2):261–274
Pinho DB, Firmino AL, Ferreira-Junior WG, Pereira OL (2012a) An efficient protocol for DNA extraction from Meliolales and the description of *Meliola centellae* sp. nov. Mycotaxon 122:333–345
Pinho DB, Pereira OL, Firmino AL, da Silva M, Ferreira-Júnior WG, Barreto RW (2012b) New Meliolaceae from the Brazilian Atlantic forest 1. Species on hosts in the families *Asteraceae*, *Burseraceae*, *Euphorbiaceae*, *Fabaceae* and *Sapindaceae*. Mycologia 104(1):121–137
Pinho DB, Firmino AL, Ferreira-Junior WG, Pereira OL (2013) New Meliolaceae from the Brazilian Atlantic Forest 2: species on host families *Annonaceae*, *Cecropiaceae*, *Meliaceae*, *Piperaceae*, *Rubiaceae*, *Rutaceae* and *Tiliaceae*. Mycologia 105(3):697–711
Pinho DB, Honorato Junior J, Firmino AL, Hora Junior B, Mizubuti ESG, Pereira OL (2014) Reappraisal of the black mildews (Meliolales) on *Hevea brasiliensis*. Trop Plant Pathol 39(1):89–94
Pons N, Sutton BC (1988) *Cercospora* and similar fungi on yams (*Dioscorea* species). Mycol Pap 160:1–78
Rodríguez JD, Piepenbring M (2007) Two new species of *Appendiculella* (Meliolaceae) from Panama. Mycologia 99(4):544–552
Rodríguez JD, Kirschner R, Piepenbring M (2015) New species and new records of *Meliolaceae* from Panama. Fungal Divers 70(1):73–84
Rodríguez-Hernández M (1985) Dos nuevas especies de Ascomycetes de la fumagina: *Chaetothyrium diversum* y *Dennisiella longispora*. Revista del Jardín Botánico Nacional Universidad de la Habana 6(1):69–73
Rodríguez-Hernández M, Camino MV (1987) New species of meliolaceous fungi from Cuba. Feddes Repertorium 98(9–10):509–513
Seifert KA, Morgan-Jones G, Gams W, Kendrick B (2011) The genera of Hyphomycetes. CBS-KNAW Fungal Biodiversity Centre Utrecht, Utrecht
Zunino M, Zullini A (2003) Biogeografía, 2. Edición. Fondo de Cultura Económica, México

# Chapter 10
# Mesofungi

**Bryce Kendrick**

The title of the book is *Microfungi*. This is a short 'bridging' chapter, in which I suggest that readers might usefully extend that term in the course of their collecting activities to include numerous but largely overlooked taxa that do not quite fit that description. Not because I do not find microfungi fascinating, indeed, after I first discovered them, I spent 60 years looking at, thinking about, investigating and writing about them, and I look forward to reading this book.

Fruiting macrofungi draw visual attention to themselves, unless they are buried in the earth (hypogeous fungi) when olfactory clues, at least for some mammals, take over. Microfungi, despite their virtual invisibility, also began to be studied centuries ago, Persoon making 1801 a banner year. Later, they were realized to be the causes of food spoilage and of many plant and animal diseases, the source of mycotoxins, important agents of food processing, and the producers of antibiotics and other pharmaceuticals that helped to stem the tide of various bacterial plagues that had previously ravaged the human race and our domestic animals. This led to a string of increasingly detailed and sophisticated compilations (e.g., Sutton 1980, Nag Raj 1993), the most recent appearing only 2 years ago (Seifert et al. 2012).

But there are fungi which fall through the cracks—neither micro nor macro—which are too small to be regularly collected by most people, of no interest to mycophagists, and in most cases of unknown biochemical potential.

A recently published book has the title *Ascomycete Fungi of North America: A Mushroom Reference Guide* (Beug et al. 2014). The first phrase is factually correct, the second not so much. It is appropriate to apply the term mushroom to such things as morels, false morels and saddle fungi and even truffles, all of which can be macroscopic and are often considered in handbooks alongside the agarics and boletes,

B. Kendrick, Ph.D., D.Sc., F.R.S.C. (✉)
8727 Lochside Drive, Sidney-by-the-Sea, BC, Canada V8L 1M8

Department of Biology, University of Waterloo, Waterloo, ON, Canada
e-mail: bryce@mycolog.com



D.-W. Li (ed.), *Biology of Microfungi*, Fungal Biology,
DOI 10.1007/978-3-319-29137-6_10

while their ecological roles and edibility and toxicity are widely pondered or experimentally investigated. But when we consider many of the other organisms illustrated in the *Ascomycete* book, the word mushroom would not spring to our lips. Indeed, most fungivorous humans would fail to see them, or would spurn or ignore them. They are just too small.

When I wrote the second edition of my mycology textbook *The Fifth Kingdom*, I knew I could not exclude the *Oomycota*. They are not true fungi, since for many valid reasons they belong to Kingdom *Chromista*, beside the brown algae and diatoms, but their hyphae are such good mimics of those produced by what we call the true fungi (Kingdom *Eumycota*), having been molded by processes of convergent evolution, that they had to be included. But the myxomycetes and mycetozoa were clearly beyond the pale. Not only are they absolutely *not* fungi at all, since for much of their life cycles they don't even have cell walls, but biologists have long since placed them among the extremely diverse Kingdom *Protozoa* (or *Protoctista*). Yet before the appearance of the third edition of my text, I was asked by many of my teaching colleagues at other universities to put them back in. Their response to my objections was that if we mycologists didn't teach them, then no one would. This argument was enough for me to bring what I now call the *Myxostelida* and other kinds of 'slime mould' back from the wilderness. This helped to broaden my mind and possibly those of some students. The spectrum of organisms awaiting study is still huge and partially unperceived.

During my doctoral studies at the University of Liverpool, I spent thousands of hours examining forest soil from various horizons, seeking out and exploring the ecology of truly microscopic fungi—in some cases, exciting new taxa—and the other organisms with which they shared that environment. In the process, I came across many small invertebrate animals—beetle mites, nematodes, enchytraeid worms and springtails—and soon found that although taxonomically disparate, they interacted in important ways with the microfungi and were placed by soil scientists in an informal but well-recognized category called the *meiofauna* or *mesofauna*.

Wikipedia now tells us that this category contains 'organisms (animals) that will pass through a 1 mm mesh, but will be retained by a 45 μm mesh'. This term is usually applied to benthic (aquatic) fauna, but is easily and conveniently transferred to a wide range of soil invertebrates.

In mycological terms, when defining mesofungi, I am including fungi that do not require a microscope to be seen, but which are small enough, or fruiting in such unexpected places or such unusual forms as to be passed over by most collectors. If unseen or ignored, they will fail to be collected and studied at the microscopic—or any other—level. I have included a number of illustrations to give a sample of such fungi and places.

Mesofungi are most likely to be seen by those looking for microfungi or macrofungi in the field, and if we are eventually to get a good hold on the full spectrum of what is fruiting out there, and injecting spores into the atmosphere, some of us must raise our sights and widen our horizons a little.

I have gone through the most recent book on *Ascomycete Fung of North America* (mentioned above) and have found that a fairly large number of the 600 taxa dealt with in detail are very small and unlikely to draw much attention. I counted about 50 genera in which this was the case. For example, O*rbilia luteorubella* has ascomata that are between

0.2 and 1.5 mm in diameter, 200–1500 μm; *Nectria cinnabarina*, 200–400 μm; *Helminthosphaeria clavariarum*, up to 300 μm; and *Roseodiscus subcarneus*, 400–1500 μm. Whether these organisms are observed or not may depend on their colour. The first is yellow, the second red, the third black and the fourth white. *Mollisia cinerea* ranges from 500 to 3000 μm, but is grey and may not be noticed or thought worth collecting even when its ascomata are present in numbers on the surface of rotting wood.

*Helminthosphaeria* is often missed altogether because it is just assumed to be a very dark specimen of its clavariaceous host. On a field trip in Algonquin Park, Ontario, many years ago, one of my former students made me think that he might have the makings of a future professional mycologist when he spotted this colour as denoting a parasitic fungus. Fortunately for our discipline, my prediction was correct.

Other mesofungi may be seen or missed because of the degree of aggregation or clustering of their ascomata, or its absence. A few ascomata of *Godronia urceolus* (500–1500 μm) would probably be missed: a large aggregation would probably draw attention to itself—but in most cases would not be collected. Many senescing leaves develop dark spots in fall, but unless there is a pattern of multiple similarly sized spots, as in species of *Rhytisma*, *Coccomyces* or *Trochila*, these may not be recognized as fungal fructifications. In the case of *Rhytisma punctatum*, which attacks *Acer macrophyllum* on the west coast, each of the spots, which represents an original single-spore infection, may be quite extensive and encompasses a fairly large number of tiny ascomata (Fig. 10.1). Less conspicuous, because each spot incorporates a single ascoma, are *Trochila ilicina* (Figs. 10.2, 10.3 and 10.4) and *Coccomyces dentatus* (Figs. 10.5 and 10.6) which will become recognizable as discomycetes only when incubated in a damp environment.

**Fig. 10.1** Colonies of *Rhytisma punctatum* producing large spots in senescent leaves of *Acer macrophyllum* (bigleaf maple). Each spot is developing numerous black apothecia which will not mature until the following year

**Fig. 10.2** A dead leaf of *Ilex aquifolium* (holly) with very numerous black apothecia of *Trochila ilicina*

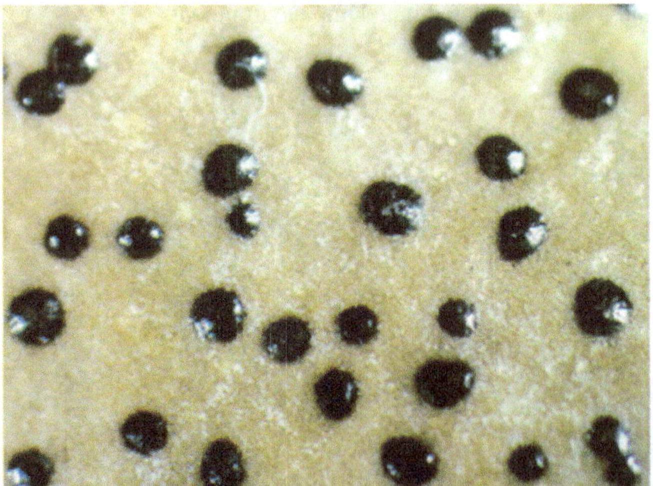

**Fig. 10.3** Close-up of developing apothecia of *Trochila ilicina* in dead holly leaf

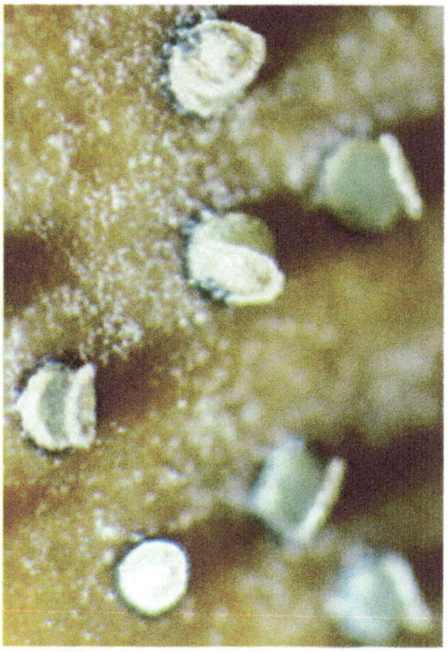

**Fig. 10.4** Open apothecia of *Trochila ilicina* in dead holly leaf, showing greyish hymenium

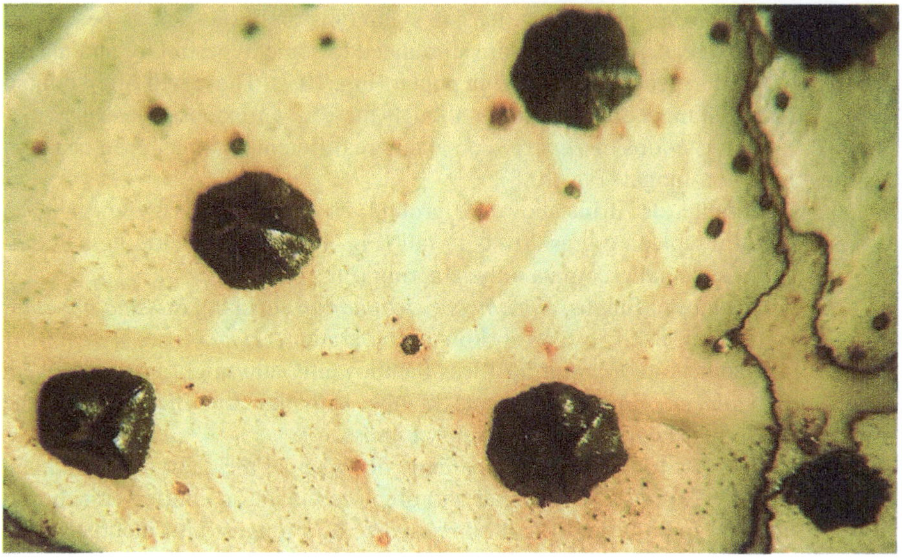

**Fig. 10.5** Black apothecia of *Coccomyces dentatus* in dead *Mahonia* (Oregon grape) leaf, triangular valves of lid just opening

**Fig. 10.6** Open apothecia of *Coccomyces dentatus* revealing pale grey hymenium

The genus *Taphrina* produces extensive symptoms on hosts such as peach and poplar, in which areas of the leaves are stimulated to some hypertrophy and distortion, and turn red in peach and yellow in poplar (Fig. 10.7). These will often remain unrecognized as fungal fructifications, though there is an extensive ascal hymenium. The manifestations of *Taphrina* on alder are different and may even be taken for something non-fungal (Fig. 10.8).

Other mesofungi also fruit above the ground, as does *Claviceps purpurea*, its stromata initially replacing grass fruits in fall (Fig. 10.9), but developing its complex stipitate perithecial ascomata only in spring (Fig. 10.10).

Although the lobster fungus, *Hypomyces lactifluorum* attacking *Russula brevipes*, draws attention to itself by its brilliant colouration, *Hypomyces luteovirens* (Fig. 10.11) blankets its *Russula* host less flamboyantly and might be overlooked. *Nectriopsis violacea* (Fig. 10.12) attacks the aethalial fructifications of the myxostelid, *Fuligo septica*, and is also easy to miss.

I also found a few fungi that had apparently flown under the radar of the authors.

For example, *Colpoma*, widespread and often dominant on conifer branches at higher elevations and a little difficult to recognize as a fungus until incubated in a damp chamber (Figs. 10.13 and 10.14), is not mentioned in the book. Nor is *Didymascella thujina*, the cause of Keithia blight of western red cedar (Fig. 10.15).

*Onygena* species (Fig. 10.16) fruit, often inconspicuously, on hoof or horn, substrates most people would not examine closely. *Diatrype* (Fig. 10.17) covers wood surfaces under bark and because of its extent and uniformity may easily be missed or misinterpreted as scorched wood. Members of the *Erysiphales*, *Thelebolales*, *Myxotrichaceae* and *Pseudeurotiaceae*, as noted in Beug et al. (2014), do not produce fruit bodies large enough to attract attention and so are obvious candidates for

**Fig. 10.7** *Taphrina populina* fruiting on leaves of *Populus* (poplar)

**Fig. 10.8** *Taphrina alni* (alder tongue) fruiting on cones of *Alnus rubra* (alder)

**Fig. 10.9** Dark sclerotia of *Claviceps purpurea* on *Leymus mollis* (Poaceae)

**Fig. 10.10** Germinated sclerotia of *Claviceps purpurea* with stalked, complex perithecial ascomata

**Fig. 10.11** *Hypomyces luteovirens* fruiting all over the gills of a species of *Russula*

**Fig. 10.12** *Nectriopsis violacea* fruiting on an aethalium of the slime mould *Fuligo septica*

**Fig. 10.13** Black, lirellate (elongate) apothecia of *Colpoma crispum* on dead conifer branch

**Fig. 10.14** Apothecia of *Colpoma crispum* opening to show whitish hymenium

**Fig. 10.15** *Didymascella thujina* (Keithia blight) on *Thuja plicata* (western red cedar)

**Fig. 10.16** Whitish mazaedia of *Onygena equina* on deer antler

**Fig. 10.17** Extensive fruiting stroma of *Diatrype stigma*. The black dots are the ostioles of perithecial ascomata. The stroma can be mistaken for scorched wood

inclusion in the mesofungi. One can certainly see the tiny dark ascomata of many *Erysiphales* occurring in gradually maturing clusters on host leaves (Figs. 10.18 and 10.19), but unless they are collected and properly examined, the beauty and information content of these teleomorphs are apt to be missed.

Another problem with mesofungi is that they may have a long ripening period and are often immature when collected, especially in northern regions, so mature asci and ascospores cannot be found. Even in the macrofungal *Geoglossaceae*, where the fruit bodies are relatively large, the ascospores are often found not to have completed their pattern of septation: 'Are they 3-septate or 7-septate? Are they mature?' Molecular techniques present a way around this problem.

Not so many years ago, we had good excuses for not studying such fungi. Their ascomata were too small and were probably immature anyway. But now with readily available and progressively less expensive sequencing techniques, we have largely lost those excuses.

Then there are the Basidiomycetes. One of the best examples I can think of is *Mucronella* (Fig. 10.20), a genus represented in the immediate vicinity of my home by at least three species. The problem with locating this genus during its extended fruiting period is that the basidiomata tend to develop inside deep cracks in the rotten wood of large Douglas fir stumps and are unlikely to be seen unless directly sought for. The elicitation of the seeking behaviour (usually one of the cognoscenti tells you where to look) suggests a certain level of prior knowledge that most people lack.

**Fig. 10.18** A group of dark ascomata of *Uncinula* (*Erysiphales*) on a living leaf

**Fig. 10.19** Ascomata of *Phyllactinia* (*Erysiphales*) on a leaf

I have spent many years looking for microfungi fruiting in their natural habitats, whether these be rotting logs, dead leaves or wallboard. But rotting wood is not reserved for microfungi, and I have inevitably come across many tiny ascomycetes and basidiomycetes, not to mention *Myxostelida*, which are not microscopic but may not be large enough or conspicuous enough to be spotted by casual observers. In many cases, they are small enough to discourage further investigation by someone already focused on other kinds of target. How can you get a spore print from an agaric with a cap 1 mm across? Recently, Oluna and Adolf Ceska (pers. comm.)

**Fig. 10.20** Icicle-like basidiomata of *Mucronella pulchra* on rotting stump of *Pseudotsuga menziesii* (Douglas fir)

found a species of agaric with a basidioma of that size, but on close inspection, they were able to find mature angular pink basidiospores typical of *Leptonia* (*Entoloma*) *cephalotrichum* (Fig. 10.21).

One of the best lab exercises I ever adopted as a teacher was for students to follow the succession of fungi colonizing and fruiting on herbivore dung. We usually looked at the digestive by-products of horses which, if the horses were being fed naturally in pastures and we were able to discourage the insect larvae, almost always gave us a diverse and interesting sequence, including zygomycetes, ascomycetes and basidiomycetes (and sometimes dictyostelids and myxobacteria). Although some of the fungi, for example, *Pilobolus* (Fig. 10.22), were large enough to see with the unaided eye (with sporangiophores over 1 cm tall), many such as *Saccobolus* (Fig. 10.23) and *Ascobolus*, though their ascomata could be picked out by the naked eye, required a microscope for any further investigation. And it was only because students were focusing on the dung that they even noticed these fungi. The sequence matured with small species of coprinoid agarics, which tended to be extremely ephemeral.

If people go out in the field and look for fungi, what are the chances that the average amateur will find *Sphaerobolus stellatus* and *Heyderia abietis* (Fig. 10.24), both of which I spotted during a recent foray by lying down on the ground and poking around assiduously in the litter.

Here are some others that qualify. *Bisporella citrina* has ascomata that are 1–3 mm in diameter. However, they are yellow and often present in large

**Fig. 10.21** Basidioma of *Leptonia cephalotrichum* with pink basidiospores

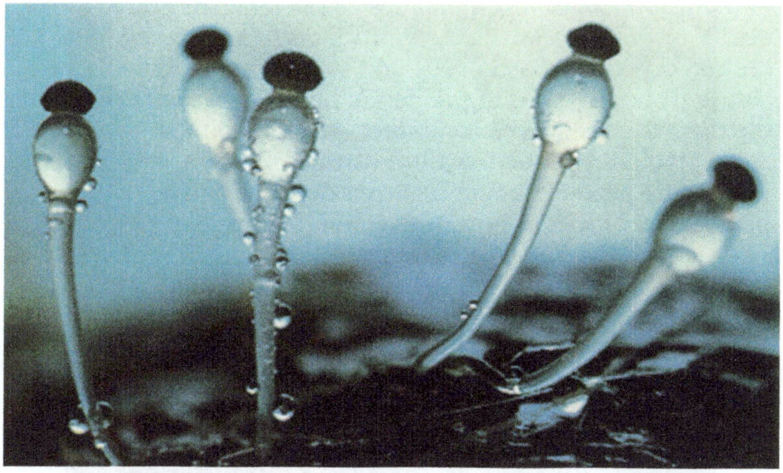

**Fig. 10.22** Erect sporangiophores of *Pilobolus* on horse dung

numbers, and so they can be easily detected. To a lesser extent, this is true of *Mollisia cinerea* which ranges from 0.5 to 3 mm—and is much less visible because it is grey.

But *Rosellinia subiculata*, with perithecia of only 0.5 mm, is less likely to be spotted, as is *Roseodiscus subcarneus*, at 0.4–1.5 mm, and *Orbilia luteorubella*, 0.2–1.5 mm (check Beug et al. 2014 for pictures). *Heyderia abietis*, already mentioned, with a head 3 × 2 mm, is very likely to be missed.

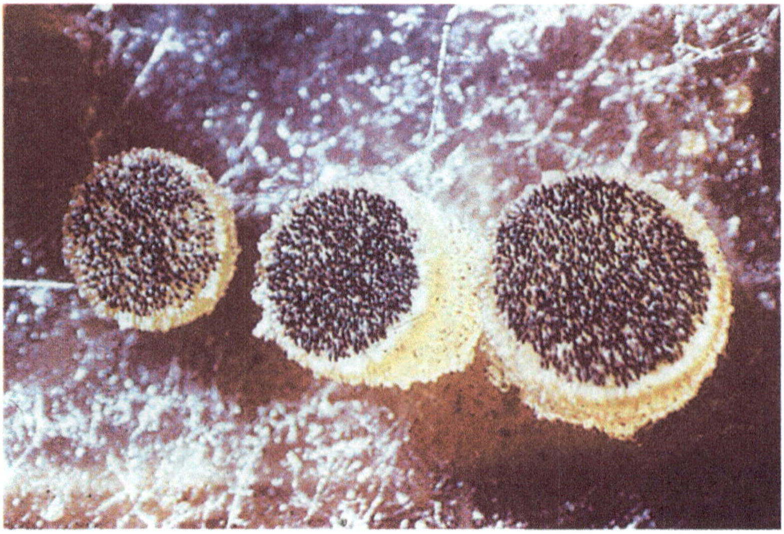

**Fig. 10.23** Hemispherical apothecia of *Saccobolus*

**Fig. 10.24** Stipitate apothecia of *Heyderia abietis*

I have merely presented a gesture sketch of the mesofungi and must leave it to those who collect microfungi in the field to consider both widening their search and their intellectual horizons in the service of their discipline.

## References

Beug MW, Bessette AE, Bessette AR (2014) Ascomycete fungi of North America. University of Texas Press, Austin, TX, 488 pp

Nag Raj TR (1993) Coelomycetous anamorphs with appendage-bearing conidia. Mycologue Publications, Sidney, 1101 pp

Persoon CH (1801) Synopsis methodica Fungorum. Gottingen

Seifert KA, Morgan-Jones G, Gams W, Kendrick B (2012) The Genera of Hyphomycetes CBS, Utrecht 997 pp

Sutton BC (1980) The coelomycetes. CABI Publishing, Wallingford, 696 pp

# Chapter 11
# Evolution of Fungi and Update on Ethnomycology

**De-Wei Li, R.F. Castañeda-Ruiz, and James LaMondia**

In the past 20 years, research on fungal evolution, archaeology, and ethnomycology has made significant advances and produced enormous results. The latest developments allow us to better understand fungi and their origins and usage of fungi by human beings in ancient times. The new results have made it possible for us to examine these topics from a new angle with a new vision. This chapter intends to cover the areas which are often not addressed in the scientific literature.

---

D.-W. Li, Ph.D. (✉)
The Connecticut Agricultural Experiment Station Valley Laboratory,
153 Cook Hill Road, Windsor, CT 06095, USA

Co-Innovation Center for Sustainable Forestry in Southern China, Nanjing Forestry University, Nanjing, Jiangsu 210037, China
e-mail: dewei.li@ct.gov

R.F. Castañeda-Ruiz, Ph.D.
Instituto de Investigaciones Fundamentales en Agricultura Tropical 'Alejandro de Humboldt' (INIFAT), Académico Titular de la Academia de Ciencias de Cuba, Calle 1 Esq. 2, Santiago de Las Vegas, C.P. 17200 La Habana, Cuba
e-mail: icastaneda@ceis.cujae.edu.cu; rfcastanedaruiz@gmail.com; rfcastaneda@inifat.co.cu

J. LaMondia, Ph.D.
The Connecticut Agricultural Experiment Station Valley Laboratory,
153 Cook Hill Road, Windsor, CT 06095, USA
e-mail: James.lamondia@ct.gov



D.-W. Li (ed.), *Biology of Microfungi*, Fungal Biology,
DOI 10.1007/978-3-319-29137-6_11

# Evolution of Fungi

## Early Evolution of Fungi

It is important to understand the origin and evolution of fungi and the diversification of major fungal lineages. Fossil records were used as direct evidence of fungal evolution for much of the past (Takamatsu 2013). Fungi were mainly preserved in two ways: amber and chert. Fungal fossils are scarce and rare due to their fragile nature and difficulty in preservation (Braun and Cook 2012). Studies on the evolution, diversity, ecology, and reproductive biology of fossil fungi and fungal-like microorganisms have lagged behind those of other fossil organisms for many years due to a number of reasons (Schwendemann et al. 2009). Indisputably, the lack of fungal fossil is one of the major reasons. Uncertainty in fossil ages and their phylogenetic positions are the two other reasons (Hipsley and Müller 2014). Molecular methods offer new ways to study fungal evolution. Using DNA sequences to date evolutionary events of fungi is gradually becoming more popular (Rutschmann 2006; Prieto and Wedin 2013), and molecular dating has become an important method to temporally determine diversification of the Tree of Life (Tamura et al. 2012).

However, there are limits in molecular dating techniques and there is no single best molecular dating method available (Rutschmann 2006). One of the limits is the lack of reliable clock calibration (Tamura et al. 2012). The limits had led to discrepancies of dating divergences for major fungal lineages in the literature (Prieto and Wedin 2013; Beimforde et al. 2014). For example, the date of the first divergence in Ascomycota was estimated from 325 to 1316 million years ago (Ma) in previous studies (Prieto and Wedin 2013; Beimforde et al. 2014; Berbee and Taylor 1993, 2007; Gueidan et al. 2011; Lücking et al. 2009). Beimforde et al. (2014) indicated that recent developments in fungal phylogenies and molecular clock models have allowed mycologists to establish increasingly more realistic models than those in previous studies for fungal evolution, contributing to the discrepancies in dating fungal diversification. It is obvious that the dates proposed for temporal origins of some fungal lineages are tentative at present. These dates need to be verified or perfected with more studies and improved methodologies. Using a molecular clock to date major fungal evolutionary events or diversification may be inaccurate without appropriate calibration (Hipsley and Müller 2014). Thus, clear justification and proper usage of molecular clock calibration are crucial to accurately determine dates in the evolutionary history of fungi (Hipsley and Müller 2014). Despite these limitations, research in this area has made significant progress in the last two decades with assistance from the new dating techniques for fossils and the advancement of molecular clock methods. These developments have provided insight into how fungi evolved. Pirozynski (1976), Taylor and Krings (2005), and Taylor et al. (2015) have published thorough reviews on fungal fossils, using early and more recent literature, respectively. In this chapter, the author will mainly cover the advances in fungal fossil discoveries made in the last decade. For molecular dating, the focus is on the studies which used either fungal fossils or geological events to calibrate their molecular clock.

## *Fossil Evidences and Molecular Clock Dating*

The divergence of animals from fungi was estimated at ca. 965 Ma (Doolittle et al. 1996). Oberwinkler (2012) recently drew a figure to show the evolutionary tree of true fungi based on a number of key literature citations published in the last 10 years. In his figure, he indicated the diversification of the fungal taxonomic groups and animals at ca. 850 Ma, calibrated using fossil records. Preceding land plants, fungi possibly colonized the land during the Cambrian (542–488.3 Ma) (Brundrett 2002; Berbee and Taylor 2010). Fungi were first traced back to the Silurian period, about 408–438 Ma in the Paleozoic era based in fossil records (Alexopoulos et al. 1996). Studies conducted in the last two decades showed that this dating was rather conservative. Fossilized hyphae and spores recovered from the Ordovician of Wisconsin (460 Ma) resemble modern-day Glomeromycota and existed at a time when the land flora likely consisted of only nonvascular bryophyte-like plants (Redecker et al. 2000). Glomeromycota is a much older evolutionary lineage than those of Ascomycota and Basidiomycota (Schüßler and Walker 2011). The first zygomycetous fungi were estimated to appear on earth during the Precambrian, ca. 1.2–1.4 Ga ago using molecular clock (Heckman et al. 2001; Blair 2009; Krings et al. 2013). However, more conservative estimates dated the divergence at ca. 800 Ma (Berbee and Taylor 2001). The conflicting results indicated that dating the divergence of zygomycetous fungi is far from resolved and needs further studies. Krings et al. (2014) recently published the first fossil record of a fungal "sporocarp" of *Mycocarpon rhyniense* from the Lower Devonian Rhynie chert ca. 410 Ma ago. The taxonomic placement of *Mycocarpon rhyniense* remains unclear, but the authors hypothesized that the fossilized fungus might belong to zygomycetous fungi and predates the oldest evidence of fungal "sporocarps" by ca. 80 Ma (Krings et al. 2014). Dating analysis by Gryganskyi et al. (2012) showed that the Entomophthoromycota (formerly considered as zygomycetous fungi) originated 405 ± 90 Ma using molecular clock models. Based on ancestral state reconstruction, the ancestor of all Entomophthoromycota was suggested to be morphologically similar to species of *Conidiobolus*. The authors suggested that entomopathogenic lineages in Entomophthoraceae probably evolved from ancestors of saprobes or facultative pathogens during or shortly after the evolutionary diversification of the arthropods. Unfortunately, fossil evidence of these fungi is extremely rare. Strullu-Derrien et al. (2014) reported two new endophytes *Palaeoglomus boullardii* (Glomeromycota) and *Palaeoendogone gwynne-vaughaniae* (Mucoromycotina) in *Horneophyton lignieri* dating back to ca. 407 million years old from the Rhynie chert. Krings et al. (2013) suggested that some "sporocarp" types from fossils represented by mantled zygosporangia and zygomycetous fungi probably were important in terrestrial paleoecosystems by the Carboniferous period (359–299 Ma). Fungal fossils were infrequent and controversial before the early Devonian (416–359 Ma), when fungal fossils including Chytridiomycota, Ascomycota, Glomeromycota, and Peronosporomycetes became abundant in the Rhynie chert, mostly as Glomeromycota (formerly reported as Zygomycota) and Chytridiomycota (Brundrett 2002; Taylor and Taylor 1997; Redecker et al. 2000; Taylor et al. 2005; Krings et al. 2007).

Considerable uncertainties exist in the timing of the origins of land-plant fungal symbioses due to the fact that paleontological data and molecular clock date estimates are often greatly different (Field et al. 2015). A recent study provided some evidence that the earliest mycorrhizal fungi are the Mucoromycotina which predate the Glomeromycota (Bidartondo et al. 2011). This report called into question the ancestral position of Glomeromycota in fungal evolution and in forming symbioses with land plants, since the study by Bidartondo et al. (2011) did not calibrate their molecular clock using any available methods. Further research will be necessary to confirm this study in the future. Strullu-Derrien et al. (2014) reported fossil evidence *Palaeoendogone gwynne-vaughaniae* (Mucoromycotina) from Rhynie chert from 407 Ma (Scotland, UK).

Whether Glomeromycota, zygomycetes, and Chytridiomycota are paraphyletic or monophyletic, phylogenetics remain to be determined (Berbee and Taylor 2007). Fossil evidence indicated that these fungi were present in the Devonian, early in the history of vascular plants (Berbee and Taylor 2007).

In the latest study, Beimforde et al. (2014) studied all of the oldest Ascomycete fossils available from amber (Albian to Miocene) and chert (Devonian and Maastrichtian) representing four major ascomycete lineages (Lecanoromycetes, Laboulbeniomycetes, Dothideomycetes, and Eurotiomycetes) and used a multi-gene data set from 145 taxa representing all main taxonomic groups of the Ascomycota to estimate divergence times for Ascomycota. The authors demonstrated that Ascomycota started its initial diversification in the Ordovician (485–443 Ma), followed by repeated diversification of lineages throughout the Phanerozoic, and this continuous diversification was not affected by mass extinctions (Beimforde et al. 2014). A recent study was conducted using molecular dating and fossil calibration to estimate the divergence times of most of the major groups of Ascomycota (Prieto and Wedin 2013). The results from this study suggested that the origin and diversification of the Pezizomycotina occurred in the Early Cambrian (531 Ma). The main lineages of lichen-forming Ascomycota originated at least as early as the Carboniferous, with successive radiations in the Jurassic and Cretaceous generating the diversity of the main modern groups (Prieto and Wedin 2013). The origin of Pezizomycotina was dated at ca. 419 Ma with a calibrated molecular clock and two fossil records (Yang et al. 2012a), while the results of Prieto and Wedin (2013) suggested it be ca. 530 Ma.

The lichen-like fossils discovered in marine phosphorite of the Doushantuo Formation from South China, involving filamentous hyphae closely associated with coccoidal cyanobacteria or algae, were dated back ca. 600 Ma ago. These fossils showed that fungi developed symbiotic relationships with photoautotrophic organisms prior to the evolution of vascular plants (Yuan et al. 2005). A fossil of a perithecial fungus with both sexual and asexual states was found from the early Devonian (400 Ma) several years ago in the cortex just beneath the epidermis of aerial stems and rhizomes of the vascular plant *Asteroxylon* (Taylor et al. 2005). A fossil of a perithecial sordariomycete was reported from the Lower Cretaceous (ca. 133 Ma) on Vancouver Island, British Columbia, Canada (Bronson et al. 2013). Lichen fossils were found from early Devonian (415 Ma) (Honegger et al. 2013),

Lower Cretaceous (ca. 133 Ma) (Matsunaga et al. 2013), and Eocene (56 to 34 Ma) (Rikkinen and Poinar 2008). The Pezizomycotina lichen fossils from the early Devonian represent the oldest known record of lichens with symbionts (Honegger et al. 2013). Takamatsu (2013) indicated that the powdery mildews originated in the Late Cretaceous (66–100 Ma) and the first diversification of the major lineages occurred at the Cretaceous/Paleogene boundary according to molecular clocks using the 28S rDNA D1/D2 and ITS and fossil records. He stated that ancestral powdery mildews may have first initiated diversification on broad-leaved deciduous trees in the high latitudes of the northern hemisphere, and the cradle of four genera, *Blumeria*, *Golovinomyces*, *Leveillula*, and *Neoerysiphe,* may have been distributed in the areas from Central/West Asia to the Mediterranean (Takamatsu 2013).

A recent study linked the spores of the ascomycete genus *Potamomyces* with the fossil form taxa *Mediaverrunites* (Schlütz and Shumilovskikh 2013). The authors concluded that the genus *Potamomyces* evolved at least 25 Ma at the inception of the younger Tertiary according to the fossil findings for *Mediaverrunites*. Schmidt et al. (2014) indicated that new fossil evidence dated the fossil record of sooty molds to ca. 100 Ma, from the early Miocene (17 Ma) to the Early Cretaceous (Albian, ca. 100–113 Ma). Mesozoic and Cenozoic ambers from different regions of the world contained sooty molds from eight northern hemisphere amber deposits. Fragments of superficial subicula identical to those produced in the Metacapnodiaceae (Capnodiales) were recorded since the Albian. The fossil fungi demonstrated that capnodialean sooty molds had occupied their specialized niches since the time when early angiosperms first appeared in the fossil record.

Berbee and Taylor (2001) estimated 500 Ma as a minimum age for Basidiomycota. The Ascomycota and Basidiomycota diverged approximately 452 Ma (Taylor and Berbee 2006). However, this dating was adjusted to 430 Ma (Berbee and Taylor 2007). The later dating was supported by fossils reported by Krings et al. (2011). Taylor and Krings (2005) stated that all major fungal lineages except the Basidiomycota were present in the Rhynie chert (ca. 410 Ma), based on the reports on fossil fungi. All modern major taxonomic groups of fungi were present by the Late Carboniferous (Pennsylvanian, 318–299 Ma) (Blackwell et al. 2012; Alexopoulos et al. 1996). Fossil records of *Palaeancistrus martinii* R. L. Dennis with clamp connections found in PA used to be the oldest direct fossil evidence to date Basidiomycota to the middle Pennsylvanian period (ca. 305 Ma) (Dennis 1970; Taylor and Krings 2005). A recent study on the fossil filicalean fern *Botryopteris antiqua* preserved in a late Visean chert (Mississippian; ca. 330 Ma) from Esnost (Autun Basin) in central France showed that this fossil contained fungi with clamp connections and predated *Palaeancistrus martinii*; consequently, it is the oldest direct fossil evidence of Basidiomycota (Krings et al. 2011). A genomic study by Padamsee et al. (2012) based on their phylogenetic analysis of 71 proteins showed that *Wallemia sebi*, a microfungus, is the earliest diverging lineage of Agaricomycotina. The time of this evolutionary diversion was not dated in their study.

Hibbett et al. (1995) reported fossilized mushrooms from mid-Cretaceous (90–94 Ma) in amber collected from New Jersey, USA, and the mushrooms were later described as *Archaeomarasmius leggetti* Hibbett et al. (1997). Fossilized basidiomata of several *Fomes* spp. had been discovered in southwest Idaho dating back to the Tertiary (66–2.6 Ma) (Brown 1940). A new fossil basidioma of *Ganodermites libycus* A. Fleischmann et al. (Polyporales, Ganodermataceae) was reported from the Lower Miocene (Neogene) (23–2.6 Ma) of Jebel Zelten in the northern part of the Libyan Sahara, North Africa, which represents the earliest convincing fossil evidence for polypores. The fossil fungus is closely related to the modern genus *Ganoderma* and showed a variety of characters present in *Ganoderma* sp. (Fleischmann et al. 2007).

The hyphomycete in the Early Eocene Indian amber (48–56 Ma) collected from the Tarkeshwar Lignite Mine of Gujarat State, Western India, is *Monotosporella doerfeltii* Sadowski et al. (Sordariomycetes), which is attached to its substrate, and is likely the degraded thallus of a cladoniform lichen (Sadowski et al. 2012). Conidiophores of *Aspergillus collembolorum* Dörfelt and A. R. Schmidt growing on a springtail were found in Eocene amber from the Baltics (50–35 Ma) (Taylor et al. 2015; Dörfelt and Schmidt 2005). The first fossil *Aspergillus* was found from Dominican amber, and it was estimated to be from Lower Miocene-Oligocene (ca 40 Ma) (Thomas and Poinar 1988).

## Evolution of Ecological Functions of Fungi

Yang et al. (2012a) revealed that fungal carnivorism diverged from saprophytism ca. 419 Ma according to molecular clock calibration with two fossil records at the same time period. Fossil evidence of mycoparasitism and hypermycoparasitism was reported in Early Cretaceous Burmese amber (100 Ma). *Palaeoagaracites antiquus* and R. Buckley (an agaric) was parasitized by another fungus, *Mycetophagites atrebora* Poinar and R. Buckley, which was parasitized by *Entropezites patricii* Poinar and R. Buckley, a hyperparasitic fungus (Poinar and Buckley 2007). *Paleoophiocordyceps coccophagus* G. H. Sung et al. from the Early Cretaceous (Upper Albian) is the oldest fossil record of a fungal parasite of an animal (Sung et al. 2008).

Krings et al. (2007) studied the Rhynie chert plant *Nothia aphylla* Lyon ex El-Saadawy and Lacy (400 Ma) and found that three fungal endophytes were present in prostrate axes of *N. aphylla* as narrow hyphae producing clusters of small spores, large spherical spores/zoosporangia, and wide aseptate hyphae that form intercellular vesicles.

Ducousso et al. (2004) suggested that a single origin for the ectomycorrhizal status of the common ancestor of Dipterocarpoideae and Sarcolaenaceae was dated prior to Madagascar's separation from the India-Seychelles block (ca. 88 Ma). Their report, thus, predated the previous oldest ectomycorrhizal fossil (50 Ma) (LePage et al. 1997). Moyersoen (2006) studied the South American ectomycorrhizal dipterocarp *Pakaraimaea* and suggested that a Gondwanaland evolution of the ectomycorrhizal habit is at least 135 Ma. Hibbett and Matheny (2009) stated that ectomycorrhizae have multiple origins. Tedersoo et al. (2010) proposed that the ectomycorrhizal habit had evolved at least 66 times. The origin of freshwater ascomycetes was dated at 390 Ma using a Bayesian relaxed-clock method (Vijaykrishna et al. 2006).

## Update on Ethnomycology

There is much that we can learn from our ancestors regarding the utilization of fungi for various purposes. In the last three decades, the science of ethnomycology has made significant advancements. A comprehensive review of information on ethnomycology was covered in a book *Conspectus of World Ethnomycology* by Dugan (2011). However, the ethnomycological studies remain incomplete in some geographic areas. A number of facts contribute to this phenomenon. One of the major hurdles is the language skill to read ancient texts, such as ancient Chinese. Another one is insufficient archaeological studies to verify and date writing records. It is crucial to conduct more ethnomycological studies in the future. If the update in this chapter can initiate more interest and research in this area, our intentions will be served.

### *Microfungi in Human History*

The utilization of fungi by human beings as food, for fermented beverages, medicine, and other purposes such as ritual ceremonies, has been traced to prehistoric and preliterate periods when human beings were hunter-gatherers. Domestication of plants and animals led to the rise of agriculture and allowed food surplus and fermentation possible for making alcoholic beverages and baking. Since there were no written records during that period, archaeological evidence and molecular clock dating are used to study the relationships between human beings and fungi.

Fungi have played and are still playing an important role in human history. We often underestimated the ability of our ancestors to utilize fungi and are often surprised to recognize the significance of fungi in the evolutionary history of human beings. The archaeological discoveries and molecular technologies made in the last three decades have greatly improved our understanding of how human beings use fungi to their benefit. In early stages, the major purpose of human use of fungi was for survival. When food became sufficient and food surpluses were possible, fungi turn to being used to improve the quality of human life.

Mildews (microfungi) of cereals and vine crops were reported in the Old Testament (c. 750 BC) (Agrios 2005). Theophrastus, the Greek philosopher (c. 300 BC), was the first one to write about diseases including rusts of cereals, legumes, and trees (Agrios 2005). This topic has been intensively covered in plant pathology books. Therefore, it will not be covered further here.

Fungi, especially yeasts, were utilized for wine making, baking, medicinal use, and other fermentation purposes in China dating back to several thousand years ago. Liquor and wine making in China goes back to 6000–7000 years ago (Yu and Zhuang 2006). Guo (1976) (aka Kuo Mo-jo) found that during the Yangshao culture period (5000–3000 BC), there are a good number of bronze wine-ware vessels excavated in China from the Zhou and later dynasties. Some of these historical relic pieces are on display in museums not only in China, but also in Europe and North America, as well as other areas. The earliest bronze wine cup, excavated in 1975, dated back to the Xia dynasty (c. 2070–c. 1600 BC) (Luoyang Museum 2014).

## *Fermented Food and Condiments*

In ancient Egypt, Egyptians considered fermentation a gift from the god Osiris, whereas ancient Romans ascribed the emergence of mushrooms and truffles to lightning bolts cast to the earth by Jupiter (Alexopoulos et al. 1996).

A number of microfungi have been used to produce fermented food and condiments. Yeasts are the most important ones. Fermented food can be found in all geographic areas and continents in the world. Different ethnic groups from different continents all make significant contributions to fermented foods or condiments.

The earliest archaeological evidence showed that plant food processing and possibly the production of flour dated to ca. 30,000 years ago in Europe (Revedin et al. 2010). Fermented foods dated to ca. 10,000 BC to the Middle Ages for preserving or salvaging food surpluses (Liu et al. 2013). Emmer wheat was domesticated in the Middle East about 8000 BC (Gornicki and Faris 2014). The development of leavened bread dated to 7000 BC (Liu et al. 2013), but the earliest archaeological evidence of leavened bread is from ancient Egypt dated to ca. 2000–1200 BC (Samuel 1996, 1999). Scanning electron microscopy detected yeast cells in several ancient Egyptian loaves and indicated that these bread loaves (emmer wheat (*Triticum dicoccum* Schübl.)) were leavened (Samuel 1996).

Shevchenko et al. (2014) reported direct evidence that the Subeixi sourdough bread excavated in Subeixi cemetery (500–300 BC) in Xinjiang, China, was made from barley and broomcorn millet leavened using baker's yeast and lactic acid bacteria based on the geLC-MS/MS proteomics analysis.

The Chinese used not only yeasts but also several microfungi for various purposes. *Aspergillus* was used for fermented pasta and *Mucor* for making preserved tofu (soy cheese); *Monascus purpureus* Went was used in an additional recipe for making wine and *Neurospora* for fermented soybean crumb. Soy sauce originated from China approximately 2000 years ago, and it is fermented in a two-step process using yeast and bacteria (Lioe et al. 2012).

Dairy products were associated with the domestication of milk-producing animals in human history. The domestication of cattle, sheep, and goats had already taken place in the Near East by 8000 BC or before, and among these animals, sheep (*Ovis aries* L.) were the first to be domesticated from mouflon (*Ovis orientalis*

Gmelin) between 11,000 and 9000 BC in Iran (Krebs 2003; Clutton-Brock 1999; Garrard et al. 1996).

Cheese making was suggested as dating to the same era as bread making (7000 BC) (Liu et al. 2013). However, that date probably is the earliest archaeological evidence for milk use reported by Evershed et al. (2008). The earliest archaeological evidence of cheese making dated back to Neolithic (5500 BC) from Kujavia, Poland, where strainers with milk fat molecules were discovered (Salque et al. 2013). Cheese making was suggested as discovered accidentally by storing milk in a container made from the stomach of an animal where the milk was turned into curd and whey. By Roman times, cheese had become a daily food. Yeasts are present in cheeses, and *Debaryomyces hansenii* (Zopf) Lodder and Kreger, *Yarrowia lipolytica* (Wick., Kurtzman, and Herman) Van der Walt and Arx, *Saccharomyces cerevisiae* Meyen ex E. C. Hansen, *Kluyveromyces lactis* (Dombr.) Van der Walt, and *K. marxianus* (E. C. Hansen) Van der Walt are the predominant ones (Gori et al. 2011). However, the exact role played by yeasts in cheese making remains unsolved. Maturation and aroma formation in Camembert were suggested by a number of researchers (Gripon 1999). *Penicillium roqueforti* Thom is crucial to blue cheese making and it was used as a secondary starter culture (Gori et al. 2011). *Penicillium camemberti* Thom (*Penicillium candidum* Link) is used to produce white mold cheeses, such as Brie and Camembert (Gori et al. 2011).

At present, cheese is one of the major agricultural products in the world, and there are over 500 varieties in the market. Worldwide production of cheese in 2010 was over 20 million tons according to the "FAO Statistical Yearbook 2013: World Food and Agriculture" (FAO 2013).

Fermented meat is suggested to relate to preservation of the meat leftover from large animals killed by our hunter-gatherer ancestors before the beginning of agriculture. However, there is no archaeological evidence dating back to that era to prove this hypothesis at present. Fermented meat is suggested to date back to ca. 1500 BC (Liu et al. 2013). *Debaryomyces hansenii* is the dominant yeast in fermented meat products (Asefa et al. 2009). Several hyphomycetes, including *Penicillium commune* Thom, *P. nalgiovense* Laxa, and *P. solitum* Westling, occur on fermented meat products and cheeses (Asefa et al. 2009; Filtenborg et al. 1996; Nout and Aidoo 2011).

The greatest varieties of fermented food are present in Asian countries (Nout and Aidoo 2011). Shurtleff and Aoyagi (2011a) indicated that Douchi, a black soybean, is the earliest known fermented or processed soyfood based on the fact that fermented black soybeans were unearthed from the Han Tomb no. 1 (dated to about 165 BC) at Mawangdui near Changsha, Hunan province, in China. Yokotsuka (1985) indicated that China is the birthplace of fermented vegetables and the use of *Aspergillus* and *Rhizopus* microfungi to make processed food. The book called *Shu Jing* or *Shu Ching* written in the Zhou dynasty in China (1121–256 BC) refers to the use of "qu" or "chu" (koji in Japanese) a fermented grain product (Yokotsuka 1985). Qu (koji) is also mentioned in the Zhouli [Rites of the Zhou dynasty] (300 B.C.E.) in China (Shurtleff and Aoyagi 2012). Qu is a culture prepared by inoculating either *Aspergillus oryzae* or *Monascus purpureus* microfungi onto cooked grains and/or soybeans in a warm, humid place. Discovery of qu making was a milestone in Chinese food technology, as it is the basis for three major fermented soyfoods: jiang/miso, soy sauce, and fermented black soybeans, grain-based wines (including Japanese sake), and jiuniang (or Japanese amazake) (Shurtleff and Aoyagi 2012).

Tempeh is a very popular fermented food that originated on the Java island in Indonesia and the only soyfood that did not originate from China or Japan (Shurtleff and Aoyagi 2007, 2011b). It is produced by fermenting dehulled cooked soybean cotyledons with *Rhizopus oligosporus*. Its history was not well documented. It was suggested that Indonesians had been making tempeh for several centuries, probably beginning in the early 1600s (Shurtleff and Aoyagi 2007, 2011b), but it has been speculated that this food may have originated over 2000 years ago (Sastroanijoyo 1971). The earliest known reference to tempeh appeared in 1815 in a classic Javanese literature, the Serat Centhini (see Volume 1, p. 295) (Shurtleff and Aoyagi 2011b) and in 1875 in European literature (Gericke and Rooda 1875). At present, it has been introduced to other countries and commercially produced in Japan, Netherlands, and the USA (Shurtleff and Aoyagi 2007). Soy tempeh contains 19.5 % protein (Shurtleff and Aoyagi 2007).

Wikipedia (2015) listed over 110 fermented foods from around the world. It is an incomplete list. These fermented foods originated from different geographic areas and countries. Some of them were fermented using microfungi. Without fermentation supplied by microfungi, our foods would be much less tasty. For additional information, *Industrial Applications, The Mycota X, 2nd edition* by Nout and Aidoo (2011) is an excellent reference.

## *Yeasts and Alcoholic Beverages*

Yeasts are phylogenetically microfungi, with 1500 species currently described (Kurtzman and Fell 2006) and estimated to be 1.5 % of all described fungal species. Yeasts have been used for thousands of years by human beings for fermenting food and beverages and preserving food (Samuel 1996). Fermentations in the Neolithic times probably originated from natural inoculation by yeasts in the nature. It remains unknown when human beings started to intentionally add selected yeast to make beer, wine, steamed buns, or bread. Samuel (1996) indicated that fermentation with selected yeasts led to speciation of new species in the *Saccharomyces* sensu stricto complex by interspecies hybridization or polyploidization.

Guerra-Doce (2014) indicated that alcoholic fermentation might have been discovered long prior to the domestication of plants and animals during the Neolithic. The earliest alcoholic drink in the world at present was fermented from wild grapes (*Vitis* sp.), or hawthorn fruit (*Crataegus* sp.), rice, and honey uncovered at the early Neolithic village of Jiahu, in the Yellow River Valley in Henan Province, China, ca. 7000–6600 BC according to chemical analysis of traces absorbed and preserved of ancient pottery jars (McGovern et al. 2004; Guerra-Doce 2014). The earliest wine making dated to 6000 BC in Georgia (Vouillamoz et al. 2006). Archaeologists have found similar evidence from residues in two ceramic vessels uncovered at the site of Hajji Firuz Tepe in the Zagros Mountains in Mesopotamia of northwestern Iran, ca. 5400–5000 BC during the Neolithic. Biochemical analysis showed that these vessels contained a resinated wine (McGovern et al. 1996). The earliest evidence of beer brewing was found from a ceramic vessel recovered at the cave of Can Sadurní, Barcelona, in the early Neolithic Europe dated to 5000 cal BC (Blasco et al. 2008). At the same time, evidence of malting, one of the key steps in beer brewing, was found on two grinding stones (Blasco et al. 2008). Archaeological evidence of alcoholic beverages was found in other areas around the world also. Evidence of alcoholic beverages was found in Babylon and dated to 5000 BC (Battcock 1998), 2000 BC in pre-Hispanic Mexico (Battcock 1998), and 1500 BC in Sudan (Dirar 1993; Pederson 1979).

Evidence of alcoholic beverages brewed with *Saccharomyces cerevisiae* was found dating from 3150 BC in Egypt (Cavalieri et al. 2003). White wine was excavated from the tomb of King Tutankhamun, an Egyptian pharaoh of the 18th dynasty (ca. 1332 BC–1323 BC) (Guasch-Jané et al. 2006). The authors sequenced ITS from the residue inside one of the earliest wine jars recovered from the tomb of King Scorpion I in Egypt, and the result of sequencing showed that the fungus responsible for alcoholic fermentation was *Saccharomyces cerevisiae*.

Archaeologists excavated unique cereal alcoholic beverages preserved as liquids inside sealed bronze vessels of the Shang and Western Zhou dynasties (ca. 1600–1046 BC and ca. 1046–722 BC, respectively), and the actual alcoholic beverages still remained in liquid state when the bronze vessels were unearthed after over 3000 years in the ground (McGovern et al. 2004; Anyang Archaeological Team 2004; Henan Institute of Cultural Relics and Archaeology 2000). These findings provide direct evidence for fermented beverages in ancient Chinese culture, which were of considerable social, religious, and medical significance, and helped elucidate their earliest descriptions in the Shang dynasty oracle inscriptions (McGovern et al. 2004).

Archaeologists unearthed a whole set of wine making gear from a tomb of the Dawenkou culture (4100–2600 BC, the Neolithic period) in Lingyinhe, Ju county, Shandong, in 1979 (Bao 2008). Wang (1987) speculated that the tomb occupant could have been a professional wine maker.

Archaeologists excavated a fully equipped winery consisting of basins for winepresses, fermentation vats, storage jars, drinking bowls, and remains of domesticated grapes from a cave complex of Areni-1, a Chalcolithic site in southeast Armenia dated to around 4000 BC.

Quantities of carbonized grape berries (*Vitis vinifera*) in pots dated to 4200 BC were unearthed from the prehistoric site of Dikili Tash in eastern Macedonia, Greece. The significance of the discovery is that the grape berries had been pressed which indicated the extraction of juice from grapes and suggested wine making (Valamoti et al. 2007).

Since Neolithic times, alcoholic drinks have not been essential to the survival of human beings. What is the reason that our human beings discovered alcoholic beverage usage prior to domestication of plants and animals? Does this behavior contribute any fitness to human beings during evolution?

## *Edible Mushrooms as Food and Risk*

Human beings collect mushrooms in the wild as food and for medicinal purposes, probably dating back to the prehistoric period. Mushrooms were used as a delicacy, food, and medication for therapeutic properties and often in religious ceremonies during the early civilizations of the Chinese, the Fertile Crescent (including Egyptians and others), Greeks, Indians, Latin Americans, and Romans (Miles and Chang 2004). Roman emperors prohibited ordinary people from eating mushrooms so that the mushrooms could be strictly reserved for nobility. Mushrooms were highly regarded as luxuries and delicacies by Romans. Recipes for mushrooms suggested by Diphilus of Siphnos dated to 300 BC (Dalby 2003; van Rossenberg 2005). Classical Greek authors considered mushrooms as famine food, similar to acorns. The discovery of a bowl of mushrooms collected from the wild in a Bronze Age house near Nola in Italy is the first evidence that mushrooms were used as food in prehistoric Europe (van Rossenberg 2005).

Chinese ancestors had started to collect edible mushrooms for the table dating back to 4000 BC based on archaeological studies which excavated mushrooms along with rice, *Crataegus hupehensis* Sarg., and other foodstuffs from an ancient site, Hemudu culture, Yuyao in Zhejiang, China, in 1977 (Hemedu Site Museum 2015).

Collecting mushrooms for direct consumption always has some risks. Mycologists often say that bad fungal taxonomy kills. In history, Charles VI, Holy Roman Emperor, was killed by eating death cap mushrooms *Amanita phalloides* (Fr.) Link in 1740 (Wasson 1972). His death led to a war and a diversion of European history. It is possible that the death of Pope Clement VII in 1534 was the result of eating this mushroom also (Wasson 1972). Every year, there are some isolated cases of human death caused by eating poisonous mushrooms due to misidentification around the world. Some cities set up hotlines for mushroom poisoning.

A mysterious epidemic occurred in Yunnan, China, in 1978. Some local villagers suddenly died from an unknown cause. It was reported that victims even dropped dead in the middle of a conversation (Stone 2010). Thus, this mysterious epidemic was referred as sudden unexplained death syndrome (SUDS). From 1978 to 2010, SUD had caused over 400 deaths from sudden cardiac arrest and several dozen nonfatal cardiac cases in the area. The victims are children, adults, and senior people (Zhou et al. 2012; Stone 2010). Prior to 2005, the Chinese CDC had sent scientists to the area to study SUDS several times, but failed to determine the cause. In 2005, a new research team was formed including cardiologists, epidemiologists, mycologists, medical examiners, and medicinal chemists. The "little white mushroom" or "nail-like mushroom" thought to be edible by locals in that area was suggested by the team to be the cause of a proportion of these deaths to local villagers after 5 years of research. SUDS caught attention around the world after (Stone 2010). Two years later, the scientists put the pieces together and determined that the little white mushroom was the cause. They found three mycotoxins including two new ones (amino acids 2R-amino-4S-hydroxy-5-hexynoic acid (1) and 2R-amino-5-hexynoic acid (2) and the known toxin g-guanidinobutyric acid (3)) extracted from fruiting bodies of the little mushroom were responsible for SUDS and found two unusual toxins that killed mice with an $LD_{50}$ of 71 and 84 mg $kg^{-1}$, respectively (Zhou et al. 2012). This little white mushroom was found to be new to science and described as *Trogia venenata* Zhu L. Yang et al. (Cantharellaceae) (Yang et al. 2012b). Hypoglycemia induced by *Trogia* toxin is suggested to be the mechanism. Is this over 30-year mystery fully deciphered? It seems not so. At present, there is a disagreement whether these toxins are the sole culprit. The exact toxin(s) produced by *Trogia venenata* and responsible for SUDS remain not fully established, and the associated mechanism has not been fully confirmed (Graeme 2014; Stone 2012). However, since the public campaign by Chinese CDC to advise local villagers of stopping consumption of *Trogia venenata* in 2009, no new SUDS cases have been reported, and as a public health threat, the case is closed (Stone 2012).

## *Microfungi as Food*

Microfungi play very important roles in fermented food and beverages. Two microfungi are used directly as vegetable/food. Both are smut fungi belonging to *Ustilago*. Jiaobai is a vegetable widely cultivated and consumed in China and several other Asian countries (Terrell and Batra 1982; Guo et al. 2007; Chung and Tzeng 2004; Zhang et al. 2012). The vegetable is *Zizania latifolia* Turcz. (Manchurian ricegrass, broad-leaved wild rice, Asian wild rice, and water oat) infected by *Ustilago esculenta* Henn., which causes host stems to hypertrophy for consumption (Chan and Thrower 1980). *Zizania latifolia* was planted as one of the six major grain crops in China since the Zhou dynasty (from 771 to 221 BC) (Guo et al. 2007). Since Jiaobai was cultivated, *Zizania latifolia* was no longer cultivated as a grain crop. This fungus is federally regulated and quarantined in the USA due to its potential risk to native wild rice (*Zizania aquatica* L.) in North America (Yamaguchi et al. 1990). Thus, the cultivation of this vegetable is banned in the USA.

Another smut-related food is huitlacoche (corn smut, *Ustilago maydis*), which is a delicacy in Mexico. Huitlacoche consumption originates from Aztec cuisine. Only immature galls are harvested for cooking purposes. Fully mature galls lose their cooking value. It is consumed as a filling in quesadillas and other tortilla-based foods and soups. Since the mid-1990s, farms in Florida and Pennsylvania have been permitted by the US Department of Agriculture (USDA) to produce huitlacoche due to demand from high-end restaurants (Pataky and Chandler 2003).

A number of microfungi, *Fusarium* spp., are notorious for producing mycotoxins in cereal crops that have detrimental effects on human and animal health if they enter the food chain as food or animal feeds. However, *Fusarium venenatum* Nirenberg is an exception. One of its strains (IMI 145425, ATCC PTA-2684) has a high protein content and does not produce mycotoxin. The potential utilization of this strain to produce mycoprotein has been studied since the 1960s and has been commercially used for the production of the single-cell protein, mycoprotein (Wiebe 2002). Its product was approved for sale as food in the UK in 1984 (Wiebe 2002).

The mycoproteins are marketed under a brand name, Quorn, and are currently sold in 13 countries including the USA with 90 products ranging from steak strips to burgers to fillets (Marlow Foods Ltd 2014).

Microfungi as potential protein producers or resources as alternatives to animal meats should be studied further.

## Medicinal Fungi

Fungi with medicinal properties are an important resource to modern medicine. Medicinal fungi have been used for a long period of time. Unfortunately, medicinal fungi are often overlooked and it leads to a deficiency in research on medicinal fungi. Only a very small number of medicinal fungi have been utilized in modern medicine (Li 2011). Li (2011) estimated that there are about 1500–2000 medicinal fungi. A majority of them are macrofungi (polypores and mushrooms) and some microfungi (Ying et al. 1987; Li 2011). In China alone, over 540 fungi have been used in Chinese medicine (Li 2011). There are over 220 fungi both having medicinal properties used as medicinal herbs and being edible consumed as food in China (Li 2011).

The discovery of Ötzi the Iceman, a well-preserved natural mummy (c. 3300 BC) discovered from a glacier in Ötztal Alps, near Hauslabjoch on the border between Austria and Italy in 1991, was big news, not only for archaeology and anthropology, but also other fields of science, such as mycology. The iceman was carrying three items of two polyspores, among other artifacts with him (Wikipedia 2014; Fowler 2001). The two polypores were later identified as *Piptoporus betulinus* (Bull.) P. Karst. (for medicinal purposes) and *Fomes fomentarius* (L.) Fr. (for fire initiation) (Peintner et al. 1998).

DNA sequencing from ancient archaeological specimens of polypores discovered by archaeologists in the early Neolithic village of "La Marmotta" (Anguillara Sabazia, Rome) (ca. 7000 years ago) showed that the fungus is *Daedaleopsis tricolor* (Bull.) Bondartsev and Singer (Bernicchia et al. 2006). The authors suggested that the fungus was collected for use during ritual functions or for their pharmacological properties. Montroux and Lundstrom-Baudais (1979) discovered *Fomes* in swampy areas of the Neolithic sites of Clairvaux and Charavines in France that dated between 2900 and 2399 BC.

*Ophiocordyceps sinensis* (Berk.) G. H. Sung et al. have been used for at least 2000 years (Shrestha et al. 2010) for its believed abilities to treat many diseases related to lungs, kidney, and erectile dysfunction (Li 2011; Ying et al. 1987). Li Shi- Zhen (1578) described over 30 medicinal fungi in his historical work on Chinese traditional medicine *Compendium of Materia Medica* including macrofungi, *Auricularia auricula* (L.) Underw., *Ganoderma lingzhi* S. H. Wu, Y. C. and Y. C. Dai (reported as *Ganoderma lucidum* (Curtis) P. Karst prior to 2012), *Morchella* spp., *Omphalia lapidescens* Schroet., *Phellinus igniarius* (L.) Quél., *Polyporus umbellatus* (Pers.) Fr., *Poria cocos* (Schw.) Wolf., *Termitomyces albu-*

*minosus* (Berk.) R. Heim., *Tremella fuciformis* Berk., *Trametes robiniophia* Murr., and microfungi, *Aecidium mori* (Barclay) Dietel, *Ophiocordyceps sinensis* (Berk.) G. H. Sung et al. (*Cordyceps sinensis*), *Tilletia hordei* Körn., *Ustilago crameri* Körn., etc. (Li 1578). Ying et al. (1987) reported 272 medicinal fungi in details from China in his monographic work *Icons of Medicinal Fungi from China*.

In addition to the medicinal microfungi described by Li (1578), there are a number of other microfungi that are often used in Chinese medicine. Use of *Monascus purpureus* dates back over a thousand years in the North Song dynasty in China (Shi and Pan 2011). In addition to use in rice wine making, food fermentation, and industrial applications, *Monascus purpureus* has been used as a medication in Chinese medicine (Ying et al. 1987). This fungus produces several secondary metabolites such as dimerumic acid, pigments, polyketide monacolins, and γ-aminobutyric acid. Studies showed that this fungus is effective for lowering blood cholesterol and the management of diabetes, blood pressure, obesity, and Alzheimer's disease and has anticancer properties (Shi and Pan 2011; Ying et al. 1987; Lee et al. 2006).

*Hypocrella bambusae* (Berk. and Broome) Sacc. is used in Chinese medicine to treat stomach problems and rheumatoid arthritis (Ying et al. 1987). Hypocrellin A produced by *H. bambusae* showed promising antifungal and antileishmanial activities (Ma et al. 2004). *Shiraia bambusicola* Henn. is a pathogen to bamboo, primarily in the southern provinces of China, and its stromata are commonly used in Chinese medicine to treat a variety of disorders in humans (Yang et al. 2009). It also produces a number of perylenequinoid pigments, including hypocrellin A and hypocrellin B, and the hypocrellins showed excellent light-induced antitumor and antiviral activities (Estey et al. 1996; Deininger et al. 2002; Zhang et al. 2004).

Several microfungi causing smut diseases have been used for their medicinal properties, such as *Ustilaginoidea virens* (Cooke) Takah., *Sphacelotheca sorghi* (Ehrenb. ex Link) G.P. Clinton (*Sporisorium sorghi* Ehrenb. ex Link), and several more species of *Ustilago*: *Ustilago esculenta* Henn., *Ustilago maydis* (DC.) Corda, *Ustilago nuda* (C.N. Jensen) Rostrup., *Sclerotinia sclerotiorum* (Lib.) de Bary, *Rhizoctonia* sp., and *Beauveria bassiana* (Bals.-Criv.) Vuill. (Ying et al. 1987; Li 2011).

*Claviceps purpurea* (Fr.) Tul., a notorious mesofungus, causes both plant diseases in a number of host grasses and human ergotism (aka *ignis sacer*, holy fire, or St. Anthony's fire). It also has some useful medicinal properties, which has been known to humans for ca. three millennia (Ainsworth 1986). However, there are conflicting reports on early fungus (ergot) description, its medicinal properties, and epidemics of ergotism. The current brief review on ergot and ergotism is to clarify or verify several contradictory reports in the literature. Among the hosts, rye, *Secale cereale* L., is very susceptible to *C. purpurea*. The fungal pathogen will develop ergot or spurred rye in the infected rye inflorescences. The earliest record of ergot is probably an inscription to Gudea on a Babylonian tablet ca. 2500 BC with a text "the women who gather noxious grasses and who were expelled from the city with the exorcists and mutterers of charms" (Wellcome 1908). Another early record possibly referring to ergot is the "noxious pustule in the ear of grain" on an Assyrian cuneiform tablet ca. 600 BC (Lapinskas 2007; Wellcome 1908; Van Dongen and de

Groot 1995). Written reports of ergotism had not appeared until 857 in the Annales Xantenses "a Great plague of swollen blisters consumed the people by a loathsome rot, so that their limbs were loosened and fell off before death" (Barger 1931). Ergotism is caused by ingesting ergot-contaminated grain products made from rye infected by *C. purpurea*. Ergotism is the earliest known mycotoxicosis in human history (Schlegel 2013). It is a well-studied fungus. A literature search using its scientific name, *Claviceps purpurea* with Google Scholar, yielded 12,800 hits and ergot, 71,800 hits, respectively.

The epidemics caused by *C. purpurea* were not only reported in the scientific literature, but also captured in fine artworks (Pokorny 2003; Musée du Louvre 2015). Pieter Bruegel the Elder (ca. 1525–1569) reflected ergotism epidemics in his several works. The most famous one is an oil painting "The Beggars (The Cripples)" in 1568, which is in the possession of the Louvre in Paris (Fig. 11.1). Richard et al. (2003) opined that this painting actually portrayed the gangrenous ergotism victims of the tragic epidemics of the era. This is the reason the painting was chosen as the cover of the book *Mycotoxins: Risks in Plant, Animal and Human Systems* (Richard et al. 2003). In the description of the painting in the webpage of the Louvre, there are two sentences that attracted our attention: "[t]he underlying significance of the scene, the meaning of the foxtails in particular, remains unclear" and "[m]any hypotheses have been put forward to interpret this painting, particularly addressing the question of what the foxtails, hanging from the garments worn by the beggars,

**Fig. 11.1** The Beggars by Pieter Bruegel the elder (ca. 1525–1569) in 1568. Copyright of Réunion des Musées Nationaux, Louvre, Paris, France

are meant to symbolize" (Musée du Louvre 2015). It is rather possible that the inflorescences hanging from garments in the painting may not be foxtails, but rye, which was a staple crop in the Middle Ages and the sixteenth century in Europe. It seems that Bruegel may have painted ergots to several of the seed heads in his painting (Fig. 11.2). The seed heads of two plants are rather similar in morphology when comparing them to the inflorescences of rye in the illustration of ergots (Secale luxurians) in Caspar Bauhin's *Theatri Botanici* in 1658 (Fig. 11.3). Secale luxurians was the pharmaceutical named by Caspar Bauhin as one of the synonyms for ergot, not for its host plant rye. However, there is no evidence to show whether Pieter Bruegel the Elder had suspected the relationship between ergot and ergotism. Caspar Bauhin's illustration was considered the first one on ergot (Ainsworth 1986; Van Dongen and de Groot 1995). However, the first illustration of ergot should actually go to Adam Lonicer (Adam Lonitzer or Adamus Lonicerus (1528–1586)), a German physician/botanist with a drawing of ergotized rye with a caption of "Hamelkorn" in Germany (Fig. 11.4) and giving a Latin name "Clavi siliginis" for ergot in "Kräuterbuch" possibly in 1569 (Lonicer 1557–1577). Lonicer is also the first one in the west to describe the medicinal properties of ergot and as a medication used by midwives for assisting childbirth (Hofmann 1978; Ainsworth 1986). Several authors had studied ergotism portrayed in a number of masterpieces of fine arts from the Middle Ages, such as "Isenheim Altarpiece" painted for the Monastery of St. Anthony by Matthias Grünewald and "St. Anthony Triptych" by Hieronymus Bosch (ca. 1450–1516) to understand the ergotism epidemics and the roles played by the monasteries of St. Anthony and apothecaries in the Middle Ages from medi-

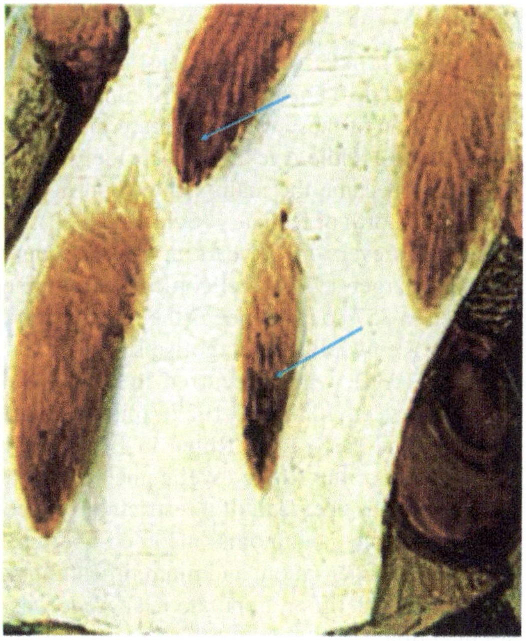

**Fig. 11.2** Detail from "The Beggars" showing possibly ergots by the *arrows*

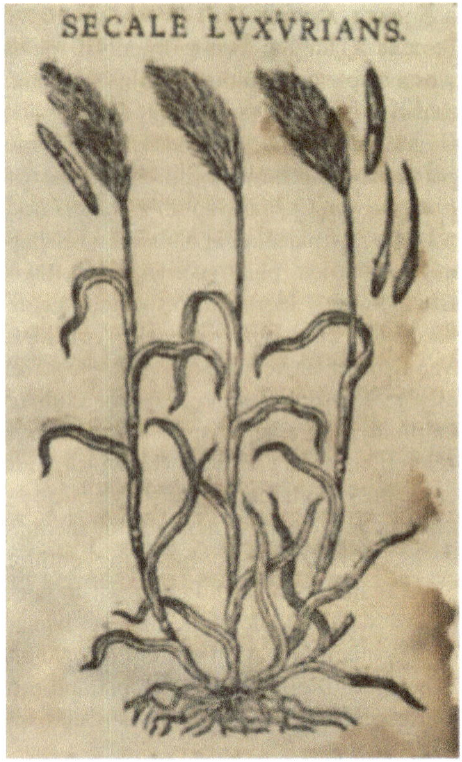


**Fig. 11.3** Drawing of ergots (Secale luxurians) and rye, the host plant by Caspar Bauhin in *Theatri Botanici* (1658)

cal, social, artistic, and religious perspectives (Kierulf 1982; Dixon 1984; de Yébenes and de Yébenes 1990; Morán Suárez 1996; Battin 2009) due to confusing interpretation of these works (Morán Suárez 1996).

Since the Middle Ages, rye was widely cultivated in Central and Eastern Europe. Rye bread became the main staple food in most areas east of the French-German border and north of Hungary (Schlegel 2013). Epidemics of ergotism were common occurrences every 5–10 years in Europe from the Middle Ages to the nineteenth century (Barger 1931). Wellcome (1908) opined that the ergotism epidemics that occurred in 857 AD in France is the first report.

There are two kinds of ergotisms: gangrenous ergotism and convulsive ergotism (Bar ger 1931), which occurred in different geographic areas in Europe (Eadie 2003). Between 1085 and 1927, epidemics of "convulsive ergotism" were widespread in the east of the Rhine in Europe, while gangrenous ergotism occurred mainly west of the Rhine (Wellcome 1908; Eadie 2003). Gangrenous ergotism—ischemia—was the result of a restriction in blood supply to tissues or vasoconstrictive effects. Its symptoms included nausea, limb pain, limbs turning black, impairment of sensation, and mummification, causing infected extremities to spontaneously break off or amputate. Gangrene was sometimes complicated by secondary infection, and the mortality rate was high (Eadie 2003). This type of ergotism is

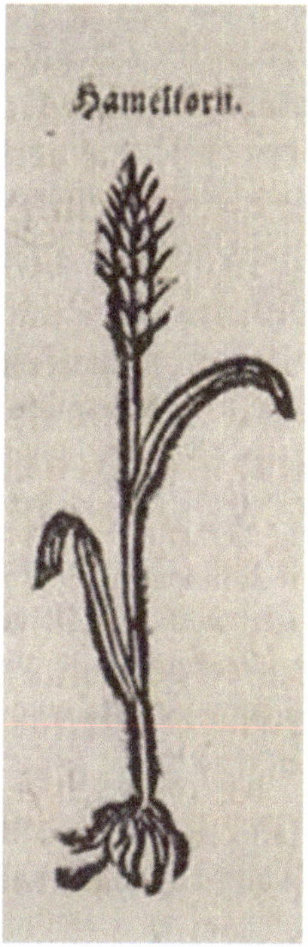


**Fig. 11.4** The earliest illustration of ergotized rye by Adam Lonicer in page 525 of "Kräuterbuch" (1557–1577). The caption is "Hamelkorn" in Germany

characterized by the burning sensation in the extremities that led to the name "ignis sacer", i.e., holy fire (Wellcome 1908). Convulsive ergotism (St. Vitus Dance) resulted in a nervous dysfunction, where the victim is twisting and contorting their body in pain, trembling and shaking, and wryneck, painful seizures, spasms, convulsions, confusions, and delusions. Hallucinations, mania, or psychosis may occur.

The discovery of the cause of gangrenous ergotism in 1630 is attributed to a French physician, Dr. Tuillier (or Thuillier, a later spelling), via his experimentation (Barger 1931). In several literature citations, this date was reported as 1670 or 1676 (Caporael 1976; Schumann 2000; Miedaner and Geiger 2015). However, the controversy about the cause of convulsive ergotism continued to 1800 in Germany (Wellcome 1908; Barger 1931). However, Garrison (1929) opined that the medical faculty at Marburg in 1597 reported the cause of ergotism to be consumption of bread made from spurred rye. It could be the first report to link the disease to the cause. At present, we are not sure whether this report was supported by any experimentation.

The epidemics of ergotism decreased with increased knowledge of the fungus and the mycotoxins it produced, implementation of regulations, and advances in milling procedures (Belser-Ehrlich et al. 2013). Since 1900, ergotism outbreaks in humans have been very infrequent, but not eradicated. Several outbreaks have occurred since then. During 1926–1927, a severe ergotism outbreak occurred in Southern Russia with 10,000 cases of convulsive ergotism (Kent and Evers 1994).

In 1927, Manchester in England suffered an outbreak with >200 cases of gangrenous ergotism (Robertson and Ashby 1928; Kent and Evers 1994). An infamous outbreak occurred in Pont St. Esprit, France, in 1951 with ca. 150 people sick and four fatalities due to muscular spasm and cardiovascular collapse for most cases and one case with moist gangrene at the toes (Gabbai 1951). However, several alternative theories were later proposed to explain this epidemic, such as mercury poisoning. At present, consensus on the true cause of the epidemics has not been reached (Belser-Ehrlich et al. 2013). India suffered a number of outbreaks caused by other members of the genus *Claviceps* with devastating results in 1956–1957 and 1975 due to consumption of *C. fusiformis* ergotized pearl millet, *Pennisetum glaucum* (L.) R.Br. (Krishnamachari and Bhat 1976; Patel et al. 1958), and sorghum (*Sorghum bicolor* (L.) Moench) caused by *Claviceps sorghi* P. Kulkarni et al. and *C. africana* Freder. et al. in 2001 (Navi et al. 2002; Pažoutová et al. 2000). Outbreaks occurred in Ethiopia in 1977 and 2001, respectively, by consuming barley containing ergotized wild oats (Demeke et al. 1979; Urga et al. 2002).

The ergot sclerotia contain high concentrations (up to 2 % of dry weight) of the alkaloid ergotamine (derivatives of 6,8-dimethylergoline and lysergic acid derivatives) with biological activities affecting circulation and neurotransmission (Eadie 2003; Tudzynski et al. 2001). The first clear account of ergot and its poisonous properties is provided by the Persian physician, Abu Mausur Muwaffak Harawi (or Abu Mansur Muwaffaq Heravi) (?950 AD) in his *Book of the Remedies* (Kitab al-abnyia 'an Haqa'iq al-adwiya) (Wellcome 1908). Since his book was written between 968 and 977 AD (Golzari et al. 2012), the date of 950 AD cited by Wellcome is questionable. Ergotamine has been prescribed for various causes of headaches, including migraines. Ergometrine is used to control postpartum hemorrhage (Mahmud et al. 2014), and ergonovine causes contraction of the uterus (Balki et al. 2015). The knowledge that the ergot could be used for the latter was known since the seventeenth century when midwives prepared extracts of ergots for child delivery. Ergotamine in extracts of ergot had first been used to treat migraine headaches in Italy in 1862 (Eadie 2004). The use of ergot as an obstetrical remedy was known to the Chinese (emperor of Zhou dynasty 1100 BC) (p. 123, reference 15) (Bove 1970; Schiff 2006). Natural ergot alkaloid and its derivatives have been used to assist child delivery and treat postpartum hemorrhage; diseases of the ears (including inner auditory tubes), eyes (including cornea), lips, nose, tongue, and skin, as well as hyperthyroid; autonomic nervous regulation; and motion sickness in China (Ying et al. 1987; Li 2011).

At present, clinical ergotism due to ingestion of contaminated cereal grains is rare (Ayarragaray 2014). However, it may occur by medications related to ergot alkaloids administered for treating other medical conditions. An HIV-positive individual was diagnosed with ergotism due to taking ergotamine tartrate for 2 weeks to treat his severe migraine headache (Fröhlich et al. 2010), and a recent case was reported for the same reason (Adam et al. 2014). Pharmaceutical properties of ergot have been used to treat several medical conditions. Its alkaloids are still in use as an oxytocic drug and used to treat migraines and Parkinson's disease (Houghton and Howes 2004; Gizzo et al. 2013; Tepper 2013). The gene for technological biosynthesis of ergot alkaloids recently attracted attention (Li and Unsöld 2006). At present, research has been conducted to identify cellular and molecular factors which determine the response of cancer cells to six ergot alkaloids and their possibility for tumor therapy (Mrusek et al. 2015). Ergoflavin isolated from *Claviceps purpurea* showed cytotoxicity against five human cancer cell lines (Deshmukh et al. 2009). Ergotamine produced by *Claviceps purpurea* was considered a potential biosecurity threat (Paterson 2006; Wilson and Ho 2015). The genome of *Claviceps purpurea* (strain 20.1) had been sequenced being 32.1 Mb in size (Schardl et al. 2013). The authors who studied the dynamic of alkaloid loci of Clavicipitaceae with multi-genome analysis found that the fungi including *C. purpurea* are under selection for alkaloid diversification.

Medicinal properties of microfungi are great sources for the modern medicine and pharmaceutical industry. Since the discovery of penicillin by Alexander Fleming (1922), a number of microfungi have been used to produce medicinal drugs, such as antibiotics, antifungals, immunosuppressants, etc. Using fungi to produce medicinal drugs has been well covered in the literature. However, the microfungi, which have been widely used in some traditional (alternative) medicine since ancient times, are less studied. Some mechanisms of medicinal properties of these fungi remain unclear.

A recent study on the endophytic fungi of 29 traditional Chinese medicinal plants found 31 fungal groups at different taxonomic levels and 73 morphospecies (Huang et al. 2008). Some phenolic compounds coexisted with certain endophytic fungi in the same hosts. The results raise a question whether the endophytic fungi and their secondary metabolites play any role in the medicinal properties of these plants. Endophytic fungi are the new sources of anticancer bioactive compounds (Shukla et al. 2014; Kumar et al. 2014; Deshmukh et al. 2014). The endophytic fungi of medicinal plants and their pharmaceutical properties should be studied in the future.

At present, we have a better understanding of when and how the major fungal phyla and some of their functions evolved. However, there are still some discrepancies in dating fungal evolutionary events. It is clear that fungi have played a significant role in human history and evolution. There is so much we can still learn from the usage of fungi by our ancestors. Fungi are great resources to the human race which should be better studied, conserved, and utilized.

**Acknowledgements** The authors are appreciative to Dr. Di Lu, Nanjing Agricultural University, and Yale University Library for their assistance to obtain some literature which otherwise will not be available to the authors.

## References

Adam G, Kurt T, Cinar C, Sariyildirim A, Resorlu M, Uysal F, Yener AÜ, Ozcan S, Sacar M, Ozdemir H (2014) Ergotamine-induced vasospastic ischemia mimicking arterial embolism: unusual case. Ulus Travma Acil Cerrahi Derg 20(4):291–294

Agrios G (2005) Plant pathology, 5th edn. Academic, New York

Ainsworth G (1986) Introduction to the history of medical and veterinary mycology. Cambridge University Press, Cambridge
Alexopoulos CJ, Mims CM, Blackwell M (1996) Introductory mycology, 4th edn. Wiley, New York
Anyang Archaeological Team (2004) Kaoguxue Jikan. Archaeol Q 15:359–390 (no. M1046:1042)
Asefa DT, Møretrø T, Gjerde RO, Langsrud S, Kure CF, Sidhu MS, Nesbakken T, Skaar I (2009) Yeast diversity and dynamics in the production processes of Norwegian dry-cured meat products. Int J Food Microbiol 133(1):135–140
Ayarragaray JEF (2014) Ergotism: a change of perspective. Ann Vasc Surg 28(1):265–268
Balki M, Erik-Soussi M, Ramachandran N, Carvalho JC (2015) The contractile effects of oxytocin, ergonovine, and carboprost and their combinations: an in vitro study on human myometrial strips. Anesth Analg 120(5):1074–1084
Bao Q-A (2008) The Dawnekou culture relic and wine making. China Brewing 178:100–101
Barger G (1931) Ergot and ergotism. Gurney and Jackson, London
Battcock M (1998) Fermented fruits and vegetables: a global perspective, vol 134. Food & Agriculture Organization, United Nations, Rome
Battin J (2009) Antony's fire (gangrenous ergotism) and medieval iconography. Bull Acad Natl Med 193(8):1925–1936
Beimforde C, Feldberg K, Nylinder S, Rikkinen J, Tuovila H, Dörfelt H, Gube M, Jackson DJ, Reitner J, Seyfullah LJ (2014) Estimating the phanerozoic history of the ascomycota lineages: combining fossil and molecular data. Mol Phylogenet Evol 78:386–98
Belser-Ehrlich S, Harper A, Hussey J, Hallock R (2013) Human and cattle ergotism since 1900: symptoms, outbreaks, and regulations. Toxicol Ind Health 29(4):307–316. doi:10.1177/0748233711432570
Berbee ML, Taylor JW (1993) Dating the evolutionary radiations of the true fungi. Can J Bot 71(8):1114–1127
Berbee ML, Taylor JW (2001) Fungal molecular evolution: gene trees and geologic time. In: McLaughlin DJ, McLaughlin EG, Lemke PA (eds) Systematics and evolution, vol 7B. Springer, p 229–245. doi:10.1007/978-3-662-10189-6_10
Berbee ML, Taylor JW (2007) Rhynie chert: a window into a lost world of complex plant–fungus interactions. New Phytol 174(3):475–479
Berbee ML, Taylor JW (2010) Dating the molecular clock in fungi–how close are we? Fungal Biol Rev 24(1):1–16
Bernicchia A, Fugazzola MA, Gemelli V, Mantovani B, Lucchetti A, Cesari M, Speroni E (2006) DNA recovered and sequenced from an almost 7000 y-old Neolithic polypore, *Daedaleopsis tricolor*. Mycol Res 110(1):14–17
Bidartondo MI, Read DJ, Trappe JM, Merckx V, Ligrone R, Duckett JG (2011) The dawn of symbiosis between plants and fungi. Biol Lett 7(4):574–577. doi:10.1098/rsbl.2010.1203
Blackwell M, Vilgalys R, James T, Taylor J (2012) Fungi. Eumycota: mushrooms, sac fungi, yeast, molds, rusts, smuts, etc. Tree of Life Web Project Retrieved 04–25
Blair JE (2009) Fungi. In: Hedges S, Kumar S (eds) The timetree of life. Oxford University Press, New York, pp 215–219
Blasco A, Edo M, Villalba MJ (2008) Evidencias de procesado y consumo de cerveza en la cueva de Can Sadurní (Begues, Barcelona) durante la Prehistoria. In: IV Congreso del Neolítico Peninsular: 27–30 de noviembre de 2006. Museo Arqueológico de Alicante-MARQ, pp 428–431
Bove FJ (1970) The story of ergot. Karger Publishers, Basel
Braun U, Cook RT (2012) Taxonomic manual of the Erysiphales (powdery mildews). CBS-KNAW Fungal Biodiversity Centre, Utrecht
Bronson AW, Klymiuk AA, Stockey RA, Tomescu AM (2013) A perithecial sordariomycete (Ascomycota, Diaporthales) from the lower cretaceous of Vancouver Island, British Columbia, Canada. Int J Plant Sci 174(3):278–292
Brown R (1940) A bracket fungus from the late Tertiary of southwest Idaho. J Wash Acad Sci 30:422–424

Brundrett MC (2002) Coevolution of roots and mycorrhizas of land plants. New Phytol 154(2):275–304
Caporael LR (1976) Ergotism: the Satan loosed in Salem. Science 192(4234):21–26
Cavalieri D, McGovern P, Hartl D, Mortimer R, Polsinelli M (2003) Evidence for *S. cerevisiae* fermentation in ancient wine. J Mol Evol 57(1):S226–S232. doi:10.1007/s00239-003-0031-2
Chan Y, Thrower L (1980) The host-parasite relationship between *Zizania latifolia* Turcz and *Ustilago esculanta* P. Henn. growth substances in the host-parasite combination. New Phytol 85:225–233
Chung K-R, Tzeng DD (2004) Nutritional requirements of the edible gall-producing fungus *Ustilago esculenta*. J Biol Sci 4(2):246–252
Clutton-Brock J (1999) A natural history of domesticated mammals. Cambridge University Press, Cambridge
Dalby A (2003) Food in the ancient world from A to Z. Psychology Press, London
de Yébenes JG, de Yébenes PG (1990) Dystonia in the painting of Matias Grunewald. Epidemic ergotism in the late middle ages. Arch Neurobiol 54(2):37–40
Deininger MH, Weinschenk T, Morgalla MH, Meyermann R, Schluesener HJ (2002) Release of regulators of angiogenesis following Hypocrellin-A and -B photodynamic therapy of human brain tumor cells. Biochem Biophys Res Commun 298(4):520–530. doi:10.1016/S0006-291X(02)02512-3
Demeke T, Kidane Y, Wuhib E (1979) Ergotism—a report on an epidemic, 1977–78. Ethiop Med J 17(4):107–113
Dennis RL (1970) A middle Pennsylvanian basidiomycete mycelium with clamp connections. Mycologia 62(3):578–584. doi:10.2307/3757529
Deshmukh SK, Mishra PD, Kulkarni-Almeida A, Verekar S, Sahoo MR, Periyasamy G, Goswami H, Khanna A, Balakrishnan A, Vishwakarma R (2009) Anti-inflammatory and anticancer activity of ergoflavin isolated from an endophytic fungus. Chem Biodivers 6(5):784–789
Deshmukh SK, Verekar SA, Bhave SV (2014) Endophytic fungi: a reservoir of antibacterials. Front Microbiol 5:715. doi:10.3389/fmicb.2014.00715
Dirar HA (1993) The indigenous fermented foods of the Sudan: a study in African food and nutrition. CAB international, Wallingford
Dixon LS (1984) Bosch's "St. Anthony Triptych"—an apothecary's apotheosis. Art J 44(2):119–131
Doolittle RF, Feng D-F, Tsang S, Cho G, Little E (1996) Determining divergence times of the major kingdoms of living organisms with a protein clock. Science 271(5248):470–477
Dörfelt H, Schmidt AR (2005) A fossil *Aspergillus* from Baltic amber. Mycol Res 109(08):956–960
Musée du Louvre (2015) The beggars. Musée du Louvre. http://www.louvre.fr/en/oeuvre-notices/beggars. Accessed 1 May 2015
Ducousso M, Béna G, Bourgeois C, Buyck B, Eyssartier G, Vincelette M, Rabevohitra R, Randrihasipara L, Dreyfus B, Prin Y (2004) The last common ancestor of Sarcolaenaceae and Asian dipterocarp trees was ectomycorrhizal before the India–Madagascar separation, about 88 million years ago. Mol Ecol 13(1):231–236
Dugan FM (2011) Conspectus of world ethnomycology: fungi in ceremonies, crafts, diets, medicines, and myths. American Phytopathological Society, St. Paul, MN
Eadie MJ (2003) Convulsive ergotism: epidemics of the serotonin syndrome? Lancet Neurol 2(7):429–434. doi:10.1016/S1474-4422(03)00439-3
Eadie MJ (2004) Ergot of rye—the first specific for migraine. J Clin Neurosci 11(1):4–7. doi:10.1016/j.jocn.2003.05.002
Estey EP, Brown K, Diwu Z, Liu J, Lown JW, Miller GG, Moore RB, Tulip J, McPhee MS (1996) Hypocrellins as photosensitizers for photodynamic therapy: a screening evaluation and pharmacokinetic study. Cancer Chemother Pharmacol 37(4):343–350
Evershed RP, Payne S, Sherratt AG, Copley MS, Coolidge J, Urem-Kotsu D, Kotsakis K, Ozdogan M, Ozdogan AE, Nieuwenhuyse O, Akkermans PMMG, Bailey D, Andeescu R-R, Campbell

S, Farid S, Hodder I, Yalman N, Ozbasaran M, Bicakci E, Garfinkel Y, Levy T, Burton MM (2008) Earliest date for milk use in the Near East and southeastern Europe linked to cattle herding. Nature 455(7212):528–531
FAO (2013) Statistical yearbook 2013: world food and agriculture. FAO (Food and Agriculture Organization of the United Nations), Rome
Field KJ, Rimington WR, Bidartondo MI, Allinson KE, Beerling DJ, Cameron DD, Duckett JG, Leake JR, Pressel S (2015) First evidence of mutualism between ancient plant lineages (Haplomitriopsida liverworts) and Mucoromycotina fungi and its response to simulated Palaeozoic changes in atmospheric CO2. New Phytol 205(2):743–756
Filtenborg O, Frisvad JC, Thrane U (1996) Moulds in food spoilage. Int J Food Microbiol 33(1):85–102
Fleischmann A, Krings M, Mayr H, Agerer R (2007) Structurally preserved polypores from the Neogene of North Africa: *Ganodermites libycus* gen. et sp. nov. (Polyporales, Ganodermataceae). Rev Palaeobot Palynol 145(1–2):159–172. doi:10.1016/j.revpalbo.2006.10.001
Fleming A (1922) On a remarkable bacteriolytic element found in tissues and secretions. Proc R Soc Lond Ser B 93:306–317
Fowler B (2001) Iceman: uncovering the life and times of a prehistoric man found in an Alpine glacier, 1st edn. University of Chicago Press, Chicago, IL
Fröhlich G, Kaplan V, Amann-Vesti B (2010) Holy fire in an HIV-positive man: a case of 21st-century ergotism. Can Med Assoc J 182(4):378–380. doi:10.1503/cmaj.091362
Gabbai L (1951) Ergot poisoning at Pont St. Esprit. Br Med J 2(4732):650–651
Garrard A, Colledge S, Martin L (1996) The emergence of crop cultivation and caprine herding in the" marginal zone" of the southern Levant. In: Harris DR (ed) The origins and spread of agriculture and pastoralism in Eurasia. University College London, London, pp 204–226
Garrison FH (1929) An introduction to the history of medicine, 4th edn. W. B Saunders Company, Philadelphia
Gericke J, Rooda T (1875) Javaansch-Nderduitsch Handwoordenboek (Javanese-Low German Concise Dictionary). Johannes Mueller, Amsterdam
Gizzo S, Patrelli TS, Di Gangi S, Carrozzini M, Saccardi C, Zambon A, Bertocco A, Fagherazzi S, D'Antona D, Nardelli GB (2013) Which uterotonic is better to prevent the postpartum hemorrhage? Latest news in terms of clinical efficacy, side effects, and contraindications a systematic review. Reprod Sci 20(9):1011–1019
Golzari SEJ, Khodadoust K, Alakbarli F, Ghabili K, Islambulchilar Z, Shoja MM, Khalili M, Abbasnejad F, Sheikholeslamzadeh N, Shahabi NM, Hosseini SF, Ansarin K (2012) Sleep paralysis in medieval Persia—the Hidayat of Akhawayni (?–983 AD). Neuropsychiatr Dis Treat 8:229–234. doi:10.2147/NDT.S28231
Gori K, Cantor M, Jakobsen M, Jespersen L (2011) Production of bread, cheese and meat. In: Hofrichter M (ed) Industrial applications, vol 10, The mycota. Springer, Berlin, pp 3–27. doi:10.1007/978-3-642-11458-8_1
Gornicki P, Faris JD (2014) Rewiring the wheat reproductive system to harness heterosis for the next wave of yield improvement. Proc Natl Acad Sci 111(25):9024–9025. doi:10.1073/pnas.1407956111
Graeme K (2014) Mycetism: a review of the recent literature. J Med Toxicol 10(2):173–189. doi:10.1007/s13181-013-0355-2
Gripon J (1999) Mould-ripened cheeses. In: Fox P (ed) Cheese: chemistry, physics and microbiology, 2nd edn. Springer, Berlin, pp 111–136
Gryganskyi AP, Humber RA, Smith ME, Miadlikovska J, Wu S, Voigt K, Walther G, Anishchenko IM, Vilgalys R (2012) Molecular phylogeny of the entomophthoromycota. Mol Phylogenet Evol 65(2):682–694. doi:10.1016/j.ympev.2012.07.026
Guasch-Jané MR, Andrés-Lacueva C, Jáuregui O, Lamuela-Raventós RM (2006) First evidence of white wine in ancient Egypt from Tutankhamun's tomb. J Archaeol Sci 33(8):1075–1080
Gueidan C, Ruibal C, De Hoog G, Schneider H (2011) Rock-inhabiting fungi originated during periods of dry climate in the late Devonian and middle Triassic. Fungal Biol 115(10):987–996

Guerra-Doce E (2014) The origins of inebriation: archaeological evidence of the consumption of fermented beverages and drugs in prehistoric Eurasia. J Archaeol Method Theory:1–32. doi:10.1007/s10816-014-9205-z
Guo M-R (1976) History of China 中国史稿. People's Publishing House, Beijing
Guo H, Li S, Peng J, Ke W (2007) *Zizania latifolia* Turcz. Cultivated in China. Genet Res Crop Evol 54(6):1211–1217
Heckman DS, Geiser DM, Eidell BR, Stauffer RL, Kardos NL, Hedges SB (2001) Molecular evidence for the early colonization of land by fungi and plants. Science 293(5532): 1129–1133
Hemedu Site Museum (2015) Hemedu site museum. http://www.hemudusite.com
Henan Institute of Cultural Relics and Archaeology (2000) Luyi Taiqinggong Changzikou mu (Zhongzhou Classical Texts, Zhengzhou, China)
Hibbett DS, Matheny PB (2009) The relative ages of ectomycorrhizal mushrooms and their plant hosts estimated using Bayesian relaxed molecular clock analyses. BMC Biol 7(1):13
Hibbett D, Grimaldi D, Donoghue M (1995) Cretaceous mushrooms in amber. Nature 377:487
Hibbett D, Grimaldi D, Donoghue M (1997) Fossil mushrooms from miocene and cretaceous ambers and the evolution of homobasidiomycetes. Am J Bot 84(7):981
Hipsley CA, Müller J (2014) Beyond fossil calibrations: realities of molecular clock practices in evolutionary biology. Front Genet 5:138. doi:10.3389/fgene.2014.00138
Hofmann A (1978) Historical view on ergot alkaloids. Pharmacology 16(Suppl 1):1–11
Honegger R, Edwards D, Axe L (2013) The earliest records of internally stratified cyanobacterial and algal lichens from the Lower Devonian of the Welsh Borderland. New Phytol 197(1):264–275. doi:10.1111/nph.12009
Houghton P, Howes M (2004) Natural products and derivatives affecting neurotransmission relevant to Alzheimer's and Parkinson's disease. Neurosignals 14(1–2):6–22
Huang W, Cai Y, Hyde K, Corke H, Sun M (2008) Biodiversity of endophytic fungi associated with 29 traditional Chinese medicinal plants. Fungal Divers 33:61–75
Kent NL, Evers A (1994) Technology of cereals: an introduction for students of food science and agriculture. Woodhead Publishing, Cambridge
Kierulf H (1982) Ergotism in Bosch and Grunewald. Nouv Presse Med 11(38):2803–2809
Krebs CA (2003) Groundbreaking scientific experiments, inventions, and discoveries of the ancient world. Greenwood Publishing Group, Westport, CT
Krings M, Taylor TN, Hass H, Kerp H, Dotzler N, Hermsen EJ (2007) Fungal endophytes in a 400-million-yr-old land plant: infection pathways, spatial distribution, and host responses. New Phytol 174(3):648–657. doi:10.1111/j.1469-8137.2007.02008.x
Krings M, Dotzler N, Galtier J, Taylor TN (2011) Oldest fossil basidiomycete clamp connections. Mycoscience 52(1):18–23
Krings M, Taylor TN, Dotzler N (2013) Fossil evidence of the zygomycetous fungi. Persoonia 30:1–10. doi:10.3767/003158513X664819
Krings M, Taylor TN, Taylor EL, Kerp H, Dotzler N (2014) First record of a fungal "sporocarp" from the Lower Devonian Rhynie chert. Palaeobiodivers Palaeoenviron 94(2):221–227
Krishnamachari K, Bhat R (1976) Poisoning by ergoty bajra (pearl millet) in man. Indian J Med Res 64(11):1624–1628
Kumar V, Rai S, Gaur P, Fatima T (2014) Endophytic fungi: novel sources of anticancer molecules. In: Verma VC, Gange AC (eds) Advances in endophytic research. Springer, New Delhi, pp 389–422
Kurtzman CP, Fell JW (2006) Yeast systematics and phylogeny—implications of molecular identification methods for studies in ecology. In: Gábor P, Rosa C (eds) Biodiversity and ecophysiology of yeasts. Springer, Berlin, pp 11–30
Lapinskas V (2007) A brief history of ergotism: from St. Anthony's fire and St. Vitus's dance until today. Medicicinos Teorija ir Praktika 13(2):202–206
Lee C-L, Tsai T-Y, Wang J-J, Pan T-M (2006) In vivo hypolipidemic effects and safety of low dosage *Monascus* powder in a hamster model of hyperlipidemia. Appl Microbiol Biotechnol 70(5):533–540

LePage B, Currah R, Stockey R, Rothwell G (1997) Fossil ectomycorrhizae from the Middle Eocene. Am J Bot 84(3):410

Li S-Z (1578) Compendium Materia Medica 本草纲目.

Li Y (2011) Fungal resource. China Agricultural Press, Beijing

Li S-M, Unsöld IA (2006) Post-genome research on the biosynthesis of ergot alkaloids. Planta Med 72(12):1117–1120

Lioe HN, Apriyantono A, Yasuda M (2012) Soy sauce: typical aspects of Japanese Shoyu and Indonesian kecap. In: Hui YH (ed) Handbook of plant-based fermented food and beverage technology, 2nd edn. CRC Press, Boca Raton, FL, p 821

Liu S, Zhang D, Chen J, Zhu Y (2013) History of solid-state fermented foods and beverages. In: Chen J, Zhu Y (eds) Solid state fermentation for foods and beverages, Fermented foods and beverages series. CRC Press, Boca Roton, FL, pp 95–118. doi:10.1201/b16054-5

Lonicer A (1557–1577) Kreuterbuch. Frankfurt, Germany

Lücking R, Huhndorf S, Pfister DH, Plata ER, Lumbsch HT (2009) Fungi evolved right on track. Mycologia 101(6):810–822

Luoyang Museum (2014) Introduction to historical relic treasures exhibit of LuoYang Museum 洛阳文物珍宝展展品介绍. http://www.lymuseum.com/?action=5&bid=3&cid=12&did=285. Accessed 23 Apr 2014

Ma G, Khan SI, Jacob MR, Tekwani BL, Li Z, Pasco DS, Walker LA, Khan IA (2004) Antimicrobial and antileishmanial activities of hypocrellins A and B. Antimicrob Agents Chemother 48(11):4450–4452

Mahmud G, Javaid K, Tasnim N, Tabassum A, Bangash KT (2014) Where does ergometrine stand in prevention of postpartum haemorrhage in caesarean section? J Pak Med Assoc 64(8):911–914

Marlow Foods Ltd (2014) Quorn. http://www.quorn.com.

Matsunaga KK, Stockey RA, Tomescu AM (2013) *Honeggeriella complexa* gen. et sp. nov., a heteromerous lichen from the lower cretaceous of Vancouver Island (British Columbia, Canada). Am J Bot 100(2):450–459. doi:10.3732/ajb.1200470

McGovern PE, Glusker DL, Exner LJ, Voigt MM (1996) Neolithic resinated wine. Nature 381(6582):480–481

McGovern PE, Zhang J, Tang J, Zhang Z, Hall GR, Moreau RA, Nuñez A, Butrym ED, Richards MP, Wang C-S (2004) Fermented beverages of pre-and proto-historic China. Proc Natl Acad Sci USA 101(51):17593–17598

Miedaner T, Geiger HH (2015) Biology, genetics, and management of ergot (*Claviceps* spp.) in rye, sorghum, and pearl millet. Toxins 7(3):659–678

Miles PG, Chang S-T (2004) Mushrooms: cultivation, nutritional value, medicinal effect, and environmental impact. CRC press, Boca Raton, FL

Montroux O, Lundstrom-Baudais K (1979) Polyporace´es des sites ne´olithiques de Chairvaux et Charavines (France). Candollea 34:153–166

Morán Suárez I (1996) The fire of Saint Anthony: a study of ergotism in the painting of Bosch. Asclepio 48(2):173–193

Moyersoen B (2006) *Pakaraimaea dipterocarpacea* is ectomycorrhizal, indicating an ancient Gondwanaland origin for the ectomycorrhizal habit in Dipterocarpaceae. New Phytol 172(4):753–762

Mrusek M, Seo E-J, Greten HJ, Simon M, Efferth T (2015) Identification of cellular and molecular factors determining the response of cancer cells to six ergot alkaloids. Invest New Drugs 33(1):32–44

Navi S, Bandyopadhyay R, Rao TN, Tooley P (2002) An outbreak of sorghum ergot in parts of Andhra Pradesh, India. Int Sorghum Millets Newsl 43:68–70

Nout MR, Aidoo KE (2011) Asian fungal fermented food. In: Hofrichter M (ed) Industrial applications, vol X, 2nd edn. Springer, Berlin, pp 29–58

Oberwinkler F (2012) Evolutionary trends in basidiomycota. Stapfia 96:45–104

Padamsee M, Kumar TA, Riley R, Binder M, Boyd A, Calvo AM, Furukawa K, Hesse C, Hohmann S, James TY (2012) The genome of the xerotolerant mold *Wallemia sebi* reveals adaptations to osmotic stress and suggests cryptic sexual reproduction. Fungal Genet Biol 49(3):217–226

Pataky JK, Chandler MA (2003) Production of huitlacoche, *Ustilago maydis*: timing inoculation and controlling pollination. Mycologia 95(6):1261–1270

Patel T, Boman T, Dalal U (1958) An epidemic of ergot poisoning through ingestion of infected bajri (*Pennisetum typhoideum*) in southern parts of Bombay State. Indian J Med Sci 12(4):257

Paterson RRM (2006) Fungi and fungal toxins as weapons. Mycol Res 110(9):1003–1010

Pažoutová S, Bandyopadhyay R, Frederickson DE, Mantle PG, Frederiksen RA (2000) Relations among sorghum ergot isolates from the Americas, Africa, India, and Australia. Plant Dis 84(4):437–442. doi:10.1094/PDIS.2000.84.4.437

Pederson CS (1979) Microbiology of food fermentations, 2nd edn. AVI Publishers, Westport, CT

Peintner U, Pöder R, Pümpel T (1998) The iceman's fungi. Mycol Res 102(10):1153–1162

Pirozynski K (1976) Fossil fungi. Annu Rev Phytopathol 14(1):237–246

Poinar GO Jr, Buckley R (2007) Evidence of mycoparasitism and hypermycoparasitism in early cretaceous amber. Mycol Res 111(4):503–506. doi:10.1016/j.mycres.2007.02.004

Pokorny E (2003) Bosch's cripples and drawings by his imitators. Master Drawings 41(3):293–304

Prieto M, Wedin M (2013) Dating the diversification of the major lineages of Ascomycota (Fungi). PLoS One 8(6), e65576

Redecker D, Kodner R, Graham LE (2000) Glomalean fungi from the Ordovician. Science 289(5486):1920–1921

Revedin A, Aranguren B, Becattini R, Longo L, Marconi E, Lippi MM, Skakun N, Sinitsyn A, Spiridonova E, Svoboda J (2010) Thirty thousand-year-old evidence of plant food processing. Proc Natl Acad Sci 107(44):18815–18819

Richard J, Payne G, Desjardins A, Maragos C, Norred W, Pestka J (eds) (2003) Mycotoxins: risks in plant, animal and human systems, vol 139. CAST Task Force Report

Rikkinen J, Poinar GO (2008) A new species of *Phyllopsora* (Lecanorales, lichen-forming Ascomycota) from Dominican amber, with remarks on the fossil history of lichens. J Exp Bot 59(5):1007–1011. doi:10.1093/jxb/ern004

Robertson J, Ashby HT (1928) Ergot poisoning among rye bread consumers. Br Med J 1(3503):302

Rutschmann F (2006) Molecular dating of phylogenetic trees: a brief review of current methods that estimate divergence times. Divers Distrib 12(1):35–48. doi:10.1111/j.1366-9516.2006.00210.x

Sadowski EM, Beimforde C, Gube M, Rikkinen J, Singh H, Seyfullah LJ, Heinrichs J, Nascimbene PC, Reitner J, Schmidt AR (2012) The anamorphic genus *Monotosporella* (Ascomycota) from Eocene amber and from modern Agathis resin. Fungal Biol 116(10):1099–1110

Salque M, Bogucki PI, Pyzel J, Sobkowiak-Tabaka I, Grygiel R, Szmyt M, Evershed RP (2013) Earliest evidence for cheese making in the sixth millennium bc in northern Europe. Nature 493(7433):522–525, http://www.nature.com/nature/journal/v493/n7433/abs/nature11698.html#supplementary-information

Samuel D (1996) Investigation of ancient Egyptian baking and brewing methods by correlative microscopy. Science 273:488–490

Samuel D (1999) A new look at old bread: ancient Egyptian baking. Archaeology Int 3:28–31

Sastroanijoyo M (1971) An answer to the world food crisis. Australian National University Reporter 2:19

Schardl CL, Young CA, Hesse U, Amyotte SG, Andreeva K, Calie PJ, Fleetwood DJ, Haws DC, Moore N, Oeser B (2013) Plant-symbiotic fungi as chemical engineers: multi-genome analysis of the Clavicipitaceae reveals dynamics of alkaloid loci. PLoS Genet 9(2), e1003323

Schiff PL (2006) Ergot and its alkaloids. Am J Pharm Educ 70(5):98. doi:10.5688/aj700598

Schlegel RH (2013) Rye: genetics, breeding, and cultivation. CRC Press, Boca Raton, FL

Schlütz F, Shumilovskikh LS (2013) On the relation of *Potamomyces armatisporus* to the fossil form-type *Mediaverrunites* and its taxonomical and ecological implications. Fungal Ecol 6(4):309–315. doi:10.1016/j.funeco.2013.03.007

Schmidt AR, Beimforde C, Seyfullah LJ, Wege S-E, Dörfelt H, Girard V, Grabenhorst H, Gube M, Heinrichs J, Nel A (2014) Amber fossils of sooty moulds. Rev Palaeobot Palynol 200:53–64

Schumann GL (2000) Ergot of rye. APS. http://www.apsnet.org/edcenter/intropp/lessons/fungi/ascomycetes/Pages/Ergot.aspx. Accessed 1 May 2015

Schüßler A, Walker C (2011) Evolution of the 'plant-symbiotic' fungal phylum, glomeromycota. In: Pöggeler S, Wöstemeyer J (eds) Evolution of fungi and fungal-like organisms. Springer, Berlin, pp 163–185

Schwendemann AB, Taylor TN, Taylor EL, Krings M, Dotzler N (2009) *Combresomyces cornifer* from the Triassic of Antarctica: evolutionary stasis in the Peronosporomycetes. Rev Palaeobot Palynol 154(1–4):1–5. doi:10.1016/j.revpalbo.2008.11.003

Shevchenko A, Yang Y, Knaust A, Thomas H, Jiang H, Lu E, Wang C, Shevchenko A (2014) Proteomics identifies the composition and manufacturing recipe of the 2500-year old sourdough bread from Subeixi cemetery in China. J Proteomics 105:363–371. doi:10.1016/j.jprot.2013.11.016

Shi Y-C, Pan T-M (2011) Beneficial effects of *Monascus purpureus* NTU 568-fermented products: a review. Appl Microbiol Biotechnol 90(4):1207–1217

Shrestha B, Zhang W, Zhang Y, Liu X (2010) What is the Chinese caterpillar fungus *Ophiocordyceps sinensis* (Ophiocordycipitaceae)? Mycology 1(4):228–236. doi:10.1080/21501203.2010.536791

Shukla S, Habbu P, Kulkarni V, Jagadish K, Pandey A, Sutariya V (2014) Endophytic microbes: a novel source for biologically/pharmacologically active secondary metabolites. Asian J Pharmacol Toxicol 2(3):1–16

Shurtleff W, Aoyagi A (2007) History of Tempeh. Soyinfo Center, Lafayette, CA

Shurtleff W, Aoyagi A (2011a) History of fermented black soybeans (165 BC to 2011). Soyinfo Center, Lafayette, CA

Shurtleff W, Aoyagi A (2011b) History of tempeh and tempeh products (1815–2011): extensively annotated bibliography and sourcebook. Soyinfo Center, Lafayette, CA

Shurtleff W, Aoyagi A (2012) History of Koji-Grains and/or soybeans enrobed with a mold culture (300 BCE to 2012): extensively annotated bibliography and sourcebook. Soyinfo Center, Lafayette, CA

Stone R (2010) Will a midsummer's nightmare return? Science 329(5988):132–134. doi:10.1126/science.329.5988.132

Stone R (2012) Heart-stopping revelation about how Chinese mushroom kills. Science 335(6074):1293

Strullu-Derrien C, Kenrick P, Pressel S, Duckett JG, Rioult J-P, Strullu D-G (2014) Fungal associations in *Horneophyton ligneri* from the Rhynie Chert (c. 407 million year old) closely resemble those in extant lower land plants: novel insights into ancestral plant–fungus symbioses. New Phytol 203(3):964–979. doi:10.1111/nph.12805

Sung G-H, Poinar GO Jr, Spatafora JW (2008) The oldest fossil evidence of animal parasitism by fungi supports a cretaceous diversification of fungal–arthropod symbioses. Mol Phylogenet Evol 49(2):495–502. doi:10.1016/j.ympev.2008.08.028

Takamatsu S (2013) Origin and evolution of the powdery mildews (Ascomycota, Erysiphales). Mycoscience 54(1):75–86. doi:10.1016/j.myc.2012.08.004

Tamura K, Battistuzzi FU, Billing-Ross P, Murillo O, Filipski A, Kumar S (2012) Estimating divergence times in large molecular phylogenies. Proc Natl Acad Sci 109(47):19333–19338

Taylor JW, Berbee ML (2006) Dating divergences in the fungal tree of life: review and new analyses. Mycologia 98(6):838–849

Taylor TN, Krings M (2005) Fossil microorganisms and land plants: associations and interactions. Symbiosis 40(3):119–135

Taylor TN, Taylor EL (1997) The distribution and interactions of some Paleozoic fungi. Rev Palaeobot Palynol 95(1):83–94

Taylor TN, Hass H, Kerp H, Krings M, Hanlin RT (2005) Perithecial ascomycetes from the 400 million year old Rhynie chert: an example of ancestral polymorphism. Mycologia 97(1):269–285. doi:10.3852/mycologia.97.1.269

Taylor TN, Krings M, Taylor EL (2015) Fungal diversity in the fossil record. In: McLaughling DJ, Spatafora JW (eds) Systematics and evolution, vol Part B. The Mycota, vol VII, Part B, 2nd edn. Springer, Heidelberg, pp 259–278

Tedersoo L, May TW, Smith ME (2010) Ectomycorrhizal lifestyle in fungi: global diversity, distribution, and evolution of phylogenetic lineages. Mycorrhiza 20(4):217–263

Tepper SJ (2013) Orally inhaled dihydroergotamine: a review. Headache J Head Face Pain 53(S2):43–53

Terrell EE, Batra LR (1982) *Zizania latifolia* and *Ustilago esculenta*, a grass-fungus association. Econ Bot 36(3):274–285. doi:10.1007/BF02858549

Thomas GM, Poinar GO Jr (1988) A fossil *Aspergillus* from Eocene Dominican amber. J Paleontol 62(1):141–143

Tudzynski P, Correia T, Keller U (2001) Biotechnology and genetics of ergot alkaloids. Appl Microbiol Biotechnol 57(5–6):593–605

Urga K, Debella A, Agata N, Bayu A, Zewdie W (2002) Laboratory studies on the outbreak of gangrenous ergotism associated with consumption of contaminated barley in Arsi, Ethiopia. Ethiop J Health Dev 16(3):317–323

Valamoti S, Mangafa M, Koukouli-Chrysanthaki C, Malamidou D (2007) Grape-pressings from northern Greece: the earliest wine in the Aegean? Antiquity 81(Pt 311):54–61

Van Dongen PW, de Groot AN (1995) History of ergot alkaloids from ergotism to ergometrine. Eur J Obstet Gynecol Reprod Biol 60(2):109–116

van Rossenberg E (2005) The discovery of an early bronze age village at Nola (Campania, Italy). The Pompeii premise put to the test. Profiel:6

Vijaykrishna D, Jeewon R, Hyde KD (2006) Molecular taxonomy, origins and evolution of freshwater ascomycetes. Fungal Divers 23(1):351–390

Vouillamoz JF, McGovern PE, Ergul A, Söylemezoğlu G, Tevzadze G, Meredith CP, Grando MS (2006) Genetic characterization and relationships of traditional grape cultivars from Transcaucasia and Anatolia. Plant Genet Res 4(02):144–158. doi:10.1079/PGR2006114

Wang S (1987) Wine brewing in the late phase of the Dawenkou culture. Chin Cuisine 9:3–5

Wasson RG (1972) The death of Claudius or mushrooms for murderers. Botanical Museum Leaflets. Harvard University, Cambridge, pp 101–128

Wellcome HS (1908) From ergot to" Ernutin": An historical Sketch. Burroughs Wellcome & Co., London

Wiebe M (2002) Myco-protein from *Fusarium venenatum*: a well-established product for human consumption. Appl Microbiol Biotechnol 58(4):421–427. doi:10.1007/s00253-002-0931-x

Wikipedia (2014) Ötzi. http://en.wikipedia.org/wiki/%C3%96tzi

Wikipedia (2015) List of fermented foods. http://en.wikipedia.org/wiki/List_of_fermented_foods

Wilson BA, Ho M (2015) The biosecurity threat posed by biological toxins. In: Gopalakrishnakone P, Balali-Mood M, Llewellyn L (eds) Biological toxins and bioterrorism, toxinology. Springer, Berlin, p 604

Yamaguchi M, Janick J, Simon J (1990) Asian vegetables. In: Advances in new crops. Proceedings of the first national symposium 'New crops: research, development, economics', Indianapolis, IN, USA, 23–26 Oct 1988. Timber Press, pp 387–390

Yang H, Xiao C, Ma W, He G (2009) The production of hypocrellin colorants by submerged cultivation of the medicinal fungus *Shiraia bambusicola*. Dyes Pigment 82(2):142–146. doi:10.1016/j.dyepig.2008.12.012

Yang E, Xu L, Yang Y, Zhang X, Xiang M, Wang C, An Z, Liu X (2012a) Origin and evolution of carnivorism in the Ascomycota (fungi). Proc Natl Acad Sci 109(27):10960–10965. doi:10.1073/pnas.1120915109

Yang ZL, Li Y, Tang L, Shi G, Zeng G (2012b) *Trogia venenata* (Agaricales), a novel poisonous species which has caused hundreds of deaths in southwestern China. Mycol Prog 11(4):937–945

Ying J, Mao X, Ma Q, Zong Y, Wen H (1987) Icons of medicinal fungi from China. Science Press, Beijing, China

Yokotsuka T (1985) Traditional fermented soybean foods. In: Moo-Young M (ed) Comprehensive biotechnology: the principles, applications, and regulations of biotechnology in industry, agriculture, and medicine. FAO, pp 395–427

Yu Y-N, Zhuang W-Y (2006) Foreward of Flora Fungorum Sinicorum. In: Zang M (ed) Flora fungorum sinicorum: 'Boletaceae'(1), vol 44. Science Press, Beijing

Yuan X, Xiao S, Taylor TN (2005) Lichen-like symbiosis 600 million years ago. Science 308(5724):1017–1020

Zhang LZ, Tang G-Q, Gao B-W, Zhang G-L (2004) Spectroscopic studies on the excited-state properties of the light-induced antiviral drug hypocrellin A loaded in the mesoporous solid. Chem Phys Lett 396(1–3):102–109. doi:10.1016/j.cplett.2004.07.124

Zhang J-Z, Chu F-Q, Guo D-P, Hyde KD, Xie G-L (2012) Cytology and ultrastructure of interactions between *Ustilago esculenta* and *Zizania latifolia*. Mycol Prog 11(2):499–508

Zhou ZY, Shi GQ, Fontaine R, Wei K, Feng T, Wang F, Wang GQ, Qu Y, Li ZH, Dong ZJ (2012) Evidence for the natural toxins from the mushroom *Trogia venenata* as a cause of sudden unexpected death in Yunnan Province, China. Angew Chem Int Ed 51(10):2368–2370

# Chapter 12
# Phylogenetic Diversity of Fungi in the Sea including the *Opisthosporidia*

**Ka-Lai Pang and E.B. Gareth Jones**

## Introduction

Kohlmeyer and Kohlmeyer (1979) stated that the ocean is a stable environment with little fluctuation of temperature and salinity and a lack of growth substrates, which exerts little selection pressure for the evolution of diverse lineages of marine fungi. This opinion may have been true prior to molecular sequencing of biodiversity, but modern techniques have revealed a much greater diversity (Nagahama et al. 2003; Arfi et al. 2012). The open ocean has more or less constant physical conditions in terms of salinity and temperature, and there is a general lack of growth substrates for fungi. However, the majority of marine fungi found so far have been reported from the terrestrial/ocean interface where the physical conditions can change considerably seasonally or even daily. For instance, salinity of mangrove water varies from 5 ‰ in Mai Po Mangrove, Hong Kong (Jones 2000), to 46 ‰ in mangroves of the Red Sea (El-Sharouny et al. 1998). In addition, plentiful organic materials are available for fungal colonization including wood; herbaceous plants such as mangrove leaves, sea grasses, and palm fronds; dead animals; and algae (Luo and Pang 2014). A recent documentation of over 1000 marine fungi (Jones et al. 2015) and an estimated number of 10,000 (Jones 2011) further suggest that there is a high fungal diversity in the sea. Marine fungi mostly

K.-L. Pang, Ph.D. (✉)
Institute of Marine Biology and Center of Excellence for the Oceans, National Taiwan Ocean University, 2 Pei-Ning Road, Keelung 20224, Taiwan (R.O.C.)
e-mail: klpang@ntou.edu.tw

E.B.G. Jones, Ph.D.
Botany and Microbiology Department, College of Science, King Saud University, Riyadh 1145, Saudi Arabia
e-mail: torperadgj@gmail.com



D.-W. Li (ed.), *Biology of Microfungi*, Fungal Biology,
DOI 10.1007/978-3-319-29137-6_12

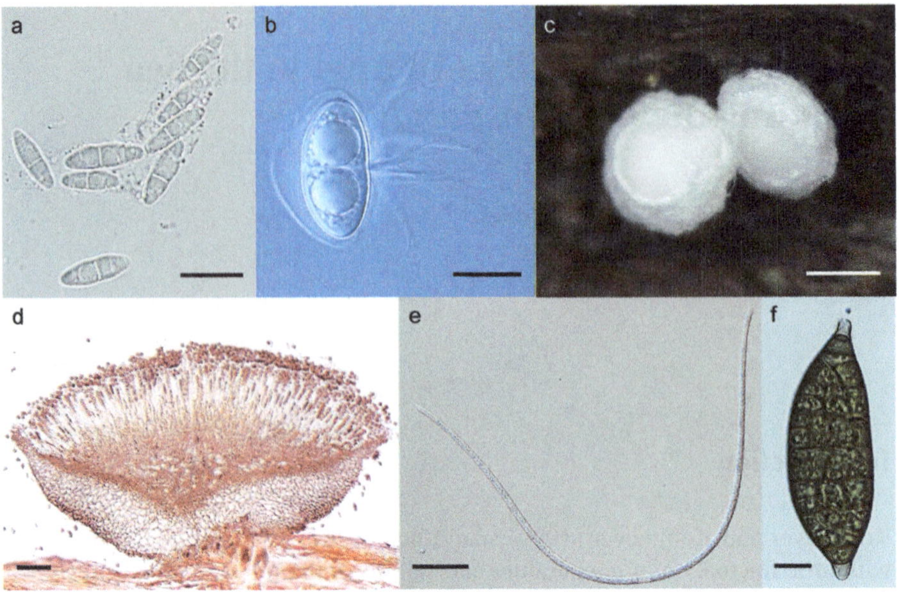

Fig. 12.1 Marine fungi. (**a**) A deliquescing ascus. (**b**) *Nimbospora effusa* with large oil globules, sheath, and appendages. (**c**) *Halocyphina villosa* on mangrove wood. (**d**) A section of the discomycete *Dactylospora vrijmoediae*. (**e**) A filamentous ascospore of *Lulworthia* sp. (**f**) A muriform ascospore of *Aigialus grandis*. Scale bar: **a**–**b**, **f** = 10 μm; **c** = 2 mm; **e** = 30 μm

belong to the *Ascomycota* and asexual forms with only a few basidiomycetes and chytrids (Jones et al. 2009a). Many marine fungi, especially in the family *Halosphaeriaceae*, have evolved special morphological features as an adaptation to a marine life, including deliquescing asci (Fig. 12.1a) for aquatic dispersal of ascospores in water, large oil globules (Fig. 12.1b) in ascospores for flotation, and elaborate ascospore appendages/sheaths (Fig. 12.1b) for attachment to substrates in the sea (Pang 2002). The diverse morphology of ascospore appendages has provided key characters for the delineation of genera and species in the *Halosphaeriaceae* and other groups (Jones 1994).

Early studies in the field of marine mycology were descriptive, with mostly diversity studies of various substrata (Kohlmeyer and Kohlmeyer 1979) and taxonomic studies to classify marine fungi based on morphology at light and electron microscopic levels (Jones and Moss 1978; Jones 1995). The taxonomic outline of marine fungi by Johnson and Sparrow (1961) listed 2 orders, 5 families, and 65 genera. Later, Kohlmeyer and Kohlmeyer (1979) classified the marine fungi into 7 orders, 17 families, and 61 genera. With the advent of phylogenetic inference using sequence data, many previously thought to be closely related taxa have been shown to be distantly related and subsequently transferred to other orders and families (Suetrong et al. 2009; Jones et al. 2014). New orders and families have also been established for taxa forming independent lineages, e.g., *Lulworthiales* (Kohlmeyer et al. 2000), *Savoryellales* (Boonyuen et al. 2011), and *Dyfrolomycetales*

(Hyde et al. 2013). A recent update by Jones et al. (2009a) documented 530 marine fungi belonging to 29 orders, 56 families, and 209 genera. This observation suggests that many morphological characters of marine fungi are possibly gained through convergent evolution and therefore are not reliable taxonomic characters. It also shows the high phylogenetic diversity of marine fungi in the sea.

In this chapter, we provide an updated account of the phylogenetic diversity of fungi in the sea. We also include current knowledge of the *Cryptomycota*, the *Microsporidia*, and the new phylum *Aphelida* in the sea, taxa which form a branch at the basal position to the rest of the fungi (Karpov et al. 2014a).

## Marine Fungi: More Evolved Branches

At least eleven major lineages have been discovered in the kingdom fungi (Blackwell 2011). Marine fungi are predominantly ascomycetes and asexual fungi with an unknown taxonomic position, while only a few basidiomycetes have been reported from the marine environment (Jones et al. 2009a). Also, few chytridiomycetes and other groups, including the *Mucoromycotina*, have been reported from marine habitats (Jones and Pang 2012). Isolation of the zygomycetes (*Mucoromycotina*, *Kickxellomycotina*), marine yeasts, and predacious fungi may require special isolation techniques, e.g., dilution plating of sediment or seawater or inclusion of nematodes for the discovery of predacious fungi (Shearer et al. 2007; Statzell-Tallman et al. 2010; Swe et al. 2011). Marine zygomycetes have been reported from sponges (Holler et al. 2000; Passarini et al. 2013), sediments (Bubnova 2010), and wood (Rämä et al. 2014) or found associated with marine arthropods (Lichtwardt 2012).

Most marine basidiomycetes belong to the *Agaricomycotina*, while *Flamingomyces ruppiae* and *Parvulago marina* are in the *Ustilaginomycotina* (Jones et al. 2009a). The few marine *Basidiomycota* are not monophyletic, suggesting individual lines of evolution from terrestrial environment to the sea (Hibbett and Binder 2001; Binder et al. 2006). *Calathella mangrovei*, *Halocyphina villosa* (Fig. 12.1c), *Nia vibrissa*, *Physalacria maipoensis*, and *Mycaureola dilseae* are in the euagaric clades; *C. mangrovei*, *H. villosa*, and *N. vibrissa* cluster in the *Nia* clade with *Lachnella* spp., *Flagelloscypha* spp., *Cyphellopsis anomala*, *Merismodes fasciculata*, and *Dendrothele acerina*, while *P. maipoensis* and *M. dilseae* form a separate clade with *Gloiocephala*, *Armillaria*, and *Flammulina velatipes* (Binder et al. 2006). Many marine *Agaricomycetes* are intertidal, with minute, enclosed basidiomata on wood, marsh plants, and red algae, and have lost ballistospory, morphological adaptations for an aquatic life (Binder et al. 2006; Jones et al. 2009a). Marine basidiomycetes, such as common mangrove species *Halocyphina villosa* and *Calathella mangrovei*, are mainly saprobic, but *Mycaureola dilseae* is parasitic on the red alga *Dilsea carnosa* (Binder et al. 2006). Marine basidiomycetous yeasts contribute further to marine fungal diversity; Jones et al. (2015) document marine yeasts: *Ascomycota*, 138 species (in 35 genera), and *Basidiomycota*, 75 species (in 26 genera), many reported from deep water including many new species of the genera

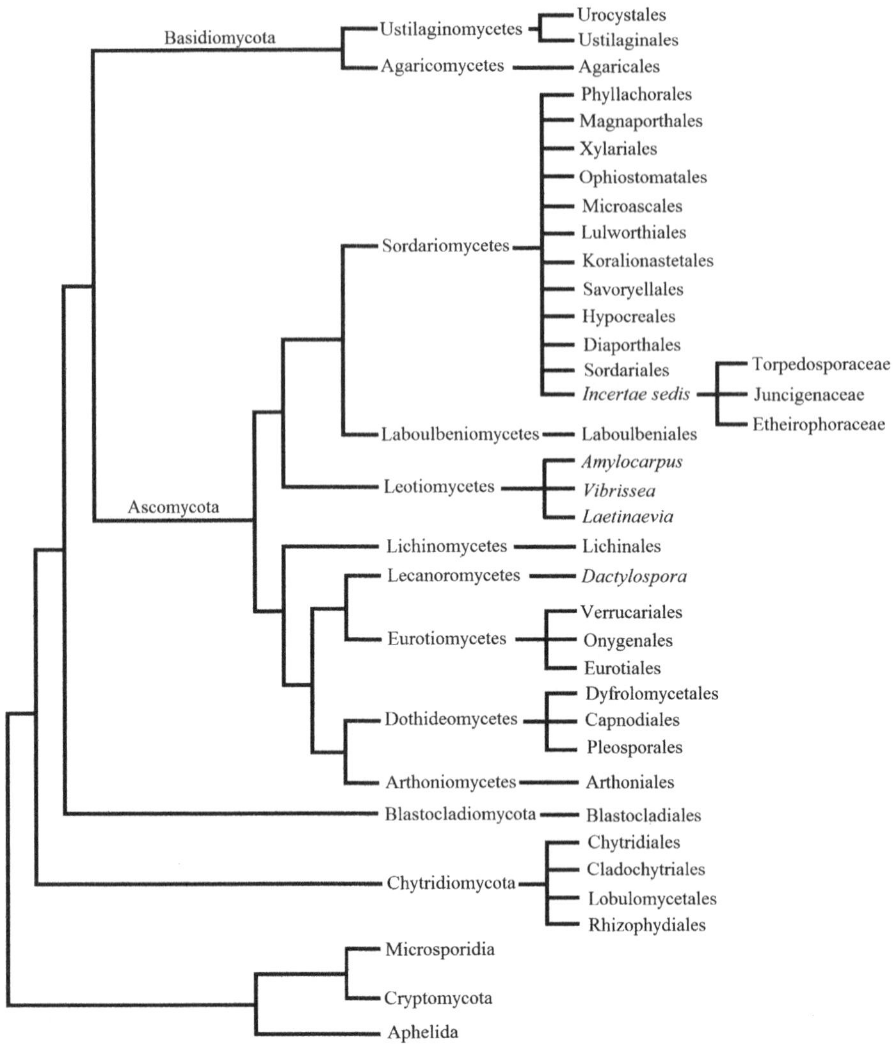


**Fig. 12.2** Phylogenetic lineages of fungi in the sea. Topology is based on Schoch et al. (2009) and Karpov et al. (2014a)

*Cryptococcus*, *Rhodosporidium*, *Rhodotorula*, and *Sporobolomyces* (Kutty and Philip 2008; Fell et al. 2011).

The *Ascomycota* constitutes roughly 80 % of the total marine mycota, and they belong to different classes including *Eurotiomycetes*, *Laboulbeniomycetes*, *Lecanoromycetes*, *Leotiomycetes*, *Lichinomycetes*, *Arthoniomycetes*, *Sordariomycetes*, and *Dothideomycetes* (Fig. 12.2), with the majority in the last two classes and many with unknown taxonomic positions, such as the genera with cleistothecial ascomata: *Biflua*, *Crinigera*, *Dryosphaera*, and *Eiona* (Jones et al. 2009a;

Jones 2011). Many of these genera have restricted distribution and were infrequently collected, thus preventing a molecular study. *Dothideomycetes* and *Sordariomycetes* were also found to be the dominant classes of fungi occurring on the mangrove trees *Avicennia marina* and *Rhizophora stylosa* using tag-encoded 454 pyrosequencing of the 18S and ITS regions of the rRNA gene (Arfi et al. 2012). With the increased mycological research activity in Asian mangroves in recent years, many (rare) marine fungi have been recollected after their original descriptions, which enabled a systematic evaluation. An example is *Manglicola guatemalensis*, a species originally described on *Rhizophora* in Guatemala (Kohlmeyer and Kohlmeyer 1971), but recollected in Thai mangroves (Suetrong et al. 2010). This enabled its classification in the *Jahnulales*, an order previously not represented in marine habitats.

*Laboulbenia marina* is the only species in the *Laboulbeniomycetes*, and it was found on a beetle in the *Laminaria* zone in France (Kohlmeyer and Volkmannn-Kohlmeyer 2003). The *Laboulbeniomycetes* are predominantly terrestrial and freshwater species, so the marine occurrence of this fungus was questioned by Jones et al. (2009a).

Marine *Dactylospora* species were classified, based on morphological observations, in the *Lecanorales* (*Lecanoromycetes*), along with their terrestrial *Dactylospora* counterparts. Recently, Schoch et al. (2009) and Diederich et al. (2013), based on sequence analyses, found that marine *Dactylospora* (*D. mangrovei*, *D. haliotrepha*) formed a separate lineage with *Sclerococcum sphaerale* in the *Eurotiomycetes* and well separated from their terrestrial counterparts and unrelated to the *Lecanoromycetes*. Pang et al. (2014) further confirmed these observations and described a new marine species, *D. vrijmoediae* (Fig. 12.1d). Other marine *Eurotiomycetes* include *Eupenicillium* and *Gymnascella*, both infrequently reported (Jones et al. 2009a), which probably will have a much wider occurrence with greater emphasis on molecular techniques.

*Amylocarpus encephaloides* is the only species sequenced in the marine *Leotiomycetes*, an assignment suggested by a phylogenetic analysis of the 18S rDNA gene (Landvik 1996). This species is likely to be related to taxa of the *Erysiphales* as suggested by a number of studies (Landvik 1996; Berbee 2001; Hambleton and Sigler 2005). No sequences are available for *Vibrissea nypicola* (but terrestrial species are assigned to the *Helotiales*, *Leotiomycetidae*, and *Leotiomycetes*) and *Laetinaevia marina*. *Lichina* (*Lichinaceae*, *Lichinomycetes*) is a lichenized genus with two marine species, *L. confinis* and *L. pygmaea* (Jones et al. 2009a), that group with *Pseudopaulia tessellata* (Schultz and Büdel 2003). *Halographis* is a marine lichenoid species in the *Arthoniomycetes* and placed in the *Roccellaceae*, *Arthoniales* (MycoBank), but no sequence is available for this species. The most specious marine lichens are members of the *Verrucariales* with the genera: *Collemopsidium* (7 species), *Hydropunctaria* (6 species), *Mastodia* (1 species), *Wahlenbergiella* (3 species), and *Verrucaria* (16 species) (Jones et al. 2015). The placement of *Melaspilea mangrovei* in the *Melaspileaceae*, *Arthoniales*, *Arthoniomycetidae*, and *Arthoniomycetes* is unresolved as it has not been sequenced. This species has been reported from mangrove wood and is probably saprobic (Vrijmoed et al. 1996).

Most marine *Sordariomycetes* belong to the *Halosphaeriaceae*, *Microascales* (Pang 2002; Jones et al. 2009a; Sakayaroj et al. 2011). This family includes predominantly marine fungi from wood, but several genera can also be found in freshwater habitats, such as *Aniptodera*, *Halosarpheia*, and *Nais* (Pang 2002). The taxonomic position of the freshwater genus *Fluviatispora* in the *Halosphaeriaceae* requires a molecular study. The same is true for *Chadefaudia* and *Trailia*, which are found on marine algae (Jones et al. 2009a). Spatafora and Blackwell (1994) were the first to include a sequence of the *Halosphaeriaceae* (*Halosphaeriopsis mediosetigera*) and its assignment to the *Microascales*. Over the last two decades, many taxa of the *Halosphaeriaceae* have been collected, isolated and sequenced, and shown to be a monophyletic group (Spatafora et al. 1998; Sakayaroj et al. 2011). However, the phylogenetic relationships between genera and species are not well resolved, and Jones (1995) suspected there has been a rapid evolutionary radiation. Spatafora et al. (1998) concluded that members of the *Halosphaeriaceae* were probably evolved from terrestrial environments over a short period of time. Adaptive features include: development of wide range of ascospore appendages for attachment to comparatively scarce substrata in the sea and deliquescing asci as a means for spore dispersal (Jones 1994, 2006). Pang et al. (2003) speculated that unfurling ascospore appendage type in genera such as *Halosarpheia* and *Saagaromyces*, which can be commonly found in mangrove environments, may be an intermediate form between unornamented species and those with complex ascospore appendage morphology and ontogeny, e.g., *Corollospora* and *Naufragella*. Sakayaroj et al. (2011) further suggested that unfurling ascospore appendages might have been gained and lost several times during evolution. Consequently, mangrove environment may be a stepping stone for oceanic taxa to evolve from terrestrial habitats.

*Kohlmeyeriella*, *Lindra*, and *Lulworthia* (Fig. 12.1e) were originally placed in the *Halosphaeriaceae* based on their morphological characteristics, although the latter two genera produce filiform ascospores (Jones 1995). Spatafora et al. (1998) first suggested that *Lindra* and *Lulworthia* were unrelated to the *Halosphaeriaceae* and formed a new lineage in the *Sordariomycetes*, the *Lulworthiaceae* (*Lulworthiales*) (Kohlmeyer et al. 2000). *Kohlmeyeriella*, *Rostrupiella*, *Haloguignardia*, and *Spathulospora* were later added to this new order (Inderbitzin et al. 2004; Campbell et al. 2005; Koch et al. 2007), although they share few morphological features in common. The validity of the *Spathulosporaceae* requires further studies as this family did not form a distinct lineage in the *Lulworthiales*. *Spathulospora* is a parasite of the red algal genus *Ballia*, near the Antarctic region, and has been rarely reported (Kohlmeyer and Kohlmeyer 1975; Jones et al. 2009a). Other members of this order are mainly saprobic on wood or algae. *Pontogeneia*, a parasitic genus on brown algae, and *Koralionaste* from corals (*Koralionastetales*) have recently been shown to form a dichotomy with the *Lulworthiales* (Campbell et al. 2009). Members of both orders differ in morphology, including the presence of paraphyses and periphyses, and the lack of apical mucus-containing chambers or gelatinous sheaths of the ascospores in the *Koralionastetales* (Campbell et al. 2009).

Other new lineages of marine fungi include the *Savoryellales* and the new families established for the taxa formerly known as the "TBM clade" (Schoch et al.

2006). *Savoryella* is an aquatic genus with freshwater, brackish, and marine species (Jones et al. 2009a) and morphologically similar to the *Halosphaeriaceae* (Read et al. 1993) and *Sordariales* (Jones and Hyde 1992). As a result, the taxonomy of *Savoryella* had been unresolved and was referred to various orders or *Ascomycota incertae sedis* (Jones et al. 2009a). Vijaykrishna et al. (2006), in a phylogenetic analysis of the 18S rRNA gene, firstly suggested that *Savoryella* is unrelated to the *Halosphaeriaceae* nor the *Sordariales*, but rather a group of taxa in the *Hypocreales*. In a multigene phylogenetic analysis (18S, 28S, 5.8S rDNA, rpb1, rpb2, tef1), Boonyuen et al. (2011) discovered that *Savoryella* is closely related to two freshwater genera *Ascotaiwania* and *Ascothailandia* and they together constitute a new lineage, the *Savoryellales*, in the *Hypocreomycetidae*.

*Swampomyces* is characterized by forming a clypeus on wood, cylindrical asci with an apical structure, septate paraphyses, and hyaline ascospores (Kohlmeyer and Volkmann-Kohlmeyer 1987). *Torpedospora radiata*, on the other hand, produces coriaceous ascomata, deliquescing asci, and ascospores with several armlike appendages (Meyers 1957). Both genera formed a monophyletic, yet unknown, lineage in the *Hypocreomycetidae*, based on a phylogenetic analysis using 18S and 28S rDNA (Sakayaroj et al. 2005). This observation was further confirmed by Schoch et al. (2006) who used a wider representation of taxa (*Juncigena adarca*, *Etheirophora* spp.) and genes (18S and 28S rDNA, rpb2) and tentatively named it as "*Torpedospora*/*Bertia*/*Melanospora* (TBM) clade" but no taxonomic change was proposed. Recently, Jones et al. (2014) demonstrated that the marine fungus *Chaetosphaeria chaetosa* and the terrestrial genus *Falcocladium* also belong to the assortment of species in the TBM clade and formally established new families: *Juncigenaceae*, *Etheirophoraceae*, *Falcocladiaceae*, and *Torpedosporaceae* to accommodate these taxa.

Marine fungi can also be found in other orders of the *Sordariomycetes*: *Hypocreales*, *Diaporthales*, *Sordariales*, *Ophiostomatales*, *Xylariales*, *Magnaporthales*, and *Phyllachorales* (Jones et al. 2009a). However, many orders are only represented by a single marine species, while many others have not been subjected to a molecular study. Xylariaceous taxa are common on wood and the brackish water palm *Nypa fruticans*, for example, *Halorosellinia oceanica* and *Nemania maritima* were confirmed to be members of the *Xylariaceae* (Smith et al. 2003; Pinnoi et al. 2010). *Pedumispora rhizophorae*, a taxon previously assigned to the *Diaporthales*, was recently found to be related to the *Diatrypaceae*, *Xylariales*, based on a phylogenetic analysis of the 28S and ITS regions of the rDNA, although the deliquescing asci and filiform ascospores with polar hooklike end cells do not fit the morphological delimitation of other diatrypaceous genera (Klaysuban et al. 2014). *Diatrypasimilis* is another member of the *Diatrypaceae* recently reported from decaying *Rhizophora* wood collected in Australian mangroves (Chalkley et al. 2010), with features more typical of the family. Marine *Hypocreales*, including *Kallichorma* spp., *Heleococcum japonense*, and *Emericellopsis maritima*, are referred to the *Bionectriaceae* (Rossman et al. 2001; Zuccaro et al. 2003), while *Sedecimiella taiwanensis* from mangroves of Taiwan and China is the sole member of the *Niessliaceae* (Pang et al. 2010).

Marine *Dothideomycetes* predominantly belong to the subclass *Pleosporomycetidae*, which have been mainly collected from mangrove substrata (Jones et al. 2009a). These fungi occupy the upper tidal zone with occasional seawater submergence and have active spore discharge mechanisms (Suetrong et al. 2009). A significant change in the taxonomic classification of a number of marine *Dothideomycetes* has been observed in the last few years with the extensive collection and isolation of mangrove fungi in Thailand and other Asian countries. As a result, a number of new orders and families have been established, including *Dyfrolomycetaceae* (*Dyfrolomycetales*), *Halojulellaceae*, *Trematosphaeriaceae*, and *Manglicolaceae*, and this may suggest the wide morphological diversity expressed by the marine *Dothideomycetes* (Hyde et al. 2013).

Few marine species in the subclass *Dothideomycetidae* have been sequenced. *Mycosphaerella eurypotami* and *Scirrhia annulata* grouped with other terrestrial *Mycosphaerella* spp. in the *Mycosphaerellaceae*, *Capnodiales* (Suetrong et al. 2009), while *M. pneumatophorae* is distantly related to this group and grouped with *Phaeotrichum benjaminii*, *Tyrannosorus pinicola*, and *Venturia inaequalis* (Schoch et al. 2009).

Considerable advancement has been made in the classification of the marine *Pleosporomycetidae* in the last decade. Suetrong et al. (2009) identified 17 families in the *Pleosporales*, with many marine taxa forming lineages outside these recognized families resulting in the introduction of four new families (Jones et al. 2012a). Fifteen families (*Aigialaceae*, *Amniculicolaceae*, *Didymellaceae*, *Didymosphaeriaceae*, *Halojulellaceae*, *Leptosphaeriaceae*, *Lenthitheciaceae*, *Lophiostomataceae*, *Massarinaceae*, *Morosphaeriaceae*, *Phaeosphaeriaceae*, *Pleosporaceae*, *Sporormiaceae*, *Testudinaceae*, *Trematosphaeriaceae*) of the *Pleosporales* include marine taxa (Suetrong et al. 2009; Jones et al. 2012a; Ariyawansa et al. 2013). Many marine taxa in these families are represented by a single/few species, further proving their terrestrial origin: *Neomassariosphaeria typhicola* predominantly from marsh plants in the *Amniculicolaceae*, a family of freshwater ascomycetes (Zhang et al. 2009a); *Didymella fucicola* from brown algae in the *Didymellaceae* (Suetrong et al. 2009); *Lentithecium phragmiticola* and *Keissleriella rarum* from saltmarsh plants in the new family *Lentitheciaceae* (Zhang et al. 2009b); *Lophiostoma scabridisporum* and *Paraliomyces lentiferus* in the *Lophiostomaceae*; *Tremateia halophile* on *Juncus roemerianus* in the *Didymosphaeriaceae*; *Loratospora aestuarii* and *Phaeosphaeria* species (*Ph. albopunctata*, *Ph. olivacea*, *Ph. spartinicola*, *Ph. typharum*) in the *Phaeosphaeriaceae* (Suetrong et al. 2009; Zhang et al. 2009b); *Paradendryphiella* spp., *Decorospora gaudefroyi*, and *Alternaria maritima* in the *Pleosporaceae* (Suetrong et al. 2009; Zhang et al. 2009b; Inderbitzin et al. 2002; Suetrong et al. 2009; Zhang et al. 2009b); *Amorosia littoralis* in the *Sporormiaceae* (Mantle et al. 2006); and *Massarina ricifera* and *Verruculina enalia* in the *Testudinaceae* (Suetrong et al. 2009). Many marine *Pleosporales* still await phylogenetic analysis before they can be properly taxonomically assigned, e.g., marine *Massarina* spp., and *Leptosphaeria* spp. (Jones et al. 2012a).

Suetrong et al. (2009) investigated the phylogeny of *Aigialus* spp. (Fig. 12.1f), *Ascocratera manglicola*, and *Rimora* (*Lophiostoma*) *mangrovei* and found they formed a robust monophyletic group in a basal position of the *Pleosporales*, separating from other known families of the order, so a new family, the *Aigialaceae*, was established for this group. These genera have similar ascomatal morphology and asci, while they mainly differ in ascospore morphology: muriform and dark-colored in *Aigialus* and hyaline in *Ascocratera* and *Rimora*. A new family, the *Morosphaeriaceae*, was proposed to accommodate *Morosphaeria velatispora*, *M. ramunculicola*, *Helicascus* spp., and *Kirschsteiniothelia elaterascus* (Suetrong et al. 2009). There is great morphological variation in this group, especially the ascospores. Subsequently, *K. elaterascus* was transferred to *Morosphaeria* (Boonmee et al. 2012). *Kirschsteiniothelia* is a polyphyletic genus with *K. maritima*, another marine species, grouping with *Mytilinidiales* (Suetrong et al. 2009). The marine mangrove fungi *Halomassarina thalassiae* and *Falciformispora lignatilis* constituted a monophyletic group with *Trematosphaeria pertusa*, the type of species of the newly proposed family *Trematosphaeriaceae* that was reinstated (Suetrong et al. 2009). However, these fungi have few morphological characters in common. Whether the marine *Trematosphaeria* species affiliate with this group will require further study.

*Julella avicenniae* was described from intertidal wood of *Avicennia* but can also be found on the more terrestrial part of the tree (Borse 1987; Hyde 1992; Jones et al. 2009a). *Julella avicenniae* differs from *J. buxi* (type species: *Didymosphaeriaceae*, *Pleosporales*) in that the ascus contains eight ascospores with an apical apparatus. Therefore, a new genus *Halojulella* was proposed for *J. avicenniae* in the new family *Halojulellaceae* based on morphological and sequence data (Ariyawansa et al. 2013; Ariyawana et al. 2014). The taxonomic placement of *J. herbatilis*, on *Juncus roemerianus*, requires sequence data to confirm its position in the *Halojulellaceae* and its relationship with *H. avicenniae*.

Asexual marine fungi are fairly common and constitute about 15 % of the total marine fungi (Jones et al. 2009a). Abdel-Wahab and Bahkali (2012) listed 143 species of asexual marine fungi belonging predominantly to the *Dothideomycetes* and the *Sordariomycetes* (*Ascomycota*). Asexual marine fungi can also be found in other classes, including the *Eurotiomycetes*, the *Leotiomycetes*, and the *Orbiliomycetes* (Abdel-Wahab and Bahkali 2012). *Halenospora varia*, a cosmopolitan species on wood, was found to be unrelated to *Zalerion*, an earlier assignment based on morphology (Bill et al. 1999; Jones et al. 2009a). Other asexual genera have also been proven to be polyphyletic, e.g., *Cirrenalia* and *Sigmoidea* (Jones et al. 2009b; Abdel-Wahab et al. 2010), which suggests that conidial morphology is unreliable in classifying asexual marine fungi. Sequences for many asexual marine fungi have been made available, allowing their formal taxonomic classification. Asexual fungi of the *Orbiliomycetes* are nematode-trapping fungi in the genera *Arthrobotrys*, *Dactylella*, and *Monacrosporium*, while *Aspergillus sydowii*, found on the sea fan *Gorgonia* spp., belongs to the *Eurotiomycetes*. Species of *Aspergillus* and *Penicillium* have been repeatedly isolated from marine sediments and marine animals

(Khudyakova et al. 2000; Zhang et al. 2012) and whether some of them are marine would require a detailed investigation at the molecular level.

Within the *Ascomycota*, most asexual marine fungi belong to the *Dothideomycetes* (*Pleosporales*) and the *Sordariomycetes* (*Halosphaeriaceae*, *Lulworthiales*), which corroborate the observation that most marine sexual fungi are in the same two classes and that teleomorphic and anamorphic phases of the same fungus occupy a similar niche (Alexopoulos et al. 1996). Many sexual/asexual relationships have been linked from culture observations, e.g., *Nereiospora cristata*/*Monodictys pelagica* and *Halosphaeriopsis mediosetigera*/*Trichocladium achrasporum* (Mouzouras and Jones 1985; Shearer 1986), while others can only be referred to a genus, e.g., *Halosigmoidea marina* and *H. parvula* (putative stage in *Corollospora*), or are unknown, especially those in the *Lulworthiales*, e.g., *Cumulospora*, *Hydea*, and *Matsusporium* (Abdel-Wahab et al. 2010; Abdel-Wahab and Bahkali 2012). Sequence data can only be used to classify asexual fungi rather than linking asexual/sexual relationships due to the incomplete representation of reference sequences in the GenBank.

Marine chytrids have been poorly studied, few have been documented in recent years, and most marine texts do not include them (Kohlmeyer and Kohlmeyer 1979; Jones et al. 2009a). However, their importance as parasites of marine animals and their retrieval from sediments in hydrothermal vents suggest that they may have an important role in the marine environment (Nagano et al. 2010; Gleason et al. 2012; Hatai 2012). Currently marine *Chytridiomycota* species include members of the orders *Chytridiales* (two families, four genera, 12 species), *Cladochytriales* (one genus, one species), *Lobulomycetales* (one family, one genus, one species), *Rhizophydiales* (two families, two genera, two species), and *incertae sedis* (five genera, 10 species) with a cumulative number of 26 species (Jones et al. 2015). Another zoosporic group of organisms the *Blastocladiomycota* is represented by one species in the genus *Catenaria* (*Blastocladiales*, *Catenariaceae*).

## Marine Fungi: Basal Lineages

The new superphylum *Opisthosporidia* was recently established to accommodate the three phyla, *Cryptomycota*, *Microsporidia*, and *Aphelida*, which have been found to be related to the kingdom Fungi but form a basal branch to other fungi (Karpov et al. 2014a). The three phyla are united by (1) primarily phagotrophic mode of nutrition, (2) formation of cysts with penetrative apparatus, and (3) presence of tubular/flat mitochondrial cristae (Karpov et al. 2014a). Marine *Microsporidia* and *Aphelida* are mainly parasites of marine organisms (Karpov et al. 2014a; Stentiford et al. 2013), while marine *Cryptomycota* have been detected in the se diment, in the form of environmental sequences of small subunit ribosomal RNA gene (van Hannen et al. 1999; Lara et al. 2010; Jones et al. 2011a).

*Pseudaphelidium drebesii* is the only marine species represented in the *Aphelida*, parasitic on the marine diatom *Thalassiosira punctigera* (Karpov et al. 2014b). The life cycle of *P. drebesii* involves attachment of zoospores to the host leading to cysts, penetration of host cells via infection tubes, and formation of a plasmodium cleaving into amoeboid cells which encyst and are released to form a new generation of zoospores (Karpov et al. 2014b).

*Microsporidia* are obligate intracellular parasites of eukaryotic hosts, which can be found in terrestrial, freshwater, and marine environments. The life cycle of the *Microsporidia* varies and it generally involves infection using a polar tube, intracellular replication as meronts, and differentiation into spores (Kearney and Gleason 2014). In marine environment, the *Microsporidia* infect a wide variety of hosts including crabs (Stentiford et al. 2014), nematodes (Sapir et al. 2014), fish (Matthews et al. 2013), copepods (Jones et al. 2012b), lobsters (Stentiford et al. 2010), shrimps (Brown and Adamson 2006), and amphipods (Dunn et al. 2001). The classification of the *Microsporidia* is based on their ecology, i.e., classes Terresporidia, Aquasporidia, and Marinosporidia for microsporidia infecting terrestrial, freshwater, and marine hosts, respectively (Vossbrinck and Debrunner-Vossbrinck 2005). Marinosporidia is polyphyletic, suggesting the unreliability of this criterion (Stentiford et al. 2013). Site of infection (tissue type) to the hosts may be more phylogenetically relevant (Stentiford et al. 2013).

*Cryptomycota* was first described as a group of sequences (LKM11) closely related to fungi derived from denaturing gradient gel electrophoresis (van Hannen et al. 1999). Subsequently, other studies also detected the presence of sequences related to LKM11 in different terrestrial, freshwater, and marine habitats (Jones et al. 2011a; Livermore and Mattes 2013). In particular, LKM11 was detected in marine anoxic sediment (Dawson and Pace 2002), deep-sea sediment (Nagano et al. 2010), deep-sea methane cold-seep sediments (Nagahama et al. 2011), and marine upper-water column (Livermore and Mattes 2013). Lara et al. (2010) was the first study to show that LKM11 was related to *Rozella allomycis* with a good support using 18S rRNA gene sequence and named the group Rozellida. Jones et al. (2011b) formally named this group the *Cryptomycota* due to the cryptic nature of the group (discovered only by environmental sequences). No culture is available for LKM11 but by using tyramide signal amplification fluorescence *in situ* hybridization (TSA-FISH) coupled with monoclonal antibody TAT116 against α-tubulin, cells of the *Cryptomycota* are ovoid, 3–5 μm in size; one or more cells are attached to diatoms; and a flagellum is present in at least one stage of their life cycle (Jones et al. 2011a). These characters can also be found in *Rozella* spp., which are parasites of Oomycota, *Blastocladiomycota*, and *Chytridiomycota* (Gleason et al. 2014). However, no chitin or cellulose was detected in cells of the *Cryptomycota* by Calcofluor white (Jones et al. 2011a); in *R. allomycis*, chitin or cellulose was present in young resting spores (James and Berbee 2012). One marine species, *R. marina*, was reported as a parasite of *Chytridium polysiphoniae*, which in turn, is parasitic on brown algae (Sparrow 1936; Glockling et al. 2012). Zoospores of *R. marina* are formed inside the sporangium of *C. polysiphoniae* (Johnson 1966).

## Concluding Remarks

The paper by Barghoorn and Linder (1944) is generally regarded as the stimulus for the study of marine fungi. However, progress was slow until the 1960s when many studies reported on marine *Sordariomycetes* growing on submerged wood and driftwood. The discovery of marine fungi, predominantly Dothideomycetes, on mangrove substrata led to a dramatic increase in our knowledge during 1980s to late 1990s, with many new taxa described from the tropics, especially Asia. From 2000, the number of species described has dropped with mycologists focusing on determining the phylogenetic affinity of taxa already described. However, the study of marine fungi has proceeded along two paths: the mycelial forms (circa 899 species) and yeasts (circa 213 species), with little interaction between the two groups of mycologists studying them. Other taxa have not been critically studied, especially the so– called marine-derived fungi and facultative marine fungi, taxa mainly documented by those interested in the discovery of compounds with new bioactivity. The last decade has seen greater awareness of the "fungal-like" organisms once referred to the *Phycomycetes*. This interest has been stimulated by the realization that many are important pathogens of marine organisms, their source of polyunsaturated fatty acids (PUFAs), and their report from deep sea hydrothermal vents. Few publications have focused on bringing these different approaches together, so that we have a better understanding of the marine mycota and related organisms (Jones and Pang 2012). Marine lineages of fungi are summarized in Fig. 12.2.

The next decade will be dominated by the development of new molecular techniques for their study, e.g., tag-encoded 454 pyrosequencing of the ribosomal rDNA and rRNA and fluorescence *in situ* hybridization (FISH). These will bring different challenges, especially dealing with the vast amount of data generated and in their interpretation. Also better documentation of sequences generated from these studies and correct identification of data deposited in world data banks enable a better understanding of the phylogenetic relationship of the different taxa. Currently, the number of marine fungi is circa 1112, while Jones (2011) predicts a figure of 11,000–12,500 species (Jones et al. 2015).

**Acknowledgments** K.L. Pang would like to thank the Ministry of Science and Technology, Taiwan, for financial support (NSC101-2621-B-019-001-MY3).

## References

Abdel-Wahab MA, Bahkali AHA (2012) Taxonomy of filamentous anamorphic marine fungi: morphology and molecular evidence. In: Jones EBG, Pang KL (eds) Marine fungi and fungal-like organisms. De Gruyter, Berlin, pp 65–90

Abdel-Wahab MA, Pang KL, Nagahama T, Abdel-Aziz FA, Jones EBG (2010) Phylogenetic evaluation of anamorphic species of *Cirrenalia* and *Cumulospora* with the description of eight new genera and four new species. Mycol Prog 9:537–558

Alexopoulos CJ, Mims CW, Blackwell M (1996) Introductory mycology. Wiley, New York

Arfi Y, Buée M, Marchand C, Levasseur A, Record E (2012) Multiple markers pyrosequencing reveals highly diverse and host-specific fungal communities on the mangrove trees *Avicennia marina* and *Rhizophora stylosa*. FEMS Microbiol Ecol 79:433–444

Ariyawana HA, Tanaka K, Thambugala KM, Phookamsak R, Tian Q, Camporesi E et al (2014) A molecular phylogenetic reappraisal of the Didymosphaeriaceae (= Montagnulaceae). Fungal Divers. doi:10.1007/s13225-014-0305-6

Ariyawansa HA, Jones EBG, Suetrong S, Alias SA, Kang JC, Hyde KD (2013) Halojulellaceae a new family of the order Pleosporales. Phytotaxa 130:14–24

Barghoorn ES, Linder DH (1944) Marine fungi: their taxonomy and biology. Farlowia 1:395–467

Berbee ML (2001) The phylogeny of plant and animal pathogens in the Ascomycota. Physiol Mol Plant Pathol 59:165–187

Bills GF, Platas G, Peelaez F, Masurekar P (1999) Reclassification of a pneumocandin producing anamorph, *Glarea lozoyensis* gen. *et* sp. nov., previously identified as *Zalerion arboricola*. Mycol Res 103:179–192

Binder M, Hibbett DS, Wang Z, Farnham WF (2006) Evolutionary relationships of *Mycaureola dilseae* (Agaricales), a basidiomycetes pathogen of a subtidal Rhodophyte. Am J Bot 93:547–556

Blackwell M (2011) The fungi: 1, 2, 3 … 5.1 million species? Am J Bot 98:426–438

Boonmee S, Ko TWK, Chukeatirote E, Hyde KD, Chen H, Cai L, McKenzie EHC, Jones EBG, Kodsueb R, Hassan BA (2012) Two new *Kirschsteiniothelia* species with *Dendryphiopsis* anamorphs cluster in Kirschsteiniotheliaceae fam. nov. Mycologia 104:698–714

Boonyuen N, Chuaseeharonnachai C, Suetrong S, Sri-indrasutdhi V, Sivichai S, Jones EBG, Pang KL (2011) Savoryellales (Hypocreomycetidae, Sordariomycetes): a novel lineage of aquatic ascomycetes inferred from multiple-gene phylogenies of the genera *Ascotaiwania*, *Ascothailandia*, and *Savoryella*. Mycologia 103:1351–1371

Borse BD (1987) Marine fungi from India—V. Curr Sci 56:1109–1111

Brown AMV, Adamson ML (2006) Phylogenetic distance of *Thelohania butleri* Johnston, Vernick, and Sprague, 1978 (Microsporidia; Thelohaniidae), a parasite of the smooth pink shrimp *Pandalus jordani*, from its congeners suggests need for major revision of the genus *Thelohania* Henneguy, 1892. J Eukaryot Microbiol 53:445–455

Bubnova EN (2010) Fungal diversity in bottom sediments of the Kara Sea. Bot Mar 53:595–600

Campbell J, Volkmann-Kohlmeyer B, Gräfenhan T, Spataofora JW, Kohlmeyer J (2005) A reevaluation of Lulworthiales: relationships based on 18S and 28S rDNA. Mycol Res 109:556–568

Campbell J, Inderbitzin P, Kohlmeyer J, Volkmann-Kohlmeyer B (2009) Koralionastetales, a new order of marine Ascomycota in the Sordariomycetes. Mycol Res 113:373–380

Chalkley DB, Suh SO, Volkmann-Kohlmeyer B, Kohlmeyer J, Zhou JJ (2010) *Diatrypasimilis australiensis*, a novel xylarialean fungus from mangrove. Mycologia 102:430–437

Dawson SC, Pace NR (2002) Novel kingdom level eukaryotic diversity in anoxic environments. Proc Natl Acad Sci USA 99:8324–8329

Diederich P, Ertz D, Lawrey JD, Sikaroodi M, Untereiner WA (2013) Molecular data place the hyphomycetous lichenicolous genus Sclerococcum close to *Dactylospora* (Eurotiomycetes) and *S. parmeliae* in *Cladophialophora* (Chaetothyriales). Fungal Divers 58:61–72

Dunn AM, Terry RS, Smith JE (2001) Transovarial transmission in the microsporidia. Adv Parasitol 48:57–100

El-Sharouny HM, Raheem AM, Abdel-Wahab MA (1998) Manglicolous fungi of the Red Sea in upper Egypt. Microbiol Res 153:81–96

Fell JW, Statzell-Tallman A, Scorzetti G, Gutiérrez MH (2011) Five new species of yeasts from freshwater and marine habitats in the Florida Everglades. Anton Leeuw 99:533–549

Gleason FH, Küpper FC, Glockling SL (2012) Zoosporic true fungi. In: Jones EBG, Pang KL (eds) Marine fungi and fungal-like organisms. De Gruyter, Berlin, pp 103–116

Gleason FH, Lilje O, Marano AV, Sime-Ngando T, Sullivan BK, Kirchmair M, Neuhauser S (2014) Ecological functions of zoosporic hyperparasites. Front Microbiol 5:244. doi:10.3389/fmicb.2014.00244

Glockling SL, Jones MDM, Gleason FH (2012) Cryptomycota (Rozellida) and Mesomycetozoea (Ichthyosporea) the super-group Opisthokonta. In: Jones EBG, Pang KL (eds) Marine fungi and fungal-like organisms. De Gruyter, Berlin, pp 153–165

Hambleton S, Sigler L (2005) *Meliniomyces*, a new anamorph for root-associated fungi with phylogenetic affinities to *Rhizoscyphus ericae* (=*Hymenoscyphus ericae*), Leotiomycetes. Stud Mycol 53:1–27

Hatai K (2012) Disease of fish and shellfish caused by marine fungi. In: Raghukumar S (ed) Biology of marine fungi. Springer, Berlin, pp 15–52

Hibbett DS, Binder M (2001) Evolution of marine mushrooms. Biol Bull 201:319–322

Holler U, Wright AD, Matthee GF, Konig GM, Draeger S, Aust HJ, Schulz B (2000) Fungi from marine sponges: diversity, biological activity and secondary metabolites. Mycol Res 104:1354–1365

Hyde KD (1992) *Julella avivenniae* (Borse) comb. nov. (Thelenellaceae) from intertidal mangrove wood and miscellaneous fungi from the North East Coast of Queensland. Mycol Res 95:939–942

Hyde KD, Jones EBG, Liu JK, Ariyawansha H, Boehm E, Boonmee S et al (2013) Families of Dothideomycetes. Fungal Divers 63:1–313

Inderbitzin P, Kohlmeyer J, Volkmann-Kohlmeyer B, Berbee ML (2002) *Decorospora*, a new genus for the marine ascomycetes *Pleospora gaudefroyi*. Mycologia 91:651–659

Inderbitzin P, Lim SR, Volkmann-Kohlmeyer B, Kohlmeyer J (2004) The phylogenetic position of *Spathulospora* based on DNA sequences from dried herbarium material. Mycol Res 108:737–748

James TY, Berbee ML (2012) No jacket required—new fungal lineage defies dress code. Bioessays 34:94–102

Johnson TW Jr (1966) *Rozella marina* in *Chytridium polysiphoniae* from Icelandic water. Mycologia 58:490–494

Johnson TW, Sparrow FK (1961) Fungi in oceans and estuaries. J. Cramer, Germany

Jones EBG (1994) Fungal adhesion. Mycol Res 98:961–981

Jones EBG (1995) Ultrastructure and taxonomy of the aquatic ascomycetous order Halosphaeriales. Can J Bot 73:S790–S801

Jones EBG (2000) Marine fungi: some factors influencing biodiversity. Fungal Divers 4:53–73

Jones EBG (2006) Form and function of fungal spore appendages. Mycoscience 47:167–183

Jones EBG (2011) Are there more marine fungi to be described? Bot Mar 54:343–354

Jones EBG, Hyde KD (1992) Taxonomic studies on *Savoryella* Jones *et* Eaton (Ascomycotina). Bot Mar 35:83–91

Jones EBG, Moss ST (1978) Ascospore appendages of marine ascomycetes: an evaluation of appendages as taxonomic criteria. Mar Biol 49:11–26

Jones EBG, Pang KL (2012) Introduction of marine fungi. In: Jones EBG, Pang KL (eds) Marine fungi and fungal-like organisms. De Gruyter, Berlin, pp 1–13

Jones EBG, Sakayaroj J, Suetrong S, Somrithipol S, Pang KL (2009a) Classification of marine Ascomycota, anamorphic taxa and Basidiomycota. Fungal Divers 35:1–187

Jones EBG, Zuccaro A, Mitchell JI, Nakagiri A, Chatmala I, Pang KL (2009b) Phylogenetic position of freshwater and marine *Sigmoidea* species: introducing a marine hyphomycete *Halosigmoidea* gen. nov. (Halosphaeriales). Bot Mar 52:349–359

Jones MDM, Forn I, Gadelha C, Egan MJ, Bass D, Massana R, Richards TA (2011a) Discovery of novel intermediate forms redefines the fungal tree of life. Nature 474:200–205

Jones MDM, Richards TA, Hawksworth DL, Bass D (2011b) Validation and justification of the phylum name Cryptomycota phyl. Nov. IMA Fungus 2:173–175

Jones EBG, Hyde KD, Pang KL, Suetrong S (2012a) Phylogeny of the Dothideomycetes and other classes of marine Ascomycota. In: Jones EBG, Pang KL (eds) Marine fungi and fungal-like organisms. De Gruyter, Berlin, pp 17–34

Jones SRM, Prosperi-Porta G, Kim E (2012b) The diversity of microsporidia in parasitic copepods (Caligidae: Siphonostomatoida) in the northeast Pacific Ocean with description of *Facilispora margolisi* n. g., n. sp. and a new family Facilisporidae n. fam. J Eukaryot Microbiol 59:206–217

Jones EBG, Suetrong S, Rungjindamai N, Somrithipol S, Sakayaroj J, Abdel-Wahab MA, Pang KL (2014) An additional fungal lineage in the Hypocreomycetidae (*Falcocladium* species) and the taxonomic re-evaluation of *Chaetosphaeria chaetosa* and *Swampomyces* species, based on morphology, ecology and phylogeny. Cryptogam Mycol 35:119–138

Jones EBG, Suetrong S, Sakayaroj J, Bahkali AH, Abdel-Wahab MA, Boekhout T, Pang KL (2015) Classification of marine Ascomycota, Basidiomycota, Blastocladiomycota and Chytridiomycota. Fungal Divers [In Submission]3:1–72

Karpov SA, Mamkaeva MA, Aleoshin VV, Nassonova E, Lilje O, Gleason FH (2014a) Morphology, phylogeny, and ecology of the aphelids (Aphelidea, Opisthokonta) and proposal for the new superphylum Opisthosporidia. Front Microbiol. doi:10.3389/fmicb.2014.00112

Karpov SA, Mamkaeva MA, Benzerara K, Moreira D, Lopez-Garcia P (2014b) Molecular phylogeny and ultrastructure of *Aphelidium* aff. *Melosirae* (Aphelida, Opisthosporidia). Protist 165:512–526

Kearney R, Gleason FH (2014) Microsporidia. In: Jones EBG, Hyde KD, Pang KL (eds) Freshwater fungi and fungal-like organisms. De Gruyter, Berlin, pp 157–176

Khudyakova YV, Pivkin MV, Kuznetsova TA, Svetashev VI (2000) Fungi in sediments of the sea of Japan and their biologically active metabolites. Microbiology 69:608–611

Klaysuban A, Sakayaroj J, Jones EBG (2014) An additional marine fungal lineage in the Diatrypaceae, Xylariales: *Pedumispora rhizophorae*. Bot Mar 57:413–420

Koch J, Pang KL, Jones EBG (2007) Rostrupiella anica gen. et sp. nov., a Lulworthia-like marine lignicolous species from Denmark and the USA. Bot Mar 50:294–301

Kohlmeyer J, Kohlmeyer E (1971) Marine fungi from tropical America and Africa. Mycologia 63:831–861

Kohlmeyer J, Kohlmeyer E (1975) Biology and geographical distribution of *Spathulospora* species. Mycologia 67:629–637

Kohlmeyer J, Kohlmeyer E (1979) Marine mycology–the higher fungi. Academic, New York

Kohlmeyer J, Volkmann-Kohlmeyer B (1987) Marine fungi from Belize with a description of two new genera of Ascomycetes. Bot Mar 30:195–204

Kohlmeyer J, Volkmannn-Kohlmeyer B (2003) Marine Ascomycetes from algae and animal hosts. Bot Mar 46:285–306

Kohlmeyer J, Spatafora JW, Volkmann-Kohlmeyer B (2000) Lulworthiales, a new order of marine Ascomycota. Mycologia 92:453–458

Kutty SN, Philip R (2008) Marine yeasts—a review. Yeast 25:465–483

Landvik S (1996) *Neolecta*, a fruit-body-producing genus of the basal ascomycetes, as shown by SSU and LSU rDNA sequences. Mycol Res 100:199–202

Lara E, Moreira D, Lopez-Garcia P (2010) The environmental clade LKM11 and Rozella form the deepest branching clade of fungi. Protist 161:116–121

Lichtwardt RW (2012) Trichomycete gut fungi from tropical regions of the world. Biodivers Conserv 21:2397–2402

Livermore JA, Mattes TE (2013) Phylogenetic detection of novel Cryptomycota in an Iowa (United States) aquifer and from previously collected marine and freshwater targeted high-throughput sequencing sets. Environ Microbiol 15:2333–2341

Luo ZH, Pang KL (2014) Fungi from substrates in marine environment. In: Misra JK, Tewari JP, Deshmukh SK, Vágvölgyi C (eds) Fungi from different substrates. CRC Press, London

Mantle PG, Hawksworth DL, Pazoutova S, Collinson LM, Rassing BR (2006) *Amorosia littorlais* gen. *et* sp. nov., a new genus and species name for the scorpinone and caffeineproducing hyphomycetes from the littoral zone in The Bahamas. Mycol Res 110:1371–1378

Matthews CGG, Richards RH, Shinn AP, Cox DI (2013) Gill pathology in Scottish farmed Atlantic salmon, Salmo salar L., associated with the microsporidian *Desmozoon lepeophtherii* Freeman *et* Sommerville, 2009. J Fish Dis 36:861–869

Meyers SP (1957) Taxonomy of marine Pyrenomycetes. Mycologia 49:475–528
Mouzouras R, Jones EBG (1985) *Monodictys pelagica*, the anamorph of *Nereiospora cristata* (Halosphaeriaceae). Can J Bot 63:2444–2447
Nagahama T, Hamamoto M, Nakase T, Takaki Y, Horikoshi K (2003) *Cryptococcus surugaensis* sp. nov., a novel yeast species from sediment collected on the deep-sea floor of Suruga Bay. Int J Syst Evol Microbiol 53:2095–2098
Nagahama T, Takahashi E, Nagano Y, Abdel-Wahab MA, Miyazaki M (2011) Molecular evidence that deep-branching fungi are major fungal components in deep-sea methane cold-seep sediments. Environ Microbiol 13:2359–2370
Nagano Y, Nagahama T, Hatada Y, Nunoura T, Takami H, Miyazaki J, Takai K, Horikoshi K (2010) Fungal diversity in deep-sea sediments-the presence of novel fungal groups. Fungal Ecol 3:316–325
Pang KL (2002) Systematics of the Halosphaeriales: which morphological characters are important? In: Hyde KD (ed) Fungi in marine environments. Fungal Diversity Press, Hong Kong, pp 35–57
Pang KL, Vrijmoed LLP, Kong RYC, Jones EBG (2003) Polyphyly of *Halosarpheia* (Halosphaeriales, Ascomycota): implications on the use of unfurling ascospore appendage as a systematic character. Nova Hedwig 77:1–18
Pang KL, Alias SA, Chiang MWL, Vrijmoed LLP, Jones EBG (2010) *Sedecimiella taiwanensis* gen. *et* sp. nov., a marine mangrove fungus in the Hypocreales (Hypocreomycetidae, Ascomycota). Bot Mar 53:493–498
Pang KL, Guo SY, Alias SA, Hafellner J, Jones EBG (2014) A new species of marine *Dactylospora* and its phylogenetic affinities within the Eurotiomycetes, Ascomycota. Bot Mar 57:315–321
Passarini MRZ, Santos C, Lima N, Berlinck RGS, Sette LD (2013) Filamentous fungi from the Atlantic marine sponge *Dragmacidon reticulatum*. Arch Microbiol 195:99–111
Pinnoi A, Phongpaichit P, Jeewon R, Tang AMC, Hyde KD, Jones EBG (2010) Phylogenetic relationships of *Astrocystis eleiodoxae* sp. nov. (Xylariaceae). Mycosphere 1:1–9
Rämä T, Norden J, Davey ML, Mathiassen GH, Spatafora JW, Kauserud H (2014) Fungi ahoy! Diversity on marine wooden substrata in the high North. Fungal Ecol 8:46–58
Read SJ, Jones EBG, Moss ST (1993) Taxonomic studies of marine Ascomycotina: ultrastructure of the asci, ascospores, and appendages of *Savoryella* species. Can J Bot 71:273–283
Rossman AY, McKemy JM, Pardo-Schultheiss RA, Schroers HJ (2001) Molecular studies of the Bionectriaceae using large subunit rDNA sequences. Mycologia 93:100–110
Sakayaroj J, Pang KL, Jones EBG, Vrijmoed LLP, Abdel-Wahab MA (2005) A systematic reassessment of marine ascomycetes *Swampomyces* and *Torpedospora*. Bot Mar 48:395–406
Sakayaroj J, Pang KL, Jones EBG (2011) Multi-gene phylogeny of the Halosphaeriaceae: its ordinal status, relationships between genera and morphological character evolution. Fungal Divers 46:87–109
Sapir A, Dillman AR, Connon SA, Grupe BM, Ingels J, Mundo-Ocampo M et al (2014) Microsporidia-nematode associations in methane seeps reveal basal fungal parasitism in the deep sea. Front Microbiol 5:43. doi:10.3389/fmicb.2014.00043
Schoch CL, Sung GH, Volkmann-Kohlmeyer B, Kohlmeyer J, Spatafora JW (2006) Marine fungal lineages in the Hypocreomycetidae. Mycol Res 110:257–263
Schoch CL, Sung GH, López-Giráldez F, Townsend JP, Miadlikowska J, Hofstetter V et al (2009) The Ascomycota tree of life: a phylum-wide phylogeny clarifies the origin and evolution of fundamental reproductive and ecological traits. Syst Biol 58:224–239
Schultz M, Büdel B (2003) On the systematic position of the lichen genus Heppia. Lichenologist 35:151–156
Shearer CA (1986) The significance of teleomorph/anamorph connections in the classification of marine Ascomycota. In: Moss ST (ed) The biology of marine fungi. Cambridge University Press, Cambridge, pp 253–262
Shearer CA, Descals E, Kohlmeyer B, Kohlmeyer J, Marvanová L, Padgett D et al (2007) Fungal biodiversity in aquatic habitats. Biodivers Conserv 16:49–67

Smith GJD, Liew ECY, Hyde KD (2003) The Xylariales: a monophyletic order containing 7 families. Fungal Divers 13:175–208

Sparrow FK Jr (1936) Biological observation on the marine fungi of Woods Hole Waters. Biol Bull 70:236–263

Spatafora JW, Blackwell M (1994) The polyphyletic origins of ophiostomatoid fungi. Mycol Res 98:1–9

Spatafora JW, Volkmann-Kohlmeyer B, Kohlmeyer J (1998) Independent terrestrial origins of the Halosphaeriales (marine Ascomycota). Am J Bot 85:1569–1580

Statzell-Tallman A, Scorzetti G, Fell JW (2010) *Candida spencermartinsiae* sp. nov., *Candida taylorii* sp. nov. and *Pseudozyma abaconensis* sp. nov., novel yeasts from mangrove and coral reef ecosystems. Int J Syst Evol Microbiol 60:1978–1984

Stentiford GD, Bateman KS, Small HJ, Moss J, Shields JD, Reece KS, Tuck I (2010) *Myospora metanephrops* (n. g., n. sp.) from marine lobsters and a proposal for erection of a new order and family (Crustaceacida; Myosporidae) in the class Marinosporidia (Phylum Microsporidia). Int J Parasitol 40:1433–1446

Stentiford GD, Feist SW, Stone DM, Bateman KS, Dunn AM (2013) Microsporidia: diverse, dynamic, and emergent pathogens in aquatic systems. Trends Parasitol 29:567–578

Stentiford GD, Bateman K, Feist SW, Oyarzun S, Uribe JC, Palacios M, Stone DM (2014) *Areospora rohanae* n.gen. n.sp. (Microsporidia; Areosporiidae n. fam.) elicits multi-nucleate giant-cell formation in southern king crab (*Lithodes santolla*). J Invertebr Pathol 118:1–11

Suetrong S, Schoch CL, Spatafora JW, Kohlmeyer J, Volkmann-Kohlmeyer B, Sakayaroj J, Phongpaichit S, Tanaka K, Hirayama K, Jones EBG (2009) Molecular systematics of the marine Dothideomycetes. Stud Mycol 64:155–173

Suetrong S, Sakayaroj S, Phongpaichit S, Jones EBG (2010) Morphological and molecular characteristics of a poorly known marine ascomycete, *Manglicola guatemalensis* (Jahnulales: Pezizomycotina; Dothideomycetes, *Incertae sedis*): new lineage of marine ascomycetes. Mycologia 102:83–92

Swe A, Li J, Zhang KQ, Pointing SB, Jeewon R, Hyde KD (2011) Nematode-Trapping fungi. Curr Res Environ Appl Mycol 1:1–26

van Hannen EJ, Mooij WM, van Agterveld MP, Gons HJ et al (1999) Detritus-dependent development of the microbial community in an experimental system: qualitative analysis by denaturing gradient gel electrophoresis. Appl Environ Microbiol 65:2478–2484

Vijaykrishna D, Jeewon R, Hyde KD (2006) Molecular taxonomy, origins and evolution of freshwater ascomycetes. Fungal Divers 23:351–390

Vossbrinck CR, Debrunner-Vossbrinck BA (2005) Molecular phylogeny of the Microsporidia: ecological, ultrastructural and taxonomic considerations. Folia Parasitol 52:131–142

Vrijmoed LLP, Hyde KD, Jones EBG (1996) *Melaspilea mangrovei* sp. nov., from Australia and Hong Kong mangroves. Mycol Res 100:291–294

Zhang Y, Schoch CL, Fournier J, Crous PW, de Gruyter J, Woudenberg JHC, Hirayama K, Tanaka K, Pointing SB, Spatafora JW, Hyde KD (2009a) Multi-locus phylogeny of Pleosporales: a taxonomic, ecological and evolutionary re-evaluation. Stud Mycol 64:85–102

Zhang Y, Crous PW, Schoch CL, Hyde KD (2009b) Pleosporales. Fungal Divers 53:1–221

Zhang XY, Bao J, Wang GH, He F, Xu XY, Qi SH (2012) Diversity and antimicrobial activity of culturable fungi isolated from six species of the South China Sea gorgonians. Microb Ecol 64:617–627

Zuccaro A, Schulz B, Mitchell JI (2003) Molecular detection of ascomycetes associated with *Fucus serratus*. Mycol Res 107:1451–1466

# Chapter 13
# Biology and Ecology of Freshwater Fungi

**Clement K.M. Tsui, Christiane Baschien, and Teik-Khiang Goh**

## Introduction

Freshwater fungi complete at least one part of their life cycle in water and distribute propagules (spores, conidia, sporangia) in or above water. It has been estimated that there are more than 3000 species of fungi occurring in the aquatic habitats (Abdel-Aziz 2008). Taxonomically, aquatic fungi comprise taxa from all fungal phyla (*Cryptomycota*, *Chytridiomycota*, *Blastocladiomycota*, *Mucoromycotina*, *Glomeromycota*, *Dikaryomycota*).

Fungal morphology in freshwater ranges from zoospores and nonmotile single cells of *Cryptomycota*, *Chytridiomycota*, *Blastocladiomycota*, yeasts, and aseptate and septate hyphae up to interwoven hyphae in more or less complex plectenchyma in higher fungi. Sporangiophores and sporangia (*Chytridiomycota*, *Blastocladiomycota*, and *Mucoromycotina*), ascomycetous and basidiomycetous

---

Our work is dedicated to John Webster (1925–2014) for his pioneer and influential work on freshwater fungi.

C.K.M. Tsui, Ph.D. (✉)
Department of Pathology and Laboratory Medicine, University of British Columbia, Vancouver, BC, Canada, V6T 1Z4
e-mail: clementsui@gmail.com

C. Baschien, Ph.D.
Department of Microbial Ecology and Diversity, Leibniz Institute DSMZ-German Collection of Microorganisms and Cell Cultures, Inhoffenstr. 7 B, 38124 Braunschweig, Germany

T.-K. Goh, Ph.D.
School of Biological Sciences, Faculty of Integrative Sciences and Technology, Quest International University Perak, No. 227, Plaza Teh Teng Seng (Level 2), Jalan Raja Permaisuri Bainun, 30250 Ipoh, Perak Darul Ridzuan, Malaysia
e-mail: teikkhiang.goh@qiup.edu.my



D.-W. Li (ed.), *Biology of Microfungi*, Fungal Biology,
DOI 10.1007/978-3-319-29137-6_13

yeasts, conidiophores, and a high diversity of more or less conspicuous conidia (*Ascomycota* and *Basidiomycota* derived) are also visible under the microscope. In addition, complex structures such as acervuli, pycnidia, and ascocarps and basidiocarps as well as lichens can be found on substrates in freshwaters. Over the past few decades, various morphological and ecological groups of water-associated fungi have been identified (Fuller and Jaworski 1987; Goh and Hyde 1996; Hyde et al. 1997; Ingold 1975; Jones 1981). These include the zoosporic fungi, the aquatic ascomycetes, the "Ingoldian fungi," the aero-aquatics, and a great diversity of mitosporic fungi (asexual *Ascomyceta*) occurring on submerged plant materials.

Worldwide, freshwater comprises diverse habitats such as groundwater, streams, rivers, canals, and lakes but also includes amphibious habitats, such as ditches, peats, and swamps (Shearer et al. 2007). Generally, most freshwater fungi are associated with organic matter derived from decaying plants and animals. However, there are many microhabitats in and adjacent to freshwater, which provide space and possibilities for different fungal life strategies. These microhabitats include roots and other parts of submerged and riparian plants, the canopy, the ambient soil, and sediments. Indeed, freshwater fungi are recorded from the tree canopy (Ando 1992; Ando and Tubaki 1984a, b; Sridhar et al. 2009), rainwater (Gönczöl and Révay 2004), and soil (Park 1974). Also leaf litter in treeholes (Gönczöl and Révay 2003), dew drops (Tubaki et al. 1985), and honey (Magyar et al. 2005) have been identified as locality for spores of freshwater fungi. Fungi have been observed within cooling towers (Eaton and Jones 1971a, b), groundwater (Krauss et al. 2005), and potable water distribution systems (Doggett 2000; Nagy and Olson 1982) including tap water (Heinrichs et al. 2013).

## Who Are the Members of Freshwater Fungi?

Traditional taxonomy and identification of fungi is mostly based on fruiting bodies and spores. Freshwater fungi are divided into various morphological and ecological groups (Shearer et al. 2007; Goh and Hyde 1996). The different groups require specialized methods to examine their biodiversity, taxonomy, distribution, population dynamics, and ecological functions. Traditionally, it has been a challenge to characterize all the different groups of freshwater fungi within a freshwater habitat.

The first group is (*I*) the aquatic hyphomycetes (more than 300 described mitosporic fungi) (Figs. 13.1 and 13.2), also known as "the Ingoldian fungi" in honor of the pioneer mycologist Prof. C. T. Ingold, which comprises conidial states of mainly *Ascomycota* and a few *Basidiomycota* (Shearer et al. 2007; Jones et al. 2014). Ingold (1942) discovered the conidia in the foam of streams and showed the connection to their associated mycelia on submerged leaves. Most aquatic hyphomycetes produce conspicuous stauroconidia (e.g., tetraradiate [*Alatospora*, *Articulospora*] or other branched forms [*Varicosporium*]) or scolecoconidia (sigmoid, curved, or straight [*Anguillospora*, *Flagellospora*]). The conidial shape is an adaptation to survival and

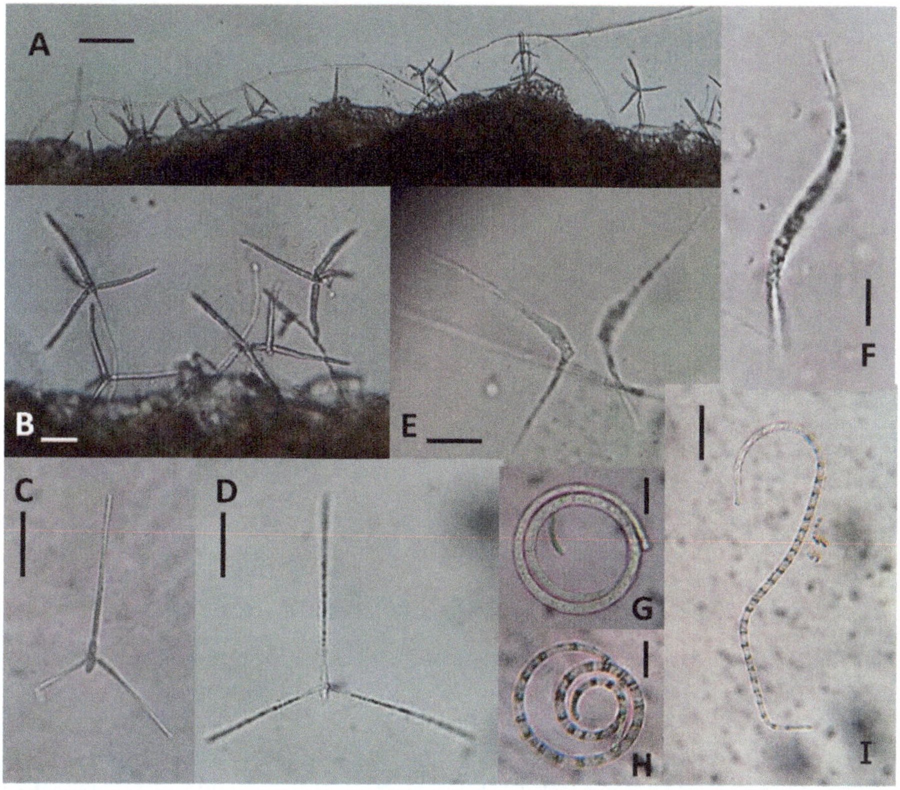


**Fig. 13.1** Ingoldian fungi. (**a**, **b**) *Triscelophorus acuminatus*. Conidia production seen at the edge of submerged leaf. Each conidium has four arms. (**c**, **d**) *Triscelophorus monosporus*. Conidia with three arms. (**e**) *Lunulospora cymbiformis*. Conidia which are sickle-shaped and distinctly bent more or less at right angle. (**f**) *Anguillospora crassa*. Typical sigmoid-shaped conidium. (**g**, **h**) *Helicomyces roseus*. Typical coiled conidia. (**i**) *Helicomyces roseus*. Conidium which has uncoiled and become more or less sigmoid shaped in water. Scale bars: **a** = 50 μm; **b** – **i** = 20 μm

dispersal in aquatic habitats discussed below (Dix and Webster 1995; Webster 1959). Aquatic hyphomycetes are known as important decomposers in the turnover of leaf litter in woodland streams [overview in (Gessner et al. 2007)]

The second group is called the (*II*) aero-aquatic fungi (mitosporic ascomycetes) (~90 species described) (Fig. 13.1g, h), which are found on submerged plant litter and wood in flat lentic waters, such as woodland ponds and ditches, which may have periodical levels of water. In contrast to aquatic hyphomycetes, the aero-aquatic fungi release their propagules aerially. Therefore, they produce conspicuous air-trapping dispersal units, which can flow on the water surface; for example, some spores are helicoid in more than one plane (Fig. 13.1g, h) (e.g., *Helicoma, Helicoon*) or spiral into a sphere or a net [e.g., *Candelabrum*, *Spirosphaera* (Voglmayr 2004, 2011; Voglmayr and Delgado-Rodriguez 2003; Voglmayr et al. 2011)]. The aero-aquatic life strategy was first described by Agathe van Beverwijk (1951).

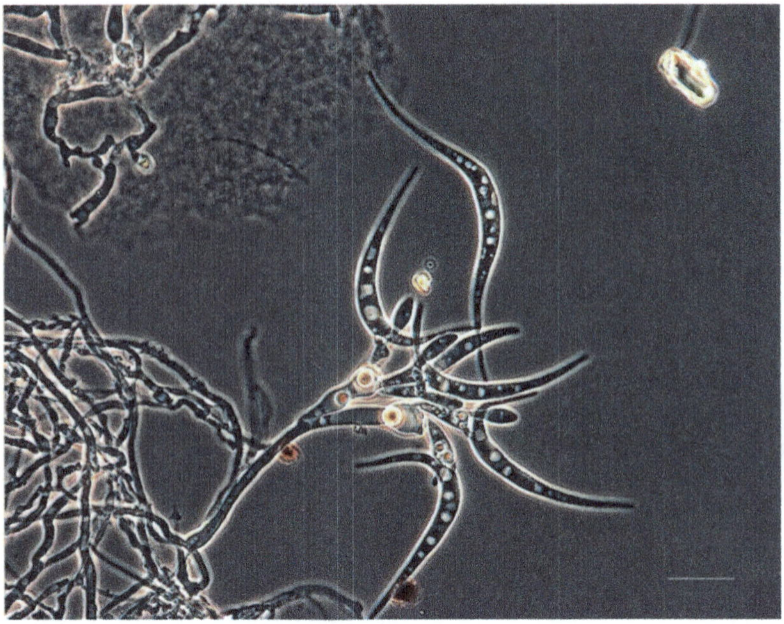

**Fig. 13.2** *Lunulospora curvula*. Conidiogenesis of crescent-shaped unbranched conidia. Bar = 25 μm

The third group is named (*III*) freshwater ascomycetes (~600 species described meiosporic fungi), which comprise the sexual states of phylogenetically heterogeneous ascomycete fungi occurring worldwide in freshwater habitats on herbaceous and woody substrates (Shearer et al. 2007, 2009; Goh and Hyde 1996; Shearer 1993a; Vijaykrishna et al. 2006). In addition, they are also collected from submerged dead macrophytes (Shearer 1993a; Fallah and Shearer 2001). The morphologies in asci and ascospores are very diverse and noticeable. Asci of freshwater ascomycetes are either deliquescent, with apical apparatus (e.g., *Massarina*, *Jahnula*), or fissitunicate (ectoascus and coiling endoascus, e.g., *Kirschsteiniothelia* (transferred to *Helicascus*)) (Shearer 1993b). Many freshwater ascomycetes produce ascospores with appendages, which facilitate attachment to substrates (Shearer et al. 2007; Shearer 1993a; Wong et al. 1998). Gelatinous gel-like sheaths and/or thick-walled hyphae (Shearer et al. 2009; Ingold 1955) are thought to enhance attachment and adhesion to plant materials (Digby and Goos 1987; Ingold and Chapman 1952). However, spores borne with gelatinous sheath and the active discharge of ascospores are also present in strictly terrestrial ascomycetes. In comparison to aquatic and aero-aquatic hyphomycetes, freshwater ascomycetes seem to be less exclusively adapted to life in aquatic habitats (Vijaykrishna et al. 2006).

The fourth group of freshwater fungi is the (*IV*) zoosporic fungi (Fig. 13.3), which produce flagellate zoospores as part of their life cycle. They are known for

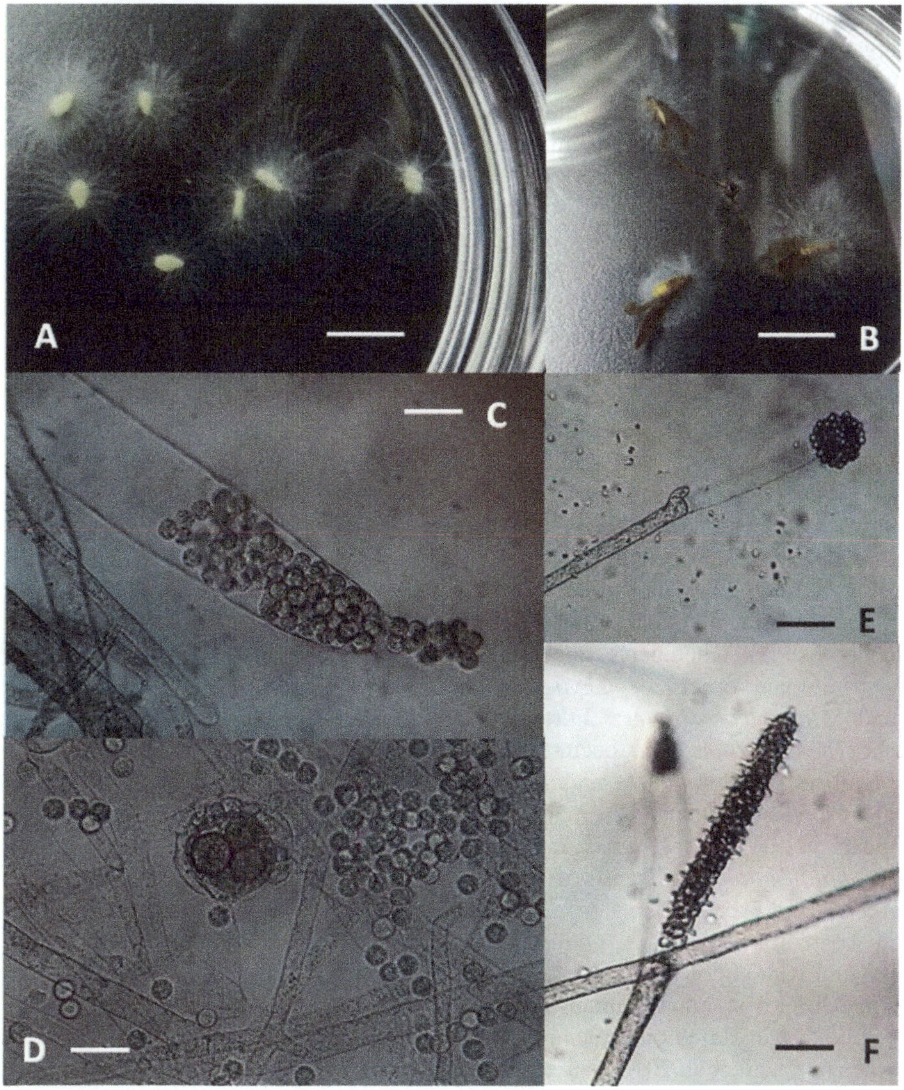


**Fig. 13.3** Oomycetes. (**a**) Baiting of oomycetes, using sesame seeds. (**b**) Baiting of oomycetes, using insect carcasses. (**c**) *Saprolegnia* sp., zoosporangium releasing zoospores. (**d**) *Saprolegnia* sp., oogonium, with attached antheridia, and many encysted primary zoospores. (**e**) *Achlya* sp., zoosporangium showing discharged and encysted zoospores at the apical pore. (**f**) *Dictyuchus* sp., a dictyoid zoosporangium in the process of zoospore discharge. Scale bars: **a**, **b** = 5 mm; **c** = 10 µm; **d–f** = 20 µm

their parasitic life strategies but they are also involved in the decomposition of organic matter (Sparrow 1960). Historically, they have been known as the members of "phycomycetes," a functionally defined group. They have also been traditionally termed the "lower fungi" which basically comprises species without a septate hyphal system. With the advent of molecular studies and recent taxonomic treatments based on phylogenetics, the "phycomycetes" is in fact heterogeneous, comprising of members from the *Eumycota* ("true fungi") and the *Chromista* (*Oomycetes* and *Labyrinthulomycetes*). The biology, taxonomy, and phylogenetic relationship of these aquatic organisms have been well documented (Fuller and Jaworski 1987; Beakes 2003; Bowman et al. 1992; Buczacki 1983; Powell 1993).

The freshwater zoosporic fungi belong mostly to the *Chytridiomycetes* and the fungal-like *Oomycetes*, which are microscopic organisms not producing any fruiting bodies visible to the naked eye. Members of the *Chytridiomycetes* are generally called chytrids, whereas those belonging to the *Oomycetes* are usually called water molds. They usually reproduce asexually by means of zoospores, but in *Oomycetes*, sexual reproduction may occur by means of oogamy, resulting in the formation of oospores, which are survival structures generally resistant to adverse environmental conditions. The chytrids produce uniflagellate haploid zoospores, whereas those of the water molds are biflagellate and diploid. These taxa are associated with dead and living plant materials but also with algae, cyanobacteria (Sonstebo and Rohrlack 2011), invertebrates, fish, and amphibians (Powell 1993; Longcore et al. 1999; Powell et al. 2013).

The main orders of *Chytridiomycetes* are *Chytridiales*, *Spizellomycetales*, and *Monoblepharidales*. Furthermore members of *Neocallimastigomycota* and *Blastocladiomycota* occur in aquatic habitats (Powell and Letcher 2014). Members of *Chytridiales* and *Blastocladiales* are more frequently encountered in freshwater systems. The main orders of *Oomycetes* found in aquatic environments are the *Saprolegniales*, *Leptomitales*, and *Peronosporales*. Common genera of *Saprolegniales* frequently isolated from the aquatic systems are *Saprolegnia*, *Achlya*, and *Dictychus*.

Apart from chytrids, *Cryptomycota* is a recently discovered (Jones et al. 2011) phylum with endoparasitic lifestyle (James et al. 2013; Lazarus and James 2015). They differ from other filamentous fungi in that they lack chitinous cell walls in the trophic stage. Based on molecular markers, they have been detected in lakes and wetlands (Ishii et al. 2015; Wurzbacher et al. 2014).

The fifth group (*V*) belongs to aquatic–terrestrial hyphomycetes (mitosporic ascomycetes) (Figs. 13.4, 13.5 and 13.6) that comprise a huge morphological diversity of conidial (mitosporic) fungi growing on decaying plant material (Shearer et al. 2007; Hu et al. 2014) and capable of sporulating underwater (Bärlocher et al. 2008; Baschien et al. 2009). They are distinguished based on the features of conidia, conidiophores, and the conidiogenesis. For example, common fungi include *Dactylaria, Dictyochaeta, Canalisporium,* and *Sporoschisma*; some of these fungi are linked with their corresponding sexual relatives using cultural or molecular studies. Some studies also demonstrate the presence of fungi of terrestrial origin, including saprobes from plants and soil, such as *Cladosporium*, *Alternaria*, and *Penicillium* species. Fungi on leaves from the canopy (Gönczöl and Révay 2006) may fall into the water.

**Fig. 13.4** Lignicolous hyphomycetes. (**a**) *Ellisembia adscendens*. Colony on submerged wood. Conidia are long and projecting upward on the wood surface. (**b**) *Phaeoisaria clementidis*. Colony on submerged wood. Conidiophores are in the form of long synnemata projecting upward on the wood surface. (**c**) *Vermiculariopsiella* sp. Colony on submerged wood. Conidiophores are in the form of dense sporodochia on the wood surface, surrounded by long setae and producing conidia in the form of white slimy mass. (**d**) *Monotosporella setosa*. Colony on submerged wood. Conidiophores are long, erect, and projecting from the wood surface. (**e–g**) *Monotosporella setosa*. Conidiophores producing conidia at their apex. (**h**) *Ellisembia adscendens*. Conidia. (**i**) *Canalisporium elegans*. Conidia, which were produced in the form of sporodochia on the surface of submerged wood. (**j**, **k**) *Dictyosporium elegans*. Conidia. Scale bars: **a–d** = 200 μm; **e–h**, **j**, **k** = 20 μm; **i** = 50 μm

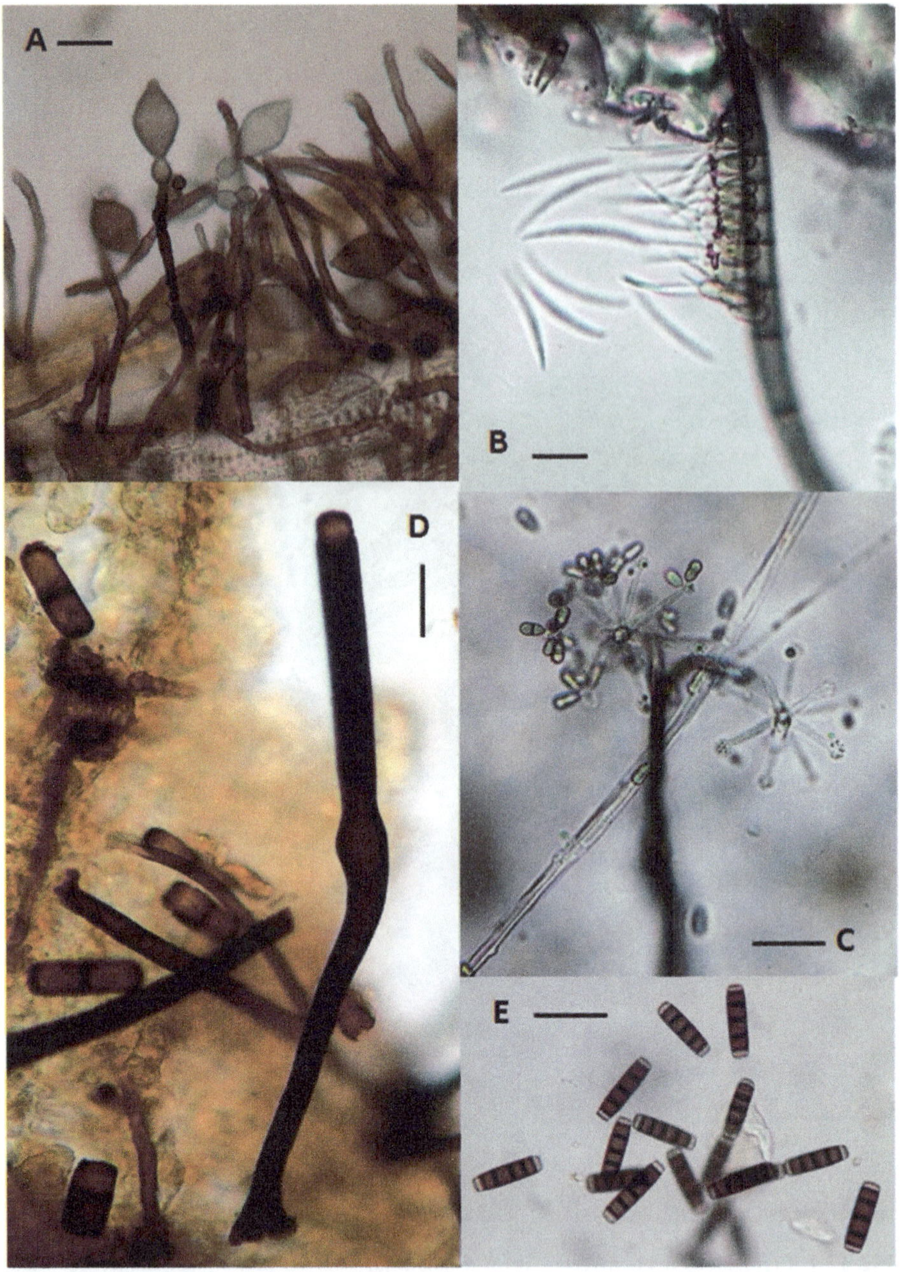


**Fig. 13.5** Lignicolous hyphomycetes. (**a**) *Beltrania africana*. Conidiophores and conidia from submerged wood. (**b**) *Cryptophialoidea secunda*. Part of setiform conidiophore showing the fertile region, bearing unilateral phialidic conidiogenous cells, producing falcate conidia. (**c**) *Pseudobotrytis terrestris*. Conidiophore bearing umbellately arranged polydenticulate conidiogenous cells, producing two-celled conidia. (**d**) *Sporoschisma uniseptatum*. Conidiophore which has a swollen venter, producing endogenous conidia, usually in basipetal chains. (**e**) *Sporoschisma mirabile*. Conidia, initially borne on chains, have become disarticulated in water mount. Scale bars: **a** = 20 μm; **b** = 10 μm; **c**, **d** = 20 μm; **e** = 50 μm

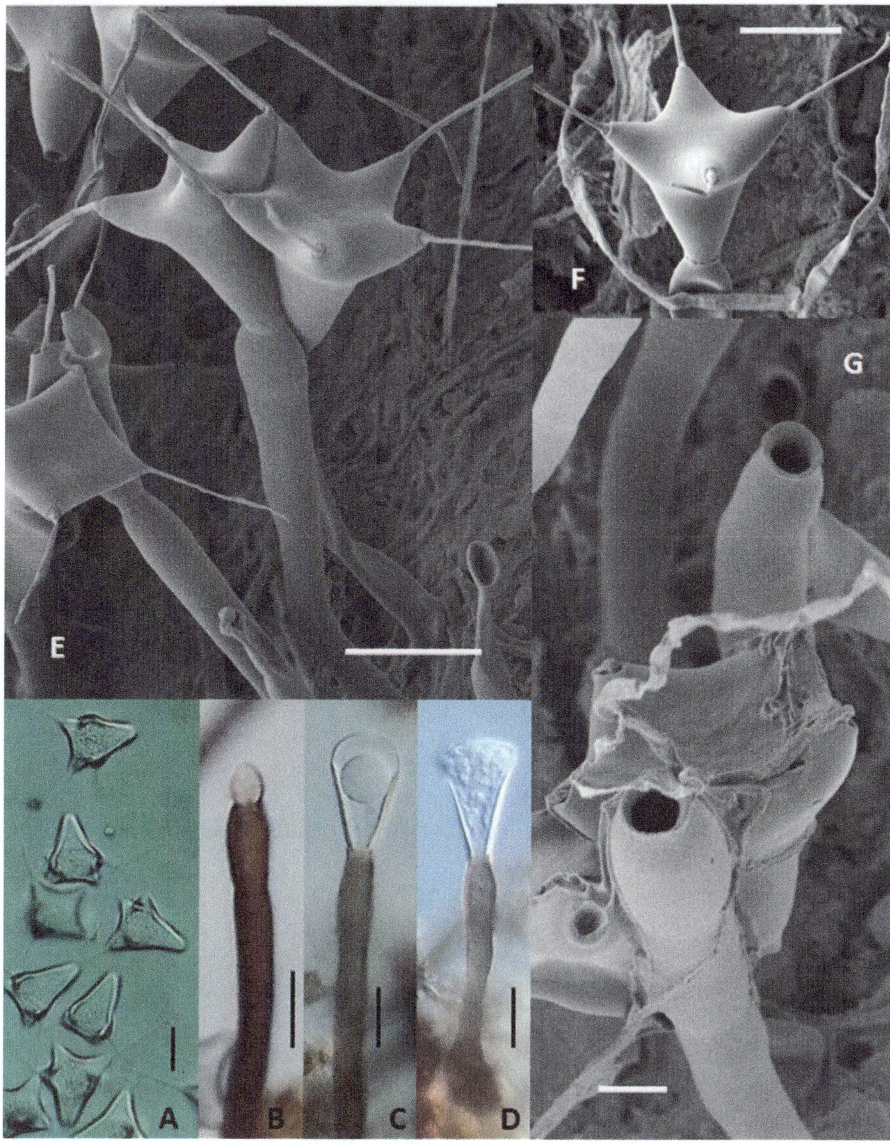


**Fig. 13.6** *Nawawia quadrisetulata*. (**a**) Conidia, bearing 4–5 setulae at the corners of the distal end. (**b–d**) Conidiophores showing sequential development of conidia. (**e**, **f**) Scanning electron micrographs of setulate conidia produced at the apex of conidiophores. (**g**) Conidiophores showing the terminal phialides. Scale bars: **a–d** = 20 μm; **g** = 5 μm

## How Are They Adapted to an Aquatic Lifestyle?

Spores are the dispersal propagules of fungi. Aquatic fungi need to produce spores that can be dispersed in water, and then become entrapped to, and subsequently colonize new substrates. The completion of the life cycle and hence the survival of the species is reliant on the spores.

The zoosporic fungi release motile spores that can actively disperse/swim in the lentic aquatic environments (e.g., ponds, lakes, ditches, pools, and swamps). They are usually chemotactic, being sensitively attracted to their host or substrates that release specific sugars and amino acids (Fuller and Jaworski 1987). Usually the zoospores swim for a few minutes to hours before they stop and encyst. At the initial stage of zoospore encystment, depending on the species, the flagella are either retracted or shed before the zoospore assumes a spherical shape. An important property of the zoospores in general is their ability to stick firmly to the surfaces of the substrates upon which they settle down to encyst. Adhesion ensures that the zoospore, upon reaching a favorable substrate, accidentally, or through specific tactic responses, would remain firmly attached. Evidently, this behavior has great ecological value for both saprobes and parasites as it establishes a permanent contact between the fungus and its potential food source. Swimming zoospores are for dispersal and thus are not adhesive but become so in the initial stage of encystment before a cyst wall is made. The adhesive phase lasts only about 30–60 s (Sing and Bartnicki-Garcia 1972). Apparently, the timing of the adhesive phase, coinciding with the change from motile to sessile form, provides obvious ecological advantage to the fungus. Once the cysts mature, i.e., after a well-defined wall is made, the cells lose their ability to attach themselves to solid surfaces. After successful adhesion and encystment of the spore, germination of the cyst leads to penetration and colonization of the substrate.

Besides the zoosporic fungi, most aquatic fungi are saprobes that grow/colonize on submerged wood or waterlogged leaves in freshwater. The spores of freshwater ascomycetes and Ingoldian fungi are not motile in the aquatic environment. These fungi usually occur in lotic habitats (e.g., streams, creeks, rivers, brooks), and thus they possess strategies to survive in turbulent water. Their propagules are released from conidiophores growing out of substrates and thus are able to detach from the surface of the substrates, disperse (float) in water, and eventually become entrapped (settled) or attached (adhere) to new substrates, which they can colonize by penetration of a germ tube. The mechanisms of fungal adhesion and the role of mucilage in fungal attachment in the aquatic environment have been well documented (Wong et al. 1998; Au et al. 1996; Ingold 1966; Jones 2006). An excellent review of the adaptations for dispersal in filamentous freshwater fungi is given by Goh and Hyde (1996).

Many freshwater ascomycetes have ascospores with various sheaths, appendages, or wall ornamentations, which probably function in dispersal and/or attachment of the spores. Shearer (1993a) provided a list of ascomycete species possessing ascospores surrounded by mucilaginous sheaths. There are ascospores that possess unfurling mucilaginous appendages, which uncoil in water to form long viscous threads. Several unique appendage types have now been shown to exist in freshwater ascomycetes, which account for their success in the aquatic environments (Jones 2006; Wong and Hyde 1999).

The Ingoldian fungi have been known for decades to have sigmoid or tetraradiate spores (Ingold 1975; Goh 1997). Sigmoid conidia often become attached at their sticky poles and straighten in the direction of the water current so that they are less likely to be washed away in a lotic habitat (Webster and Davey 1984). Tetraradiate and variously branched spores act as an anchor and allow their entrapment to the substrates or in surface foam (Ingold 1966). Tetraradiate spores can also effectively attach to the substrate with three "legs" forming a strongly adhesive tripod (Webster and Descals 1981). Adhesive mucilaginous material is also produced at each arm of the tetraradiate spore in contact with a surface and attaches them firmly to the substrate.

The aero-aquatic hyphomycetes, such as *Beverwykella*, *Cancellidium*, and *Clathrosphaerina*, are normally found on decaying plant materials in slow-flowing streams and stagnant ponds (Gessner et al. 2007; Voglmayr and Delgado-Rodriguez 2001, 2003; Voglmayr and Krisai-Greilhuber 1997). Their propagules are beautiful, extraordinary, and interesting as they possess a special floatation device, usually composed of intricate branching hyphal network and thus air trapping, enabling these fungi to be dispersed from one static water habitat to another. Conidia of *Helicoon* species are composed of tightly coiled filaments, which assume a barrel shape, and can trap air in water for floating (Goh and Hyde 1996). For further discussion of the adaptation strategies in the aero-aquatic fungi, see Fisher (Fisher et al. 1977) and Webster and Descals (1981).

A great diversity of dematiaceous hyphomycetes have been discovered from wood submerged in freshwater (Goh and Tsui 2003). Among these lignicolous hyphomycetes, many possess long mononematous stiptate conidiophores, which stand erect from the submerged substrates and bear masses of conidia at the apices. Examples of such hyphomycetes are *Acrogenospora*, *Cryptophiale*, and *Spadicoides*. Others produce erect synnemata (e.g., *Bactrodesmium longisporum*, *Nawawia dendroidea*, and *Phaeoisaria clematidis*) or occur as sporodochia (e.g., *Dictyosporium*, *Canalisporium*, and *Yinmingella*). Producing easily detachable conidia at the apex of long, erect conidiophores, or having conidia produced in dense masses from sporodochia, may be adaptation strategies conducive to effective spore dispersal in the aquatic habitats. Moreover, certain features observed in some of the commonly found hyphomycetes from submerged wood are worth mentioning, because they may represent special adaptation features for dispersal in the aquatic habitats. For example, *Chalara*, *Sporoschisma*, and *Sporoschismopsis* (Goh et al. 1997) produce long chains of conidia from erect conidiophores, which eventually disarticulated in water for ease of dispersal. The muriform conidia of *Canalisporium* have distinct pores or canals in the septa (Goh et al. 1998) which enable air to trap inside the conidia for floating. The various spore types of these lignicolous hyphomycetes in freshwater also are provided with modified appendages, mucilaginous sheaths, setulae, or arms (Goh 1997), and these are functionally comparable to those of aquatic ascomycetes and Ingoldian fungi.

## How Did They Evolve?

The absolute age of fungal phyla is an area of active research, and only broad estimates exist for the oldest clades (Taylor and Berbee 2006). Currently, the divergence of younger phyla from the *Cryptomycota*, *Chytridiomycota*, and *Blastocladiomycota* clades—and the loss of the zoospore stage—is thought to have occurred during the Neoproterozoic, approximately 600–800 Myr (Stajich et al. 2009). The first divergence of terrestrial fungi (*Endogonales* and *Glomales*) has been estimated to have occurred about 600 Myr or even earlier, while *Ascomycota* and *Basidiomycota* separated 500 Myr (Berbee and Taylor 2001). The origin and radiation of *Euascomycetes* (whose groups commonly produce conidia adapted in aquatic habitats) may have taken place in the Mesozoic, about 240 Myr. This puts a lower limit to the earliest appearance of aquatic ascomycetes and relatives. Did the majority of the taxa appear in a brief burst of rapid radiation, or was this spread out over an extended period of time? Are new species still invading running waters today (Schlütz and Shumilovskikh 2013)?

Fungi in freshwater are grouped ecologically and morphologically, and it is not surprising that they have evolved independently from multiple lineages. Shearer (Shearer 1993a) first proposed the multiple origins of freshwater ascomycetes, which are present as endophytes, pathogens, or saprobes on plants, and have become adapted to aquatic environment when these plants invaded water. 18S rRNA data (Vijaykrishna et al. 2006; Baschien et al. 2006; Belliveau and Bärlocher 2005) showed that freshwater fungi (sexual and asexual ascomycetes) evolved from terrestrial fungi in (more than) four different classes, *Sordariomycetes*, *Dothideomycetes*, *Leotiomycetes*, and *Orbiliomycetes*. Many freshwater fungi have terrestrial relatives, supporting the fact of secondary adaptation to the freshwater environment. A comprehensive study of 84 fungi of described and undescribed freshwater *Dothideomycetes* and 85 additional ascomycetes representative of the major orders and families of *Dothideomycetes* based on ribosomal genes also confirmed the polyphyletic origins of freshwater ascomycetes (Shearer et al. 2009). Apart from these, molecular studies of fungi on submerged leaves using denaturing gradient gel electrophoresis [DGGE; (Kelly et al. 2010; Nikolcheva et al. 2003)] and the analysis of terminal restriction fragment length polymorphism [TRFLP; (Nikolcheva et al. 2003; Nikolcheva and Bärlocher 2005)] and of clone libraries (Clivot et al. 2014; Harrop et al. 2009) provided evidence for the presence of fungi of terrestrial origin.

Many genera of freshwater fungi (ascomycetes and their asexual relatives) are also not monophyletic. The polyphyletic origin of aquatic hyphomycetes has been reinforced by additional studies (Baschien et al. 2006, 2013; Campbell et al. 2006, 2009). For example, Ingoldian fungi are assigned to four classes: *Sordariomycetes* (~11 spp.), *Dothideomycetes* (~10 spp.), *Pezizomycetes* (1 sp.), *Orbiliomycetes* (3–5 spp.), and *Leotiomycetes* (>75 spp.). The morphology of tetraradiate and sigmoid conidial shape has been recognized as convergent development in unrelated aquatic hyphomycete taxa (Ingold 1966; Webster 1980). Molecular studies of ribosomal genes also reassured the convergence of the conidial shape (Baschien et al. 2006, 2013; Campbell et al. 2006; Belliveau and Bärlocher 2005). Also, the helicosporous aero-aquatic fungi have evolved from multiple lineages within the ascomycetes (Tsui and Berbee 2006).

The molecular studies mentioned above showed that many ascomycetous and basidiomycetous freshwater fungi are closely related to terrestrial fungi. A few species of the genera *Varicosporium*, *Tetracladium*, *Filosporella*, and *Anguillospora* have been isolated as endophytes from aquatic or terrestrial roots (Fisher et al. 1991; Kohout et al. 2012; Nemec 1969; Sati and Belwal 2005; Watanabe 1975). The step back from terrestrial to aquatic life cycles could have been aided by an endophytic lifestyle (Selosse et al. 2008). Furthermore, the localization of freshwater fungi inside roots or other amphibious plant parts may function as temporary reservoir during different stages within a life cycle. The production of different conidial shapes and synanamorphs may also be an adaptation to shifts between aquatic, semiaquatic, and terrestrial habitats (Baschien et al. 2013). Kohout and co-workers (Kohout et al. 2013) proposed the scenario that terrestrial ancestors of recent aquatic plants interacted with different root-associated fungi (RAF). The aquatic hyphomycetes could have evolved from non-mycorrhizal RAF that once entered aquatic habitats together with their host plants.

## What Are They Doing in Freshwater Habitats?

The primary ecological role of fungi in aquatic habitats is to decompose dead plant material—both woody and herbaceous debris. When Kaushik and Hynes (1971) demonstrated the crucial role of fungi in the decomposition of plant materials to detritus in streams, limnologists recognized the important ecological function of aquatic hyphomycetes. Bärlocher and Kendrick (1974) showed that aquatic hyphomycetes condition leaf material for aquatic invertebrates (Plecoptera, Trichoptera, Coleoptera, Crustacea, Gastropoda) by increasing the palatability using exoenzymes [cellulases, pectinases, laccases; (Abdel-Raheem and Shearer 2002; Chamier 1985)]. Most aquatic fungi have the ability to decompose a wide range of organic substrates, although a few species may be limited to one or a few types of substrates. For example, Gulis (2001) showed that wood/twig substrates bear fungal communities distinct from those on leaves. In general, aquatic ascomycetes and basidiomycetes are thought to be responsible for the decomposition of woody debris while Ingoldian fungi decompose either leaves or herbaceous debris. Through decomposition, freshwater fungi can facilitate the transfer of nutrients and energy between trophic levels in the food web (Gessner et al. 2007).

Wood is a complex substance and its major chemical constituents contain cellulose, hemicellulose, and lignin. The degradation mechanisms of wood are well known in terrestrial fungi, and it is assumed that similar mechanisms are present in freshwater fungi. Cellulose hydrolysis is achieved by endoglucanases and cellobiohydrolases, collectively termed cellulases (Eaton and Hale 1993). Hydrolysis of hemicellulose, a mixed polymer, occurs via the action of hydrolytic xylanases, mannanases, and possibly other hydrolases with broad substrate specificity (Eaton and Hale 1993). The degradation of lignin involves two peroxidases, lignin peroxidase and Mn-dependent peroxidase, and a polyphenol oxidase, laccase, known as lignin-modifying enzymes (LMEs) (Pointing 2001). The vast majority of freshwater ascomycetes, regardless of eco-climate, habitat, and substrate distributions, are capable of breaking down cellobiose, hemicellulose, and xylan and starch, which are important carbon compounds in plant-based debris (Bucher et al. 2004; Simonis et al. 2008; Yuen et al. 1998).

Three fungal wood decay types are recognized on the basis of whether or not they can degrade cellulose and lignin or just cellulose alone, namely, soft rot, white rot, and brown rot (Eaton and Hale 1993). Soft rots occur in wood that has an unusually high level of moisture, which is often the case for woody substrates in aquatic environments. Most of the freshwater fungi are capable to form soft rot decay and to produce cellulases (Bucher et al. 2004; Abdel-Raheem and Shearer 2002). White-rot fungi normally have the enzymes to degrade both cellulose and lignin simultaneously. The residual material that is left behind has a somewhat fibrous appearance and is very pale in color, looking as if it had been bleached. In contrast, members of brown-rot fungi can degrade cellulose. After decomposition, wood is brown in color and tends to be broken up into cubical fragments that quickly disintegrate into a powdery brown residue. It has been questionable whether or not freshwater ascomycetes can form white rot or brown rot because the breakdown of lignin in woody debris is primarily accomplished by basidiomycetes that do not seem common in the aquatic habitat. However, Junghanns et al. (2005) showed that the aquatic hyphomycete *Clavariopsis aquatica* produced laccases. Five putative laccase genes (*lcc1* to *lcc5*) identified in *C. aquatica* were differentially expressed in response to the fungal growth stage and potential laccase inducers (Sole et al. 2012). Recently, Kerr and co-workers (2013) showed the oxidation of lignin in leaf litter by undetermined aquatic fungi using high spatial resolution infrared microspectroscopy. The modification of lignin by freshwater fungi degrades lignin into carbohydrate-depleted recalcitrant carbon, which may influence the carbon pool in aquatic environments.

Leaves and other nonwoody plant parts (e.g., fruits and seeds) represent a different type of substrate than wood and bark. In general, angiosperm leaves are readily decomposed. When a leaf falls from a tree from the riparian zone, it would be colonized by soil-inhabiting fungi—those already present in the layer of litter at the soil surface or various aquatic fungi when the leaf becomes submerged. Most aquatic hyphomycetes can degrade cellulose, various hemicelluloses, and pectin (Chamier 1985; Abdel-Raheem and Ali 2004; Chandrashekar and Kaveriappa 1991; Zemek et al. 1985). Many studies have examined the enzymatic activities of freshwater fungi and reported amylase, $\beta$-glucosidase, $\beta$-xylosidase, endoglucanase, endoxylanase, lipase, pectinase, and protease activity (Chamier 1985).

To which extent aquatic—terrestrial—hyphomycetes play an important role in aquatic systems is still unclear, although they are common on leaves and wood. It is generally assumed that the so-called aquatic–terrestrial fungi are not able to compete against aquatic hyphomycetes and hence are removed after the first 2–3 days of the decomposition process (Harrop et al. 2009; Bärlocher and Kendrick 1974; Nikolcheva et al. 2005; Perez et al. 2012). However, aquatic–terrestrial fungi are present on longer-exposed leaves in streams [e.g., (Hameed et al. 2008; Smither-Kopperl et al. 1998)]. Kelly et al. (2010) studied fungal communities on decomposing maple and aspen leaves, and his group reported that the majority of operational taxonomic units (OTUs) represented terrestrial *Cladosporium* species, whereas aquatic hyphomycete sequences were not observed. Indeed, several aquatic–terrestrial hyphomycetes are able of decomposing leaf litter (Bucher et al. 2004; Singh et al. 2014). For example, the terrestrial leaf litter ascomycete *Torula herbarum*, isolated from a tropical stream, has been found to be able to break down lignin (Bucher et al. 2004).

Few aquatic fungi form a symbiotic mycorrhizal relationship with the roots of trees and other plants and macrophytes. This association is mutually beneficial to both the plant and the fungus. The fungus enables the plant to take up nutrients that are unavailable, and the plant provides nutrition for the fungus. There are two fundamentally different types of mycorrhizal associations—ectomycorrhizal (usually involving a basidiomycete) and endomycorrhizal (most often involving a member of the *Glomeromycota*). In the former, the fungus produces a covering of hyphae (called a sheath, Hartig net, or mantle) around the outside of smaller rootlets of the host plant. Other hyphae invade the cortex of the rootlet but do not disrupt the individual cells. In endomycorrhizal associations, no sheath is formed and hyphae of the fungus actually invade cells of the cortex of the rootlet. Perhaps 80 % of all vascular plants form mycorrhizal associations with fungi. First aquatic vesicular–arbuscular mycorrhiza was discovered in *Littorella uniflora*, *Lobelia dortmanna*, and *Isoetes lacustris* by Sondergaard and Laegaard (1977). Vesicular–arbuscular mycorrhizas (VAM) in *Isoetes* plants were later also observed by Sudova et al. (2011).

Furthermore, dark septate endophytes (DSE) and fungal root associates (RAF) from aquatic habitats were reported (Kohout et al. 2012; Seena et al. 2008). Mycorrhiza and some DSE are known to be mutualistic. On the other hand, endophytism can be the balance between rather antagonistic relationships between fungus and host which may even develop into a pathogenic outcome (Schulz and Boyle 2005). Several species of aquatic hyphomycetes were isolated as endophytes from aquatic or terrestrial plants (Sati and Belwal 2005). Some aquatic hyphomycetes were also found in roots of terrestrial habitats [e.g., (Tedersoo et al. 2007)]. However, there are few reports about the transmission of freshwater fungi either vertical to the next generation of the same host or horizontal between hosts. It is not known if endophytic aquatic hyphomycetes have a harmful or beneficial effect on the hosts. Further thorough studies are needed to elucidate the aspects of endophytic life of aquatic hyphomycetes.

For instance, all the described *Minutisphaera* spp. (*M. fimbriatispora*, *M. japonica*, *M. aspera*, and *M. parafimbriatispora*) have been isolated from submerged wood in freshwater habitats so far (Ferrer et al. 2011; Raja et al. 2013), suggesting that they play an ecological role in nutrient cycling and organic matter decomposition in freshwater habitats (Pointing 2001). A recent BLAST search (Altschul et al. 1990) of newly sequenced ITS strains of *M. aspera* and *M. parafimbriatispora* in GenBank identified two ITS sequences from endophytes ("*Pleosporales* sp. 39 g," JX244063, and "*Didymosphaeria* sp. TS_04_050," HQ713763) as the top BLAST matches with high percent identity values and coverage. Based on the uncorrected p-distances calculated in PAUP*, the two ITS sequences from endophytes were identical to each other and shared 99 % sequence similarity with ITS sequences of *M. aspera*. The high ITS sequence similarity between these fungal endophytes and *M. aspera* could imply that *Minutisphaera* may have a dual mode of lifestyle as saprobes on submerged wood as well as fungal endophytes inside the roots of trees. However, additional studies are warranted to test this ecological hypothesis (Selosse et al. 2008).

In contrast to the aquatic ascomycetes (meiosporic and mitosporic fungi), which generally colonize organic matter, which are comparatively larger in size, such as fallen leaves, submerged wood, or aquatic plants, the zoosporic fungi are colonizers of smaller substrates. They are particularly fond of substrates which contain chitin, keratin, or cellulose (Wong et al. 1998). In case of small particles such as algae, pollen grains, seeds, and zooplankton carcasses, and other temporarily available substrates decomposition is achieved by the much smaller chytrids (*Chytridiomycetes*) and water molds (*Oomycetes*), rather than the aquatic hyphomycetes. This is because the zoosporic fungi are relatively simpler in structures. They do not depend on macroscale hyphal networks and thus are capable of very fast responses to changes in their environment. Being actively motile, the zoospores of these aquatic fungi actively search for adequate substrates using chemotaxis. Once a suitable substrate has been reached, the zoospore encysts. An appressorium or a penetration tube is formed and the food particle (substrate) is invaded by tiny rhizoids tapping the internal nutrient reservoirs. Zoosporogenesis can occur in a short time, producing a prolific number of zoospores from zoosporangia. Their life cycle can be completed in days, either endobiotic, epibiotic, endophytic, or ectophytic, depending on the relationship of the thallus with the host or organic substrate.

Like other aquatic fungi, the chytrids and water molds are heterotrophs. Most of the species are benign saprobes, but they often exist as parasites, sometimes as symbionts, and of course as decomposers. The aquatic systems harbor a wealth of organisms that can serve as suitable hosts for the parasitic zoosporic fungi: algae from different phyla, cyanobacteria, protists, zooplankton, fish, birds, mussels, eggs of liver flukes, nematodes, crayfish, mites, insect larvae, amphibians, mammals, plants, and other aquatic fungi. For the decomposers, resources of organic matters derived from animals include fish scales, fish eggs, carcasses, feathers, and hair, while plant-derived resources include pollen, spores, seeds, small fruits, and plant debris (Cole et al. 1990).

Chytrids are surprisingly abundant on filamentous algae, phytoplanktons, and diatoms, and some species are known to severely deplete local populations of their algal hosts. The abundance of chytrids in aquatic systems are considered much higher than traditionally thought (Kagami et al. 2014). The phytoplanktons infected with chytrid zoospores could become an excellent food source for zooplanktons in terms of size, shape, and nutritional quality, and the nutrients from within the planktons can be transferred to the zooplanktons through the "mycoloop" pathway in the aquatic ecosystems (Kagami et al. 2014). Also pollen deposited in lakes could be consumed by saprotrophic chytrids, rather than parasitic relatives (Masclaux et al. 2013).

Ibelings and co-workers studied the host–parasite interactions between freshwater phytoplankton and chytrid fungi and found that algal population was naturally regulated by the parasitism of these fungi (Ibelings et al. 2004). Encounters with these parasitic chytrids can be fatal to algae, particularly if their defense mechanism is breached by the fungal parasites. When the alga is attacked by the fungus, it undergoes a "suicide" response, which is a controlled hypersensitive reaction. This hypersensitivity is regarded as a common defense mechanism in algae. If this controlled "suicide" progress is initiated at the right moment during fungal infection, it results in the successful interruption of the fungal infection cycle, because the parasite's ability to reproduce via spore production is inhibited. This mechanism is conducive to maintaining a healthy algal population because it reduces the abundance of the deadly fungal parasite in the environment. If such control is unsuccessful, however, the parasitic chytrid prevails and thus resulting in mass mortality of the algal species. The ecological relevance of this negative interaction between the two parties is obvious. The failure of the algal "suicide" mechanism in response to fungal infection can lead to shifts in the algal community composition in a given aquatic system.

In rare, but important cases, some zoosporic fungi cause severe damage to larger aquatic organisms. They infect frogs, shrimps, fishes, or fish eggs (Chukanhom and Hatai 2004; Noga 1993) and thereby exert strong population pressure. Such damage is of great importance for aquaculture and often demands antifungal treatments. Some oomycetes, especially species of *Aphanomyces* and *Saprolegnia*, are aggressive pathogens of fish and crustacea. For example, *Aphanomyces astaci* causing the crayfish plague has driven the European crayfish population to the edge of extinction (Reynolds 1988). The most notorious parasitic chytrid is *Batrachochytrium dendrobatidis*, which cause worldwide extinction of several known and unknown species of frogs (Berger et al. 1998; Skerratt et al. 2007). Aquatic plants are also greatly affected by some oomycetes. For example, *Pythium phragmites* has been found to cause reed decline (Nechwatal et al. 2005). For more discussion on diseases of freshwater fishes caused by zoosporic fungi, see Willoughby (2003).

## Distribution and Biodiversity

### *Distribution Pattern*

Regarding the distribution pattern of freshwater fungi, some may be restricted to tropical, temperate, or cold water habitats, while others are cosmopolitan. The geographical distribution of Ingoldian mitosporic ascomycetes (=anamorph or asexual fungi) are relatively well studied compared to those of the freshwater ascomycetes. The Ingoldian fungi most commonly occur on shed leaves in streams and rivers, and they are documented by stream biologists. Some are cosmopolitan and some are restricted in distribution. Also the phylogeographic pattern varies among and within a species. Molecular barcoding (ITS) of 130 isolates of six Ingoldian fungi revealed significant genetic differentiation between continents within a single fungal species (Duarte et al. 2012). The knowledge on the distribution pattern of freshwater ascomycetes is accumulating even though most investigations are concentrating in tropical and subtropical Asia, North America, as well as the neotropics.

Some studies have reported shifts in fungal community composition by latitude and temperature (Arnold and Lutzoni 2007). Such spatial shifts/turnover in community is also expected in freshwater fungi. Wood-Eggenschwiler and Bärlocher (1985) used distribution data obtained from the literature (Webster and Descals 1981) for over 150 species of Ingoldian mitosporic fungi and they concluded, "on a worldwide scale, temperature together with its influence on vegetation in different climatic regions is the major factor in determining distribution patterns of Ingoldian mitosporic fungi." Wood-Eggenschwiler and Bärlocher (1985) discovered that there was a higher similarity in species composition of Ingoldian fungi between geographically distinct tropical locations (South America, West Africa) than between tropical and temperate regions that were located on the same continent, either African or North and South American. Raja et al. (2009) also reported a change in species composition of freshwater ascomycetes along the temperate–subtropical latitudinal ecotone in Florida, USA.

Apart from the macroclimatic factors, microenvironmental factors also affect the distribution and abundance of freshwater fungi. Chauvet (1991) studied the distribution of Ingoldian mitosporic fungi at 27 stations in France, and he concluded that the most important environmental factors are altitude, pH, temperature, and season, although the relationship between species composition and each environmental factor is hard to establish. Longitudinal distribution patterns in freshwater fungi along a river and stream are reported for both leaf litter and woody substrates (Gönczöl 1989; Shearer and Webster 1985a, 1991; Tsui et al. 2001a). Tsui et al. (2001a) reported changes in fungal communities and taxonomic compositions from upstream to downstream in responses to salinity and riparian vegetation. Shearer and Webster (1985a) reported that Ingoldian mitosporic fungi communities in headwater streams were distinctly different from the downstream communities in the River Teign. Using water filtration, leaf pack baiting, and collection of naturally occurring substrates, lower species diversity with a lower frequency occurrence of species was observed in the headwaters (Shearer and Webster 1985a). Using molecular data of DGGE, Miura and Urabe (2014) also demonstrated that taxonomic composition and richness of epilithic fungal assemblages change along the longitudinal gradient of the river, according to the water temperature, and the spatial variation in abundance and composition of dissolved organic matter and nutrients. While species diversity could change spatially, the genetic variability within a species does not vary locally. Using eight microsatellite markers, Anderson and Shearer (2011) revealed small genetic differentiations among populations of *Tetracladium marchalianum* from Wisconsin and Illinois, USA. They concluded that the fungal populations may be highly connected in local habitats.

## *Substrate Preference*

Freshwater fungi demonstrate substrate specialization, even though they are saprobes. For instance, of the 548 species of freshwater ascomycetes reported up to 2009 (http://fungi.life.uiuc.edu/), 60 % are reported only from submerged woody debris and about 30 % are reported only from herbaceous substrates, while only about 10 % species are reported from both submerged wood and herbaceous substrates [reviewed in (Raja et al. 2009)]. During the substrate distribution pattern investigation of freshwater ascomycetes in the Florida Peninsula (Raja et al. 2009), the results implied substrate preference among freshwater fungi. Of the 132 fungal taxa collected in freshwater habitats, 100 were reported only on woody debris, 14 species occurred exclusively on herbaceous debris, and 18 species were found on both woody and herbaceous debris (Raja et al. 2009). Cai et al. (2003) also reported substrate preferences in freshwater fungi during an investigation on the biodiversity of freshwater fungi on submerged bamboo and submerged wood in Liput River in the Philippines. Fifty-eight and 38 fungal taxa were collected on bamboo and wood, respectively, but only 16 among them were in common on both substrates (Cai et al. 2003).

Thomas et al. (1992) observed *Alatospora acuminata* more frequently on *Acacia* leaves, while *Tetrachaetum elegans* was more common on *Eucalyptus* leaves. The authors suggested five possible reasons: First, different substrates have different nutrients, favoring the growth of some fungi over others. Second, different substrates contain different inhibitory chemicals, for example, tannin, impacting sporulation and spore germination. Third, variations in the gross physical structure of substrates affect differentially the impaction efficiency of various fungal spores. Fourth, variations in fine physical structure of substrates affect penetration and colonization by fungi. Fifth, different substrates vary in their decay rate—durable substrate has a much longer exposure time to the spores and to fungal colonization. Gulis (2001) investigated five different substrate types from 92 watercourses of Belarus for aquatic hyphomycete colonization (52 species). He found specific fungal assemblages correlating with leaf litter types which suggests possible substrate preferences of aquatic hyphomycetes triggered in particular by lignin content.

## *Seasonal Variation*

Seasonal occurrence with fluctuations in conidial numbers has been reported in many parts of the world (Thomas et al. 1989). Although most species can be collected throughout the year, their relative abundance (measured in terms of conidial production) is influenced by the seasonal availability of substrates, which is in agreement with seasonal input of deciduous tree litter in temperate regions (Bärlocher 1992). For instance, the conidial peak in summer in Australia was also highly correlated with the leaf fall of eucalypt forests in summer (Thomas et al. 1989), and the timing of conidial maxima in New Zealand streams was well correlated with the prevailing litter fall patterns (Aimer and Segedin 1985). Recent studies employing DGGE illustrated seasonal changes in fungal communities in a lake in Japan (Ishii et al. 2015).

### *Human Disturbance*

Most freshwater habitats are vulnerable to human disturbance such as agriculture, urbanization, and industrialization. Any perturbation around the riparian environment affects significantly the in-stream fungal communities and the biogeochemical cycles through runoff processes. Organic pollution caused by the mass discharged of domestic or agricultural wastes reduces the amount of oxygen in the water. Most freshwater fungi cannot survive in such anoxic and polluted environments. For example, organic pollution reduced substantially the diversity of aquatic hyphomycetes (Raviraja et al. 1998) and ascomycetes (Tsui et al. 2001b). Toxic metal entering the rivers as a result of mining and industrialization also impact negatively the fungal communities. Spore production, biomass, and fungal diversity are severely depleted under high concentrations of coal, copper, zinc, and cadmium (Krauss et al. 2001; Niyogi et al. 2009; Sridhar et al. 2005). Similarly, the application of chemical pesticides can also change the fungal communities. For example, the antigen and biomass production of *Neonectria* (*Heliscus*) *lugdunensis* was influenced by the herbicide mecoprop (Bermingham et al. 1998). Recent pyrosequencing (metagenomics) data also showed declining fungal diversity in most eutrophic streams (Duarte et al. 2014). However, previous studies demonstrated the opposite—anthropogenic nutrients stimulated fungal spore production and mycelial biomass on leaves (Gulis and Suberkropp 2003).

## Functional Biodiversity/Outlook: Who Is Doing What?

The most prominent ecological function of freshwater fungi is the decomposition of allochthonous organic matter in aquatic systems. The research of freshwater fungal ecology has been focused on stream-inhabiting aquatic hyphomycetes mostly in temperate low-order streams (Bärlocher 2010). Properties of ecosystem functions, such as fungal biomass, fungal productivity, and fungal impact on the decomposition process (Lecerf and Richardson 2010a), have been investigated in hundreds of field and laboratory (microcosm) studies [meta-analyzed in (Ferreira et al. 2014)]. These studies were accompanied by inventory biodiversity (morphological and molecular) approaches of aquatic hyphomycetes [e.g., (Nikolcheva et al. 2003; Seena et al. 2008; Duarte et al. 2014; Casas et al. 2011; Pascoal et al. 2005; Shearer and Webster 1985b)]. However, the results of studies investigating the functional consequences of biodiversity changes often remained unpredictable or inconsistent [summarized in (Graça et al. 2015)]. Within the last decade, it has become apparent that species traits and functional diversity may be better correlated with ecosystem function than taxonomic identity (Gessner et al. 2010; Lecerf and Richardson 2010b).

Nowadays we have the methods at hand to even more thoroughly elucidate species assemblages and, furthermore, investigate the possible and active traits of freshwater fungi. The progress in sequencing and annotating fungal genomes will soon shed light on the genetic diversity and metabolic potential of freshwater fungi from all fungal phyla. Transcriptomes [e.g., using RNA-seq; (Wang et al. 2009)] from single cultures or microcosm studies of freshwater fungi will enable us to study the metabolism during decomposition and/or degradation processes. For the identification of metabolic pathways, the quality of genome annotation is crucial (Kuske et al. 2015). The optimal study design is to have very well-annotated reference genomes, the transcriptome (the expressed genes), and the corresponding proteome (the produced enzymes) as shown by Hori and co-workers (2014). Freshwater fungal communities are complex assemblages of mostly filamentous fungi, single-celled chytrids, and yeasts. They are members of food webs and mediators of biogeochemical pathways for the energy transfer between different trophy levels. Future methodical improvements will hopefully ease the challenges of meta-analyses to understand the "why-is-who-doing-what" in the ecology of freshwater fungi.

**Acknowledgments** The authors wish to thank the following students of Professor T.K. Goh, for providing the pictures of freshwater fungi used in this chapter: Bao Yee Loh (*Oomycetes*), Huey Kei Lee (Ingoldian fungi), and Wee Kee Sim (lignicolous hyphomycetes). We are grateful to Dr. Huzefa Raja for comments and suggestions. The authors thank the master student of Christiane Baschien, Jens Schmidt for the picture of Lunulospora curvula conidiogenesis.

## References

Abdel-Aziz FA (2008) Diversity of aquatic fungi on *Phragmites australis* at Lake Manzala, Egypt. Sydowia 60(1):1–14, PubMed PMID: WOS:000257909100001

Abdel-Raheem AM, Ali EH (2004) Lignocellulolytic enzyme production by aquatic hyphomycetes species isolated from the Nile's delta region. Mycopathologia 157:277–286

Abdel-Raheem A, Shearer C (2002) Extracellular enzyme production by freshwater ascomycetes. Fungal Divers 11:1–19

Aimer RD, Segedin BP (1985) Some aquatic hyphomycetes from New Zealand streams. N Z J Bot 23:273–299

Altschul SF, Gish W, Miller W, Myers EW, Lipman DJ (1990) Basic local alignment search tool. J Mol Evol 215:403–410

Anderson JL, Shearer CA (2011) Population genetics of the aquatic fungus *Tetracladium marchalianum* over space and time. PLoS One 6(1):10. doi:10.1371/journal.pone.0015908, PubMed PMID: WOS:000286516500015

Ando K (1992) A study of terrestrial aquatic hyphomycetes. Trans Mycol Soc Jpn 33:415–425

Ando K, Tubaki K (1984a) Some undescribed hyphomycetes in rainwater draining from intact trees. Trans Mycol Soc Jpn 25:39–47

Ando K, Tubaki K (1984b) Some undescribed hyphomycetes in the rain drops from intact leaf-surface. Trans Mycol Soc Jpn 25:21–37

Arnold AE, Lutzoni F (2007) Diversity and host range of foliar fungal endophytes: are tropical leaves biodiversity hotspots? Ecology 88:541–549

Au DWT, Jones EBG, Moss ST (1996) Spore attachment and extracellular mucilage of aquatic hyphomycetes. Biofouling 10:123–140

Bärlocher F (1992) The ecology of aquatic hyphomycetes. Springer, Berlin

Bärlocher F (2010) Molecular approaches promise a deeper and broader understanding of the evolutionary ecology of aquatic hyphomycetes. J N Am Benthol Soc 29(3):1027–1041. doi:10.1899/09-081.1, PubMed PMID: WOS:000280692400021

Bärlocher F, Kendrick B (1974) Dynamics of the fungal population on leaves in a stream. J Ecol 62:761–791

Bärlocher F, Seena S, Wilson KP, Williams DD (2008) Raised water temperature lowers diversity of hyporheic aquatic hyphomycetes. Freshw Biol 53(2):368–379. doi:10.1111/j.1365-2427.2007.01899.x, PubMed PMID: WOS:000252393800013

Baschien C, Marvanová L, Szewzyk U (2006) Phylogeny of selected aquatic hyphomycetes based on morphological and molecular data. Nova Hedwigia 83(3–4):311–352. doi:10.1127/0029-5035/2006/0083-0311, PubMed PMID: WOS:000242664300003

Baschien C, Rode G, Böckelmann U, Götz F, Szewzyk U (2009) Interactions between hyphosphere-associated bacteria and the fungus *Cladosporium herbarum* on aquatic leaf litter. Microb Ecol 58(3):642–650. doi:10.1007/s00248-009-9528-6, PubMed PMID: WOS:000269928300020

Baschien C, Tsui CKM, Gulis V, Szewzyk U, Marvanová L (2013) The molecular phylogeny of aquatic hyphomycetes with affinity to the Leotiomycetes. Fungal Biol 117(9):660–672. doi:10.1016/j.funbio.2013.07.004, PubMed PMID: WOS:000324899800008

Beakes GW (2003) Lower fungi: a review of microscopial techniques for the taxonomic and ecological study of zoosporic freshwater fungi. In: Tsui CKM, Hyde KD (eds) Fungal diversity research series 10 freshwater mycology -a practical approach. Fungal Diversity Press, Hong Kong, pp 51–79

Belliveau MJ-R, Bärlocher F (2005) Molecular evidence confirms multiple origins of aquatic hyphomycetes. Mycol Res 109(12):1407–1417, PubMed

Berbee ML, Taylor JW (2001) Fungal molecular evolution: gene trees and geologic time. In: Mc Laughlin DJ and Mc Laughlin EJ, Lemke PA (eds) The mycota VII part b systematics and evolution. Springer, Berlin

Berger L, Speare R, Daszak P, Green DE, Cunningham AA, Goggin CL et al (1998) Chytridiomycosis causes amphibian mortality associated with population declines in the rain forests of Australia and Central America. Proc Natl Acad Sci USA 95(15):9031–9036. doi:10.1073/pnas.95.15.9031, PubMed PMID: WOS:000075143900110

Bermingham S, Fisher PJ, Martin A, Marriott M, Lappin-Scott H (1998) The effect of the herbicide Mecoprop on *Heliscus lugdunensis* and its influence on the preferential feeding of *Gammarus pseudolimnaeus*. Microb Ecol 35:199–204

Beverwijk AL (1951) *Candelabrum spinulosum*, a new fungus species. Anton Leeuw Int J Gen Mol Microbiol 17:9–12

Bowman BH, Taylor JW, Brown AG, Lee J, Lu SD, White TJ (1992) Molecular evolution of the fungi relationship of the basidiomycetes, ascomycetes and chytridiomycetes. Mol Biol Evol 9:285–296

Bucher VVC, Hyde KD, Pointing SB, Reddy CA (2004) Production of wood decay enzymes, mass loss and lignin solubilization in wood by marine ascomycetes and their anamorphs. Fungal Divers 15:1–14

Buczacki ST (1983) Zoosporic plant pathogens: a modern perspective. Academic, London

Cai L, Zhang KQ, McKenzie EHC, Hyde KD (2003) New species of *Dictyosporium* and *Digitodesmium* from submerged wood in Yunnan, China. Sydowia 55(2):129–135, PubMed PMID: WOS:000188231200001

Campbell J, Shearer C, Marvanová L (2006) Evolutionary relationships among aquatic anamorphs and teleomorphs: *Lemonniera*, *Margaritispora*, and *Goniopila*. Mycol Res 110:1025–1033. doi:10.1016/j.mycres.2006.04.012, PubMed PMID: WOS:000241957700003

Campbell J, Marvanová L, Gulis V (2009) Evolutionary relationships between aquatic anamorphs and teleomorphs: *Tricladium* and *Varicosporium*. Mycol Res 113:1322–1334. doi:10.1016/j.mycres.2009.09.003, PubMed PMID: WOS:000272811000009

Casas JJ, Gessner MO, Lopez D, Descals E (2011) Leaf-litter colonisation and breakdown in relation to stream typology: insights from Mediterranean low-order streams. Freshw Biol 56(12):2594–2608. doi:10.1111/j.1365-2427.2011.02686.x, PubMed PMID: WOS:000296502000014

Chamier A (1985) Cell-wall-degrading enzymes of aquatic hyphomycetes: a review. Bot J Linn Soc 91:67–81

Chandrashekar KR, Kaveriappa KM (1991) Production of extracellular cellulase by *Lunulospora curvula* and *Flagellospora penicillioides*. Folia Microbiol 36:249–255

Chauvet E (1991) Aquatic hyphomycete distribution in South-Western France. J Biogeogr 18:699–706

Chukanhom K, Hatai K (2004) Freshwater fungi isolated from eggs of the common carp (*Cyprinus carpio*) in Thailand. Mycoscience 45:42–48

Clivot H, Cornut J, Chauvet E, Elger A, Poupin P, Guerold F et al (2014) Leaf-associated fungal diversity in acidified streams: insights from combining traditional and molecular approaches. Environ Microbiol 16(7):2145–2156. doi:10.1111/1462-2920.12245, PubMed PMID: WOS:000338983600013

Cole JJ, Caraco NF, Likens GE (1990) Short-range atmospheric transport—a significant source of phosphorus to an Oligotrophic Lake. Limnol Oceanogr 35(6):1230–1237, PubMed PMID: WOS:A1990EP94300002

Digby S, Goos RD (1987) Morphology, development and taxonomy of *Loramyces*. Mycologia 79:821–831

Dix NJ, Webster J (1995) Fungal ecology. Springer, Berlin

Doggett MS (2000) Characterization of fungal biofilms within a municipal water distribution system. Appl Environ Microbiol 66(3):1249–1251. doi:10.1128/aem.66.3.1249-1251.2000, PubMed PMID: WOS:000085604800059

Duarte S, Seena S, Baerlocher F, Cassio F, Pascoal C (2012) Preliminary insights into the phylogeography of six aquatic hyphomycete species. PLoS One 7(9), e45289. doi:10.1371/journal.pone.0045289, PubMed PMID: WOS:000311313900114

Duarte S, Baerlocher F, Cassio F, Pascoal C (2014) Current status of DNA barcoding of aquatic hyphomycetes. Sydowia 66(2):191–202, PubMed PMID: WOS:000347436300003

Eaton RA, Jones EBG (1971a) The biodeterioration of timber in water-cooling towers. II. Fungi growing on wood in different positions in a water cooling system. Mater Org 6:81–92

Eaton RA, Jones EGB (1971b) The biodeterioration of timber in water cooling towers. I. Fungal ecology and the decay of wood at Connah's Quay and Ince. Mater Org 6:51–80

Eaton RA, Hale MDC (1993) Wood: decay, pests and protection. Chapman & Hall, London

Fallah PM, Shearer CA (2001) Freshwater ascomycetes: new or noteworthy species from north temperate lakes in Wisconsin. Mycologia 93(3):566–602. doi:10.2307/3761741, PubMed PMID: WOS:000168935100016

Ferreira V, Castagneyrol B, Koricheva J, Gulis V, Chauvet E, Graça MA (2014) A meta-analysis of the effects of nutrient enrichment on litter decomposition in streams. Biol Rev. doi:10.1111/brv.12125

Ferrer A, Miller AN, Shearer CA (2011) *Minutisphaera* and *Natipusilla*: two new genera of freshwater Dothideomycetes. Mycologia 103(2):411–423. doi:10.3852/10-177, PubMed PMID: WOS:000288887400017

Fisher PJ, Sharma PD, Webster J (1977) Cellulolytic ability of aero-aquatic hyphomycetes. Trans Br Mycol Soc 69:495–496, PubMed PMID: WOS:A1977EF05300018

Fisher PJ, Petrini O, Webster J (1991) Aquatic hyphomycetes and other fungi in living aquatic and terrestrial roots of *Alnus glutinosa*. Mycol Res 95:543–547

Fuller MS, Jaworski A (1987) Zoosporic fungi in teaching and research. Southeastern Publishing Corporation, Athen, GA

Gessner MO, Gulis V, Kuehn KA, Chauvet E, Suberkropp K (2007) 17 Fungal decomposers of plant litter in aquatic ecosystems. In: Christian PK, Druzhinina IS (eds) Environmental and microbial relationships, 4th edn. Springer, Berlin

Gessner MO, Swan CM, Dang CK, McKie BG, Bardgett RD, Wall DH et al (2010) Diversity meets decomposition. Trends Ecol Evol 25(6):372–380, PubMed PMID: WOS:000278682500008

Goh TK (1997) Tropical freshwater hyphomycetes. In: Hyde KD (ed) Biodiversity of tropical microfungi. Hong Kong University Press, Hong Kong

Goh T, Hyde K (1996a) Biodiversity of freshwater fungi. J Ind Microbiol 17(5–6):328–345

Goh TK, Tsui CKM (2003) Key to common dematiaceous hyphomycetes from freshwater. In: Tsui CKM, Hyde KD (eds) Freshwater mycology. Fungal Diversity Press, Hong Kong

Goh TK, Ho WH, Hyde KD, Umali TE (1997) New records and species of *Sporoschisma* and *Sporoschismopsis* from submerged wood in the tropics. Mycol Res 101:1295–1307

Goh TK, Ho WH, Hyde KD, Whitton SR, Umali TE (1998) New records and species of *Canalisporium* (Hyphomycetes), with a revision of the genus. Can J Bot-Revue Canadienne De Botanique 76(1):142–152, PubMed PMID: WOS:000073021500017

Gönczöl J (1989) Longitudinal distribution patterns of aquatic hyphomycetes in a mountain stream in Hungary—experiments with leaf packs. Nova Hedwigia 48(3–4):391–404, PubMed PMID: WOS:A1989AD12000009

Gönczöl J, Révay A (2003) Treehole fungal communities: aquatic, aero-aquatic and dematiaceous hyphomycetes. Fungal Divers 12:19–34, PubMed PMID: WOS:000181503800003

Gönczöl J, Révay A (2004) Fungal spores in rainwater: stemflow, throughfall and gutter conidial assemblages. Fungal Divers 16:67–86, PubMed PMID: WOS:000222743900006

Gönczöl J, Révay A (2006) Species diversity of rainborne hyphomycete conidia from living trees. Fungal Divers 22:37–54, PubMed PMID: WOS:000239781700003

Graça MAS, Ferreira V, Canhoto C, Encalada AC, Guerrero-Bolano F, Wantzen KM et al (2015) A conceptual model of litter breakdown in low order streams. Int Rev Hydrobiol 100(1):1–12. doi:10.1002/iroh.201401757, PubMed PMID: WOS:000347777600001

Gulis V (2001) Are there any substrate preferences in aquatic hyphomycetes? Mycol Res 105:1088–1093. doi:10.1016/s0953-7562(08)61971-1, PubMed PMID: WOS:000171808900009

Gulis V, Suberkropp K (2003) Leaf litter decomposition and microbial activity in nutrient-enriched and unaltered reaches of a headwater stream. Freshw Biol 48(1):123–134. doi:10.1046/j.1365-2427.2003.00985.x, PubMed PMID: WOS:000179693900011

Hameed AAA, El Hawarry S, Kamel MM (2008) Prevalence and distribution of airborne and waterborne fungi and actinomycetes in the Nile river. Aerobiologia 24(4):231–240. doi:10.1007/s10453-008-9101-7, PubMed PMID: WOS:000260611900006

Harrop BL, Marks JC, Watwood ME (2009) Early bacterial and fungal colonization of leaf litter in Fossil Creek, Arizona. J N Am Benthol Soc 28(2):383–396. doi:10.1899/08-068.1, PubMed PMID: WOS:000266645700011

Heinrichs G, Hübner I, Schmidt CK, de Hoog GS, Haase G (2013) Analysis of black fungal biofilms occurring at domestic water taps. I: compositional analysis using Taq-Encoded FLX Amplicon Pyrosequencing. Mycopathologia 175:387–397

Hori C, Ishida T, Igarashi K, Samejima M, Suzuki H, Master E et al (2014) Analysis of the *Phlebiopsis gigantea* genome, transcriptome and secretome provides insight into its pioneer colonization strategies of wood. PLoS Genet 10, e1004759

Hu DM, Cai L, Jones EBG, Zhang H, Boonyuen N, Hyde KD (2014) Taxonomy of filamentous asexual fungi from freshwater habitats, links to sexual morphs and their phylogeny. In: Hyde KD, Jones EBG, Pang KL (eds) Freshwater fungi: and fungal-like organisms. Walter de Gruyter GmbH & Co KG, Berlin

Hyde KD, Wong SW, Jones EBG (1997) Freshwater ascomycetes. In: Hyde KD (ed) Biodiversity of tropical microfungi Hong Kong. Hong Kong University Press, Hong Kong

Ibelings BW, De Bruin A, Kagami M, Rijkeboer M, Brehm M, van Donk E (2004) Host parasite interactions between freshwater phytoplankton and chytrid fungi (Chytridiomycota). J Phycol 40(3):437–453. doi:10.1111/j.1529-8817.2004.03117.x, PubMed PMID: WOS: 000221644400001

Ingold CT (1942) Aquatic hyphomycetes of decaying alder leaves. Trans Br Mycol Soc 25:339–417

Ingold CT (1955) Aquatic Ascomycetes: further species from the English Lake District. Trans Br Mycol Soc 38:157–168

Ingold CT (1966) The tetraradiate aquatic fungal spore. Mycologia 58(1):43–56. doi:10.2307/3756987, PubMed PMID: WOS:A19667371900002

Ingold CT (1975) Hooker lecture 1974: convergent evolution in aquatic fungi: the tetraradiate spore. Biol J Linn Soc 7(1):1–25. doi:10.1111/j.1095-8312.1975.tb00731.x

Ingold CT, Chapman B (1952) Aquatic Ascomycetes: *Loramyces juncicola* Weston and *L. macrospora* n. sp. Trans Br Mycol Soc 35:269–272

Ishii N, Ishida S, Kagami M (2015) PCR primers for assessing community structure of aquatic fungi including Chytridiomycota and Cryptomycota. Fungal Ecol 13:33–43

James TY, Pelin A, Bonen L, Ahrendt S, Sain D, Corradi N et al (2013) Shared signatures of parasitism and phylogenomics unite Cryptomycota and Microsporidia. Curr Biol 23(16):1548–1553. doi:10.1016/j.cub.2013.06.057, PubMed PMID: WOS:000323401100019

Jones EBG (1981) Observations on the ecology of lignicolous aquatic hyphomycetes. In: Wicklow DT, Carroll GC (eds) The fungal community. Marcel Dekker, New York, pp 731–742

Jones EBG (2006) Form and function of fungal spore appendages. Mycoscience 47:167–183

Jones MDM, Forn I, Gadelha C, Egan MJ, Bass D, Massana R et al (2011) Discovery of novel intermediate forms redefines the fungal tree of life. Nature 474(7350):200–203. doi:10.1038/nature09984, PubMed PMID: WOS:000291397800047

Jones EBG, Hyde KD, Pang KL (2014) Freshwater fungi and fungal-like organisms. Walter de Gruyter GmbH, Berlin

Junghanns C, Moeder M, Krauss G, Martin C, Schlosser D (2005) Degradation of the xenoestrogen nonylphenol by aquatic fungi and their laccases. Microbiology 151:45–57. doi:10.1099/mic.0.27431-0, PubMed PMID: WOS:000226352800006

Kagami M, Miki T, Takimoto G (2014) Mycoloop: chytrids in aquatic food webs. Front Microbiol 5:9. doi:10.3389/fmicb.2014.00166, PubMed PMID: WOS:000334664400001

Kaushik NK, Hynes HBN (1971) The fate of the dead leaves that fall into streams. Archiv Fur Hydrobiologie 68:465–515

Kelly JJ, Bansal A, Winkelman J, Janus LR, Hell S, Wencel M et al (2010) Alteration of microbial communities colonizing leaf litter in a temperate woodland stream by growth of trees under conditions of elevated atmospheric $CO_2$. Appl Environ Microbiol 76(15):4950–4959. doi:10.1128/aem.00221-10, PubMed PMID: WOS:000280266200005

Kerr JL, Baldwin DS, Tobin MJ, Puskar L, Kappen P, Rees GN et al (2013) High spatial resolution infrared micro-spectroscopy reveals the mechanism of leaf lignin decomposition by aquatic fungi. PLoS One 8, e60857

Kohout P, Sykorova Z, Ctvrtlikova M, Rydlova J, Suda J, Vohnik M et al (2012) Surprising spectra of root-associated fungi in submerged aquatic plants. FEMS Microbiol Ecol 80(1):216–235. doi:10.1111/j.1574-6941.2011.01291.x, PubMed PMID: WOS:000301051600019

Kohout P, Tesitelova T, Roy M, Vohnik M, Jersakova J (2013) A diverse fungal community associated with *Pseudorchis albida* (Orchidaceae) roots. Fungal Ecol 6(1):50–64. doi:10.1016/j.funeco.2012.08.005, PubMed PMID: WOS:000314256500007

Krauss G, Bärlocher F, Schreck P, Wennrich R, Glasser W, Krauss GJ (2001) Aquatic hyphomycetes occur in hyperpolluted waters in Central Germany. Nova Hedwigia 72(3–4):419–428, PubMed PMID: WOS:000169242400010

Krauss G, Sridhar KR, Bärlocher F (2005) Aquatic hyphomycetes and leaf decomposition in contaminated groundwater wells in Central Germany. Archiv Fur Hydrobiologie 162(3):417–429. doi:10.1127/003-9136/2005/0162/0417, PubMed PMID: WOS:000228391300009

Kuske CR, Hesse CN, Challacombe JF, Cullen D, Herr JR, Mueller RC et al (2015) Prospects and challenges for fungal metatranscriptomics of complex communities. Fungal Ecol 14:133–137

Lazarus KL, James TY (2015) Surveying the biodiversity of the Cryptomycota using a targeted PCR approach. Fungal Ecol 14:62–70

Lecerf A, Richardson JS (2010a) Biodiversity-ecosystem function research: insights gained from streams. River Res Appl 26(1):45–54. doi:10.1002/rra.1286, PubMed PMID: WOS:000274310500006

Lecerf A, Richardson JS (2010b) Litter decomposition can detect effects of high and moderate levels of forest disturbance on stream condition. For Ecol Manag 259(12):2433–2443. doi:10.1016/j.foreco.2010.03.022, PubMed PMID: WOS:000278303700022

Longcore JE, Pessier AP, Nichols DK (1999) *Batrachochytrium dendrobatidis* gen et sp. nov., a chytrid pathogenic to amphibians. Mycologia 91(2):219–227. doi:10.2307/3761366, PubMed PMID: WOS:000079317100001

Magyar D, Gönczöl J, Révay A, Grillenzoni F, Seijo-Coello MDC (2005) Stauro- and scolecoconidia in floral and honeydew honeys. Fungal Divers 20:103–120, PubMed PMID: WOS:000234463100008

Masclaux H, Perga ME, Kagami M, Desvilettes C, Bourdier G, Bec A (2013) How pollen organic matter enters freshwater food webs. Limnol Oceanogr 58:1185–1195

Miura A, Urabe J (2014) Spatial and seasonal changes in species diversity of epilithic fungi along environmental gradients of a river. Freshw Biol 60:673–685

Nagy LA, Olson BH (1982) The occurrence of filamentous fungi in drinking-water distribution-systems. Can J Microbiol 28(6):667–671, PubMed PMID: WOS:A1982NV58700017

Nechwatal J, Wielgoss A, Mendgen K (2005) *Pythium phragmitis* sp. nov., a new species close to *P. arrhenomanes* as a pathogen of common reed (*Phragmites australis*). Mycol Res 109:1337–1346. doi:10.1017/s0953756205003990, PubMed PMID: WOS:000233734000003

Nemec S (1969) Sporulation and identification of fungi isolated from root rot-diseased strawberry plants. Phytopathology 59(10):1552–1553, PubMed PMID: WOS:A1969E429700058

Nikolcheva LG, Bärlocher F (2005) Seasonal and substrate preferences of fungi colonizing leaves in streams: traditional versus molecular evidence. Environ Microbiol 7(2):270–280. doi:10.1111/j.1462-2920.2004.00709.x, PubMed PMID: WOS:000226376800013

Nikolcheva LG, Cockshutt AM, Bärlocher F (2003) Determining diversity of freshwater fungi on decaying leaves: comparison of traditional and molecular approaches. Appl Environ Microbiol 69(5):2548–2554. doi:10.1128/aem/69.5.2548.2554.2003, PubMed PMID: WOS:000182808300015

Nikolcheva LG, Bourque T, Bärlocher F (2005) Fungal diversity during initial stages of leaf decomposition in a stream. Mycol Res 109:246–253. doi:10.1017/s0953756204001698

Niyogi DK, Cheatham CA, Thomson WH, Christiansen JM (2009) Litter breakdown and fungal diversity in a stream affected by mine drainage. Fundam Appl Limnol 175(1):39–48. doi:10.1127/1863-9135/2009/0175-0039, PubMed PMID: WOS:000269326500003

Noga EJ (1993) Fungal diseases of marine and estuarine fishes. In: Couch JA, Fournie JW (eds) Pathology of marine and estuarine organisms. CRC Press, Boca Raton, FL, pp 85–110

Park D (1974) Aquatic hyphomycetes in non-aquatic habitats. Trans Br Mycol Soc 63(1): 183–187

Pascoal C, Cassio F, Marvanová L (2005) Anthropogenic stress may affect aquatic hyphomycete diversity more than leaf decomposition in a low-order stream. Archiv Fur Hydrobiologie 162(4):481–496. doi:10.1127/0003-9136/2005/0162-0481, PubMed PMID: WOS:000229072700004

Pérez J, Descals E, Pozo J (2012) Aquatic Hyphomycete communities associated with decomposing alder leaf litter in reference headwater streams of the Basque Country (northern Spain). Microb Ecol 64(2):279–290. doi:10.1007/s00248-012-0022-1, PubMed PMID: WOS:000306174700001

Pointing SB (2001) Feasibility of bioremediation by white-rot fungi. Appl Microbiol Biotechnol 57:20–33

Powell MJ (1993) Looking at mycology with a Janus face: a glimpse at Chytridiomycetes active in the environment. Mycologia 85:1–20

Powell MJ, Letcher PM (2014) 6 Chytridiomycota, Monoblepharidomycota, and Neocallimastigomycota. In: McLaughlin DJ, Spatafora JW (eds) Systematics and evolution. 7A, 2nd edn. Springer, Berlin, pp 141–175

Powell MJ, Letcher PM, Blackwell WH (2013) A new aquatic cellulose-degrading chytrid in the Chytridiales. Phytopathology 103(6):115, PubMed PMID: WOS:000322799500634

Raja HA, Schmit JP, Shearer CA (2009) Latitudinal, habitat and substrate distribution patterns of freshwater ascomycetes in the Florida Peninsula. Biodivers Conserv 18(2):419–455. doi:10.1007/s10531-008-9500-7, PubMed PMID: WOS:000262965400011

Raja HA, Oberlies NH, Figueroa M, Tanaka K, Hirayama K, Hashimoto A et al (2013) Freshwater Ascomycetes: *Minutisphaera* (Dothideomycetes) revisited, including one new species from Japan. Mycologia 105(4):959–976. doi:10.3852/12-313, PubMed PMID: WOS:000322849500014

Raviraja NS, Sridhar KR, Bärlocher F (1998) Breakdown of *Ficus* and *Eucalyptus* leaves in an organically polluted river in India: fungal diversity and ecological functions. Freshw Biol 39:537–545

Reynolds JD (1988) Crayfish extinctions and crayfish plague in central Ireland. Biol Conserv 45(4):279–285. doi:10.1016/0006-3207(88)90059-6, PubMed PMID: WOS:A1988P839500004

Sati SC, Belwal M (2005) Aquatic hyphomycetes as endophytes of riparian plant roots. Mycologia 97(1):45–49. doi:10.3852/mycologia.97.1.45, PubMed PMID: WOS:000229365800005

Schlütz F, Shumilovskikh LS (2013) On the relation of *Potamomyces armatisporus* to the fossil form-type *Mediaverrunites* and its taxonomical and ecological implications. Fungal Ecol 6(4):309–315

Schulz B, Boyle C (2005) The endophytic continuum. Mycol Res 109:661–686. doi:10.1017/s095375620500273x, PubMed PMID: WOS:000230687600001

Seena S, Wynberg N, Baerlocher F (2008) Fungal diversity during leaf decomposition in a stream assessed through clone libraries. Fungal Divers 30:1–14, PubMed PMID: WOS:000258548800001

Selosse MA, Vohnik M, Chauvet E (2008) Out of the rivers: are some aquatic hyphomycetes plant endophytes? New Phytol 178(1):3–7. doi:10.1111/j.1469-8137.2008.02390.x, PubMed PMID: WOS:000253711800002

Shearer CA (1993a) The fresh-water Ascomycetes. Nova Hedwigia 56(1–2):1–33, PubMed PMID: WOS:A1993KT94300001

Shearer CA (1993b) A new species of *Kirschsteiniothelia* (Pleosporales) with an unusual fissitunicate ascus. Mycologia 85:963–969

Shearer CA, Webster J (1985a) Aquatic hyphomycete communities in the River Teign. 1. Longitudinal distribution patterns. Trans Br Mycol Soc 84:489–501, PubMed PMID: WOS:A1985AHE9100011

Shearer CA, Webster J (1985b) Aquatic hyphomycete communities in the River Teign. 3. Comparison of sampling techniques. Trans Br Mycol Soc 84:509–518, PubMed PMID: WOS:A1985AHE9100013

Shearer CA, Webster J (1991) Aquatic hyphomycete communities in the River Teign. 4. Twig colonization. Mycol Res 95:413–420, PubMed PMID: WOS:A1991FL70400006

Shearer CA, Descals E, Kohlmeyer B, Kohlmeyer J, Marvanová L, Padgett D et al (2007) Fungal biodiversity in aquatic habitats. Biodivers Conserv 16(1):49–67. doi:10.1007/s10531-006-9120-z, PubMed PMID: WOS:000244185900004

Shearer CA, Raja HA, Miller AN, Nelson P, Tanaka K, Hirayama K et al (2009) The molecular phylogeny of freshwater Dothideomycetes. Stud Mycol 64:145–153. doi:10.3114/sim.2009.64.08, PubMed PMID: WOS:000274365900009

Simonis JL, Raja HA, Shearer CA (2008) Extracellular enzymes and soft rot decay: are ascomycetes important degraders in freshwater? Fungal Divers 31:135–146

Sing VO, Bartnicki-Garcia S (1972) Adhesion of zoospores of *Phytophthora palmivora* to solid surface. Phytopathology 62:790

Singh S, Harms H, Schlosser D (2014) Screening of ecologically diverse fungi for their potential to pretreat lignocellulosic bioenergy feedstock. Appl Microbiol Biotechnol 98(7):3355–3370. doi:10.1007/s00253-014-5563-4, PubMed PMID: WOS:000334167900045

Skerratt LF, Berger L, Speare R, Cashins S, McDonald KR, Phillott AD et al (2007) Spread of chytridiomycosis has caused the rapid global decline and extinction of frogs. EcoHealth 4(2):125–134. doi:10.1007/s10393-007-0093-5, PubMed PMID: WOS:000248231900004

Smither-Kopperl ML, Charudattan R, Berger RD (1998) Dispersal of spores of *Fusarium culmorum* in aquatic systems. Phytopathology 88(5):382–388. doi:10.1094/phyto.1998.88.5.382, PubMed PMID: WOS:000073487300002

Sole M, Muller I, Pecyna MJ, Fetzer I, Harms H, Schlosser D (2012) Differential regulation by organic compounds and heavy metals of multiple laccase genes in the aquatic Hyphomycete *Clavariopsis aquatica*. Appl Environ Microbiol 78(13):4732–4739. doi:10.1128/aem.00635-12, PubMed PMID: WOS:000305376600023

Sondergaard M, Laegaard S (1977) Vesicular-arbuscular mycorrhiza in some aquatic vascular plants. Nature 268(5617):232–233. doi:10.1038/268232a0, PubMed PMID: WOS: A1977DN97200031

Sonstebo JH, Rohrlack T (2011) Possible implications of chytrid parasitism for population subdivision in freshwater cyanobacteria of the genus *Planktothrix*. Appl Environ Microbiol 77(4):1344–1351. doi:10.1128/aem.02153-10, PubMed PMID: WOS:000287078100023

Sparrow FK (1960) Aquatic phycomycetes. University of Michigan Press, Ann Arbor, MI, 1187 p

Sridhar KR, Bärlocher F, Krauss GJ, Krauss G (2005) Response of aquatic hyphomycete communities to changes in heavy metal exposure. Int Rev Hydrobiol 90(1):21–32. doi:10.1002/iroh.200410739, PubMed PMID: WOS:000227722800002

Sridhar KR, Duarte S, Cassio F, Pascoal C (2009) The role of early fungal colonizers in leaf-litter decomposition in Portuguese streams impacted by agricultural runoff. Int Rev Hydrobiol 94(4):399–409. doi:10.1002/iroh.200811154, PubMed PMID: WOS:000269832100005

Stajich JEBM, Blackwell M, Hibbett DS, James TY, Spatafora JW, Taylor JW (2009) The fungi. Curr Biol 19:R840–R845

Sudova R, Rydlova J, Ctvrtlikova M, Havranek P, Adamec L (2011) The incidence of arbuscular mycorrhiza in two submerged Isoetes species. Aquat Bot 94(4):183–187. doi:10.1016/j.aquabot.2011.02.003, PubMed PMID: WOS:000290502100006

Taylor JW, Berbee ML (2006) Dating divergences in the fungal tree of life: review and new analyses. Mycologia 98:838–849

Tedersoo L, Pellet P, Kõljalg U, Selosse M-A (2007) Parallel evolutionary paths to mycoheterotrophy in understorey Ericaceae and Orchidaceae: ecological evidence for mixotrophy in Pyroleae. Oecologia 151(2):206–217

Thomas K, Chilvers GA, Norris RH (1989) Seasonal occurrence of conidia of aquatic hyphomycetes (fungi) in Lees Creek, Australian Capital Territory. Aust J Mar Freshwat Res 40:11–23

Thomas K, Chilvers GA, Norris RH (1992) Aquatic hyphomycetes from different substrates; substrate preference and seasonal occurrence. Aust J Mar Freshwat Res 43:491–509

Tsui CKM, Berbee ML (2006) Phylogenetic relationships and convergence of helicosporous fungi inferred from ribosomal DNA sequences. Mol Phylogenet Evol 39(3):587–597. doi:10.1016/j.ympev.2006.01.025, PubMed PMID: WOS:000238155300001

Tsui CKM, Hyde KD, Hodgkiss IJ (2001a) Longitudinal and temporal distribution of freshwater ascomycetes and dematiaceous hyphomycetes on submerged wood in the Lam Tsuen River, Hong Kong. J N Am Benthol Soc 20(4):533–549. doi:10.2307/1468086, PubMed PMID: WOS:000172517400003

Tsui CKM, Hyde KD, Hodgkiss IJ (2001b) Colonization patterns of wood-inhabiting fungi on baits in Hong Kong rivers, with reference to the effects of organic pollution. Anton Leeuw Int J Gen Mol Microbiol 79(1):33–38. doi:10.1023/a:1010210631215, PubMed PMID: WOS:000167920200004

Tubaki K, Tokumasu S, Ando K (1985) Morning dew and *Tripospermum* (Hyphomycetes). Bot J Linn Soc 91:45–50

Vijaykrishna D, Jeewon R, Hyde KD (2006) Molecular taxonomy, origins and evolution of freshwater ascomycetes. Fungal Divers 23:351–390, PubMed PMID: WOS:000243414100015

Voglmayr H (2004) *Spirosphaera cupreorufescens* sp. nov., a rare aeroaquatic fungus. Stud Mycol 50(1):221–228

Voglmayr H (2011) Phylogenetic relationships and reclassification of *Spirosphaera lignicola*, an enigmatic aeroaquatic fungus. Mycotaxon 116:191–202

Voglmayr H, Delgado-Rodriguez G (2001) *Dendroclathra caeruleofusca* gen.nov. et sp.nov., an aeroaquatic hyphomycete from Cuba. Can J Bot-Revue Canadienne De Botanique 79(9):995–1000, PubMed PMID: WOS:000171070800001

Voglmayr H, Delgado-Rodriguez G (2003) New species, notes and key to the aeroaquatic genera *Beverwykella* and *Ramicephala* gen. nov. Mycol Res 107:236–244. doi:10.1017/s0953756203007172, PubMed PMID: WOS:000182647300015

Voglmayr H, Krisai-Greilhuber I (1997) *Pseudoclathrosphaerina evamariae* gen. et sp. nov. and *Sympodioclathra globosa* gen. et sp. nov., two aeroaquatic fungi similar to *Clathrosphaerina*. Mycologia 89(6):942–951. doi:10.2307/3761115, PubMed PMID: WOS:A1997YK93200014

Voglmayr H, Park MJ, Shin HD (2011) *Spiroplana centripeta* gen. & sp. nov., a leaf parasite of *Philadelphus* and *Deutzia* with a remarkable aeroaquatic conidium morphology. Mycotaxon 116:203–216

Wang Z, Gerstein M, Snyder M (2009) RNA-Seq: a revolutionary tool for transcriptomics. Nat Rev Genet 10:57–63

Watanabe T (1975) *Tetracladium setigerum* an aquatic hyphomycetes associated with gentian and strawberry roots. Trans Mycol Soc Jpn 16:348–350

Webster J (1959) Experiments with spores of aquatic hyphomycetes I: sedimentation and impaction on smooth surfaces. Ann Bot 23:595–611

Webster J (1980) Introduction to fungi, 2nd edn. Cambridge University Press, Cambridge

Webster J, Davey RA (1984) Sigmoid conidial shape in aquatic fungi. Trans Br Mycol Soc 83:43–52, PubMed PMID: WOS:A1984TH62300005

Webster J, Descals E (1981) Morphology, distribution, and ecology of conidial fungi in freshwater habitats. In: Cole GC, Kendrick B (eds) Biology of conidial fungi. Academic, London, pp 295–335

Willoughby LG (2003) Diseases of freshwater fishes. In: Tsui CKM, Hyde KD (eds) Freshwater mycology. Fungal Diversity Press, Hong Kong, pp 111–126

Wong SW, Hyde KD (1999) *Proboscispora aquatica* gen. et sp. nov., from wood submerged in freshwater. Mycol Res 103:81–87. doi:10.1017/s0953756298006820, PubMed PMID: WOS:000078869900013

Wong MKM, Goh TK, Hodgkiss IJ, Hyde KD, Ranghoo VM, Tsui CKM et al (1998) Role of fungi in freshwater ecosystems. Biodivers Conserv 7(9):1187–1206. doi:10.1023/a:1008883716975, PubMed PMID: WOS:000078129900005

Wood-Eggenschwiler S, Bärlocher F (1985) Geographical distribution of Ingoldian fungi. Verhandlungen Internationaler nereinigung fuer Theoretische und Angewanot d Liminologie 22:2780–2785

Wurzbacher C, Rosel S, Rychla A, Grossart H-P (2014) Importance of saprotrophic freshwater fungi for pollen degradation. PLoS One 9(4), e94643

Yuen TK, Hyde KD, Hodgkiss IJ (1998) Physiological growth parameters and enzyme production in tropical freshwater fungi. Mater Org 32(1):1–16, PubMed PMID: WOS:000079120700001

Zemek J, Marvanov L, Kuniak L, Kadlecikova B (1985) Hydrolytic enzymes in aquatic hyphomycetes. Folia Microbiol 30:363–372

# Chapter 14
# Dispersal Strategies of Microfungi

**Donát Magyar, Máté Vass, and De-Wei Li**

*"Fungi cannot walk or run, but some can swim, most can soar, a few can jump, and some must be carried" (Kendrick 1985).*

## Introduction

Spore dispersal is an extensively studied field in microfungi biology. Even in the early 1900s, the knowledge accumulated as in some volumes of Buller's book (Buller 1909). Since that time many books and review papers have been dedicated to this topic, and most informative papers were published in the 1960s. Among them Ingold's *Fungal Spores* is the most comprehensive work, in which all the four major branches—fungal dispersal with air, water, plants, and animals—are covered. In the last 40 years, however, no similar work was published—only from the studies of aquatic hyphomycete dispersal and airborne fungi tended to become a distinct lineage of mycology. Others, e.g., entomophilous dispersal, have been only

D. Magyar, Ph.D. (✉)
Department of Air Hygiene and Aerobiology, National Public Health Center, Budapest, Hungary
e-mail: magyar.donat@gmail.com

M. Vass
Deptartment of Ecology and Genetics, Limnology, Uppsala University, Uppsala, Sweden
e-mail: mate.vass@ebc.uu.se

D.-W. Li, Ph.D.
The Connecticut Agricultural Experiment Station Valley Laboratory,
153 Cook Hill Road, Windsor, CT 06095, USA

Co-Innovation Center for Sustainable Forestry in Southern China, Nanjing Forestry University, Nanjing, Jiangsu 210037, China
e-mail: Dewei.Li@ct.gov



D.-W. Li (ed.), *Biology of Microfungi*, Fungal Biology,
DOI 10.1007/978-3-319-29137-6_14

sporadically studied. In recent years, fungal spore dispersal has been studied with increasing intensity due to their allergological, phytopathological, and ecological potential, and a wealth of new information has been gained. Although some fundamentals in understanding spore dispersal should be provided, the aim of this chapter is to review the last 40 year's discoveries, updating the knowledge of the dispersal of microfungi according to Ingold's original concept, and show the most important results and directions.

## *The Spore Dispersal Pathway*

Many fungal species developed different strategies—adaptations or mechanisms to aid dispersal, mainly by means of air currents, water, or animals. Most widely, airborne dispersal of spores is studied. The concept of "aerobiology pathway" involves three stages (Edmonds 1979): (1) liberation, (2) transport, and (3) deposition. The journey of the spores does not always finish here: a subsequent phase, resurfacing (4) may also occur. Multiple resurfacing in airborne dispersal (re-aerosolization) is important, giving a "second chance" to find the host. These steps are also present in other dispersal strategies. Thus, it is reasonable expanding this concept to create the new term "spore dispersal pathway." The pathway and the dispersal strategies are summarized in Fig. 14.1 [detailed description of the steps and mechanisms shown in this figure exceeds the frame of this chapter; for further details, see (Gregory 1961; Ingold 1971; Kendrick 1990; Madelin 1994; Lacey 1996; Deacon 2006; Lacey and West 2006; Elbert et al. 2007; Money and Fischer 2009)]. Dispersal steps are often combined, e.g., airborne transport and deposition of the spore in the canopy are often followed by stemflow transport.

The number of spores produced in and emitted from a fungal colony is on a wide scale (from one big spore of *Valdensinia heterodoxa* to 400,000,000 conidia of a colony of *Penicillium* sp. 2.5 cm in diameter or even 5.4 trillion spores of *Ganoderma applanatum*) (Ingold 1971). Even the world's human population—approximately 7 billion—is insignificant as a number when compared to the number of spores that just a single fungus may produce within its perennial fruiting body over a 6-month period. This is also an important factor in fungal dispersal, concerning the K- or r-strategy of the fungi. The r-selected organisms produce many spores, but with only a low chance of reproductive success (e.g., *Cladosporium*, *Rhizopus*), while K-selected organisms produce few spores, but with significant chance of establishment in a new environment (Andrews 1992). Concentration dynamics of the spores are strongly influenced by each stage of the pathway (Ingold 1971; Lacey 1981, 1996).

Spores can also be classified as primary and secondary inocula. Primary inoculum consists of propagules of a pathogenic fungus that start the disease cycle in a new growing season of a crop. Secondary inoculum distributes the pathogen within the main growing season, and this usually leads to the development of epidemics (McGee 2003).

Spores are dispersed in different space and time scales. The microscale dispersal is limited to rather short time and small space scales on the order of < an hour and

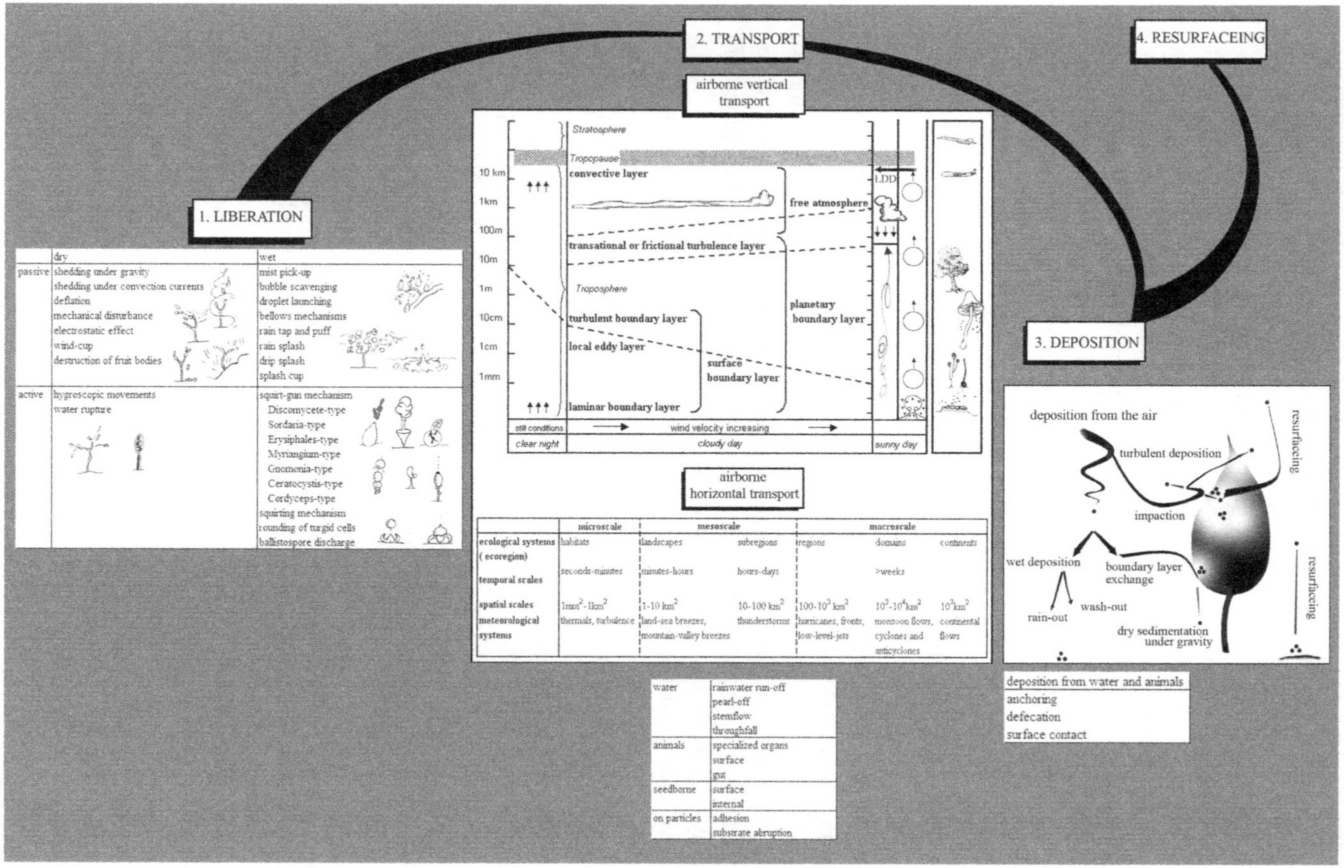


**Fig. 14.1** The aerobiology pathway (liberation, transport, deposition, and resurfacing) modified after Edmonds' concept (Edmonds 1979), including Lacey's (1991) classification of liberation mechanisms, vertical transport after Gregory (1961), horizontal transport after Gage (1999)

space scales up to several hundred meters. For mesoscale transport, the time and space scales are days and several hundred km. Macroscale transport is the greatest in time and space scales including global circulation patterns (Magyar 2007).

In recent years the importance of understanding the detailed dispersal strategy has been enhanced by an unprecedented number of fungal diseases, which are considered a worldwide threat to food security and a cause of extinctions of wild plant species (Fisher et al. 2012). Successfully controlling these emerging diseases depends on whether their propagates can be controlled, thus it is important to know the dispersal strategies of fungi. Many fungi have developed different active and passive strategies to release, disperse, and colonize new environments (Incagnone et al. 2014). In this chapter, these mechanisms are presented and discussed.

## Dispersal Units of Fungi

Fungi may be present as propagules (spores, hyphal fragments, partial conidiophores, pieces of mycelium, sclerotia) or attached to substrata (Fig. 14.2), but spores are the main units of dispersal (Bärlocher 1992a). The transportation of each of these forms results in the fungi widespread occurrence. Fungal spores occur in a fantastic scale of shapes from the simple globose to the exceptionally complex. Morphology of spores may be indicative of its ecology, dispersal, or sporogenesis as well.

Thick-walled, pigmented spores have higher resistance against drying and UV radiation during aerial transport than thin-walled and hyaline ones. Spore size and roughness affect the dispersal of spores by the rate at which spores fall through the air and their ability to impact on stems and other obstacles or surfaces. The large-spored leaf- and stem-pathogenic fungi appear to be specialized to impact on the host plants (impactors), while the spheroid, minute spored molds (penetrators) appear to be specialized to penetrate deeper into the vegetation (Gregory 1961). Spores could be grouped according to their wettability, into hydrophobic dry spores (xerospores) and readily wettable slime spores (gloiospores) (Fig. 14.1). Gloiospores are carried within rain droplets, but xerospores are found on their surface. Certain fungi have two or more morphologically different spore types, and each of them is adapted for different dispersal circumstances. It is often observed in the anamorph–teleomorph pairs (e.g., orbiliaceous fungi have penetrator-type ascospores and radiate, anchoring conidia). Spores with appendages and mucilage on their surface are usually dispersed by rainwater or animals. The function of these structures in *Coelomycetes* fungi are summarized by Nag Raj (1993). The role of some structures on the spores is still unclear, e.g., hornlike hyaline cells on some bark fissure-inhabiting fungi (*Excipularia*, *Oncopodiella*, and *Oncopodium*). Cytoplasmic viscosity and the presence of ergosterol of airborne and waterborne fungal spores are significantly different between the two spore types and correlated with spore survival. The mode of spore dispersal is related to cellular composition (Van Leeuwen et al. 2010).

The different spore forms of rust fungi have different modes of dispersal: pycniospores are released from their conidiogenous cells into a viscous liquid and locally allocated by insects, splashing water, and contact among host plant organs (Littlefield

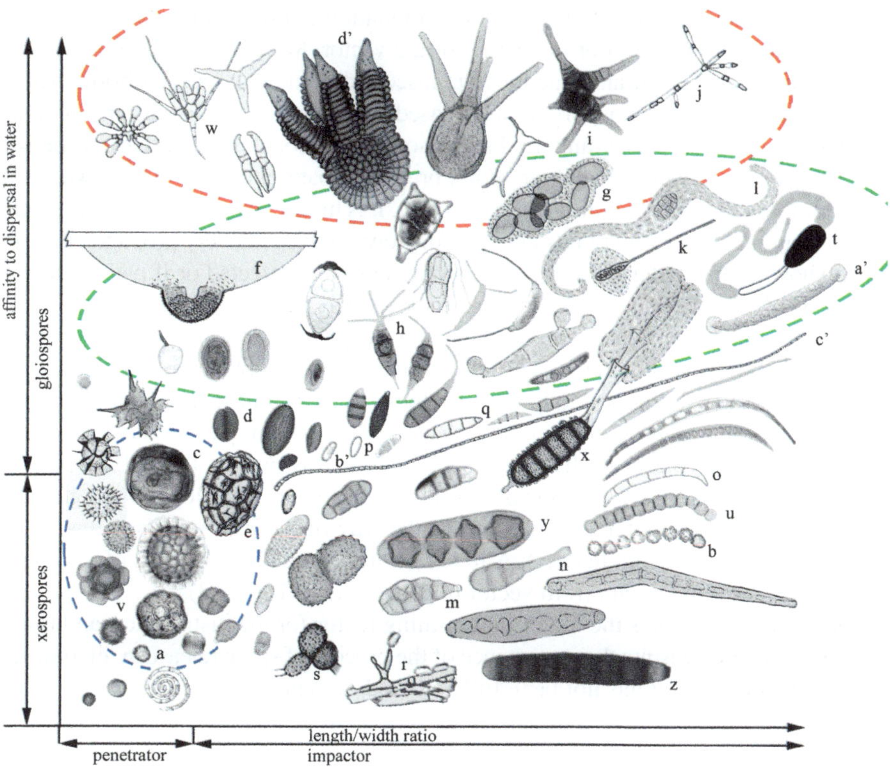


**Fig. 14.2** Fungal dispersal units (spores). Drawn at different scales. Group outlined with a *blue line*: dry spores with rough surface. Group outlined with a *green line*: spores with appendages or associated with mucilage. Group outlined with a *red line*: radiate spores. a *Aspergillus* conidia; b *Aspergillus* conidia in chain; c fungal mycelia inside an airborne pollen grain; d two xylariaceous ascospores adhered on their flat side form a sphere minimizing air drag; e ascospore of the subterranean fungus *Tuber mesentericum*; f *Pilobolus*, a discharged sporangium just landed on a glass slide; g *Ascobolus immersus* (reported as *Dasyobolus immerses*) ascospore octade; h *Pestalotiopsis* conidia, i *Oncopodiella* sp. germinated conidia from stemflow rainwater; j *Metschnikowia reukaufii* yeast cells from floral honey; k *Loramyces juncicola* ascospore; l *Pleospora scirpicola* ascospore; m–n *Alternaria* sp. m with short rostrum and n with long rostrum; o–s *Fusarium* sp., o macroconidium, p microconidium, q mesoconidium, r mycelia and conidia on airborne plant debris, s chlamydospore; t *Podospora fimicola*; u *Torula herbarum*; v *Urocystis* spores surrounded by sterile cells that possibly aid wind dispersal; w *Tetracladium* conidium from streamwater foam; x *Phragmidium mucronatum* teliospore; y *Aglaospora profusa* ascospore; z *Dendryphion nanum* conidium from the burrows of earthworms; a' *Leptosphaeria maculans* ascospore and b' conidia of its anamorph, *Phoma lingam*; c' *Cordyceps militaris*. d' Cross-shaped conidia of *Valdensinia heterodoxa* turn to have a lacrymiform shape when discharged, which aerodynamically should make the conidium travel further. g, f, l, k, t, x, and c' were redrawn after Ingold (Ingold 1971, 1978) and d' after Zhao and Shamoun (2010)

1981). Aeciospores are produced in tightly packed chains, released by the dissolution of intercalary cells, and aerially disseminated (Littlefield and Heath 1979), while the teliospores may remain attached to the host organ on which they were produced (Littlefield 1981).

The distribution of infected plants in a population is useful information pointing out the source of inoculum or the nature of the vector. Random distribution is rather rare. It can be a result of airborne inocula, insect-borne inocula or seed-borne pathogens, via a long distant transportation. Diseased plants aggregated in clusters are more common, and it indicates a random distribution of the inoculum. A disease dispersed by aphids or a root fungal pathogen might produce a pattern of aggregation when the inoculum spreads from the initial infected point. It is highly unusual to observe a regular distribution of diseased plants in the field. However, vegetatively propagated crops could show such a pattern, if all planting materials were infected or if pathogens left in the soil by a previous crop with even space were able to infect the next crop. Soilborne pathogens, such as root fungal pathogens or nematodes, or diseases carried by soil-inhabiting vectors show a patch distribution. Soil organisms usually spread slowly, as do the diseases transported by these vectors, and consequently lead to patchy distribution. A very common disease distribution pattern is a gradient. Gradients typically indicate that the inoculum source is outside the field crop. The gradient steepness is relative to the proximity of the source. The slope of the gradient can be influenced by other factors, such as the way of the pathogen is spread (e.g., flying insects vs. crawling insects). A much greater effect on the spread of disease is from vector movement than from vector population (McGee 2003).

Dispersal of spores is the major determining factor for the distribution and range of fungi in nature. Despite the importance of the process of spore release in allergology and plant pathology, it has not been fully studied in many fungi.

## Liberation

Liberation mechanisms of spores vary greatly among different groups of fungi. These mechanisms have been reviewed by Gregory (1961) and Ingold (1971). Based on these works, the mechanisms are summarized and illustrated (Fig. 14.2), but the present descriptions are focused on the knowledge gained in the past years.

At the interface of the solid objects, the air is stationary. Airflows around the objects are moving slowly due to the drag created by the object's surface. This is the laminar boundary layer which surrounds all surfaces at a variable depth between 0.1 mm and 9.0 mm on a leaf surface (depending upon leaf size and wind speed), to a meter or more that often exist on a woodland floor on calm days (Nobel 1991).

Above this boundary layer, the spores could be dispersed because the air becomes progressively more turbulent in local eddies, until there is a net movement of the air mass. In order to become airborne, spores must cross the laminar boundary layer to reach a new site. This is the essential feature of spore dispersal, so microfungi require different strategies to accomplish it. Release mechanisms could be classified as active (utilizing the energy generated by the fungus) and passive (utilizing the energy from the environment). Fungi that grow on more rigidly supported surfaces can release the spores by active processes, and others involve adaptations of the spore-bearing structures rather than of the spores themselves. Airborne dispersal needs different strategies under dry and rainy conditions, and many fungi have adapted to both circumstances (e.g., the *Pleospora* is liberated in wet, while its anamorph, *Alternaria*, in dry weather) (Magyar 2005). Molecular techniques are major alternatives or supplementary to morphological identification of fungi. However, the latter is still raison d'etre in dispersal research, according to the differences in the ecological adaptations of different spore forms.

In windless conditions, gravity alone may overcome adhesive forces attaching the spore, when it is elevated on a tall sporophore, stem or leaf. Mushrooms and some myxomycetes are examples for this strategy. When the stalk (stipe) of a mushroom elongates, the cap (pileus) reaches the layer of the turbulent air. The basidiospores fall from the gills or pores vertically into turbulent air and carried away by wind. In calm days when higher temperature differences are present, spores could be lifted up from cultures to the top of 10–12-cm glass cylinders by convection alone (e.g., at 10 °C differences for *Botrytis cinerea* and *Chrysonilia sitophila*) (Lynch and Poole 1984).

Wind plays a major role in spore release. Moderate wind speed [0.4–2.0 m $s^{-1}$ (Gregory 1961)] is enough to detach spores from the colony of many fungal species, but the maximum number of the removed spores varies among species (e.g., between 0.4 and 1.0 m $s^{-1}$, *Alternaria alternata* (reported as *Alternaria tenuis*) (Rotem 1994); between 0.5 and 1.0 m $s^{-1}$, *Erysiphe graminis* (Aylor et al. 1981); at 0.5 m $s^{-1}$, *Aspergillus fumigatus* (Pasanen et al. 1991); between 1.8 and 2.3 m $s^{-1}$, *Puccinia recondita* f. sp. *tritici*; between 1.8 and 2.8 m $s^{-1}$, *Puccinia striiformis* (Geagea et al. 1997); between 3.0 and 5.0 m $s^{-1}$, *Spilocaea pomi* (sexual state, *Venturia inaequalis* (Wiseman 1932). Fungal spores may be deflated by a higher wind speed from the ground (3.0–5.4 m $s^{-1}$) than from the phyllosphere (0.5–2.0 m $s^{-1}$) (Jones and Harrison 2004). The efficiency of deflation should be reduced when spores are slimy or the substrate is wetted (Jones and Harrison 2004; Ward and Manners 1974). In this case, spores are cemented to each other and to the surface (e.g., Zoberi 1961; Pady et al. 1969; Cohen and Rotem 1970; Rotem 1994; Geagea et al. 1997). It has been shown that especially intermittent wind and vibration is effective (Aylor 1990, 1993; Górny et al. 2001), because it provides a small but continuous stream of viable airborne spores in the air, removing the highly mobile (dry and detached) part of spore mass produced in the fungal colony. Gloiospores (e.g., *Stachybotrys chartarum*), however, remain attached on the surface and could not be removed by low airspeeds [0.3–1.6 ms (Tucker et al. 2007)].

Dispersal of a high proportion of lichen-forming fungi is efficiently carried out by vegetative symbiotic propagules: soredia, isidia, blastidia, and thallus fragments (Honegger 2009). The predominant mode in reindeer lichens (*Cladonia* spp.) is thallus fragments, which cover thousands of $km^2$ area of arctic tundras. Soredia, powdery propagules, are composed of fungal hyphae wrapped around green algae or cyanobacteria. Large amounts of mycobiont-derived crystalline secondary metabolites are carried by the soredia of a high number of lichens at their surfaces and are highly hydrophobic. This phenomenon facilitates wind dispersal.

Colonies of some microfungi growing on leaf surfaces sometimes produce chains of spores from a basal cell so that the mature spores are pushed upwards through the boundary layer as more spores are produced at the base of the chain (e.g., *Blumeria graminis*). The spores are then removed by air currents or, sometimes more effectively, by mist-laden air (e.g., *Cladosporium*) (Harvey 1970). Wadia et al. (1998) showed that spore counts of *Passalora personata* were decreased by steady wind, but intermittent wind gusts caused a high concentration. Continuous wind in 3–6 days exhausted the colonies. Stronger wind gusts tear off the strongly attached mycoparticles, sclerotia, and immature conidia as well. Spores should not be produced until new conidiophores develop and replace the broken ones (Langenberg et al. 1977). During evolution, two groups of *Alternaria* spp. adapted different strategies through the boundary layer by developing a beak, a unique conidiogenous apparatus on conidia. The dispersal of small-spored *Alternaria* involved the catenate proliferation of conidia emerging from the beak or the secondary conidiophore of the precedent conidium. For the other group, the liberation of filament-beaked *Alternaria* was facilitated by the elongated filamentous beak. Both strategies result in the elevation of conidia through the boundary layer (Chou and Wu 2002). The erosion of fungal colonies is stronger on the upper part of the plant canopy than near the ground (Rotem 1994; Aylor 1990). Some spores hardly became airborne, because their dispersal is not primary anemophilous, like coprophilic fungi. Staurosporae, "giant spores" (*Bipolaris*, *Podosphaera*, *Phragmidium*, and *Tetraploa*), fruiting bodies, sclerotia, and fungal particles aggregated with pollens and plant debris could be also aerosolized by strong wind. Rotem (1964) observed immature conidia of *Alternaria solani* in the windstorms of Negev desert and some of them were attached on plant debris. Meredith (1966) mentioned that conidiophores with immature conidia, hyphal fragments, and plant debris with spores were collected by a Hirst-type air sampler. Hyphal fragments of phylloplane fungi should be dispersed due to scrubbing action of host plants which takes place during wind currents (Tilak and Pande 2005). Airborne plant debris could carry the oospores of *Peronosclerospora sorghi* (Bock et al. 1997).

Ballistospore discharge is a feature of basidiomycetes. The process of ballistospory and the role of the droplets observed on the spores were poorly understood until recently, because this process occurs so rapidly that the launch of the ballistospore has never been visualized. Analyses show that ballistospores catapult into the air at initial accelerations in excess of 10,000 g. Recent technological development of ultrahigh-speed video cameras allowed to capture the fast motions during spore discharge at camera speeds of up to 100,000 frames $s^{-1}$ (Pringle et al. 2005; Noblin et al. 2009). According to the records, just few seconds before discharge, fluid begins to condense on the spore surface at two locations (Fig. 14.3; McLaughlin

**Fig. 14.3** (continued) complex. This force puts the hilum under tension, which provides a counteracting force that cannot exceed the fracture force (FB). Fourth, the hilum is fractured, thus releasing the spore. (**b**) The corresponding stages in jumping. First, the center of mass is lowered to allow the legs to do work on the substratum. At this stage, the gravitational force (Fg) and the ground reaction force (FR) are balanced. Second, as the legs unfold, the moments at the joints (M) are resisted by the substratum, thus providing the impulse (I) necessary to accelerate the center of mass. Third, late in the jump, the fast-moving upper body starts to entrain the legs, which to this point were moving slowing upward. Fourth, after takeoff all body parts are moving at similar speeds and only gravity acts on the body (**c**) A typical basidium with four spores. (**d**) Structure of the lower half of the spore. (**e**) Spore ejection in *Auricularia auricula* [**a**–**c**, **e**, pictures courtesy from Xavier Noblin and Jacques Dumais; **d**, based on McLaughlin et al. (1985)]

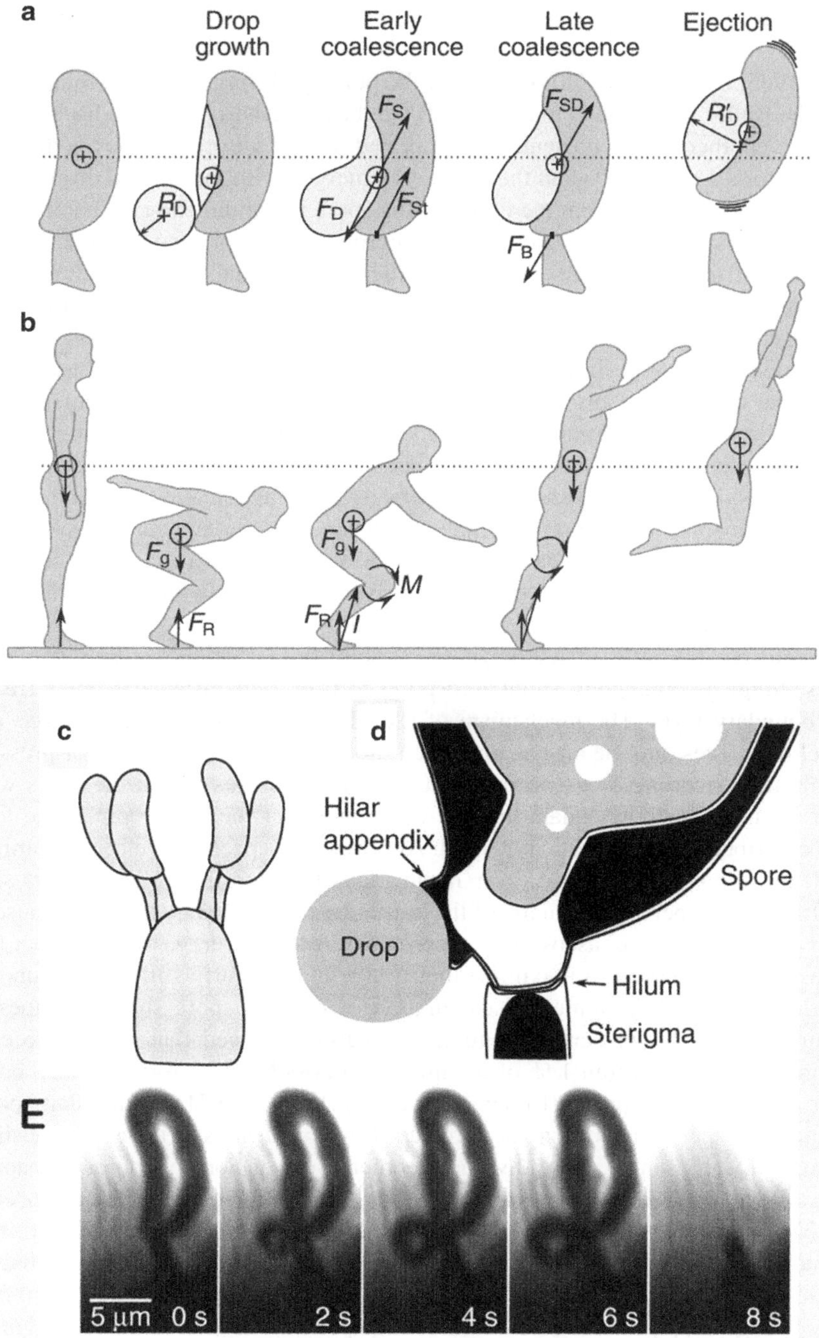


**Fig. 14.3** The four stages of ballistospore ejection. (**a**) First, the growth of the drop brings the center of mass of the spore–drop complex closer to the end of the sterigma. Second, at the start of the coalescence process, the drop and spore exert on each other forces of equal magnitude but opposite direction (FD and FS). The expected downward displacement of the spore is prevented by the presence of the sterigma, giving rise to a reaction force FSt acting at the hilum. Third, in late coalescence, the momentum of the drop is transferred to the spore, which was immobile until then. The transfer of momentum is equivalent to a force FSD applied at the center of mass of the spore–drop

et al. 1985; Noblin et al. 2009). (1) Over the punctum lacrymans of the hilar appendix, the Buller's drop starts to bubble. (2) On the adaxial drop on the adjacent spore surface, another drop is present. Once initiated, Buller's drop increases in diameter rapidly for a few seconds and then suddenly merged with the adaxial drop. As the drop redistribute mass from the hilum of the spore in the direction of the other end of the spores, the spore is catapulted. Recordings demonstrated that coalescence may result from the directed collapse of Buller's drop onto the spore, but it may also involve the movement of the spore toward the drop. The energy of spore discharge is derived from the rapid, surface tension powered movement of Buller's drop onto the spore surface, to push spore off from the sterigma. The release of surface tension energy at coalescence provides the kinetic energy and directional momentum to launch the spore away from the fungus. This mechanism is responsible for launching basidiospores from mushrooms, but microfungi also utilize it in their dispersal, such as basidiomycetous yeasts, rusts, and smuts. It was calculated that the relative sizes of the spore and drop determine the distance of the launch. Increase in the radius of the drop tends to catapult the spore over a greater distance (Money and Fischer 2009). The results revealed a surprising similarity with the mechanics of jumping in animals (Fig. 14.3b; Noblin et al. 2009).

Explosive discharge of Ascomycota is one of the fastest movements in nature. These fungi, as small guns, shoot ascospores to 1- or 2-cm distance to break free of the boundary layer. The mechanism of asci that act as small water cannons and expel the spores into the air has long been thought to be driven by turgor pressure within the extending ascus. In recent studies the pressures within the ascus were measured. Such studies quantified the components of the ascus epiplasmic fluid that contribute to the osmotic potential. Although few species have been examined in detail, the results indicate diversity in ascus function that reflects ascus size, fruiting body type, and the niche of the particular species (Trail 2007). Because of their microscopic size, ascospores experience great fluid drag. It was shown that ascospores are shaped to maximize their range in the nearly still air surrounding fruiting bodies using numerical calculation of optimal spore shapes (shapes of minimum drag for prescribed volumes). Analysis showed that spores are constrained to remain within 1 % of the minimum possible drag for their size, being near their physical optima (Roper et al. 2008). Ingold (1971) opined that spores would be shaped to maximize the force used by apical rings of the asci to push on them. But, surprisingly, the individual geometric dimensions of spores and apical ring critical to these hypotheses are either very weakly or not correlated (Fritz et al. 2013). There is a mechanism avoiding energy losses during spore's ejection through the apical ring. This mechanism is based on a physical principle discovered 50 years ago in the study of elastomeric seals and O-rings used to control fluid flow in engines, pipes, and other engineering applications (Dowson and Higginson 1959). The apical ring is an elastic seal and distorts significantly when the spore, which is lubricated by a thin fluid layer, passes through it. Some apothecial fungi, e.g., the plant-pathogenic *Sclerotinia sclerotiorum*, are dispersed by synchronizing the ejection of thousands of ascospores, and this creates a blast of air that carries spores through the laminar boundary layer (Fig. 14.4). High-speed imaging showed

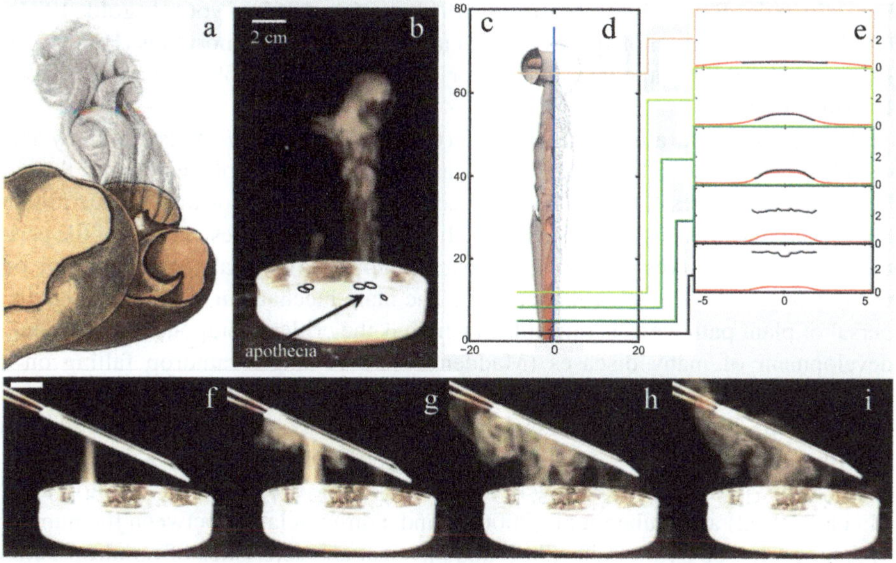


**Fig. 14.4** Synchronized ejection of ascospores creates a coherent jet of air, enhancing dispersal in open and crowded environments. (**a**) Early documentation of synchronous spore release in a drawing of *Otidea cochleata* (Bulliard 1791). (**b**) Spore jets originating from *Sclerotinia sclerotiorum* apothecia travel much farther than individually ejected spores. (**c–e**) Mathematical modeling shows how spores mobilize the air around the ascocarp. (**c**) Simulated jet profile for synchronous, random ejection showing both (**e**) The flow of air that carries the spores and (**d**) spore motions. (**e**) Average spore (*black curves*) and air (*red curves*) speeds on cross sections of the jet show how the spores initially push upon the still air, and as the jet develops the spores are pulled upward by the air that they set into motion. (**f–i**) If the spore jet impacts upon an obstacle, such as a glass slide, then pressure gradients within the jet displace spores out and around the obstacle. Images are taken at t=0, 0.11, 0.32, 0.55 s after initial impact [pictures courtesy from Roper et al. (2010)]

that synchronization is self-organized and likely triggered by mechanical stresses. Spores "cooperate" maximally to generate the favorable airflow with optimal geometry (Roper et al. 2010). Some quantity of moisture is usually required for spore discharge in different Ascomycetes. Consequently, the peak of ascospore concentration could be measured some minutes after the start of rainfall or 5–10 h later. Sometimes fog is enough to generate spore discharge (Gregory and Stedman 1958; Carter and Moller 1961; Hirst and Stedman 1962; Arseniuk et al. 1997). Asci may be exhausted in rainfalls lasting for many days (Gregory and Stedman 1958; Turott and Levetin 2001).

Drying-induced hygroscopic movements exist in some fungi. When drying out, their spore-bearing structures twist, leading to the release of the spores (e.g., *Botrytis*, *Phytophthora infestans*, and *Peronospora*). The reduction of R.H. is triggered by various factors (temperature/solar radiation, wind). Results of several laboratory experiments on spore discharge of fungal species under controlled climatic conditions showed active spore discharged in case of decreasing relative humidity (de Bary 1887; Meredith 1961, 1962, 1965, 1966; Ingold 1971). Outdoor concentration of spores reaches its maximum at the mornings (Hirst 1953; Panzer et al. 1957; Waggoner and Taylor 1958; Hamilton 1959; Sreeramulu 1959; Dutzmann 1985).

Airborne concentration of ascospores (*Chaetomium*, *Xylariaceae*, etc.) may also correlate positively with drying. In this case ascospores are not ejected into the air, but a mass of spores (cirrus) is formed covering the surface of the fungal fruiting body. After the cirrus is dried, spores are liberated. Incipient desiccation of the perithecium of some *ascomycetes* (e.g., *Claviceps purpurea*) may also exert pressure on its content so rate of discharge increases. The main mechanism involved in the dispersal of plant pathogens is rain splash which is the critical component in epidemic development of many diseases (Madden 1992). A single raindrop falling on a *Botrytis* colony in a leaf lesion could contaminate an area of 2.5 $m^2$ (Weston and Taylor 1948). Rain splash occurs when a raindrop falls onto a surface covered by a thin film of water, forming a crater, with many (100–5000) secondary droplets produced at its periphery (see Edgerton's iconic "milk drop coronet" photograph). Madden (1992) and Huber et al. (2006) found a direct relation between the number of spores released by the rain splash and physical characteristics of raindrops, e.g., kinetic energy, velocity, and Weber number. Weber number is a dimensionless, pertinent parameter for the description of splash droplet formation after impact. The bigger the Weber number is the more the water drop interface is deformed and potentially breaks into splash droplets (Saint-Jean et al. 2006). The literature of hydrodynamic modeling and physics of splash dispersal was vast in the last decades, especially in plant pathogens, like *Botryosphaeria* (Ahimera et al. 2004), *Colletotrichum* (Peres et al. 2005), *Fusarium*/*Gibberella* (Hörberg 2002; Paul et al. 2004), and rusts (Sache 2000). Experiments were also conducted, such as splash from single water drop impactions, spore transport with simulated rain over small areas, and disease spread in the field with naturally occurring rain. Studies showed that with increasing size of impacting raindrops (and hence increasing velocity and kinetic energy), the number of splash droplets produced also increased, as well as the number of spores disseminated and flight distance of the splash droplets. It was found that transport distance is very short, generally <15 cm in each splash event. This indicates that the deposition of spores on a potential infection site is a result of continual re-splashing of droplets and spores in them (Madden 1997).

Raindrops or hail can also release dry spores on a surface, by "puff" and "tap" mechanisms. Spores can be discharged by raindrops when the raindrops hit lesions at terminal velocity after an unimpeded free fall from a height of several meters or as secondary droplets at lower-velocity dripping from vegetation (Herwitz 2006). A raindrop splotches sideways, when it falls on a rigidly supported surface, and the consequential gust of air disturbs the boundary layer, leading to the dry spores to become airborne. The first large drops of an intensive rainstorm increase the concentration of dry fungal spores (Hirst 1953; Ainsworth 1952; Harvey 1967; Rantio-Lehtimäki 1977; Hjelmroos 1993; Kurkela 1997; Venables et al. 1997; Allitt 2000). A study by Hirst and Stedman (1963) showed the possibility of the dry dispersal of fungal spores by incident raindrops in case of some rust, smut, and conidial fungi.

The mature basidiospores of some puffballs (Basidiomycota) are enclosed in a papery basidioma with an apical pore, so that raindrops "puff" the spores into the air, like bellows. Ascospores of *Venturia inaequalis* are released from the lesions on leaves following the impact of raindrops during rain. Dampened periodic vibrations were induced by raindrops on leaf surfaces. Vibrations, linear, and/or oscillatory flows of water on the leaves induced spore releases at low energy thresholds. The fungal perceptibility of kinetic energy facilitated selective discharge of spores when environmental conditions are most favorable to survival. Determining kinetic effects of rain which reveal start, thresholds, and proportional distribution is significant for research on disease prognosis, and at the same time, it may serve as a model for a common fungal sensing mechanism (Alt and Kollar 2010).

## Airborne Transport

It is unequivocal that wind plays an important role in transportation of fungal spores. Airborne spores are primarily present in the troposphere. Dispersal of airborne spores is closely related to the size, shape, roughness, density, and electrostatic charging of individual spores and air viscosity, convection, layering, wind, turbulence, wind gradients close to the ground, and patterns of atmospheric circulation. When mixing of air masses is low, populations of spores should increase in still air. Daily maximum of spore population should evolve in certain geographical areas, where diurnal high winds occur on a regular basis (Gregory 1961; Ogden et al. 1969). Spore free winds or high wind velocity reduces the number of spores hovering above the source (Hamilton 1959; Lopez and Salvaggio 1983; Giner et al. 2001). According to the theories of eddy diffusion, turbulence dilutes clouds of spores. Eddies dilute spore cloud as it travels downwind, spreading it both horizontally and vertically, or physically move it. Air movements vary from small turbulent eddies to large frontal systems (several thousand km in length and several hundred km in width) and jet streams in the stratosphere and in the upper troposphere where spores are speedily transported. Gregory (1961) classified distinct zones of troposphere with regard to the dispersal of airborne fungi.

The laminar boundary layer is thickened at nighttime. The local eddy layer is eddies resulted from small cup-shaped depressions or roughness on the surface. In the turbulent boundary layer, diurnal changes are less pronounced. The transitional or frictional turbulence layer attains up to 500–1000 m in height, in which turbulence weakens with altitude. The top of this layer is the limit to which fungal spores may be transported upwards by turbulence. The convective layer is a stratum in which airborne particulates can be transported only by convection. During summer, when the sun heats the ground, warm air bubbles carrying a large number of spores may form in areas of 1.25 $km^2$ every 6–15 min.

Small propagules like fungal spores may be dispersed by the wind over distances of thousands of kilometers. Long-distance dispersal of fungi was proved by spore-trapping devices mounted on the outsides of aircrafts. One of the approaches toward understanding the long-distance dispersal is trajectory analysis based on circulation patterns as derived from worldwide synoptic observations. Long-distance dispersal can be highly significant for plant disease epidemiology, especially when new pathogenic races or fungicide-resistant strains develop and are spread across or between continents. In some fungi, asexual reproductive structures (mitospores) have been reported to disperse within [e.g., continental Europe to Great Britain (Brown et al. 1991)] and between continents [e.g., from South Africa to Australia (Watson and De Sousa 1983), Fig. 14.5]. Many other examples of long-distance dispersal (LDD) are in evidence (Aylor et al. 1982; Brown and HovmØller 2002; Frank et al. 2008), e.g., in India (Nagarajan and Singh 1976), between Paraguay and Argentina (Waller 1979), from Angola to Brazil (Bowden et al. 1971), across the Tasman Sea from Australia to New Zealand (Viljanen-Rollinson and Cromey 2002), and between North Africa and Southeast Europe (Tuboly and Vörös 1962). For additional examples, see (Taylor et al. 2006).

Meteorological data support the suggestion that *Hemileia vastatrix* urediospores developed in a coffee rust epidemic in Angola in 1966 were transported across the Atlantic at an altitude of 1500–2000 m in 5–7 days and deposited over the coffee plantations in Bahia, Brazil (Bowden et al. 1971). Similarly, urediospores could be easily transported from South Africa to Australia in <5 days traveling at 12,000-m altitude (Nagarajan and Singh 1990).

Long-distance dispersal is an important ecological process. It can significantly increase the range in which pathogen epidemics spread across a landscape, result in rapid spread of a disease to previously disease free areas, and influence the spatial distribution of pathogen populations in fragmented landscapes. Spores released during the hottest part of the day are shown to be more likely to undergo long-distance dispersal than those released at other times (Savage et al. 2012).

A number of features of spores are important for long-distance dispersal: resistance to desiccation resulted from hydrophobins in the walls, resistance to ultraviolet radiation conferred by pigments in cell walls, or the shape and size of spores. These characters may differentially influence long-distance dispersal among taxa (Finlay 2002). For instance, *Glomeromycota*, which develop relatively large, normally non-wind-dispersed asexual spores, had the lowest average geographical range (Tedersoo et al. 2014). The relatively delicate conidia of aquatic hyphomycetes are not suitable for longer-distance dispersal (Duarte et al. 2012). Generally region-based distribution patterns of fungi are, to some extent, incompatible with clustering of plants and animals, where Holarctic lineages are deeply nested within larger tropical groups (Sanmartín and Ronquist 2004). Overall, fungi have broader geographical distribution than those of flowering plants, but even if an accepted hypothesis on the distribution patterns of species on continents is present, it is a huge challenge to elucidate these patterns on geographical or biological islands, especially on the most remote and isolated ones (Incagnone et al. 2014; Whittaker and Fernández-Palacios 2007).

Long-distance dispersal is a stochastic occurrence which may be exceptionally significant in determining a population (Wingen et al. 2013), since the dispersal pat-

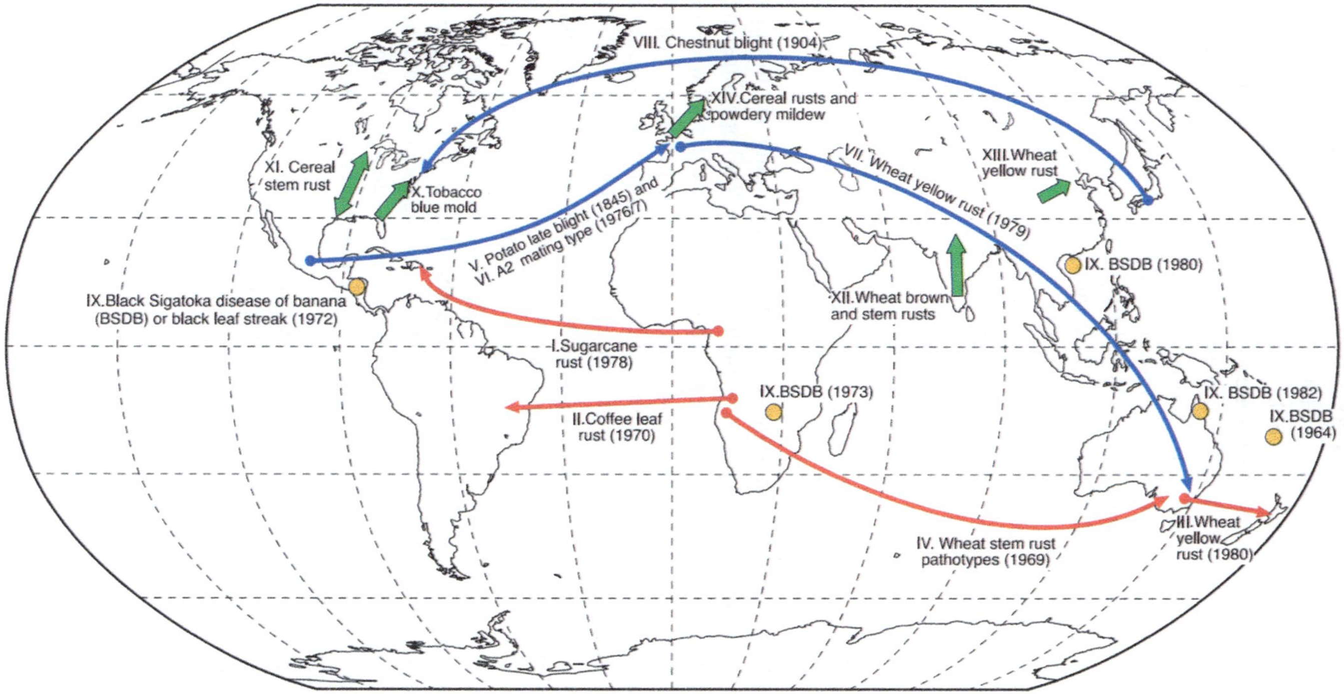


**Fig. 14.5** Selected dispersal events of fungal pathogens. *Red* and *blue arrows* indicate invasions of new territories (first year recorded in brackets). *Red arrows* indicate dispersal that probably occurred by direct movement of airborne spores. *Blue arrows* indicate pathogens that were probably transported to the new territory in infected plant material or by people and spread thereafter as airborne spores. *Orange circles* indicate the worldwide spread of black Sigatoka disease of banana; the first outbreak recorded on each continent is marked. *Green arrows* indicate periodic migrations of airborne spores in extinction–recolonization cycles (from Fig. 14.1 from James K. M. Brown and Mogens S. Hovmøller, SCIENCE 297:537 (2002). Background World Map: © Luckinbeal, Southern Connecticut State University, New Haven, Ct. Reprinted with permission from AAAS)

terns of fungal spores influence population structure, gene flow, and fungal community structure (Lilleskov and Bruns 2005) and play an important role in most conceptual models of community assembly. However, it is difficult to directly measure dispersal of fungal spores across a whole community at ecologically relevant spatial scales. The passively liberated spores are less efficient on long distance than actively liberated ones, and their dispersal distances are not associated with the mass of the propagules (Jenkins et al. 2007). Microorganisms generally have a cosmopolitan distribution (Incagnone et al. 2014). In the words of O'Malley (2008): "Everything is everywhere, but the environment selects." Wood-Eggenschwiler and Bärlocher (1985) concluded that geological barriers or distance rarely restricts species occurrence. In contrast, a recent study showed that fungi largely exhibit strong biogeographic patterns that appear to be driven by dispersal limitation (Tedersoo et al. 2014). Temperature and its influence on vegetation are considered as the main factor determining fungal distribution worldwide.

## Spore Dispersal by Indoor Air

Air movement outdoors influences air movement indoors forcing air through cracks on the windward side and sucking it out on the leeward. Wall openings and gaps at structural joints (electrical outlets, cable boxes, etc.) are 1 mm wide in average. Thus, through-wall gaps may serve as a route for dispersal of fungal spores. Within a building, thermal rise or chimney effect transports spores upward via elevator pits, staircases, or floor openings. Air circulates indoors as a result of thermal convection. Thermal gradients between heaters, walls, and windows generate mixing (Daws 1967). Convection may very efficiently carry fungal spores from the first- to fourth-floor halls in five min and into rooms in 20 min (Christensen 1950). Aerosols are translocated by heat sources, even by heat from a lamp. The heat from the human body may be adequate to change the local convection system moving air up from floor level (Lewis et al. 1969). Occupants generate turbulence via their activities. Spores may be dispersed by bed-making, sweeping, and building repair work, resulting in spore concentrations 10–17 times greater than before disturbance (Maunsell 1954; Reponen et al. 1992).

Ventilation systems could rapidly circulate spores or transport them from any point source throughout entire buildings (Philips 1965). Concentration and viability of the airborne spores could also be reduced by filtering and UV lamps installed in air ducts (Kowalski 2006; Burroughs and Hansen 2008). Spore cloud may be diluted, but not completely removed by artificial ventilation. Even without any mechanical ventilation, dispersal of a spore cloud may be distributed all over a large facility in a short period of time.

In grain warehouses, postharvest and plant-pathogenic fungi are common because spores from the stored grain become airborne. It was suggested that *Tilletia* teliospores spread readily inside grain warehouses, possibly by the activities of transportation systems. Teliospore dispersal takes place rapidly between rooms and floors via wall openings and spouts. Spouts are used for loading and emptying the storage room to another room at another floor, generating airborne dust and turbulent air movements (Halász et al. 2014).

## Deposition

Airborne spores can be removed from the air in three key ways: sedimentation, washout, or impaction. Spores are heavier than air and so tend to sediment, but upward movements by convection and by turbulence impede this process. Gravity certainly determines the range of the spore dispersal, even after high-speed launches, but viscous drag from the air acts as a far greater brake, causing a rapid deceleration. The heavier (larger) spores settle faster than lighter (smaller) spores. Stokes' law describes the relationships between terminal velocity ($V_t$) and size of smooth spheres 1–50 μm in diameter in viscous fluids. The relevant equation is $V_t = 0.0121r$. $V_t$ is the terminal velocity (cm $s^{-1}$) and r, the spore radius (μm). The simplified equation is appropriate to determine the $V_t$ of many spores in the air. The sedimentation rates agree closely with Stokes' law for perfect spheres of unit density (1.0). Value of $V_t$ varies depending on spore shape (deviation from the ideal sphere), drag (by spore surface roughening), degree of hydration, and aggregation of spores (with trapped air between cells). The sedimentation rates for spores with unusual shapes are adjusted with correction factors.

The effect of the direction of the launch is insignificant on the distance that a spore is shot. The effect of gravity is indiscernible until the spore is slowed down by air viscosity. When spores are launched horizontally, their typical traveling path is a "Wile E. Coyote trajectory," similar to the tragic coyote featured in Warner Brothers cartoons falling off a cliff (Money and Fischer 2009).

Rain removes spores from air by impaction on raindrops, by capturing on cloud droplets, or even by forming cloud nuclei. Impaction of spores on the wet surfaces is increased by wind speed. In saturated air, spores gaining weight accelerate sedimentation (Weinhold 1955). Madelin and Johnson (1992) showed that spores had larger aerodynamic diameters when aerosolized at 95–98 % R.H. than the ones at 40 %.

In prolonged rain, droplets wash off the xerospore from air (Hamilton 1959; Ho et al. 1995; Katial et al. 1997; Fernández et al. 1998; Lim et al. 1998). Gloiospores (being wettable) become incorporated in the raindrops and spread as a film across a wettable surface or drip from a nonwettable one. While nonwettable spores covered with rodlets of hydrophobins remain on the surface of the raindrop, if this rolls across a nonwettable surface, such as a leaf cuticle, it will leave a trail of spores behind (Talbot 1997).

Dense vegetation efficiently facilitates spore removal from the air by both impaction and reducing air current speed to allow sedimentation. An individual tree can filter the 66–80 % of the airborne particulates passing through its canopy and amount of dust removed may reach several hundred kg (Kovács 1985). Tree leaves after rain can recuperate their original filtering capacity. The deposition of fluorescent powder dispersed as an aerosol on different plant leaves (*Ilex*, *Poa*, *Raphanus*, and *Solanum*) was studied by Hirst and Stedman (Hirst and Stedman 1971) in a small wind tunnel. Deposition was greatest on the small plant parts, such as petioles, stems, and longer spines of the holly leaves (*Ilex*). There was much less deposition on the larger blades of the leaves, particularly when these leaves were oriented nearly parallel to wind direction. Similar results were reported from experiments with fungal spores. Impaction is most efficient for large spores blown fast toward small objects. When airborne spores move toward an object (or vice versa), the air is deflected around the object and inclines to bring spores with it. The momentum (mass × velocity) of a spore inclines to take it along its current path for at least some distance. Three points are associated with this phenomenon: (a) larger spores have a better chance of impacting than smaller spores at any given air speed; (b) as the air speed increments, so gradually smaller spores can impact; and (c) as the receiving object size increments, so the air deflection is greater, and this decreases the probability of impaction (Carter 1965). Small fungal spores (4–5 μm in diameter) fail to impact on objects only 1 mm in diameter at a wind speed of 2 m $s^{-1}$ (the typical maximal wind speed in vegetation). Filtering efficiency was higher in *Rosa rugosa*, and the efficiencies of other studied plants are listed in a descending order: *Acer campestre*, *Carpinus betulus*, *Lonicera* sp., and *Ligustrum vulgare* (Kovács 1985).

Carter (1965) found that spore deposition of *Eutypa armeniacae* is high on thin stems of apricot. The spore clusters impacted best on the narrow leaf petioles of ca. 1–2 mm in diameter, less well on the thicker stems of young apricot, and worse on the broader leaf blades at all wind speeds. It might be considered that the impaction efficiencies were rather low, <3 % in all cases. Spores were present on the adaxial surface of leaves, when sedimentation occurred under gravity. Spores impacted on both abaxial and adaxial surfaces of the leaf surfaces when the impaction was due to airflow. Wet and sticky surfaces covered by rainwater, plant sap, honeydew, etc. act as excellent natural spore traps (Chamberlain and Chadwick 1972). Waxy cuticle and trichomes on leaf surfaces may enhance the trapping efficiency and assist retention of certain types of spores (Forster 1977), Hydrophobic characteristics of the leaves reduce the efficiency of wet deposition of spores (lotus effect). Spores incline to occur in preferred patterns associated with leaf anatomy. The most active zone of the trees for spore impaction and resurfacing is the top of the canopies of these woody plants (Aylor 1978).

## Resurfacing

A large number of spores could resuspend into the air when adhesion diminished by drying airflow. Not all spores that land become securely attached. Some are resuspended again by the agitation of strong wind or by rain splash. The demarcation of the resurfacing phase is important to introduce the concept of multiple-stage dispersal of spores. Observations of Carter (1965) in *Eutypa armeniacae* become the classic description of multiple-stage dispersal. This fungus is a pathogen of apricot trees and grapevines, naturally releases its spores as clusters of eight ascospores held together in mucilage—a relatively large propagule with sufficient momentum to impact on twigs at relatively low wind speeds. After impaction, the fugal pathogen depends on secondary spread by water, either rain or irrigation water, which wash off the mucilage and carries the separate ascospores down to any wound sites.

Ascospores of *Pleospora herbarum* discharged from the perithecia in wet weather can be rediscovered in the air samples several months later, during dust storms, when deposited spores are resuspended into the airstream. The possibility of resurfacing of spores is scarcely considered; however, such events may be hazardous and unexpected.

## Dispersal in Water

Almost two-thirds of Earth's surface is covered by oceans, rivers and streams and reservoirs which represent a substantial amount of water. Despite of this huge water mass, fungi with primary (*Chytridiomycetes*) or secondary adaption (aquatic hyphomycetes, yeasts) to aquatic environments (Ingold 1971; Jakucs and Vajna 2003; Wurzbacher et al. 2010) account for only less than 2 % of the overall described approximately 3000 taxa (Shearer et al. 2007). The main ontogenic stages of their whole dispersal in water are similar to that of nonaquatic fungi: spore production, liberation, transport, and deposition. The boundary layer is also present in water, due to the flow of water around solids, but such layers are much thicker. However, due to the different media, dispersal strategies are somewhat different from those of terrestrial (aerial) ones (Ingold 1971) presented in the previous section of this chapter. Furthermore, dispersal strategies are highly influenced by the two major types of the aquatic environment (Sigee 2005): standing (lentic systems) and flowing waters (lotic systems).

In flowing waters passive movements are predominantly unidirectional (Bärlocher 1992a; Hynes 1970). Particles, including fungal propagules, suspended in the water will be transported downstream for a certain distance before resettling. This process is referred to as "spiraling" (Minshall et al. 1985; Mulholland et al. 1985) and has a great significance for the dispersal pattern of fungi and also the availability of their substrates in such environments. What are the convincing mechanisms that permit aquatic organisms with little or no active movements to persist in a given river reach? If water current were the only effective mechanism, it would be inevitable for their gradual depletion and ultimate elimination from rivers. The explanation to their sustained presence in rivers must, thus, recline in processes that annul the effects of streamflow (Bärlocher 1992a).

Organic particles [litter, feces, precipitates of dissolved organic matter (Lock 1981; Suberkropp and Klug 1980)] transported by streamflow within the "spiraling" process provide degradable energy sources for saprobic fungi and also for other microorganisms attached to these particles. The major part of these particles derives from leaves (Abelho and Graca 1996; Grigg and Mulligan 1999) which represent the major energy source in streams (Kovács 2012). Before a leaf falls and blows into the stream, the litter is already colonized by terrestrial fungi and yeasts (Bärlocher 1992a) which may contribute to decomposition, but they decline steadily and are replaced by aquatic fungi, the main decomposer (Suberkropp 1992), aquatic hyphomycetes (Ingold 1942), Ingoldian fungi (Bärlocher 1992b), or amphibious hyphomycetes. Some studies concluded that the average distance that leaves travel from a starting point in streams was 200 m, and within 1000 m all leaves were generally entrained (Young et al. 1978). Shorter distances (19 and 91 m) were found by Prochazka et al. (1991). Since fungal asexual spores (conidia) of aquatic hyphomycetes are nonmotile, all of these pioneer studies clearly indicated the importance of this transporting pathway (water movements) in dispersal of fungal propagules attached to drifting matters, since conidia seem ill suited for longer-distance dispersal (Duarte et al. 2012). Independent conidia of aquatic hyphomycetes released into flowing water may be carried from a few hundred meters to a few kilometers (Thomas et al. 1990) and can maintain their ability to germinate for several days (Iqbal and Webster 1973; Sridhar and Bärlocher 1994). The distances that the drifting particles (with attached fungal propagules) travel are also relevant for microbial movements, strongly determining the possible dispersal of fungi (Bärlocher 1992a) at a local scale. However, leaves and their fragments quickly develop less suitable for fungal growth and reproduction (Bärlocher 1982) and compel the fungal species to passively and very rapidly move from substrate to substrate after leaves secede and drop (Bärlocher 1992a). Against the risk of being drifted from their target (substrate) and swept away from the favorable upper reaches of a stream, Bärlocher suggested that the shapes and sizes of conidia of aquatic hyphomycetes (Bärlocher 1992a) and being carried by leaf-eating animals, shredders [i.e., *Gammarus pulex* (Bärlocher 1981)], attempt to minimize this impact. Many species have branched spores (e.g., tetraradiate, consisting of four arms diverging from a point) or sigmoid/filiform spores (with a curvature lying in more than one plane), but ovoid or spherical shapes also occur in some species (Ingold 1942, 1971). In Webster's (Webster 1959) experiment, a horizontal water tunnel apparatus was used to determine the efficiency of impaction of some common spores of aquatic hyphomycetes. He found that the tetraradiate spores (e.g., *Articulospora tetracladia*, *Tetracladium marchalianum*) have higher trapping efficiency than those with longer, sigmoid (*Anguillospora longissima*) or other types of spores (*Heliscus lugdunensis*). Results of Dang et al. (2007) led to a similar conclusion in their microcosm experiment with three aquatic hyphomycetes (*Heliscus lugdunensis*, *Flagellospora curvula*, and *Tetrachaetum elegans*). In contrast, Cornut et al. (2014) found that leaf litter buried by streambed sediments (hyporheic zone) is colonized more efficiently by filiform spore species than by species with tetraradiate (branched) spores due to the filtering effect of sedimentary matrix. Conidia are not likely to have long-distance aquatic dispersal (Bärlocher 2009), since conidia are comparatively fragile and quickly lose their germinability even under favored conditions (Sridhar and Bärlocher 1994).

Several studies focused on the exchanges with the surrounding coastal/terrestrial habitats try to reveal more detailed image about the dispersal pathways of typical aquatic fungi (Bärlocher 1992a). The exchange between groundwater and streamwater (Hynes 1983) is well known and raises the question whether this hydrological pathway occurs in both directions. Kuehn and Koehn (1988) proved subterranean origin of terrestrial hyphomycetes and zygomycetous fungi, but not for aquatic hyphomycetes in a study on artesian. Although Bärlocher and Murdoch (Bärlocher and Murdoch 1989) demonstrated that multi-radiate and sigmoid conidia can penetrate quite deeply into the sediment, but in that form they did not reach the groundwater bases. Overall, it might be concluded that those aquatic fungal propagules which cannot overcome the currents of flowing waters (drifting to the lower reaches of the stream) also cannot be returned to the upper favorable reaches of a stream by the exchanges with currents of the groundwater.

It is widely accepted that occurrence of aquatic fungi is not restricted to the lotic systems itself. There must be some unexplored transport mechanism between flowing waters and terrestrial habitats, thereby offering reservoirs for these aquatic biotas where they may survive in terrestrial sexual state (Bärlocher 1992a). There are many records evidencing that many aquatic hyphomycetes can occur and survive in soils (Bandoni 1972; Park 1974; Waid 1954); in ephemeral aquatic micro-ecosystems, like tree holes (Gönczöl 1976; Gönczöl and Révay 2003; Karamchand and Sridhar 2008; Vass et al. unpublished data); on roofs of several buildings (Czeczuga and Orlowska 1999) and gutters (Gönczöl and Révay 2004); and even in the tree canopies (Czeczuga and Orlowska 1994; Sridhar 2009; Sridhar and Karamchand 2009). Thus, it seems that they survive in these habitats not as conidia rather as mycelia or dormant structures associated with plant detritus (Bärlocher 2009). Shearer (Shearer 1992) described the wood as a potential source of inoculum of aquatic hyphomycetes, but their transport mechanisms into the previously listed terrestrial habitats have not been explored yet. Even though, Selosse et al. (2008) in their letter have raised the idea of "flying from water to plants" by wind or aerosols as a potential process, but without aerobiological observations this hypothesis is not supported firmly. For tetraradiate spores, one potential explanation might be the aqueous film theory described by Bandoni (1974) and Bandoni and Koske (1974). It may explain the movement of spores in the water film on leaves and also on tree barks (Karamchand and Sridhar 2008). As Webster and Descals (1981) stated, one of the most interesting aspects of the aquatic fungi, especially of the aquatic hyphomycetes, is their successful worldwide distribution, which does not show any consistent inter- and intracontinental phylogeographic structure (Duarte et al. 2012). The observation of identical morphospecies on geologically young islands a long way from mainlands, such as the Hawaiian Islands (Ranzoni 1979), points to efficient long-distance transportation of viable inocula. Existence of unexplored pathway(s) has been the greatest challenge in the field of aquatic mycology, and to go beyond speculations molecular investigations are essential in the future. One possible explanation of the paradox of the worldwide distribution of freshwater fungi is that such fungi are not dispersed in their staurosporous form for long distances but rather with their teleomorphs. Regarding many aquatic hyphomycetes (anamorph) have a sexual state, teleomorph (Webster 1992). Many teleomorphs of Ingoldian fungi do not need to be submerged in water in order to discharge their spores, but do so freely in air. These airborne meiospores are penetrator types and permit dispersal over large distances (Bärlocher 2009).

The main difference of lentic systems (standing waters) to those discussed above is the lack of permanent, strong, unidirectional water currents. Regarding the spatial heterogeneity of standing waters, we can distinguish three main zones: profundal, littoral, and pelagic zones (Hutchinson 1967). Due to the high diversity of lakes, we cannot ultimately characterize these zones by main types of their energy sources (allochthonous or autochthonous), but there are habitats permanently or temporarily under dominant influence of allochthonous or in situ produced (primary production by algae) autochthonous material (Pieczynska 1990). In the former case, the main part of the external input consists of leaf litter like flowing waters in general. Therefore, leaves colonized by terrestrial fungi often enter lakes via wind and inflowing streams. Moreover, the high loads of airborne propagules by dry or wet deposition also occur (Smirnov 1964) (see subchapters above). Reaching a lake does not mean unambiguously that all transported species from lotic systems are able to continue their activities in standing waters but represent merely an additional energy source for local decomposer biota.

Pollen grains as allochthonous sources for lentic systems could be already colonized by pollen-degrading fungi in air or on bark tissues, but their entering the lakes results in a fungal replacement by aquatic fungi (like chytrids) (Skvarla and Anderegg 1972). Wurzbacher et al. (2014) provided a detailed analysis on pollen degradation by aquatic fungi in which the major fungal part consists of *Chytridiomycota* beside the three other phyla (Ascomycota, Basidiomycota, and Cryptomycota) with a majority of lower zoosporic fungi. The zoosporic fungi [*Chytridiomycota*, Oomycota, and plasmodiophorids; only *Chytridiomycota* being true fungi (Deacon 2006)] seem to have an aquatic evolutionary origin, developing motile spores with a posterior flagellum of the whiplash type which liberate themselves from the sporangium due to their own motility in water (Ingold 1971; Money and Fischer 2009) and dispersal capacity in the water column (Sparrow 1968). Zoospore-like motile form is also present in Myxogastria (myxomycetes). Depending on the environmental conditions, either a myxoflagellate or myxamoebae bud from the spore. Myxoflagellates develop two flagella. One is usually shorter than the other and occasionally only remnant. Myxamoebae move like amoebae—that is, crawling on the substrate. The flagella function for locomotion to assist to move close to food sources (Ingold 1971). Certain features of zoospores are important from the point of view of dispersal: their lifetime, swimming amplitude, and ability in the selection of suitable substrates. In average, zoosporic fungi may swim from several hours to several days (Ingold 1971; Deacon 2006) with a speed of about 0.25 cm $min^{-1}$ (Ingold 1971) to 1.0 cm $min^{-1}$ (Royle and Hickman 1964). However, Deacon (2006) mentioned lower rates (0.006 cm $min^{-1}$) for zoospores of Oomycota. These numbers allow estimating that these organisms may colonize a minimum of 3–4 m new environments during their lifetimes. Therefore, biological advantage of motility might focus on the mechanism of escape spores from the zoosporangium, and also there are some major advantages in selecting a substratum, but comparatively little benefit in actual dispersal. As Ingold thought, the zoospores of water molds are chemotactic based on the observation of *Saprolegnia* spp. (Fischer and Werner 1958), which are highly attracted to small quantities of amino acids and salts released by dead animals as their potential nutrient sources.

In the littoral zone of lakes, benthic community dominated by extended algal mats could be very beneficial for saprobic and parasitic fungi, and their detached, floating parts could be heavily colonized by fungi (Zopf 1884). Such biotic surfaces like algal mats are often used as microhabitats for predatory fungi to capture nematodes, rotifers, or tardigrades (Sommerstorff 1911).

According to available knowledge, we may conclude that dispersal of aquatic fungi in lakes is, similarly to the fungal dispersal strategies in lotic systems, restricted to habitats where decomposing materials are concentrated [in lakes it means littoral region (Pieczynska 1990)]. We may interpret this as "edge effect"—an increase of biota abundances in ecotonal habitats (di Castri et al. 1988).

Despite cell numbers or biomasses of aquatic fungi are much lower in the open water (Wurzbacher et al. 2010) than in littoral habitats, their role in matter and energy transport might be important and described as mycoloop (Kagami et al. 2007) of aquatic food web (Sigee 2005). The role of saprobic or parasitic fungi with zoospores establishes an association between filter-feeding zooplankton and large, "inedible" algae (e.g., *Asterionella*) (Kagami et al. 2004). In this case, the individual motility of zoosporic fungi allows fulfilling this ecological function in aquatic ecosystems along with the successful fungi distribution in the open water (pelagic zone). Therefore, there are important pathways in fungal dispersal carried out by other biotas, focusing on parasitic fungi.

Beyond freshwaters, in marine zones approximately 500 fungal species have been observed. Most of these species belong to Ascomycota or *Mastigomycotina* and species from Basidiomycota are represented with low numbers. Their classification as true aquatic (marine) fungi is strongly questionable. In groups of fungi occurring in marine zones, similar tetraradiate spores could be detected as in freshwaters (Jakucs and Vajna 2003), but in contrast, the ascospores often have flakes of wall material or mucilaginous appendages, which must be of functional significance influencing sinking properties as the shapes give high surface tension to stay afloat (Padisák et al. 2003; Rees 1980) in marine conditions (Deacon 2006). Most of them are saprobic, but parasitic and symbiotic species have also been found. For example, the spores of *Gloeosporidina cecidii* can easily colonize brown algae species of *Sargassum* (Kohlmeyer and Demoulin 1981), but the parasitic fungus (*Cytospora rhizophorae*) of red mangrove (Kohlmeyer and Kohlmeyer 1971; Wier et al. 2000) and the *Rhabdospora avicenniae* living on black mangrove (Kohlmeyer and Kohlmeyer 1971) are documented, and the mycological observation (Ananda et al. 1998) of fungi growing on different animal substrates (like shells, fish "bones") in coastal zones is also worth to mention. There are detected species in deep sea as well: *Allescheriella bathygena* in 1722 m and *Periconia abyssa* in 3975 and 5315 m of depths (Kohlmeyer 1977). Moreover, the surprisingly high fungal activities in deep-sea hydrothermal ecosystems and its composition (in which members are considered key organisms of terrestrial habitats) (Le Calvez et al. 2009) further confuse the patterns that we gained when trying generalize dispersal mechanisms of fungi. However, the ocean currents probably are the major transporting pathways in the case of microfungi as well as in marine systems.

Freshwater environments include all those sites where freshwater serves as the main external medium, either in the liquid or frozen state. Frozen aquatic environments have long been considered as microbiological deserts, but studies found that this is contrary (Sigee 2005). Ekelöf (1907) reported fungi from the Antarctica for the first time. The Antarctic subcontinent, for example, is now known to be rich in microorganisms (Vincent 1988). More than 20 species of microfungi including *Stachybotrys chartarum* had been isolated from freshwater on the King George Island, Antarctica (Kong and Qi 1991; Yu and Wang 1995). Most Antarctic environments get microbial propagules (including fungi) via air as result of spore traps showed and reports of the microflora in Antarctic snow and ice or at geothermal sites, with a high occurrence of cosmopolitans in most habitats (Pearce and Galand 2008; Ruisi et al. 2007). A majority of fungi reported from Antarctica are asexual microfungi including both endemic and indigenous ones (Ruisi et al. 2007). Besides the other biota, fungi are often locally abundant and interacting within highly structured communities. However, their dispersal patterns and origins have remained largely unexplored.

Dispersal of fungi in tap water via plumbing system is also an emerging issue. A wide variety of fungi is known to be common in wet indoor environments, as well as in the drinking water resources [i.e., (Hageskal et al. 2007)]. Formation of tenacious and massive black biofilms was occasionally observed at the water–air interphase of water taps and in associated habitats at several locations in Germany. *Exophiala lecanii-corni* is the dominant component of these biofilms. A retrograde route of contamination in case of *E. lecanii-corni* can be assumed (Heinrichs et al. 2013). Meanwhile in a German survey (Göttlich et al. 2002), the fungal flora was dominated by species of *Acremonium*, *Exophiala*, *Penicillium*, and particularly *Phialophora*. Some of them occurred throughout the entire drinking water system and constitute a resident fungal flora in water supplies. It has been also concluded (Short et al. 2011) that the plumbing systems serve as a significant environmental reservoir of human-pathogenic fungi, like *Fusarium. Fusarium oxysporum* f. sp. *cucumerinum* was detected with biofilm-forming capacity in a recent study as well (Peiqian et al. 2014). Picioreanu et al. (2001) considered the detachment process is due to internal stress created by moving liquid past the biofilm. Two biofilm detachment processes, sloughing (large-biomass-particle removal) and erosion (small-particle loss), were modeled. Simulations showed that erosion resulted in smoother biofilm surface, while sloughing made biofilm surface rougher. Under similar hydrodynamic conditions and biofilm strength, faster-growing biofilms lead to a quicker rate of detachment than slow-growing biofilms. The experimental data showed that detachment depends on both shear and microbial growth rates. Instability in biofilm accumulation and rapid biomass loss (sloughing) were triggered by high growth rates. High liquid shear, combined with low biomass growth rates, can avoid massive sloughing. Biofilm patches filled the entire cavity where they commence their growth, but they are not able to spread out the carrier peaks and to completely colonize the substrate. Although most of the knowledge of biofilm dispersal regards to bacteria, similar processes may distribute biofilm-forming fungi in indoor wet environments.

Apparently some pathogenic fungi, especially fusaria, are able to disperse more efficiently in other fluids than water as well, like sap inside living plants or animal blood. Some pathogenic fungi are introduced into plants through lesions in stems, twigs (Sieber et al. 1995), or roots. Small hyphal parts or spores might then be transported with the transpiration stream from root lesions to the leaves. *Fusarium oxysporum* infects healthy plants penetrating the plant's root tips, root wounds, or lateral roots by means of mycelia or by germinating spores. The mycelium grows intracellularly through the root cortex and into the xylem. Once in the xylem, the mycelium remains exclusively in the xylem vessels and produces microconidia. The microconidia enter into the sap stream and are transported upward where the flow of the sap stops and the microconidia germinate (Agrios 2006).

*Fusarium verticillioides* is a dimorphic fungus (Szécsi and Magyar 2011). In the nature or in axenic cultures at 25–30 °C, it has a mold-like appearance while changing to a yeastlike form in rich culture media at 37 °C. There are several reports that this fungus has been isolated from human and animal bodies associated with diseases (Nucci and Anaissie 2007). In many cases, fungemia—the presence of a fungus in the blood—is the only manifestation of the infection. The yeastlike form diffuses more readily in human and animal bodies through blood and lymph systems than mold-like cells (Wang et al. 1999).

## Dispersal by Plants and Substrate Particles

Plant seeds are efficiently dispersed. Therefore fungal particles on or in the seeds will be distributed with equal effectiveness. Seed-borne fungi are of particular interest to plant pathologists, as contaminated seeds are responsible for some plant disease outbreaks. Humans also have a huge influence over seed dispersal through a number of effective and largely generalist dispersal vectors (Auffret et al. 2014).

Smuts, e.g., *Tilletia* and *Ustilago* species, infect the developing caryopsis ("seed") of grasses (Poaceae). Sori (bunt balls) are formed in the ovary of the host plants, where hyphae of the pathogen replace the tissue of the young ovary, which is then converted into a powdery, brownish-black mass of teliospores. Each diseased "seed" (bunted kernel) can produce thousands to millions of teliospores. Teliospores are often released when bunt balls are ruptured during harvest. However, some of the bunt balls remain intact and are found among the harvested grains. Teliospores are released from these grains during maintenance or processing. Interestingly, teliospores of *Tilletia caries* are combustible, so explosions and fires result from their ignition by sparks from mechanical harvesters (Wiese 1998). Such teliospores are easily dispersed by wind. If impacting on the host plant and reaching the sterigma of a healthy flower, the teliospores germinate and infect the ovary. Because large-scale seed treatments as well as resistant cultivars are widely applied, some smut species are less damaging nowadays than 50 years ago (Wiese 1998).

Transmission of endophytes is through seeds, or other vegetative propagules are also known (see subchapter below regarding man-made dispersal).

Orchids have been found to associate with saprobic basidiomycetes of several lineages, collectively named *Rhizoctonia* (as anamorph). These fungi may strongly impact seed germination patterns of orchids (Oros et al. 2014). Orchid seedlings are fully dependent on nutrients supplied by a mycorrhizal fungus. Its recruitment success will be strongly influenced by the availability of a suitable fungal strain. Mycorrhizal fungi have an aggregated distribution within the habitats (Jersáková and Malinová 2007). However, in certain circumstances the mycorrhizal fungi are likely to be distributed independently of the orchids (Feuerherdt et al. 2005). Co-distribution model of the mycorrhizal fungi (*Rhizoctonia solani*) and orchid species (i.e., *Pterostylis acuminata*) was observed in a laboratory experiment by Perkins and McGee (1995).

Spores and mycelia adhering on detached particles of their substrate can be transported as "hitchhikers." Host plants or its tissues, fruits, etc. usually contain a wide variety of free spores of different species, including pathogens. *Puccinia psidii* (one of the most important pathogens of *Eucalyptus* and other Myrtaceae) can be spread mainly by windblown fungal spores, but infected rooted cuttings, seedlings, pollen, and other host tissues increase the risk of its introduction to rust-free countries as well as of spreading new pathogen lineages (Lana et al. 2012). Strong disturbance increments the drift of colonized substrate particulates as vectors (Pasanen et al. 1991). Plant debris carrying *F. verticillioides* are abundantly aerosolized during the operation of harvesters (Magyar et al. 2011). Once airborne, large particles may be carried by the wind to a distance of 300–400 km from the source (Ooka and Kommendahl 1977).

Joint dispersal is prevalent in diverse symbioses and a predominantly common reproductive mode in lichens (Wornik and Grube 2010). Lichenized algae can be dispersed together with their fungal partners in asexually produced organs, either produced directly on the vegetative thallus or in specialized regions. It is an efficient approach to retain successful associations and to circumvent low symbiont availability.

## Dispersal by Animals

In most cases, the dispersal agents are animals, from arthropods and gastropods to mammals and birds (Castellano et al. 2004). There are numerous fungus–vector associations, ranging from almost incidental associations to highly evolved mutualism (Deacon 2006). Associations between fungi and insects are among the most species rich, diverse, and complex in terrestrial ecosystems. Spores dispersed by insects are recognized in many groups of fungi, including Ascomycota, Basidiomycota, anamorphic fungi, and zygomycetous fungi (Kendrick 1985; Ingold 1953), as well as in the myxomycetes (Stephenson and Stempen 1994). Nevertheless, the role of fungivorous insects as spore vectors has been poorly documented (Schigel 2012). The adaptations in various fungal groups as a result of selection for arthropod dispersal are discussed in Abott's article (Abbott 2002). The significance of insects as vectors of fungal phytopathogens has frequently been undervalued (Agrios 1980). Compared to the general postulation that wind transport is the principal mode of dispersal of fungi, Kluth et al. (2002) showed that the occurrence of spores dispersed by insects may also effectively spread the pathogen to isolated weed stands (alternative host) (Morrison et al. 1998). The unique adaptations and dispersal abilities of both fungi and insects result in a variety of interactions (Schigel 2012). Interactions between animals and fungi, such as attraction, mycophagy, mutualism, and parasitism, are occasionally documented, but some of them have high ecological or economic impact. Specialized entomophilic species are far less common in the kingdom of *Fungi* than in plants. Spores with adhesive or echinate surface can spontaneously attach onto or trapped into animal hairs.

Many lichens are dispersed by animals since soredia adhere electrostatically to the cuticle of invertebrates [insects, spiders, mites; (Stubbs 1995)] or to the feathers, fur, and extremities of vertebrates. Short- and long-distance transport is also provided by animals which use lichen fragments to camouflage either their own body (insects) or their nests [birds, squirrels, (Gressitt 1965; Richardson 1975; Gerson and Seaward 1977; Scharf 1978; Seyd and Seaward 1984; Brodo et al. 2001; Allgaier 2007)].

Masses of spores and sticky sap droplet exuded form pycnidial and perithecial cirrhi or often on the top of an elongated stalk are carried away by arthropods passing through colonies. Microfungi such as *Gliocladium*, *Graphium*, *Leptographium*, *Myrothecium*, *Pesotum*, *Stilbella*, and *Stachybotrys* develop complex conidiophores and utilize this dispersal mode (Ingold 1953; Abbott 2000; Seifert 1985; Upadhyay 1981; Wingfield et al. 1993). These relatively large, complex structures develop vertically from the substrates and are tall enough to hitchhike onto large insects moving over the surface.

Other microfungi, including *Acremonium*, *Fusarium*, *Gliomastix*, *Trichoderma*, and *Verticillium*, produce large numbers of droplets at the apex of conidiophores (Ingold 1953; Carmichael et al. 1980). The conidiophores are developed in different orientations throughout the mycelium and are particularly efficient to touch small insects traveling through a mycelial mass. The genus *Cephalotrichum* develops its asexual spores in a dry head at the apex of a synnema. These synnemata are frequently up to 1 mm high and are produced at a right angle to the substrate surface. The fused hyphae of the synnemata offer resilience and spring back into the upright position when gently manipulated in the laboratory (Abbott 2000). As the insects are in motion through a sporulating colony resembling miniature forest and brush against the synnemata, small clouds of conidia are released, effectively dusting the insect with conidia. This differs from the "paintbrush" method used by the slimy-spored synnemata of fungi, such as *Graphium*. Some myxomycetes (slime molds) may develop spores in similar stalked structures (e.g., *Stemonitis*) and are dispersed by tiny slime mold beetles (e.g., *Anisotoma*, *Agathidium*), which feed on the fructifications of the slime molds (Stephenson and Stempen 1994). Ing (1967) reports that woodlice distribute myxomycete spores. The sporocarps of *Enteridium*, *Tubifera*, and *Fuligo* commonly provide habitat for beetles and several other biotas (members of Anisotomidae, Leiodidae, or Agathidiidae). Their spores can be found in the fecal pellets of these animals. Therefore, they play a major role in the dispersal mechanism of myxomycete fungi. A number of fungi have developed a strategy to attract animals when spores are mature and lead the animals to consume the spores. The attractants are volatile organic compounds, including pheromones (Claus et al. 1981).

Komonen (2008) studied the abilities of Ciidae, the minute tree fungus beetles to colonize *Trametes*, and the results show that the beetles are able to fly up to 1.5 km toward the odors of fruit bodies of *Trametes*. The fungal spores are often covered by sugary matter with attractive odors to attract insects (Webster 1980). For *Phallus* and *Mutinus*, a strong fetid odor is produced to attract flies to feed on the glebal surface of the fungi. The basidiospores adhere to the legs and bodies of the flies, and the insects may deplete the entire slimy basidiospore layer in several hours (Abbott unpublished). The spores are transported to adjacent sites by the flies and are excreted, relatively intact (Ingold 1965). The insects are vectors for some fungi as they carry spores in the wound site. Some fungi from leaves eaten by invertebrates survive passage through the digestive tract (Bärlocher 1981). Endophytic fungi colonize plant tissues without causing symptoms of disease and live cryptically in all higher plants. *Endophytes* which are once deposited on the leaf surface might easily infect the plant through openings that are caused by herbivores. There are indications that endophytes can be dispersed by herbivores. For instance, when the infected herbivore migrates to new feeding places, endophytes such as the entomopathogenic *Beauveria bassiana* could be transferred (Vega et al. 2008). Grasshoppers or other phytophagous insects could serve as an agent for the dispersal of non-grass fungal endophytes (e.g., *Colletotrichum gloeosporioides*) in plant communities such as tropical forests. It was shown that endophytic fungi can pass the gut of grasshoppers without being destroyed and can be dispersed in this way (Devarajan and Suryanarayanan 2006).

In plant-pathogenic fungi, insect transmission of spores is extensively studied. It is necessary to possess some knowledge of vector behavior for reliable epidemiological predictions (Dye 1986). For instance, patchy distribution of hosts and aggregation of vectors within patches cause an increase in parasite's basic reproductive rate which is predicted by random host–vector contact models. Conditions for steady coexistence between parasites and hosts, as well as the equilibrium frequency of infection in host populations, are dependent on the degree of feeding preference of vectors for either healthy or diseased hosts. Different vector species may exhibit different efficacies of transmission due to differences in morphology or behavior (Webber 1990). Therefore, in order to better understand the epidemiological consequence of the dispersal of a vector-borne pathogenic fungus, it is important to investigate the mechanisms of transmission efficiency and the interactions among vectors, pathogens, and hosts (Shykoff and Bucheli 1995).

*Microbotryum violaceum* (Pers.) G. Deml and Oberw. is a causative agent of the anther smut disease common in members of the pink family (Caryophyllaceae). Flower visitors that usually serve to pollinate flowers can transfer infectious spores from diseased to healthy plants (Fig. 14.6a). The female flower of the infected dioecious plant develops anther-like structures filled with spores instead of pollen grains.

**Fig. 14.6** *Silene acaulis* infected by *Microbotryum violaceum*. (**a**) *Purple spores* are produced in anthers (photo courtesy from Michael Hood). (**b**) Insect-attracting pseudoflowers on *Arabis* sp., caused by *Puccinia monoica* (photo courtesy from Michael Wood)

Flowers of infected plants last significantly longer than those of healthy plants, probably because the infection strengthened floral organs, such as the flower base and the anther filaments (Uchida et al. 2003). Pollinators prefer plants with large floral displays and also prefer males to females and healthy to diseased plants (Shykoff and Bucheli 1995). Male plants consistently produced nectar with higher sugar concentration, thereby offering higher-quality floral rewards than either females or diseased plants. It was suggested that more attractive plants may be predisposed to infection since pollinating insects also serve as vectors for this fungal disease. In a field experiment, those plants that became infected produce larger numbers of flowers than those healthy individuals (Thrall and Jarosz 1994). Pollinators visit more healthy plants after an "accidental" visit to a diseased plant than they might if visiting plants at random (Real and Rathke 1991).

*Puccinia monoica* Arthur is a rust fungus, which inhibits flowering in its host plant, *Arabis*. The fungus causes the plant to produce pseudoflowers, which attract insects to spread the rust spores. These pseudoflowers attract insects not only by visual means but also by producing a pungent fragrance and exuding a sugar-rich solution (Raguso and Roy 1998).

The dispersal of plant pathogens by adult shore flies was found to be aerial vectors by an experiment for three microfungal pathogens: *Fusarium oxysporum* f. sp. *basilica*, *Thielaviopsis basicola*, and *Verticillium dahliae*. Adult shore flies are attracted to sporulating cultures and infected plant tissues. Shore flies acquired fungal conidia and other structures both by feeding and physical contact. The minimum acquisition time for soilborne fungi was 10–20 min. Acquisition incremented with time to reach 100 % frass deposits infestation following 2 h of acquisition. Pathogens dispersed by adult shore flies were fast over time at 2.21 $cm^2$/h/insect. The dispersal area by adult shore flies incremented with the lengthening in exposure time (El Hamalawi 2008).

*Claviceps* conidia were found on a range of insects, including moths, flies, leafhoppers, *Orius* spp., beetles, midge flies, head bugs, hymenopterous insects, and thrips. Such conidia are carried by up to 100 % of moths and 75 % of flies collected from some fields (Butler et al. 2001). However, this fungus uses an arsenal of strategies to provide successful spore dispersal. The anamorph of these species (*Sphacelia*) produces three different types of single-celled, hyaline spores (micro-, macro-, secondary conidia). An ergot kernel develops when a spore of *Claviceps* sp. infects a floret of flowering grass or cereal. The infection process is similar to a pollen grain growing into an ovary during fertilization. Infection requires that the spores of the fungal pathogen have access to the stigma, and as a result plants infected by *Claviceps* are primarily outcrossing species with open flowers. The growing mycelium subsequently destroys the host ovary and connects with the vascular bundle originally developed for seed nutrition. The first stage of ergot infection develops a white soft tissue (known as sphacelia) exuding sugary honeydew, which contains millions of macro- and microconidia (Fig. 14.7). Macroconidia are immersed in a nearly clear sugary liquid which takes on an opaque, yellow-brown, or orange to pink color because of the conidia. The syrup-like honeydew is exuded onto the surface of the sphacelia and continues to drip down across the sorghum head and onto other plant parts and the soil surface. The macroconidia have limited capacity to spread because of their occurrence in a sticky liquid environment. When exposed to high relative humidity (>90 %), the hygroscopic honeydew droplets acquire an increased water content, which stimulates germination of the macroconidia from which germ tubes extend and penetrate the honeydew surface that function as sterigmata, ending in apical secondary conidia that are easily disseminated by wind (long-distance spread) and/or fall onto soil where secondary conidia can also infect plants and act as primary inoculum for disease initiation. Moreover, honeydew containing conidia may adhere to farm personnel and equipment leading to the dispersal of the fungi from field to field. Honeydew in contaminated seed lots is also easily transferred, causing rapid uncontrollable spread. Although insects as vector of the conidia of *C. purpurea* is known, they may not play a significant role as water splash or wind in spreading sorghum ergot. The teleomorphs of *Claviceps* species produce long, filiform ascospores, which utilize passive dispersal pathways by wind and rain (Bandyopadhyay et al. 1998).

*Thrips obscuratus* is capable of carrying conidia of *Botrytis cinerea* on its body. Adult thrips artificially contaminated with *B. cinerea* had most conidia on the head, thorax, legs, and the abdominal distal segments; few were found on the wings. The conidia were observed most frequently in sculptured areas and intersegmental regions or trapped under setae, but very few on the smooth areas. This distribution pattern suggests that adhesion is mechanical. Up to 17 % of adult thrips sampled were naturally contaminated by the fungus, and up to 12 propagules per contaminated thrips were found. Field infestation of kiwifruit by *T. obscuratus* was shown to increase the susceptibility of kiwifruit petals to *B. cinerea* (Fermaud and Gaunt 1995). Fungal pathogens of insects (e.g., Entomophthorales) have also adapted to dispersal from one individual to others within an insect community (Pirozynski and Hawksowrth 1988)

**Fig. 14.7** (**a**) High severity of ergot (*Claviceps africana*) with abundant production of honeydew containing large amounts of conidia. Opacity of honeydew is due to the high content of macroconidia. (**b**) Secondary conidia production as the whitish area at the upper ends of the ergot droplets. (**c**) White secondary sporulation on the surface of honeydew that dripped onto the sorghum leaf from sphacelia in the sorghum head above. (**d**) Sporulation of *C. africana* on soil surface where honeydew dripped. Courtesy from G. Odvody. Reproduced with permission from APS Feature (Odvody et al. 1998)

Some birds and mammals (mouse and shrew) are also proved to be vectors of plant-pathogenic fungi, e.g., in transporting *Cryphonectria parasitica* (chestnut blight pathogen) in mixed hardwood forests (Scharf and DePalma 1981). Birds are visually attracted to some fungi. Mycophagy by birds is also observed many times and well reviewed (Bailey 1904; Simpson 1998, 2000). However, Simpson (2000) marveled the reason why the fungi, regarding to its wide availability and nutrient benefits

**Fig. 14.8** (**a**) Douglas squirrel (*Tamiasciurus douglasii*) feeding on and transporting a truffle (photo courtesy from Bernie Krausse). (**b**) *Paurocotylis pila* imitating fruits of *Podocarpus* spp. that are consumed by birds (photo courtesy from Clive Shirley)

(Cork and Kenagy 1989), are not utilized by more birds than observed in total. In New Zealand, which lacks native mammals, birds appear to be important vectors of sequestrate fungi. *Paurocotylis pila* develops a scarlet peridium. As ascomata of this fungus enlarge, the ascomata are elevated to the surface of the humus, occasionally detaching completely and positing loose on the surface. Their size and color imitate fruits of nearby *Podocarpus* spp. and other plants that are consumed by birds (Fig. 14.8b). In addition, *P. pila* matures at the same time as the fruits of *Podocarpus*. *Paurocotylis* ascomata on the forest floor among podocarp fruits are almost surely ingested by birds (Castellano et al. 1989). *Picoa lefebvrei* in deserts of the Arabian Peninsula and North Africa produces different visual signals. Several to greater than a dozen small ascomata congregate to form a distinct hump on an otherwise flat desert floor. Birds find these humps, scrape away the covering soil, and ingest the ascomata (Alsheikh and Trappe 1983). All sterile fungal tissues are digested, but the ascospores pass through the digestive systems undamaged. As feces containing ascospores deteriorate with age, the ascospores are released into the soil and potentially contact feeder roots of receptive mycorrhizal hosts (Trappe and Maser 1977). According to the first two present authors' unpublished observations, the songbirds have high importance in dispersal of tree bark-inhabiting and plant-pathogenic fungal spores and mycelia by their movements (Vass and Magyar unpublished).

## *Hypogeous Fungi*

Besides insects and other arthropods, mammals are also often utilized, especially by hypogeous fungi to move their spores from the site of growth and production to new substrata for colonization. Such fungi lack mechanisms of spore discharge to the air (Fogel and Trappe 1978). Loss of forcible discharge of spores to air must be accompanied by mutations that adapt sequestrate fungi to other spore dispersal tactics. Spore dispersal of hypogeous fungi including gourmet fungi, truffles, is dependent exclusively on animals. The role of mycophagous small mammals to disperse hypogeous fungi is well established (Urban et al. 2012).

Such fungi (and other macrofungi) are part of the diet of small rodents (Rodentia), such as mice (Murinae, Muridae, Myomorpha), dormice (Gliridae, Sciuromorpha), voles (Arvicolinae, Cricetidae, Myomorpha), and squirrels (Sciuridae, Sciuromorpha) (Maser et al. 2008) and insectivorous shrews (*Sorex* spp., Soricidae, Eulipotyphla) frequently feed on hypogeous fungi (Kataržytė and Kutorga 2011; Schickmann et al. 2012). However, at present no information is available on whether spore dispersal by small mammals contributing to productivity is analogous to the relevance of pollination in fruit orchards (Urban et al. 2012). Mycophagy can also affect fungal diversity within their home ranges by making sure the unremitting and efficient dispersal of spores from one site to others. Movement of bush rats (*Rattus fuscipes*) transports fungal spores among different fungal communities. These animals have the potential to affect the structures of vegetation communities via spore dispersal (Vernes and Dunn 2009). The significance of small mammals as consumers and dispersal agents of mycorrhizal fungal spores in tropical and temperate ecosystems had been reported in several studies. A study strongly suggests that the subterranean rodent *Ctenomys* cf. *knighti* may play an important role as a dispersal agent of arbuscular mycorrhizae and pigmented septate endophytic fungi (Fracchia et al. 2011). Rodent dispersal of fungal spores also promotes seedling establishment away from mycorrhizal networks in host plants (Frank et al. 2009).

Mycophagous mammals excavate and consume fruiting bodies of hypogeous ectomycorrhizal fungi and discharge excrement which contains numerous spores. However, the spores passing through vertebrates' digestive systems can influence the viability and activity of the spores consumed (Schickmann et al. 2012; Castillo-Guevara et al. 2011, 2012). Passage through the digestive system of flying squirrels (*Glaucomys sabrinus*) may improve germination and inoculation potential of spores, while active mycelia in forest soils may be the most important and efficient way for seedlings to develop mycorrhizae under natural environments (Caldwell et al. 2005).

The hypogeous and epigeous fungi ingested by red squirrels (*Sciurus vulgaris*) in subalpine conifer forests in the Alps were studied to determine the presence and taxa of fungal spores in excrement samples. Almost all live-trapped squirrels had fed on fruit bodies of hypogeous fungi in summer and fall, but only some red squirrels had ingested epigeous fungi, *Boletus* and/or *Laccaria*. It seems that fruit bodies of hypogeous and, to a lesser extent, epigeous fungi provide an important seasonal food resource for red squirrels in conifer forests in the Alps. Squirrels which have a large home-range size and long dispersal distances perhaps play a key role to disperse spores for hypogeous fungi (Bertolino et al. 2004; Teron and Hutchison 2013, Fig. 14.8a).

Piattoni et al. (2012) investigated the significant role of wild boars in dissemination of hypogeous fungal spores. They have found poor abundance of spores in boars' feces which suggested that the wild boar can be considered an opportunistic mycophagist, but due to the movements of wild boar during seasonal migrations, wild boar could have an important role though in truffle long-distance dispersal.

Mycophagists can indirectly affect vegetation succession by dispersing spores and other reproductive structures of mycorrhizal fungi, thus helping the distribution and regeneration of mycorrhizal plant species (Bruns 1995; Wiemken and Boller 2006). Maser et al. (2008) postulate that growth, regeneration, and adaptation of the mycorrhizal fungi–tree network would be greatly impaired, if not impossible, without dispersal of spores of hypogeous fungi by animals. The distributions of spores of microfungi were studied in soil litter and on the fur of small mammals (*Sorex araneus, S. caecutiens, S. minutus*, and *Clethrionomys glareolus*). There are 156 species of microfungi reported on the animal fur. Fungal spores were observed to attach to hairs with mucus or warty walls under electronic microscopy. The composition of spores isolated from the fur of the mammals and the soil litter was significantly different and had a seasonal variation. The microfungi peculiar to the fur were differentiated. All the studied species of small mammals are different in the composition of microfungi on their fur. The spore composition can be used as a marker for studying ecological niches of animals and fungal dispersal. The differences in the spore compositions on the fur increment with the distance between habitats of the animals (Shchipanov et al. 2006).

Transportation of an immense number of *Gorgomyces* by nematodes in soil litter was observed using apparently specialized mucilaginous appendages on the spores [(Gönczöl and Révay 1985), Fig. 14.13].

A number of Basidiomycota produce resupinate (crust-like) basidiomata buried in the soil. The basidiospores of these fungi are actively discharged, but it seems that they are often not well adapted for aerial dispersal. *Tomenlella sublilacina*, a widespread ectomycorrhizal fungus, sporulates in the soil organic horizon and can establish from the spore bank in a short time following disturbance. Gut contents of centipede and feces of centipede and salamander contained many apparently intact spores of this fungus. These results showed the potential for *T. sublilacina* spore dispersal by arthropods (mites, springtails, millipedes, beetles, fly larvae) and their predators (centipedes, salamanders) in soil food webs and might assist to elucidate the widespread distribution of this fungus. It is possible that this is a common mechanism of dispersal for fungi developing resupinate basidiomata in the soils, signifying a necessity to better understanding of the linkages of soil food webs and spore dispersal (Lilleskov and Bruns 2005) by arthropods and vertebrate predators.

According to their scanning electron microscopic (SEM) studies, the role of some beetle species (*Leiodes cinnamomea, Agaricophagus cephalotes, Colenis immunda, Agaricophagus reitteri*) in spreading of the fungus, burgundy truffles (*Tuber uncinatum*), was documented. Large numbers of fungal spores were found adhering to the underparts and legs of adult beetles feeding on truffles. Additionally, SEM revealed no spores on flies hatched from the fruit bodies of truffles, thus excluding their role in spreading the fungus (Bratek et al. 2010).

Each sequestrate (truffle-like) species produces its unique range of aromas, often a mixture of a number of compounds (Marin et al. 1984). Immature fruiting bodies have little or indistinctive odor. When spores start to mature, the attractant compounds are developed, and as more spores mature, the aroma increments in pungency and intensity (Trappe and Maser 1977).

The results of Varga and Naár (2002) supported the assumption that Collembola species have an important role in the distribution of fungi living in bryophytes as well as in the distribution of the ones living in the soil. According to the results of Dromph (2001) who had observed the dispersal mechanism of insect-pathogenic fungi by three different Collembola species (*Folsomia fimetaria*, *Hypogastrura assimilis*, and *Proisotoma minuta*), the fungal propagules in the feces of their carrier species could preserve their viability. Thus, their dissemination is supported by the consumer (Collembola) biota, establishing these fungi in new habitats (Williets et al. 1989).

## *Mutualism*

Mutualistic relations between fungi and insects also exist. Leaf-cutter ants (e.g., *Atta* and *Acromyrmex*) cultivate fungus gardens in subtropical and tropical Americas. The ants carry and maintain a selected fungal species (e.g., *Leucoagaricus*) to inoculate the piles of harvested leaf pieces to provide a food source for their larvae (Fisher et al. 1994; Wheeler 1907). Other fungi (e.g., *Termitomyces*) are associated with termites (e.g., *Termes*) in Africa and Asia (Wheeler 1907). *Ophiocordyceps* species known as "zombie ant fungi" control the ant hosts by inducing a behavior of anchoring biting (Fig. 14.9; Hughes et al. 2011). The fungi emit a mixture of behavior-controlling chemicals when encountering the brain of its natural target host, but not when infecting other ant species (de Bekker et al. 2014). These fungi infect many insects, and the species that infect ants induce hosts to die attached by their mandibles to plant material to provide a platform from which the fungus can

**Fig. 14.9** The "zombie ant" dispersing pathogenic *Ophiocordyceps* spores. Photo: David Hughes, with permission

develop and actively discharge spores to infect other ants. The jet ant, *Lasius fuliginosus*, cultivates *Cladosporium myrmecophilum*, which functions to provide stability to the walls of its carton nest (Fig. 14.10). The spores of this fungus (and those of many other species) are carried inside the mouth of the ant and ingestion is prevented by a filter-like organ (Magyar and Babinszkyné Nagy unpublished).

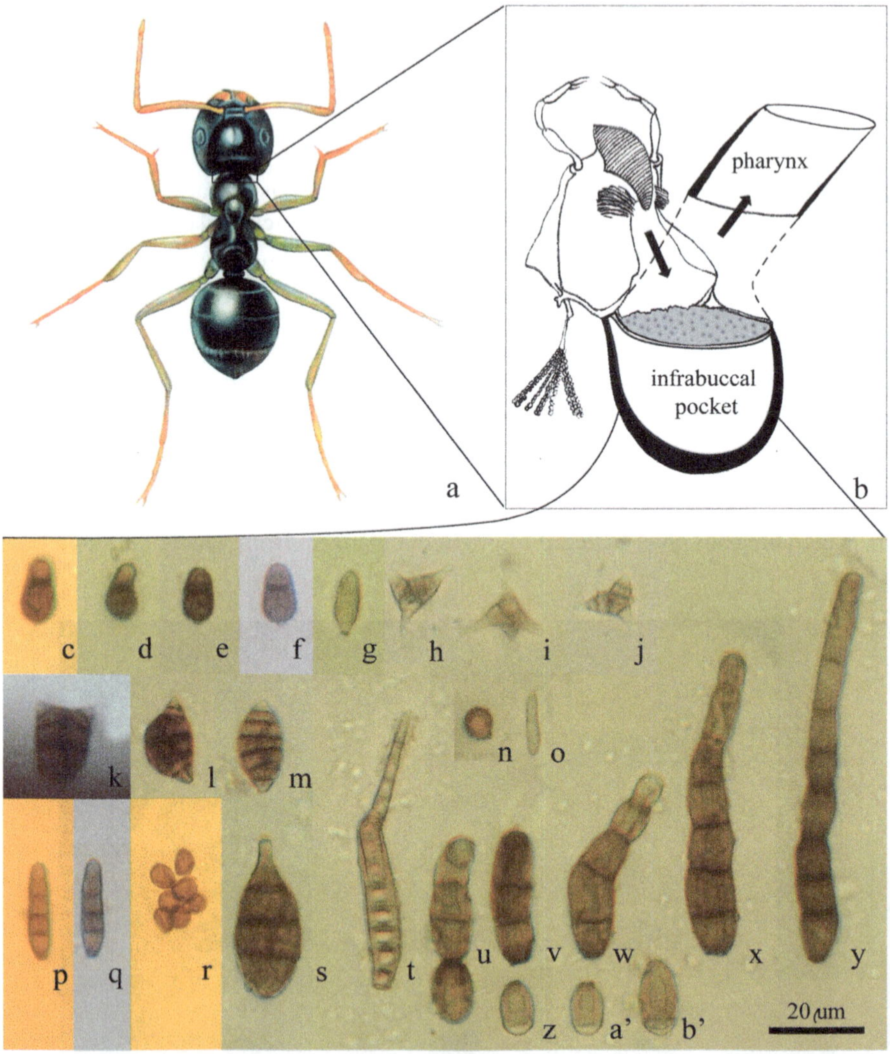


**Fig. 14.10** a: Jet ant (*Lasius fuliginosus*, picture courtesy from Kenneth Ervik), b: infrabuccal pocket of the ant where fungal spores are transported [redrawn after Gotwald (1969)], c–b': spores found in the infrabuccal pocket (Magyar and Babinszkyné Nagy unpublished). c–f: *Cladosporium myrmecophilum* (the ant's carton nest fungus), g: *Cladosporium* sp., h–j: *Oncopodiella guamensis*, k–m: *Oncopodiella trigonella*, n: unknown basidiospore, o–r: unknown species, s: *Alternaria* sp., t: *Ellisembia* sp., u–b': *Corynespora* sp.

*Xyleborus* spp. (ambrosia beetles), bark borers, cultivate *Ambrosiella* spp., the ambrosia fungi to feed their larvae and adults. The ambrosia fungi are transported in mycangia (specialized pockets) and are an essential part of beetle brood galleries (Wheeler 1907; Cassar and Blackwell 1996). Most blue-stain fungi or lumber molds (*Ophiostomatales*) are well documented to be dispersed by insects, and such dispersals associate with bark or ambrosia beetles. Ophiostomatoid fungi frequently colonize the wood in the galleries of bark beetles. Ascospores are developed in wet spore masses at the apices of the perithecial necks in *Ophiostoma*, *Ceratocystis*, and *Sphaeronaemella*. The long-necked ascomata and long-stalked conidiophores stick out into the insect passing routes and efficiently force the insects to touch the spore masses as they pass through the restricted spaces and to pick up spores (Upadhyay 1981; Wingfield et al. 1993).

Bark beetles, *Dendroctonus* and *Ips*, are well documented for their roles in dispersal of the ascospores and the conidia of blue-stain fungi (e.g., *Leptographium* and *Pesotum*) (Upadhyay 1981; Wingfield et al. 1993). One classic example is the dispersal of Dutch elm disease caused by *Ophiostoma ulmi* and *O. novo-ulmi* by specialized bark beetle vectors (*Scolytus* and *Hylurgopinus*). These pathogens enter the plant via wounds chewed by bark beetles and then translocated in the xylem vessels by developing in a yeastlike budding phase. It results in reactions in the xylem vessels to lead to obstruction and death of part or all of the xylem. The disease cycle commences when pathogen-carrying beetles emerge from the bark of dead or dying elms in early spring, fly to adjacent healthy trees, and eat the bark of the young shoots. The bark beetles damage the xylem during feeding, so to infect the elms with the pathogens. Then the fungal pathogen spreads in the xylem, resulting in the death of the whole tree or some of its major branches. The bark of the newly died elms is then targeted by the female beetles to oviposit. The female beetle bores into the inner bark and chews out a channel, oviposits along its length of the bark beetle gallery. The eggs hatch and the young larvae make a series of radiating channels by feeding prior to pupating for overwintering. At the same time, the pathogen develops from the xylem into the bark and produces spores in the tunnels. In the following spring, the adult beetles emerge from the pupae and are covered by spores. The adult beetles fly away from the bark and search for new elms. The disease cycle repeats.

The southern pine beetle (*Dendroctonus frontalis*) has two mycangia, separated by a sclerotized mycangial bridge, suggesting that each mycangium functions independently. Mycangia are surrounded by abundant tracheoles connecting the structures to outside via openings within the prothorax. It was hypothesized that these openings may play roles in determining the species of fungi entering to and growing in the mycangium (Fig. 14.11) (Yuceer et al. 2011).

It has been reported that myxomycete spores are stored in the pits in the mandibles of Sphindidae, an entirely myxomycophagous family, and the venter of slime mold-feeding latridiid species.

There is a hypothesis that a loose mutualistic association is present between dust mites and fungi. It is beneficial to the mites for feeding on animal (or human) skin scales colonized by fungi, and in return it benefits the fungi since spores are dis-

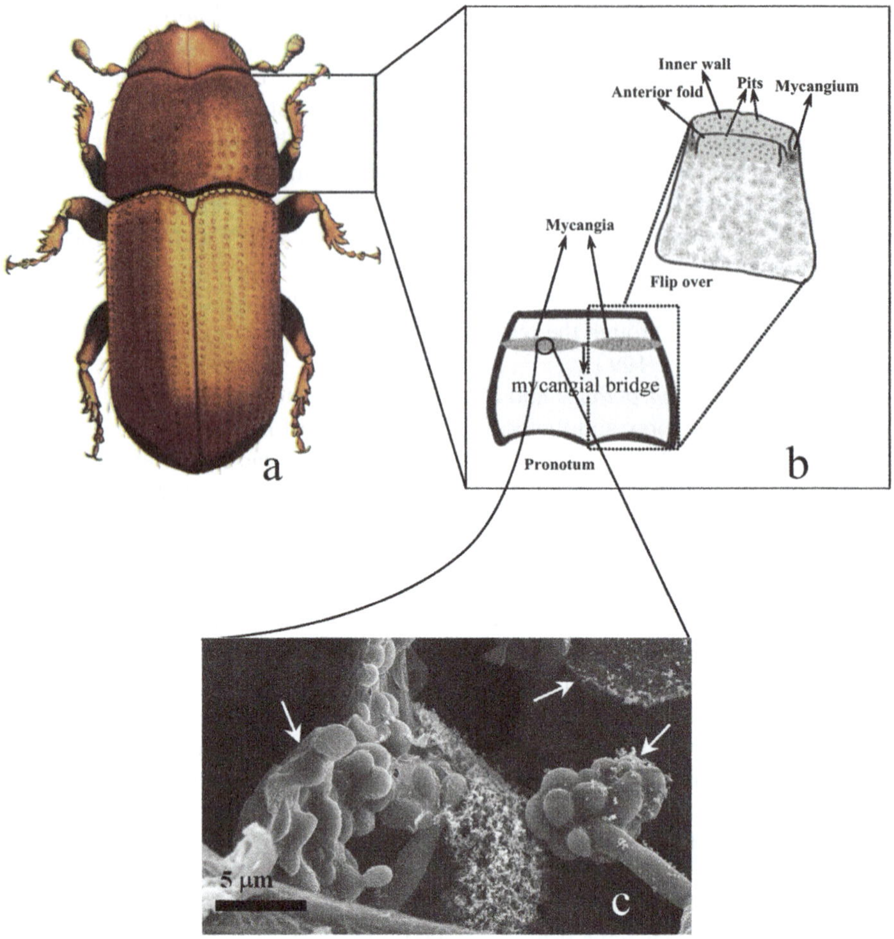


**Fig. 14.11** (**a**) The southern pine beetle (*Dendroctonus frontalis*, picture courtesy from the USDA Forest Service Archive, USDA Forest Service, Bugwood.org). (**b**) Locations of the mycangia. (**c**) Spore-like fungal cells (*arrows*) in the mycangium (**b**, **c** courtesy from Cetin Yuceer)

persed by the mites. Conidia are moved on the mite cuticle, but the spores are concentrated within the feces of the mites as well (Fig. 14.12). Spores survive passage through the digestive tracts and able to germinate in the fecal pellets, which serve as a sufficient substrate to the fungi to develop hyphae and conidiophores without any other food sources (Colloff 2009). A mutualistic relationship between *Acarus siro* and the molds and yeasts that colonize the wheat endosperm is hypothesized by Levinson et al. (Levinson et al. 1991). The mite is apt not to feed on the endosperm unless it becomes moldy. The fungi produce ammonia from broken and sprouted wheat as a metabolite. Ammonia is a rather efficient attractant for colonizing mites (kairomone effect). Also, the guanine in the mite feces is solubilized by ammonia to

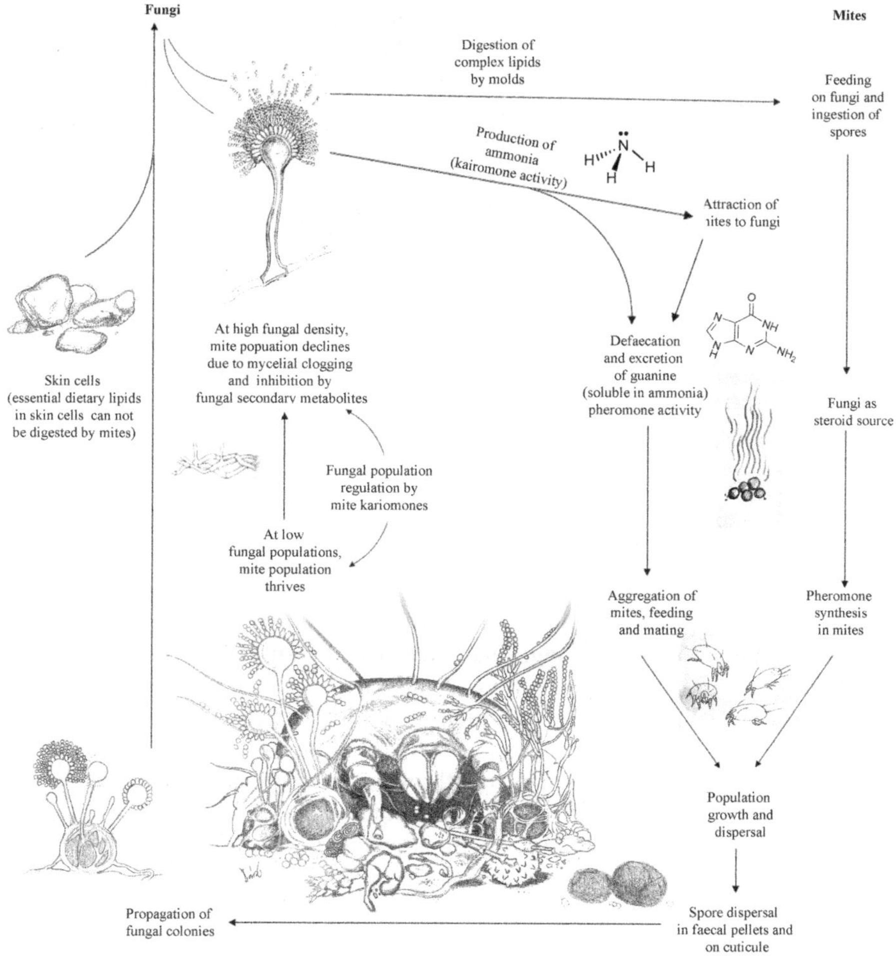


**Fig. 14.12** The dispersal of mold spores by feeding dust mites [redrawn by Magyar after Colloff with permission (Colloff 2009)]

attract opposite sex for copulation (pheromone effect). The mites mate and disperse, distributing the fungal spores via their digestive tracts. Shed human dander is colonized by *Aspergillus glaucus* group and their lipids are broken down to form free fatty acids. One species of *Fungitarsonemus* (Prostigmata, Eleutherengonides), common foliar mites in tropical areas, releases an adherent layer to allow spores and pollen grains to adhere as an additional layer (perhaps tactile or visual camouflage). However, their role in fungal dispersal remains poorly understood. Mites are also involved in dispersal of a broad range of coprophilous fungi (Malloch and Blackwell 1992) and several myxomycetes (Keller and Smith 1978). Fungal mites (e.g., *Tyrophagus*) ingest the mycelia and spores. Fecal materials of insects and mites often contain fungal spores, and these spores often appear intact and undamaged by passing through the digestive systems of arthropods (Abbott, unpublished). The effects of panphytophagous oribatid mites on the recovery of the microbial community in partially decomposed litter layer from forests after strong disturbance (freezing and heating) were investigated in a laboratory microcosm experiment. Generally, oribatid mites enhanced the recovery of the disturbed systems by accelerating the recolonization of litter materials by fungal species and the recovery of the microbial metabolism by dispersal of spores and by grazing on microbial populations (Maraun et al. 1998).

Mutualism between snails and microfungi has also been observed. Snails (*Littoraria irrorata*) graze grass primarily not to feed but to prepare substrate for growth of *Phaeosphaeria* and *Mycosphaerella* spp. and consume these invasive fungi. However, unlike ants and termites, snails do not seem to carry spores to inoculate prepared substrate to initiate fungal growth (Silliman and Newel 2003).

## *Coprophilous Fungi*

Coprophilous fungi use the dung of herbivores as substrates. The mechanisms of spore dispersal of these fungi guarantee that the spores are propelled from the dung onto the nearby vegetation (mostly by ballistic mechanisms), where the fungi will be fed on along with plant materials and pass through an animal digestive system to repeat the cycle. The coprophilous ascospores are dispersed by a number of dispersal mechanisms from the dung to the nearby vegetation. The ascospores are often covered by mucilage or have gelatinous appendages for attaching to the plant parts on which they land without difficulties (Wicklow 1981). Herbivores eat the ascospores that frequently are darkly pigmented and well protected against both gastric juices and the UV light from the sun. Spore germination can even be triggered by gastric juices of the herbivores (Webster 1970). *Basidiobolus ranarum* (*Entomophthorales*) growing on the feces of lizards and frogs employs a different discharge mechanism. The sporangium of this species is developed on a subsporangial vesicle, when the vesicle ruptures at its base, jetting the sap backwards to propel the sporangium forwards, like a rocket. *Pilobolus*, the hat-thrower fungus, is known for their explosive spore dispersal. *Pilobolus* has to pass through the digestive systems of grazing animals in their life cycle. Since the animals circumvent foraging near their dungs, *Pilobolus* uses a phototropic (light-following) squirt gun mechanism to propel their sporangia up to 3 m away onto uninfested vegetation. The ballistic discharge of *Pilobolus* may reach speeds of up to 90 km/h due to the release of highly pressurized fluids from the sporangiophore stalk after the rupture of the subsporangial vesicle. The sporangia have a sticky mucilaginous ring for adhering to vegetation when wetted by the propelling fluids. Once ingested, the spores pass through the digestive systems undamaged and are deposited into a fresh pile of dung, thus

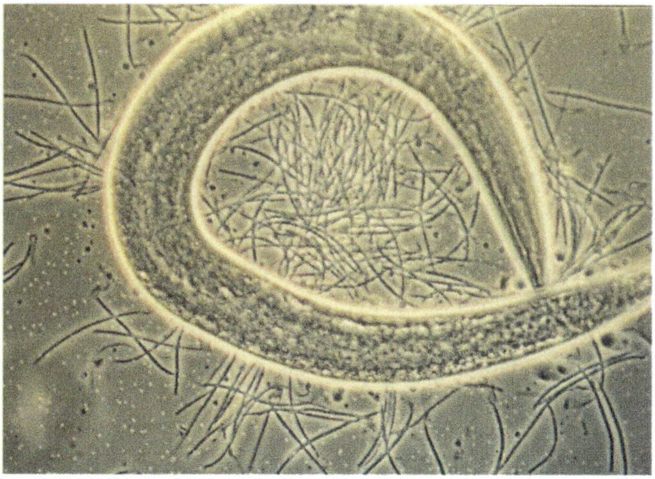

**Fig. 14.13** Nematode carrying *Gorgomyces* conidia. Photo by Ágnes Révay, with permission

continuing the asexual life cycle. *Sphaerobolus stellatus*, the cannonball fungus (Basidiomycota), develops basidiospores in a large ball-like structure within a cup-shaped basidioma. At maturity, the inner layer of the cup separates from the outer layer and suddenly inverts, like a trampoline, forcibly ejecting the spores containing the ball into the air (Ingold 1971).

There are more than 40 species of bioluminescent fungi (Weitz 2004). This phenomenon may raise the role of luminescence for attracting invertebrates to assist fungal spore dispersal (Sivinski 1981, 1998).

Aquatic animals also represent main keys in dispersal of aquatic fungi (amphibians—chytridiomycosis caused by *Batrachochytrium dendrobatidis*; fish—some species of the fungus-like organism called Oomycetes; zooplankton—Microsporidia species on copepods) in lakes.

## Man-Made Spore Dispersal

Human activity also plays an important role in spore dispersal. High amount of fungi are aerosolized worldwide by man-made kinetic energy, e.g., by harvesting (Magyar et al. 2012; Skjøth et al. 2012), or demolition of moldy buildings (Bouza et al. 2002). Spore dispersal by man-made thermal energy, especially in heat production by industrial composting, is also remarkable (Sebők et al. unpublished). The effect of these human activities can be intensive and drastic and often higher than energy requirement for the natural fungal spore liberation, either passive or active.

An example for the unintentional dispersal of a pathogen by humans is the global emergence and spread of the pathogenic, virulent, and highly transmissible fungus *Batrachochytrium dendrobatidis*, belonging to zoosporic aquatic fungi. It has caused the decline or extinction of up to about 200 species of frogs, resulting in the disease chytridiomycosis (Fisher et al. 2009; Skerratt et al. 2007). In the distribution of this fungus, human activities like visiting waterbodies one by one (regional scale) or human transport (continent scale) also have huge relevance and impact (Skerratt et al. 2007).

Spores of microfungi are also intentionally dispersed by human activity, e.g., in fermentation and food industry, research laboratories, and some fields of mycotechnology (Magyar 2007; Samson 2010). Open-field cultivation of microfungi is rarely in practice, since inoculation may pose a risk in agriculture. Some exceptions are corn smut, ergot, and Jiaobai smut [*Ustilago esculenta* infecting *Zizania latifolia* as a vegetable (Chan and Thrower 1980; Terrell and Batra 1982)] cultivation.

Corn smut is a delicacy in Mexico and is known as huitlacoche there. It is preserved and sold for a considerably higher price than uninfected corn. Corn smut consumption originates from Aztec cuisine. For culinary use, the immature galls are harvested. As fully mature galls are dry and almost entirely full of spores. In the mid-1990s, due to demand from high-end restaurants, farms in Florida and Pennsylvania were permitted by the United States Department of Agriculture (USDA) to purposely infect corn with huitlacoche. The initiative is still under way (Pataky and Chandler 2003).

Jiaobai (or gau sun) has been widely cultivated as a vegetable for several centuries in China (Terrell and Batra 1982). When *Zizania latifolia* is infected by *Ustilago esculenta*, its cum apex becomes swollen forming a juicy stem gall and contains smut mycelia (Chan and Thrower 1980). The galled stems are harvested as vegetable at the stage when no smut spores are developed (Chung and Tzeng 2004). This vegetable is produced by vegetatively segregating infected plants.

Endophytic fungi are common in leaves and stems of grasses (Szécsi et al. 2013). Their systemic infection also produces infected seeds. These fungi do not cause any disease in the grasses, but under most circumstances they are beneficial to the growth and survival of infected plants, enhancing drought tolerance, summer survival, and insect resistance. In recognition of the beneficial effects, turf grass breeders now offer seeds of a variety of "endophyte-enhanced" cultivars (Grewal and Richmond 2004). This man-made seed-borne dispersal of microfungi has an emerging agricultural and commercial potential. Fungal spores could also be intentionally dispersed as biological weapons. Karnal bunt (*Tilletia indica*) spores were produced in Iraq as an anticrop agent. It must have been intended for use in the war against Iran, whose main crop is wheat (Whitby 2002; Suffert et al. 2009). The M115 anticrop bomb also referred as the feather bomb or the E73 bomb was a US biological cluster bomb aimed to deliver *Puccinia graminis* f. sp. *tritici*, wheat stem rust (Wheelis et al. 2006). Wheat stem rust bomb production includes a dry particulate matter (rust spores) which was adhered to a lightweight vector, usually feathers.

## Applications of Dispersal Strategies of Microfungi: Future Trends

Research of dispersal strategies of microfungi lead for new applications and inventions in surprisingly different fields of science. Analysis of the spore discharge process using high-speed video is proved to be a useful modern tool to understand micromechanical processes. The ballistospory biomechanics were analogous to the recent development of a surface tension motor (Pringle et al. 2005). The production of this nanoscale machine driven by the coalescence of liquid droplets demonstrates that processes powered by surface tension may have significant applications in engineering (Regan et al. 2005). Investigations of other forcible spore discharge mechanisms and the discharge of basidiospore may offer a natural model for this kind of devices, and the findings have had major significance for future development of machines that operate on the micrometer and nanometer scales.

The practical use of insect dispersal is in the biological control. Western honeybee, *Apis mellifera* L., was used to deliver conidia of *Clonostachys rosea* to control *Botrytis cinerea* on strawberry (*Fragaria × ananassa*) (Peng et al. 1992). Bumblebee, *Bombus impatiens*, was studied as a delivery vehicle of *C. rosea* (reported as *G. roseum* and *G. catenulatum*) to control *Botrytis cinerea* on raspberry (*Rubus idaeus*) and rabbiteye blueberry (*Vaccinium virgatum*) (Yu and Sutton 1997; Smith et al. 2012). Promising research results showed that the western honeybee (*Apis mellifera*) could be used to disperse *Trichoderma* spp. to reduce disease incidence of sunflower head rot caused by *Sclerotinia sclerotiorum* (Escande et al. 2002).

Free fungal spores can be studied concerning their biodiversity in a sample. Spore composition is characteristic, and their analysis is a useful tool to identify the origin of samples, indicating vegetation (air samples from a known trajectory, stemflow and honeydew samples, forensic, archeological, or geological sediments). It was suggested that the possibility of identification of origin of honeys could obviously be tested with the multivariate analysis of fungal diversity data (Magyar et al. 2005).

Plasmodium movements of slime molds are driven by the grainy cytoplasm, which flows by unidirectional pulsation in a cell. The cell attains a speed of up to 1 mm/s (Nowotny 2000). Such movements of slime molds are used to imitate the formation of the motorway transport network (Adamatzky et al. 2013) or the migrations of human population (Adamatzky and Martinez 2013) in laboratory conditions. Plasmodium movement will be used in future designs of self-growing wetware circuits and devices and integration of slime mold electronics into unconventional biohybrid systems (Adamatzky 2013).

**Acknowledgments** We thank Ágnes Révay, Zoltán Bratek, and Gyula Oros for their excellent suggestions and providing literature.

## References


Abbott SP (2000) Holomorph studies of the Microascaceae. Ph.D. Dissertation, University of Alberta

Abbott SP (2002) Insects and other arthropods as agents of vector-dispersal in fungi. Unpublished. Online version: http://www.thermapure.com/pdf/AbbottInsectdispersal.pdf

Abelho M, Graca MAS (1996) Effects of eucalyptus afforestation on leaf litter dynamics and macroinvertebrate community structure of streams in Central Portugal. Hydrobiologia 324:195–204

Adamatzky A (2013) Physarum wires: self-growing self-repairing smart wires made from slime mould. Biomed Eng Lett 3(4):232–241

Adamatzky A, Martinez GJ (2013) Bio-imitation of Mexican migration routes to the USA with slime mould on 3D terrains. J Bionic Eng 10:242–250

Adamatzky A, Lees M, Sloot P (2013) Bio-development of motorway network in the Netherlands: a slime mould approach. Adv Complex Syst 16(02n03)

Agrios GN (1980) Insect involvement in the transmission of fungal pathogens. In: Harris KF, Maramorosch K (eds) Vectors of plant pathogens. Academic, New York, pp 293–323

Agrios GN (2006) Plant pathology, 5th edn. Academic, New York

Ahimera N, Gisler S, Morgan DP, Michailides TJ (2004) Effects of single-drop impactions and natural and simulated rains on the dispersal of *Botryosphaeria dothidea* conidia. Phytopathology 94:1189–1197

Ainsworth GC (1952) The incidence of air-borne *Cladosporium* spores in the London region. J Gen Microbiol 7:358–361

Allgaier C (2007) Active camouflage with lichens in a terrestrial snail, Napaeus (N.) barquini Alonso and Ibáñez, 2006 (Gastropoda, Pulmonata, Enidae). Zool Sci 24:869–876

Allitt U (2000) Airborne fungal spores and the thunderstorm of 24 June 1994. Aerobiologia 16:397–406

Alsheikh AM, Trappe JM (1983) Taxonomy of *Phaeangium lefebvrei*, a desert truffle eaten by birds. Can J Bot 61:1919–1925

Alt S, Kollar A (2010) Hydrodynamics of raindrop impact stimulate ascospore discharge of *Venturia inaequalis*. Fungal Biol 114(4):320–324

Ananda K, Prasannarai K, Sridhar KR (1998) Occurrence of higher marine fungi on marine animal substrates of some beaches along the west coast of India. Indian J Mar Sci 27(2):233–236

Andrews JH (1992) Fungal life-history strategies. In: Carroll CG, Wicklow DT (eds) The fungal community, 2nd edn. Marcel Dekker, New York, pp 119–145

Arseniuk E, Goral T, Scharen AL (1997) Seasonal patterns of spore dispersal of *Phaeosphaeria* spp. and *Stagonospora* spp. Plant Dis 82:187–194

Auffret AG, Berg J, Cousins SA (2014) The geography of human-mediated dispersal. Divers Distrib 20(12):1450–1456

Aylor DE (1978) Dispersal in time and space: aerial pathogens. In: Horsfall JG, Cowling EB (eds) Plant disease—an advanced treatise. Academic Press, New York, pp 159–180

Aylor DE (1990) The role of intermittent wind in the dispersal of fungal pathogens. Annu Rev Phytopathol 28:73–92

Aylor DE (1993) Relative collection efficiency of Rotorod and Burkard spore samplers for airborne *Venturia inequalis* ascospores. Phytopathology 83:1116–1119

Aylor DE, McCartney HA, Brainbridge A (1981) Deposition of particles liberated in gusts of wind. J Appl Meteorol 20:1212–1221

Aylor DE, Taylor GS, Raynor GS (1982) Long-range transport of tobacco blue mold spores. Agric Meteorol 27(3):217–232

Bailey FM (1904) Additional notes on the birds of the upper Pecos. Auk 21(3):349–363

Bandoni RJ (1972) Terrestrial occurrence of some aquatic hyphomycetes. Can J Bot 50(11):2283–2288

Bandoni RJ (1974) Mycological observations on the aqueous films covering decaying leaves and other litter. Trans Mycol Soc Jpn 15:309–315
Bandoni RJ, Koske RE (1974) Monolayers and microbial dispersal. Science 183:1079–1081
Bandyopadhyay R, Frederickson DE, McLaren NW, Odvody GN, Ryley MJ (1998) Ergot: a new disease threat to sorghum in the Americas and Australia. Plant Dis 82:356–367
Bärlocher F (1981) Fungi on the food and in the faeces of *Gammarus pulex*. Trans Br Mycol Soc 76(1):160–165
Bärlocher F (1982) Conidium production from leaves and needles in four streams. Can J Bot 60:1487–1494
Bärlocher F (1992) Recent developments in stream ecology and their relevance to aquatic mycology. In: Bärlocher F (ed) The ecology of aquatic hyphomycetes. Springer-Verlag, Berlin (Ecol Stud 94:16–37)
Bärlocher F (1992) Research on aquatic hyphomycetes: historical background and overview. In: Bärlocher F (ed) The ecology of aquatic hyphomycetes. Springer, Berlin (Ecol Stud 94:1–15)
Bärlocher F (2009) Reproduction and dispersal in aquatic hyphomycetes. Mycoscience 50:3–8
Bärlocher F, Murdoch JH (1989) Hyporheic biofilms—a potential food source for interstitial animals. Hydrobiologia 184:61–67
Bertolino S, Vizzini A, Wauters LA, Tosi G (2004) Consumption of hypogeous and epigeous fungi by the red squirrel (*Sciurus vulgaris*) in subalpine conifer forests. Forest Ecol Manage 202(1):227–233
Bock CH, Jeger MJ, Fitt BDL, Sherrington J (1997) Effect of wind on the dispersal of oospores of *Peronosclerospora sorghi* from sorghum. Plant Pathol 46:439–449
Bouza E, Peláez T, Pérez-Molina J, Marín M, Alcalá L, Padilla B, Muñoz P, Adán P, Bové B, Bueno MJ, Grande F, Puente D, Rodríguez MP, Rodríguez-Créixems M, Vigil D, Cuevas O, Aspergillus Study Team (2002) Demolition of a hospital building by controlled explosion: the impact on filamentous fungal load in internal and external air. J Hosp Infect 52:234–42
Bowden J, Gregory PH, Johnson CG (1971) Possible wind transport of coffee leaf rust across the Atlantic Ocean. Nature 229:500–501
Bratek Z, Merenyi Z, Illies Z, Laslo P, Anton A, Papp L, Merkl O, Garay J, Vikor J, Brandt S (2010) Studies on the ecophysiology of *Tuber aestivum* populations in the Carpatho-Pannonian region. Österreichische Zeitschrift für Pilzkunde 19:221–226
Brodo IM, Sharnoff SD, Sharnoff S (2001) Lichens of North America. Yale University Press, New Haven
Brown JKM, HovmØller MS (2002) Aerial dispersal of pathogens on the global and continental scales and its impact on plant disease. Science 297:537–541
Brown JKM, Jessop AC, Rezanoor HN (1991) Genetic uniformity in barley and its powdery mildew pathogen. Proc R Soc B 246:83–90
Bruns TD (1995) Thoughts on the processes that maintain local species diversity of ectomycorrhizal fungi. Plant Soil 170:63–73
Buller AHR (1909) Researches on fungi I. Longmans, Green and Co, London, 192p
Bulliard JB (1791) Histoire des champignons de la France. l'auteur, Paris
Burroughs HE, Hansen SJ (2008) Managing indoor air quality. The Fairmont Press, Lilburn
Butler MD, Alderman SC, Hammond PC, Berry RE (2001) Association of insects and ergot in Kentucky bluegrass seed production fields. J Econ Entomol 94(6):1471–1476
Caldwell IR, Vernes K, Barlocher F (2005) The northern flying squirrel (*Glaucomys sabrinus*) as a vector for inoculation of red spruce (*Picea rubens*) seedlings with ectomycorrhizal fungi. Sydowia 57(2):166–178
Carmichael JW, Kencrick WB, Conners IL, Sigler L (1980) Genera of hyphomycetes. University of Alberta Press, Edmonton, AB
Carter MV (1965) Ascospore deposition of *Eutypa armeniacae*. Aust JAgr Res 16:825–836
Carter MV, Moller WJ (1961) Factors affecting the survival and disseminations of *Mycosphaerella pinodes* (Berk. & Blox.) Vestergr. in South Australian irrigated pea fields. Aust J Agr Res 12:879–888
Cassar S, Blackwell M (1996) Convergent origins of ambrosia fungi. Mycologia 88:596–601

Castellano MA, Trappe JM, Maser Z, Maser C (1989) Key to spores of the genera of hypogeous fungi of north temperate forests with special reference to animal mycophagy. Mad River Press, Arcata, CA

Castellano MA, Trappe JM, Luoma DL (2004) Sequestrate fungi. In: Foster MS, Mueller GM, Bills GF (eds) Biodiversity of fungi: inventory and monitoring methods. Academic, Burlington, pp 197–213

Castillo-Guevara C, Sierrra J, Galindo-Flores G, Cuautle M, Lara C (2011) Gut passage of epigeous ectomycorrhizal fungi by two opportunistic mycophagous rodents. Curr Zool 57(3):293–299

Castillo-Guevara C, Lara C, Pérez G (2012) Micofagia por roedores en un bosque templado del centro de México. Revista Mexicana de Biodiversidad 83(3):772–777

Chamberlain AC, Chadwick RC (1972) Deposition of spores and other particles on vegetation and soil. Ann Appl Biol 71:141–158

Chan YS, Thrower L (1980) The host-parasite relationship between *Zizania caduciflora* Turcz. and *Ustilago esculenta* P. Henn. I. Structure and development of the host and host–parasite combination. New Phytol 85(2):201–207

Chou H-Hung WW-S (2002) Phylogenetic analysis of internal transcribed spacer regions of the genus *Alternaria*, and the significance of filament-beaked conidia. Mycol Res 106(2):164–169

Christensen CM (1950) Intramural dissemination of spores of *Hormodendrum resinae*. J Allergy 21:409–413

Chung K, Tzeng DD (2004) Nutritional requirements of the edible gall-producing fungus *Ustilago esculenta*. J Biol Sci 4(2):246–252

Claus R, Hoppen HO, Karg H (1981) The secret of truffles: a steroidal pheromone? Experientia 37:1178–1179

Cohen Y, Rotem J (1970) The relationship of sporulation to photosynthesis in some obligatory and facultative parasites. Phytopathology 60:1600–1604

Colloff MJ (2009) Dust mites. Springer, Dordrecht

Cork SJ, Kenagy GJ (1989) Nutritional value of a hypogeous fungus for a forest-dwelling ground squirrel. Ecology 70:577–586

Cornut J, Chauvet E, Florian M-B, Assemat F, Elger A (2014) Aquatic hyphomycetes species are screened by the hyporheic zone of woodland streams. Appl Environ Microbiol 80(6):1949–1960

Czeczuga B, Orlowska M (1994) Some aquatic fungi of hyphomycetes on tree leaves. Annales Academiae Medicae Bialostocensis 39:86–92

Czeczuga B, Orlowska M (1999) Hyphomycetes in rain water, melting snow and ice. Acta Mycol 34:181–200

Dang C, Gessner M, Chauvet E (2007) Influence of conidial traits and leaf structure on attachment success of aquatic hyphomycetes on leaf litter. Mycologia 99(1):24–32

Daws LF (1967) Movement of air streams indoors. In: Gregory PH, Monteith JL (eds) Airborne microbes. Cambridge University Press, Cambridge, pp 31–59

de Bary A (1887) Comparative morphology and biology of the fungi, mycetozoa and bacteria. Clarendon Press, Oxford, p 525

de Bekker C, Quevillon L, Smith PB, Fleming K, Ghosh D, Patterson AD, Hughes DP (2014) Species-specific ant brain manipulation by a specialized fungal parasite. BMC Evol Biol 14(1):166

Deacon J (2006) Fungal biology, 4th edn. Blackwell Publishing, Malden, MA

Devarajan PT, Suryanarayanan TS (2006) Evidence for the role of phytophagous insects in dispersal of non-grass fungal endophytes. Fungal Divers 23:111–119

di Castri F, Hansen AJ, Holland MM (eds) (1988) A new look at ecotones: emerging international projects on landscape boundaries. Biology International, Special Issue 17. IUBS, Paris

Dowson D, Higginson GR (1959) A numerical solution to the elasto-hydrodynamic problem. J Mech Eng Sci 1:5–15

Dromph KM (2001) Dispersal of entomophagic fungi by collembolan. Soil Biol Biochem 33:2047–2051

Duarte S, Seena S, Bärlocher F, Cássio F, Pascoal C (2012) Preliminary insights into the phylogeography of six aquatic hyphomycete species. PLoS One 7(9), e45289

Dutzmann S (1985) Zur Analyse der Beziehung zwinschen Klimadaten und Sporenproduktion sowie Sporenverbreitung von *Erysiphe graminis* f. sp. *hordei*. Zeitschrift für Pflanzenkrankheiten und Pflanzenschutz 92:629–642

Dye C (1986) Vectorial capacity: must we measure all the components? Parasitol Today 2:203–209

Edmonds RL (1979) Aerobiology—the ecological systems approach. Dowden, Hutchinson & Ross Inc, Stroudsburg, 386p

Ekelöf E (1907) Studien uber den Bakteriengehalt der Luft und des Erdbodens der antarktischen Gegenden, ausgefuhrt wahrend der schwedischen Sudpolar-Eypedition 1901–1903. Z Hyg Infekt Krankh 56:344–370

El Hamalawi ZA (2008) Attraction, acquisition, retention and spatiotemporal distribution of soilborne plant pathogenic fungi by shore flies. Ann Appl Biol 152(2):169–177

Elbert W, Taylor PE, Andreae MO, Poschl U (2007) Contribution of fungi to primary biogenic aerosols in the atmosphere: wet and dry discharged spores, carbohydrates, and inorganic ions. Atmos Chem Phys 7:4569–4588

Escande AR, Laich FS, Pedraza MV (2002) Field testing of honeybee-dispersed *Trichoderma* spp. to manage sunflower head rot (*Sclerotinia sclerotiorum*). Plant Pathol 51(3):346–351

Fermaud M, Gaunt RE. (1995) *Thrips obscuratus* as a potential vector of *Botrytis cinerea* in kiwifruit. Mycol Res 99(3):267–273

Fernández D, Valencia MV, Molnár T, Vega A, Sagues E (1998) Daily and seasonal variations of *Alternaria* and *Cladosporium* airborne spores in León (North-West, Spain). Aerobiologia 14:215–220

Feuerherdt L, Petit S, Jusaitis M (2005) Distribution of mycorrhizal fungus associated with the endangered pink-lipped spider orchid (*Arachnorchis* (syn. *Caladenia*) *behrii*) at Warren Conservation Park in South Australia. N Z J Bot 43:367–371

Finlay BJ (2002) Global dispersal of free-living microbial eukaryote species. Science 296:1061–1063

Fischer FG, Werner G (1958) Die chemotaxis der schwarmsporen von wasserpilzen (Saprolegniaceen). Hoppe-Seylers Z Physiol Chem 310:65–91

Fisher PJ, Stradling DJ, Pegler DN (1994) Leaf cutting ants, their fungus gardens and the formation of basidiomata of *Leucoagaricus gongylophorus*. Mycologist 8(3):128–131

Fisher MC, Garner TW, Walker SF (2009) Global emergence of *Batrachochytrium dendrobatidis* and amphibian chytridiomycosis in space, time, and host. Annu Rev Microbiol 63:291–310

Fisher MC, Henk DA, Briggs CJ, Brownstein JS, Madoff LC, McCraw SL, Gurr SJ (2012) Emerging fungal threats to animal, plant and ecosystem health. Nature 484:186–194

Fogel R, Trappe JM (1978) Fungus consumption (mycophagy) by small animals. Northwest Sci 52:1–31

Forster GF (1977) Effect of leaf-surface wax on the deposition of airborne propagules. Trans Br Mycol Soc 68:245–250

Fracchia S, Krapovickas L, Aranda-Rickert A, Valentinuzzi VS (2011) Dispersal of arbuscular mycorrhizal fungi and dark septate endophytes by *Ctenomys* cf. *knighti* (Rodentia) in the northern Monte Desert of Argentina. J Arid Environ 75(11):1016–1023

Frank KL, Geils BW, Kalkstein LS, Thistle HW Jr (2008) Synoptic climatology of the long-distance dispersal of white pine blister rust II. Combination of surface and upper-level conditions. Int J Biometeorol 52(7):653–666

Frank JL, Anglin S, Carrington EM, Taylor DS, Viratos B, Southworth D (2009) Rodent dispersal of fungal spores promotes seedling establishment away from mycorrhizal networks on *Quercus garryana*. Botany 87(9):821–829

Fritz JA, Seminara A, Roper M, Pringle A, Brenner MP (2013) A natural O-ring optimizes the dispersal of fungal spores. J R Soc Interface 10(85):20130187. doi:10.1098/rsif.2013.0187

Gage SH, Isard SA, Colunga-G M (1999) Ecological scaling of aerobiological processes. Agr Forest Meteorol 97:249–261

Geagea LW, Huber L, Sache I (1997) Removal of urediniospores of brown (*Puccinia recondita* f. sp. *tritici*) and yellow (*Puccinia striiformis*) rusts of wheat from infected leaves submitted to a mechanical stress. Eur J Plant Pathol 103:785–793
Gerson U, Seaward MRD (1977) Lichen–invertebrate associations. In: Seaward MRD (ed) Lichen ecology. Academic, London, pp 69–119
Giner MM, Garcia JSC, Camacho CN (2001) Airborne *Alternaria* spores in SE Spain (1993–98)-occurrence patterns, relationship with weather variables and prediction models. Grana 40(3):111–118
Gönczöl J (1976) Ecological observations on the aquatic hyphomycetes of Hungary II. Acta Bot Acad Sci Hung 22:51–60
Gönczöl J, Révay Á (1985) *Gorgomyces*, gen.nov. an unusual Hyphomycete from terrestrial litter of Hungary. Nova Hedwig 41:453–461
Gönczöl J, Révay Á (2003) Treehole fungal communities: aquatic, aero-aquatic and dematiaceous hyphomycetes. Fungal Divers 12:19–34
Gönczöl J, Révay Á (2004) Fungal spores in rainwater: stemflow, throughfall and gutter conidial assemblages. Fungal Divers 16:67–86
Górny RL, Reponen T, Grinshpun SA, Willeke K (2001) Source strength of fungal spore aerosolization from moldy building material. Atmos Environ 35:4853–4862
Göttlich E, van der Lubbe W, Lange B et al (2002) Fungal flora in groundwater-derived public drinking water. Int J Hyg Environ Health 205(4):269–279
Gotwald WH (1969) Comparative morphological studies of the ants, with particular reference to the mouthparts (Hymenoptera: Formicidae). Memoirs of the Cornell University Agricultural Experiment Station, no. 408, 150p
Gregory PH (1961) The microbiology of the atmosphere. Intersience Publishers Inc., London, 252p
Gregory PH, Stedman OJ (1958) Spore dispersal in *Ophiobolus graminis* and other fungi of cereal foot rots. Trans Br Mycol Soc 41(4):449–456
Gressitt JL (1965) Flora and fauna on backs of large Papuan moss-forest weevils. Science 150:1833–1835
Grewal PS, Richmond DS (2004) New benefits of endophyte-infected grasses emerge. Turfgrass Trends 2004:86–91
Grigg AH, Mulligan DR (1999) Litterfall from two eucalypt woodlands in central Queensland. Aust J Ecol 24:662–664
Hageskal G, Gaustad P, Heier BT, Skaar I (2007) Occurrence of moulds in drinking water. J Appl Microbiol 102(3):774–780
Halász Á, Magyar D, Bobvos J (2014) Aerobiological aspects of quarantine risks in grain warehouses—a study on bunt (*Tilletia* spp.) dispersal. Aerobiologia 30(2):161–171
Hamilton ED (1959) Studies on the air spora. Acta Allergol 13:143–175
Harvey R (1967) Air-spora studies at Cardiff I. *Cladosporium*. Trans Br Mycol Soc 50(3):479–495
Harvey R (1970) Spore productivity in Cladosporium. Mycopathologia et mycologia applicata 41(3–4):251–256
Heinrichs G, Hübner I, Schmidt CK, de Hoog GS, Haase G (2013) Analysis of black fungal biofilms occurring at domestic water taps (II): potential routes of entry. Mycopathologia 175(5–6):399–412
Herwitz SR (2006) Raindrop impact and water flow on the vegetative surfaces of trees and the effects on stemflow and throughfall generation. Earth Surf Process Landf 12:425–432
Hirst JM (1953) Changes in atmospheric spore content: diurnal periodicity and the effects of weather. Trans Br Mycol Soc 36:375–393
Hirst JM, Stedman OJ (1962) The epidemiology of apple scab (*Venturia inaequalis* (Cke.) Wint.). III. The supply of ascospores. Ann Appl Biol 50:551–567
Hirst JM, Stedman OJ (1963) Dry liberation of fungus spores by raindrops. J Gen Microbiol 33:335–344

Hirst JM, Stedman OJ (1971) Patterns of spore dispersal in crops. In: Preece TF, Dickinson CH (eds) Ecology of leaf surface micro-organisms. Academic Press, London, p 640
Hjelmroos M (1993) Relationship between airborne fungal spore presence and weather variables. Grana 32:40–47
Ho TM, Tan BH, Ismail S, Bujang MK (1995) Seasonal prevalence of air-borne pollen and spores in Kuala Lumpur, Malaysia. Asian Pac J Allergy Clin Immunol 13:17–22
Honegger R (2009) Lichen-forming fungi and their photobionts. In: Deisin HB (ed) Plant relationships. Springer, Berlin, pp 307–333
Hörberg HM (2002) Patterns of splash dispersed conidia of *Fusarium poae* and *Fusarium culmorum*. Eur J Plant Pathol 108:73–80
Huber L, Madden LV, Fitt BDL (2006) Environmental biophysics applied to the dispersal of fungal spores by rain-splash. In: The epidemiology of plant diseases, 2nd edn. Springer, Dordrecht
Hughes DP, Andersen SB, Hywel-Jones NL, Himaman W, Billen J, Boomsma JJ (2011) Behavioral mechanisms and morphological symptoms of zombie ants dying from fungal infection. BMC Ecol 11(1):13
Hutchinson GE (1967) A treatise on limnology, vol II: introduction to lake biology and the limnoplankton. Wiley, New York
Hynes HBN (1970) The ecology of running waters. University of Toronto Press, Toronto
Hynes HBN (1983) Groundwater and stream ecology. Hydrobiologia 100:93–99
Incagnone G, Marrone F, Barone R, Robba L, Naselli-Flores L (2014) How do freshwater organisms cross the "dry ocean"? A review on passive dispersal and colonization processes with a special focus on temporary ponds. Hydrobiologia:1–21
Ing B (1967) Myxomycetes as food for other organisms. Proceedings of the South London Entomological and Natural History Society 1967:18–23
Ingold CT (1942) Aquatic hyphomycetes of decaying alder leaves. Trans Br Mycol Soc 25:339–417
Ingold CT (1953) Dispersal in fungi. Oxford University Press, London
Ingold CT (1965) Spore liberation. Oxford University Press, London
Ingold CT (1971) Fungal spores: their liberation and dispersal. Clarendon, Oxford, 302 p
Ingold CT (1978) Role of mucilage in dispersal of certain fungi. Trans Br Mycol Soc 70(1):137–140
Iqbal SH, Webster J (1973) The trapping of aquatic hyphomycete spores by air bubbles. Trans Br Mycol Soc 60:37–48
Jakucs E, Vajna L (2003) Mikológia. Agroinform Press, Budapest, p 478
Jenkins DG, Brescacin CR, Duxbury CV, Elliott JA, Evans JA, Grablow KR, Hillegass M, Lyon BN, Metzger GA, Olandese ML, Pepe D, Silvers GA, Suresch HN, Thompson TN, Trexler CM, Williams GE, Williams NC, Williams SE (2007) Does size matter for dispersal distance? Global Ecol Biogeogr 16:415–425
Jersáková J, Malinová T (2007) Spatial aspects of seed dispersal and seedling recruitment in orchids. New Phytol 176:237–241
Jones AM, Harrison RM (2004) The effects of meteorological factors on atmospheric bioaerosol concentrations-a review. Sci Total Environ 326:151–180
Kagami M, van Donk E, de Bruin A, Rijkeboer M, Ibelings B (2004) Daphnia can protect diatoms from fungal parasitism. Limnol Oceanogr 49:680–685
Kagami M, de Bruin A, Ibelings B, van Donk E (2007) Parasitic chytrids: their effects on phytoplankton communities and food-web dynamics. Hydrobiologia 578:113–129
Karamchand KS, Sridhar KR (2008) Water-borne conidial fungi inhabiting tree-holes of the west coast and Western Ghats of India. Czech Mycol 60:63–74
Kataržytė M, Kutorga E (2011) Small mammal mycophagy in hemiboreal forest communities of Lithuania. Cent Eur J Biol 6:446–456
Katial RK, Zhang Y, Jones RH, Dyer PD (1997) Atmospheric mould spore counts in relation to meteorological parameters. Int J Biometeorol 41:17–22

Keller HW, Smith DM (1978) Dissemination of myxomycete spores through the feeding activities (ingestion/defecation) of an acarid mite. Mycologia 70:1239–1241

Kendrick B (1985) The fifth kingdom. Mycologue Publications, Waterloo, ON

Kendrick B (1990) Fungal allergens. In: Smith EG (ed) Sampling and identifying allergenic pollens and moulds. Blewstone Press, San Antonio, pp 41–165

Kluth S, Kruess A, Tscharntke T (2002) Insects as vectors of plant pathogens: mutualistic and antagonistic interactions. Oecologia 133(2):193–199

Kohlmeyer J (1977) New genera and species of higher fungi from the deep sea (1615–5315m). Revue de mycologie 41:189–206

Kohlmeyer J, Demoulin V (1981) Parasitic and symbiotic fungi on marine algae. Bot Mar 24(1):9–18

Kohlmeyer J, Kohlmeyer E (1971) Marine fungi from tropical America and Africa. Mycologia 63(4):831–861

Komonen A (2008) Colonization experiment of fungivorous beetles (Ciidae) in a lake–island system. Entomol Tidskr 129:141–145

Kong H-Z, Qi Z-T (1991) Some filamentous fungi isolated from the mate rials of King George Island in Antarctica. Mycosystema 10:59–62

Kovács M (1985) A nagyvárosok környezete. Gondolat Press, Budapest, p 108

Kovács K (2012) Avarlebontási kísérletek dombvidéki kisvízfolyásokon (Litter-decomposition experiments in woodland streams). Dissertation, University of Pannonia

Kowalski WJ (2006) Aerobiological engineering handbook: airborne disease and control technologies. McGraw-Hill, New York

Kuehn K, Koehn R (1988) A mycofloral survey of an artesian community within the Edwards Aquifer of central Texas. Mycologia 80:646–652

Kurkela T (1997) The number of *Cladosporium* conidia in the air in different weather conditions. Grana 36:54–61

Lacey J (1981) The aerobiology of conidial fungi. In: Cole GT, Kendrick WB (eds) The biology of conidial fungi, vol 1. Academic, New York, pp 373–415

Lacey J (1991) Aggregation of spores and its effect on aerodynamic behaviour. Grana 30:347–445

Lacey J (1996) Spore dispersal– its role in ecology and disease: the British contribution to fungal aerobiology. Mycol Res 100(6):641–660

Lacey M, West J (2006) The air spora: a manual for catching and identifying airborne biological particles. Springer, Dordrecht

Lana VM, Mafia RG, Ferreira MA, Sartório RC, Zauza EAV, Mounteer AH, Alfenas AC (2012) Survival and dispersal of *Puccinia psidii* spores in eucalypt wood products. Australas Plant Pathol 41(3):229–238

Langenberg WJ, Sutton JC, Gillespie TJ (1977) Relation of weather variables and periodicities of airborne spores of *Alternaria dauci*. Phytopathology 67:879–883

Le Calvez T, Burgaud G, Mahé S, Barbier G, Vandenkoornhuyse P (2009) Fungal diversity in deep-sea hydrothermal ecosystems. Appl Environ Microbiol 75(20):6415–6421

Levinson HZ, Levinson AR, Muller K (1991) The adaptive function of ammonia and guanine in the biocenotic association between Ascomycetes and flour mites. Naturwissenschaften 78:174–176

Lewis HE, Foster AR, Mullan BJ, Cox RN, Clark RP (1969) Aerodynamics of the human microenvironment. Lancet 1:1273–1277

Lilleskov EA, Bruns TD (2005) Spore dispersal of a resupinate ectomycorrhizal fungus, *Tomentella sublilacina*, via soil food webs. Mycologia 97(4):762–769

Lim SH, Chew FT, Dali SDBM, Tan HTW, Lee BW, Tan TK (1998) Outdoor airborne fungal spores in Singapore. Grana 37(4):246–252

Littlefield LJ (1981) Biology of the plant rusts. An introduction. Iowa State University Press, Ames, IA

Littlefield LJ, Heath MC (1979) Ultrastructure of rust fungi. Academic, New York

Lock MA (1981) River epilithon—a light and organic energy transducer. In: Lock MA, Williams DD (eds) Perspectives in running water ecology. Plenum Press, New York, pp 3–40
Lopez M, Salvaggio EJ (1983) Climate-weather-air pollution 1203–1214. In: Middleton E, Reed CE, Ellis EF (eds) Allergy: principles and practice. C.V. Mosby St. Louis, Toronto, p 1440
Lynch JM, Poole NJ (1984) Aerial dispersal and the development of microbial communities. In: Lynch JM, Poole NJ (eds) Microbial ecology: a conceptual approach. Blackwell, Oxford, pp 140–170
Madden LV (1992) Rainfall and the dispersal of fungal spores. Adv Plant Pathol 8:40–79
Madden LV (1997) Effects of rain on splash dispersal of fungal pathogens. Can J Plant Pathol 19(2):225–230
Madelin TM (1994) Fungal aerosols: a review. J Aerosol Sci 25:1405–1412
Madelin TM, Johnson HE (1992) Fungal and actinomycete spore aerosols measured at different humidities with an aerodynamic particle sizer. J Appl Bacteriol 72:400–409
Magyar D (2005) Aerobiological studies on mycobiota. PhD thesis, St. István University
Magyar D (2007) Aeromycological aspects of mycotechnology. In: Rai MK (ed) Mycotechnology: current trends and future prospects. I.K. International Publishing House, New Delhi, pp 226–263
Magyar D, Gönczöl J, Révay Á, Grillenzoni F, Seijo-Coello MDC (2005) Stauro- and scolecoconidia in floral and honeydew honeys. Fungal Divers 20:103–120
Magyar D, Eszéki ER, Gy O, Szécsi Á, Kredics L, Hatvani L, Körmöczi P (2011) The air spora of an orchid greenhouse. Aerobiologia 27(2):121–134
Magyar D, Szőke C, Koncz Z, Szécsi Á, Bobvos J (2012) Identification of airborne propagules of the *Gibberella fujikuroi* species complex during maize production. Aerobiologia 28(2):263–271
Malloch D, Blackwell M (1992) Dispersal of fungal diaspores. In: Carroll GC, Wicklow DT (eds) The fungal community: its organization and role in the ecosystem, 2nd edn. Marcel Dekker, New York, pp 147–171
Maraun M, Visser S, Scheu S (1998) Oribatid mites enhance the recovery of the microbial community after a strong disturbance. Appl Soil Ecol 9(1):175–181
Marin AB, Libbey LM, Morgan ME (1984) Truffles: on the scent of buried treasure. McIlvainea 6:34–38
Maser CA, Claridge AW, Trappe JM (2008) Trees, truffles, and beasts: how forests function. Rutgers University Press, New Brunswick
Maunsell K (1954) Concentration of airborne spores in dwelling under normal conditions and under repair. Int Arch Allergy Appl Immunol 5:373–376
McGee PA (2003) Pathogen survival and dispersal of plant parasites. University of Sydney. http://bugs.bio.usyd.edu.au/learning/resources/PlantPathology/survival_dispersal/survivalDispersal.html. Accessed 7 Jan 2015
McLaughlin D, Beckett A, Yoon K (1985) Ultrastructure and evolution of ballistosporic basidiospores. Bot J Linn Soc 91:253–271
Meredith DS (1961) Spore discharge in *Deightoniella torulosa* (Syd.) Ellis. Ann Bot 25:271–278
Meredith DS (1962) Some components of the air-spora in Jamaican banana plantations. Ann Appl Biol 50:577–592
Meredith DS (1965) Violent spore release in *Helminthosporium turcicum*. Phytopathology 55:1099–1102
Meredith DS (1966) Spore dispersal in *Alternaria porri* (Ellis) Neerg. on onion in Nebraska. Ann Appl Biol 57:67–73
Minshall GW, Cummins KW, Peterson RC, Cushing CE, Bruns DA, Sedell JR, Vannote RL (1985) Developments in stream ecosystem theory. Can J Fish Aquat Sci 42:1045–1055
Money NP, Fischer MVF (2009) Biomechanics in spore release in phytopathogens. In: Deisig H (ed) The mycota V. Plant relationships. Springer-Verlag, Berlin
Morrison KD, Reekie EG, Jensen KIN (1998) Biocontrol of common St Johnswort (*Hypericum perforatum*) with *Chrysolina hyperici* and a host-specific *Colletotrichum gloeosporioides*. Weed Tech 12:426–435

Mulholland PJ, Newbold JD, Elwood JW, Webster JR (1985) Phosphorus spiraling in a woodland stream; seasonal variations. Ecology 66:1012–1023

Nag Raj TR (1993) Coelomycetous anamorphs with appendage-bearing conidia. Mycologue Publications, Waterloo

Nagarajan S, Singh H (1976) Preliminary studies on forecasting wheat stem rust appearance. Agric Meteorol 17:281–289

Nagarajan S, Singh DV (1990) Long-distance dispersion of rust pathogens. Annu Rev Phytopathol 28:139–153

Nobel PS (1991) Physiochemical and environmental plant physiology. Academic, New York

Noblin X, Yang S, Dumais J (2009) Surface tension propulsion of fungal spores. J Exp Biol 212(17):2835–2843

Nowotny W (2000) Myxomyceten (Schleimpilze) und Mycetozoa (Pilztiere)–Lebensformen zwischen Pflanze und Tier. Stapfia 73:7–38

Nucci M, Anaissie E (2007) *Fusarium* infections in immunocompromised patients. Clin Microbiol Rev 20:695–704

O'Malley MA (2008) 'Everything is everywhere but the environment selects': ubiquitous distribution and ecological determinism in microbial biogeography. Stud Hist Philos Biol Biomed Sci 39(3):314–325

Odvody G, Bandyopadhyay R, Frederiksen RA, Isakeit T, Frederickson D, Kaufman H, Dahlberg J, Velasquex R, Torres H (1998) Sorghum ergot goes global in less than three years. APS Feature. www.apsnet.org/publications/apsnetfeatures/Pages/Ergot.aspx

Ogden EC, Hayes JV, Raynor GS (1969) Diurnal patterns of pollen emission in *Ambrosia*, *Phleum*, *Zea* and *Ricinus*. Am J Bot 56:16–21

Ooka JJ, Kommendahl T (1977) Wind and rain dispersal of *Fusarium moniliforme* in corn fields. Phytopathology 67:1023–1026

Oros GR, Eszéki E, Naár Z, Magyar D (2014) Phytopatological properties of symbiotic *Rhizoctonia solani* strains associated to orchids. Acta Agraria Debreceniensis 62:65–71

Padisák J, Soróczki-Pintér É, Rezner Z (2003) Sinking properties of some phytoplankton shapes and the relation of form resistance to morphological diversity of plankton—an experimental study. In Aquatic Biodiversity (pp. 243–257). Springer Netherlands

Pady SM, Kramer CL, Clary R (1969) Periodicity in spore release in *Cladosporium*. Mycologia 60(1):87–98

Panzer JD, Tullis EC, van Arsdel EP (1957) A simple 24-hour slide spore collector. Phytopathology 47:512–514

Park D (1974) Aquatic hyphomycetes in non-aquatic habitats. Trans Br Mycol Soc 63(1):183–187, IN30–IN31

Pasanen AL, Pasanen P, Jantunen MJ, Kalliokoski P (1991) Significance of air humidity and air velocity for fungal spore release into the air. Atmos Environ 25:459–462

Pataky JK, Chandler MA (2003) Production of huitlacoche, *Ustilago maydis*: timing inoculation and controlling pollination. Mycologia 95(6):1261–1270

Paul PA, El-Allaf SM, Lipps PE, Madden LV (2004) Rain splash dispersal of *Gibberella zeae* within wheat canopies in Ohio. Phytopathology 94:1342–1349

Pearce DA, Galand PE (2008) Microbial biodiversity and biogeography. Polar lakes and rivers: limnology of arctic and antarctic aquatic ecosystems, p 213–230

Peiqian L, Xiaoming P, Huifang S, Jingxin Z, Ning H, Birun L (2014) Biofilm formation by *Fusarium oxysporum* f. sp. *cucumerinum* and susceptibility to environmental stress. FEMS Microbiol Lett 350(2):138–145

Peng G, Sutton JC, Kevan PG (1992) Effectiveness of honey bees for applying the biocontrol agent *Gliocladium roseum* to strawberry flowers to suppress *Botrytis cinerea*. Can J Plant Pathol 14(2):117–129

Peres NA, Timmer LW, Adaskaveg JE, Correll JC (2005) Lifestyles of *Colletotrichum acutatum*. Plant Dis 89:784–796

Perkins AJ, McGee PA (1995) Distribution of the orchid mycorrhizal fungus, *Rhizoctonia solani*, in relation to its host, *Pterostylis acuminata*, in the field. Aust J Bot 43(6):565–575

Philips GB (1965) Safety in the chemical laboratory. XIII. Microbiological hazards in the laboratory, I. J Chem Educ 42:A43–44, A46–A48
Piattoni F, Ori F, Morara M, Iotti M, Zambonelli A (2012) The role of wild boars in spore dispersal of hypogeous fungi. Acta Mycol 47(2):145–153
Picioreanu C, van Loosdrecht MC, Heijnen JJ (2001) Two-dimensional model of biofilm detachment caused by internal stress from liquid flow. Biotechnol Bioeng 72(2):205–218
Pieczynska E (1990) Lentic aquatic-terrestrial ecotones: their structure, functions, and importance. In: Décamps H, Naiman RJ (eds) The ecology and management of aquatic-terrestrial ecotones, vol 4. CRC Press, Boca Raton
Pirozynski KA, Hawksowrth DL (eds) (1988) Coevolution of fungi with plants and animals. Academic, London
Pringle A, Patek SN, Fischer M, Stolze J, Money NP (2005) The captured launch of a ballistospore. Mycologia 97(4):866–871
Prochazka K, Stewart BA, Davies BR (1991) Leaf litter retention and its implications for shredder distribution in two headwater streams. Arch Hydrobiol 120:315–325
Raguso RA, Roy BA (1998) 'Floral' scent production by *Puccinia* rust fungi that mimic flowers. Mol Ecol 7(9):1127–1136
Rantio-Lehtimäki A (1977) Research on airborne fungus spores in Finland. Grana 16:163–165
Ranzoni FV (1979) Aquatic hyphomycetes from Hawaii. Mycologia 71:786–795
Real LA, Rathke BJ (1991) Individual variation in nectar production and its effect on fitness in *Kalmia latifolia*. Ecology 72:149–155
Rees G (1980) Factors affecting the sedimentation rate of marine fungal spores. Bot Mar 23:375–385
Regan BC, Aloni S, Jensen K, Zettl A (2005) Surface-tension-driven nanoelectromechanical relaxation oscillator. Appl Phys Lett 86(123119):1–3
Reponen T, Lehtonen M, Raunemaa T, Nevalainen A (1992) Effect of indoor sources on fungal spore concentrations and size distributions. J Aerosol Sci 23:663–666
Richardson DHS (1975) The vanishing lichens. David and Charles, Newton
Roper M, Pepper RE, Brenner MP, Pringle A (2008) Explosively launched spores of ascomycete fungi have drag-minimizing shapes. Proc Natl Acad Sci USA 105(52):20583–20588
Roper M, Seminara A, Bandi MM, Cobb A, Dillard HR, Pringle A (2010) Dispersal of fungal spores on a cooperatively generated wind. Proc Natl Acad Sci USA 107(41):17474–17479
Rotem J (1964) The effect of weather on dispersal of *Alternaria* spores in a semi-arid region of Israel. Phytopathology 54:628–632
Rotem J (1994) The genus *Alternaria*. APS Press, St Paul, MN, p 326
Royle DJ, Hickman CJ (1964) Analysis of factor governing in vitro accumulation of zoospores of *Pythium aphanidermatum* on roots. I. Behaviour of zoospores. Can J Microbiol 10:151–162
Ruisi S, Barreca D, Selbmann L, Zucconi L, Onofri S (2007) Fungi in Antarctica. Rev Environ Sci Biotechnol 6(1–3):127–141
Sache I (2000) Short-distance dispersal of wheat rust spores by wind and rain. Agronomie 20:757–767
Saint-Jean S, Testa A, Madden LV, Huber L (2006) Relationship between pathogen splash dispersal gradient and Weber number of impacting drops. Agr Forest Meteorol 141:257–262
Samson RA (2010) Food and indoor fungi, CBS laboratory manual series. CBS-KNAW Fungal Biodiversity Centre, Utrecht
Sanmartín OI, Ronquist F (2004) Southern hemisphere biogeography inferred by event-based models: plant versus animal patterns. Syst Biol 53:216–243
Savage D, Barbetti MJ, MacLeod WJ, Salam MU, Renton M (2012) Seasonal and diurnal patterns of spore release can significantly affect the proportion of spores expected to undergo long-distance dispersal. Microb Ecol 63(3):578–585
Scharf CS (1978) Birds and mammals as passive transporters for algae found in lichens. Can Field Nat 92:70–71
Scharf SC, DePalma KN (1981) Birds and mammals as vectors of the chestnut blight fungus (*Endothia parasitica*). Can J Zool 59(9):1647–1650

Schickmann S, Urban A, Kräutler K, Nopp-Mayr U, Hackländer K (2012) The interrelationship of mycophagous small mammals and ectomycorrhizal fungi in primeval, disturbed and managed Central European mountainous forests. Oecologia 170(2):395–409
Schigel DS (2012) Fungivory and host associations of Coleoptera: a bibliography and review of research approaches. Mycology 3(4):258–272
Seifert KA (1985) A monograph of *Stilbella* and some allied hyphomycetes. Stud Mycol 27:1–235
Selosse M-A, Vohník M, Chauvet E (2008) Out of the rivers: are some aquatic hyphomycetes plant endophytes? New Phytol 178:3–7
Seyd EL, Seaward MRD (1984) The association of oribatid mites with lichens. Zool J Linn Soc 80:369–420
Shchipanov NA, Aleksandrov DY, Aleksandrova AV (2006) Dispersal of micromycete spores by small mammals. Zool Zhurnal 851:101–113
Shearer CA (1992) The role of woody debris. In: Bärlocher F (ed) The ecology of aquatic hyphomycetes. Springer, Berlin (Ecol Stud 94:77–98)
Shearer CA, Descals E, Kohlmeyer B, Kohlmeyer J, Marvanová L, Padgett D, Padgett D, Porter D, Raja HA, Schmit JP, Thorton HA, Voglymayr H (2007) Fungal biodiversity in aquatic habitats. Biodivers Conserv 16(1):49–67
Short DP, O'Donnell K, Zhang N, Juba JH, Geiser DM (2011) Wide spread occurrence of diverse human pathogenic types of the fungus *Fusarium* detected in plumbing drains. J Clin Microbiol 49(12):4264–4272
Shykoff JA, Bucheli E (1995) Pollinator visitation patterns, floral rewards and the probability of transmission of *Microbotryum violaceum*, a veneral disease of plants. J Ecol 83(2):189–198
Sieber TN, Kowalski T, Holdenrieder O (1995) Fungal assemblages in stem and twig lesions of *Quercus robur* in Switzerland. Mycol Res 99(5):534–538
Sigee DC (2005) Freshwater microbiology: biodiversity and dynamic interactions of microorganisms in the aquatic environment. Wiley, Chichester, p 544
Silliman BR, Newel SY (2003) Fungal farming in a snail. PNAS 100(26):15643–15648
Simpson JA (1998) Why don't more birds eat more fungi? Australas Mycol Newsl 17:67–68
Simpson JA (2000) More on mycophagous birds. Australas Mycol 19(2):49–51
Sivinski J (1981) Arthropods attracted to luminous fungi. Psyche 88:383–390
Sivinski JM (1998) Phototropism, bioluminescence, and the diptera. Fla Entomol 81:282–292
Skerratt LF, BergerL SR, Cashins S, McDonald KR, Phillott AD, Hines HB, Kenyon N (2007) Spread of chytridiomycosis has caused the rapid global decline and extinction of frogs. EcoHealth 4(2):125–134
Skjøth CA, Sommer J, Frederiksen L, Gosewinkel Karlson U (2012) Crop harvest in Denmark and Central Europe contributes to the local load of airborne *Alternaria* spore concentrations in Copenhagen. Atmos Chem Phys 12(22):11107–11123
Skvarla JJ, Anderegg DE (1972) Infestation of cedar pollen by Rhizophidium (Chytridiomycetes). Grana 12(1):47–51
Smirnov NN (1964) On the quantity of allochthonous pollen and spores received by the Rybinsk Reservoir. Hydrobiologia 24:421–429
Smith BJ, Sampson BJ, Walter M (2012) Efficacy of bumble bee disseminated biological control agents for control of *Botrytis* blossom blight of rabbiteye blueberry. Int J Fruit Sci 12(1–3):156–168. doi:10.1080/15538362.2011.619359
Sommerstorff H (1911) Ein Tiere fangender Pilz. Plant Syst Evol 61:361–373
Sparrow F (1968) Ecology of freshwater fungi. In: Ainsworth G, Sussman A (eds) The fungi—an advanced treatise, the fungal population, vol III. Academic, New York, pp 41–93
Sreeramulu T (1959) The diurnal and seasonal periodicity of spores of certain plant pathogens in the air. Trans Br Mycol Soc 42(1):177–184
Sridhar KR (2009) Fungi in the tree canopy—an appraisal. In: Rai M, Bridge P (eds) Applied mycology. CAB International, New York, pp 73–91
Sridhar KR, Bärlocher F (1994) Viability of aquatic hyphomycete conidia in foam. Can J Bot 72:106–110

Sridhar KR, Karamchand KS (2009) Diversity of water-borne fungi in throughfall and stemflow of tree canopies in India. Sydowia 61:327–344

Stephenson SL, Stempen H (1994) Myxomycetes: a handbook of slime molds. Timber Press, Portland, OR

Stubbs CS (1995) Dispersal of soredia by the oribatid mite, *Humerobates arborea*. Mycologia 87:454–458

Suberkropp K (1992) Aquatic hyphomycetes communities. In: Caroll GC, Wicklow DT (eds) The fungal community: its organization and role in the ecosystem, 2nd edn. Marcel Dekker, New York, pp 729–747

Suberkropp K, Klug MJ (1980) The maceration of deciduous leaf litter by aquatic hyphomycetes. Can J Bot 58:1025–1031

Suffert F, Latxague É, Sache I (2009) Plant pathogens as agroterrorist weapons: assessment of the threat for European agriculture and forestry. Food Secur 2:221–232

Szécsi Á, Magyar D (2011) Yeast-mycelium dimorphism in fumonisin-producing *Fusarium verticillioides*. Acta Phytopathologica et Entomologica Hungarica 46(2):185–190

Szécsi Á, Magyar D, Tóth S, Szőke C (2013) Poaceae: a rich source of endophytic fusaria. Acta Phytopathologica et Entomologica Hungarica 48(1):19–32

Talbot NJ (1997) Fungal biology: growing into the air. Curr Biol 7(2):R78–R81

Taylor JW, Turner E, Townsend JP, Dettman JR, Jacobson D (2006) Eukaryotic microbes, species recognition and the geographic limits of species: examples from the kingdom Fungi. Philos Trans R Soc B 361:1947–1963

Tedersoo L, Bahram M, Põlme S, Kõljalg U, Yorou NS, Wijesundera R, Ruiz LV, Vasco-Palacios AM, Thu PQ, Suija A, Smith ME, Sharp C, Saluveer E, Saitta A, Rosas M, Riit T, Ratkowsky D, Pritsch K, Põldmaa K, Piepenbring M, Phosri C, Peterson M, Parts K, Pärtel K, Otsing E, Nouhra E, Njouonkou AL, Nilsson RH, Morgado LN, Mayor J, May TW, Majuakim L, Lodge DJ, Lee SS, Larsson K-H, Kohout P, Hosaka K, Hiiesalu I, Henkel TW, Harend H, Guo L-d, Greslebin A, Grelet G, Geml J, Gates G, Dunstan W, Dunk C, Drenkhan R, Dearnaley J, De Kesel A, Dang T, Chen X, Buegger F, Brearley FQ, Bonito G, Anslan S, Abell S, Abarenkov K (2014) Global diversity and geography of soil fungi. Science 346(6213):1256688

Teron JN, Hutchison LJ (2013) Consumption of truffles and other fungi by the American Red Squirrel (*Tamiasciurus hudsonicus*) and the Eastern Chipmunk (*Tamias striatus*) (Sciuridae) in northwestern Ontario. Can Field-Nat 127(1):57–59

Terrell EE, Batra LR (1982) *Zizania latifolia* and *Ustilago esculenta*, a grass-fungus association. Econ Bot 36(3):274–285

Thomas K, Chilvers GA, Norris RH (1990) A dynamic model of fungal spora in a freshwater stream. Mycol Res 95:184–188

Thrall PH, Jarosz AM (1994) Host-pathogen dynamics in experimental populations of *Silene alba* and *Ustilago violacea*. I. Ecological and genetic determinants of disease spread. J Ecol 82:549–560

Tilak ST, Pande BN (2005) Current trends in aeromycological research. In: Rai MK, Deshmukh SK (eds) Fungi: diversity and biotechnology. Scientific Publishers, Jodhpur, pp 281–510

Trail F (2007) Fungal cannons: explosive spore discharge in the Ascomycota. FEMS Microbiol Lett 276(1):12–18

Trappe JM, Maser C (1977) Ectomycorrhizal fungi: interactions of mushrooms and truffles with beasts and trees. In: Walters T (ed) Mushrooms and man, an interdisciplinary approach to mycology. Linn-Benton Community College, Albany, OR, pp 65–179

Tuboly L, Vörös J (1962) A dohányperonoszpóra és leküzdésének lehetőségei. Mezőgazdasági világirodalom 4:46–53

Tucker K, Stolze JL, Kennedy AH, Money NP (2007) Biomechanics of conidial dispersal in the toxic mold *Stachybotrys chartarum*. Fungal Genet Biol 44:641–647

Turott C, Levetin E (2001) Correlation of spring spore concentrations and meteorological conditions in Tulsa, Oklahoma. Int J Biometeorol 45:64–74

Uchida W, Matsunaga S, Sugiyama R, Kazama Y, Kawano S (2003) Morphological development of anthers induced by the dimorphic smut fungus *Microbotryum violaceum* in female flowers of the dioecious plant *Silene latifolia*. Planta 218(2):240–248
Upadhyay HP (1981) A monograph of *Ceratocystis* and *Ceratocystiopsis*. University of Georgia Press, Athens, GA
Urban A, Kataržytė M, Schickmann S, Kräutler K, Pla T (2012) Is small mammal mycophagy relevant for truffle cultivation? Acta Mycol 47(2):139–143
Van Leeuwen MR, Van Doorn TM, Golovina EA, Stark J, Dijksterhuis J (2010) Water-and air-distributed conidia differ in sterol content and cytoplasmic microviscosity. Appl Environ Microbiol 76(1):366–369
Varga J, Naár Z (2002) A collembolák mint lehetséges terjesztõi a mohákban élõ gombáknak. Folia Hist Nat Mus Matraensis 26:115–120
Vega FE, Posada F, Aime MC, Pava-Ripoll M, Infante F, Rehner SA (2008) Entomopathogenic fungal endophytes. Biol Control 46(1):72–82
Venables KM, Allitt U, Collier CG, Emberlin J, Greig JB, Hardaker PJ, Highham JH, Laing-Morton T, Maynard RL, Murray V, Strachan D, Tee RD (1997) Thunderstorm-related asthma-the epidemic of 24/25 June 1994. Clin Exp Allergy 27:725–736
Vernes K, Dunn L (2009) Mammal mycophagy and fungal spore dispersal across a steep environmental gradient in eastern Australia. Austral Ecol 34(1):69–76
Viljanen-Rollinson SLH, Cromey MG (2002) Pathways of entry and spread of rust pathogens: implications for New Zealand's biosecurity. N Z Plant Prot 55:42–48
Vincent W (1988) Microbial ecosystems in Antarctica. Cambridge University Press, Cambridge
Wadia KDR, McCartney HA, Butler DR (1998) Dispersal of *Passalora personata* conidia from groundnut by wind and rain. Mycol Res 102:355–360
Waggoner PE, Taylor GS (1958) Dissemination by atmospheric turbulence: spores of *Peronospora tabacina*. Phytopathology 48:46–51
Waid JS (1954) Occurrence of aquatic hyphomycetes upon the root surfaces of beech grown in woodland soils. Trans Br Mycol Soc 37(4):420–421
Waller JM (1979) The recent spread of coffee rust (*Hemileia vastatrix*) and attempts to control it. In: Ebbels DL, King JE (eds) Plant health. Blackwell, Oxford, pp 275–283
Wang ZG, Wang ZG, Cong LM, Liu XM, Tong Z, Cheng SY, Ge SJ (1999) Dimorphic fungus characteristics of fumonisin-producing strains of *Fusarium moniliforme* from Zhejiang. Mycopathology 144:165–167
Ward SV, Manners JG (1974) Environmental effects on the quantity and viability of conidia produced by *Erysiphe graminis*. Trans Br Mycol Soc 62:119–128
Watson I, De Sousa C (1983) Long distance transport of spores of *Puccinia graminis tritici* in the southern hemisphere. Proc Linnean Soc NSW 106:311–321
Webber JF (1990) Relative effectiveness of *Scolytus scolytus*, *S. multistriatus* and *S. kirschi* as vectors of Dutch elm disease. Eur J For Pathol 20:184–192
Webster J (1959) Experiments with spores of aquatic hyphomycetes. I. Sedimentation and impaction on smooth surfaces. Ann Bot 23:595–611
Webster J (1970) Coprophilous fungi. Trans Br Mycol Soc 54:161–180
Webster J (1980) Introduction to fungi, 2nd edn. Cambridge University Press, Cambridge
Webster J (1992) Anamorph-teleomorph relationships. In: Bärlocher F (ed) The ecology of aquatic hyphomycetes. Springer, Berlin, pp 99–117
Webster J, Descals E (1981) Morphology, distribution, and ecology of conidial fungi in freshwater habitats. In: Cole GC, Kendrick B (eds) Biology of conidial fungi, vol 1. Academic, New York
Weinhold AR (1955) Rate of fall of uredospores of *Puccinia graminis tritici* Erikss & Henn. as affected by humidity and temperature. Technical Report of Botany and Plant Pathology, Department Colorado, p 104
Weitz WHJ (2004) Naturally bioluminescent fungi. Mycologist 18(01):4–5
Weston WD, Taylor RE (1948) The plant in health and disease. Crosby Lockwood, London
Wheeler WM (1907) The fungus growing ants of North America. In: Wingfield MJ, Seifert KA, Weber JF (eds) (1993) *Ceratocysitis* and *Ophiostoma* taxonomy. Dover Reprint, New York

Wheelis M, Rózsa L, Dando M (eds) (2006) Deadly cultures: biological weapons since 1945. Harvard University Press, Cambridge, MA
Whitby SM (2002) Biological warfare against crops. Palgrave, New York
Whittaker RJ, Fernández-Palacios JM (2007) Island biogeography: ecology, evolution, and conservation. Oxford University Press, Oxford
Wicklow DT (1981) Biogeography and conidial fungi. In: Cole GT, Kendrick B (eds) The biology of conidial fungi, vol 1. Academic, New York, pp 417–447
Wiemken V, Boller T (2006) Delayed succession from alpine grassland to savannah with upright pine: limitation by ectomycorrhiza formation? For Ecol Manage 237:492–502
Wier AM, Tattar TA, Klekowski EJ (2000) Disease of Red Mangrove (*Rhizophora mangley*) in Southwest Puerto Rico Caused by *Cytospora rhizophorae1*. Biotropica 32(2):299–306
Wiese MV (1998) Smuts. Common Bunt (Stinking Smut); Dwarf Bunt. In: Compendium of wheat diseases, 2nd edn. APS Press, St Paul, MN, pp 18–21
Williets HM, Thomson AR, Morris G (1989) Laboratory studies on the selection of *Pythium ultinum*, plant-pathogenic fungus by *Onychiurus auranticus* (Collembola). Asp Appl Biol 23:373–378
Wingen LU, Shaw MW, Brown JKM (2013) Long-distance dispersal and its influence on adaptation to host resistance in a heterogeneous landscape. Plant Pathol 62(1):9–20
Wingfield MJ, Seifert KA, Weber JF (eds) (1993) *Ceratocystis* and *Ophiostoma* taxonomy, ecology, and pathogenicity. APS Press, St. Paul, MN
Wiseman R (1932) Untersuchungen über die überwinterung des Apfelschorfpilzes *Fusicladium dendriticum* (Wallr.) Fckl. im tötem Blatt sowie die Ausbreitung der Sommersporen (konidien) des Apfelschorfes. Landwirtschaftliches Jahrbuch der Schweiz 46:619–679
Wood-Eggenschwiler S, Bärlocher F (1985) Geographical distribution of Ingoldian fungi. Verh Internat Verein Limnol 22:2780–2785
Wornik S, Grube M (2010) Joint dispersal does not imply maintenance of partnerships in lichen symbioses. Microb Ecol 59(1):150–157
Wurzbacher CM, Bärlocher F, Grossart HP (2010) Fungi in lake ecosystems. Aquat Microb Ecol 59:125–149
Wurzbacher C, Rösel S, Rychła A, Grossart HP (2014) Importance of saprotrophic freshwater fungi for pollen degradation. PloS one 9(4):e94643
Young SA, Kovalek WP, del Signore KA (1978) Distances travelled by autumn-shed leaves introduced into a woodland stream. Am Midl Nat 100:217–222
Yu H, Sutton JC (1997) Effectiveness of bumblebees and honeybees for delivering inoculum of *Gliocladium roseum* to raspberry flowers to control *Botrytis cinerea*. Biol Control 10(2):113–122
Yu Y, Wang Y (1995) Notes on some fungi isolated from freshwater in Antarctica. Acta Mycol Sin 15(2):81–85
Yuceer C, Hsu C-Y, Erbilgin N, Klepzig KD (2011) Ultrastructure of the mycangium of the southern pine beetle, *Dendroctonus frontalis* (Coleoptera: Curculionidae, Scolytinae): complex morphology for complex interactions. Acta Zool 92(3):216–224
Zhao S, Shamoun SF (2010) Conidial morphology, structure and development in Valdensinia heterodoxa. Mycology 1:113–120
Zoberi MH (1961) Take-off of mould spores in relation to wind speed and humidity. Ann Bot 25:53–64
Zopf W (1884) Zur Kenntniss der Phycomyceten—Zur Morphologie und Biologie der Ancylisteen und Chytridiaceen zugleich ein Beitrag zur Phytopathologie, vol 47. Nova Acta der Kaiserlichen Leopoldino-Carolinischen Deutschen Akademie der Naturforscher, Halle

# Chapter 15
# Microfungi in Indoor Environments: What Is Known and What Is Not

**Chin Yang, Sepideh Pakpour, John Klironomos, and De-Wei Li**

## Introduction

Indoor molds refer to the microfungi occurring in indoor environments. A majority of indoor fungi belong to the microfungi. A few of them are members of Basidiomycota, such as wood-decaying fungi and species of *Coprinus* s.l. A member of Ascomycota, *Peziza domiciliana* is occasionally found in indoor environments. The World Health Organization (WHO) recognized that "[h]ealthy indoor air is recognized as a basic right" (WHO 2009). The indoor bio-deterioration and health effects associated with fungal infestation in indoor environments have become a research priority (Green et al. 2011). More and more studies in the last decade demonstrated the association between indoor fungi and their detrimental effects on human health. "WHO guidelines for indoor air quality: dampness and mould" concluded that "the most important effects are increased prevalence of

---

C. Yang, Ph.D.
Prestige EnviroMicrobiology, Inc, 242 Terrace Boulevard Suite B-1, Voorhees, NJ 08043, USA
e-mail: chins.yang@prestige-em.com; csyang1000@gmail.com

S. Pakpour • J. Klironomos, Ph.D., F.R.S.C.
Department of Biology, University of British Columbia,
Okanagan campus, Kelowna, BC, Canada V1V 1V7

I.K. Barber School of Arts and Sciences, The University of British Columbia,
Okanagan campus, ASC 402, 3333 University Way, Kelowna, BC, Canada V1Y 4S8
e-mail: sepideh.pakpour@ubc.ca; john.klironomos@ubc.ca

D.-W. Li, Ph.D. (✉)
The Connecticut Agricultural Experiment Station Valley Laboratory,
153 Cook Hill Road, Windsor, CT 06095, USA

Co-Innovation Center for Sustainable Forestry in Southern China, Nanjing Forestry University, Nanjing, Jiangsu 210037, China
e-mail: dewei.li@ct.gov



D.-W. Li (ed.), *Biology of Microfungi*, Fungal Biology,
DOI 10.1007/978-3-319-29137-6_15

respiratory symptoms, allergies and asthma as well as perturbation of the immunological system" (WHO 2009). Molds associated with poor indoor air quality are important risk factors for human health in both low-income and middle- and high-income countries (WHO 2009).

Part of the difficulty is that several hundred indoor microfungi have been reported in indoor environments. This number of microfungi is significantly underestimated based on authors' observations. It is common to find more than two dozen fungi coexisting in a residence from a single investigation of indoor molds. Frequently, the indoor fungi are only identified to genus level due to the costs and a lack of well-trained mycologists. It poses a huge challenge to segregate multiple factors that determine which fungus or fungi are responsible for the adverse effects on health, since the occupants are exposed to multiple agents simultaneously. Also, other biological agents, such as dust mites, pollen, bacteria, viruses, and pets, are present in conjunction with indoor molds. These confounding agents/factors make the determination of causal agents even more difficult. The other challenge is to accurately estimate exposure level and to link it to relevant symptoms and health outcomes due to the exposure (WHO 2009). It is understandable why at present the causal relationship between indoor microfungi and some associated health problems, such as sick building syndrome, is still subject to debate.

It is indisputable that significant development has been made in indoor molds research. It has led to the publication of "WHO guidelines for indoor air quality: dampness and mould" in 2009. In 2012, the American Society of Heating, Refrigerating and Air-Conditioning Engineers (ASHRAE) changed their position and recognized "the association between damp buildings and the potential for adverse health effects" and advised its membership that the association of molds with health risks is real (Harriman 2012).

Since several books and review papers on indoor molds have been published (Straus 2004; Prezant et al. 2008; Yang and Heinsohn 2007; Jaakkola et al. 2013; Quansah et al. 2012; Nevalainen et al. 2015), this chapter mainly focuses on the latest developments and new directions of research in the last decade.

## Fungal Systematics and Indoor Microfungi Research

Systematic studies on a number of indoor microfungi-related genera, such as *Acremonium*, *Aspergillus*, *Cladosporium*, *Penicillium*, and *Stachybotrys*, using DNA sequence data had led to the redelineation of a number of these genera in the last decade (Bensch et al. 2010; Zalar et al. 2007; Summerbell et al. 2011).

Scott et al. (2004) studied indoor *Penicillium chrysogenum* in Ontario, Canada. The results showed that *P. chrysogenum* is the most common fungus indoors (Scott et al. 2004). It is also the most common fungus in water-damaged buildings in Denmark and Greenland (Andersen et al. 2011). The isolates collected during the study by Scott et al. (2004) formed four clades. The first clade contained more than 5.6 % of house isolates clustered with the ex type strains of *P. chrysogenum* and

*P. notatum*. The 4th clade clustered with Fleming's strain and was the dominant one which included more than 90 % of house isolates (Scott et al. 2004). A further study *P. chrysogenum* was conducted by Houbraken et al. (2011). Their results showed that Fleming's strain is not *Penicillium chrysogenum*, but *P. rubens* which is the most common fungi (clade 4 in Scott et al. study) in indoor environments.

*Cladosporium cladosporioides* s.l. was studied by Bensch et al. (2010) and 22 species were segregated and described as new species according to phylogenetic analysis using sequencing data and cryptic morphological differences (Bensch et al. 2010). *Cladosporium sphaerospermum* s.l. was also redelineated. *Cladosporium dominicanum*, *C. halotolerans*, *C. psychrotolerans*, *C. spinulosum*, and *C. velox* were described, all with globoid conidia (Zalar et al. 2007). The subtle morphological differences among these newly described species result in a major challenge to identify them morphologically.

According to the study by Summerbell et al. (2011), *Acremonium strictum*, a very common indoor microfungus, was combined as *Sarocladium strictum* and *Acremonium kiliense*, another indoor microfungus, was changed to Sarocladium kiliense.

A number of new species and records have been described or reported from indoor environments, such as the new species, *Balaniopsis triangularis* D. W. Li and W. B. Kendrick, and a new record in Canada and the USA, *Triadelphia australiensis* B. Sutton (Li et al. 2008, 2013). Several noteworthy fungi isolated from indoor environments in the USA were described and reported for the first time: *Ascotricha erinacea* Zambett., *Sporoschisma saccardoi* Mason & Hughes apud Huges, *Stachybotrys microspora* (Mathur & Sankhla) Jong & Davis, *S. nephrospora* Hansford, and *Zygosporium masonii* Hughes (Li and Yang 2004b). (Li and Yang 2004b). In fact, this was the first report of *Ascotricha erinacea* since it was described by Zambettakis in 1955 (Zambettakis 1955). *Stachybotrys elegans* (Pidopl.) W. Gams and *Parascedosporium putredinis* (Corda) Lackner & de Hoog were reported from indoor environments for the first time (Li et al. 2013). Recently, *Stachybotrys elegans* was reported from intensive care units in Assiut University Hospitals in Egypt (Aboul-Nasr et al. 2014). *Stachybotrys chartarum* and *S. elegans* are species complexes and further studies are necessary to delineate the two species complexes in the future (Wang et al. 2015).

*Wallemia sebi*, a xerophilic and mycotoxigenic microfungus, was sometimes reported from indoor environments (Samson et al. 2004). *Wallemia* was previously considered to be Ascomycetes (Kirk et al. 2001). Its dolipore/parenthesome septa were reported over four decades ago (Terracina 1974). Moore (1986) opined that *W. sebi* may be a basidiomycete and a teleomorph. Moore's former opinion was confirmed by a phylogenetic study not long ago. The genus *Wallemia* indeed belongs to Basidiomycota and a new class Wallemiomycetes and a new order Wallemiales were erected to accommodate it (Zalar et al. 2005). A genomic study by Padamsee et al. (2012) suggested that *W. sebi* has cryptic sexual reproduction. Moore's latter opinion may not be far off.

*Tritirachium* spp., a hyphomycete genus occasionally found indoor environments, were formerly placed in Pezizomycotina, Ascomycota (Kirk et al. 2008).

However, a recent study by Schell et al. (2011) found that this genus is a new lineage in Basidiomycota. Subsequently, subphylum Pucciniomycotina, class Tritirachiomycetes, order Tritirachiales, and family Tritirachiaceae were proposed to accommodate this genus (Schell et al. 2011).

Since the passage of the International Code of Nomenclature for algae, fungi, and plants (Melbourne Code) or (ICN) adopted by the 18th International Botanical Congress Melbourne, Australia, July 2011 (McNeill et al. 2012), the principle one fungus = one name has been in effect. Dual name systems (asexual and sexual names) for fungi were terminated. Only one name is used for each species. It has led to name changes for many fungi. Name changes for some taxa are pending, such as *Aspergillus*, *Penicillium*, *Fusarium*, and their teleomorphs (Geiser et al. 2013; Pitt and Taylor 2014). This change of the Melbourne Code has significant implications for indoor microfungi research and investigation. Both states of *Aspergillus* and its teleomorphs, *Eurotium* and *Emericella*, as well as *Penicillium* and its teleomorphs, *Eupenicillium* and *Talaromyces*, are common in indoor environments or frequently found on culture media. The availability of the genus name of *Paecilomyces* remains to be determined. *Bipolaris* and *Curvularia* are linked to teleomorphic *Cochliobolus*. A recent study placed *Bipolaris* and *Curvularia* into two separate clades (Manamgoda et al. 2012) and may solve the problem and allow us to continue using these two names. The name *Tetraploa*, the conidia of this genus sometimes found in the air, is linked to its sexual stage, *Massarina*, which is polyphyletic including two other genera, *Acrocalymma* and *Ceratophoma*. Unfortunately, many more names related to indoor microfungi can be listed. Uncertainty of name changes for these common indoor fungi is leading to an increasing concern to indoor mold researchers and investigators. To better understand the changes to article 59 of the Melbourne Code and its implication to indoor mold research, please refer to Chap. 2 by Walter Gams for a detailed discussion.

## Fungal Fragments

Airborne fungal fragments as a component of bioaerosols have been overlooked in aeromycological studies in the past. Samir et al. (2014) indicated that hyphal fragments and hyphae as aeroallergens are underestimated. A significant number of studies did not include fungal fragments as one of the fungal categories/groups/taxa. Fungal fragments include hyphal and conidial fragments and partial conidiophores, ascomata and basidiomata, and other partial fungal structures. Hyphal fragments were reported in various aeromycological studies and laboratory reports in the past (Li and Kendrick 1995a, b; Li and Yang 2004a), but not other fungal fragments. The difficulties in identifying and enumerating these fungal fragments are implicating scientific understanding of the significance of fungal exposure in indoor environments and associated health effects. Exposure to fungal fragments has attracted attention in the last decade. There is an increasing interest in aerosolized fungal fragments and their association with asthma severity (Green et al. 2006), especially the submicron fungal fragments. The health effects of exposure to respirable sized fungal fragments in indoor and occupational environments infested by fungi remain less clear and need further studies in the future.

Green et al. (2011) coined two new terms for fungal particulates: gonomorphic and non-gonomorphic. Gonomorphic refers to the fungal structures which are differentiated as separable, dispersive reproductive structures (sexual and asexual spores). Non-gonomorphic particles are defined to be mechanically severed from the parent mycelium but were not programmatically differentiated as separable (Green et al. 2011). Non-gonomorphic particles include hyphal fragments (<100 μm), chlamydospores, partial multicellular conidia, and subcellular fragments of hyphae and conidia (Eduard et al. 2012). A number of researches have focused on it. Immunodiagnostic methods demonstrated that non-gonomorphic particles contain antigens as well as allergens (Eduard et al. 2012). These preliminary studies have initiated collaborative studies into the occurrence and possible health effects associated with personal exposure to non-gonomorphic particles.

Particle fragmentation can be facilitated by several biotic (fungal autolysis, hyphal vacuolation, shizolytic/rhexolytic separation, as well as prokaryote, protozoan, and microarthropod comminution) or abiotic processes (wind, vibration, anthropogenic, and mechanical disturbances). In some environments, larger non-gonomorphic particles (>2.5 mm) may represent a significant proportion of the fungal bioaerosol load (ca. 56 %) and are derived from species within the orders Capnodiales, Eurotiales, and Pleosporales (Green et al. 2011).

Fungal fragments (< spores) were found to be released from fungal colonies in air chamber studies (Eduard et al. 2012). Hyphal fragments are non-gonomorphic particles that tend to be primarily derived from outdoor sources during the summer and from indoor sources during the winter (Li and Kendrick 1996). Hyphal fragments in the air were often within 7–100 μm in length and viable (Pady and Kramer 1960; Pady and Gregory 1963). Meirer and Lindbergh (1935) reported airborne hyphal fragments near the Arctic Circle.

Green et al. (2005) demonstrated that airborne fungal fragments (hyphal fragments and fragmented conidia) and conidia of a number of previously undocumented genera are new aeroallergen sources. Green et al. (2006) indicated that respiratory deposition models suggest that submicron fragments of *Stachybotrys chartarum* may be deposited in 230–250-fold higher numbers than spores. Their studies showed that the concentrations of total airborne hyphae were frequently significantly higher than those of conidia of individual allergenic genera. Approximately, 25 % of all hyphal fragments expressed detectable allergens and conidia of previously uncharacterized genera were sources of allergens (Green et al. 2006).

Samir et al. (2014) demonstrated that immunostaining of fungal hyphae was heterogeneous, and ~27 % of all fungal fragments and hyphae expressed detectable allergens compared with nonstained hyphae. Their results showed that fungal hyphae and fragments are underestimated sources of airborne allergens since positively immunostained hyphal fragments were detected in all samples and the quantity of the detected fungal hyphae in any of the individual protein-binding membranes was significantly higher than the spore counts of *Alternaria* spp., *Aspergillus* spp., and *Cladosporium* spp. (Samir et al. 2014).

Madsen (2012) found that aerosolization of conidia and fungal fragments of *Botrytis cinerea* occurred under an airflow of 1.5 m $s^{-1}$ or 0.5 m $s^{-1}$. She determined that the size of the respirable fraction of the aerosolized particles was dependent on the RH. At high RH, about 30 % of the aerosolized particles were of respirable size, while at low RH, about 70 % were of respirable size. Under low RH, more fungal (1 → 3)-β-D-glucan and chitinase became airborne than under high RH.

Size-selective studies on exposure to fungal spores and fragments were not well studied until a decade ago. Reponen et al. (2007) stated that fungal fragments with smaller sizes (<1 μm) may contribute to mold-related health effects. A number of laboratory studies showed that large numbers of submicrometer-sized fungal fragments (30 nm–1 μm) were liberated along with unbroken spores from infested surfaces. The number of released fungal fragments was constantly higher, up to 500 times, than the number of unfragmented spores, and the number of fragments is not correlated with spores (Górny et al. 2002; Cho et al. 2005). Reponen et al. (2007) studied airborne fungal particles in three sized groups: (a) >2.25 μm (spores), (b) 1.05–2.25 μm (mixture), and (c) <1.0 μm (submicrometer-sized fragments). The authors concluded that the real contribution of fungal fragments to the overall exposure may be very high, even much higher than those estimates from previous laboratory studies.

A recent size-selective study on personal exposure of agricultural workers to airborne fungi and fungal fragments showed that *Alternaria* and *Botrytis* were highly correlated with (1 → 3)-β-D-glucan at the aerodynamic size < 1 μm, which was much smaller than the expected spore sizes (the average aerodynamic sizes were 18.5 μm for *Alternaria* and 6.1 μm for *Botrytis*). Thus, the assumption is that *Alternaria* and *Botrytis* might release small fragments into the air and these aerosolized submicron fungal particulates could enter the deep lung and cause respiratory diseases (Lee and Liao 2014).

Cho et al. (2005) found that the released fungal fragments of *Stachybotrys chartarum* were 380 particles $cm^{-3}$, which was approximately 514 times higher than those of conidia, and *Aspergillus versicolor* released comparable numbers of spores and fragments. Their modeling showed that *S. chartarum* fragments had 230–250-fold higher respiratory deposition than its conidia and *A. versicolor* had a comparable result.

Adhikari et al. (2013) studied particulates of fungal fragments in three aerodynamic size fractions: <1.0, 1.0–1.8, and >1.8 μm and used N-acetylhexosaminidase (NAHA) as a marker of fungal cell biomass. Significant relationships were found between the amounts of NAHA in the total amount and in the size fraction >1.8 μm but not in the smaller fractions.

Seo et al. (2014) found that the concentration of (1,3)-β-D-glucan in submicron fungal fragments in indoor air was 2 times higher in homes with asthmatic children (50.9 pg/$m^3$) compared to homes with non-asthmatic children (26.7 pg/$m^3$) in South Korea and that relative humidity was negatively correlated with the concentration of (1,3)-β-D-glucan in submicron fungal fragments. At present, a number of studies have concluded that fungal hyphae and fragments are underestimated sources of aeroallergens (Green et al. 2006; Samir et al. 2014). A number of studies are starting to focus on using nanotechnology in indoor molds (Gong et al. 2014; Filip 2009; Rácová et al. 2013).

## Mold, Dampness, and Water Damage

Water damage/intrusion and dampness in buildings without immediate corrective actions are the determining factor which allows indoor molds to grow, eventually leading to fungal infestations indoors. Dampness/high humidity and long-term water damage lead to different fungal compositions and fungal succession. The difference is determined by water activity ($a_w$) in the substrates/building materials. Exposure to molds associated with dampness and water damage in indoor environments has received significant attention in recent research (Kennedy and Grimes 2013). For an assessment of dampness in indoor environments, the review paper by Kennedy and Grimes (2013) can be consulted.

Dampness due to high relative humidity (RH) in the air is an often overlooked problem which results in fungal infestation. High RH allows fungi, especially xerophilic taxa, to develop in indoor environments. A number of fungi (*Alternaria* spp., *Cladosporium* spp., *Acremonium* sp., and *Ulocladium* sp., etc.) and visible fungal colonies developed on ceiling tiles in a school with inadequate operation of the A/C system during the summer in a coastal area of Connecticut (Li unpublished observation).

Karvala et al. (2014) studied patients with occupational asthma (OA) or work-exacerbated asthma (WEA) and concluded that adverse work ability outcomes are associated with asthma in relation to workplace dampness. They opined that there was a need to put effective preventive measures in place to assist workers with indoor air symptoms to sustain their work ability. Haverinen-Shaughnessy et al. (2012) found that moisture problems (water damage, dampness, and molds) are rather common in schools in the Netherlands, Spain, and Finland. The occurrence and severity may show a discrepancy across geographical areas, which can be partly explained by building characteristics.

Black Aspergilli (*Aspergillus* section Nigri) were studied in indoor environments in six countries: Algeria, Croatia, Hungary, the Netherlands, Thailand, and Turkey recently (Varga et al. 2014). The highest species diversity (seven species including a new species) was observed in indoor samples from Thailand, while the lowest (two species) was found in Algeria. *A. niger*, *A. tubingensis*, *A. luchuensis*, and *A. welwitschiae* were identified in all three temperate European countries, while *A. tubingensis*, *A. luchuensis*, and *A. welwitschiae* were also detected in Turkey, but not *A. niger* (Varga et al. 2014).

A recent study with 13,335 participants on associations between home dampness and asthma and related symptoms in 4- to 6-year-old children in Shanghai found that home dampness was strongly and significantly associated with dry cough, wheeze, and rhinitis symptoms. Children exposed to visible mold spots had a 32 % higher risk of asthma (adjusted OR, 95 % CI: 1.32, 1.07–1.64) (Hu et al. 2014).

## Mites

House dust mites and other indoor mites and their association with indoor molds are little studied, although they have been observed and reported on a few occasions (Yang 2007; Samson and Houbraken 2011). Mites can be serious contaminants in laboratories by feeding on fungal colonies and carrying fungal spores around. House dust mites (HDM) are major allergens indoors and most studies were focused on allergens of HDM and associated allergies in the past. However, HDM are often present on building materials, especially on drywall infested by indoor molds. The relationship between HDM and indoor molds has been overlooked in the past. Part of the reason is lack of expertise in mites. Very few people working on indoor molds have the acarological background to identify mites to its taxon. Very few acarologists are available to assist identifying mites collected from indoor environments. The ecological roles played by house dust mites associated with fungi in indoor environments are unclear, such as feeding behaviors on fungal colonies and spreading fungal infestations by transporting fungal spores on their bodies. In addition, little is known concerning the viability and allergenicity of fungal spores and fragments in mite fecal matter.

It is common knowledge that dust mites are the predominant producers of inhalation allergens in many regions in the world (Cui 2014). The most common mite species that produce allergens are *Dermatophagoides pteronyssinus*, the European house dust mite, and *Dermatophagoides farinae*, the American house dust mite (WHO 2009). These two species are not confined to Europe or North America as their common names imply. Another species, *Euroglyphus maynei,* also has a wide distribution. Eleven other dust mite species have been reported indoors (Cui 2014). Most research has focused on *D. pteronyssinus* and *D. farina* and their allergenicities and health effects. The major allergens produced by *D. pteronyssinus* are proteases (Der p I and Der p II), which are reported in large amounts in fecal materials, while *D. farinae* produced the major allergen Der f I (Institute of Medicine 2000). Increased levels of these allergens have been reported in house dust, mattress dust bedding, and upholstery (Van Strien et al. 2004; Cui 2014).

Recently, Naegele et al. (2013) studied feeding preferences of *Dermatophagoides farinae* to 16 indoor fungi. They found that *D. farinae* preferred to feed on *Alternaria alternata*, *Cladosporium sphaerospermum*, and *Wallemia sebi. Penicillium chrysogenum*, *Aspergillus versicolor*, and *Stachybotrys chartarum* were at the bottom of its preferential list. Naegele et al. (2013) also suggested that the food preferences of *D. farinae* may play dual roles in indoor mold infestation: a decrease in spore numbers by feeding on fungal structures and spreading fungal spores its bodies carry. However, it is necessary to study the factors which determine the feeding preferences of mites associated with indoor molds. MVOC emitted by fungi, mycotoxins/secondary metabolites in fungal structures, or water activities in building materials should be further studied.

## Indoor Substrates and Indoor Molds

There are a number of materials present in indoor environments that can serve as substrates for fungal growth such as drywall, wooden structures, furniture, carpet, paints, books, paper products, leather products (shoes, jackets, couches), upholstery, and wallpaper, etc.

### *Wood*

Wood is highly susceptible to fungal colonization by fungi such as *Cladosporium*, *Penicillium*, and *Aspergillus* (Bjurman 1994). Wood drying, specifically kiln drying, results in a higher amount of nitrogen and low molecular carbohydrates on the wood surface which makes it more susceptible to mold growth (Thelander et al. 1993; Viitanen 1997). Other studies illustrated that some engineered wood products, such as oriented strand boards (OSB), plywood, and medium-density fiberboard (MDF), are more susceptible to growth of *Aspergillus*, *Trichoderma*, and *Penicillium* than solid wood, particleboard (Chung et al. 1999), acylated wood (Suttie et al. 1998), and wood polyethylene composites (Mankowski and Morrell 2000).

### *Wallpaper and Plasters*

Both paper and glue are very good media for many indoor fungi and thus wallpapers are very susceptible to mold growth (Bissett 1987; Grant et al. 1989). Synthetic polymers such as synthetic rubber and plastic can also be degraded by different species of fungi such as *Aspergillus niger*, *Aspergillus flavus*, *Aureobasidium pullulans*, *Chaetomium* spp., *Penicillium funiculosum*, *Penicillium luteum*, and *Trichoderma* spp. (Flannigan and Miller 2001).

Recently, prefabricated gypsum plaster paper boards (drywall) are commonly used in new buildings. The boards are highly susceptible to mold growth, mainly by the cellulolytic *S. chartarum* because of the paper used to reinforce the material. In addition, the gypsum itself can support fungal growth due to its nutrient contents and additives that make it more hygroscopic at lower humidity levels (Nielsen et al. 1998; Eppley and Bailey 1973; Andersen et al. 2002).

### *Plastic and Glass*

Fungi can also grow on polyethylene and polyvinyl chloride (PVC) as they can degrade most plasticizers including common organic acid esters such as dioctyl phthalate (DOP) and dioctyl adipate (DOA) (Webb et al. 2000). Glass-reinforced plastic (GRP), popularly known as fiberglass, and fiberglass ceiling tiles which contain 10 % ureaphenol-formaldehyde resin are other susceptible materials that can support fungal growth, especially *A. versicolor* and *Penicillium* spp. (Steyn and Vleggaar 1976).

## *Paint*

Fungi can also grow on water-based or solvent-based paints; however, it is not clear whether molds found on the surface are using the paint components or taking nutrients from dirt also found on the surface (Allsopp et al. 2004). In general, paints can either enhance or decrease the susceptibility of a given base material to fungal growth. For example, paints prevent the growth of *Aureobasidium pullulans*, while *Penicillium* and *Aspergillus* species can grow rapidly on the same paints (Nielsen and Thrane 2002).

Cheng et al. (2014) showed that the surface water ratio and moisture content of mortars, brick, and tiles were affected by the pore size and distribution of these materials and different environments also showed significant effects on the surface water ratio of the building materials. The surface water ratio was a major factor affecting mold growth on these building materials (Cheng et al. 2014).

Chunduri (2014) reported *Aspergillus niger*, *Ascotricha chartarum*, *Fusarium solani*, *Aspergillus* spp., *Penicillium* spp., *Cladosporium* spp., *Phoma* spp., *Stachybotrys* spp. *Ascospora* spp., *Curvularia* spp., and *Alternaria* spp. from indoor cement walls of residential/commercial construction with water damage/intrusion. However, these results need to be verified, since cement itself may not support the growth of some of these fungi, such as *Stachybotrys* spp. and *Ascotricha chartarum* without organic dust or wooden furniture against the cement walls.

## *Other Substrates*

Fungal infestation in buildings is rather complex and facilitated by a number of factors, such as water damage, dampness, indoor environmental conditions, substrates, exposure time, building maintenance, building design, etc. The physical properties of building materials, e.g., surface structure, can be a key factor for fungal spore germination and fungal growth (Ryu and Moon 2014). Ryu and Moon (2014) inoculated *Aspergillus versicolor*, *Penicillium chrysogenum*, and *Stachybotrys chartarum* (reported as *S. atra*) on four kinds of wallpapers and their results showed that the wallpapers with irregular patterns or a flat surface structure delayed fungal growth. Wallpapers with higher moisture absorption capability prevented the development of fungi. A clear correlation between growth rates of the fungi and hygroscopic properties of the wallpapers was determined in the study also. *Aspergillus versicolor* grew faster than *P. chrysogenum* and *S. chartarum* on the wallpapers.

Mattresses are recognized as reservoirs for dust mites, but are also important reservoirs for molds (Verhoeff et al. 1994). The concentrations of molds in mattresses were reported to be $10^3$–$10^7$ spores/g of dust (Verhoeff et al. 1994).

## Natural Disasters and Indoor Molds

One of the effects of global warming is the increase in numbers of severe tropical storms (hurricanes and typhoons) and floods throughout the world (Mousavi et al. 2011). Hurricanes, superstorms, and floods cause tremendous damage to residences and other properties. In addition to physical damage, mold infestation and elevated mold and other particulate exposure of home owners, residents, first responders, and restoration workers in the damaged buildings lead to an increasing health concern for healthcare providers and disaster medicine throughout the world (Johanning et al. 2014). In the last 10 years, two major hurricanes, Katrina in 2005 and Sandy in 2012, made landfall on the east coast and caused catastrophic damage. A number of studies on indoor molds associated with these two major hurricanes have been published.

Rao et al. (2007) found that culturable fungi were significantly higher in the moderately/heavily water-damaged houses (67,000 CFU/m$^3$) than in the mildly water-damaged houses (3700 CFU/m$^3$) ($p=0.02$) 1 month after hurricanes Katrina and Rita and the predominant fungi in those houses were *Aspergillus niger*, *Penicillium* spp., *Trichoderma*, and *Paecilomyces*. At the same time, the fungal taxa and their concentrations were different from those previously reported from buildings without water damage in the southeastern USA. Fungi, fungal glucans, and endotoxins in the environment after Hurricanes Katrina and Rita in New Orleans were detected at concentrations that have been associated with health effects (Rao et al. 2007).

A retrospective, cross-sectional study of flooding, mold exposure, remediation, and respiratory symptom prevalence was conducted in Bound Brook, New Jersey, after Hurricane Floyd in September 1999 (Jones et al. 2013). The results showed that flood damage was a strong predictor of mold growth ($p<0.001$), and flooding was strongly associated with physician-confirmed respiratory symptoms in the aftermath of the flood (28 of 29 cases vs. 10 of 18 referents; $p<0.001$). Individuals involved in cleanup work without adequate personal protection were inclined to report five or more symptoms ($p<0.002$). The study concluded that exposure to molds during cleanup of moldy materials was a significant contributor to symptoms (Jones et al. 2013).

A study on total and respirable dust exposure for restoration work activities (demolition, mold remediation, trash and debris management, landscape restoration, and sewer/waterline repair) was conducted from 2005 to 2012 after Hurricane Katrina devastated the city of New Orleans in 2005 (Rando et al. 2013). The results showed that the most significant exposures were for demolition work, with average respirable dust exposures in 2005 above the action level of 2.5 mg/m$^3$ and 17.6 %

of exposures exceeding the permissible exposure limit (PEL) (5 mg/m$^3$). Average exposures to endotoxin and microbial glucan in 2005 were as high as 256 EU/m$^3$ and 118 μg/m$^3$, respectively. The results of this study support the conclusion that there is an association between respiratory illness and exposure during post-Katrina restoration work in the years immediately after the hurricane (Rando et al. 2013).

Studies on indoor molds after major natural disasters do not always support each other. The results showed some discrepancies. Grimsley et al. (2012) reported indoor and outdoor airborne mold levels were 501 and 3958 spores/m$^3$, respectively, in the homes of HEAL children that had been damaged by rain, flooding, or both. *Alternaria* antigen was reported in dust from 98 % of the homes, with 58 % having concentrations > 10 μg/g, and Mus m 1, Der p 1, and Bla g 1 were found in 60 %, 35 %, and 20 % of homes, respectively, at low concentrations. This study concluded that except for *Alternaria* antigen in dust, concentrations of airborne mold (ratio of indoor to outdoor mold) and dust allergens in the homes of HEAL children were lower than measurements found in other studies. They speculated that extensive post-Katrina mold remediation and renovations, or because children moved into cleaner homes upon returning to New Orleans, might be the reason (Grimsley et al. 2012).

Barbeau et al. (2010) evaluated the levels of indoor and outdoor molds in the months following hurricanes Katrina and Rita and found the homes with greater flood damage, especially those with >3 ft of indoor flooding, had higher levels of mold growth in comparison with homes with little or no flooding. Water intrusion originating from roof damage was also associated with mold growth. However, they did not find an increase in the occurrence of adverse health outcomes from published reports to date (Barbeau et al. 2010).

Sato et al. (2014) conducted a study on fungi from paper-based cultural objects damaged by seawater due to the tsunami in March 2011. *Stachybotrys chartarum* was found to be one of the key species causing the black alterations on tsunami-damaged, paper-based cultural items that remained wet for several months after the tsunami. *Chaetomium* and *Cladosporium* were observed on blackened documents also. *Myxotrichum deflexum* and *Streptomyces* sp. were isolated from the red-altered paper samples (Sato et al. 2014). The results showed that NaCl tolerance and cellulose-utilizing capacity were important for the microfungi that outcompeted the fungal taxa intolerant to NaCl on paper and paper products following the tsunami. A number of species of *Aspergillus* and *Penicillium* are very tolerant to NaCl, up to 20 %, while *S. chartarum* was tolerant up to 10 %. This may explain why recent studies reported *S. chartarum* was isolated from sponge in marine environments (Ma et al. 2013).

At present, there are several projects supported by NIOSH to train professionals who are involved in restoration or demolition work following Hurricane Sandy to avoid or reduce their exposure to indoor molds and other particulates when handling fungal infested/hazardous materials. Clearly, more studies are necessary in this field in the future.

There is an increasing concern by medical professionals worldwide over illnesses and health issues experienced by unprotected first responders, workers, home owners, and volunteers in recovery, restoration, and demolition of moldy indoor environments after hurricanes, typhoons, tropical storms, and floods. Avoiding and minimizing unnecessary fungal exposure are recommended and appropriate personal protective equipment (PPE) in disaster response and recovery work should be used (Johanning et al. 2014).

## Dust

House dust is a repository, concentrator, and a long-term reservoir of indoor fungi. Continuous elutriation of organic and inorganic airborne particles, originating from various indoor and outdoor sources, forms dust (Scott 2001). Dust is a complex mixture of organic and inorganic material whose composition varies depending on a given building type and use and major particle sources (Macher 2001; Chew et al. 2003). Dust samples are frequently collected during indoor fungi inspections and research to evaluate indoor environments and cumulative exposure of occupants to fungi as well as fungal composition.

The first global study on indoor fungi using dust samples collected from indoor environments on different continents found that indoor fungal composition is geographically patterned and more diverse in temperate zones than in the tropics (Amend et al. 2010). This result deviates from the previous understanding that indoor mold diversity in tropical areas is higher. Amend et al. (2010) reported that among 4473 operational taxonomic units (OTUs) defined at 97 % ITS sequence identity, only 31 OTUs were reported in more than half of the samples. The same authors also found that all but 3 of these 31 indoor fungi belonged to the phylum Ascomycota, and among these common indoor fungi over 80 % (25 of 31) were in the class Dothideomycetes. A recent study using polymerase chain reaction (PCR)–denaturing gradient gel electrophoresis (DGGE) and culture methods was conducted on household dust. It found that fungal diversity varied between 9 and 56 operational taxonomic units. Culture methods demonstrated that the most abundant genus was *Aspergillus*, followed by *Penicillium, Mucor*, and *Rhizomucor. Trichoderma, Chrysosporium, Fusarium, Rhizopus*, and *Stachybotrys*, which were present in a limited number of houses in central Portugal (Sousa et al. 2014).

There are some conflicting results about the health effects of indoor fungi. A recent study showed that culturable fungi, (1–3, 1–6)-β-D-glucan, and ergosterol concentrations in dust collected from residences in Sweden were not associated with asthma, rhinitis, or eczema diagnoses in children (Choi et al. 2014). However, the validity of this conclusion was questioned by Rylander (2014). No doubt, the disagreement will continue.

Adams et al. (2013b) adopted a simple, passive, innovative sampling device modified from the dustfall collector published by Würtz et al. (2005) for their recent study in student housing in Berkeley, CA. With this method, a Petri-dish sampler suspends a sterile, empty, 9-cm disposable Petri dish, 0.3 m from the ceiling to collect settling airborne dust for a defined period of time at a sampling site. The main advantage of this method is to collect dust samples with clearly defined starting and finishing points. When a dust sample is collected from an indoor surface with traditional methods, it is very difficult to know the time frame the dust sample represents. The study of Adams et al. (2013b) produced some very interesting results using pyrosequencing (ITS region 1). The most abundant fungus is *Cryptococcus victoriae*. Does this result suggest that yeasts and yeast-like fungi are significantly under-evaluated or its dominancy represent a geographic difference in fungal diversity? Among the 25 most abundant microfungi, two fungi could not be identified by BLASTing against the fungal taxa in GenBank. It indicated that some indoor fungi are overlooked and even not described previously. They found that richness of individual indoor samples ranged from 17 to 271 OTUs and 643 OTUs were reported in both winter and summer across seasons and 619 OTUs were reported in both indoor and outdoor samples across localities (Adams et al. (2013b).

## Health Effects

Exposures to indoor fungi may lead to a range of diseases and symptoms, both respiratory and non-respiratory. Respiratory diseases and symptoms that originated from exposure to indoor fungi include asthma, hypersensitivity pneumonitis, cough, wheeze, dyspnea (shortness of breath), nasal and throat symptoms, respiratory infections, rhinosinusitis, and sarcoidosis (Park and Cox-Ganser 2011).

*Stachybotrys chartarum* is notorious as an indoor microfungus and mycotoxin producer that can cause mycotoxicosis (stachybotrytoxicosis) in animals and humans. This fungus has been suggested to cause various medical conditions in humans, such as acute infant pulmonary hemorrhage, asthma, adult nasal and tracheal bleeding, allergies, asthma-like symptoms, inflammation, and lung injury (Etzel et al. 1998; Vesper et al. 2001; Vesper and Vesper 2002; Al-Ahmad et al. 2010; Bhan et al. 2011; Yike and Dearborn 2011; Pieckova et al. 2009). Infant pulmonary hemorrhage incidents in Cleveland caught the attention of the media, the public, and the medical community to this toxigenic fungus and other indoor molds in the last 15 years. The causal relationship between *S. chartarum* and pulmonary hemorrhage and the risk of exposure to this species to public health is still subject to debate (Pestka et al. 2008; Yike and Dearborn 2011). Cases of infant pulmonary hemorrhage associated with *S. chartarum* in patients' residences in Cleveland keep increasing from nine cases in 1990s to 52 cases in 2011. Of the cases investigated, 91 % of patients were living in residences infested by *S. chartarum* (Yike and Dearborn 2011). Pulmonary hemorrhage in acute animal models of instillation of *S. chartarum* conidia into rodent airways had been reported in more than 10 papers (Yike and Dearborn 2011). One of the early studies isolated *Stachybotrys chartarum* (reported as *S. atra*) from bronchoalveolar lavage fluid of a child with pulmonary hemorrhage (Elidemir et al. 1999). Another one found *S. chartarum* exposure in an infant that developed laryngeal spasm and hemorrhage during general anesthesia (Tripi et al. 2000). Nagayoshi et al. (2011) demonstrated for the first time that inhalation exposure to the conidia of *S. chartarum* led to the remodeling of pulmonary arteries in mice. Yike and Dearborn (2011) opined that this study made a significant contribution to our understanding of the pathologic effects of *S. chartarum*. Recent studies showed that pulmonary hypertension and evoked pulmonary arterial remodeling in mice were caused by repeated inhalation of *S. chartarum* conidia (Nagayoshi et al. 2011).

Rakkestad et al. (2010) indicated that cell death (apoptosis) was induced by heat-treated conidia of *S. chartarum* within 3–6 h due to DNA damage. Pollard et al. (2013) demonstrated that the conidial extracts of *S. chartarum* altered surfactant protein expression. Exposure to the conidial extracts of *S. chartarum* resulted in significantly reduced viability of fetal rat lung epithelial cell and human A549 cells and their ability to produce pulmonary surfactant. This study has demonstrated that inhaling mycotoxins produced by *S. chartarum* may induce potential damage to surfactant production and function (Pollard et al. 2013).

Bhan et al. (2011) demonstrated that hypersensitivity pneumonitis was induced by *S. chartarum*. Although a low rate of IgE mediated allergic responses to the toxigenic fungus *Stachybotrys chartarum* has been reported in some studies, it is unlikely a strong allergen in the clinical setting and the toxic-irritant effects appear to be more important (Barnes et al. 2002). After comparing the allergenicity of *S. chartarum* to house dust mite extracts in a mouse model, Chung et al. (2010) developed a suggested threshold dose (10 μg) for *S. chartarum* allergy induction. For a susceptible population in damp water-damaged environments, the authors' conclusion is that exposure to *S. chartarum* might be easily over the sensitization threshold. A high prevalence of pulmonary diseases was reported among office workers of Florida court buildings following prolonged indoor exposure to *S. chartarum* and *Aspergillus versicolor* (Hodgson et al. 1998).

Gareis and Gottschalk (2014) studied guttation droplets of *Stachybotrys* spp. and found that all 15 strains of *S. chartarum* and *S. chlorohalonata* but one developed various amounts of guttation droplets. Among the isolates, five are toxigenic isolates and produced highly toxic guttation droplets. These toxin containing guttation droplets were confirmed by all methods: ELISA, effect-based bioassay (MTT cell culture test), and tandem mass spectrometry (LC-MS/MS). The concentrations of macrocyclic trichothecenes, satratoxin G and H, are reported to be between the Limit of detection (LOD) and 7160 ng/mL exudate and 280 and 4610 ng/mL, respectively. Roridins and verrucarins were also reported from guttation droplets. The study demonstrated that *S. chartarum* were able to produce toxic exudates. It suggests that the toxic exudates may be significant in understanding its toxic potential in indoor environments (Gareis and Gottschalk 2014), exposure, and its health effects. The genome of *S. chartarum* has been sequenced (Semeiks et al. 2014). The genomic information of this species will assist us to better understand the mechanisms of its medical implication to human beings and the production of secondary metabolites including mycotoxins and more.

Satratoxin-G (SG) is a trichothecene mycotoxin produced by *Stachybotrys chartarum*. Islam et al. (2006) showed that intranasal exposure to SG induced apoptosis of olfactory sensory neurons and acute inflammation in the nose and brain of mice.

Carey et al. (2012) demonstrated that SG induced acute rhinitis, atrophy of the olfactory epithelium, and apoptosis of olfactory sensory neurons occurred in both groups in monkeys. These studies shed some light on the potential risk of nasal airway injury and neurotoxicity caused by exposure to mycotoxigenic molds in water-damaged buildings.

Dannemiller et al. (2014) showed that early-life exposure to low fungal diversity in dust collected from residences was associated with increased risk for later childhood asthma development using DNA sequencing data. Tischer et al. (2011) stated that exposure to visible mold and/or dampness was associated with an increased risk of developing asthma during the first 2 years of life. The authors found that meta-analyses showed a significant association with early asthma symptoms in four cohorts and with asthma later in childhood in six cohorts and 3–10 years. There was a significant association in six cohorts with symptoms of allergic rhinitis at school age 6–8 years and at any ages between 3 and 10 years (Tischer et al. 2011).

*Wallemia sebi*, a basidiomycetous hyphomycete and xerophilic fungus, is also very common in house dust (Desroches et al. 2014; Zalar et al. 2005). Desroches et al. (2014) demonstrated that isolates of *Wallemia sebi* from indoor environments produced several secondary metabolites including the known compound walleminone and a new compound 1-benzylhexahydroimidazo [1,5-α] pyridine-3,5-dione (wallimidione). Desroches et al. (2014) stated that wallimidione is likely the most toxic mycotoxin produced by *W. sebi*.

Exposure to mycotoxigenic molds has been recognized as a significant health risk for the last 20 years. Research has shown that mycotoxins are possible causes of human disease in water-damaged buildings (Brewer et al. 2013). However, the health effects from mycotoxins carried in airborne fungal structures remain hotly debated at present.

Brewer et al. (2013) tested urine of patients with a prior diagnosis of chronic fatigue syndrome for aflatoxins, ochratoxin A, and macrocyclic trichothecenes using Enzyme-Linked Immunosorbent Assays (ELISA). Urine specimens from 104 of 112 patients (93 %) were positive for at least one mycotoxin and nearly 30 % of the cases were positive for more than one mycotoxin. The results from a healthy control population without exposure to indoor molds were negative.

Sercombe et al. (2014) found that the predominant fungal conidia that bound IgE were derived from common environmental genera including *Cladosporium* and other fungi that produce 1-celled spores and inhalable fungal aerosols are the predominant aeroallergen sources in the homes in Sydney, Australia, in comparison with Der p 1, Fel d 1, and Bla g 1 allergen particles. Simoni et al. (2011) indicated that school children in Italy, Denmark, Sweden, Norway, and France exposed to viable molds ≥300 CFU/m$^3$ showed a higher risk than those exposed to lower levels for dry cough at night in the past year (odds ratio, OR: 3.10, 95 % confidence interval, CI: 1.61–5.98) and rhinitis (OR: 2.86, 95 % CI: 1.65–4.95), as well as for persistent cough (OR: 3.79, 95 % CI: 2.40–5.60). *Aspergillus*/*Penicillium* DNA showed significant and positive association with wheeze, and *Aspergillus versicolor* DNA with wheeze, rhinitis, and cough. Significant inverse associations of *Aspergillus versicolor* DNA were found with forced vitality capacity (Simoni et al. 2011).

*Alternaria* population was associated with wheezing symptoms for children with maternal mold sensitization [OR=9.16; (1.37–61.22)], but not for those without maternal mold sensitization [OR=1.32; (0.79–2.20)] (Behbod et al. 2013). The evidence indicated that dampness and molds in the home were determining factors for the development of asthma. The association of visible molds, and especially mold odor (MVOCs) to the risk of asthma development, implicated mold-related causal factors (Quansah et al. 2012).

Hsu et al. (2010) indicated a statistically significant dose-dependent relationship between total serum IgE levels and severity of indoor visible mold growth. However, further analysis of specific IgE to commonly examined fungal allergens failed to substantiate the correlation (Hsu et al. 2010). Holme et al. (2010) failed to find an association between the spore concentration indoors and moldy odor as well as signs of visible dampness in the homes. They did not determine an association between the airborne spore concentration indoors and asthma/allergy in the children (Holme et al. 2010). A recent study found that exposure to mold odor was related to reduced lung function among non-asthmatic individuals, particularly among women (Hernberg et al. 2014). Airborne fungi were found to be the second most frequent allergen after mites (Ceylan et al. 2013). The number of household residents was found to positively correlate with the populations of airborne fungi. However, the authors did not find an association between the airborne fungi and dosage of inhaler corticosteroids used or symptom levels in asthmatics (Ceylan et al. 2013).

Racial/ethnic, neighborhood, and socioeconomic factors are found to link to asthma, but few studies have been conducted on the relationship between these factors and indoor allergens (Camacho-Rivera et al. 2014). Home owners were less likely to report the presence of mice, cockroaches, and mold within their households in comparison with renters in Los Angeles. At the neighborhood level, their results showed that neighborhood-level racial/ethnic and socioeconomic influences on indoor allergen exposure exist (Camacho-Rivera et al. 2014).

A recent study focused on the interaction of genes with environment. The result showed that there are obvious combined effects between IL-4 promoter (CT, CI, TT) and mold exposure on both additive and multiplicative scales (Hwang et al. 2012). The risk of asthma was found to be significantly associated with children carrying the CT genotype and visible mold exposure in comparison with those carrying the TT genotype without any exposure indicator. There is a similar tendency for children who were exposed to mold odor and carried CT genotype. The authors opined that gene–environment interactions between the IL-4 promoter and an indoor mold infestation may play a significant role in childhood asthma (Hwang et al. 2012).

Slime molds (Myxomycetes) are polyphyletic. Common slime molds can be encountered indoors and outdoors, such as *Physarum*, *Fuligo*, and *Stemonitis* (Fiore-Donno et al. 2012; Stephenson 2011). *Stemonitis* species have been observed growing on water-damaged wooden structures in indoor environments (Li and Yang, pers. obser.). Airborne and indoor slime molds and their potential health effects are often overlooked and inadequately studied due to the difficulties in identifying their spores from the air and inability to isolate these slime molds with regular culture media.

Lierl (2013) recently extracted allergens from nine species of slime molds. Her results of allergy prick testing in 69 subjects with symptoms typical of seasonal allergic rhinitis found that 42 % of the subjects were positive for at least 1 slime mold extract, with 9 % to 22 % reacting to each extract. Lierl (2013) further opined that the spores of these slime molds in the air might be significant aeroallergens.

For the association between indoor mold exposure and exacerbation of asthma, the conclusion is still subject to debate. In the latest review by Kanchongkittiphon et al. (2014) on literature published since 2000, the authors concluded that there is sufficient evidence of a causal association between outdoor culturable fungal exposure and exacerbation in asthmatics sensitized to fungi. However, they opined that evidence of an association between indoor culturable *Penicillium* as well as total culturable fungal exposure and exacerbation in asthmatic children is limited or suggestive (Kanchongkittiphon et al. 2014). They attributed their conclusion to the sampling methods using 1-min air samples. In their opinion, the short sampling time led to highly unreliable assessments of fungal concentrations in the air and temporal variability of fungal spores in the air is high.

## MVOC

Musty odor is often the first indication that catches our attention when we walk into a moldy property. The odors we smell are volatile organic compounds (MVOCs) emitted from microfungi growing in the moldy property. Some fungi, though not all, produce noticeable MVOCs. In the last decade, the health effects of MVOCs started to attract scientists' attention and their importance has been gradually recognized (Piechulla and Degenhardt 2014). A recent study concluded that some MVOCs may be a risk factor for sick building syndrome and certain MVOCs were slightly higher in homes with reported dampness and mold (Sahlberg et al. 2013). Exposure to mold odor (MVOCs) was related to lower lung function levels among non-asthmatic individuals, especially among female adults (Hernberg et al. 2014).

*S. chartarum* was found to emit MVOCs on gypsum wallboard and ceiling tiles (Betancourt et al. 2013). Most of the MVOCs emitted by *S. chartarum* were alcohols, ketones, ethers, and esters. Anisole (methoxybenzene) was emitted from all strains of *S. chartarum* with a maximal concentration observed when the strains were incubated for 7 days. Betancourt et al. (2013) suggested that MVOCs are suitable markers for fungal identification and could be used for early detection of hidden molds because MVOCs easily diffuse through weak barriers, such as wallpaper (Betancourt et al. 2013).

Polizzi et al. (2012) studied MVOCs collected in water-damaged buildings to evaluate their use as possible indicators of indoor fungal growth using two sampling methods. Their results showed that dynamic headspace absorption using the Tenax method is more sensitive than Solid-Phase Microextraction. The detected MVOC profiles allowed them to identify *Aspergillus versicolor* group, *Aspergillus ustus*, and *Eurotium amstelodami* based on species-specific MVOCs or MVOC patterns. MVOCs produced by *Chaetomium* spp. and *Epicoccum* spp. were different from 76 fungal strains of various genera. This study indicated that chemotaxonomic discrimination of fungal species may be used as a supplementary method to the classical morphological and molecular identification techniques for research on indoor microfungi (Polizzi et al. 2012).

An animal study found that two fungal volatiles (E)-2-octenal and oct-1-en-3-ol had cytotoxic effects on murine bone marrow stromal cells and exposure to both (E)-2-octenal and oct-1-en-3-ol led to a shift to unsaturated fatty acids and lower cholesterol levels in the cell membrane (Hokeness et al. 2013). The results indicated that the MVOCs increased membrane fluidity. These changes to the cell membrane contribute to the failure of normal cell function. Considering the importance of bone marrow stromal cells to the appropriate development and activation of immune cells, the results of this study provide crucial information to understand the mechanism at a cellular level for how exposure to MVOCs may lead to immune-related disease conditions (Hokeness et al. 2013).

Matysik et al. (2008) studied MVOCs emitted from six microfungal species *Aspergillus versicolor*, *A. fumigatus*, *A. niter*, *Penicillium expansum*, *P. chrysogenum*, and *Cladosporium cladosporioides* on DG 18 and wet wallpaper. There are differences in MVOCs emitted by the fungi growing on wet wallpaper and DG18. Fungi growing on wet wallpaper had changed MVOC patterns with less signals and significantly reduced emission rates. The authors also demonstrated that some MVOCs are species specific on wallpaper, such as 2,4-pentandione for *A. fumigatus* and 1,3-dimethoxybenzene for *A. versicolor* (Matysik et al. 2008).

Bingley et al. (2012) detected more than 150 volatile compounds from 16 fungal strains (mainly *Aspergillus* spp. and *Penicillium* spp., also including *Stachybotrys chartarum*, *Trichoderma*, *Alternaria* sp., and *Cladosporium* sp.) isolated from cinematographic film using headspace solid-phase micro-extraction coupled with Gas Chromatography–Mass Spectrometry. Three MVOCs, 1-octen-3ol from 13 of the isolates, 3-octanone from 10 of the isolates, and 3-octanol from 4 isolates, are indicative of viable fungal growth on the cinematographic film. The study is significant in preventing unnecessary discarding of valuable historical film due to health and safety concerns regarding spore inhalation and would allow safe handling (Bingley et al. 2012).

A new online MVOC database (http://bioinformatics.charite.de/mvoc/) was designed for microbial volatiles and their emitting organisms (Lemfack et al. 2014). Lemfack et al. (2014) reported 846 compounds (5431 synonyms), which are produced by 349 bacterial and 69 fungi species. Many fungi compiled in the MVOC database are species of *Aspergillus*, *Penicillium*, and *Trichoderma* (mVOC 2015). This database is a significant start for MVOC research in the future, despite the fact the newly published MVOCs studies may not be found in the MVOC database.

## Infections Caused by Indoor Fungi

A wide variety of fungi have been observed by the authors or reported in the literature growing in the indoor environment (Samson 1999; Li and Yang 2004a; Li et al. 2013; Samson et al. 2011). They include members of the Myxogastria (formerly known as Myxomycota) (e.g., *Stemonitis* spp.), the Zygomycetous fungi (e.g., *Rhizopus stolonifer*, *Mucor* spp., *Syncephalastrum racemosum*), the Ascomycota (e.g., *Ascotricha erinacea*, *Ascotricha chartarum*, *Chaetomium* spp., *Emericella* spp., *Eurotium* spp., and *Peziza domiciliana*), the Basidiomycota (e.g., *Asterostroma cervicolor*, *Coprinus* spp., *Gloeophyllum* spp., *Sistotrema brinkmanii*, *Schizophyllum commune*, and *Serpula lacrymans*), and many species belonging to asexual states of Ascomycota (Li and Yang 2004b; Li et al. 2013; Samson et al. 2011; Schmidt 2007). Many of these indoor fungi have been associated with allergic diseases, asthma, hypersensitivities, occupational respiratory diseases, and other syndromes (Li and Yang 2004a; Yang and Johanning 2007; Ellis and Day 2011; Hodgson and Flannigan 2011). More importantly, many indoor fungi are also known human pathogens. A table of pathogenic fungi compiled from several different sources was provided (Li and Yang 2004a). A table of infectious fungi is available in the article titled "Respiratory tract infections caused by indoor fungi" (Summerbell 2011). The focus of this section is on the importance of infections by these fungi in the indoor environment.

Although the health impacts of damp indoor environments and resulting fungal growth have been extensively reviewed (Favata et al. 2000; Li and Yang 2004a; Yang and Johanning 2007; Ellis and Day 2011; Hodgson and Flannigan 2011; Storey et al. 2004; WHO 2009; IOM and Health 2004) and guidelines for control of infectious agents, including fungal pathogens, in healthcare facilities are available (CDC 2003), there has been little concern expressed regarding the home and work environments where sensitive individuals or immune-deficient patients are likely to spend significant amounts of their time. With the sensitive and immune-deficient populations at almost 20 % of the US population in 1996, these populations were expected to increase significantly because of increases in life span and the number of immunecompromised individuals (Gerba et al. 1996).

Summerbell (2011) described three categories of fungal respiratory tract infections caused by virulent fungal pathogens, *Pneumocystis* (a genus of yeast-like, atypical fungi), and opportunistic fungal pathogens. The first category includes *Blastomyces dermatitidis*, *Histoplasma capsulatum*, *Paracoccidioides brasiliensis*, *Coccidioides immitis* and *C. posadasii*, *Cryptococcus neoformans* and *C. gattii*, *Sporothrix schenckii*, and *Penicillium marneffei*. The second category includes *Pneumocystis* species that are contagious and cause respiratory infections, usually in immunocompromised individuals. The third category includes many opportunistic species, primarily a few species in the genus *Aspergillus*.

Among the first category, nine fungi are considered virulent and capable of infecting healthy individuals. Their number is relatively small and they are usually limited in their geographical area of distribution or grow in some unique environments, such as avian excreta or bat guano. However, there have been few reports confirming these fungi in fact growing in typical indoor environments. Because their distributions are usually endemic to certain geographical areas, this indicates that they may have certain growth requirements that are unique to their environment for their presence. Their detection and isolation may be limited or restricted by such requirements.

Among the first category, the only species that is considered to have a wide distribution is *Sporothrix schenckii*, which is often associated with vegetative debris (including dried straw, dried reeds, or dried *Sphagnum* peat moss). The infections by *S. schenckii* are mostly subcutaneous abscesses but rarely respiratory or through airborne spores (Summerbell 2011). The other eight species are either endemic within certain geographic regions, associated with avian excreta (Tille 2014) or with certain animals (e.g., *Penicillium marneffei*). Because of their saprobic nature, all nine species are potentially capable of growing indoors when their environmental needs are present. However, a few cases of infections caused by these fungi due to their growth indoors have been reported. Cases of infections due to the introduction of spores from outdoors or other sources were discussed by Summerbell (2011). An unusual case of blastomycosis was reported in Ontario, Canada. *Blastomyces dermatitidis* was isolated from a petroleum filtering shed laden with diatomaceous earth (Bakerspigel et al. 1986). The petroleum filtering shed is not a typical indoor living or work environment. However, this does suggest that buildings within the geographical distribution areas of the fungus may harbor it if the indoor environment allows the presence or even the proliferation of this fungus. *Histoplasma capsulatum, Cryptococcus neoformans*, and *C. gattii* are the three species that have been confirmed capable of growing indoors, particularly in association with roosting birds and bats (Summerbell 2011).

In the second category, *Pneumocystis* is a genus currently including at least two species, *P. jirovecii* [reported as *P. jiroveci* prior to 2009 (Stringer et al. 2009)] and *P. carinii. P. jirovecii* is reserved for the isolates causing human *Pneumocystis carinii* pneumonia (PCP), while *P. carinii* is applied to those causing PCP in rodents (Stringer et al. 2002; Tille 2014). *Pneumocystis* is difficult to grow in culture outside the lung (Stringer et al. 2002; Tille 2014). There has been no report of *Pneumocystis* species growing outside of a host or in water-damaged environments.

In the third category, there are many species that are known or documented to grow indoors. A list of fungi, other than endomycetous yeasts which have been confirmed as causing respiratory tract infection, sinusitis, or disseminated infections potentially invading the lung in humans, includes species in the Zygomycetous fungi, the Ascomycota and its anamorphs, and the Basidiomycota and its anamorphs (Summerbell 2011). The list includes many species that are well documented as capable of growing indoors. The most important taxon in this group is the genus *Aspergillus*.

Although the genus *Aspergillus* has been extensively studied by mycologists (Thom and Church 1926; Thom and Raper 1945; Raper and Fennell 1965; Klich 2002; Samson and Pitt 1990; Pitt and Hocking 2009; Samson and Varga 2007; Bennett 2010), many taxa within the genus were still not well delineated until the last 10 years. The most medically important species in the genus, *Aspergillus fumigatus*, was generally regarded as a very variable species in morphology (Pringle et al. 2005; Hong et al. 2005). In fact, many common species in the genus, such as *A. niger* and *A. terreus*, are variable and often treated as a complex (Geiser et al. 2007). Significant efforts have been made to better define the species taxa in the genus *Aspergillus* using molecular techniques in an integrated taxonomic approach (Hong et al. 2005; Geiser et al. 2007; Samson and Varga 2007).

The genus *Aspergillus* comprises approximately 285 species (Seifert et al. 2011), of which 34 have been associated with human disease. Historically, *A. fumigatus* caused 90 % of aspergillosis cases (Barnes and Marr 2006). *Aspergillus fumigatus* is considered both a primary and opportunistic pathogen (Nierman et al. 2005). In immunocompromised individuals, the incidence of invasive infection can be as high as 50 % and the mortality rate is often about 50 % (Denning 1998). Increasingly, aspergillosis is caused by non-fumigatus species (Barnes and Marr 2006). *Aspergillus* spp. were reportedly associated with infections occurring after hematopoietic stem cell transplantation (HSCT) included *A. fumigatus* (56 % of cases), *A. flavus* (18.7 %), *A. terreus* (16 %), *A. niger* (8 %), and *A. versicolor* (1.3 %) (Barnes and Marr 2006).

Within the genus *Aspergillus*, *A. fumigatus*, *A. flavus*, *A. terreus*, *A. niger*, and *A. nidulans* are the primary causative agents of human infections (Dagenais and Keller 2009). Four species, *A. fumigatus*, *A. flavus*, *A. terreus*, and *A. nidulans*, accounted for the majority of *Aspergillus* infections (Summerbell 2011). One of the common traits among these species is that they are rapid growers at a body temperature of 37 °C (Pitt and Hocking 2009). Such a trait has been shown to correlate with pathogenicity (Dagenais and Keller 2009).

Of particular concern is that some aspergilli have been found to develop resistance to antifungal treatments. *A. terreus* is resistant to amphotericin in vitro. In addition, other species with variable susceptibilities to antifungal agents are being described (Barnes and Marr 2006; 2006). Denning and Perlin (2011) reported increasing resistance of *Aspergillus fumigatus* to an azole drug, itraconazole. Snelders et al. (2012) reported that resistance of *Aspergillus fumigatus* to medical triazoles was due to cross-resistance from triazole fungicides.

*Aspergillus fumigatus* is found in composting vegetation, wood chips, garbage, and other materials at above ambient temperatures. Its spores are common in outdoor air and in carpet and mattress dusts. It can grow on warm, wet building, and finishing materials such as wallpaper (Samson et al. 2011). It has been reported from food products, such as oilseeds, soybeans, vegetables, nuts, coffee beans, cereals, meat products, and cheeses (Pitt and Hocking 2009). However, its confirmed growth in the indoor environment has been very limited, probably because there is usually very little decaying vegetation indoors.

Although *A. fumigatus* is one of the most important opportunistic microfungi to infect humans, little is known about its ecology, its dissemination and dispersal, and its periodicity. It was indicated that all humans will inhale at least several hundred airborne spores of *Aspergillus fumigatus* per day (Latgé 1999). However, the suggestion was based on three references of limited studies. In review of the three references cited in the report, there are major discrepancies of the claim. Chazalet et al. (1998) reported 0 to 3 conidia/m$^3$ of *A. fumigatus* using impactors on Sabouraud medium. Goodley et al. (1994) conducted a 1-year study to monitor the frequency of spores of *Aspergillus* spp. in a hospital ground in the London area. The study involved once a week air sampling using a Surface Air Sampler for 2–3 min between 2 and 4 pm. On a monthly basis, airborne spores of *A. fumigatus* varied from 0 to 273 CFU/m$^3$, with the highest concentration in March. Hospenthal et al. (1998) used an Andersen single-stage impactor and Czapek-Dox agar to monitor airborne spores of *Aspergillus* species in the Washington DC area. The sampling was carried out at an average of three times per month over a 54-week period. Results of *A. fumigatus* and *A. flavus* were grouped together. Weekly concentrations varied from 0 to 12 CFU/m$^3$. An average of 2 CFU/m$^3$ were recovered for *A. fumigatus* and *A. flavus* combined. In addition, there was no seasonal variation observed; variations in concentrations from different locations were in the graphic presentations. Because all three studies were based on limited scopes and intermittent sampling, it is very difficult to suggest or to predict an average airborne concentration of *A. fumigatus*. This raises the important issue of how much do we know about the ecology of this important opportunistic human pathogen.

Mullins et al. (1984) conducted intermittent sampling of airborne *Aspergillus fumigatus* spores using Andersen samplers on a 3 days/week schedule over a 12-month period at Cardiff, UK, and St Louis, USA. On average, *A. fumigatus* spore concentrations were 15 CFU/m$^3$ in St Louis and 11 CFU/m$^3$ in Cardiff. Seasonal variations were observed with highest concentrations in October for St Louis and in November for Cardiff.

*Aspergillus flavus* is most likely from an agricultural setting and rarely growing on building materials (Samson et al. 2011). It has only been detected growing in the indoor environment on a couple of occasions by the authors. *Aspergillus terreus* is common in tropical and subtropical regions and uncommon on building materials (Klich 2002; Samson et al. 2011). There is little indication that it grows indoors but has been isolated from indoor dust (Samson et al. 2011). *Aspergillus niger* is reportedly not generally associated with contaminated building materials (Samson et al. 2011) but has been observed on occasions to grow on water-damaged books and sheetrock wallboards. *Emericella (Aspergillus) nidulans* is common in tropical and subtropical regions (Klich 2002; Samson et al. 2011). It has been observed on occasion growing on water-damaged building materials, e.g., sheetrock wallboard, by its cleistothecia, globose to subglobose Hulle cells, and red and bi-flanged ascospores.

In addition to causing respiratory infections and diseases, fungal contaminations in the indoor environment can impact human health in other ways. Reports of fungal infections caused by contaminated injectable medicines, surgical wounds, or contact lens solutions have been increasing. Faulk and Lesher (1995) reported that a patient receiving injections of prednisone corticosteroids developed draining lesions. Biopsies and isolations were made. An atypical mycobacterium, *Mycobacterium fortuitum*, and a dematiaceous fungus, *Phialophora verrucosa*, were recovered from the lesions and identified (Faulk and Lesher 1995). Two cases of *Exophiala* infections of the subcutaneous tissues in organ transplant patients were reported by Gold et al. (1994). Outbreaks of fungal infections were reported in the USA due to fungal contaminations in contact lens solutions and injectable drugs. These outbreaks have caused many cases of fungal infections, including fungal meningitis and fungal keratitis.

In 2002, five cases of systemic fungal infection caused by *Exophiala* (*Wangiella*) *dermatitidis* occurred in patients receiving injectable steroids were reported (Casadevall and Pirofski 2013). In one case, the infection was not evident until 152 days after injection of the contaminated solution. Four cases developed into meningitis, while the fifth case was sacroiliitis. The outbreak was traced back to a single compounding pharmacy, which was later found by a governmental agency to have inadequate quality control for sterility (CDC 2002).

*Exophiala* (*Wangiella*) *dermatitidis* is cosmopolitan and commonly known as a black yeast which is an inhabitant of cool moist soils but has been found in sinks, drains, and steam baths. It has also been isolated from steam baths and hot tubs (Matos et al. 2002; Summerbell 2011). It can cause fatal infections of the central nervous system in otherwise healthy, mainly adolescent individuals in East Asia (Samson et al. 2011). The source of E. dermatitidis contamination was not reported.

In 2006, the US CDC reported an outbreak of fungal keratitis among 164 confirmed patients in 33 states and 1 US territory. Microbial corneal infection is a rare but is a serious complication that can lead to permanent vision loss or the need for corneal transplantation. In this outbreak, corneal transplantation was required or planned in 55 cases (34 %). One hundred fifty-four (94 %) of the confirmed patients wore soft contact lenses. Use of a specific brand of contact lens solution was implicated because case patients were found to be significantly more likely than controls to use the brand solution. *Fusarium* was not recovered from the factory, warehouse, solution filtrate, or unopened solution bottles. Implicated solution was not clustered in lots or in time. Among 39 isolates tested, at least 10 different *Fusarium* species were identified, comprising 19 genotypes (Chang et al. 2006). *Fusarium* spores are not uncommon but usually found at low densities in outdoor air during growing and Fall seasons. In a Canadian study, using a high-throughput jet sampler *Fusarium avenaceum*, *F. graminearum*, and *F. sporotrichioides* were found to account for 93.9 % of the total *Fusarium* airborne spores populations during the sampling period. Nine other *Fusarium* species at low frequencies were also detected in the study (Martin 1988). In another Canadian study, the researchers detected spores of several *Fusarium* species, including *F. graminearum*, *F. crookwellense*, *F. sporotrichioides*, *F. moniliforme*, *F. equiseti*, *F. subglutinans*, and *F. culmorum* within a 20-day sampling period in July, 1994 (Fernando et al. 2000). These findings plus the identifications of 10 different *Fusarium* species and 19 genotypes strongly suggest that the source of contaminations was more likely from airborne spores than systematic contamination.

Another outbreak caused by fungi was reported in 2012 from patients receiving contaminated steroid methylprednisolone acetate (MPA) injections. The outbreak was first reported in Tennessee where sixty-six cases were reported with a total of

22 patients that had laboratory confirmation of *Exserohilum rostratum* infection (21 patients) or *Aspergillus fumigatus* infection (1 patient) (Kainer et al. 2012). The CDC and Food and Drug Administration (FDA) announced that *E. rostratum* had been identified in two lots of unopened vials of methylprednisolone.

The CDC expanded the investigation to other affected States and reported the outbreak of fungal meningitis and other infections among patients who received the contaminated preservative-free MPA injections from the New England Compounding Center in Framingham, Massachusetts. Chiller et al. (2013) reported clinical findings of 328 patients without peripheral-joint infection. It found that 265 (81 %) had central nervous system (CNS) infection and 63 (19 %) had non-CNS infections. Laboratory evidence of *E. rostratum* was detected in samples from 96 of 268 patients (36 %). For those with CNS infections, strokes were associated with an increased severity of abnormalities in cerebrospinal fluid ($P<0.001$). Non-CNS infections were more frequent later in the course of the outbreak (median interval from last injection to diagnosis, 39 days for epidural abscess and 21 days for stroke; $P<0.001$), and such infections developed in patients with and without meningitis. The authors warned that fungal infections caused by epidural and paraspinal injection of a contaminated glucocorticoid product can result in a broad spectrum of clinical disease (Chiller et al. 2013).

The final report of the CDC's investigation included not only fungal meningitis, but also localized spinal or paraspinal infections, and infections associated with injections in a peripheral joint space, such as a knee, shoulder, or ankle. The index case had *Aspergillus fumigatus* meningitis. Subsequent reports showed that the primary fungus identified in patients is *Exserohilum rostratum* (Lockhart et al. 2013). *Exserohilum rostratum* is considered an extremely rare cause of human fungal disease (Summerbell 2011). The final statistics shows that the outbreak infected 751 people in 20 states and caused 64 deaths. The injections were found to be contaminated with *Exserohilum rostratum* (CDC 2013).

Although ubiquitous, members of the genus *Exserohilum* are rarely pathogenic to humans. Only three species have been shown to parasitize humans: *E. rostratum*, *E. longirostratum*, and *E. macginnisii*. The most common type of infections are sinusitis and skin infections, though also a few cases of cerebral abscesses, keratitis, osteomyelitis, prosthetic valve endocarditis, and disseminated infection have been described. They described a child undergoing treatment for acute lymphoblastic leukemia (ALL) that was infected by *Exserohilum rostratum*, causing cutaneous phaeohyphomycosis. The infection was traced to contaminated intravenous dressings. Treatment of the infection is primarily based on aggressive surgical removal combined with antifungal therapy (Saint-Jean et al. 2007).

In the case of the outbreak caused by Exserohilum rostratum, presumably from an environmental source, the fungus was not on the list provided by Li and Yang (2004a, b). It was listed in Summerbell's table as "opportunistic" and "rare" in causing respiratory tract infection. The fungus was reported as common on grass and many other plants, substrates, and soil (Ellis 1971; Sivanesan 1987). In fact, it is one of the three known pathogenic species in the genus (Saint-Jean et al. 2007). It appears that a rare, opportunistic fungus can cause a major outbreak if given the opportunity and under an unfortunate situation.

## New Technologies

New technologies often provide better methods or tools to assist us to better study indoor microfungi and associated health effects. No doubt, the proper application of newly emerged technologies is beneficial to indoor fungi and aeromycological studies.

Raman Microspectroscopy-Based Identification of Individual Fungal Spores has been used to develop a reference library of Raman spectra from a number of microfungi typically associated with damp indoor environments. The acquired reference spectral library has subsequently been utilized to identify individual spores of microfungi via direct comparison of the spore Raman spectra with the reference spectral signatures in the library. In addition, the distinct peak structures of Raman spectra provide detailed understanding of the overall chemical composition of spores. Potential application of this novel methodology is anticipated in the fields of public health, forensic sciences, and environmental microbiology (Ghosal et al. 2012).

Nanotechnology–Effectiveness of antimicrobial nanometals has been extensively studied recently (Yu et al. 2013). Some of the studies explored the potential of using nanotechnology for management of indoor molds. Yu et al. (2013) showed that Ag concentration to inhibit the germination and growth of *Aspergillus niger* conidia of 5 wt% nano Ag catalyst was 65 mg/mL and ozone has a synergetic effect on nanometals' antifungal efficacy.

High-throughput sequencing technology has been applied to study compositions of indoor molds. It is a very good supplementary method to traditional culturable methods for indoor mold studies, since many fungi cannot be cultured or grow poorly on artificial media. A number of studies have used this technology to study indoor fungi (Adams et al. 2013a; Amend et al. 2010).

Every method has its pros and cons. There is no exception for high-throughput sequencing. Adams et al. (2013a) found that the method may lead to bias in community richness and composition when high abundance of a few fungal taxa of the dust samples differs to a large degree. To address this shortcoming, Adams et al. (2013c) used UPARSE, a new method aimed at clustering globally trimmed sequences into operational taxonomic units (OTUs) with a focus on reducing OTU inflation. With this method the number of OTUs was dropped from 1305 taxa in QIIME (Protocol S1) to 966 in UPARSE for indoor fungi.

Their results showed the composition detected on residents' foreheads had a surprising richness of nonresident fungi, including plant pathogens such as *Claviceps purpurea,* ergot. However, they opined that it is unlikely the majority of the fungi would grow on the indoor surfaces which seems to be more or less passive collectors of airborne fungi of putative outdoor origin, while some fungi did grow on typical household surfaces, particularly on drains and skin (Adams et al. (2013c).

## Future Perspectives

The research, especially the report by Amend et al. (2010), in the last decade raises some important questions. Is it possible that some indoor fungal species evolve independently from their outdoor populations of the same taxon? If so, can we determine the evolutionary rate at which the indoor species evolve? How do indoor microfungi evolve? Is there any difference between the evolutionary directions of fungi indoors and the ones on natural substrates outdoors? Should we still emphasize that all indoor fungi have origins of outdoor sources?

Morphology-based methods and gene-based methods are supplementary. Each has their advantages and disadvantages. One should not exclude the other. After all, morphology-based or classic fungal taxonomy is the basis of gene-based fungal systematics.

The exact number of microfungi in indoor environments is severely underestimated and remains unclear. Systematic studies on indoor microfungi should be carried on in local areas and worldwide.

### *House Dust Mites and Indoor Molds*

Do we really know the number of species of indoor mites, except for the three known species which have been studied for their allergenicity in the past and 11 other species reported from indoor environments? Our personal observation has suggested that some soil mite species may occur on building materials with water damage. Soil ecological studies showed that mites feed on organic matter which was partially decomposed by fungi. Is this relationship present on building materials indoors also? What are the ecological roles of the mites in indoor environments? What are the exact relationship and interaction between domestic mites and microfungi in indoor environments? It seems that mites are attracted to fungal colonies. Are the mites indoors chemotaxic? Are MVOCs, mycotoxin, or chitin involved in chemotaxis?

Nanoparticles are particles between 1 and 100 nm in size. They are also able to pass through cell membranes in organisms, and their interactions with biological systems are relatively unknown. Their sizes are similar to protein (SCENIHR 2015). Nano-sized fungal fragments should be studied to understand whether they still carry allergens and mycotoxins and what kinds of health effects result from exposure to them.

## *Nematodes, Termites, and Carpenter Ants*

Nematodes are very common in soils. Nematodes were observed on fungal colonies on wet dry walls damaged by a flood. Some nematodes can feed on fungi. Colonies of *Botrytis cinerea* have been used as a food source to cultivate pine wood nematodes for studying pine wood nematode wilt (Futai 2013; Wu et al. 2013). A number of fungi, such as *Arthrobotrys oligospora*, *Dactylaria candida*, *Monacrosporium cionopagum*, and *Nematoctonus geogenius*, can trap or parasitize nematodes (Barron 1977; Xie et al. 2010). Many residences have a crawl space with a bare soil surface. It provides nematodes access to indoor building materials with fungal infestation. It is not clear whether nematodes play any roles in indoor mold infestation and their interaction with indoor molds.

## *Slime Molds*

Quantitative real-time polymerase chain reaction (QPCR) for detecting indoor fungi and Environmental Relative Moldiness Index (ERMI) to screen indoor environments for molds were DNA-based methods developed in the late 1990s and 2000s, respectively (Haugland et al. 1999; Vesper et al. 2007). Although QPCR primers and probes have been developed for detecting 120 fungal species, commercial service of QPCR is normally limited to a panel of 35 microfungal species. This method is accepted in indoor mold research and investigations. ERMI was developed using ca. 1000 dust samples collected from US houses to screen indoor environments for molds (Vesper et al. 2007). However, ERMI is still subject to debate and further evaluation and validation are necessary. It is questionable whether ERMI should be applied to nonresidential environments or not. These two methods provided alternatives to morphology-based methods for indoor microfungi research and investigation. Fungal systematics is developing and some indoor fungi-related genera may be redelineated, such as *Cladosporium cladosporioides* sensu lato (species complex) and *Penicillium chrysogenum* sensu lato. Thus, molecular methods should be fine-tuned or updated following the advancement of fungal systematics. It is necessary to verify the specificities of the primers and probes of qPCR for detecting indoor fungi based on the latest development of fungal systematics. It will advance our understanding of indoor molds and their effects on public health.

No method is perfect. The same principle is also applicable to DNA-based methods. DNA-based methods, comparable to morphology-based methods and other methods, have demonstrated great advantages but also showed their limitations. Morphology- and DNA-based methods are supplementary. Pitkaranta et al. (2008) conducted a study comparing DNA sequence, cultivation, and quantitative PCR methods. The results indicated that the three methods were complementary to each other and combined results from the three methods depicted a more comprehensive picture of diversity and population of indoor fungi than the one of any individual method produced. Will and Rubinoff (2004) pointed out that "DNA sequence data are an important and powerful part of taxonomy and systematics. Molecular data have an indisputable role in the analysis of biodiversity. However, DNA-based data should not be seen as a substitute for understanding and studying whole organisms when determining identities or systematic relationships." It is important to follow the latest development of fungal taxonomy and understand the advantages and inadequacies of all the methods available for indoor fungi investigation and research. Up-to-date knowledge will help us to choose proper methods suitable to our objectives and to develop new hypotheses for future research.

## *Fungal Infections*

With the increase of sensitive populations, fungal infections, whether primary or opportunistic, are much frequently reported. Although many cases of infections are respiratory, more are reported from non-respiratory routes of infections such as wound or injections. The fungal spores that can germinate and grow at 37 °C can also grow in the human body. Monitoring and testing for indoor fungal pathogens must take into consideration such a characteristic.

*Aspergillus fumigatus* is considered the most important opportunistic fungal pathogen. It has major health impacts on the health of immunecomprimised individuals, such as cystic fibrosis patients (Latgé 1999; LiPuma 2010). However, there has been no continuous, systematic, long-term study of its airborne spores: their geographical distributions, seasonality and periodicity, yearly variations, and temporal and spatial fluctuations. Without such information, hospitals and other healthcare facilities will find it very difficult to control infiltration of *A. fumigatus* spores from outdoors.

With frequent outbreaks of fungal infections associated with injectable medicines and eye care solutions, environmental control and hygiene of manufacturing facilities as well as users' education are critical to avoid future outbreaks.

**Acknowledgment** We are very grateful to Dr. James LaMondia, The Connecticut Agricultural Experiment Station, USA, for reviewing the manuscript.

## References

Aboul-Nasr MB, Zohri A-NA, Amer EM (2014) Indoor surveillance of airborne fungi contaminating intensive care units and operation rooms in Assiut University Hospitals, Egypt. J Health Sci 2:20–27

Adams RI, Amend AS, Taylor JW, Bruns TD (2013a) A unique signal distorts the perception of species richness and composition in high-throughput sequencing surveys of microbial communities: a case study of fungi in indoor dust. Microb Ecol 66(4):735–741

Adams RI, Miletto M, Taylor JW, Bruns TD (2013b) Dispersal in microbes: fungi in indoor air are dominated by outdoor air and show dispersal limitation at short distances. ISME J 7(7): 1262–1273. doi:10.1038/ismej.2013.28

Adams RI, Miletto M, Taylor JW, Bruns TD (2013c) The diversity and distribution of fungi on residential surfaces. PLoS One 8(11), e78866. doi:10.1371/journal.pone.0078866
Adhikari A, Reponen T, Rylander R (2013) Airborne fungal cell fragments in homes in relation to total fungal biomass. Indoor Air 23(2):142–147
Al-Ahmad M, Manno M, Ng V, Ribeiro M, Liss G, Tarlo S (2010) Symptoms after mould exposure including *Stachybotrys chartarum*, and comparison with darkroom disease. Allergy 65(2):245–255
Allsopp D, Seal KJ, Gaylarde CC (2004) Introduction to biodeterioration. Cambridge University Press, Cambridge
Amend AS, Seifert KA, Samson R, Bruns TD (2010) Indoor fungal composition is geographically patterned and more diverse in temperate zones than in the tropics. Proc Natl Acad Sci 107(31):13748–13753. doi:10.1073/pnas.1000454107
Andersen B, Nielsen KF, Jarvis BB (2002) Characterization of *Stachybotrys* from water-damaged buildings based on morphology, growth, and metabolite production. Mycologia 94(3):392–403
Andersen B, Frisvad JC, Søndergaard I, Rasmussen IS, Larsen LS (2011) Associations between fungal species and water-damaged building materials. Appl Environ Microbiol 77(12): 4180–4188
Bakerspigel A, Kane J, Schaus D (1986) Isolation of *Blastomyces dermatitidis* from an earthen floor in southwestern Ontario, Canada. J Clin Microbiol 24(5):890–891
Barbeau DN, Grimsley LF, White LE, El-Dahr JM, Lichtveld M (2010) Mold exposure and health effects following hurricanes Katrina and Rita. Annu Rev Public Health 31:165–178
Barnes PD, Marr KA (2006) Aspergillosis: spectrum of disease, diagnosis, and treatment. Infect Dis Clin North Am 20(3):545–561
Barnes C, Buckley S, Pacheco F, Portnoy J (2002) IgE-reactive proteins from *Stachybotrys chartarum*. Ann Allergy Asthma Immunol 89(1):29–33
Barron GL (1977) The nematode-destroying fungi. Canadian Biological Publications Ltd, Canada
Behbod B, Sordillo J, Hoffman E, Datta S, Muilenberg M, Scott J, Chew G, Platts-Mills T, Schwartz J, Burge H (2013) Wheeze in infancy: protection associated with yeasts in house dust contrasts with increased risk associated with yeasts in indoor air and other fungal taxa. Allergy 68(11):1410–1418
Bennett JW (2010) An overview of the genus *Aspergillus*. In: Machida M, Gomi K (eds) *Aspergillus* molecular biology and genomics. Caiser Academic Press, Portland, OR, pp 1–17
Bensch K, Groenewald JZ, Dijksterhuis J, Starink-Willemse M, Andersen B, Summerell BA, Shin H-D, Dugan FM, Schroers H-J, Braun U, Crous PW (2010) Species and ecological diversity within the *Cladosporium cladosporioides* complex (Davidiellaceae, Capnodiales). Stud Mycol 67(1):1–94. doi:10.3114/sim.2010.67.01
Betancourt DA, Krebs K, Moore SA, Martin SM (2013) Microbial volatile organic compound emissions from *Stachybotrys chartarum* growing on gypsum wallboard and ceiling tile. BMC Microbiol 13(1):283. doi:10.1186/1471-2180-13-283
Bhan U, Newstead MJ, Zeng X, Ballinger MN, Standiford LR, Standiford TJ (2011) *Stachybotrys chartarum*-induced hypersensitivity pneumonitis is TLR9 dependent. Am J Pathol 179(6):2779–2787. doi:10.1016/j.ajpath.2011.08.019
Bingley GD, Verran J, Munro LJ, Banks CE (2012) Identification of microbial volatile organic compounds (MVOCs) emitted from fungal isolates found on cinematographic film. Anal Methods 4(5):1265–1271
Bissett J (1987) Fungi associated with urea-formaldehyde foam insulation in Canada. Mycopathologia 99(1):47–56
Bjurman J (1994) Ergosterol as an indicator of mould growth on wood in relation to culture age, humidity stress and nutrient level. Int Biodeter Biodegr 33(4):355–368
Brewer JH, Thrasher JD, Straus DC, Madison RA, Hooper D (2013) Detection of mycotoxins in patients with chronic fatigue syndrome. Toxins 5(4):605–617
Camacho-Rivera M, Kawachi I, Bennett GG, Subramanian S (2014) Associations of neighborhood concentrated poverty, neighborhood racial/ethnic composition, and indoor allergen exposures: a cross-sectional analysis of Los Angeles households, 2006–2008. J Urban Health 91(4):1–16

Carey SA, Plopper CG, Hyde DM, Islam Z, Pestka JJ, Harkema JR (2012) Satratoxin-G from the black mold *Stachybotrys chartarum* induces rhinitis and apoptosis of olfactory sensory neurons in the nasal airways of rhesus monkeys. Toxicol Pathol 40(6):887–898

Casadevall A, L-A P (2013) *Exserohilum rostratum* fungal meningitis associated with methylprednisolone injections. Future Microbiol 8(2):135–137

CDC (2002) *Exophiala* infection from contaminated injectable steroids prepared by a compounding pharmacy—United States, July–November 2002. MMWR Morb Mortal Wkly Rep 51(49):1109

CDC (2003) Guidelines for environmental infection control in health-care facilities. Centers for Disease Control and Prevention (CDC), Atlanta, GA. http://www.cdc.gov/hicpac/pdf/guidelines/eic_in_HCF_03.pdf

CDC (2013) Multistate outbreak of fungal meningitis and other infections. Centers for Disease Control Prevention

Ceylan E, Doruk S, Genc S, Ozkutuk AA, Karadag F, Ergor G, Itil BO, Cımrın AH (2013) The role of molds in the relation between indoor environment and atopy in asthma patients. J Res Med Sci 18(12):1067

Chang DC, Grant GB, O'Donnell K, Wannemuehler KA, Noble-Wang J, Rao CY, Jacobson LM, Crowell CS, Sneed RS, Lewis FM (2006) Multistate outbreak of *Fusarium* keratitis associated with use of a contact lens solution. JAMA 296(8):953–963

Chazalet V, Debeaupuis J-P, Sarfati J, Lortholary J, Ribaud P, Shah P, Cornet M, Thien HV, Gluckman E, Brücker G (1998) Molecular typing of environmental and patient isolates of *Aspergillus fumigatus* from various hospital settings. J Clin Microbiol 36(6):1494–1500

Cheng A, Hsin Y, Lin W-T (2014) Effects of mold growth on building materials by different environments in Taiwan. KSCE J Civ Eng 18(4):1083–1090

Chew G, Rogers C, Burge H, Muilenberg M, Gold D (2003) Dustborne and airborne fungal propagules represent a different spectrum of fungi with differing relations to home characteristics. Allergy 58(1):13–20

Chiller TM, Roy M, Nguyen D, Guh A, Malani AN, Latham R, Peglow S, Kerkering T, Kaufman D, McFadden J (2013) Clinical findings for fungal infections caused by methylprednisolone injections. N Engl J Med 369(17):1610–1619

Cho S-H, Seo S-C, Schmechel D, Grinshpun SA, Reponen T (2005) Aerodynamic characteristics and respiratory deposition of fungal fragments. Atmos Environ 39(30):5454–5465. doi:10.1016/j.atmosenv.2005.05.042

Choi H, Byrne S, Larsen LS, Sigsgaard T, Thorne PS, Larsson L, Sebastian A, Bornehag CG (2014) Residential culturable fungi, (1–3, 1–6)-β-D-glucan, and ergosterol concentrations in dust are not associated with asthma, rhinitis, or eczema diagnoses in children. Indoor Air 24(2):158–170. doi:10.1111/ina.12068

Chunduri JR (2014) Indoor fungal populations inhabiting cement structures-remedial measures. IOSR J Environ Sci Toxicol Food Technol 8(4):2319–2399

Chung W-Y, Wi S-G, Bae H-J, Park B-D (1999) Microscopic observation of wood-based composites exposed to fungal deterioration. J Wood Sci 45(1):64–68

Chung YJ, Copeland LB, Doerfler DL, Ward MDW (2010) The relative allergenicity of *Stachybotrys chartarum* compared to house dust mite extracts in a mouse model. Inhal Toxicol 22(6):460–468

Cui Y (2014) When mites attack: domestic mites are not just allergens. Parasit Vectors 7(1):411. doi:10.1186/1756-3305-7-411

Dagenais TR, Keller NP (2009) Pathogenesis of *Aspergillus fumigatus* in invasive aspergillosis. Clin Microbiol Rev 22(3):447–465

Dannemiller KC, Mendell MJ, Macher JM, Kumagai K, Bradman A, Holland N, Harley K, Eskenazi B, Peccia J (2014) Next-generation DNA sequencing reveals that low fungal diversity in house dust is associated with childhood asthma development. Indoor Air 24(3):236–247

Denning DW (1998) Invasive aspergillosis. Clin Infect Dis 26(4):781–803

Denning DW, Perlin DS (2011) Azole resistance in *Aspergillus*: a growing public health menace. Future Microbiol 6(11):1229–1232
Desroches TC, McMullin DR, Miller JD (2014) Extrolites of *Wallemia sebi*, a very common fungus in the built environment. Indoor Air 24(5):533–542
Eduard W, Heederik D, Duchaine C, Green BJ (2012) Bioaerosol exposure assessment in the workplace: the past, present and recent advances. J Environ Monit 14(2):334–339
Elidemir O, Colasurdo GN, Rossmann SN, Fan LL (1999) Isolation of *Stachybotrys* from the lung of a child with pulmonary hemosiderosis. Pediatrics 104(4):964–966
Ellis MB (1971) Dematiaceous hyphomycetes. CABI Publishing, Wallingford
Ellis A, Day J (2011) Allergenic microorganisms and hypersensitivity. In: Flannigan B, Samson RA, Miller JD (eds) Microorganisms in home and indoor work environments, diversity, health impacts, investigation and control. CRC Press, Boca Raton, FL, pp 147–182
Eppley RM, Bailey WJ (1973) 12,13-Epoxy-delta 9-trichothecenes as the probable mycotoxins responsible for stachybotryotoxicosis. Science 181(101):758–760
Etzel R, Montaña E, Sorenson WG, Kullman GJ, Allan TM, Dearborn DG (1998) Acute pulmonary hemorrhage in infants associated with exposure to stachybotrys atra and other fungi. Arch Pediatr Adolesc Med 152(8):757–762, doi:10-1001/pubs.Pediatr Adolesc Med.-ISSN-1072-4710-152-8-poa8148
Faulk CT, Lesher JL Jr (1995) Phaeohyphomycosis and *Mycobacterium fortuitum* abscesses in a patient receiving corticosteroids for sarcoidosis. J Am Acad Dermatol 33(2):309–311
Favata E, Neill H, Yang C (2000) Emerging microbial diseases of the indoor environment. In: Couturier AJ (ed) Occupational and environmental infectious diseases. OEM Press, Beverly Farms, MA, pp 697–716
Fernando WG, Miller J, Seaman W, Seifert K, Paulitz T (2000) Daily and seasonal dynamics of airborne spores of *Fusarium graminearum* and other *Fusarium* species sampled over wheat plots. Can J Bot 78(4):497–505
Filip Z (2009) Environmental microbiology, soil science, nanoscience: a planetary view. In: Simeonov LI, Hassanien MA (eds) Exposure and risk assessment of chemical pollution—contemporary methodology. Springer, Dordrecht, pp 259–270
Fiore-Donno AM, Kamono A, Meyer M, Schnittler M, Fukui M, Cavalier-Smith T (2012) 18S rDNA phylogeny of *Lamproderma* and allied genera (Stemonitales, Myxomycetes, Amoebozoa). PLoS One 7(4), e35359
Flannigan B, Miller J (2001) Microbial growth in indoor environments. In: Flannigan B, Samson R, Miller J (eds) Microorganisms in home and indoor work environments, vol 36. Taylor & Francis, New York, p 67
Futai K (2013) Pine wood nematode, *Bursaphelenchus xylophilus*. Annu Rev Phytopathol 51:61–83
Gareis M, Gottschalk C (2014) *Stachybotrys* spp. and the guttation phenomenon. Mycotoxin Res 30:151–159
Geiser D, Klich M, Frisvad JC, Peterson S, Varga J, Samson RA (2007) The current status of species recognition and identification in *Aspergillus*. Stud Mycol 59:1–10
Geiser DM, Aoki T, Bacon CW, Baker SE, Bhattacharyya MK, Brandt ME, Brown DW, Burgess LW, Chulze S, Coleman JJ (2013) One fungus, one name: defining the genus *Fusarium* in a scientifically robust way that preserves longstanding use. Phytopathology 103(5):400–408
Gerba CP, Rose JB, Haas CN (1996) Sensitive populations: who is at the greatest risk? Int J Food Microbiol 30(1):113–123
Ghosal S, Macher JM, Ahmed K (2012) Raman microspectroscopy-based identification of individual fungal spores as potential indicators of indoor contamination and moisture-related building damage. Environ Sci Technol 46(11):6088–6095
Gold WL, Vellend H, Salit IE, Campbell I, Summerbell R, Rinaldi M, Simor AE (1994) Successful treatment of systemic and local infections due to *Exophiala* species. Clin Infect Dis 19(2):339–341
Gong J-Y, Chen Y-C, Huang Y-T, Tsai M-C, Yu K-P (2014) For inactivation of mold spores by UVC radiation, TiO2 nanoparticles may act as a "sun block" better than as a photocatalytic disinfectant, and ozone acts as a promoter. Photochem Photobiol Sci 13:1305–1310

Goodley J, Clayton Y, Hay R (1994) Environmental sampling for aspergilli during building construction on a hospital site. J Hosp Infect 26(1):27–35

Górny RL, Reponen T, Willeke K, Schmechel D, Robine E, Boissier M, Grinshpun SA (2002) Fungal fragments as indoor air biocontaminants. Appl Environ Microbiol 68(7):3522–3531. doi:10.1128/aem.68.7.3522-3531.2002

Grant C, Hunter C, Flannigan B, Bravery A (1989) The moisture requirements of moulds isolated from domestic dwellings. Int Biodeterior 25(4):259–284

Green BJ, Sercombe JK, Tovey ER (2005) Fungal fragments and undocumented conidia function as new aeroallergen sources. J Allergy Clin Immunol 115(5):1043–1048. doi:10.1016/j.jaci.2005.02.009

Green BJ, Tovey ER, Sercombe JK, Blachere FM, Beezhold DH, Schmechel D (2006) Airborne fungal fragments and allergenicity. Med Mycol 44(Supplement 1):S245–S255. doi:10.1080/13693780600776308

Green BJ, Schmechel D, Summerbell RC (2011) Aerosolized fungal fragments. In: Adnan O, Samson RA (eds) Fundamentals of mold growth in indoor environments and strategies for healthy living. Wageningen Academic Publishers, Amsterdam, pp 211–243

Grimsley L, Chulada PC, Kennedy S, White L, Wildfire J, Cohn RD, Mitchell H, Thornton E, El-Dahr J, Mvula MM (2012) Indoor environmental exposures for children with asthma enrolled in the HEAL study, post-Katrina New Orleans. Environ Health Perspect 120(1):1600–1606

Harriman L (2012) Mold and dampness. ASHRAE J 54(12):132–133

Haugland RA, Heckman JL, Wymer LJ (1999) Evaluation of different methods for the extraction of DNA from fungal conidia by quantitative competitive PCR analysis. J Microbiol Methods 37(2):165–176

Haverinen-Shaughnessy U, Borras-Santos A, Turunen M, Zock JP, Jacobs J, Krop EJM, Casas L, Shaughnessy R, Täubel M, Heederik D, Hyvärinen A, Pekkanen J, Nevalainen A (2012) Occurrence of moisture problems in schools in three countries from different climatic regions of Europe based on questionnaires and building inspections—the HITEA study. Indoor Air 22(6):457–466. doi:10.1111/j.1600-0668.2012.00780.x

Hernberg S, Sripaiboonkij P, Quansah R, Jaakkola JJ, Jaakkola MS (2014) Indoor molds and lung function in healthy adults. Respir Med 108(5):677–684

Hodgson M, Flannigan B (2011) Occupational respiratory disease: hypersensitivity pneumonitis and other forms of interstitial lung disease. In: Flannigan B, Samson RA, Miller JD (eds) Microorganisms in home and indoor work environments, diversity, health impacts, investigation and control. CRC Press, Boca Raton, FL, pp 183–203

Hodgson MJ, Morey P, Leung WY, Morrow L, Miller D, Jarvis BB, Robbins H, Halsey JF, Storey E (1998) Building-associated pulmonary disease from exposure to *Stachybotrys chartarum* and *Aspergillus versicolor*. J Occup Environ Med 40(3):241–249

Hokeness K, Kratch J, Nadolny C, Aicardi K, Reid CW (2013) The effects of fungal volatile organic compounds on bone marrow stromal cells. Can J Microbiol 60(1):1–4. doi:10.1139/cjm-2013-0708

Holme J, Hägerhed-Engman L, Mattsson J, Sundell J, Bornehag CG (2010) Culturable mold in indoor air and its association with moisture—related problems and asthma and allergy among Swedish children. Indoor Air 20(4):329–340

Hong S-B, Go S-J, Shin H-D, Frisvad JC, Samson RA (2005) Polyphasic taxonomy of *Aspergillus fumigatus* and related species. Mycologia 97(6):1316–1329

Hospenthal D, Kwon-Chung K, Bennett J (1998) Concentrations of airborne *Aspergillus* compared to the incidence of invasive aspergillosis: lack of correlation. Med Mycol 36(3):165–168

Houbraken J, Frisvad JC, Samson RA (2011) Fleming's penicillin producing strain is not *Penicillium chrysogenum* but *P. rubens*. IMA Fungus 2(1):87–95. doi:10.5598/imafungus.2011.02.01.12

Hsu NY, Wang JY, Su HJ (2010) A dose-dependent relationship between the severity of visible mold growth and IgE levels of pre-school-aged resident children in Taiwan. Indoor Air 20(5):392–398. doi:10.1111/j.1600-0668.2010.00663.x

Hu Y, Liu W, Huang C, Zou Z, Zhao Z, Shen L, Sundell J (2014) Home dampness, childhood asthma, hay fever, and airway symptoms in Shanghai, China: associations, dose—response relationships, and lifestyle's influences. Indoor Air 24(5):450–463

Hwang B-F, Liu I, Huang T-P (2012) Gene–environment interaction between interleukin-4 promoter and molds in childhood asthma. Ann Epidemiol 22(4):250–256

Institute of Medicine (2000) Clearing the air: asthma and indoor air exposures. National Academies Press, Washington, DC

IOM IoMCoDIS, Health (2004) Damp indoor spaces and health. National Academies Press, Washington DC

Islam Z, Harkema JR, Pestka JJ (2006) Satratoxin G from the black mold Stachybotrys chartarum evokes olfactory sensory neuron loss and inflammation in the murine nose and brain. Environ Health Perspect 114(7):1099–1107

Jaakkola MS, Quansah R, Hugg TT, Heikkinen SA, Jaakkola JJ (2013) Association of indoor dampness and molds with rhinitis risk: a systematic review and meta-analysis. J Allergy Clin Immunol 132(5):1099–1110, e1018

Johanning E, Auger P, Morey P, Yang C, Olmsted E (2014) Review of health hazards and prevention measures for response and recovery workers and volunteers after natural disasters, flooding, and water damage: mold and dampness. Environ Health Prev Med 19(2):93–99. doi:10.1007/s12199-013-0368-0

Jones EK, Sumner KG, Gochfeld M (2013) Residential flood damage after hurricane Floyd, mold, household remediation, and respiratory health. Remediat J 24(1):107–120. doi:10.1002/rem.21381

Kainer MA, Reagan DR, Nguyen DB, Wiese AD, Wise ME, Ward J, Park BJ, Kanago ML, Baumblatt J, Schaefer MK (2012) Fungal infections associated with contaminated methylprednisolone in Tennessee. N Engl J Med 367(23):2194–2203

Kanchongkittiphon W, Mendell MJ, Gaffin JM, Wang G, Phipatanakul W (2014) Indoor environmental exposures and exacerbation of asthma: an update to the 2000 review by the Institute of Medicine. Environ Health Perspect 123(1):6–20. doi:10.1289/ehp.1307922

Karvala K, Nordman H, Luukkonen R, Uitti J (2014) Asthma related to workplace dampness and impaired work ability. Int Arch Occup Environ Health 87(1):1–11. doi:10.1007/s00420-012-0830-0

Kennedy K, Grimes C (2013) Indoor water and dampness and the health effects on children: a review. Curr Allergy Asthma Rep 13(6):672–680

Kirk P, Cannon P, David J, Stalpers J (2001) Ainsworth & Bisby's dictionary of the fungi, 9th edn. CAB International, Wallingford

Kirk PM, Cannon PF, Minter D, Stalpers JA (2008) Dictionary of the fungi. CABI, Wallingford

Klich MA (2002) Identification of common *Aspergillus* species. Centraalbureau voor Schimmelcultures, Utrecht

Latgé J-P (1999) *Aspergillus fumigatus* and aspergillosis. Clin Microbiol Rev 12(2):310–350

Lee S-A, Liao C-H (2014) Size-selective assessment of agricultural workers' personal exposure to airborne fungi and fungal fragments. Sci Total Environ 466–467:725–732. doi:10.1016/j.scitotenv.2013.07.104

Lemfack MC, Nickel J, Dunkel M, Preissner R, Piechulla B (2014) mVOC: a database of microbial volatiles. Nucleic Acids Res 42(D1):D744–D748

Li D-W, Kendrick B (1995a) Indoor aeromycota in relation to residential characteristics and allergic symptoms. Mycopathologia 131(3):149–157. doi:10.1007/BF01102894

Li D-W, Kendrick B (1995b) A year-round comparison of fungal spores in indoor and outdoor air. Mycologia 87(2):190–195

Li D-W, Kendrick B (1996) Functional and causal relationships between indoor and outdoor airborne fungi. Can J Bot 74(2):194–209

Li D-W, Yang CS (2004a) Fungal contamination as a major contributor to sick building syndrome. In: Straus D (ed) Sick building syndrome, vol 55, Advances in applied microbiology. Elsevier-Academic Press, San Diego, pp 31–112

Li D-W, Yang CS (2004b) Notes on indoor fungi I: new records and noteworthy fungi from indoor environments. Mycotaxon 89:473–488

Li D-W, Kendrick B, Spero D, Macdonald C (2008) *Balaniopsis triangularis* sp. nov. from indoor environments. Mycotaxon 105:105–110

Li D-W, Zhao G, Yang C, Jalsrai A, Kerin B (2013) Four noteworthy hyphomycetes from indoor environments. Mycotaxon 125(1):111–121

Lierl MB (2013) Myxomycete (slime mold) spores: unrecognized aeroallergens? Ann Allergy Asthma Immunol 111 (6):537-541. e532. doi:10.1016/j.anai.2013.08.007

LiPuma JJ (2010) The changing microbial epidemiology in cystic fibrosis. Clin Microbiol Rev 23(2):299–323

Lockhart SR, Pham CD, Gade L, Iqbal N, Scheel CM, Cleveland AA, Whitney AM, Noble-Wang J, Chiller TM, Park BJ (2013) Preliminary laboratory report of fungal infections associated with contaminated methylprednisolone injections. J Clin Microbiol 51(8):2654–2661

Ma X, Li L, Zhu T, Ba M, Li G, Gu Q, Guo Y, Li D (2013) Phenylspirodrimanes with anti-HIV activity from the sponge-derived fungus *Stachybotrys chartarum* MXH-X73. J Nat Prod 76(12):2298–2306

Macher J (2001) Review of methods to collect settled dust and isolate culturable microorganisms. Indoor Air 11(2):99–110

Madsen AM (2012) Effects of airflow and changing humidity on the aerosolization of respirable fungal fragments and conidia of Botrytis cinerea. Appl Environ Microbiol 78(11):3999–4007

Manamgoda DS, Cai L, McKenzie EH, Crous PW, Madrid H, Chukeatirote E, Shivas RG, Tan YP, Hyde KD (2012) A phylogenetic and taxonomic re-evaluation of the *Bipolaris-Cochliobolus-Curvularia* complex. Fungal Divers 56(1):131–144

Mankowski M, Morrell JJ (2000) Patterns of fungal attack in wood-plastic composites following exposure in a soil block test. Wood Fiber Sci 32(3):340–345

Martin R (1988) Use of a high-through-put jet sampler for monitoring viable airborne propagules of *Fusarium* in wheat. Can J Plant Pathol 10(4):359–360

Matos T, De Hoog G, De Boer A, De Crom I, Haase G (2002) High prevalence of the neurotrope *Exophiala dermatitidis* and related oligotrophic black yeasts in sauna facilities. Mycoses 45(9-10):373–377

Matysik S, Herbarth O, Mueller A (2008) Determination of volatile metabolites originating from mould growth on wall paper and synthetic media. J Microbiol Methods 75(2):182–187

McNeill J, Barrie F, Buck W, Demoulin V, Greuter W, Hawksworth D, Herendeen P, Knapp S, Marhold K, Prado J (2012) International Code of Nomenclature for algae, fungi, and plants (Melbourne Code). Koeltz Scientific Books, Germany

Meirer FC, Lindbergh CA (1935) Collecting microorganisms in the artic atmosphere with field notes and material. Sci Mon 40:5–20

Moore R (1986) A note on Wallemia sebi. Antonie Van Leeuwenhoek 52(2):183–187

Mousavi ME, Irish JL, Frey AE, Olivera F, Edge BL (2011) Global warming and hurricanes: the potential impact of hurricane intensification and sea level rise on coastal flooding. Clim Change 104(3–4):575–597

Mullins J, Hutcheson PS, Slavin RG (1984) *Aspergillus fumigatus* spore concentration in outside air: Cardiff and St Louis compared. Clin Exp Allergy 14(4):351–354. doi:10.1111/j.1365-2222.1984.tb02215.x

mVOC (2015) mVOC: microbial volatile organic compound database. http://bioinformatics.charite.de/mvoc/index.php?site=home. Accessed 12 Jan 2015

Naegele A, Reboux G, Scherer E, Roussel S, Millon L (2013) Fungal food choices of *Dermatophagoides farinae* affect indoor fungi selection and dispersal. Int J Environ Health Res 23(2):91–95

Nagayoshi M, Tada Y, West J, Ochiai E, Watanabe A, Toyotome T, Tanabe N, Takiguchi Y, Shigeta A, Yasuda T (2011) Inhalation of *Stachybotrys chartarum* evokes pulmonary arterial remodeling in mice, attenuated by Rho-Kinase Inhibitor. Mycopathologia 172(1):5–15

Nevalainen A, Taubel M, Hyvarinen A (2015) Indoor fungi: companions and contaminants. Indoor Air 25(2):125–156. doi:10.1111/ina.12182

Nielsen KF, Thrane U (2002) Mould growth on building materials: secondary matabolites, mycotoxins and biomarkers. Technical University of Denmark (Danmarks Tekniske Universitet), Department of Systems Biology (Institut for Systembiologi),

Nielsen KF, Thrane U, Larsen TO, Nielsen P, Gravesen S (1998) Production of mycotoxins on artificially inoculated building materials. Int Biodeter Biodegr 42(1):9–16

Nierman WC, Pain A, Anderson MJ, Wortman JR, Kim HS, Arroyo J, Berriman M, Abe K, Archer DB, Bermejo C (2005) Genomic sequence of the pathogenic and allergenic filamentous fungus *Aspergillus fumigatus*. Nature 438(7071):1151–1156

Padamsee M, Kumar TA, Riley R, Binder M, Boyd A, Calvo AM, Furukawa K, Hesse C, Hohmann S, James TY (2012) The genome of the xerotolerant mold *Wallemia sebi* reveals adaptations to osmotic stress and suggests cryptic sexual reproduction. Fungal Genet Biol 49(3):217–226

Pady S, Gregory P (1963) Numbers and viability of airborne hyphal fragments in England. Trans Br Mycol Soc 46(4):609–613

Pady S, Kramer C (1960) Kansas aeromycology VI: hyphal fragments. Mycologia 52:681–687

Park J-H, Cox-Ganser JM (2011) Mold exposure and respiratory health in damp indoor environments. Front Biosci (Elite Ed) 3:757–771

Pestka JJ, Yike I, Dearborn DG, Ward MDW, Harkema JR (2008) *Stachybotrys chartarum*, trichothecene mycotoxins, and damp building–related illness: new insights into a public health enigma. Toxicol Sci 104(1):4–26

Piechulla B, Degenhardt J (2014) The emerging importance of microbial volatile organic compounds. Plant Cell Environ 37(4):811–812

Pieckova E, Hurbankova M, Cerna S, Liskova A, Kovacikova Z, Kollarikova Z, Wimmerova S (2009) Inflammatory and haematotoxic potential of indoor *Stachybotrys chartarum* (Ehrenb.) Hughes metabolites. Arch Ind Hyg Toxicol (Arhiv Za Higijenu Rada i Toksikologiju) 60(4):401–409

Pitkaranta M, Meklin T, Hyvarinen A, Paulin L, Auvinen P, Nevalainen A, Rintala H (2008) Analysis of fungal flora in indoor dust by ribosomal DNA sequence analysis, quantitative PCR, and culture. Appl Environ Microbiol 74(1):233–244. doi:10.1128/aem.00692-07

Pitt JI, Hocking AD (2009) *Aspergillus* and related teleomorphs. In: Pitt JI, Hocking AD (eds) Fungi and food spoilage. Springer, New York, pp 275–337

Pitt JI, Taylor JW (2014) *Aspergillus*, its sexual states, and the new International Code of Nomenclature. Mycologia 106(5):1051–1062. doi:10.3852/14-060

Polizzi V, Adams A, Malysheva SV, De Saeger S, Van Peteghem C, Moretti A, Picco AM, De Kimpe N (2012) Identification of volatile markers for indoor fungal growth and chemotaxonomic classification of *Aspergillus* species. Fungal Biol 116(9):941–953. doi:10.1016/j.funbio.2012.06.001

Pollard GF, Shaw A, Sowa M, Rand T, Thliveris JA, Scott JE (2013) *Stachybotrys chartarum* (*atra*) spore extract alters surfactant protein expression and surfactant function in isolated fetal rat lung epithelial cells, fibroblasts and human A549 cells. Open J Pediatr 3(3):243–256

Prezant B, Weekes DM, Miller JD (2008) Recognition, evaluation, and control of indoor mold. American Industrial Hygiene Association, Fairfax, VA

Pringle A, Baker DM, Platt JL, Wares JP, Latge JP, Taylor JW (2005) Cryptic speciation in the cosmopolitan and clonal human pathogenic fungus *Aspergillus fumigatus*. Evolution 59(9):1886–1899

Quansah R, Jaakkola MS, Hugg TT, Heikkinen SAM, Jaakkola JJK (2012) Residential dampness and molds and the risk of developing asthma: A systematic review and meta-analysis. PLoS One 7(11):1–9. doi:10.1371/journal.pone.0047526

Rácová Z, Wasserbauer R, Ryparová P (2013) Microscopic filamentous fungi in buildings, preventing their occurrence and their remediation using nanofibers. Adv Mater Res 649:89–92

Rakkestad KE, Skaar I, Ansteinsson VE, Solhaug A, Holme JA, Pestka JJ, Samuelsen JT, Dahlman HJ, Hongslo JK, Becher R (2010) DNA damage and DNA damage Responses in THP-1 monocytes after exposure to spores of either *Stachybotrys chartarum* or *Aspergillus versicolor* or to T-2 toxin. Toxicol Sci 115(1):140–155

Rando RJ, Kwon C-W, Lefante JJ (2013) Exposures to thoracic particulate matter, endotoxin, and glucan during post-hurricane Katrina Restoration Work, New Orleans 2005–2012. J Occup Environ Hyg 11(1):9–18. doi:10.1080/15459624.2013.839879

Rao CY, Riggs MA, Chew GL, Muilenberg ML, Thorne PS, Van Sickle D, Dunn KH, Brown C (2007) Characterization of airborne molds, endotoxins, and glucans in homes in New Orleans after Hurricanes Katrina and Rita. Appl Environ Microbiol 73(5):1630–1634

Raper KB, Fennell DI (1965) The genus *Aspergillus*. Williams & Wilkins, Philadelphia, PA

Reponen T, Seo S-C, Grimsley F, Lee T, Crawford C, Grinshpun SA (2007) Fungal fragments in moldy houses: a field study in homes in New Orleans and Southern Ohio. Atmos Environ 41(37):8140–8149. doi:10.1016/j.atmosenv.2007.06.027

Rylander R (2014) Fungi in homes—how do we measure? Indoor Air 24(2):221–222. doi:10.1111/ina.12075

Ryu SH, Moon HJ (2014) Mould germination and the growth rate on wallpapers with different physical properties and the surface structures. Indoor Built Environ 23(1):171–179

Sahlberg B, Gunnbjörnsdottir M, Soon A, Jogi R, Gislason T, Wieslander G, Janson C, Norback D (2013) Airborne molds and bacteria, microbial volatile organic compounds (MVOC), plasticizers and formaldehyde in dwellings in three North European cities in relation to sick building syndrome (SBS). Sci Total Environ 444:433–440

Saint-Jean M, St-Germain G, Laferrière C, Tapiero B (2007) Hospital-acquired phaeohyphomycosis due to *Exserohilum rostratum* in a child with leukemia. Can J Infect Dis Med Microbiol 18(3):200

Samir H, Wageh W, Abd-Elaziz Emam M (2014) Demonstration of aeroallergenicity of fungal hyphae and hyphal fragments among allergic rhinitis patients using a novel immunostaining technique. Egypt J Otolaryngol 30(1):17–22. doi:10.4103/1012-5574.127186

Samson R (1999) Ecology, detection and identification problems of moulds in indoor environments. In: Johanning E (ed) Bioaerosols, fungi and mycotoxins: health effects, assessment, prevention and control. Eastern New York Occupational and Environmental Health Center, Albany, NY, pp 33–37

Samson R, Houbraken J (2011) Isolation and identification of fungi. In: Flannigan B, Samson RA, Miller JD (eds) Microorganisms in home and indoor work environments, diversity, health impacts, investigation and control. CRC Press, Boca Raton, pp 265–278

Samson RA, Pitt JI (1990) Modern concepts in *Penicillium* and *Aspergillus* classification. Plenum Publishing Corporation, New York

Samson RA, Varga J (2007) *Aspergillus* systematics in the genomic era, vol 59. CBS Fungal Biodiversity Centre, Utrecht

Samson RA, Hoekstra ES, Frisvad JC (2004) Introduction to food-and airborne fungi, 7th edn. Centraalbureau voor Schimmelcultures (CBS), Utrecht

Samson R, Houbraken J, Summerbell R, Flannigan R, Miller J (2011) Common and important species of fungi and actinomycetes in indoor environments. In: Flannigan B, Samson RA, Miller JD (eds) Microorganisms in home and indoor work environments, diversity, health impacts, investigation and control. CRC Press, Boca Raton, pp 321–513

Sato Y, Aoki M, Kigawa R (2014) Microbial deterioration of tsunami-affected paper-based objects: a case study. Int Biodeter Biodegr 88:142–149. doi:10.1016/j.ibiod.2013.12.007

SCENIHR (2015) Nanotechnologies. http://ec.europa.eu/health/scientific_committees/opinions_layman/en/nanotechnologies/l-2/6-health-effects-nanoparticles.htm. Accessed 14 Jan 2015

Schell WA, Lee AG, Aime MC (2011) A new lineage in Pucciniomycotina: class Tritirachiomycetes, order Tritirachiales, family Tritirachiaceae. Mycologia 103(6):1331–1340. doi:10.3852/10-333

Schmidt O (2007) Indoor wood-decay basidiomycetes: damage, causal fungi, physiology, identification and characterization, prevention and control. Mycol Prog 6(4):261–279. doi:10.1007/s11557-007-0534-0

Scott JA (2001) Studies on indoor fungi. University of Toronto, Toronto

Scott J, Untereiner WA, Wong B, Straus NA, Malloch D (2004) Genotypic variation in *Penicillium chysogenum* from indoor environments. Mycologia 96(5):1095–1105

Seifert KA, Morgan-Jones G, Gams W, Kendrick B (2011) The genera of hyphomycetes. CBS-KNAW Fungal Biodiversity Centre, Utrecht

Semeiks J, Borek D, Otwinowski Z, Grishin NV (2014) Comparative genome sequencing reveals chemotype-specific gene clusters in the toxigenic black mold *Stachybotrys*. BMC Genomics 15(1):590. doi:10.1186/1471-2164-15-590

Seo S, Choung JT, Cehn BT, Lindsley WG, Kim KY (2014) The level of submicron fungal fragments in homes with asthmatic children. Environ Res 131:71–76

Sercombe JK, Liu-Brennan D, McKay KO, Green BJ, Tovey ER (2014) Domestic exposure to fungal allergenic particles determined by halogen immunoassay using subject's serum versus particles carrying three non-fungal allergens determined by allergen-specific HIA. Indoor Air 24(4):438–445. doi:10.1111/ina.12087

Simoni M, Cai GH, Norback D, Annesi-Maesano I, Lavaud F, Sigsgaard T, Wieslander G, Nystad W, Canciani M, Viegi G (2011) Total viable molds and fungal DNA in classrooms and association with respiratory health and pulmonary function of European schoolchildren. Pediatr Allergy Immunol 22(8):843–852

Sivanesan A (1987) Graminicolous species of *Bipolaris*, *Curvularia*, *Drechslera*, *Exserohilum* and their teleomorphs. CAB International, Wallingford

Snelders E, Camps SM, Karawajczyk A, Schaftenaar G, Kema GH, Van der Lee HA, Klaassen CH, Melchers WJ, Verweij PE (2012) Triazole fungicides can induce cross-resistance to medical triazoles in *Aspergillus fumigatus*. PLoS One 7(3), e31801

Sousa ACA, Almeida JRSL, Pereira CC, Ramiro Pastorinho M, Pereira ÂMC, Nogueira AJA, Taborda-Barata L, Teixeira JP, Correia ACM, Alves A (2014) Characterization of fungal communities in house dust samples collected from central Portugal—a preliminary survey. J Toxicol Environ Health A 77(14–16):972–982. doi:10.1080/15287394.2014.911137

Stephenson SL (2011) From morphological to molecular: studies of myxomycetes since the publication of the Martin and Alexopoulos (1969) monograph. Fungal Divers 50(1):21–34

Steyn PS, Vleggaar R (1976) The structure of dihydrodeoxy-8-epi-austdiol and the absolute configuration of the azaphilones. J Chem Soc Perkin Trans 1(2):204–206

Storey E, Dangman K, Schenck P, DeBernardo R, Yang C, Bracker A, Hodgson M (2004) Guidance for clinicians on the recognition and management of health effects related to mold exposure and moisture indoors. University of Connecticut Health Center, Division of Occupational and Environmental Medicine, Center for Indoor Environments and Health, Storrs, CT

Straus DC (2004) Sick building syndrome, vol 55. Elsevier Academic Press, Amsterdam

Stringer J, Beard C, Miller R, Wakefield A (2002) A new name (*Pneumocystis jiroveci*) for *Pneumocystis* from humans. Emerg Infect Dis 8:891–896

Stringer JR, Beard CB, Miller RF (2009) Spelling *Pneumocystis jirovecii*. Emerg Infect Dis 15(3):506

Summerbell R (2011) Respiratory tract infections caused by indoor fungi. In: Flannigan B, Samson RA, Miller JD (eds) Microorganisms in home and indoor work environments, diversity, health impacts, investigation and control. CRC Press, Boca Raton, pp 205–226

Summerbell RC, Gueidan C, Schroers HJ, De Hoog GS, Starink M, Rosete YA, Guarro J, Scott JA (2011) *Acremonium* phylogenetic overview and revision of *Gliomastix*, *Sarocladium*, and *Trichothecium*. Stud Mycol 68:139–162

Suttie E, Hill C, Jones D, Orsler R (1998) Chemically modified solid wood—I. Resistance to fungal attack. Mater Org 32:159–182

Terracina FC (1974) Fine structure of the septum in *Wallemia sebi*. Can J Bot 52(12):2587–2590

Thelander O, Bjurman J, Boutelje J (1993) Increase in the content of low molecular carbon hydrates at lumber suracces during drying and correlation with nitrogen content, yellowing and mould growth. Wood Sci Technol 27:381–389

Thom C, Church MB (1926) The Aspergilli. Williams & Wilkins, Baltimore, MD

Thom C, Raper KB (1945) A manual of the Aspergilli, vol 60. Soil science, vol 60. Williams & Wilkins Company, Baltimore, MD

Tille P (2014) Opportunistic atypical fungus: *pneumocystis jiroveci*. Bailey & Scott's diagnostic microbiology, 13th edn. Mosby, Inc., an affiliate of Elsevier Inc,

Tischer C, Hohmann C, Thiering E, Herbarth O, Müller A, Henderson J, Granell R, Fantini M, Luciano L, Bergström A (2011) Meta—analysis of mould and dampness exposure on asthma and allergy in eight European birth cohorts: an ENRIECO initiative. Allergy 66(12):1570–1579

Tripi PA, Modlin S, Sorenson W, Dearborn DG (2000) Acute pulmonary haemorrhage in an infant during induction of general anaesthesia. Pediatr Anesth 10(1):92–94

Van Strien R, Gehring U, Belanger K, Triche E, Gent J, Bracken M, Leaderer B (2004) The influence of air conditioning, humidity, temperature and other household characteristics on mite allergen concentrations in the northeastern United States. Allergy 59(6):645–652

Varga J, Kocsubé S, Szigeti G, Baranyi N, Vágvölgyi C, Despot DJ, Magyar D, Meijer M, Samson RA, Klarić MŠ (2014) Occurrence of black Aspergilli in indoor environments of six countries. Arh Hig Rada Toksikol 65(2):219–23

Verhoeff A, Ev R-H, Samson R, Brunekreef P, Jv W (1994) Fungal propagules in house dust. I Allergy 49(7):533–539

Vesper SJ, Vesper MJ (2002) Stachylysin may be a cause of hemorrhaging in humans exposed to *Stachybotrys chartarum*. Infect Immun 70(4):2065–2069

Vesper SJ, Magnuson ML, Dearborn DG, Yike I, Haugland RA (2001) Initial characterization of the hemolysin stachylysin from *Stachybotrys chartarum*. Infect Immun 69(2):912–916

Vesper S, McKinstry C, Haugland R, Wymer L, Bradham K, Ashley P, Cox D, Dewalt G, Friedman W (2007) Development of an environmental relative moldiness index for US homes. J Occup Environ Med 49(8):829–833

Viitanen HA (1997) Modelling the time factor in the development of mould fungi-the effect of critical humidity and temperature conditions on pine and spruce sapwood. Holzforschung-International J Biol Chem Phys Technol Wood 51(1):6–14

Wang Y, Hyde KD, McKenzie EHC, Jiang Y-L, Li D-W, Zhao D-G (2015) Overview of *Stachybotrys* (*Memnoniella*) and current species status. Fungal Divers 71(1):17–83. doi:10.1007/s13225-014-0319-0

Webb JS, Nixon M, Eastwood IM, Greenhalgh M, Robson GD, Handley PS (2000) Fungal colonization and biodeterioration of plasticized polyvinyl chloride. Appl Environ Microbiol 66(8):3194–3200

WHO (2009) WHO guidelines for indoor air quality: dampness and mould. World Health Organization—Regional Office for Europe, DK-2100 Copenhagen O, Denmark.

Will KW, Rubinoff D (2004) Myth of the molecule: DNA barcodes for species cannot replace morphology for identification and classification. Cladistics 20(1):47–55. doi:10.1111/j.1096-0031.2003.00008.x

Wu X-Q, Yuan W-M, Tian X-J, Fan B, Fang X, Ye J-R, Ding X-L (2013) Specific and functional diversity of endophytic bacteria from pine wood nematode *Bursaphelenchus xylophilus* with different virulence. Int J Biol Sci 9(1):34

Würtz H, Sigsgaard T, Valbjørn O, Doekes G, Meyer HW (2005) The dustfall collector-a simple passive tool for long-term collection of airborne dust: a project under the Danish Mould in Buildings program (DAMIB). Indoor Air 15(s9):33–40

Xie H, Aminuzzaman F, Xu L, Lai Y, Li F, Liu X (2010) Trap induction and trapping in eight nematode-trapping fungi (Orbiliaceae) as affected by juvenile stage of *Caenorhabditis elegans*. Mycopathologia 169(6):467–473

Yang CS (2007) A retrospective and forensic approach to assessment of fungal growth in the indoor environment. In: Yang Chin S, Heinsohn P (eds) Sampling and analysis of indoor microorganisms. Wiley, Hoboken, NJ, pp 215–229

Yang CS, Heinsohn PA (2007) Sampling and analysis of indoor microorganisms. Wiley, Hoboken, NJ

Yang C, Johanning E (2007) Airborne fungi and mycotoxins. In: Hurst CJ, Crawford RL, Garland JL, Lipson DA, Mills AL, Stetzenbach LD (eds) Manual of environmental microbiology, 3rd edn. American Society for Microbiology, Washington, DC, pp 972–988

Yike I, Dearborn D (2011) Guest editorial—novel insights into the pathology of *Stachybotrys chartarum*. Mycopathologia 172(1):1–3

Yu K-P, Huang Y-T, Yang S-C (2013) The antifungal efficacy of nano-metals supported $TiO_2$ and ozone on the resistant *Aspergillus niger* spore. J Hazard Mater 261:155–162

Zalar P, de Hoog GS, Schroers H-J, Frank JM, Gunde-Cimerman N (2005) Taxonomy and phylogeny of the xerophilic genus *Wallemia* (Wallemiomycetes and Wallemiales, cl. et ord. nov.). Antonie Van Leeuwenhoek 87(4):311–328

Zalar P, de Hoog GS, Schroers H-J, Crous PW, Groenewald JZ, Gunde-Cimerman N (2007) Phylogeny and ecology of the ubiquitous saprobe *Cladosporium sphaerospermum*, with descriptions of seven new species from hypersaline environments. Stud Mycol 58(1):157–183. doi:10.3114/sim.2007.58.06

Zambettakis C (1955) Un nouveau champignon papyricole [*Ascotricha erinacea*]. Bulletin de la Société Botanique de France 102:219–225

# Chapter 16
# Biology of the Whiskey Fungus

James A. Scott and Richard C. Summerbell

## Introduction

Since the industrial revolution, sooty discoloration has become an ever-apparent feature of the exterior surfaces of buildings and monuments worldwide. Discoloration arises both from the deposition of airborne, carbonaceous combustion products and from the growth of melanized microbes on surfaces. The development of dark discoloration on outdoor surfaces such as concrete, brick, and wood as well as many other construction materials exposed to low levels of ethyl alcohol (ethanol) vapour has been associated with the aging of spirits, particularly in areas of damp climate, for over a century. Known as "warehouse staining", this phenomenon differs little from disfigurement caused by the soot that billowed copiously from factories during the industrial revolution. But unlike the sooty accumulations arising from coal combustion, warehouse staining is caused by a microbial biofilm consisting of thick, black, confluent, crust-like colonies that extend over large, exposed outdoor surfaces. The substrates for warehouse staining encompass a wide range of materials, including man-made items such as construction materials, fences, road signs, outdoor furniture, and vehicles. Natural materials such as vegetation and rock are also affected. Although this biofilm probably exists in nature sporadically as isolated

J.A. Scott, Ph.D. (✉)
Division of Occupational and Environmental Health, Dalla Lana School of Public Health, University of Toronto, 223 College Street, Toronto, ON, Canada M5T 1R4

Sporometrics Inc., 219 Dufferin Street, Suite 20-C, Toronto, ON, Canada M6K 1Y9
e-mail: james.scott@utoronto.ca

R.C. Summerbell, Ph.D.
Sporometrics Inc., 219 Dufferin Street, Suite 20-C, Toronto, ON, Canada M6K 1Y9

Dalla Lana School of Public Health, Gage Occupational and Environmental Health Unit, University of Toronto, 223 College Street, Toronto, ON, Canada M5T 1R4
e-mail: rsummerbell@sporometrics.com



D.-W. Li (ed.), *Biology of Microfungi*, Fungal Biology,
DOI 10.1007/978-3-319-29137-6_16

microcolonies where its growth may be engendered by natural ethanolic vapours, it is only in the unnatural flux of ethanol emissions from industry that the stains flourish prominently.

Even though warehouse staining has been recognized for many years, the phenomenon received sparse study until recently. We trace the history of study on warehouse staining, provide a summary of the current state of knowledge on this interesting phenomenon, and identify areas for future research.

## *Nineteenth Century*

In 1872, Antonin Baudoin, director of the French Distillers' Association, noticed the accumulation of a soot-like blackening of walls around distilleries along the Charente River in Cognac, France. This "plague of soot", he observed, obliged the distillery proprietors to engage in regular cleaning. As he described it, the town of Cognac seemed as though it were covered in mourning crape—a reference to the crinkled black silk fabric of the Victorian era costumes worn by grief-stricken widows (Roumeguère 1881). Curiously, distilleries downstream in the historical province of Aunis to the northwest as well as those in the southern countryside were unaffected. Knowledgeable in biology through his training as a pharmacist, Baudoin judged the growth to represent a species of the fungus *Xenodochus*; he provided specimens to Durrieu de Maisonneuve (Richon and Petit 1881). The latter, in consultation with Roumeguère, concurred in his diagnosis and considered naming the fungus "*X. baudoini*" to honour Baudoin (Roumeguère 1881). A formal description was never published. However, Roumeguère circulated specimens from the collection in his Fungi Gallici exsiccati, Centurie XVI, No. 1695. As outlined by Scott et al. (2007), Roumeguère's description referred the specimen to *Torula conglutinata* var. *compniacensis*. His collection labels, however, erroneously listed the specimens as "*T. conglomerata* * *compniacensis*" (DAOM 66898, Fig. 16.1).

**Fig. 16.1** Packet containing Centurie XVI, No. 1695 of Roumeguère's Fungi Gallici exsiccati (DAOM 66898). The name written on the packet is "*Torula conglomerata* * *compniacensis*" despite that the name published by Roumeguère (1881) was "*Torula conglutinata* Corda var. *Compniacensis* Richon"

The transcription error on the packet was carried forward by Saccardo (1886) and Crane (2001).

In 1878, following a detailed study of specimens observed under higher power magnification, Baudoin reconsidered his characterization of the causal agent and decided the organism was cyanobacterial rather than fungal, a species of the genus *Nostoc* (Richon and Petit 1881). Nearly a decade after his initial discovery, Baudoin published a pamphlet outlining his theories on the blackening phenomenon, attributing the growth to *Nostoc* (Baudoin 1893).

Baudoin was not alone in his early observation of warehouse staining. While presiding over the June 14, 1878, meeting of the French Botanical Society, physician-mycologist Gaspard Adolphe Chatin presented to the society examples of tile and stone collected near spirit maturation warehouses in Cognac. The fragments were blackened, he believed, by an unknown cryptogam. No further information was provided in the meeting transcript (Chatin 1878). Richon and Petit (1881) later reasoned that Chatin's discovery could only have been the fungus they formally described as *Torula compniacensis*.

## *Twentieth Century*

In the century following the work of Richon and Petit (1881) and Roumeguère (1881), there were few sporadic reports in the scientific literature relating to warehouse staining. In discussing anthropophilic fungi, Moreau (1954) referred to the warehouse staining fungus ("*Torula conglutinata* var. *compniacensis*") as an example of a saprotroph—in this case, one responsible for blackening tiles of houses in the village of Cognac that were exposed to the alcoholic vapours emitted from its famous distilleries. Although he was clearly familiar with the fungus and earlier work done by Baudoin, Richon, and Petit, Moreau did not contribute any original observations.

An elegant study done by Annelisa Kjøller (1961) at the University of Copenhagen reported the presence of a dense, confluent, black fungal colonization on asbestos-cement roofing at the Heering Distillery in Dalby, Sjælland, Denmark. Direct microscopic observations of mounts from these materials revealed a predominant form that Kjøller considered to be a single entity, although she noted the absence of typical hyphae or differentiated conidiophores. Kjøller (1961) identified the fungus as *Torula compniacensis*, and the identification was verified at the time by Dr. S.J. Hughes in Canada, after he had compared the collections to authentic material representing this taxon from Roumeguère's Fungi Gallici exsiccati, Cent. XVI, No. 1695 "*Torula conglomerata * compniacensis*" (DAOM 66898, Fig. 16.1). Later examinations of two specimens from Kjøller's study by Scott et al. (2007), one a stone chip from the factory roof (DAOM 6687) and the other a fragment of asbestos tile (DAOM 109435), confirmed Kjøller's conclusions and supported the placement of her collections in the genus *Baudoinia* J.A. Scott and Unter. as *Baudoinia compniacensis* (Richon) J.A. Scott and Unter.

In addition to her microscopic observations, Kjøller (1961) prepared cultures by spreading cells scraped from dried specimens on malt agar, incubating the cultures at 14 °C, and examining the agar plates directly under the microscope after 1 week. Using this method she successfully germinated cells that had been stored dry for up to a year. Although Kjöller is undoubtedly the first scientist to have successfully cultured the warehouse staining fungus, lamentably no cultures remain from this work for verification by modern methods.

In her studies of germinating "chlamydospores", Kjøller (1961) observed considerable variation and distinguished three basic patterns of germination and maturation. She found that adulteration of the isolation medium with alcohol, sherry, or prune juice failed to alter the germination or growth pattern. Kjøller's first pattern was characterized by initial swelling of the chlamydospore and bursting of the ornamented wall followed by the production of elongating germ tubes associated with greenish-yellow mucilage (e.g. Fig. 16.2). The structures matured after 3 weeks to become roughened, monilioid chlamydospores similar in appearance to those she observed on the roofing material. Her illustrations of germinating chlamydospores (Kjøller 1961, Fig. 16.1) bear close resemblance to the description and illustrations of Scott et al. (2007, Figs. 16.4 and 16.5). These collections likely correspond to *Baudoinia*. The second pattern involved the direct germination of chlamydospores to a budding, yeast-like phase that ultimately matured into dark, smooth, muriform chlamydospores. These structures connote the genus *Aureobasidium*, which contains ubiquitous, unspecialized yeast-like fungi encountered commonly in the mature warehouse staining mycobiota (Ewaze et al. 2008b; Scott et al. 2007; Watson et al. 1984). In her third pattern, chlamydospores germinated to produce hyaline germ tubes that became moniliform, and dark, smooth-walled muriform chlamydospores at maturity. Kjøller interpreted these to be arthroconidial chains. While the identity of this fungus is unknown, it is unlikely to be affiliated with the genus *Baudoinia*.

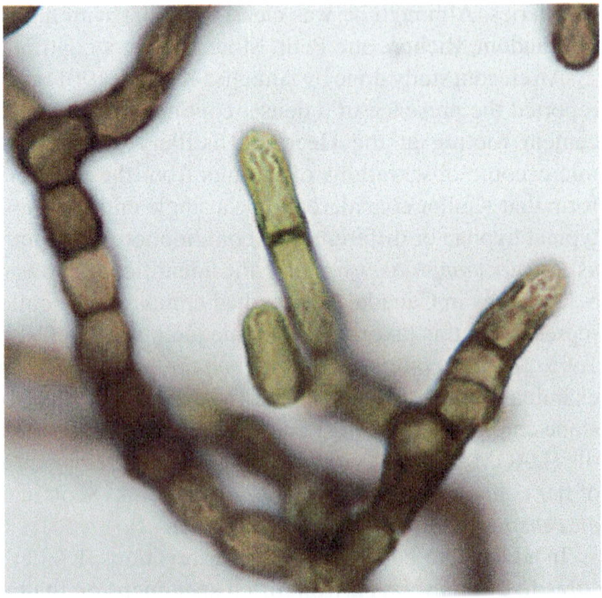

**Fig. 16.2** Germ tube of *Baudoinia* photographed in transmitted light microscopy demonstrating the greenish coloration of the nascent cell wall (SCCM 2488428) (2000 ×)

Several years after the publication of Kjøller's work, Auger-Barreau (1966) published preliminary findings on the identity of the agent of warehouse staining from Cognac, France, as a basis for later investigations of allergy-related health effects in communities affected by the fungus. Auger-Barreau was initially sent a specimen from a Cognac cellar which she examined microscopically and attempted to culture. She described the thick, brown, felt-like growth as macroscopically resembling the velvety interior context of the "amadou" fungus, *Fomes fomentarius* (L.) Fr. However, her microscopic and culture examinations suggested that it consisted of multiple species. When Auger-Barreau visited the warehouse to collect more specimens, she observed two distinctive morphological patterns of growth. One pattern, seen on the interior walls of cellars, conformed to her observations of the specimen she was initially sent. The other, a black and soot-like growth, was restricted to building exteriors (Auger-Barreau 1966). Auger-Barreau (1966) reasonably concluded that the outdoor fungus was probably Richon's *T. compniacensis*, whereas the indoor fungus more closely resembled a species of *Cladosporium*. In all likelihood, the latter collection represented the cellar fungus, *Zasmidium cellare* (Pers.) Fr. (Chlebicki and Majewska 2010; Tribe et al. 2006).

Two decades after the work of Kjøller and Auger-Barreau, Watson et al. 1984 reported on the warehouse staining phenomenon from bond warehouses in Scotland. They subjected scraping samples to culture and reported a suite of non-specialist fungi that included *Alternaria tenuissima* (Kunze: Fr.) Wiltshire, *Ascochyta* sp., *Aspergillus terreus* Thom, *Aureobasidium pullulans* (deBary) Arnaud, *Cladosporium tenuissimum* Booke, *Curvularia* sp., *Epicoccum nigrum* Link, *Phoma capitulatum* Pawar Mathur and Thirum., *P. nebulosa* (Pers.) Berk., *P. tropica* Schneider & Boerema, and *Pseudodiplodia* sp. They likened the black staining to a polymicrobial biofilm named "*Fumago vagans*" that has been known to develop on aphid honeydews.

Watson et al. (1984) provided a series of calculations on ethanol losses from the bond warehouses they examined, projecting a 2 % loss in the first year of aging and a 1 % loss in the second and subsequent years. Based on approximately 2.25 million litres of whiskey in storage, they calculated the loss to correspond to about 90,000 l over the first few years of maturation. They proposed a relationship between the ethanol emission from bond warehouses and the development of black fungal growth on surfaces, observing that the "growth is denser and extends further from the warehouses following the direction of the prevailing wind". They further noted that growth was absent near unused warehouses and only initiated once the warehouses were filled with whiskey barrels. Although they were clearly aware of the association of the cellar fungus with fugitive emissions of ethanol, Watson et al. (1984) were unfamiliar with prior work on the whiskey fungus.

## *Twenty-First Century*

In 2007, the warehouse staining fungus was again rediscovered by Scott et al. (2007) during a survey commissioned by a Canadian distiller, of fungal colonists of outdoor surfaces near spirit maturing warehouses. They compared their environmental samples against those of Roumeguère (1881) and Kjøller (1961), concluding that all specimens consisted predominantly of what appeared to be a single taxon conforming to the description of *Torula compniacensis*. Using a technique similar to that of Kjøller (1961), Scott et al (2007) successfully germinated single cells, isolating them in pure culture. Scott et al. incorporated 5 ppm ethanol in their primary isolation medium to enhance recovery of the whiskey fungus by reducing the proliferation of contaminant moulds. Microscopic examination of the resulting isolates showed darkly pigmented, extremely slow-growing colonies that resembled the first predominant pattern described by Kjøller (1961).

Scott et al. (2007) examined 13 specimens and 8 cultures collected from a range of geographic localities. Phylogenetic analysis of several isolates based on nucSSU rRNA gene sequences supported the erection of a new genus in the order Capnodiales. They named the genus *Baudoinia* in recognition of Antonin Baudoin's early contributions to the study of the organism (Scott et al. 2007). They provided a synonymy partly based on Crane (2001), although as Illana-Esteban (2013) later noted that Crane (2001) and Scott et al. (2007) had incorrectly cited the journal reference for *T. compniacensis* Richon. Firstly, both authors listed the journal title *Rev. Mycol. (Paris)*. The journal abbreviation *Rev. Mycol. (Paris)* refers to *Revue de Mycologie*, published from 1936 to 1979; by contrast, *Rev. Mycol. (Toulouse)* denotes *Revue Mycologique* whose publication ran from 1878 to 1906 and included the article in question. Illana-Esteban (2013) also observed that *T. compniacensis* Richon was actually initially published by Richon and Petit (1881) in the third volume of another journal, Brebissonia, in February of 1881. Roumeguère then reprinted Richon's diagnosis in the *Revue Mycologique* which appeared in July of the same year. Thus, the basionym of *B. compniacensis* is *T. compniacensis* Richon in Richon and Petit, *Brebissonia* 3(8): 115 (1881).

Illana-Esteban (2013) incorrectly attributed authorship of Richon's reprinted diagnosis to Baudoin. As published, the last sentence of Roumeguère's article appears to read: "Je distribue la nouvelle variété dans ma centurie XVI. M. Baudoin et M.", implying that Baudoin and an unnamed individual, "M.", authored the article ("centurie XVI" here referred to the 16th set of 100 specimens distributed in Roumeguère's Fungi Gallici exsiccati in which the specimen of interest was circulated as No. 1695 and incorrectly labelled as *T. conglomerata* * *compniacensis*). However, the actual concluding sentence of the article was truncated, with the remainder erroneously transposed to the head of the following page. The final sentence of the article thus correctly reads: "M. Baudoin et M. Paul Brunaud après lui, ont bien voulu m'en approvisionner, C.R." ("Mr. Baudoin, and Mr. Paul Brunaud after him have generously supplied it to me, C.R."), confirming Roumeguère as the article's author. The synonym *T. conglutinata* Corda var. *compniacensis* (Richon) Sacc. in Roum., *Rev. Mycol. (Toulouse)* 3(11):17 (1881), was created in a footnote.

Scott et al. (2007) designated a representative of Roumeguère's exsiccata, centurie XVI, No. 1695 (DAOM 238773) as the lectotype of *B. compniacensis* and selected a living culture obtained during their studies near the type locality, Cognac, France, as epitype (UAMH 10808) from which DNA sequence and cultural characters were derived. The genus *Baudoinia* was shown by Scott et al. (2007) to be a member of the Capnodiales on the basis of nucSSU sequence analysis. Although no sexual state has been demonstrated, Crous (2009) recognized *Baudoinia* as an anamorph linked to the genus *Mycosphaerella*. Using nucLSU sequence analysis, Crous et al. (2009) treated the genus in the Teratosphaeriaceae, and this determination was later accepted by Wijayawardene et al. (2012).

Based on sequencing of multiple gene regions in 16 collections of *Baudoinia* originating from a range of geographic locations, Scott et al. (2016) described four additional species in the genus: *B. antilliensis*, *B. caladoniensis*, *B. orientalis* and *B. panamericana*. They indicated that the global ranges of these taxa appeared disjunct at a continental scale but proposed that international trade in used, colonized aging barrels may promote global redistribution of whiskey fungi.

## Physiology

As its unique habitat purports, *B. compniacensis* is able to metabolize a range of carbon sources including acetate and ethanol in addition to the simple sugar, glucose. Glucose is used optimally in the range of 1–10 mM, and acetate (in the form of ammonium acetate) is used optimally at a concentration of 1 mM (Ewaze et al. 2007). Ethanol is used optimally at a continuous concentration in the range of 40–50 mM (0.25 % v/v) in the presence of 5–10 mM of a variety of nitrogen sources, both inorganic (e.g. ammonium chloride, ammonium nitrate, potassium nitrate, sodium nitrate) and organic (alanine, asparagine, glutamate, glutamine, casamino acids). Urea, however, was not utilized (Ewaze et al. 2007). Other simple alcohols were used poorly or not at all (Ewaze et al. 2007).

Higher concentrations of ethanol were associated with sharply attenuated growth. For example, the biomass yield of cultures was substantially diminished at a continuous ethanol concentration of 172 mM (1 % v/v) and was completely arrested at 860 mM (5 % v/v) (Ewaze et al. 2008a). Concentrations up to 2410 mM (14 % v/v), while not permitting growth, were tolerated for a brief period (Ewaze et al. 2008a). Concentrations of 3440 mM (20 % v/v) produced complete cell death for all exposure time periods investigated (Ewaze et al. 2008a).

Brief exposure of dormant cells to a very low concentration of ethanol in liquid form (5 mM; 290 ppm v/v), while suboptimal for vegetative growth, resulted in optimal germination and colony formation (Ewaze et al. 2008a). A similar pattern of optimized germination activation was noted for dormant cells exposed briefly to ethanol vapour at a concentration of 10 ppm (19 mg $m^{-3}$ at 25 °C) (Ewaze et al. 2008a). At this ethanol concentration, germination exceeded the non-exposed control by 200 %. Similarly, an airborne concentration of 1 ppm ethanol increased germination by 140 %, and 100 ppb resulted in an increase in germination of 60 %. Enhancement of germination of the fungus in the presence of alcohol, while suspected, was not observed by Kjøller (1961).

Protein electrophoresis studies confirmed that both alcohol dehydrogenase II and acetaldehyde dehydrogenase were produced when isolates of the fungus were grown on ethanol as a sole carbon source, but were absent when the isolates were grown on glucose (Ewaze et al. 2008a). Both the phosphorylated (active) and non-phosphorylated (inactive) forms of the TCA cycle enzyme isocitrate dehydrogenase were found when isolates of the fungus were grown on glucose. This enzyme is responsible for the oxidative decarboxylation of isocitrate to oxalosuccinate, and its presence in glucose-grown cultures confirms the normal functionality of the TCA cycle. When cultures were grown on acetate, however, both forms were absent. In their place, Ewaze et al. (2008a) confirmed the presence of isocitrate lyase and malate synthase, two glyoxylate pathway enzymes that support the TCA cycle-mediated conversion of acetyl-CoA to oxaloacetate, when the fungus is grown on ethanol or acetate in the absence of glucose. Greene et al. (2014) identified a gene sequence of one isocitrate lyase isoform closely linked with a gene cluster supporting L-tyrosine degradation.

## *Stress Responses*

Ewaze et al. (2007) reported that hydrated cells of *B. compniacensis* were readily killed at 52 °C. Preconditioning of hydrated cells at 37 °C prior to transfer to 52 °C resulted in significantly improved survival (Scott et al. 2007) and coincided with the production of several putative heat shock proteins (Ewaze et al. 2007). A comparable pattern of altered protein expression was stimulated following brief exposure to 7 % ethanol followed by high temperature incubation (Ewaze et al. 2007), suggesting a role for ethanol in the induction of stress response systems. Production of the stress-protective disaccharide trehalose was similarly stimulated by both sub-lethal heat shock and ethanol exposure (Al-Naama et al. 2009). As expected, dried cells demonstrated much greater tolerance of high temperature than hydrated cells, retaining viability after 21 d at 70 °C but exhibiting complete cell death at 75 °C (Ewaze et al. 2007).

## *Nuclear Genome*

The nuclear genome of *Baudoinia compniacensis* is the smallest that has been documented for a member of the class, Dothideomycetes, consisting of only 21.88 mB (Ohm et al. 2012). The genome is almost entirely non-redundant (0.4 % repetitive content) and contains roughly 10,500 predicted proteins. It also features fewer putative coding genes, fewer genes with introns, and shorter distances between coding genes than are found in other Dothidiomycetes (Ohm et al. 2012). Only 428 putative pathogenesis-associated genes were found in *B. compniacensis*, including small secreted proteins (67), proteins involved in secondary metabolism (12), carbohydrate active enzymes (283), secreted peptidases (28), and lipases (38) (Ohm et al. 2012). Two polyketide synthases, two non-ribosomal peptide synthetases, and eight terpene synthase genes were predicted (Ohm et al. 2012). Several putative mycotoxin-like genes were also identified including orthologues of *Ds-Nor-1*, *Ds-AvnA*, *Ds-AdhA* (dothistromin), and *Af-verA*, *Af-omtB* (sterigmatocystin) (Ohm et al. 2012). Leducq (2014) interpreted the relatively small genome and paucity of pathogenicity-related genes in *Baudoinia* as signifying a saprotrophic ecology.

Degradation of the amino acid L-tyrosine by fungi is accomplished by two alternate pathways that diverge from the intermediate, homogenisate. The first of these parallels the AKU pathway found in mammals and yields fumarate and acetoacetate following two additional enzymatic steps. The second yields pyruvate and fumarate after four additional enzymatic steps. Considerable variability is present across the fungal phylogeny with respect to the presence of one or both tyrosine degradation pathways. When present, each is typically supported by a tightly linked gene cluster whose pattern of inheritance closely reflects phylogeny. Using comparative genomic analysis, Greene et al. (2014) demonstrated the presence of only the former pathway in *B. compniacensis*. It comprised a cluster of eight putative genes, four of which shared high homology with genes found in the phylogenetically distant *Exophiala dermatitidis*, although gene arrangement and orientation were not conserved. Two additional putative homologous genes encoding the glyoxylate pathway enzyme isocitrate lyase and a membrane transport protein were likewise common to the L-tyrosine degradation linkage clusters in both taxa but absent from other taxa surveyed. These findings were interpreted as evidence of horizontal gene transfer from an ancestor of the extremophilic, melanized fungus *Exophiala dermatitidis* (Chaetothyriales) to the *Baudoinia* clade (Wisecaver and Rokas 2015).

## Mitochondrial Genome

The mitochondrial genome of *B. compniacensis* is similarly comparatively small in size (26.0 kB) and encodes the 14 coding genes normally found in mycelial fungi (ATP synthase subunits atp6, atp8, and atp9, cytochrome b, cytochrome oxidase subunits cox1–cox3, and the NADH dehydrogenase subunits nad1–nad4, nad4L, nad5–nad6) (Goodwin et al. 2014). A single additional unique open reading frame (ORF) was identified that coded for an unknown protein. All 20 amino acids were represented by 26 putative tRNA genes variously oriented in the forward and reverse direction (Goodwin et al. 2014). The compact nature of the mitochondrial genome reflects the absence of introns and superfluous ORFs.

## Mating System

Riley et al. (2014) examined the nuclear genome sequence of *B. compniacensis* in an effort to evaluate the mating system of the species. They identified a putative sequence corresponding to the *MAT1-2-1* gene encoding a high mobility group

(HMG) domain. The solitary presence of the *MAT1-2-1* idiomorph implies a heterothallic mating system, although a sexual state has so far not been recorded.

## Ecology

*Baudoinia* forms a thick, soot-like crust on ethanol-exposed outdoor surfaces near distillery aging warehouses and commercial bakeries (Scott et al. 2007). The surfaces most affected tend to be highly sun exposed and periodically subject to condensing conditions. Although the growth appears indiscriminately on a wide range of surfaces ranging from stainless steel and paint to masonry and vegetation, certain materials remain predictably devoid of growth. For example, Watson et al. (1984) remarked on the absence of fungal disfigurement in the drainage plane of copper or zinc-clad surfaces, indicating the inhibition of growth by low concentrations of ions of these metals (Fig. 16.3). Likewise, anecdotal reports have suggested a low tolerance to salt based on observations that oceanfront elevations of bond warehouses manifest less growth than landward-facing walls. Ewaze et al. (2007), however, showed isolates of *Baudoinia* to withstand sodium chloride concentrations up to 2 M.

The genus is widely distributed geographically and is known so far from the Americas, Europe, and Asia (Scott et al. 2007, 2016). The range of localities where the fungus has been collected indicates a proclivity for damp riparian and coastal climates. Although the fungus remains unknown from non-industrial habitats, the

**Fig. 16.3** Exterior wall panel finished in *white* vinyl siding. The portion of panel to the left is situated beneath a cap of metallic copper flashing, whereas copper flashing is absent at the top of the panel section at right. The area of siding located in the drainage plan beneath the copper flashing shows substantially less colonization of *Baudoinia compniacensis* than the unflashed portion. This is consistent with the observation of Watson et al. (1984) and their suggestion that copper and zinc may have antimicrobial potential in the management of warehouse staining (Bar = 10 cm)

existence of a natural, non-industrial habitat is highly probable given that the *Baudoinia* clade diverged within the Capnodiales between 55 and 75 million years ago (Ohm et al. 2012). Spirit production and maturation, by contrast, has only been practised for the past few centuries (Rogers 2011, 2014). Scott et al. (2007) speculated that the clade may have arisen in non-anthropogenic, ethanol-exposed environments such as fruit drops and other naturally occurring composts.

## *Ethanol Vapour as a Habitat Determinant*

Environmental ethanol emissions arise naturally in terrestrial environments from the aerial parts of plants that are subject to drought and other stresses. Typical airborne ethanol vapour levels in North America have been reported around 5 ppb (roughly 9 μg $m^{-3}$) (Farmer and Dawson 1982; Snider and Dawson 1985; Millet et al. 2005). The failure of molecular-based environmental surveys to recover *Baudoinia* sequences from environments lacking industrial ethanol emissions suggests that the naturally occurring background levels of ethanol vapour common in most terrestrial environments alone may be insufficient to engender the organism's ecological success on a scale sufficient to be noticeable. Clarification of this theory awaits systematic study.

While *Baudoinia* colonizes surfaces exposed to a wide range of airborne ethanol concentrations, laboratory studies have shown that it can grow entirely in the absence of ethanol. Relatively high levels of ethanol vapour, such as those found inside spirit aging warehouses coupled with the low levels of available moisture, probably inhibit the growth of the fungus indoors or to halt its growth altogether. It is noteworthy that the interiors of maturation warehouses typically lack any sign of the fungus, yet the fungus will be found growing immediately outside the warehouses. Moderate ethanol levels, in the range of 5–10 ppm (9.5–19 mg $m^{-3}$ at 25 °C), are optimally promotive of growth. The stimulation diminishes at successively lower ethanol concentrations but remains perceptible at least as low as 100 ppb (190 μg $m^{-3}$ at 25 °C).

The relationship between *B. compniacensis* and ethanol is complex. The fungus can use ethanol as a nutrient. In addition, ethanol serves as a germination activator and an inducer of cellular stress responses. Unlike the indoor-dwelling cellar fungus *Zasmidium cellare* (Tribe et al. 2006), *B. compniacensis* grows outdoors, colonizing ethanol-exposed substrates that experience extreme daily fluctuations in temperature and moisture (Scott et al. 2007). In Fig. 16.4, we propose a provisional model for the role of ethanol in the diurnal cycle of *Baudoinia*. In this model, desiccated cells are subject to moist air and cool substrates during the evening into the early morning. The formation of dew on substrates serves to absorb airborne ethanolic vapour concentrating the alcohol up to 10-fold. Hydrated, ethanol-exposed cells are stimulated to germinate, and ethanol is then metabolized as a nutrient. At the same time, the ethanol induces the production of cellular stress response mechanisms including the synthesis of trehalose heat shock proteins. As sunlight warms the

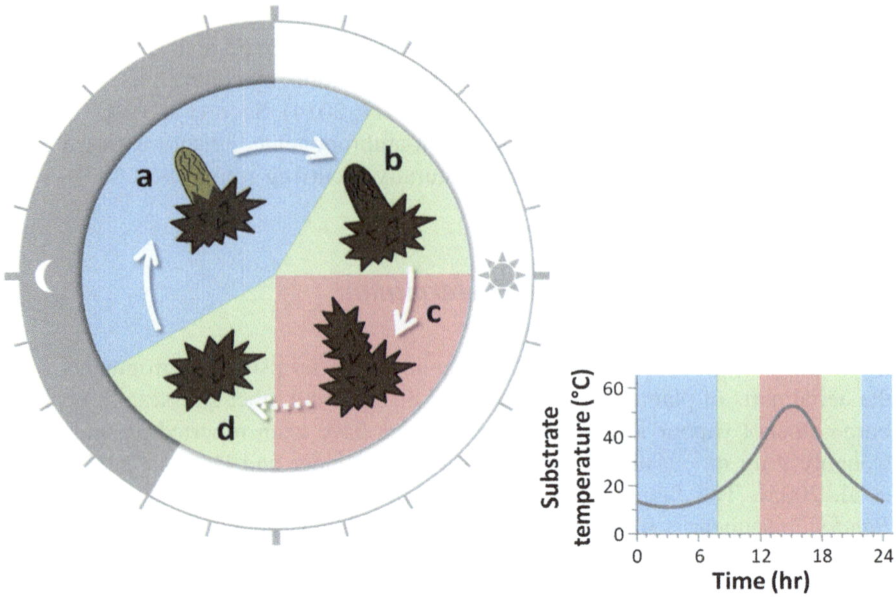


**Fig. 16.4** (**a–d**) Diagrammatic representation of the proposed generalized growth cycle of *Baudoinia*, highlighting the role of ethanol vapour, moisture, and heat (*left*). The *outer circle* indicates time of day divided into 24 hourly increments. *Thickened hatch lines* denote 0:00, 6:00, 12:00, and 18:00 h (clockwise from the left). The *inner circle* is sectioned according to the graph (inset at lower right), depicting a prototypical temperature profile of an exposed outdoor substrate during the growing season in a typical northern temperate region. (**a**) During the evening and early morning, substrates bearing cells of the fungus become cool (*blue*, < 18 °C), leading to the formation of condensation near midnight (moon symbol). Ethanol, having a greater affinity for liquid phase over vapour phase, readily dissolves in dew droplets, achieving an ethanol concentration up to tenfold greater than that of the surrounding air (Ewaze et al. 2008a). Ethanol stimulates the germination of dormant cells, serves as a nutrient for growth, and initiates cellular stress protective physiology. (**b**) Sunlight exposure warms the substrate to average ambient temperature (*green*, 18–37 °C), evaporating the dew, dehydrating cells, and reinstating dormancy. (**c**) By midday (sun symbol), the temperature of the sun-exposed outdoor substrate climbs towards its peaks (*red*, > 37 °C) (Scott et al. 2007). (**d**) Dry dormant cells may remain adherent to the substrate or be subject to fragmentation (*dashed arrow*). The substrate cools (*green*) as the sun sets and the cycle repeats

substrate, the ethanol-rich dew evaporates and the fungal cells desiccate. As the substrate warms further and the sun reaches its peak zenith angle, the dried, preconditioned cells are subjected to peak midday temperatures. Then, as the sun sets, the substrate temperature cools and the cycle repeats.

## *Dispersal*

Despite the friable nature of dried colonies of *Baudoinia*, Scott et al. (2007) remarked on the relative rarity of cells resembling *B. compniacensis* in air samples retrieved around distillery aging facilities. This observation implies that mechanisms other

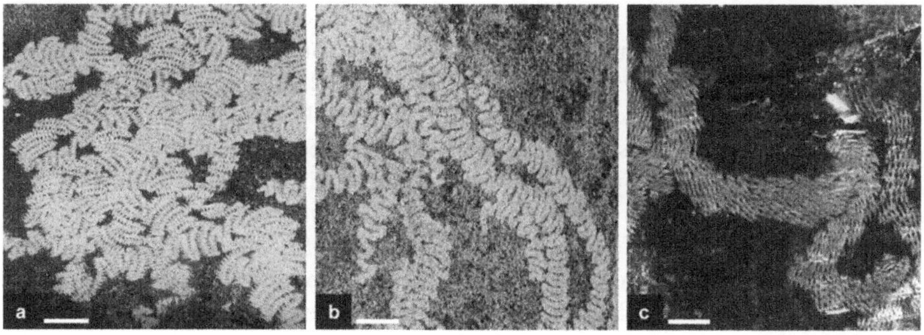


**Fig. 16.5** (**a–c**) Molluscan grazing trails on colonies of *Baudoinia* on the surface of (**a**) discarded stainless steel tank in an outdoor storage area (southwestern Ontario) and (**b**) precast concrete lamp standard on a riverside boardwalk (Cognac, France). (**c**) Trails left by gastropods observed grazing *B. compniacensis* mycelium on the surface of a truck of *Betula* sp. (Cognac, France) (Bars: **a**=30 mm, **b**=25 mm, **c**=10 mm)

than aerosol propagation may contribute to dispersal. Some degree of vector-mediated dispersal cannot be discounted. For example, during their study, Scott et al. (2007) noted colonies in several geographic locations that bore signs of terrestrial gastropod grazing where fungal material had been scraped from the substrate in patterns resembling Brownian trails (Fig. 16.5a–c).

## *Taxonomic Composition*

Early anecdotal reports attempted to characterize warehouse staining by culture methods, recovering a number of common phyloplane fungi (e.g. Kjøller 1961; Watson et al. 1984). Watson et al. (1984) did not connect their observations on warehouse staining with earlier work on *T. compniacensis*. These authors misattributed the black growth to a suite of common environmental fungi. However, they correctly recognized the characteristic ecological features of the whiskey fungus, including its close association with fugitive ethanol vapour emissions and its proclivity to colonize plant surfaces such as twigs, branches, and leaves to an extent causing severe damage and death.

Scott et al. (2007) thought that the environmentally common fungi reported by Kjøller (1961) and Watson et al. (1984) were unlikely to be responsible for the highly characteristic, sooty appearance of warehouse staining. Direct microscopic examination of freshly collected and archival samples of the growth from multiple geographic sites revealed the strong predominance of what they interpreted to be a single organism conforming to *T. compniacensis* (Scott et al. 2007). Despite lacking the differentiated conidial structures typical of many ascomycetous asexual states, *Baudoinia* has micromorphology that is distinctive and that cannot be confused with the ubiquitous genera that Watson et al. (1984) reported from their culture studies. Using a combination of a semiselective primary cultivation medium and careful isolation techniques, Scott et al. (2007) obtained pure cultures that manifested the same morphological characteristics as those represented in the descriptions and illustrations of *T. compniacensis* by Richon and Petit (1881) and Kjøller (1961).

The presence of the taxa reported by Kjøller (1961) and Watson et al. (1984) remains interesting. Because the taxa are all extremely common in outdoor air, Scott et al. (2007) reasoned that passive deposition on colonized surfaces over time gave rise to these allochthonous taxa as opposed to *Baudoinia* which they demonstrated to be a true autochthonous colonist chiefly responsible for warehouse staining.

Studies carried out in recent years have further supported *Baudoinia* to be the principal agent in the warehouse staining community (e.g. Scott et al. 2007; Ewaze et al. 2007, 2008a, b; Al-Naama et al. 2009). Ewaze et al (2008b) found that the incorporation of ethanol into a defined primary isolation at a concentration of 3 % (520 mM) strongly inhibited rapidly growing mould contaminants and still allowed the outgrowth of *Baudoinia* colonies. Al-Naama et al. (2009) acknowledged the multiorganismal nature of warehouse staining biofilms but referred to species of *Baudoinia* as "founding colonists" of the growth.

## Conclusions and Future Directions

As public awareness of the whiskey fungus grows and consumer demand for aged spirits increases, the need for information on the whiskey fungus by both the public and the spirits industry is likely to intensify. There remain a number of unanswered questions in relation to the lavish proliferation of the whiskey fungus in human communities, and several of these address fundamental aspects of its biology that remain unresolved. For example, what is the broader compositional nature and complexity of the warehouse staining biofilm? In addition to *Baudoinia*, are there other regularly occurring, autochthonous microbial participants? Despite intensive study over the past few years, the genus *Baudoinia* remains mysterious. Still unknown is the mechanism of ethanol sensing in the fungus remains uncertain.

Other outstanding questions are more of a practical nature. People living in areas affected with heavy colonization often express concern about the potential for negative health effects. Although plausible, this aspect of warehouse staining broadly, and of the whiskey fungus specifically, has not been evaluated. Likewise there is strong interest both from communities and the spirits industry to determine safe and sustainable remedial measures and preventive strategies in relation to the whiskey fungus.

## References

Al-Naama M, Green BJ, Scott JA (2009) Trehalose and heat resistance in *Baudoinia*. Int Biodeter Biodegrad 63:765–768

Auger-Barreau R (1966) Note sur *Torula compniacensis* Richon: Hyphomycetae. Actes de la Société Linnéenne Bordeaux Ser A 103(12):3 7

Baudoin A (1893) Les eaux-de-vie et la fabrication du Cognac. Librairie J.-B. Baillière et fils, Paris

Chatin GA (1878) Séance du 14 juin 1878. Bull Soc Bot Fr 25:205

Chlebicki A, Majewska M (2010) *Zasmidium cellare* in Poland. Acta Mycol 45(1):121–124

Crane JL (2001) A nomenclator of *Torula* Pers.: Fr. and *Torula* sensu Turpin. Mycotaxon 80:109–162

Crous PW (2009) Taxonomy and phylogeny of the genus *Mycosphaerella* and its anamorphs. Fungal Divers 38:1–24

Crous PW, Summerell BA, Carnegie AJ, Wingfield MJ, Hunter GC, Burgess TI, Andjic V, Barber PA, Groenewald JZ (2009) Unravelling *Mycosphaerella*: do you believe in genera? Persoonia 23:99–118

Ewaze JO, Summerbell RC, Scott JA (2007) Physiological studies of the warehouse staining fungus, *Baudoinia compniacensis*. Mycol Res 111:1422–1430

Ewaze JO, Summerbell RC, Scott JA (2008a) Ethanol physiology in the warehouse staining fungus, *Baudoinia compniacensis*. Mycol Res 112:1373–1380

Ewaze JO, Summerbell RC, Scott JA (2008b) Semiselective isolation of the ethanol imbibing sooty mould of distillery aging warehouses, *Baudoinia*. Can J Microbiol 54:331–333

Farmer JC, Dawson GA (1982) Condensation sampling of soluble atmospheric trace gases. J Geophys Res 87(C11):8931–8942

Goodwin S, McCorison C, Baker S, Scott JA, Grigoriev I (2014) The compact mitochondrial genomes of the ethanol-metabolizing fungi *Baudoinia compniacensis* and *Zasmidium cellare*. 12th European conference on fungal genetics, Seville, Spain, 23–27 Mar 2014

Greene GH, McGary KL, Rokas A, Slot JC (2014) Ecology drives the distribution of specialized tyrosine metabolism modules in fungi. Genome Biol Evol 6(1):121–132

Illana-Esteban C (2013) *Baudoinia compniacensis*, el moho negro de Cognac. Bol Soc Micol Madrid 37:225–228

Kjøller A (1961) Morphological and biological studies on *Torula compniacensis*. Bot Tidskr 57:47–52

Leducq JB (2014) Ecological genomics of adaptation and speciation in fungi. In: Landry CR, Aubin-Horth N (eds) Ecological genomics: ecology and the evolution of genes and genomes, pp 49 ff (Adv Exp Med Biol 781)

Millet DB, Donahue NM, Pandis SN, Polidori A, Stanier CO, Turpin BJ, Goldstein AH (2005) Atmospheric VOC measurements during the Pittsburgh air quality study: results, interpretation and quantification of primary and secondary contributions. J Geophys Res 110, D07S07

Moreau F (1954) Les champignons, vol 2. Lechevalier, Paris

Ohm RA, Feau N, Henrissat B, Schoch CL, Horwitz BA, Barry KW, Condon BJ, Copeland AC, Dhillon B, Glaser F, Hesse CN, Kosti I, LaButti K, Lindquist EA, Lucas S, Salamov AA, Bradshaw RE, Ciuffetti L, Hamelin RC, Kema GH, Lawrence C, Scott JA, Spatafora JW, Turgeon BG, de Wit PJ, Zhong S, Goodwin SB, Grigoriev IV (2012) Diverse lifestyles and strategies of plant pathogenesis encoded in the genomes of eighteen Dothideomycetes fungi. PLoS Pathog 8(12), e1003037

Richon MM, Petit P (1881) Note sur la plante cryptogame des murs de Cognac (*Torula compniacensis* sp. n.). Brebissonia 3(8):113–116

Riley R, Charron P, Idnurm A, Farinelli L, Dalpé Y, Martin F, Corradi N (2014) Extreme diversification of the mating type–high–mobility group (*MATA-HMG*) gene family in a plant-associated arbuscular mycorrhizal fungus. New Phytologist 201(1):254–268

Rogers A (2011) The angels' share. Wired 19(6):153–159, 167

Rogers A (2014) Proof: the science of booze. Houghton Mifflin Harcourt, New York, 227 pp

Roumeguère C (1881) La plante cryptogame des murs de Cognac. Rev Mycol (Toulouse) 3(11):16–18

Saccardo PA (1886) Sylloge Fungorum 4:262–263

Scott JA, Ewaze JO, Summerbell RC, Arocha-Rosete Y, Maharaj A, Guardiola Y, Saleh M, Wong B, Bogale M, O'Hara MJ, Untereiner WA (2016) Multilocus DNA sequencing of the whiskey fungus reveals a continental-scale speciation pattern. Persoonia 37:13–20

Scott JA, Untereiner WA, Ewaze EO, Wong B, Doyle D (2007) *Baudoinia*, a new genus to accommodate *Torula compniacensis*. Mycologia 99:592–601

Snider JR, Dawson GA (1985) Tropospheric light alcohols, carbonyls, and acetonitrile: concentrations in the southwest United States and Henry's Law data. J Geophys Res 90(D2): 3797–3805

Tribe HT, Thines E, Weber RWS (2006) Moulds that should be better known: the wine cellar mould, *Racodium cellare* Persoon. Mycologist 20:171–175

Watson RD, Minter DW, McKelvie AD (1984) Dense growth of Deuteromycetes on and around bonded distillery warehouses in Scotland. Bull Br Mycol Soc 18(1):57–58

Wijayawardene DNN, McKenzie EHC, Hyde KD (2012) Towards incorporating anamorphic fungi in a natural classification—checklist and notes for 2011. Mycosphere 3(2):157–228

Wisecaver JH, Rokas A (2015) Fungal metabolic gene clusters—caravans traveling across genomes and environments. Front Microbiol 6(161):1–11

# Chapter 17
# Allergenic Microfungi and Human Health: A Review on Exposure, Sensitization, and Sequencing Allergenic Proteins

**Mercedes Amado and Charles Barnes**

## Historical Association of Allergy and Mold

Charles Harrison Blackley (1820–1900) is generally credited with the initial understanding that pollen causes symptoms related to hay fever or allergic rhinitis. And, from the very beginning he included fungi in the list of organisms related to allergy and asthma (Blackly 1873). The Dutch pharmacologist and researcher Willem Storm van Leeuwen (1882–1933) described the high frequency (50 %) of Dutch asthmatics whose skin tested positive to molds (mostly *Mucor*, *Penicillium*, and *Aspergillus* spp.). He also identified housing conditions related to mold sensitivity and asthma. He also developed an "asthma-free room" as a form of environmental control (Van Leeuwen et al. 1925; Denning et al. 2006). In 1935, Mott and Kesten reported ophthalmic hypersensitivity reactions in rabbits using extracts of a yeast fungus (*Monilia psilosis*, current name: *Candida albicans*) (Mott and Kesten 1931). Other pioneering fungal investigations by F.T. Cadham (1924), Jiminez-Diaz (1932), and HA Hyde (1949) expanded the understanding of fungal spores as allergens. And SM Feinberg and OC Durham solidified fungi as part of the allergen repertoire in the 1946 "Allergy in Practice Yearbook." Yet, they probably missed 90 % of the relevant taxa because they only identified easily visually differentiated organisms or those that grew readily on agar plates.

---

M. Amado, M.D.
Pediatric Allergy and Immunology, Allergy Asthma and Immunology, Children's Mercy Hospitals and Clinics, 2401 Gillham Road, Kansas City, MO 64108, USA

University of Missouri-Kansas City School of Medicine, Kansas City, MO, USA
e-mail: mamado@cmh.edu

C. Barnes, Ph.D. (✉)
Division of Allergy/Immunology Laboratory, Children's Mercy Hospitals and Clinics, 2401 Gillham Road, Kansas City, MO 64108, USA

University of Missouri-Kansas City School of Medicine, Kansas City, MO, USA
e-mail: cbarnes@cmh.edu



D.-W. Li (ed.), *Biology of Microfungi*, Fungal Biology,
DOI 10.1007/978-3-319-29137-6_17

## *Fungal Extracts and Problems*

A recent clinical comentary review on environmental fungi describes fungi as: "*True fungi have cell walls that contain beta-(1–3) and beta-(1–6) glucans and chitin as structural components* (Flannigan and Miller 2011) *Fungi may be unicellular when they are yeast forms, but typically have a thread-like or tube-like body composed of hyphae, which range from 2 to 10 μm in diameter. Hyphae grow at their tips and frequently branch resulting in an interconnected network of hyphae called a mycelium. Fungi reproduce by producing spores, the majority of which are adapted for airborne dispersal. Spores may be produced by asexual processes or may be the products of sexual reproduction.*" In a more practical vein, the American Society for Testing and Materials International references *The Fifth Kingdom* by Bryce Kendrick and defines fungi as: "*eukaryotic, heterotrophic, absorptive organisms that usually develop a rather diffuse, branched, tubular body (i.e., network of hyphae) and usually reproduce by means of spores*" (Levetin et al. 2015). The terms "mold" and "mildew" though not scientific terms are frequently used by laypersons when referring to various fungal growths (Kendrick 2008). One specific factor concerning fungi that has come to light in recent years is the realization that fungal surfaces contain a wide array of molecular patterns that can be important targets for recognition by the immune system. In addition to beta glucans, they contain chitin, mannans, and mannoproteins and galactomannans.

Medical use of fungal extracts is plagued by problems related to extracts for immunotherapy and fungal sensitivity testing. These problems include the absence of validated, standardized extracts recognized by the US Food and Drug Administration, the number of different fungal species that can induce allergic reactions, the extraction time- and culture condition-dependent production of allergenic molecules (Portnoy et al. 1993), the complex number of different allergen structures produced by fungi, and the presence of highly cross-reactive allergens in multiple fungal taxa (Crameri et al. 2009).

## *Extracts Available*

The fungal derived injectable preparations for humans are essential for the diagnosis and treatment of allergic disease only. Fungi are not typically adapted for growing at the warm temperatures found in the human body, and fungal infections of healthy humans are found mostly in the peripheral cooler parts such as the scalp and feet (Flint and Cain 2014). Failure to mount adequate immune response to invasive thermophilic fungi is devastating (Fleming et al. 2014). Table 17.1 lists the major fungal preparations available for diagnostic purposes either as extracts for skin testing or as immobilized material for in vitro specific IgE testing. Since many of these preparations are approved for use, at least in the USA, through their historic or sometimes obsolete names, the table sometime lists more than one name for an organism. And,

**Table 17.1** Microfungi extracts available for diagnostic purposes (Greer Laboratories 2015)

| Traditional scientific name | Other name | Current nomenclature | | | |
|---|---|---|---|---|---|
| | | Genus | Species | In vivo (skin test) | In vitro (immunocap) |
| *Acremonium strictum* | *Cephalosporium acremonium* | *Acremonium* | *strictum* | Yes | |
| *Acremonium kiliense* | *Cephalosporium acremonium* | *Acremonium* | *kiliense* | | Yes |
| *Aureobasidium pullulans* | *Pullularia pullulans* | *Aureobasidium* | *pullulans* | Yes | Yes |
| *Alternaria alternata* | *Alternaria tenuis* | *Alternaria* | *alternata* | Yes | Yes |
| *Aspergillus flavus* | | *Aspergillus* | *flavus* | Yes | Yes |
| *Aspergillus fumigatus* | *Neosartorya fumigata* | *Aspergillus* | *fumigatus* | Yes | Yes |
| *Aspergillus niger* | *Aspergillus brasiliensis* | *Aspergillus* | *niger* | Yes | Yes |
| *Aspergillus terreus* | *Aspergillus terrestris* | *Aspergillus* | *terreus* | | Yes |
| *Botrytis cinerea* | | *Botrytis* | *cinerea* | Yes | Yes |
| *Candida albicans* | | *Candida* | *albicans* | Yes | Yes |
| *Chaetomium globosum* | | *Chaetomium* | *globosum* | Yes | Yes |
| *Cladosporium herbarum* | *Byssus herbarum* | *Cladosporium* | *herbarum* | Yes | Yes |
| *Cladosporium sphaerospermum* | | *Cladosporium* | *sphaerospermum* | Yes | |
| *Curvularia lunata* | *Cochliobolus lunatus* | *Curvularia* | *lunata* | | Yes |
| *Curvularia spicifera* | *Cochliobolus spicifer* | *Curvularia* | *spicifera* | Yes | |
| *Epicoccum nigrum* | *Epicoccum purpurascens* | *Epicoccum* | *purpurascens* | Yes | |
| *Epicoccum purpurascens* | *Epicoccum nigrum* | *Epicoccum* | *purpurascens* | | Yes |
| *Epidermophyton floccosum* | | *Epidermophyton* | *floccosum* | Yes | |
| *Fusarium moniliforme* | *Gibberella fujikuroi* | *Fusarium* | *proliferatum* | Yes | Yes |
| *Fusarium solani* | *Haematonectria haematococca* | *Fusarium* | *solani* | Yes | |
| *Geotrichum candidum* | | *Geotrichum* | *candidum* | Yes | |
| *Gliocladium viride* | *Gliocladium deliquescens* | *Gliocladium* | *viride* | Yes | |
| *Helminthosporium halodes* | *Setomelanomma rostrata* | *Helminthosporium* | *halodes* | | Yes |
| *Helminthosporium solani* | *Spondylocladium atrovirens* | *Helminthosporium* | *solani* | Yes | |
| *Malassezia orbiculare* | *Pityrosporum orbiculare* | *Malassezia* | *orbiculare* | Yes | Yes |

(continued)

**Table 17.1** (continued)

| Traditional scientific name | Other name | Current nomenclature | | In vivo (skin test) | In vitro (immunocap) |
|---|---|---|---|---|---|
| | | Genus | Species | | |
| *Mucor plumbeus* | | *Mucor* | *plumbeus* | Yes | |
| *Mucor racemosus* | *Mucor circinelloides* | *Mucor* | *racemosus* | Yes | Yes |
| *Penicillium chrysogenum* | *Penicillium notatum* | *Penicillium* | *chrysogenum* | Yes | Yes |
| *Penicillium glabrum* | | *Penicillium* | *glabrum* | | Yes |
| *Phoma betae* | *Pleospora betae* | *Phoma* | *betae* | Yes | Yes |
| *Rhizopus nigricans* | *Rhizopus stolonifer* | *Rhizopus* | *nigricans* | Yes | Yes |
| *Rhizopus oryzae* | *Rhizopus arrhizus* | *Rhizopus* | *oryzae* | Yes | |
| *Rhodotorula mucilaginosa* | *Rhodotorula rubra* | *Rhodotorula* | *mucilaginosa* | Yes | |
| *Saccharomyces cerevisiae* | | *Saccharomyces* | *cerevisiae* | Yes | |
| *Stachybotrys chartarum* | *Stachybotrys atra* | *Stachybotrys* | *chartarum* | Yes | |
| *Stemphylium herbarum* | *Stemphylium botryosum* | *Pleospora* | *herbarum* | | Yes |
| *Stemphylium solani* | | *Stemphylium* | *solani* | Yes | |
| *Trichophyton mentagrophytes* | | *Trichophyton* | *mentagrophytes* | Yes | Yes |
| *Trichophyton rubrum* | | *Trichophyton* | *rubrum* | Yes | Yes |
| *Trichosporon pullulans* | *Guehomyces pullulans* | *Guehomyces* | *pullulans* | | Yes |
| *Trichoderma harzianum* | *Trichoderma viride* | *Trichoderma* | *viride* | Yes | Yes |
| *Trichoderma viride* | *Hypocrea rufa (teleomorph)* | *Trichoderma* | *harzianum* | Yes | Yes |
| *Ulocladium chartarum* | | *Ulocladium* | *chartarum* | | Yes |

often the new or old nomenclature is not accepted by one group or another. It can be confusing. For example, *Wallemia sebi* is a Basidiomycota related to *Ustilago* frequently identified in house dust samples by DNA-based methods. Yet, *Wallemia* was not a relevant taxon in early outdoor collections and an extract for testing sensitivity to this fungus is not available at least in the USA. Also, for some previously recognized fungi, the taxonomy of the entire genus has undergone reorganization. For example, in *Rhodotorula* with three commonly recognized species (*R. glutinis*, *R. minuta*, and *R. mucilaginosa*), the formerly recognized *R. rubra* has been reclassified. And, the extract formerly available under that name is actually a strain of *R. mucilaginosa*. An extract is currently available under the name *R. mucilaginosa*, but it is the same extract that was formerly available under the name *R. rubra*. However, the use of multiple names is not unique in medical and scientific literature and just like for cytokines and clusters of differentiation those who wish to be knowledgeable in the field must master the multiple names.

## Extracts Approved in the USA for Testing and Treatment

There is currently a move by the FDA in the USA to remove from the US pharmacopeia those fungal extracts that do not have evidence of safety and efficacy in the medical literature (Slater et al. 2012). Also, there is limited evidence supporting the efficacy of immunotherapy for fungi other than *Alternaria* and *Cladosporium* (Portnoy et al. 2015). Therefore, several currently available extracts are scheduled to be withdrawn from the marketplace at least in the USA. And, if the major extract providers cease producing these fungal preparations, it is unclear if the current in vivo testing catalog will remain intact (Portnoy 2012).

## Allergenic Proteins from Fungi and Cross-Reactivity

The international efforts to identify and characterize major fungal allergenic proteins have made major progress over the past 15 years. Allergome (http://www.allergome.org/), probably the most inclusive of the allergen databases, lists 258 allergenic fungal taxa containing 594 allergenic proteins. The International Union of Immunological Societies (IUIS) database that is perhaps the most exclusive (http://www.allergen.org/) lists 29 taxa and 100 allergenic proteins. Nearly all allergy practitioners will agree that many fungal allergenic proteins are not unique to individual fungal species. Fungal cross-reactivity has been observed since the early days of allergen skin testing (Crameri et al. 2009; Dobrey 1962). Significant fungal cross-reactivity has been demonstrated by clinical history of exposure, epicutaneous skin sensitivity testing, and RAST inhibition assays (O'Neil et al. 1990; De Zubiria et al. 1990; Bisht et al. 2002). Recent high-throughput sequence work with fungal allergens indicates that fungi produce complex repertoires of species-specific and cross-reactive allergens (Crameri et al. 2006).

## *Families of Fungal Allergic Proteins*

A significant amount of sequencing work has been performed on fungal allergens and currently they are divided into several basic families of proteins. A search of the AllFam database (http://www.meduniwien.ac.at/allergens/allfam/) indicates that for fungi there are 66 recognized protein families of which 28 families have been identified in more than one fungal species. The family with the greatest number of individual fungi contributing at least one protein is the subtilisin-like serine protease family. At least 6 *Aspergillus* species, 2 *Cladosporium* species, a *Curvulata* species, and a *Penicillium* species have proteins identified as members of this family. Other major families of fungal proteins include peroxisomal membrane proteins (Asp f 3, Pen c 3, Mal f2, Mal f3), acid ribosomal proteins (Alt a 12, Clad h12, Pen b 26, Alt a 5, Cla h 5, Asp f 8, and others), cyclophilins (Asp f 11, Asp f 27), heat shock proteins (Alt a 3, Clad a 12, Pen c 19, Asp f 12), enolase (Alt a 6, Clad h 6, Asp f 22, Pen c 22), aldehyde dehydrogenase (Alt h10, Cla h10), glutathione S transferase (Pen c24), and Catalase (Pen c30, Asp n30). In addition, there are many fungal allergens where the function or family has not been identified yet. The most notable of these being Alt a 1.

## *Fungal Protein Cross-Reactivity (Theoretical and Practical)*

To examine cross-reactivity in fungi, specific IgE against four fungi, *Alternaria alternata, Aspergillus fumigatus, Cladosporium herbarum*, and *Penicillium chrysogenum*, was retrieved from clinical laboratory records. Specific IgE testing results from 6565 individuals were reviewed. A total of 2607 individuals had at least one positive test. The highest number of positive tests was to *Alternaria* (2156) and the fungus with the fewest positive tests was *Cladosporium* (1508). Most significantly, of the individuals having at least one positive test nearly half (1208) tested positive for all four fungi (Amado et al. 2014). In skin testing results, multiple-fungus sensitized patients reacted in descending order to *Alternaria*, *Candida*, *Cladosporium*, *Aspergillus*, *Saccharomyces*, *Penicillium*, and *Trichophyton* (Mari et al. 2003).

It is accepted that IgE cross-reactivity is an important aspect of fungal allergy and contributes significantly to fungal allergenicity (Crameri et al. 2009). And, elegant scientific investigations have been conducted using allergens produced through recombinant protein technology (Chou et al. 2011). Because of noted extensive fungal allergen cross-reactivity (Gutierrez-Rodriguez et al. 2011), it has been suggested that a limited number of skin tests could be used to identify patients with fungal sensitization. Once fungal sensitivity is identified, an allergist could then use an extensive panel of fungi to identify specific species sensitization if necessary. This might be done if immunotherapy is being considered or if an association between exposure and symptoms due to fungi identified in a patient's environment is suspected (Portnoy and Jara 2015).

Certain fungal allergen cross-reactivities are well documented. Alt a 1 has been quantified in culture filtrates of *Stemphylium botryosum, Ulocladium botrytis, Curvularia lunata, Alternaria tenuissima, C. herbarum, Penicillium chrysogenum,* and *Asp. fumigatus*. And, immunoblotting experiments using culture filtrate extracts from the same fungi probed with rabbit anti-recombinant Alt a 1 serum have shown significant amounts of Alt a 1. It has also been demonstrated (Crameri 2011) by inhibition experiments using serum from mold-sensitized patients and *A. alternata* extract in solid phase that IgE binding can be inhibited both by extracts closely related (*Curvularia lunata, Stemphylium botryosum*, and *Ulocladium botrytis*) and up to 70 % by extracts of more distant fungi (*Cladosporium herbarum* or *Aspergillus fumigatus*). Therefore, it has become generally recognized that this major allergen of *Alternaria alternata* (Alt a 1) is expressed in other members of the Pleosporaceae family (Saenz-de-Santamaria et al. 2006). Also, in the Pleosporaceae family it has been documented that antigenic components ranging from 14 to 20 kDa strongly react with specific serum raised against taxonomically related species (Saenz-de-Santamaria et al. 2006). Yet, in spite of the extensive cross-reactivity, there are many unique fungal allergens. Using purely bioinformatic approaches comparing structural motifs and BLAST searches for proteins, it is estimated that the size of the allergen repertoire necessary to represent the full range of fungal allergens could be as high as 5000 (Crameri 2011). Also, using sequencing data and literature searches at least 60 documented inter-phyla fungal cross-reactivities and 69 cross-reactivities have been documented between differing fungal phyla. There are even a remarkable number of documented cross-reactivities with non-fungal phyla (Simon-Nobbe et al. 2008).

## *Exposure and Sensitization*

As far as fungal exposure and sensitization are concerned, the most common prick skin test sensitivity in many studies is *Alternaria* (12.6 % of atopics, 61 % of fungal sensitized). Surprisingly, the next most sensitizing fungi was *Candida* (8.5 % of atopics, 13 % of fungal sensitized) (Mari et al. 2003). In patients who were sensitized to only 1 or 2 fungi, the most common consistently reactive genera were *Alternaria* and *Candida*. The extensive *Candida* sensitization likely results from cross-reacting allergens since *Candida* is not a common aeroallergen (Mari et al. 2003). Among asthmatics fungal sensitivity has been reported as high as 80 % (Lopez and Salvaggio 1985). A good synopsis of fungal sensitivities can be found in the 2008 review by Simon-Nobbe et al. (2008).

## *Human Exposure and Fungal Ecology*

Ecology is the study of living things, their environment, and the relation between the two. It can be extended to cover the indoor and outdoor places where airborne fungal spores are found and where humans can come into contact with fungi.

From an allergen point of view, human fungal interaction can cause sensitization, fungal related hypersensitivity disease, and the development of fungal related allergy symptoms. It is very rare to find an environment devoid of fungi and their spores. Outdoor airborne fungal spore estimations can range up to 50,000 per cubic meter of air in the warm months of the year and even in the depths of winter a few hundred spores per cubic meter of air can be found on most days (NAB 2000). Individual weather conditions can have a direct and cumulative impact on outdoor airborne spore exposure (Weber 2003). Precipitation and humidity have a tendency to elevate airborne spore levels and snow does a marvelous job of clearing the air of particles in general. Wind speed is important in bringing especially heavy spores into the air, but even light winds are adequate for small very aerodynamic spores to travel miles from their source. Thunderstorms, especially in summer, provide a complex set of conditions that typically tend to increase airborne spore levels (Nasser and Pulimood 2009), and dispersal of mold spores is strongly linked to precipitation and humidity. Sexual stage ascospore and basidiospore levels can rise strongly in relation to rainfall, and release of many taxa is triggered by high humidity, whereas asexual stage ascospores are often content to grow rapidly during periods of adequate moisture only to produce spores that become airborne during dry periods between rainfall events (Levetin 2014).

In the absence of facilitating factors that cause undue indoor amplification of fungi, indoor airborne spore levels are controlled by outdoor airborne fungal levels (Burge 2002). However, when facilitative factors are present in a house, elevated indoor spore levels can result in excessive human exposure. The extent of any disease caused by this exposure has been the subject of disagreement (Bush et al. 2006). The forces arrayed on either side of the issue leave little room for a consensus middle ground and often do battle in the law courts (Chapman et al. 2003). Consultants and experts are sometimes financially conflicted and their conflicted opinions frequently leach into the scientific literature (Bardana et al. 2003).

Facilitating factors typically involve the presence of moisture resulting from conditions as roof leaks, basement leaks, sewage leaks, plumbing leaks, high humidity, condensation, and inadequate insulation all associated with generally poor housing. Typical soil fungi easily adapt for growth on a wide variety of building materials. The only essential elements are a source of carbon and water. The carbon source can be the paper facing on dry wall or the acrylates in paint (Ito et al. 2005). Different fungal species have adapted to prefer specific moisture conditions, barely damp (xerophilic) to very wet (hydrophilic). Traditional plaster and solid wood are more resistant to fungal growth and chemical treatments have been used to add additional resistance to fungal growth. However, many of the most toxic chemicals have been justifiably removed from the market. Between 1960 and 1980 in North America, the composition of interior building walls was shifted away from plaster to surfaced gypsum board or drywall. As the name implies it must remain dry, for when moist this material is susceptible to fungal growth (Flannigan and Miller 2011). Drywall along with other changes like wall-to-wall carpeting, reduced ventilation rates, indoor appliances, and even plastic wrapping resulted in North American housing, and that of many other parts of the world, often becoming characterized as damp buildings. Estimates are that up to 30 % of homes in the USA and Canada have dampness problems. Reports indicate that 10 % of US homes had water damage from exterior leaks and 8 % had water damage from interior leaks. The population-weighted average prevalence of dampness or mold in housing was reported as 47 % in the USA (Park and Cox-Ganser 2011; Dales et al. 2008). In large Canadian studies, damp spots, visible mold or mildew, water damage, or flooding was reported by 38 % of respondents (Dales et al. 1991).

The association of damp building conditions and respiratory disease was first strongly made in the 2004 Institution of Medicine publication Damp Indoor Spaces and Health (Medicine Io 2004). This publication was bolstered by epidemiology studies in Europe, Canada, and the USA that consistently show building/house dampness and mold have been associated with increased risks for respiratory symptoms, asthma, and respiratory infections (Park and Cox-Ganser 2011; Mendell et al. 2011; Quansah et al. 2012). Subsequent intervention studies (Krieger et al. 2010) continue to support these conclusions (Maheswaran et al. 2014). And, the relative risk for increased asthma and other conditions remains after controlling for confounding exposures (Dales and Miller 1999). Mechanistic studies concerning the innate immune system and fungal associated molecular patterns have shed additional light on this relationship. Damp building exposure includes fragments of fungi (both micronic and sub-micronic), fungal allergens, bacterial endotoxins, fungal glucans, mannans, and low-molecular-weight metabolites as well as other allergens (e.g., pets, dust mites, roaches, etc.). It is important to realize that fungal exposure does not occur in a vacuum but is nearly always part of a milieu of exposures associated with building dampness as a facilitating factor for numerous allergen sources.

In addition to allergic rhinitis and asthma, there are several other medical conditions recognized to be associated with mold sensitivity. Allergic bronchopulmonary aspergillosis (ABPA) is diagnosed in patients with asthma or cystic fibrosis. It is characterized by pulmonary infiltrates, mucus containing *Aspergillus fumigatus* hyphae, elevations of total serum IgE, specific precipitating IgG antibodies, and eosinophils in the sputum (Greenberger 2012). ABPA is the most common form of allergic bronchopulmonary mycosis (ABPM). It results from an allergic inflammatory response to the colonization by *Aspergillus* in the airways although colonization by other fungi is reported (Natarajan and Subramanian 2014; Glancy et al. 1981). Allergic bronchopulmonary mycosis (ABPM) is also reported as a hypersensitivity-mediated disease associated with fungi other than *Aspergillus*. A survey reported the commonest etiologic agents in this group to be *Candida albicans* (60 % of cases), *Bipolaris* species (13 %), *Schizophyllum commune* (11 %), *Curvularia* species (8 %), *Pseudallescheria boydii* (3 %) and, rarely, *Alternaria alternata*, *Fusarium vasinfectum*, *Penicillium* species, *Cladosporium cladosporioides*, *Stemphylium lanuginosum*, *Rhizopus oryzae*, *Ca. glabrata*, *Saccharomyces cerevisiae*, and *Trichosporon beigelii* (Chowdhary et al. 2014). The characteristics of ABPM include severe asthma, eosinophilia, markedly increased total IgE and specific IgE levels, specific precipitating IgG antibodies, central bronchiectasis, and mold colonization of the airways. The term severe asthma associated with fungal sensitization (SAFS) has been suggested to illustrate the high rate of fungal sensitivity in patients with persistent severe asthma and improvement with antifungal treatment. Regardless of which acronym is used, the understanding of these diseases is just now being developed (Knutsen et al. 2012). However, studies into these conditions coupled with genetic studies involving mice with critical elements of the immune system knocked out are starting to yield additional insight into the complexity of the innate immune system and its response to fungi.

## *Allergic Fungal Sinusitis*

Similar in pathogenicity to ABPA or allergic fungal pulmonary mycosis is allergic fungal sinusitis in which an immunocompetent person presents with nasal polyposis, chronic pansinusitis, and noninvasive fungal hyphae and allergic mucin on histopathological examination. Grossly, the allergic mucous appears tan to brown and thick in consistency and has been described as peanut butter like. And, on microscopic exam, eosinophils, Charcot–Leyden crystals, and elevated levels of eosinophilic cationic protein are also found. Patients have IgE and IgG precipitating antibodies to fungi, typically *Aspergillus* although other dematiaceous fungal species (*Alternaria, Bipolaris, Curvularia*) have been identified. Treatment consists of surgical debridement and topical and oral corticosteroids, and antifungal therapy is advocated by some specialists. Recurrence of disease is common. In contrast, invasive fungal sinusitis occurs in immunocompromised individuals such as persons undergoing chemotherapy for malignancy. An Aspergilloma or *Aspergillus* fungus ball presents usually in only one sinus (maxillary or sphenoid) without allergic mucin. And surgical removal is curative.

Hypersensitivity pneumonitis or extrinsic allergic alveolitis is a pulmonary inflammatory disease with several forms including acute, subacute, or intermittent and chronic progressive disease. It is caused by inhalation of organic dusts including molds such as farmer's lung (*Micropolyspora faeni*), cheese worker's lung (*Penicillium casei*), and malt worker's lung (*Aspergillus clavatus*). Fungi are not the only cause of hypersensitivity pneumonitis and persons may present with symptoms relating to inhaling antigens of non-fungal origin such as avian proteins (bird fanciers), trimellitic anhydride, diisocyanate, and methylene diisocyanate. Unfortunately, iatrogenic causes of invasive fungal meningitis have been recently reported secondary to unsterile compounding pharmacies with devastating morbidity and mortality (Pettit et al. 2012; Smith et al. 2013; Kainer et al. 2012).

There are other diseases caused by inhalation of fungal spores in the soil. Although not directly associated with allergic microfungi they must nonetheless be mentioned here. Most persons who inhale fungal spores are asymptomatic or present with mild pulmonary symptoms such as cough similar to an upper respiratory tract infection. However, those with weakened immune systems may have disseminated disease and need treatment with antifungal medication. In the Southeastern USA, parts of Mexico, and Central America, coccidiomycosis occurs due to inhalation of fungal spores of *Coccidioides*. In the USA, this is known as Valley Fever. In the Ohio and Mississippi valley of the USA, histoplasmosis due to the inhalation of fungal spores associated with bird and bat droppings occurs. Blastomycosis occurs secondary to the inhalation of *Blastomyces dermatitidis*. This fungus lives in moist soil and is associated with decomposing organic matter such as wood and leaves. Blastomycosis occurs chiefly in the east central part of the USA.

## *Fungi and Innate Immunity*

The earliest reference to "innate immunity" that is easily found is by Miller and Watson when they used it in the title of a 1965 article in Medical Clinics of North America (Miller and Watson 1965). At its maximum in 2012, interest in the subject generated 1531 review articles according to PubMed. Innate immunity, being defined as the protection against infection that relies on mechanisms that exist before infection, can encompass a great variety of functions (Lichtman et al. 2007). This discussion will consider only those recently discovered elements of innate immunity that sense molecular patterns common to fungi. Generally that means a subset of the pattern recognition receptors (PRRs), specifically PRRs that recognize pathogen-associated molecular patterns (PAMPs) and damage-associated molecular patterns (DAMPS) that are related to fungal presence. The major players are Toll-like receptors (TLRs) and C-type Lectin Receptors (CLRs) that are directly involved in fungal recognition and modulation of the innate immune response. Also associated are the nucleotide-binding oligomerization domain (NOD)-like receptors (NLRs) and the cytosolic dsDNA sensors (CDSs) that sense mostly DNA and RNA elements some of which are fungal in origin and the retinoic acid-inducible protein 1 (RIG) like receptors (RLRs) that play a major role in virus detection.

The chief molecular patterns found in fungi are components of the fungal cell wall (Hardison and Brown 2012). These components include mannan (mannosylated proteins), β-glucan, and chitin. The typical picture involves an outermost layer of proteins that are heavily modified by multiple mannose containing structures. This outer layer is underlain by a rigid carbohydrate polymer of beta glucan supported by an even more rigid layer of polymerized n-acetyl glucosamine or chitin. However it should be recognized that the fungal cell wall can be a dynamic structure. It varies in composition among the numerous fungal species. And, it can change and adapt during the various morphological transitions fungi undergo. During these transitions and developmental stages the various elements of the fungal cell wall can be exposed to the surface. Therefore different fungi, or even the same fungus in differing developmental stages or grown in differing media can present a differing immunological picture (Amarsaikhan et al. 2014).

The CLRs are a large family of transmembrane receptors that bind to carbohydrates and recognize many glucan and mannin structures from fungi. CLRs include Dectin-1, Mincle (macrophage-inducible C-type lectin), DC-SIGN (dendritic cell-specific ICAM3-grabbing nonintegrin), DNGR-1 (DC NK lectin group receptor-1), and MBL (mannose-binding lectin). They basically recognize fungi and modulate innate immune response. Often they are divided into type 1, type 2, and soluble CLRs. Type I includes DEC 205 and MMR (macrophage mannose receptor), and type 2 includes Dectin-1, Dectin-2, Mincle, DC-SIGN, and DNGR-1. Soluble CLRs include MBL. CLRs are expressed by many cell types including macrophages and dendritic cells.

Dectin-1 is a specific receptor for β-glucans (Brown et al. 2003) in the cell walls of fungi. Dectin-1 has extracellular domain connected to a transmembrane region, followed by a cytoplasmic region with an Immunoreceptor Tyrosine-based Activation Motif (ITAM). Dectin-1 modulates expression of cytokines by inducing Nuclear Factor of Activated T Cells (NFAT) through the $Ca^{2+}$–calcineurin–NFAT pathway (Goodridge et al. 2007). Dectin-1 signaling has been shown to collaborate with TLR2 signaling to enhance the responses triggered by each receptor (Gantner et al. 2003). Dectin-1 triggers phagocytosis and activation of Src and Syk kinases, through its ITAM-like motif leading to the production of reactive oxygen species (ROS), activation of NF-κB, and subsequent secretion of pro-inflammatory cytokines (Gross et al. 2006; Dennehy and Brown 2007). ROS has a direct microbiocidal role in phagocytic pathways and can also affect other pathways (Kankkunen et al. 2010).

Dectin-2 is a major PRR for fungal infection and the induction of Th17 type immunity. It binds carbohydrates with high mannose content and is the functional receptor for α-mannans, a major fungal cell wall component (Drummond and Brown 2011, 2013). Like Dectin-1, Dectin-2 is part of the group of CLRs that links pathogen recognition and adaptive immunity. Also, activation of Dectin-2 triggers ROS leading to inflammasome activation (Ifrim et al. 2014). Mincle is another PRR involved in the recognition of fungi (Brown 2008). Mincle binds fungal α-mannose among other molecules (Yamasaki et al. 2009) and interacts with the Fc receptor common γ-chain (FcRγ), triggering signaling leading to NF-κB and calcineurin-NFAT activation. DC-SIGN is involved in the recognition of *Candida* species. It activates the Raf-1-acetylation-dependent pathway and modulates TLR signaling (Brown 2010; den Dunnen et al. 2009). Mannose-binding lectin (MBL) is a soluble C-type lectin that has a role in innate immunity against yeast through enhanced complement activation and uptake of polymorphonuclear cells (van Asbeck et al. 2008). MBL binds to repetitive mannose and/or N-acetylglucosamine residues on microorganisms, producing opsonization and complement pathway activation (Bidula et al. 2013).

Toll-like receptor (TLR) was initially identified as essential for fruit fly immunity to fungi. When TLR4 was subsequently identified as an LPS receptor (Wagner 2012), it became the focus of intensive study. Up to 15 TLRs have been appreciated in Mammals, but they are not all expressed in humans. TLRs generally form heterodimers that bind to PAMPS and through many subsequent steps (>25 protein intermediates including MYD88, TRAF, IRAK) activate the immune system through interferon regulatory factor 3 (IRF-3) and nuclear factor kappa-light-chain enhancer of activated B cells (NF kappaB) (Villena et al. 2014). TLRs control reactions to fungal-specific PAMPs. The best example is TLR 2 which heterodimerizes with TLR 1, TOLLIP (Toll interacting protein) to recognize fungal beta-glucan (Netea et al. 2006), and phospholipo-mannans (PLMs) unique to *C. albicans* (Jouault et al. 2009). Also, TLR2/1 heterodimers recognize *A. fumigatus* both in humans and mice (Rubino et al. 2012). And, TLR2/TLR1and TLR2/TLR6 heterodimers recognize mannans from *Cryptococcus neoformans* (Netea et al. 2006; Roeder et al. 2004). TLR4 is activated upon ligation of O-linked mannans in *C. albicans* and *C. neoformans* (Netea et al. 2006) (Shoham and Levitz 2005). Other TLRs (TLR3, 9, and 7) recognize fungal RNA and DNA. The presence of small nuclear polymorphisms (SNPs) in some TLRs raises the risk of human fungal infection (Skevaki et al. 2015), but these specific genetic problems are only starting to be identified.

## *DNA Allows Better Identification and Characterization of Fungi and their Allergens*

The advances in DNA sequencing technology have not only caused alterations in the taxonomy of fungi, but they have also had an increasingly dramatic impact on research involving medical aspects of fungal exposure. The advent of rapid and high capacity genomic DNA sequencing allows much better identification and characterization of fungi not only in infectious processes but also in environmental exposures. As the biome of the trachea and lungs is being studied, new organisms are being identified in unexpected places (Willger et al. 2014; Kolwijck and van de Veerdonk 2014; Dickson et al. 2014). Fungal biome evaluations are also being used to study the complexities of the hygiene hypothesis often with conflicting results (Daley 2014).

## *Recombinant Allergens*

The identification and characterization of fungal allergens has generally lagged behind the characterization of plant and animal allergens. Reasons for this are numerous including difficulty in growing fungi, the presence of materials like melatonin (Paris et al. 1990) and chitin that complicate protein identification and purification methods that were developed in animal tissue, and genetic changes in fungi during the growing process. Several fungal allergens including Alt a1 a have been produced through recombinant protein technology (Asturias et al. 2005). Some of these have been found to be sufficient for diagnostic purposes. But depending on how they are grown and the cell types used for production of these proteins (Maier et al. 2014), they are different from the native allergens and may not always be suitable for diagnostic purposes. An investigation of the in vitro diagnostic value of recombinant *Alternaria* allergens for the diagnosis of sensitization to Pleosporaceae concluded that these recombinant allergens are not sufficient for a highly sensitive diagnosis of *A. alternata* allergy (Postigo et al. 2011). And, although several recombinant allergens have been suggested as potential effective agents for immunotherapy (Akdis 2014), no recombinant fungal allergens have been approved at least in the USA.

## *Sequencing Allergenic Proteins*

In spite of the difficulties involved, numerous fungal allergens have been sequenced. The sequencing process typically has involved identification of the allergic protein through immunoblotting followed by the determination of a portion of the sequence through N-terminal or tryptic digest analysis. From knowledge of a portion of the sequence, the utilization of DNA-based methods can quickly lead to the nucleotide sequence and the deduced protein sequence. Alternatively, fungal DNA has been "shot gun" cloned into plasmid vectors and an array of fungal proteins produced. These proteins are then screened with human sera containing specific IgE against fungal proteins. Once allergenic protein producing clones are identified they are expanded and sequenced. Sequences can then be compared to large DNA databases to identify similarities with known allergen protein families and further characterize the protein allergens. Over 1000 fungal protein sequences are available on Allergome (http://www.allergome.org/) and more are being added monthly. Although many of these are redundant and represent the same protein in different organisms, it is still a testimony to the usefulness of protein allergen sequences.

The diversity of fungal taxa in differing eco-niches has been the subject of much attention using high-throughput DNA sequencing methods. The two main points of focus for this fungal biome research have been house dust and areas of the human respiratory system including both sputum and lung biopsy specimens. However, at least in house dust, the determined microbiome was dependent on the DNA extraction method. Three commonly used DNA extraction methodologies (UltraClean Soil kit, High Pure PCR Template kit, and EluQuik/DNeasy kit) were evaluated for sensitivity and susceptibility to PCR inhibitors in dust for three common fungi, *Aspergillus versicolor, Rhizopus microsporus*, and *Wallemia sebi*. The extraction methods differed in their ability to extract DNA from particular species. In addition, the soil DNA extraction kit showed the greatest ability to remove PCR inhibitors from dust samples. Most importantly, the determined species composition from the sequenced clone libraries generated varied with the different DNA extraction kits. And as has often been the case, sequencing methods produced information concerning additional fungal species not seen in solid culture (Rittenour et al. 2012).

In one recent sequencing study, the diversity of airborne and dust-borne fungi in homes of asthmatic children in Kansas City, Missouri, was determined by sequencing. Sequence results were also compared to data obtained using traditional viable and nonviable fungal exposure assessment methods. Sequencing frequently identified organisms from the Ascomycota in both air (68 %) and dust (92 %) and less frequently organisms from the Basidiomycota and zygomycetous fungi (formerly Zygomycota). The majority of Ascomycota clones belonged to the Pleosporales, Eurotiales, Capnodiales, and Dothideales. Sequencing also revealed a number of unusual Pleosporales fungi present in the samples. As seen in other studies, sequencing results identified a broader diversity of Ascomycota and Basidiomycota in indoor environments than traditional methods (Rittenour et al. 2013).

Another study performed analysis on the indoor bacterial microbiome using PhyloChip Array methods. They reported 1746 operational taxonomic units (OTUs) in each sample with few genus differences between asthmatic and non-asthmatic homes. Major phyla identified were Cyanobacteria, Firmicutes, Actinobacteria, Proteobacteria, and Bacteroidetes. These all were more abundant in the asthmatic homes ($p=0.001$–$p=7.2 \times 10–6$) (Ciaccio et al. 2014). Several studies have been performed on induced sputum samples. One case-control study identified 90 species more common in asthma patients and 46 species more common in control subjects. *Psathyrella candolleana, Malassezia pachydermatis, Termitomyces clypeatus*, and *Grifola sordulenta* showed a higher frequency in the sputum of asthma patients and *Eremothecium sinecaudum, Systenostrema alba, Cladosporium cladosporioides*, and *Vanderwaltozyma polyspora* showed a higher frequency in the sputa of control subjects. This study produced evidence for the widespread nature of fungi in the sputa of healthy and asthmatic individuals, it did not answer the question whether the organisms were colonized in the sputa or appeared there as a result of deposition from the air (van Woerden et al. 2013). Another study of the microbiome of induced sputa in asthmatic and non-asthmatic adults analyzed induced sputum samples obtained from 10 non-asthmatic subjects and 10 patients with mild active asthma (8/10 were not using inhaled corticosteroids). Proteobacteria were present in higher proportions in asthmatic patients and firmicutes and actinobacteria were present more frequently in samples from non-asthmatic subjects (Marri et al. 2013).

## *Increased Allergic and Non-allergic Fungal Disease*

Both allergic and non-allergic fungi will have increased medical importance especially as more people with altered or depressed immune systems live longer and are integrated into the general population. A good example is allergic or acute bronchopulmonary aspergillosis (ABPA). There are estimated to be in excess of four million patients affected worldwide (Agarwal et al. 2013). And, a PubMed search indicates that publication activity related to ABPA has increased steadily since the early 1970s. Even though ABPA is the most common form of allergic bronchopulmonary mycosis (ABPM), other fungi, including *Penicillium* and *Candida*, have been associated (Denning et al. 2006). A higher incidence of ABPA has been reported in compost workers, and it has been recommended that commercial compost operations where high levels of *A. fumigatus* often occur routinely screen workers for asthma, *Aspergillus* sensitivity, cystic fibrosis, bronchiectasis, and immunodeficiency (Poole and Wong 2013). ABPA and other fungal related lung conditions are an increasing problem in CF (Eickmeier et al. 2013). Also, if we recognize ABPA as a specific subtype of asthma (Greenberger et al. 2014), as the prevalence of asthma increases around the world, it is probably safe to speculate that the prevalence of ABPA will also increase.

Evaluating the prevalence of morbidity and mortality of fungal infections is very difficult. It is estimated that only about half of these infections are diagnosed before the death of an individual (Dignani 2014). Invasive fungal infections are major causes of morbidity and mortality for the immunocompromised. And those infections are especially prevalent in a growing segment of the patient population, namely, those with oncohematological disorders, recipients of stem cell or solid organ transplantation, and recipients of immunosuppressive agents including high-dose steroids, antitumor necrosis factor therapies (infliximab, adalimumab, etc.), or anti-inflammatory antibodies (Belatacept) (Ordas et al. 2012).

The use of autopsy provides probably the best indication of the prevalence of fungal infections. Cumulative studies, in seven countries studying mainly four types of patient populations (general, oncohematology, stem cell transplant, and acquired immunodeficiency syndrome) (Dignani 2014), indicated that the median prevalence of infection is 8.7 per 100 autopsies. They also found that in about half of the cases (4.4 per 100 autopsies), there were two or more concomitant fungal organisms. Fungal infection rates in 4 large autopsy studies of general populations ranged from 1.4 to 8.7 % (Lehrnbecher et al. 2010; Kume et al. 2011; Colombo et al. 2012; Shimodaira et al. 2012). Whereas in reports for 5 studies of oncology and immunosuppressed populations percentage of autopsies with fungal infection were between 18.2 % and 30.6 % (Donhuijsen et al. 2008; Lewis et al. 2013; Alsharif et al. 2009; Antinori et al. 2009; Sinko et al. 2008). Clearly, as more patients undergo immunosuppressive therapy due to transplant and oncology-related treatment and more immunosuppressed individuals survive infancy, the overall number of fungal infections will increase.

## References

Agarwal R, Chakrabarti A, Shah A, Gupta D, Meis JF, Guleria R et al (2013) Allergic bronchopulmonary aspergillosis: review of literature and proposal of new diagnostic and classification criteria. Clin Exp Allergy 43:850–873

Akdis M (2014) New treatments for allergen immunotherapy. World Allergy Organ J 7:23

Alsharif M, Cameron SE, Young JA, Savik K, Henriksen JC, Gulbahce HE et al (2009) Time trends in fungal infections as a cause of death in hematopoietic stem cell transplant recipients: an autopsy study. Am J Clin Pathol 132:746–755

Amado M, Portnoy J, Barnes CM (2014) Fungal cross-allergenicity in specific ige testing. J Allergy Clin Immunol 133:AB92

Amarsaikhan N, O'Dea EM, Tsoggerel A, Owegi H, Gillenwater J, Templeton SP et al (2014) Isolate-dependent growth, virulence, and cell wall composition in the human pathogen *Aspergillus fumigatus*. PLoS One 9, e100430

Antinori S, Nebuloni M, Magni C, Fasan M, Adorni F, Viola A et al (2009) Trends in the postmortem diagnosis of opportunistic invasive fungal infections in patients with AIDS: a retrospective study of 1,630 autopsies performed between 1984 and 2002. Am J Clin Pathol 132:221–227

Asturias JA, Ibarrola I, Ferrer A, Andreu C, Lopez-Pascual E, Quiralte J et al (2005) Diagnosis of *Alternaria alternata* sensitization with natural and recombinant Alt a 1 allergens. J Allergy Clin Immunol 115:1210–1217

Bardana EJ, Chapman JA, Charlesworth EN, Jacobs RL, Terr AL (2003) Crossing over to the dark side of the mold issue: a dissenting view. Ann Allergy Asthma Immunol 91:212–213, author reply 213–215

Bidula S, Kenawy H, Ali YM, Sexton D, Schwaeble WJ, Schelenz S et al (2013) Role of ficolin-A and lectin complement pathway in the innate defense against pathogenic *Aspergillus* species. Infect Immun 81:1730–1740

Bisht V, Singh BP, Arora N, Gaur SN, Sridhara S (2002) Antigenic and allergenic cross-reactivity of *Epicoccum nigrum* with other fungi. Ann Allergy Asthma Immunol 89:285–291

Blackly C (1873) Experimental researches on the causes and nature of catarrher aestirus (Hay fever and Hay asthma). Balliere, Tindall and Cox, London

Brown GD (2008) Sensing necrosis with Mincle. Nat Immunol 9:1099–1100

Brown GD (2010) How fungi have shaped our understanding of mammalian immunology. Cell Host Microbe 7:9–11

Brown GD, Herre J, Williams DL, Willment JA, Marshall AS, Gordon S et al (2003) Dectin-1 mediates the biological effects of beta-glucans. J Exp Med 197:1119–1124

Burge HA (2002) An update on pollen and fungal spore aerobiology. J Allergy Clin Immunol 110:544–552

Bush RK, Portnoy JM, Saxon A, Terr AI, Wood RA (2006) The medical effects of mold exposure. J Allergy Clin Immunol 117:326–333

Cadham F (1924) Asthma due to grain rusts. JAMA 83:27

Chapman JA, Terr AI, Jacobs RL, Charlesworth EN, Bardana EJ Jr (2003) Toxic mold: phantom risk vs science. Ann Allergy Asthma Immunol 91:222–232

Chou H, Tam MF, Chiang CH, Chou CT, Tai HY, Shen HD et al (2011) Transaldolases are novel and immunoglobulin E cross-reacting fungal allergens. Clin Exp Allergy 41:739–749

Chowdhary A, Agarwal K, Kathuria S, Gaur SN, Randhawa HS, Meis JF et al (2014) Allergic bronchopulmonary mycosis due to fungi other than *Aspergillus*: a global overview. Crit Rev Microbiol 40:30–48

Ciaccio C, Kennedy K, Barnes C, Portnoy JM, Rosenwasser LJ (2014) The home microbiome and childhood asthma. **133**:AB70

Colombo TE, Soares MM, D'Avilla SC, Nogueira MC, de Almeida MT (2012) Identification of fungal diseases at necropsy. Pathol Res Pract 208:549–552

Crameri R (2011) The problem of cross-reactivity in the diagnosis of fungal allergy. Clin Exp Allergy 41:302–304

Crameri R, Weichel M, Fluckiger S, Glaser AG, Rhyner C (2006) Fungal allergies: a yet unsolved problem. Chem Immunol Allergy 91:121–133

Crameri R, Zeller S, Glaser AG, Vilhelmsson M, Rhyner C (2009) Cross-reactivity among fungal allergens: a clinically relevant phenomenon? Mycoses 52:99–106

Dales RE, Miller D (1999) Residential fungal contamination and health: microbial cohabitants as covariates. Environ Health Perspect 107(Suppl 3):481–483

Dales RE, Burnett R, Zwanenburg H (1991) Adverse health effects among adults exposed to home dampness and molds. Am Rev Respir Dis 143:505–509

Dales R, Liu L, Wheeler AJ, Gilbert NL (2008) Quality of indoor residential air and health. CMAJ 179:147–152

Daley D (2014) The evolution of the hygiene hypothesis: the role of early-life exposures to viruses and microbes and their relationship to asthma and allergic diseases. Curr Opin Allergy Clin Immunol 14:390–396

De Zubiria A, Horner WE, Lehrer SB (1990) Evidence for cross-reactive allergens among basidiomycetes: immunoprint-inhibition studies. J Allergy Clin Immunol 86:26–33

den Dunnen J, Gringhuis SI, Geijtenbeek TB (2009) Innate signaling by the C-type lectin DC-SIGN dictates immune responses. Cancer Immunol Immunother 58:1149–1157

Dennehy KM, Brown GD (2007) The role of the beta-glucan receptor dectin-1 in control of fungal infection. J Leukoc Biol 82:253–258

Denning DW, O'Driscoll BR, Hogaboam CM, Bowyer P, Niven RM (2006) The link between fungi and severe asthma: a summary of the evidence. Eur Respir J 27:615–626

Dickson RP, Martinez FJ, Huffnagle GB (2014) The role of the microbiome in exacerbations of chronic lung diseases. Lancet 384:691–702

Dignani MC (2014) Epidemiology of invasive fungal diseases on the basis of autopsy reports. F1000Prime Rep 6:81

Dobrey AN (1962) Mold sensitivity in the allergic respiratory diseases. Can Med Assoc J 87:1310–1313

Donhuijsen K, Petersen P, Schmid WK (2008) Trend reversal in the frequency of mycoses in hematological neoplasias: autopsy results from 1976 to 2005. Dtsch Arztebl Int 105:501–506

Drummond RA, Brown GD (2011) The role of dectin-1 in the host defence against fungal infections. Curr Opin Microbiol 14:392–399

Drummond RA, Brown GD (2013) Signalling C-type lectins in antimicrobial immunity. PLoS Pathog 9, e1003417

Eickmeier O, Rieber N, Eckrich J, Hector A, Graeppler-Mainka U, Hartl D et al (2013) Immune response, diagnosis and treatment of allergic bronchopulmonary aspergillosis in cystic fibrosis lung disease. Curr Pharm Des 19:3669–3678

Flannigan B, Miller J (2011) Microbial growth in indoor environments. In: Flannigan B, Samson R, Miller J (eds) Microorganisms in home and indoor work environments: diversity, health impacts, investigation and control, 2nd edn. Taylor & Francis, New York, pp 57–107

Fleming S, Yannakou CK, Haeusler GM, Clark J, Grigg A, Heath CH et al (2014) Consensus guidelines for antifungal prophylaxis in haematological malignancy and haemopoietic stem cell transplantation, 2014. Intern Med J 44:1283–1297

Flint WW, Cain JD (2014) Nail and skin disorders of the foot. Med Clin North Am 98:213–225

Gantner BN, Simmons RM, Canavera SJ, Akira S, Underhill DM (2003) Collaborative induction of inflammatory responses by dectin-1 and toll-like receptor 2. J Exp Med 197:1107–1117

Glancy JJ, Elder JL, McAleer R (1981) Allergic bronchopulmonary fungal disease without clinical asthma. Thorax 36:345–349

Goodridge HS, Simmons RM, Underhill DM (2007) Dectin-1 stimulation by *Candida albicans* yeast or zymosan triggers NFAT activation in macrophages and dendritic cells. J Immunol 178:3107–3115

Greenberger PA (2012) Chapter 18: Allergic bronchopulmonary aspergillosis. Allergy Asthma Proc 33(Suppl 1):S61–S63

Greenberger PA, Bush RK, Demain JG, Luong A, Slavin RG, Knutsen AP et al (2014) Allergic bronchopulmonary aspergillosis. J Allergy Clin Immunol Pract 2:703–708

Greer Laboratories G (2015) Human allergy products and services. In: Greer Laboratories G (ed) Lenoir, NC 28645. Greer Laboratories, p 10

Gross O, Gewies A, Finger K, Schafer M, Sparwasser T, Peschel C, Forster I et al (2006) Card9 controls a non-TLR signalling pathway for innate anti-fungal immunity. Nature 442:651–656

Gutierrez-Rodriguez A, Postigo I, Guisantes JA, Sunen E, Martinez J (2011) Identification of allergens homologous to Alt a 1 from *Stemphylium botryosum* and *Ulocladium botrytis*. Med Mycol 49:892–896

Hardison SE, Brown GD (2012) C-type lectin receptors orchestrate antifungal immunity. Nat Immunol 13:817–822

Hyde HA (1949) Phenology of British hay-fever plants and its significance to allergists. Br Med J 1:897

Ifrim DC, Bain JM, Reid DM, Oosting M, Verschueren I, Gow NA et al (2014) Role of dectin-2 for host defense against systemic infection with *Candida glabrata*. Infect Immun 82:1064–1073

Ito S, Okada Y, Hirai HH, Nishida T (2005) Degradation of poly(acrylic acid)s with phenolic side-chains by manganese peroxidase from white rot fungi. J Polym Environ 13:357–363

Jimenez-Diaz C, Sanchez Cuenca B, Puig J (1932) Climatic asthmas. J Allergy (Cairo) 3:396–403

Jouault T, Sarazin A, Martinez-Esparza M, Fradin C, Sendid B, Poulain D et al (2009) Host responses to a versatile commensal: PAMPs and PRRs interplay leading to tolerance or infection by *Candida albicans*. Cell Microbiol 11:1007–1015

Kainer MA, Reagan DR, Nguyen DB, Wiese AD, Wise ME, Ward J et al (2012) Fungal infections associated with contaminated methylprednisolone in Tennessee. N Engl J Med 367: 2194–2203

Kankkunen P, Teirila L, Rintahaka J, Alenius H, Wolff H, Matikainen S et al (2010) (1,3)-beta-glucans activate both dectin-1 and NLRP3 inflammasome in human macrophages. J Immunol 184:6335–6342

Kendrick B (2008) The fifth kingdom. Focus Publishing/R Pullins & Co, Newburyport, MA

Knutsen AP, Bush RK, Demain JG, Denning DW, Dixit A, Fairs A, Greenberger PA et al (2012) Fungi and allergic lower respiratory tract diseases. J Allergy Clin Immunol 129:280–291, quiz 292–283

Kolwijck E, van de Veerdonk FL (2014) The potential impact of the pulmonary microbiome on immunopathogenesis of *Aspergillus*-related lung disease. Eur J Immunol 44:3156–3165

Krieger J, Jacobs DE, Ashley PJ, Baeder A, Chew GL, Dearborn D et al (2010) Housing interventions and control of asthma-related indoor biologic agents: a review of the evidence. J Public Health Manag Pract 16:S11–S20

Kume H, Yamazaki T, Togano T, Abe M, Tanuma H, Kawana S et al (2011) Epidemiology of visceral mycoses in autopsy cases in Japan: comparison of the data from 1989, 1993, 1997, 2001, 2005 and 2007 in Annual of Pathological Autopsy Cases in Japan. Med Mycol J 52:117–127

Lehrnbecher T, Frank C, Engels K, Kriener S, Groll AH, Schwabe D et al (2010) Trends in the postmortem epidemiology of invasive fungal infections at a university hospital. J Infect 61:259–265

Levetin E (2014) An introduction to fungal aerobiology. American College of Allergy, Asthma & Immunology 2014 Annual Meeting, Atlanta, GA

Levetin E. Horner WE. Scott JA. Environmental Allergens Workgroup. Taxonomy of Allergenic Fungi. J Allergy Clin Immunol Pract. 2015 Dec 24. pii: S2213-2198(15)00576-0. doi: 10.1016/j.jaip.2015.10.012. [Epub ahead of print]

Lewis RE, Cahyame-Zuniga RE, Leventakos L, Chamilos K, Ben-Ami G, Tamboli P et al (2013) Epidemiology and sites of involvement of invasive fungal infections in patients with haematological malignancies: a 20-year autopsy study. Mycoses 56:638–645

Lichtman AH, Pillai S, Abbas AK (2007) Cellular and molecular immunology. Saunders\Elsevier, Philadelphia

Lopez M, Salvaggio JE (1985) Mold-sensitive asthma. Clin Rev Allergy 3:183–196

Maheswaran D, Zeng Y, Chan-Yeung M, Scott J, Osornio-Vargas A, Becker AB et al (2014) Exposure to Beta-(1,3)-d-glucan in house dust at age 7–10 is associated with airway hyperresponsiveness and atopic asthma by age 11–14. PLoS One 9, e98878

Maier S, Chung CH, Morse M, Platts-Mills T, Townes L, Mukhopadhyay P, et al (2015) A retrospective analysis of cross-reacting cetuximab IgE antibody and its association with severe infusion reactions. Cancer Med Jan;4(1):36-42. doi: 10.1002/cam4.333. Epub 2014 Oct 9

Mari A, Schneider P, Wally V, Breitenbach M, Simon-Nobbe B (2003) Sensitization to fungi: epidemiology, comparative skin tests, and IgE reactivity of fungal extracts. Clin Exp Allergy 33:1429–1438

Marri PR, Stern DA, Wright AL, Billheimer D, Martinez FD (2013) Asthma-associated differences in microbial composition of induced sputum. J Allergy Clin Immunol 131(346–352):e341–e343

Medicine Io (2004) Damp indoor spaces and health. Washington, DC: National Academy of Sciences, Board on Health Promotion and Disease Prevention

Mendell MJ, Mirer AG, Cheung K, Tong M, Douwes J (2011) Respiratory and allergic health effects of dampness, mold, and dampness-related agents: a review of the epidemiologic evidence. Environ Health Perspect 119:748–756

Miller TE, Watson DW (1965) Innate immunity. Med Clin North Am 49:1489–1504

Mott E, Kesten HD (1931) Hypersensitiveness to soluble specific substances from yeast-like fungi: Ii. eye hypersensitivity. J Exp Med 53:815–819

NAB (2000) National Allergy Bureau. Year 2000 Pollen and Spore Reports. Milwaukee: American Academy for Allergy, Asthma and Immunology
Nasser SM, Pulimood TB (2009) Allergens and thunderstorm asthma. Curr Allergy Asthma Rep 9:384–390
Natarajan S, Subramanian P (2014) Allergic bronchopulmonary aspergillosis: a clinical review of 24 patients: are we right in frequent serologic monitoring? Ann Thorac Med 9:216–220
Netea MG, Ferwerda G, van der Graaf CA, Van der Meer JW, Kullberg BJ (2006) Recognition of fungal pathogens by toll-like receptors. Curr Pharm Des 12:4195–4201
O'Neil CE, Horner WE, Reed MA, Lopez M, Lehrer SB (1990) Evaluation of basidiomycete and deuteromycete (Fungi Imperfecti) extracts for shared allergenic determinants. Clin Exp Allergy 20:533–538
Ordas I, Mould DR, Feagan BG, Sandborn WJ (2012) Anti-TNF monoclonal antibodies in inflammatory bowel disease: pharmacokinetics-based dosing paradigms. Clin Pharmacol Ther 91:635–646
Paris S, Fitting C, Ramirez E, Latge JP, David B (1990) Comparison of different extraction methods of Alternaria allergens. J Allergy Clin Immunol 85:941–948
Park JH, Cox-Ganser JM (2011) Mold exposure and respiratory health in damp indoor environments. Front Biosci (Elite Ed) 3:757–771
Pettit AC, Kropski JA, Castilho JL, Schmitz JE, Rauch CA, Mobley BC et al (2012) The index case for the fungal meningitis outbreak in the United States. N Engl J Med 367:2119–2125
Poole CJ, Wong M (2013) Allergic bronchopulmonary aspergillosis in garden waste (compost) collectors—occupational implications. Occup Med (Lond) 63:517–519
Portnoy JM (2012) It's about time for allergenic extracts to undergo review. Ann Allergy Asthma Immunol 109:3–4
Portnoy J, Jara D (2015) Mold allergy revisited. current allergy reports. Ann Allergy Asthma Immunol 114(2):83–89
Portnoy J, Pacheco F, Ballam Y, Barnes C (1993) The effect of time and extraction buffers on residual protein and allergen content of extracts derived from four strains of Alternaria. J Allergy Clin Immunol 91:930–938
Portnoy J, Barnes C, Larenas-Lineman D, Miler JD, Wiliams PB, Chew G, et al (2015) Environmental assessment and exposure control of fungi: a practice parameter. In: Joint Task Force on Practice Parameters American Academy of Asthma and Immunology, The American College of Allergy, Asthma and Immunology, The Joint Council of Allergy, Asthma and Immunology (eds) Allergy practice parameters. AAAAI. https://www.aaaai.org/Aaaai/media/MediaLibrary/PDF%20Documents/Announcements/Fungi-23-0.pdf
Postigo I, Gutierrez-Rodriguez A, Fernandez J, Guisantes JA, Sunen E, Martinez J et al (2011) Diagnostic value of Alt a 1, fungal enolase and manganese-dependent superoxide dismutase in the component-resolved diagnosis of allergy to Pleosporaceae. Clin Exp Allergy 41:443–451
Quansah R, Jaakkola MS, Hugg TT, Heikkinen SA, Jaakkola JJ (2012) Residential dampness and molds and the risk of developing asthma: a systematic review and meta-analysis. PLoS One 7, e47526
Rittenour WR, Park JH, Cox-Ganser JM, Beezhold DH, Green BJ (2012) Comparison of DNA extraction methodologies used for assessing fungal diversity via ITS sequencing. J Environ Monit 14:766–774
Rittenour WR, Ciaccio CE, Barnes CS, Kashon ML, Lemons AR, Beezhold DH, Green BJ (2013) Internal transcribed spacer rRNA gene sequencing analysis of fungal diversity in Kansas City indoor environments. Environ Sci Process Impacts 2014 Jan;16(1):33–43
Roeder A, Kirschning CJ, Rupec RA, Schaller M, Weindl G, Korting HC et al (2004) Toll-like receptors as key mediators in innate antifungal immunity. Med Mycol 42:485–498
Rubino SJ, Selvanantham T, Girardin SE, Philpott DJ (2012) Nod-like receptors in the control of intestinal inflammation. Curr Opin Immunol 24:398–404
Saenz-de-Santamaria M, Postigo I, Gutierrez-Rodriguez A, Cardona G, Guisantes JA, Asturias JA, Martinez J et al (2006) The major allergen of *Alternaria alternata* (Alt a 1) is expressed in other members of the Pleosporaceae family. Mycoses 49:91–95

Shimodaira K, Okubo Y, Nakayama H, Wakayama M, Shinozaki M, Ishiwatari T et al (2012) Trends in the prevalence of invasive fungal infections from an analysis of annual records of autopsy cases of Toho University. Mycoses 55:435–443
Shoham S, Levitz SM (2005) The immune response to fungal infections. Br J Haematol 129:569–582
Simon-Nobbe B, Denk U, Poll V, Rid R, Breitenbach M (2008) The spectrum of fungal allergy. Int Arch Allergy Immunol 145:58–86
Sinko J, Csomor J, Nikolova R, Lueff S, Krivan G, Remenyi P et al (2008) Invasive fungal disease in allogeneic hematopoietic stem cell transplant recipients: an autopsy-driven survey. Transpl Infect Dis 10:106–109
Skevaki C, Pararas M, Kostelidou K, Tsakris A, Routsias JG (2015) Single nucleotide polymorphisms of Toll-like receptors and susceptibility to infectious diseases. Clin Exp Immunol 180(2):165–177
Slater JE, Menzies SL, Bridgewater J, Mosquera A, Zinderman CE, Ou AC et al (2012) The US food and drug administration review of the safety and effectiveness of nonstandardized allergen extracts. J Allergy Clin Immunol 129:1014–1019
Smith RM, Mk S, Kainer MA, Wise M, Finks J, Duwve J, Fontaine E et al (2013) Fungal infections associated with contaminated methylprednisolone injections. N Engl J Med 369:1598–1609
van Asbeck EC, Hoepelman AI, Scharringa J, Herpers BL, Verhoef J (2008) Mannose binding lectin plays a crucial role in innate immunity against yeast by enhanced complement activation and enhanced uptake of polymorphonuclear cells. BMC Microbiol 8:229
Van Leeuwen WS, Nien Z, Kremer W, Varenkamp H (1925) Ztschr Immunitatsforsch 44:1
van Woerden HC, Gregory C, Brown R, Marchesi JR, Hoogendoorn B, Matthews IP et al (2013) Differences in fungi present in induced sputum samples from asthma patients and non-atopic controls: a community based case control study. BMC Infect Dis 13:69
Villena J, Aso H, Kitazawa H (2014) Regulation of toll-like receptors-mediated inflammation by immunobiotics in bovine intestinal epitheliocytes: role of signaling pathways and negative regulators. Front Immunol 5:421
Wagner H (2012) Innate immunity's path to the Nobel Prize 2011 and beyond. Eur J Immunol 42:1089–1092
Weber RW (2003) Meteorologic variables in aerobiology. Immunol Allergy Clin North Am 23:411–422
Willger SD, Grim SL, Dolben EL, Shipunova A, Hampton TH, Morrison HG et al (2014) Characterization and quantification of the fungal microbiome in serial samples from individuals with cystic fibrosis. Microbiome 2:40
Yamasaki S, Matsumoto M, Takeuchi O, Matsuzawa T, Ishikawa E, Sakuma M et al (2009) C-type lectin Mincle is an activating receptor for pathogenic fungus, *Malassezia*. Proc Natl Acad Sci USA 106:1897–1902

# Chapter 18
# What's Old is New: Recognition of New Fungal Pathogens in the Era of Phylogenetics and Changing Taxonomy and Implications for Medical Mycology

**Nathan P. Wiederhold and Deanna A. Sutton**

## Introduction

Invasive fungal infections are associated with significant morbidity and mortality as these are often difficult to diagnosis and treat. Fungi historically associated with invasive disease in humans include the yeast within the genera *Candida, Cryptococcus,* and *Trichosporon*, the dimorphic fungi *Blastomyces dermatitidis, Coccidioides immitis/posadasii,* and *Histoplasma capsulatum,* and the molds, including limited species within the genera *Aspergillus*, *Fusarium*, and *Scedosporium*, and certain members of the Order *Mucorales.* Over the last two decades, there has been a significant increase in the number of fungal species associated with invasive disease in humans. Factors that have contributed to this increase include an increase in the number of immunocompromised patients at high risk for invasive fungal infections, including HIV-AIDS patients, those receiving immunosuppressive chemotherapy for malignancies, and solid organ transplant recipients, improvements in diagnostic assays and the clinical recognition of patients with risk factors for such infections, as well as improvements in the tools used to identify fungal species. Unfortunately, the recognition of new etiologic agents of invasive mycoses has surpassed the development of new diagnostic assays and treatment strategies against these infections. The increase in the number of etiologic agents of invasive mycoses is also method driven. Taxonomic changes due to phylogenetic analysis have led to the reclassification of many previously recognized fungi. This has led to a concern of nomenclature instability in medical mycology, and the clinical relevance of many of the newly reclassified species is unknown (de Hoog et al.

N.P. Wiederhold, Pharm.D. (✉) • D.A. Sutton, Ph.D.
Fungus Testing Laboratory, Department of Pathology,
University of Texas Health Science Center at San Antonio,
7703 Floyd Curl Drive, MC7750, San Antonio, TX 78229-3900, USA
e-mail: wiederholdn@uthscsa.edu; suttond@uthscsa.edu



D.-W. Li (ed.), *Biology of Microfungi*, Fungal Biology,
DOI 10.1007/978-3-319-29137-6_18

2013, 2014). In this chapter, we review the identification of fungi, primarily filamentous organisms, in the clinical setting, and provide examples of known and emerging causes of invasive fungal infections in humans and changes in taxonomy and fungal nomenclature that have occurred and are ongoing.

## Fungal Identification in the Clinical Setting

The identification of fungi in the clinical laboratory has historically relied on morphologic characteristics and physiologic traits. The description of the colony appearance and the microscopic features of the organism, including the reproductive structures, has been the hallmark for fungal identification for many years. Certain phenotypic/physiologic traits are also combined with the morphologic features to obtain the identities of fungal isolates. For molds these include, but are not limited to, the ability of the organism to grow at certain temperatures, tolerance to cycloheximide and benomyl, nitrate assimilation, tolerance to different concentrations of sodium chloride, growth on bromcresol purple agar, growth on trichophyton agar, and growth on urea agar (Nelson et al. 1983; Pincus et al. 1988; Kane et al. 1997; Summerbell 1993). Many of these phenotypic/physiologic assays were and still are used to identify the organism to the genus and possibly species level in clinical microbiology and reference mycology laboratories. Identification to the species level is clinically important as it provides the clinician with information that may be useful in the management of patients and help guide antifungal therapy. Indeed, early identification and the initiation of appropriate therapy have been shown to influence patient outcomes while delaying appropriate therapy can be detrimental (Morrell et al. 2005; Greene et al. 2007; Garey et al. 2006; Chamilos et al. 2008). Identification to the species level is important in helping to guide appropriate therapy, as some fungi are intrinsically resistant to certain drugs. Furthermore, some species within the same species complex may have different antifungal susceptibility profiles and this can influence the choice of treatment that is used (Balajee et al. 2005a; Gilgado et al. 2006; Lackner et al. 2012). However, identification by morphologic/physiologic characteristics alone can be time-consuming, and results may not be available in a timely fashion for clinical decisions. Morphologic identification can also be fraught with errors if done by those without proper training and experience. In addition, the morphologic features of fungi may be variable (Balajee et al. 2006, 2007). Different factors can affect these features, including the media used for subculturing and exposure to external stressors, such as antifungal agents prior to recovery from clinical specimens, which can often occur in patient groups at high risk for invasive fungal infections in which empiric or preemptive antifungal therapy is often used.

The introduction of molecular and proteomic tools, such as DNA sequence analysis and matrix-assisted light desorption ionization time-of-flight (MALDI-TOF), a relatively new diagnostic tool in the clinical microbiology laboratory, has dramatically changed how fungi are identified. These methods can reduce the amount of

**Table 18.1** DNA targets used for molecular sequence identification of fungi and examples of genera these targets may be used to identify

| Targets | Genera |
|---|---|
| Internal transcribed spacer (ITS) | All genera |
| 28S rDNA large subunit (D1/D2) | All genera |
| Beta-tubulin / Calmodulin | *Aspergillus* |
| Calmodulin | *Scedosporium* |
| Translation elongation factor | *Fusarium* |
| RNA polymerase | *Penicillium* |
| Glyceraldehyde-3-phosphate dehydrogenase | *Curvularia* |

time needed to determine the identity of an organism and reduce errors associated with morphologic variability. However, these methods have their own limitations and have not eliminated the need for the morphologic evaluation of fungi in the clinical laboratory. For both DNA sequence and protein spectrum analysis, the results that are obtained must be compared to those deposited in databases from known organisms in order for an identity to be obtained. For fungi, publicly available databases for DNA sequence comparisons are available, including those at the National Center for Biotechnology Information (GenBank; www.ncbi.nlm.nih.gov/genbank/), the Centraalbureau voor Schimmelcultures Fungal Biodiversity Centre in the Netherlands (CBS-KNAW; www.cbs.knaw.nl), the International Society of Human and Animal Mycology ITS database (ISHAM; its.mycologylab.org), and the Fusarium-ID database (http://isolate.fusariumdb.org). Reference laboratories or clinical microbiology laboratories may also have their own databases. The use of sequence results can be extremely useful when compared with credible deposits. However, not all fungal deposits within databases have been confirmed to be from accurately identified organisms (Bridge et al. 2003; Deckert et al. 2002; Crous 2002). This can lead to erroneous results and the misidentification of the cultured specimen. In addition, the choice of the target sequence can be critical for the proper identification of fungi. Although the internal transcribed spacer region (ITS) has been put forth as a universal barcode for the identification of fungi (Schoch et al. 2012; Seifert 2009; Petti 2007), this target cannot be used alone to discriminate between closely related fungi. Several other DNA targets may be used to identify fungi in the clinical setting (Table 18.1), and the choice of targets is dependent on the organism. Thus, an assessment of the morphology of the organism prior to sequence analysis can provide useful information as to what DNA targets to use for identification.

## Changes in Fungal Nomenclature

Under Article 59 of the International Code for Botanical Nomenclature, polymorphic fungi were allowed to have multiple names. These names described either the sexual (teleomorph) or asexual (anamorph) stages of the organism's life cycle. The dual nomenclature system served a purpose when fungal identification was based upon the observed morphologic features. However, it became obsolete with the introduction of molecular tools into the field of mycology (de Hoog et al. 2014; Hawksworth 2011). Under the newly named International Code of Nomenclature of algae, fungi, and plants, the dual nomenclature system under Article 59 is abolished, and as of January 1, 2013, all fungi are now to have only one correct name (Norvell 2011). Regardless of the life history stage of the type, all legitimate fungal names are now to be treated equally for the purpose of establishing priority. While the abandonment of the dual nomenclature system was an important first step, mycologists are now charged with implementing this change and contributing to the production of lists of accepted and suppressed names for fungi.

The abandonment of the dual nomenclature system and the increased use of molecular tools have implications for the field of medical mycology. Phylogenetic studies have demonstrated that fungi are much more molecularly diverse than previously recognized, and this has led to an increase in the number of clinically recognized fungi, as new species have been described and other pathogens have been reclassified (de Hoog et al. 2013). Some examples of recent nomenclature changes that have occurred for clinically relevant fungi are shown in Table 18.2. However, molecular diversity does not necessarily equate to clinical diversity, and the true clinical relevance of newly discovered species or newly reclassified organisms may be unknown. It has been suggested that this increase in recognized fungal species, combined with the abandonment of the dual nomenclature system, may be compromising the stability of the nomenclature of medically important fungi (de Hoog et al. 2013, 2014). Potentially detrimental aspects may be the confusion of clinicians who do not closely follow taxonomic changes but are responsible for the care of patients and the impediment of navigating the literature to find clinically useful

**Table 18.2** Examples of changes in fungal nomenclature of clinically relevant fungi

| New species name | Previous species name | Invasive disease |
|---|---|---|
| *Curvularia spicifera* | *Bipolaris spicifera* | Fungal sinusitis, cerebral phaeohyphomycosis |
| *Curvularia hawaiiensis* | *Bipolaris hawaiiensis* | |
| *Curvularia australiensis* | *Bipolaris australiensis* | |
| *Lichtheimia corymbifera* | *Absidia corymbifera* | Mucormycosis |
| *Lomentospora prolificans* | *Scedosporium prolificans* | Pulmonary infection (scedosporiosis) |
| *Rasamsonia argillacea* | *Geosmithia argillacea* | Pulmonary infection |
| *Sarocladium kiliense* | *Acremonium kiliense* | Mycetoma, keratitis in immunocompetent hosts, invasive infections in immunocompromised individuals |
| *Talaromyces marneffei* | *Penicillium marneffei* | Disseminated infection in HIV+ individuals |
| *Verruconis gallopava* | *Ochroconis gallopava* | Cerebral phaeohyphomycosis |

information about fungal pathogens and invasive mycoses due to confusion about the name of the organism or the diseases the fungi cause. Thus, it has been proposed that in clinical practice medical mycologists be allowed to follow taxonomic changes and implement these more gradually (de Hoog et al. 2013). At the species level, once the clinical relevance of molecular sibling species is determined, novel nomenclature could be adopted. However, until such evidence becomes available, cryptic species can be referred to as species complexes in medical practice. A potential drawback of this approach is the hindrance of gaining knowledge about the clinicopathology of a particular organism. In order to establish a body of literature necessary to gain an understanding of the clinical relevance of a particular fungal species in relation to disease and its response to therapy, the true identity of the infecting organism must be known. The rate at which such knowledge is accumulated may be slowed if such similar species are lumped into species complexes without further delineation.

## Examples of Clinically Relevant/Emerging Fungi and Changes in Fungal Taxonomy and Nomenclature

### *Invasive Aspergillosis and* Aspergillus *Section* Fumigati

Invasive aspergillosis is a major invasive fungal infection and significant cause of morbidity and mortality in immunocompromised hosts. Groups at highest risk for the development of invasive aspergillosis include highly immunocompromised individuals. Such groups have historically included solid organ transplant recipients, those undergoing hematopoietic stem cell transplantation, and patients receiving highly immunosuppressive chemotherapies (Wiederhold et al. 2003). Results from the PATH Alliance Registry from 2004 to 2007, which collects information on patients with invasive fungal infections at medical centers in the USA, demonstrated that invasive aspergillosis was the most frequent invasive fungal infections in patients undergoing hematopoietic stem cell transplantation ahead of invasive candidiasis, mucormycosis, and other invasive fungal infections (Neofytos et al. 2009). Allogeneic stem cell transplant recipients who receive prolonged course of corticosteroids for the treatment of graft versus host disease are at further risk for invasive aspergillosis (Wiederhold et al. 2003; Baddley et al. 2010; Kontoyiannis et al. 2010). Of the solid organ transplant recipients, individuals who receive lung or dual heart/lung transplants are at higher risk for invasive pulmonary aspergillosis as the primary route of entry into the body is via inhalation into the lungs (Minari et al. 2002). Although pulmonary involvement is a major component of invasive aspergillosis due to the route of entry into the body via the lungs, dissemination to other organs can occur, as invasive disease has been reported in all organ systems. The mortality rate of this disease is exceptionally high with dissemination to the central nervous system in the setting of continued immunosuppression. In addition to these patient populations, invasive aspergillosis has also become more important in critically ill patients not traditionally considered at high risk, including those with acute chronic obstructive pulmonary disease and those receiving corticosteroids (Meersseman et al. 2004; Garnacho-Montero et al. 2005). Chronic pulmonary aspergillosis is also a significant problem in patients with structure damage to the lungs, such as those who have had tuberculosis or sarcoidosis (Smith and Denning 2011; Denning 2001). The prevalence of chronic pulmonary aspergillosis is estimated to be approximately 3 million patients worldwide (Denning et al. 2011, 2013a, b). Treatment of these patients often involves prolonged courses of azole therapy, which predisposes individuals to the adverse effects and drug interactions associated with these agents and the potential development of drug-resistant organisms (Howard et al. 2009).

One of the most challenging aspects of this disease is the ability to make a timely and accurate diagnosis. The diagnosis of invasive aspergillosis involves the incorporation of clinical, radiological, serological, and histopathological findings. Although studies have demonstrated the usefulness of radiographic studies, such as chest computed tomography, in patients with risk factors for invasive aspergillosis, the images obtained cannot conclusively rule in or rule out this fungal infection as other pulmonary fungal infections can show similar signs (Walsh et al. 2008). Rapid diagnosis of this disease has focused on the detection of surrogate markers of infection, including components of the cell wall within normally sterile biologic fluids. One particular strategy that is clinically used is the detection of galactomannan, a component of the cell wall released during growth of the organism (Latge et al. 1994). A commercially available assay, the Platelia *Aspergillus* ELISA kit (Bio-Rad), uses a rat monoclonal antibody (EB-A2) directed against tetra (1 → 5)-β-D-galactofuranoside, the immunodominant epitope in galactomannan (Stynen et al. 1992, 1995; Morelle et al. 2005). This assay has proven to be useful for the diagnosis of invasive aspergillosis with a high specificity (≥85 %) in patients with hematologic malignancies at high risk for this opportunistic disease (Pfeiffer et al. 2006). The detection of galactomannan using this assay now fulfills part of the diagnostic criteria for probable invasive aspergillosis (Walsh et al. 2008). Other assays that are clinically used to detect invasive fungal infections detect another component of the cell wall of many pathogenic fungi, (1 → 3)-β-D-glucan. The chromogenic assay available for clinical use (Fungitell, Associates of Cape Cod) is based on the activation of the horseshoe crab coagulation cascade and uses amebocyte enzymes from *Limulus polyphemus* (Fungitell 2008). The prompt diagnosis of invasive aspergillosis, including the use of these surrogate diagnostic markers, can have significant effects on patient outcomes. Early diagnosis and initiation of antifungal therapy have been shown to reduce mortality in patients with invasive fungal infections including invasive aspergillosis (Greene et al. 2007; Garey et al. 2006; Caillot et al. 1997). This has led to the strategy of preemptive therapy in which antifungal agents are initiated upon the first signs of a potential infection as suggested by high-resolution computer tomography and serial screening of surrogate markers. However, these assays are not without their limitations. (1 → 3)-β-D-glucan is a pan-fungal target, since many clinically relevant species contain this polysaccharide within their cell walls (Odabasi et al. 2006). Thus, a positive result can be seen in patients with infections caused by a variety of fungi including *Aspergillus*, *Candida*, *Fusarium*, *Acremonium*, *Trichosporon*, *Sporothrix*, *Histoplasma*, *Coccidioides*, and *Blastomyces* (Odabasi et al. 2006; Senn et al. 2008; Pickering et al. 2005; Ostrosky-Zeichner et al. 2005). In addition, several substances can result in false-positive test results, primarily due to glucan content. This can occur in patients receiving hemodialysis with cellulose membranes, those receiving immunoglobulin products and albumin, as well as following serous exposure to gauze, which can occur in surgical patients (Odabasi et al. 2004, 2006). Thus, while a positive assay result does provide evidence of an invasive mycosis, it does provide information on the causative organism, which is important for making decisions regarding appropriate therapy. The galactomannan assay is more specific for *Aspergillus* than those for $(1 \rightarrow 3)$-β-D-glucan. However, the sensitivity of serum galactomannan may be reduced in patients with antifungal exposure with a high degree of variability among different patient populations (Marr et al. 2005). While the specificity of this assay is consistently above 85 %, the sensitivity may vary considerably between patient populations with rates reported in the literature ranging from 29 to 100 % (Pfeiffer et al. 2006; Verweij 2005). Furthermore, reports of cross-reactivity with this assay have been reported with other fungi, including *Fusarium* and *Trichosporon* species (Fekkar et al. 2009; Mikulska et al. 2012; Tortorano et al. 2012). The galactomannan assay also will not provide information on the species of *Aspergillus* that is causing infection, which is also important for making treatment decision in patients with invasive aspergillosis. For example, the galactomannan assay is not able to distinguish between *A. fumigatus* and *A. terreus*, the latter demonstrating resistance to amphotericin B, a widely used antifungal agent (Steinbach et al. 2004).

Although there are over 200 individual *Aspergillus* species, common causes of invasive aspergillosis in humans include *A. fumigatus*, *A. flavus, A. niger, A. nidulans,* and *A. terreus*. Of these, *A. fumigatus* is the major etiologic agent of invasive disease at most institutions (Morgan et al. 2005). This species is usually readily distinguishable from the other common causes of aspergillosis based on its morphology (Balajee et al. 2006, 2007). However, the morphology of *A. fumigatus* is unstable and by phylogenetic analysis several distinct species are now recognized within the section *Fumigati* (Sugui et al. 2014). Currently, this section consists of 51 phylogenetically separate species, 15 of which have been reported to cause clinical disease in humans (Sugui et al. 2014). These include *A. felis*, *A. fumigatiaffinis*, *A. fumigatus*, *A. fumisynnematus*, *A. lentulus*, *A. novofumigatus*, *A. parafelis*, *A. pseudofelis*, *A. pseudoviridinutans*, *A. viridinutans*, *A. fischeri*, *A. hiratsukae*, *A. laciniosus*, *A. thermomutatus*, and *A. udagawae*. Infections that have been reported caused by these species include invasive pulmonary aspergillosis, osteomyelitis, peritonitis, cerebral aspergillosis, and invasive otitis (Balajee et al. 2005b; Matsumoto et al. 2002; Jarv et al. 2004; Zbinden et al. 2012; Ghebremedhin et al. 2009).

Species identification within this section is a multifaceted approach, requiring morphologic, physiologic, and molecular sequence results. Although the morphology and phenotypic features within this section are variable, some members are not easily distinguished from *A. fumigatus* without the use of molecular sequencing.

The ITS region, although recognized as the universal barcode for fungi, is not capable of discriminating between members of section *Fumigati* (Samson et al. 2014). For this purpose, sequencing of the beta-tubulin region is recommended. Thus, here is the potential for misidentification of members of this section in clinical laboratories that do not routinely use sequence analysis for identification. The need for correct identification of the species causing infection is important since several species within this section are refractory to antifungal therapy and may cause more chronic infections than *A. fumigatus* (Zbinden et al. 2012; Barrs et al. 2013; Vinh et al. 2009a). These include *A. lentulus*, *A. felis*, *A. parafelis*, *A. pseudofelis*, *A. pseudoviridinutans*, *N. pseudofischeri*, and *N. udagawae* (Balajee et al. 2005a; Sugui et al. 2010, 2014; Vinh et al. 2009b; Khare et al. 2014). Clinical failures have occurred in patients treated with antifungals for infections caused by these species that were misidentified as *A. fumigatus* (Balajee et al. 2005a, b, 2007; Zbinden et al. 2012; Vinh et al. 2009a; Khare et al. 2014).

## *Scedosporiosis and* Pseudallescheria/Scedosporium

Another group of fungi recognized to cause invasive infections for which there has been significant taxonomic change is the genus *Scedosporium*. Members of this genus are ubiquitous ascomycetes found in soil, polluted water, sewage, and manure and are capable of causing many different types of infections in humans (Cortez et al. 2008; Walsh et al. 2004; Guarro et al. 2006). Invasive infections have been reported primarily in immunocompromised hosts, and these fungi are recognized as the second most common fungal colonizers in cystic fibrosis patients behind *Aspergillus* species (Blyth et al. 2010). Because the route of entry is similar to that of *Aspergillus* species, *Scedosporium* species can cause sinopulmonary disease that is difficult to distinguish from disease caused by *Aspergillus* and other molds that may be more amendable to antifungal therapy (Walsh et al. 2004; Bouza and Munoz 2004). Such infections can be especially devastating in lung transplant recipients (Johnson et al. 2014), and because these fungi can so adversely affect patient outcomes, colonization of the lungs may be a contraindication to lung transplantation (Raj and Frost 2002; Morio et al. 2010). *Scedosporium* species have also been recognized as causes of breakthrough infections in persistently neutropenic and/or lymphopenic patients receiving antifungal therapy (Lamaris et al. 2006; Nucci 2003). In patients whose immune system fails to recover disseminated infections may occur, which portends a poor prognosis. Infections can also occur in immunocompetent individuals. Near-drowning victims can develop pulmonary infections with dissemination to the brain, which are difficult to treat and associated with significant morbidity and mortality (Guarro et al. 2006; Buzina et al. 2006; Nakamura et al. 2013). Mycetomas, chronic, tumor-like infections of subcutaneous tissue and contiguous bone with draining sinuses, can also occur in otherwise healthy hosts secondary to traumatic inoculation (Cortez et al. 2008; Walsh et al. 2004; Guarro et al. 2006).

Over the last decade, the taxonomy of *Scedosporium* has changed markedly. Due to the ability to develop sexual structures on route culture media, members of this genus were identified by clinical microbiology laboratories as *Pseudallescheria boydii* when the teleomorph was present and as *Scedosporium apiospermum* when only the anamorph was found. However, with the use of molecular phylogeny it was subsequently determined that *P. boydii* (anamorph *Scedosporium boydii*) and *Pseudallescheria apiosperma* (anamorph *S. apiospermum)* were indeed separate species (Gilgado et al. 2008, 2010). Other species that have been discovered through the use of molecular phylogenetics, but which are morphologically identical to these sibling species, include *S. aurantiacum, P. minutispora, P. desertorum,* and *S. dehoogii* (Gilgado et al. 2005, 2008; Lackner et al. 2014a). Recently, due to the abolishment of Article 59 of the Code of Botanical Nomenclature of algae, fungi, and plants, it has been proposed that *Pseudallescheria* should be treated as a synonym of *Scedosporium* and that *Scedosporium* be given precedence as it is the oldest valid generic name (Lackner et al. 2014a). The morphologically distinct species *Scedosporium prolificans*, which in contrast to other members of the *Scedosporium* genus, is a phaeoid mold with inflated versus tubular conidiogenous cells and has been renamed *Lomentospora prolificans* based on significant phylogenetic differences (Lackner et al. 2014a). The distinction between this species and *Scedosporium* species is clinically relevant, as *L. prolificans* is highly resistant to multiple antifungal agents (Lackner et al. 2012, 2014b; Cortez et al. 2008; Walsh et al. 2004; Lewis et al. 2005; Wiederhold and Lewis 2009), and infections caused by this organism are extremely difficult to treat and are associated with poor patient outcomes (Cortez et al. 2008).

For the members of the genus *Scedosporium,* it has been suggested that for the routine identification in clinical microbiology laboratories, these fungi might be identified as members of the *Scedosporium apiospermum* complex since the sibling species *S. apiospermum* and *S. boydii* are without medically relevant differences (Lackner et al. 2014a). However, there is some evidence that differentiation among the members of this complex may be clinically important, and differences in antifungal susceptibility have been reported among these species. For example, *S. apiospermum* isolates have been reported to be less susceptible to posaconazole than those of *S. boydii* (Lackner et al. 2012). In addition, several studies that have evaluated the in vitro potency of clinically available antifungals have demonstrated that *S. aurantiacum* isolates are resistant to these agents with the exception of the triazole voriconazole (Lackner et al. 2012, 2014b; Tintelnot et al. 2009; Alastruey-Izquierdo et al. 2007), which is currently the drug of choice for the treatment of infections caused by *Scedosporium* species (Tortorano et al. 2014). Since many of the *S. aurantiacum* isolates included in these studies have been of clinical origin, this may be of clinical significance. Some of the species within this genus do have reduced susceptibility to voriconazole, including *S. dehoogii*, and the recently renamed *Lomentospora prolificans* (formerly *S. prolificans*), mentioned earlier, which is resistant to all clinically available antifungals (Lackner et al. 2012). While the clinical relevance of reduced susceptibility to antifungal agents is not fully understood, with the exception of the resistance observed with *L. prolificans*, further insights into the clinicopathology of infections caused by different *Scedosporium* species may be difficult to ascertain if these fungi are routinely lumped into a species complex.

## *Mucormycosis and the Order* Mucorales

Pathogenic fungi of the Order *Mucorales* are capable of causing invasive infections termed mucormycosis. Organisms that have been associated with infections in humans include members of the genera *Rhizopus*, *Rhizomucor*, *Mucor*, *Cunninghamella*, *Lichtheimia* (formerly *Absidia*), *Saksenaea*, and *Apophysomyces* (Kontoyiannis and Lewis 2006; Mendoza et al. 2014; Kwon-Chung 2012; Petrikkos et al. 2012). Mucormycosis is a highly aggressive angioinvasive fungal infection that primarily occurs in immunocompromised hosts, including solid organ transplant recipients, hematopoietic stem cell transplant recipients, and hematologic malignancy patients receiving immunosuppressive chemotherapy (Kontoyiannis and Lewis 2006; Petrikkos et al. 2012). In addition, diabetic patients with poorly controlled disease have also been shown to be at risk for infections caused by members of the Order *Mucorales*. Mucormycosis has also been reported in otherwise healthy individuals following traumatic inoculation (Hospenthal et al. 2011; Neblett Fanfair et al. 2012). Infections caused by these fungi include rhino-orbital and rhino-cerebral disease, pulmonary, gastrointestinal, and cutaneous infections. Disseminated infections can also occur and are associated with high mortality rates (Kontoyiannis and Lewis 2006; Petrikkos et al. 2012). Aggressive treatment is needed in patients with mucormycosis, and this often involves multiple modalities including high-dose antifungal therapy and surgery when possible to remove infected and necrotic tissue (Kontoyiannis and Lewis 2006). However, in the setting of continued immunosuppression, clinical outcomes may be poor even with aggressive treatment. Furthermore, these species are also resistant to several antifungals, including voriconazole, the azole frequently used to treat other invasive mold infections such as invasive aspergillosis and scedosporiosis, and the echinocandins (Kontoyiannis and Lewis 2006; Almyroudis et al. 2007).

The name used to describe an infectious disease caused by members of the Order *Mucorales* has also been subject to change. In 1885, the first well-documented case was published by the German pathologist Paltauf, who used the term mycosis mucorina to describe a systemic infection with rhino-cerebral and gastrointestinal involvement (Kwon-Chung 2012; Paltauf 1885). The use of mucormycosis as the disease name was first used by Baker to describe an infection caused by members of the Order *Mucorales* in the 1950s (Baker 1956, 1957). Ajello et al. subsequently proposed the term zygomycosis to include infections caused by species from two separate orders: (1) *Mucorales*, including infections caused by species within the genera *Rhizomucor*, *Rhizopus*, *Mucor*, *Lichtheimia* (*Absidia*), *Apophysomyces*, and *Saksenaea*, and (2) *Entomophthorales*, due to *Conidiobolus* and *Basidiobolus* species (Ajello et al. 1976). Until recently, zygomycosis was the term frequently used synonymously in the clinical literature with mucormycosis to describe infections caused by members of the Order *Mucorales*, in animal models of invasive infections caused by them, as well as in vitro antifungal susceptibility results against these organisms. However, there are major differences between species classified in the Orders *Entomophthorales* and *Mucorales*, including morphologic and epidemiologic differences, and in the types of infections that they cause (Mendoza et al. 2014; Petrikkos et al. 2012). While the pathogenic species within *Mucorales* cause acute invasive infections in immunocompromised hosts, chronic subcutaneous infections are often caused by the *Entomophthorales* and can often occur in immunocompetent individuals (Mendoza et al. 2014; Kwon-Chung 2012). In addition, the phylum *Zygomycota* has been eliminated, as it was found to be polyphyletic and had also not been validly described (Hibbett et al. 2007). Thus, mucormycosis and entomophthoromycosis have again been put forth as the proper terms used to describe infections caused by these different groups of fungi (Kwon-Chung 1994).

Phylogenetic studies within the Order *Mucorales* have resulted in several name changes in pathogenic species. Some examples of these include *Lichtheimia corymbifera* (formerly *Absidia corymbifera*), *Mucor circinelloides* f. *janssenii* (formerly *Mucor velutinosus*), *Mucor ardhlaengiktus* (formerly *Mucor ellipsoideus*), and *Mucor irregularis* (formerly *Rhizomucor variabilis*) (Walther et al. 2013; Alastruey-Izquierdo et al. 2010). Some clinical laboratories that do not routinely use sequence analysis for identification of fungi attempt to discriminate between species within the genera *Rhizopus* and *Mucor* by the presence or absence of rhizoids. However, these morphologic features are not specific to *Rhizopus* species and can be found in several *Mucor* species, including the clinically relevant species *M. circinelloides*. There is also some controversy regarding the name of one of the most prevalent causes of mucormycosis, *Rhizopus arrhizus* or *Rhizopus oryzae*. Although the names are synonymous, some have suggested that *R. arrhizus* is the correct species name since it was used previous to that of *R. oryzae* (1892 vs. 1895) (Dolatabadi et al. 2014). The name that is used may be of consequence as *R. oryzae* is more frequently used in the medical literature, and there is some suggestion that there may be differences in in vitro potency between the triazoles isavuconazole and posaconazole (Verweij et al. 2009; Gonzalez 2009; Thompson and Wiederhold 2010; Chowdhary et al. 2014). This is significant as both agents can be orally administered, thus possibly avoiding the need for long-term intravenous therapy with the nephrotoxic agent amphotericin B in patients with mucormycosis caused by this species. There is also some controversy as to whether or not the similar species *Rhizopus delemar* is a separate species or a variety of *R. arrhizus*. By molecular analysis using ITS alone and with multiple loci, several authors had demonstrated that *R. delemar* is indeed a phylogenetically distinct species (Dolatabadi et al. 2014; Abe et al. 2006, 2007). In addition, phenotypically *R. delemar* lacks the ldhA gene and is unable to produce lactic acid and instead forms fumaric and malic acid, while *R. arrhizus* contains two genes for lactate dehydrogenase and is therefore able to produce lactic acid (Abe et al. 2007; Saito et al. 2004). The genome sequence of a *R. delemar* strain previously classified as *R. oryzae* that was obtained from a fatal case of mucormycosis is also available (http://www.ncbi.nlm.nih.gov/

bioproject/13066) (Ma et al. 2009). However, zygospore formation has been reported in crosses between *R. delemar* and *R. arrhizus* strains, which has led some to suggest that these are the same species, even though the viability of the progeny could not be demonstrated (Dolatabadi et al. 2014). It is currently unknown if there is a clinical difference between *R. delemar* and *R. arrhizus*.

## Conclusion

The introduction of molecular tools for the identification and classification of fungi has led to significant changes in fungal taxonomy and nomenclature. These changes have major implications for the field of medical mycology, which may be both beneficial and detrimental in the clinical setting. The correct identification of the species that is causing infection is important and can help guide therapy and ensure the use of appropriate antifungal agents. However, the rapid changes in fungal taxonomy may also lead to nomenclature instability in the clinical setting, and there is concern that this may impede access to clinically relevant literature. The changes in taxonomy and nomenclature are also currently outpacing our understanding of the clinical significance of newly classified cryptic species as well as how to effectively manage patients with infections caused by these organisms.

## References


Abe A, Oda Y, Asano K, Sone T (2006) The molecular phylogeny of the genus *Rhizopus* based on rDNA sequences. Biosci Biotechnol Biochem 70:2387–2393

Abe A, Oda Y, Asano K, Sone T (2007) *Rhizopus delemar* is the proper name for *Rhizopus oryzae* fumaric-malic acid producers. Mycologia 99:714–722

Ajello L, Dean DF, Irwin RS (1976) The zygomycete *Saksenaea vasiformis* as a pathogen of humans with a critical review of the etiology of zygomycosis. Mycologia 68:52–62

Alastruey-Izquierdo A, Cuenca-Estrella M, Monzon A, Rodriguez-Tudela JL (2007) Prevalence and susceptibility testing of new species of *Pseudallescheria* and *Scedosporium* in a collection of clinical mold isolates. Antimicrob Agents Chemother 51:748–751

Alastruey-Izquierdo A, Hoffmann K, de Hoog GS, Rodriguez-Tudela JL, Voigt K et al (2010) Species recognition and clinical relevance of the zygomycetous genus *Lichtheimia* (syn. *Absidia pro parte, Mycocladus*). J Clin Microbiol 48:2154–2170

Almyroudis NG, Sutton DA, Fothergill AW, Rinaldi MG, Kusne S (2007) In vitro susceptibilities of 217 clinical isolates of zygomycetes to conventional and new antifungal agents. Antimicrob Agents Chemother 51:2587–2590

Baddley JW, Andes DR, Marr KA, Kontoyiannis DP, Alexander BD et al (2010) Factors associated with mortality in transplant patients with invasive aspergillosis. Clin Infect Dis 50:1559–1567

Baker RD (1956) Pulmonary mucormycosis. Am J Pathol 32:287–313

Baker RD (1957) Mucormycosis; a new disease? J Am Med Assoc 163:805–808

Balajee SA, Gribskov JL, Hanley E, Nickle D, Marr KA (2005a) *Aspergillus lentulus* sp. nov., a new sibling species of *A. fumigatus*. Eukaryot Cell 4:625–632

Balajee SA, Gribskov J, Brandt M, Ito J, Fothergill A et al (2005b) Mistaken identity: *Neosartorya pseudofischeri* and its anamorph masquerading as *Aspergillus fumigatus*. J Clin Microbiol 43:5996–5999

Balajee SA, Nickle D, Varga J, Marr KA (2006) Molecular studies reveal frequent misidentification of *Aspergillus fumigatus* by morphotyping. Eukaryot Cell 5:1705–1712

Balajee SA, Houbraken J, Verweij PE, Hong SB, Yaghuchi T et al (2007) *Aspergillus* species identification in the clinical setting. Stud Mycol 59:39–46

Barrs VR, van Doorn TM, Houbraken J, Kidd SE, Martin P et al (2013) *Aspergillus felis* sp. nov., an emerging agent of invasive aspergillosis in humans, cats, and dogs. PLoS One 8, e64871

Blyth CC, Middleton PG, Harun A, Sorrell TC, Meyer W et al (2010) Clinical associations and prevalence of *Scedosporium* spp. in Australian cystic fibrosis patients: identification of novel risk factors? Med Mycol 48(Suppl 1):S37–S44

Bouza E, Munoz P (2004) Invasive infections caused by *Blastoschizomyces capitatus* and *Scedosporium* spp. Clin Microbiol Infect 10(Suppl 1):76–85

Bridge PD, Roberts PJ, Spooner BM, Panchal G (2003) On the unreliability of published DNA sequences. New Phytol 160:43–48

Buzina W, Feierl G, Haas D, Reinthaler FF, Holl A et al (2006) Lethal brain abscess due to the fungus *Scedosporium apiospermum* (teleomorph *Pseudallescheria boydii*) after a near-drowning incident: case report and review of the literature. Med Mycol 44:473–477

Caillot D, Casasnovas O, Bernard A, Couaillier JF, Durand C et al (1997) Improved management of invasive pulmonary aspergillosis in neutropenic patients using early thoracic computed tomographic scan and surgery. J Clin Oncol 15:139–147

Chamilos G, Lewis RE, Kontoyiannis DP (2008) Delaying amphotericin B-based frontline therapy significantly increases mortality among patients with hematologic malignancy who have zygomycosis. Clin Infect Dis 47:503–509

Chowdhary A, Kathuria S, Singh PK, Sharma B, Dolatabadi S et al (2014) Molecular characterization and in vitro antifungal susceptibility of 80 clinical isolates of mucormycetes in Delhi, India. Mycoses 57(Suppl 3):97–107

Cortez KJ, Roilides E, Quiroz-Telles F, Meletiadis J, Antachopoulos C et al (2008) Infections caused by *Scedosporium* spp. Clin Microbiol Rev 21:157–197

Crous PW (2002) Adhering to good cultural practice (GCP). Mycol Res 106:1378–1379

de Hoog GS, Haase G, Chaturvedi V, Walsh TJ, Meyer W et al (2013) Taxonomy of medically important fungi in the molecular era. Lancet Infect Dis 13:385–386

de Hoog GS, Chaturvedi V, Denning DW, Dyer PS, Frisvad JC et al (2014) Name changes in medically important fungi and their implications for clinical practice. J Clin Microbiol. doi:10.1128/JCM.02016-14

Deckert RJ, Hsiang T, Peterson RL (2002) Genetic relationships of endophytic *Lophodermium nitens* isolates from needles of Pinus strobus. Mycol Res 106:305–313

Denning DW (2001) Chronic forms of pulmonary aspergillosis. Clin Microbiol Infect 7(Suppl 2):25–31

Denning DW, Pleuvry A, Cole DC (2011) Global burden of chronic pulmonary aspergillosis as a sequel to pulmonary tuberculosis. Bull World Health Organ 89:864–872

Denning DW, Pleuvry A, Cole DC (2013a) Global burden of chronic pulmonary aspergillosis complicating sarcoidosis. Eur Respir J 41:621–626

Denning DW, Pleuvry A, Cole DC (2013b) Global burden of allergic bronchopulmonary aspergillosis with asthma and its complication chronic pulmonary aspergillosis in adults. Med Mycol 51:361–370

Dolatabadi S, de Hoog GS, Meis JF, Walther G (2014) Species boundaries and nomenclature of *Rhizopus arrhizus* (syn. *R. oryzae*). Mycoses 57(Suppl 3):108–127

Fekkar A, Brun S, D'Ussel M, Uzunov M, Cracco C et al (2009) Serum cross-reactivity with *Aspergillus* galactomannan and cryptococcal antigen during fatal disseminated Trichosporon dermatis infection. Clin Infect Dis 49:1457–1458

Fungitell (2008) Assay for (1,3)-beta-D-glucan in serum. Associates of Cape Cod Incorporated, E. Falmouth, MA

Garey KW, Rege M, Pai MP, Mingo DE, Suda KJ et al (2006) Time to initiation of fluconazole therapy impacts mortality in patients with candidemia: a multi-institutional study. Clin Infect Dis 43:25–31

Garnacho-Montero J, Amaya-Villar R, Ortiz-Leyba C, Leon C, Alvarez-Lerma F et al (2005) Isolation of *Aspergillus* spp. from the respiratory tract in critically ill patients: risk factors, clinical presentation and outcome. Crit Care 9:R191–R199

Ghebremedhin B, Bluemel A, Neumann KH, Koenig B, Koenig W (2009) Peritonitis due to *Neosartorya pseudofischeri* in an elderly patient undergoing peritoneal dialysis successfully treated with voriconazole. J Med Microbiol 58:678–682

Gilgado F, Cano J, Gene J, Guarro J (2005) Molecular phylogeny of the *Pseudallescheria boydii* species complex: proposal of two new species. J Clin Microbiol 43:4930–4942

Gilgado F, Serena C, Cano J, Gene J, Guarro J (2006) Antifungal susceptibilities of the species of the *Pseudallescheria boydii* complex. Antimicrob Agents Chemother 50:4211–4213

Gilgado F, Cano J, Gene J, Sutton DA, Guarro J (2008) Molecular and phenotypic data supporting distinct species statuses for *Scedosporium apiospermum* and *Pseudallescheria boydii* and the proposed new species *Scedosporium dehoogii*. J Clin Microbiol 46:766–771

Gilgado F, Gene J, Cano J, Guarro J (2010) Heterothallism in *Scedosporium apiospermum* and description of its teleomorph *Pseudallescheria apiosperma* sp. nov. Med Mycol 48:122–128

Gonzalez GM (2009) In vitro activities of isavuconazole against opportunistic filamentous and dimorphic fungi. Med Mycol 47:71–76

Greene RE, Schlamm HT, Oestmann JW, Stark P, Durand C et al (2007) Imaging findings in acute invasive pulmonary aspergillosis: clinical significance of the halo sign. Clin Infect Dis 44:373–379

Guarro J, Kantarcioglu AS, Horre R, Rodriguez-Tudela JL, Cuenca Estrella M et al (2006) *Scedosporium apiospermum*: changing clinical spectrum of a therapy-refractory opportunist. Med Mycol 44:295–327

Hawksworth DL (2011) A new dawn for the naming of fungi: impacts of decisions made in Melbourne in July 2011 on the future publication and regulation of fungal names. IMA Fungus 2:155–162

Hibbett DS, Binder M, Bischoff JF, Blackwell M, Cannon PF et al (2007) A higher-level phylogenetic classification of the Fungi. Mycol Res 111:509–547

Hospenthal DR, Chung KK, Lairet K, Thompson EH, Guarro J et al (2011) *Saksenaea erythrospora* infection following combat trauma. J Clin Microbiol 49:3707–3709

Howard SJ, Cerar D, Anderson MJ, Albarrag A, Fisher MC et al (2009) Frequency and evolution of Azole resistance in *Aspergillus fumigatus* associated with treatment failure. Emerg Infect Dis 15:1068–1076

Jarv H, Lehtmaa J, Summerbell RC, Hoekstra ES, Samson RA et al (2004) Isolation of *Neosartorya pseudofischeri* from blood: first hint of pulmonary Aspergillosis. J Clin Microbiol 42:925–928

Johnson LS, Shields RK, Clancy CJ (2014) Epidemiology, clinical manifestations, and outcomes of *Scedosporium* infections among solid organ transplant recipients. Transpl Infect Dis 16:578–587

Kane J, Summerbell R, Sigler L, Krajden S, Land G (1997) Physiological and other special tests for identifying dermatophytes. Laboratory handbook of dermatophytes. Star Publishing Co, Belmont, CA, pp 45–79

Khare R, Gupta S, Arif S, Jentoft ME, Deziel PJ et al (2014) Misidentification of *Neosartorya pseudofischeri* as *Aspergillus fumigatus* in a lung transplant patient. J Clin Microbiol 52:2722–2725

Kontoyiannis DP, Lewis RE (2006) Invasive zygomycosis: update on pathogenesis, clinical manifestations, and management. Infect Dis Clin North Am 20:581–607, vi

Kontoyiannis DP, Marr KA, Park BJ, Alexander BD, Anaissie EJ et al (2010) Prospective surveillance for invasive fungal infections in hematopoietic stem cell transplant recipients, 2001–2006: overview of the Transplant-Associated Infection Surveillance Network (TRANSNET) Database. Clin Infect Dis 50:1091–1100

Kwon-Chung KJ (1994) Phylogenetic spectrum of fungi that are pathogenic to humans. Clin Infect Dis 19(Suppl 1):S1–S7

Kwon-Chung KJ (2012) Taxonomy of fungi causing mucormycosis and entomophthoramycosis (zygomycosis) and nomenclature of the disease: molecular mycologic perspectives. Clin Infect Dis 54(Suppl 1):S8–S15

Lackner M, de Hoog GS, Verweij PE, Najafzadeh MJ, Curfs-Breuker I et al (2012) Species-specific antifungal susceptibility patterns of *Scedosporium* and *Pseudallescheria* species. Antimicrob Agents Chemother 56:2635–2642

Lackner M, de Hoog GS, Yang L, Moreno LF, Ahmed SA et al (2014a) Proposed nomenclature for *Pseudallescheria*, *Scedosporium* and related genera. Fungal Divers 67:1–10

Lackner M, Hagen F, Meis JF, Gerrits van den Ende AH, Vu D et al (2014b) Susceptibility and diversity in the therapy-refractory genus *Scedosporium*. Antimicrob Agents Chemother 58:5877–5885

Lamaris GA, Chamilos G, Lewis RE, Safdar A, Raad II et al (2006) *Scedosporium* infection in a tertiary care cancer center: a review of 25 cases from 1989–2006. Clin Infect Dis 43:1580–1584

Latge JP, Kobayashi H, Debeaupuis JP, Diaquin M, Sarfati J et al (1994) Chemical and immunological characterization of the extracellular galactomannan of *Aspergillus fumigatus*. Infect Immun 62:5424–5433

Lewis RE, Wiederhold NP, Klepser ME (2005) In vitro pharmacodynamics of amphotericin B, itraconazole, and voriconazole against *Aspergillus*, *Fusarium*, and *Scedosporium* spp. Antimicrob Agents Chemother 49:945–951

Ma LJ, Ibrahim AS, Skory C, Grabherr MG, Burger G et al (2009) Genomic analysis of the basal lineage fungus *Rhizopus oryzae* reveals a whole-genome duplication. PLoS Genet 5, e1000549

Marr KA, Laverdiere M, Gugel A, Leisenring W (2005) Antifungal therapy decreases sensitivity of the *Aspergillus* galactomannan enzyme immunoassay. Clin Infect Dis 40:1762–1769

Matsumoto N, Shiraga H, Takahashi K, Kikuchi K, Ito K (2002) Successful treatment of *Aspergillus* peritonitis in a peritoneal dialysis patient. Pediatr Nephrol 17:243–245

Meersseman W, Vandecasteele SJ, Wilmer A, Verbeken E, Peetermans WE et al (2004) Invasive aspergillosis in critically ill patients without malignancy. Am J Respir Crit Care Med 170:621–625

Mendoza L, Vilela R, Voelz K, Ibrahim AS, Voigt K et al (2014) Human Fungal Pathogens of *Mucorales* and *Entomophthorales*. Cold Spring Harb Perspect Med 5(4)

Mikulska M, Furfaro E, Del Bono V, Gualandi F, Raiola AM et al (2012) Galactomannan testing might be useful for early diagnosis of fusariosis. Diagn Microbiol Infect Dis 72:367–369

Minari A, Husni R, Avery RK, Longworth DL, DeCamp M et al (2002) The incidence of invasive aspergillosis among solid organ transplant recipients and implications for prophylaxis in lung transplants. Transpl Infect Dis 4:195–200

Morelle W, Bernard M, Debeaupuis JP, Buitrago M, Tabouret M et al (2005) Galactomannoproteins of *Aspergillus fumigatus*. Eukaryot Cell 4:1308–1316

Morgan J, Wannemuehler KA, Marr KA, Hadley S, Kontoyiannis DP et al (2005) Incidence of invasive aspergillosis following hematopoietic stem cell and solid organ transplantation: interim results of a prospective multicenter surveillance program. Med Mycol 43(Suppl 1):S49–S58

Morio F, Horeau-Langlard D, Gay-Andrieu F, Talarmin JP, Haloun A et al (2010) Disseminated *Scedosporium/Pseudallescheria* infection after double-lung transplantation in patients with cystic fibrosis. J Clin Microbiol 48:1978–1982

Morrell M, Fraser VJ, Kollef MH (2005) Delaying the empiric treatment of *Candida* bloodstream infection until positive blood culture results are obtained: a potential risk factor for hospital mortality. Antimicrob Agents Chemother 49:3640–3645

Nakamura Y, Suzuki N, Nakajima Y, Utsumi Y, Murata O et al (2013) *Scedosporium aurantiacum* brain abscess after near-drowning in a survivor of a tsunami in Japan. Respir Investig 51:207–211

Neblett Fanfair R, Benedict K, Bos J, Bennett SD, Lo YC et al (2012) Necrotizing cutaneous mucormycosis after a tornado in Joplin, Missouri, in 2011. N Engl J Med 367:2214–2225

Nelson PD, Toussoun TA, Marasas WFO (1983) Growing *Fusarium* species for identification. *Fusarium* species: an illustrated manual for identification. Pennsylvania State University Press, University Park, PA, pp 13–14

Neofytos D, Horn D, Anaissie E, Steinbach W, Olyaei A et al (2009) Epidemiology and outcome of invasive fungal infection in adult hematopoietic stem cell transplant recipients: analysis of Multicenter Prospective Antifungal Therapy (PATH) Alliance registry. Clin Infect Dis 48:265–273

Norvell LL (2011) Melbourne approves a new CODE. Mycotaxon 116:481–490

Nucci M (2003) Emerging moulds: *Fusarium*, *Scedosporium* and Zygomycetes in transplant recipients. Curr Opin Infect Dis 16:607–612

Odabasi Z, Mattiuzzi G, Estey E, Kantarjian H, Saeki F et al (2004) Beta-D-glucan as a diagnostic adjunct for invasive fungal infections: validation, cutoff development, and performance in patients with acute myelogenous leukemia and myelodysplastic syndrome. Clin Infect Dis 39:199–205

Odabasi Z, Paetznick VL, Rodriguez JR, Chen E, McGinnis MR et al (2006) Differences in beta-glucan levels in culture supernatants of a variety of fungi. Med Mycol 44:267–272

Ostrosky-Zeichner L, Alexander BD, Kett DH, Vazquez J, Pappas PG et al (2005) Multicenter clinical evaluation of the (1->3) beta-D-glucan assay as an aid to diagnosis of fungal infections in humans. Clin Infect Dis 41:654–659

Paltauf A (1885) Mycosis mucorina: ein Beitrag zur Kenntnis der menschilchen Fadenpiltzerkrankungen. Virchows Arch Pathol Anat 102:543–564

Petrikkos G, Skiada A, Lortholary O, Roilides E, Walsh TJ et al (2012) Epidemiology and clinical manifestations of mucormycosis. Clin Infect Dis 54(Suppl 1):S23–S34

Petti CA (2007) Detection and identification of microorganisms by gene amplification and sequencing. Clin Infect Dis 44:1108–1114

Pfeiffer CD, Fine JP, Safdar N (2006) Diagnosis of invasive aspergillosis using a galactomannan assay: a meta-analysis. Clin Infect Dis 42:1417–1427

Pickering JW, Sant HW, Bowles CA, Roberts WL, Woods GL (2005) Evaluation of a (1->3)-beta-D-glucan assay for diagnosis of invasive fungal infections. J Clin Microbiol 43:5957–5962

Pincus DH, Salkin IF, Hurd NJ, Levy IL, Kemna MA (1988) Modification of potassium nitrate assimilation test for identification of clinically important yeasts. J Clin Microbiol 26:366–368

Raj R, Frost AE (2002) *Scedosporium apiospermum* fungemia in a lung transplant recipient. Chest 121:1714–1716

Saito K, Saito A, Ohnishi M, Oda Y (2004) Genetic diversity in *Rhizopus oryzae* strains as revealed by the sequence of lactate dehydrogenase genes. Arch Microbiol 182:30–36

Samson RA, Visagie CM, Houbraken J, Hong SB, Hubka V et al (2014) Phylogeny, identification and nomenclature of the genus *Aspergillus*. Stud Mycol 78:141–173

Schoch CL, Seifert KA, Huhndorf S, Robert V, Spouge JL et al (2012) Nuclear ribosomal internal transcribed spacer (ITS) region as a universal DNA barcode marker for Fungi. Proc Natl Acad Sci USA 109:6241–6246

Seifert KA (2009) Progress towards DNA barcoding of fungi. Mol Ecol Resour 9(Suppl s1):83–89

Senn L, Robinson JO, Schmidt S, Knaup M, Asahi N et al (2008) 1,3-Beta-D-glucan antigenemia for early diagnosis of invasive fungal infections in neutropenic patients with acute leukemia. Clin Infect Dis 46:878–885

Smith NL, Denning DW (2011) Underlying conditions in chronic pulmonary aspergillosis including simple aspergilloma. Eur Respir J 37:865–872

Steinbach WJ, Benjamin DK Jr, Kontoyiannis DP, Perfect JR, Lutsar I et al (2004) Infections due to *Aspergillus terreus*: a multicenter retrospective analysis of 83 cases. Clin Infect Dis 39:192–198

Stynen D, Sarfati J, Goris A, Prevost MC, Lesourd M et al (1992) Rat monoclonal antibodies against *Aspergillus* galactomannan. Infect Immun 60:2237–2245

Stynen D, Goris A, Sarfati J, Latge JP (1995) A new sensitive sandwich enzyme-linked immunosorbent assay to detect galactofuran in patients with invasive aspergillosis. J Clin Microbiol 33:497–500

Sugui JA, Vinh DC, Nardone G, Shea YR, Chang YC et al (2010) *Neosartorya udagawae* (*Aspergillus udagawae*), an emerging agent of aspergillosis: how different is it from *Aspergillus fumigatus*? J Clin Microbiol 48:220–228

Sugui JA, Peterson SW, Figat A, Hansen B, Samson RA et al (2014) Genetic relatedness versus biological compatibility between *Aspergillus fumigatus* and related species. J Clin Microbiol 52:3707–3721

Summerbell RC (1993) The benomyl test as a fundamental diagnostic method for medical mycology. J Clin Microbiol 31:572–577

Thompson GR III, Wiederhold NP (2010) Isavuconazole: a comprehensive review of spectrum of activity of a new triazole. Mycopathologia 170:291–313

Tintelnot K, Just-Nubling G, Horre R, Graf B, Sobottka I et al (2009) A review of German *Scedosporium prolificans* cases from 1993 to 2007. Med Mycol 47:351–358

Tortorano AM, Esposto MC, Prigitano A, Grancini A, Ossi C et al (2012) Cross-reactivity of *Fusarium* spp. in the *Aspergillus* Galactomannan enzyme-linked immunosorbent assay. J Clin Microbiol 50:1051–1053

Tortorano AM, Richardson M, Roilides E, van Diepeningen A, Caira M et al (2014) ESCMID and ECMM joint guidelines on diagnosis and management of hyalohyphomycosis: *Fusarium* spp., *Scedosporium* spp. and others. Clin Microbiol Infect 20(Suppl 3):27–46

Verweij PE (2005) Advances in diagnostic testing. Med Mycol 43(Suppl 1):S121–S124

Verweij PE, Gonzalez GM, Wiedrhold NP, Lass-Florl C, Warn P et al (2009) In vitro antifungal activity of isavuconazole against 345 mucorales isolates collected at study centers in eight countries. J Chemother 21:272–281

Vinh DC, Shea YR, Jones PA, Freeman AF, Zelazny A et al (2009a) Chronic invasive aspergillosis caused by *Aspergillus viridinutans*. Emerg Infect Dis 15:1292–1294

Vinh DC, Shea YR, Sugui JA, Parrilla-Castellar ER, Freeman AF et al (2009b) Invasive aspergillosis due to *Neosartorya udagawae*. Clin Infect Dis 49:102–111

Walsh TJ, Groll A, Hiemenz J, Fleming R, Roilides E et al (2004) Infections due to emerging and uncommon medically important fungal pathogens. Clin Microbiol Infect 10(Suppl 1):48–66

Walsh TJ, Anaissie EJ, Denning DW, Herbrecht R, Kontoyiannis DP et al (2008) Treatment of aspergillosis: clinical practice guidelines of the Infectious Diseases Society of America. Clin Infect Dis 46:327–360

Walther G, Pawlowska J, Alastruey-Izquierdo A, Wrzosek M, Rodriguez-Tudela JL et al (2013) DNA barcoding in *Mucorales*: an inventory of biodiversity. Persoonia 30:11–47

Wiederhold NP, Lewis RE (2009) Antifungal activity against *Scedosporium* species and novel assays to assess antifungal pharmacodynamics against filamentous fungi. Med Mycol 47:422–432

Wiederhold NP, Lewis RE, Kontoyiannis DP (2003) Invasive aspergillosis in patients with hematologic malignancies. Pharmacotherapy 23:1592–1610

Zbinden A, Imhof A, Wilhelm MJ, Ruschitzka F, Wild P et al (2012) Fatal outcome after heart transplantation caused by *Aspergillus lentulus*. Transpl Infect Dis 14:E60–E63

# Chapter 19
# Mycotoxins in Food and Feed: A Challenge for the Twenty-First Century

**J. David Miller**

By the turn of the nineteenth century, it was recognized that microfungi produced compounds deleterious to animal and human health, although much of this literature was from the USSR and Japan and not accessible to the English-speaking world (Ceigler and Bennett 1980). From the first use in the literature (circa 1960), the term "mycotoxin" refers to secondary metabolites produced by microfungi that are capable of causing disease and death in humans and other animals. By convention, this excluded mushroom toxins (Bennett and Klich 2003; Miller 1995). When the conditions favor the growth of toxigenic species on crops or food, it is an invariable and unfortunate rule that one or more of the compounds for which the fungus has the genetic potential are produced. A great deal is known about the impact of the agriculturally important mycotoxins on public health and the economics of farming (Pitt et al. 2012; Miller et al. 2014).

Mycotoxins have affected human populations since the beginning of organized crop production. Ergot of rye is mentioned several times in various Biblical texts (Schiff 2006). Epidemics of ergotism were reported in Western Europe from about 800 AD. The screams of the victims, the stench of rotting flesh, extremities falling off, and death all feature in the descriptions of the disease (Van Rensburg and Altenkirk 1974). In 1630, Dr. Thuillier was the first to prove that consumption of ergoty rye caused ergotism. He observed that the intensity of the malady was in proportion to the amount of ergoty grain consumed and that people with more diverse diets suffered less or not at all. He fed ergot sclerotia to chickens, geese, and pigs and they all died. Unfortunately, he did not publish his results. It was left to his son, also a physician, working with the Paris Academy of Sciences to repeat the experiments (and publish). More than a century after this, L'Abbé Tessier proposed cultivation of potatoes instead of rye, improved drainage, and the enforced cleaning of grains. About the same time, Johann Taube eliminated ergotism in patients by

J.D. Miller, Ph.D. (✉)
Department of Chemistry, Carleton University, Ottawa, ON, Canada K1S 5B6
e-mail: david.miller@carleton.ca



D.-W. Li (ed.), *Biology of Microfungi*, Fungal Biology,
DOI 10.1007/978-3-319-29137-6_19

controlling the quality of bread in hospitals in Gottingen (Stroup 1990; van Dongen and de Groot 1995).

There are only five agriculturally important mycotoxins with worldwide distribution: aflatoxin, deoxynivalenol, fumonisin, zearalenone, and ochratoxin. In cool temperate and generally wet areas, T-2 toxin /HT-2 toxin can be a problem. Similarly, although ergot sclerotia are efficiently removed during milling, ergot alkaloids can occur in processed cereals in affected regions. Excepting in the case of mycotoxins in corn that is nixtamalized (De La Campa et al. 2004), all of these toxins are stable in the processes typical of food and feed processing whether in cereals or nuts (Bullerman and Bianchini 2007; Kabak 2009; Park 2004). Pork products can be a minor dietary source of ochratoxin. For the remaining toxins, animal sources are not important under normal circumstances (Perši et al. 2014; Prelusky 1994). Milk can contain aflatoxin M-1 a mammalian metabolite of aflatoxin B1 albeit with much lower potency (IARC 2012).

Additionally, there are a number of *Penicillium* toxins that affect animal health that occur in silage (Nielsen et al. 2006; Rasmussen et al. 2010; Sumarah et al. 2005; Tangni et al. 2013), mold-damaged recycled food (Rundberget et al. 2004), and sometimes corn left in the field after harvest (Sumarah et al. 2005).

## Deoxynivalenol, Nivalenol, and Zearalenone

"Red mold poisoning" was reported in rural Japan throughout the 1950s. Eventually, deoxynivalenol (DON) was discovered by Japanese researches from grain that had made humans ill (Morooka et al. 1972). The same chemical was subsequently re-reported as "vomitoxin" from *Fusarium graminearum*-contaminated corn in 1973 (Vesonder et al. 1973). Large-scale acute human toxicoses from deoxynivalenol have occurred in modern times in India (Bhat et al. 1989), China, and Korea among other countries (Beardall and Miller 1994; Miller 2008). Corn contaminated by *F. graminearum* was associated with estrogenic symptoms particularly in swine from the 1920s. An active fraction was isolated from corn and ultimately the chemical structure of zearalenone was reported in 1966 (Caldwell et al. 1970).

### *The Fungi*

These toxins occur when wheat, barley, corn, and sometimes oats and rye are infected by *Fusarium graminearum* and *F. culmorum* or *F. asiaticum*. Incidence of Fusarium Head Blight is most affected by moisture at anthesis under warm conditions (Schaafsma and Hooker 2007; Sutton 1982). *F. culmorum* is associated with cooler growing conditions (Miller 1994). The same fungi cause a similar disease in corn called Gibberella or pink ear rot. Disease incidence is most affected by moisture at flowering/silk emergence. In the USA and Canada and southern China,

Fusarium head blight is typically caused by *F. graminearum*, which is common in wheat grown in North America and China. In the 1980s, *F. culmorum* was the dominant species in cooler wheat-growing areas, such as the UK, Finland, France, Poland, and The Netherlands (Daamen et al. 1991; Miller 1994), but in recent years, warmer summers have resulted in dominance by *F. graminearum* (van der Fels-Klerx et al. 2012; Xu et al. 2005).

*F. graminearum* comprises a number of closely related clades (O'Donnell et al. 2004), but *F. graminearum* sensu stricto and *F. asiaticum* appear most common. Morphologically identical isolates of *F. graminearum* sensu lato and *F. culmorum* can produce either deoxynivalenol (DON) and zearalenone or nivalenol and zearalenone as the principal toxic metabolites that accumulate in grain. Within the former group, some strains produce DON by the 3-acetylated precursor and others make the 15 acetylated precursor. Historically, DON-producing strains with the 15-acetylated precursor dominated in Canada, the USA, Mexico, and South America (Miller et al. 1983a, 1991; Mirocha et al. 1989; Piñeiro et al. 1995). The Asian and New World strains are genetically distinct (O'Donnell et al. 2000). In China and Japan, strains that produced DON generally did so via 3-ADON (Miller et al. 1991; Yoshizawa and Jin 1995). There are few older data from Europe, but these suggest that 3ADON chemotypes were more common (Logrieco et al. 1988; Miller et al. 1991). Recent surveys in Europe indicate that most (80 %) isolates of *F. graminearum* produce DON via 15-ADON, but most isolates of *F. culmorum* produce nivalenol and those that produce DON do so via 3-ADON (Somma et al. 2014; Vogelgsang et al. 2009). So far, nivalenol-producing strains of *F. graminearum* are rare in Canada (Tanaka et al. 1988) and the USA (Gale et al. 2011; Schmale et al. 2011). The toxin is not found in Canadian grain (Tittlemier et al. 2013). In addition, nivalenol- producing *F. asiaticum* occurs in the southeast USA. This had hitherto only been known in Japan and China (Gale et al. 2011).

The Asian chemotypes were probably introduced to eastern Canada from wheat cultivars imported from Europe to the Maritimes in the 1970s and 1980s. The first 3-ADON strain in the Canadian National Mycological Herbarium (DAOM) was deposited in 1969. Another source of the strains was from breeding material brought in from China and Europe. By 2007, nearly 100 % of the strains in Atlantic Canada were Asian as opposed to the native 15-ADON-producing strains. A similar situation has emerged in Manitoba and Saskatchewan. As of 2007, the 3-ADON genotype accounted for ca. 70 % of a large number of strains examined (Kelly et al. 2015). In this study, 5.8 % of the strains tested were from Ontario and, where given, were mainly from soft wheat. Where the county of harvest was provided (Kelly et al. 2015), these are soft wheat, shorter season areas. *F. graminearum* strains were isolated in a survey in 2010 of 24 fields in Essex, Chatham-Kent, Hamilton-Wentworth, Brant, Elgin, Middlesex, Huron, Perth, Wellington, and Lambton Counties. Almost all (97 %) of 155 isolates of *F. graminearum* were of the 15-ADON chemotype; the remainder were the 3-ADON genotype (Tamburic-Ilincic et al. 2015). In upstate New York, the 15-ADON genotype also predominated, but there were important differences: some 92 % (652/709) at near Lake Ontario, 78 % (332/379) somewhat inland in New York, and 53 % (167/319) at a location near Lake Champlain.

The authors suggested that “regional populations may be differentiated based on selection associated with climatic or landscape features not currently identified” (Kuhnem et al. 2015a). A companion study indicated that host (corn, wheat) “did not appear to structure the populations” examined (Kuhnem et al. 2015b).

Isolates from the 3-ADON chemotype produce significantly more DON + 3-ADON and are more fecund and have higher growth rates than isolates from the 15-ADON chemotype (Ward et al. 2008). These conclusions have been confirmed in a number of field studies in wheat in Canada (Amarasinghe et al. 2015: Gilbert et al. 2002, 2010; Tamburic-Ilincic et al. 2008; von der Ohe et al. 2010) and in the adjacent USA (Schmale et al. 2011). Clear et al. (2013) reported studies that may provide some information of the impact of this in practical terms. They inoculated a barley field with strains of the two chemotypes. Three years later, the prevalence of the two chemotypes changed such that the 3-ADON chemotype dominated. Further, the highest kernel DON concentrations were associated with the increased frequency of the 3-ADON chemotype. In general though, the aggressiveness between the two strains appears to be similar (Spolti et al. 2014; Kuhnem et al. 2015b). Thus, it appears that modest differences in temperature and perhaps rainfall play a role in the distribution of the two important clades (Backhouse 2014; Del Ponte et al. 2015; Zhang et al. 2012), but this remains unclear (e.g., Gilbert et al. 2014; Kuhnem et al. 2015a; Panthi et al. 2014).

However, genetic changes are also occurring. When the two chemotypes were recognized in 1983 (Miller et al. 1983a, 1991), the two populations were genetically homogeneous (O’Donnell et al. 2000); however, a shift has taken place towards a population of genetically divergent strains (Gale et al. 2007; Mishra et al. 2009). In China, it appears that this is manifest by a rapid shift under way to more virulent populations of *F. graminearum* sensu lato (Zhang et al. 2012). Strains that are crosses of the two ADON chemotypes but produce 15 ADON are more virulent than either parent (Foroud et al. 2012). In addition, the so-called Northland population originally isolated and first described in the Midwest USA does not produce DON or nivalenol but a related trichothecene, 3α-acetoxy, 7α,15-dihydroxy-12,13-epoxytrichothec-9-ene (Varga et al. 2015). So far, this population appears uncommon, with isolates collected in Minnesota, North Dakota, and South Dakota (Liang et al. 2014) as well as Québec, Ontario, PEI, and Saskatchewan (Kelly et al. 2015).

## The Toxins

*Deoxynivalenol (DON)* Because there can be high human exposure to this toxin, it has been a serious global challenge for four decades. Consequently, it has been necessary to carefully understand the toxicity of DON. In relevant animal models, DON is excreted rapidly in the urine, and depending on dose, a small amount may be excreted in feces after conjugated. DON is not found in meat, milk, or eggs (Miller 2008; Pestka 2010). The Joint Expert Committee on Food Additives and Contaminants of the FAO and WHO (JECFA) Provisional Maximum Tolerable

Daily Intake (PMTDI) is based on weight reduction in a 2-year study in male and female B6C3F1 mice (Iverson et al. 1995; JECFA 2011). At high doses, DON results in emesis and anorexia in humans, swine, and mink (Miller 2008). The minimum oral dose required for emesis in swine is in the order of 100 mg kg/bw (Pestka et al. 1987). The emetic response in dogs appears to occur at a similar dose (Ueno 1983). The mechanism for the emetic effect appears to be mediated by the effect of DON on peptide YY3-36 and 5-hydroxytryptamine (Wu et al. 2013). There are a number of mechanisms that result in DON-induced anorexia. DON modulates the insulin-like growth factor acid-labile subunit expression (Amuzie and Pestka 2010). Additionally, DON results in neuroendocrine signaling at the enteric and central levels (Pestka 2010). DON crosses the blood–brain barrier resulting in nausea resulting from inflammation and resulting cytokine upregulation in the appetite center (Bonnet et al. 2012). Chronic exposure above the PMTDI results in loss of intestine cell wall integrity, mucosal immune function, and immunological impairment (Pestka 2010; Pinton and Oswald 2014).

The established provisional maximum tolerable daily intake limit (PMTDI) for DON is 1 μg/kg body weight/per day on the basis of the NOAEL of 100 μg/kg bw per day in the Canadian 2-year feeding study of mice and a safety factor of 100 (JECFA 2001). This was modified in 2010 to include both acetates (3- and 15-ADON) for a group PMTDI (JECFA 2011). In 1993, IARC classified DON as a category 3, which is not classifiable as to its carcinogenicity to humans, and no data have emerged to change this determination (IARC 1993; JECFA 2001, 2011).

*Nivalenol* Much less is known about the toxicity of nivalenol, although it is broadly assumed to be similar to that of DON (Pestka 2010; Sugita-Konishi and Nakajima 2010). Its emetic potential is ca. 1/10th that of DON (Wu et al. 2013). There are a number of chronic studies in mice that have been conducted that have provided lowest observed adverse effect levels (LOEAL) but not NOAELs. EFSA and the Food Safety Commission of Japan set TDIs for nivalenol 1.2 μg/kg bw/day and 0.4 μg/kg, respectively (FSCJ 2010; EFSA 2013).

*Zearalenone* This compound is an estrogen analogue and causes hyperestrinism in female pigs and is a potent reproductive toxicant; the dietary no effect level is < 0.1 mg/kg bw/day. Cows and sheep are also sensitive to the estrogenic effects of this toxin with depressed ovulation and lower lambing percentages. The dietary NOAELs in domestic animals other than pigs are not clearly known (Fink-Gremmels and Malekinejad 2007; Metzler et al. 2010; Prelusky et al. 1994). Zearalenone efficiently binds to both mammalian estrogen receptors in a manner similar to 17β-estradiol (Takemura et al. 2007). Zearalenone has been implicated in several incidents of precocious pubertal changes in girls in Europe and South America (Falkay et al. 1993; Schiefer 1990). There is some evidence that zearalenone exposure affects reproductive health in young women. Massart et al. (2008) found that girls living in a part of Italy where corn can be contaminated by zearalenone had precocious puberty. In a small cohort, this was related to plausible serum concentrations of zearalenone in their serum (based on the pharmacokinetics and contamination levels in the region). There were other possible factors identified in the study, but a role of zearalenone could not be ruled out (Massart and Saggese 2010). Similarly, a role for zearalenone in the same phenomenon could not be excluded from a subsequent study in China (Deng et al. 2012). EFSA (2011a) concluded that there were a number of reports linking zearalenone to human disease that are biologically plausible, but the data are inconclusive. The PMTDI for zearalenone is 0.5 μg/kg bw based on the NOEAL of a 15-day study in pigs and a safety factor of 100 (JECFA 2000). The EU Scientific Committee on Food (SCF) established a temporary TDI of 0.2 μg/kg per day (EFSA 2011a).

Aside from the regulated toxins, DON, or nivalenol and zearalenone, *F. graminearum* and related species produce a wide variety of metabolites from several biosynthetic families (Miller et al. 1991). This includes other apotrichothecenes (Greenhalgh et al. 1989), calonectrins (Greenhalgh et al. 1985, 1986), sambucinol, sambucoin (Greenhalgh et al. 1986), and culmorins (Kasitu et al. 1992) and butenolide.

## Fumonisin

### *The Fungi*

Discovered only in 1988, fumonisins (Marasas 2001) are produced by *Fusarium verticillioides* (formerly *F. moniliforme*), *F. proliferatum,* and *F. fujikuroi* as well as some uncommon species, *F. anthophilum*, *F. dlamini*, *F. napiforme*, and *F. thapsinum* (Rheeder et al. 2002; Suga et al. 2014). Fumonisin can be a contaminant of rice resulting from infection by *F. proliferatum* and *F. fujikuroi* (Uegaki et al. 2014). However, most fumonisin exposure results from consumption of corn affected by the disease Fusarium Kernel Rot. *Fusarium verticillioides* or *F. proliferatum* occurs systemically in leaves, stems, roots, and kernels and can be recovered from virtually all maize kernels worldwide including those that are healthy (Miller 2001, 2008). Fumonisin can only accumulate in stressed or senescing kernel tissue under warm conditions and dry conditions between silking and early grain fill (Miller 2001). Insect damage is associated with increased fumonisin concentrations under permissive conditions. There is a strong, consistent relationship between insect damage and Fusarium kernel rot, although the insect species can vary by location (De La Campa et al. 2005; Parsons and Munkvold 2010, 2012). Maize genotypes containing the Bt protein have reduced amounts of fumonisin compared to non-Bt genotypes (De La Campa et al. 2005; Hammond et al. 2004). There are a number of naturally occurring fumonisins. In order of decreasing concentration in affected corn, these are fumonisin B1, FB2, FB3, and FB4 (Pitt et al. 2012).

Fumonisin B2 and B4 are produced by *Aspergillus niger*. These toxins have been found in grapes (Knudsen et al. 2010) and other dried fruits (Somma et al. 2012), notably figs (Heperkan et al. 2012; Moretti et al. 2010), and in wine (Logrieco et al. 2010; Mogensen et al. 2010).

## *The Toxins*

Consumption of corn contaminated by *F. verticillioides* (previously *F. moniliforme*) by horses was long associated with equine leukoencephalomalacia (ELEM), a liquefactive necrosis of the brain leading to death. It was not until the discovery of fumonisin that the cause was identified (Marasas 2001; Wilson et al. 1990). Aside from ELEM, fumonisin causes porcine pulmonary edema in pigs (Haschek et al. 2001). FB1 is toxic to the liver in all species and the kidney in a range of laboratory and farm animal species, causing apoptosis followed by mitosis in the affected tissues. FB1 is also toxic to the cardiovascular system in pigs and horses. FB1 and other fumonisins inhibit ceramide synthase in all species including laboratory and farm animals and disrupt sphingolipid metabolism, a process underlying the mechanism of toxicity and pathogenesis of fumonisin-related diseases (Voss et al. 2007; Voss and Riley 2013). In most assays, FB1 is most potent followed by FB2, FB3, and FB4.

The discovery of fumonisin resulted from studies of an area of South African with high corn consumption and very high rates of esophageal cancer. Fumonisin proved to be a potent cancer promoter and is a rodent carcinogen. The linkage between fumonisin and human cancer remains unclear (Bulder et al. 2012; Marasas 2001). Fumonisin has also been associated with neural tube birth defects in several populations mainly in Africa, China, as well as parts of Latin America and Mexico. Based on studies in two strains of rats and two non-human primate species, fumonisin does not cross the placenta (IPCS 2000). There is, however, limited epidemiological evidence suggesting that fumonisin exposure during early gestation could be involved in the increased incidence of neural tube defects (NTD) in areas of the world where maize is consumed in large amounts and diets are likely to be deficient in critical micronutrients and vitamins necessary for normal neural tube closure (Gelineau-van Waes et al. 2009; Marasas et al. 2004; Suarez et al. 2012). These epidemiological data are supported by studies in mice which have provided a partial mechanistic basis. Low folate is a risk factor for neural tube birth defects. Fumonisin in the serum prior to the development of the placenta blocks the transport of folate into the cells of the developing embryo and affects the neural crest cells (Gelineau-van Waes et al. 2005, 2012; Marasas et al. 2004; Voss et al. 2014). However, at the time of writing, there is no direct evidence that fumonisin causes NTD in humans.

Two studies from Tanzania suggest that fumonisin exposure may be associated with stunting in children. In this prospective cohort study, infants were enrolled at 6 months of age and followed until 12 months of age. Exposure was categorized as high or low, using the JECFA provisional maximal tolerable dietary intake (PMTDI). By 1 year, the highly exposed infants were significantly shorter and lighter than the 105 infants with low exposure, after controlling for total energy and protein intakes, gender, and village (Kimanya et al. 2010). Shirima et al. (2014) measured fumonisin and aflatoxin exposure. They reported that there was an inverse association between urinary FB1 and growth after 1 year. In this study, the aflatoxin exposures did not result in stunting. Fumonisin is known to affect gut cell function (Bouhet and Oswald 2007), which may provide support for these epidemiological data.

In 2012, the WHO/FAO Joint Expert Committee on Food Additives (Bulder et al. 2012) evaluated the existing human epidemiology studies linking fumonisin exposure to NTDs and concluded that the results in combination with what is known about the toxicology of fumonisin "…indicates that fumonisin exposure in pregnant women may be a contributing factor to increased NTD risk in their babies." Fumonisin B1 (FB1) has been classified as a Group 2B carcinogen, possibly carcinogenic to humans (IARC 2002). The established provisional maximum tolerable daily intake limit for the fumonisins (FB1, FB2, and FB3) from all sources is 2 μg/kg body weight/per day on the basis of the NOEL of 0.2 mg/kg bw per day and a safety factor of 100 (JECFA 2011).

Fumonisin and aflatoxin frequently co-occur and concurrent exposures can be very high in parts of the world dependent on maize as a staple food including for example Nigeria (Adetuniji et al. 2014), Kenya (Mutiga et al. 2014), and Guatemala (Torres et al. 2014). There are a number of implications of this noted following.

## Aflatoxin

### *The Fungi*

Aflatoxin is produced by 13 species of fungi, most of which are of no relevance to agriculture. This list currently includes *Aspergillus pseudotamarii*, *A. nomius*, *A. bombycis*, *A. parvisclerotigenus*, *A. minisclerotigenes*, *A. arachidicola*, *A. ochraceoroseus*, *A. rambellii*, *Emericella astellata*, *E. venezuelensis*, and *E. olivicola* (Varga et al. 2009). *Aspergillus flavus* was first recognized to cause aflatoxicosis in domestic animals (Wogan 1966) and is the most important species. *A. flavus* produces aflatoxin B1 and B2 and is a problem in many commodities, but most human exposure comes from contaminated corn, peanuts, and rice. The second important producer of aflatoxin, *A. parasiticus*, produces aflatoxin B1, B2, G1, and G2 and is primarily associated with peanuts in the Americas but also can occur on corn, figs, and pistachios (Horn 2003).

*A. flavus* infection of corn or peanuts results from (1) airborne or insect-transmitted conidia that contaminate the silks and grow into the ear when the maize is under high temperature stress and (2) insect or bird damaged kernels that become colonized with the fungus and accumulate aflatoxin. Drought-, nutrient-, or temperature-stressed corn or peanut plants are more susceptible to colonization by *A. flavus* or *A. parasiticus* (Guo et al. 2008; Horn 2007; Pitt et al. 2012; Wu et al. 2011). Studies in the USA Corn Belt have shown that, typically, *A. flavus*-infested kernels are randomly distributed in ears resulting mainly by insect damage (Payne and Widstrom 1992; Smart et al. 1990). The kernels that were the sites of initial infection have very high aflatoxin contents. In contrast, the ecology of *A. flavus* in corn in subtropical regions of Asia and the USA is different from that in the Midwest USA. In subtropical areas, apparently healthy kernels are infected (Miller 1995; Pitt et al. 2012).

It has been long suggested that many if not most species of *Aspergillus* have a tropical to subtropical distribution or more precisely abundance peaks in the subtropics (Christensen and Tuthill 1985). A systematic review of the prevalence of 52 species from plating data indicated that 30 occurred above expected frequencies in the tropical latitudes and hence were less common in higher latitudes. The data were separately considered by ecozone: forest, grassland, desert, and cultivated land. Of these, grassland had the lowest diversity (Klich 2002).

The genetics of *Aspergillus flavus* and *A. parasiticus* has been studied in great detail. In the past few years, this information has enabled informative studies on the ecology of these species. Studies of mating type distribution, female sterility, and aflatoxigenicity in *A. flavus* and *A. parasiticus* done on a global level revealed that in cooler wet areas, atoxigenic strains predominate in clonal *A. flavus* populations and toxigenic strains predominate in warm dry areas (Olarte et al. 2012). In fertile strains, there is both qualitative and quantitative variation in AF chemotypes ($B_1$, $B_2$, $G_1$, $G_2$, *o* methyl sterigmatocystin, and cyclopiazonic acid; Moore et al. 2013; Olarte et al. 2012, 2015).

This modern evidence of the high prevalence of *A. flavus* and *A. parasiticus* in warm and dry areas supports older data resulting from traditional plating methods. Propagule densities are much higher in cultivated fields and desert ecosystems are than in forests and prairies (Horn 2003; Klich 2002). In a study in Missouri (39° N), the maximum number of *A. flavus* and *A. parasiticus* propagules was found in soil cropped to a rotation of wheat, red clover, and corn using conventional tillage practices. No isolates were observed in a virgin prairie soil (Angle et al. 1982). Apart from that, the prevalence of aflatoxigenic strains is known to be associated with latitude. Both the occurrence of the two species and the percentage that produced aflatoxin were examined in a transect from northern Japan to Indonesia. A modest percentage of strains of the two species were isolated in soil from the north of Japan (43° N) and, of these, a negligible percentage produced aflatoxin. In contrast in Indonesia (6.6° S), a high percentage of soil fungi were *A. flavus* and almost all were good producers of aflatoxin (Manabe and Tsuruta 1978). In warm areas, as fields are converted to corn and peanuts, the prevalence of aflatoxigenic species increases (Horn 2003).

Aflatoxin is not a problem in colder corn-growing areas as, for example, Canada. There are few data on the occurrence of *A. flavus* in Canadian soils. Bisby et al. (1933) reported the isolation of kojic acid-producing strains from Manitoba soils. A study of soil fungi in eastern Ontario from fields planted to alfalfa and old agricultural soils did not report *A. flavus* (Keller and Bidochka (1998). Soil fungi were plated from a growing corn field in Ottawa (see Miller et al. 1983b) and *A. flavus* was found as an occasional component of the mycoflora (Miller unpublished data). Corn feed samples collected in seven midwest states in 1988–1989 were analyzed for fungi and incubated under adverse storage conditions followed by analysis for mycotoxins. Corn from Michigan did not contain aflatoxin initially or after storage as above. Samples were positive in two states just south of Ontario, Ohio, and Minnesota (Russell et al. 1991).

## *The Toxins*

There are many detailed reviews of the toxicology of aflatoxin including that of IARC (Pitt et al. 2012). Aflatoxin B1, the most toxic of the aflatoxins, causes a variety of adverse effects in different animal species, especially chickens. In poultry, these include liver damage, impaired productivity and reproductive efficiency, decreased egg production in hens, inferior egg-shell quality, inferior carcass quality, and increased susceptibility to disease (Wyatt 1991). Swine are somewhat less sensitive than poultry species with the $LD_{50}$ being perhaps half of that of chickens. Aflatoxin is hepatotoxic and its acute and chronic effects in swine are largely attributable to liver damage (Armbrecht 1978). In cattle, the primary symptom is reduced weight gain as well as liver and kidney damage. Milk production is reduced (Keyl 1978). Aflatoxin is also immunotoxic in domestic and laboratory animals with oral exposures in the ppm range. Cell-mediated immunity (lymphocytes, phagocytes, mast cells, and basophils) is more affected than humeral immunity (antibodies and complement; Bondy and Pestka 2000). The effects of aflatoxin on laboratory animals have been exhaustively reviewed by IARC (Pitt et al. 2012). Cyclopiazonic acid is also produced by most strains of *A. flavus* and some related species and accumulates in the crop along with aflatoxin. This compound is toxic and immune suppressive in various strains of mice and rats as well as swine and poultry (Burdock and Flamm 2000; de Waal 2002; King et al. 2011).

Naturally occurring mixtures of aflatoxins were classified as class 1 human carcinogens and aflatoxin B1 is also a class 1 human carcinogen. There is inadequate evidence of the human carcinogenicity of aflatoxin M1, the metabolite of aflatoxin B1 found in human and animal milk (IARC 2012). Aflatoxin exposure explains approximately 25 % of liver cancer globally notably in Africa (40 %) and Asia (27 %; Liu and Wu 2010). Many people in developing countries are seropositive for hepatitis B and C which are also liver carcinogens. Although aflatoxin is a potent chemical carcinogen, its ability to alter response to the hepatocarcinogenic viruses is perhaps of greater importance. The relative rates of liver cancer in hepatitis B positive populations are an order of magnitude greater (~60×) when exposed to aflatoxin. This is because the toxin interferes with the processing of the virus. Thus, reducing aflatoxin exposure to non-detectable levels could reduce HCC cases in high-risk areas by ~25 % (Liu et al. 2012).

Aside from the carcinogenicity of aflatoxin, exposure may be associated with child stunting. Two studies have been reported involving 680 children living in West Africa. Height and weight for age were lower in a dose-dependent fashion for increasing aflatoxin exposures as measured by the aflatoxin–albumin adduct (AF-alb) in serum (Gong et al. 2002). In a longitudinal study, over a period of 8 months, children with the highest aflatoxin exposures had the smallest gains in height (Gong et al. 2004).

As noted above, in Africa and parts of Latin America, co-exposure to aflatoxin and fumonisin at multiples of the acceptable limits is common. AFB1 is a potent mutagen and DNA-reactive carcinogen, while FB1 is an effective cancer promoter (Gelderblom et al. 1988) with a non-genotoxic mechanism of action (Bulder et al. 2012). It is, therefore, reasonable to believe that co-exposure is likely to enhance hepatoxicity and hepatocarcinogenicity in humans.

## Ochratoxin

### *The Fungi*

In cereals, *Penicillium verrucosum* is the sole producer of ochratoxin (OTA). *P. nordicum* is known to produce OTA on dried salted meats (Cabañes et al. 2010; Sonjak et al. 2011. Many species of *Aspergillus* produce OTA. In the section Circumdati, some 13 species are good producers of ochratoxin A: *A. affinis*, *A. cretensis*, *A. fresenii*, *A. muricatus*, *A. occultus*, *A. ochraceopetaliformis*, *A. ochraceus*, *A. pseudoelegans*, *A. pulvericola*, *A. roseoglobulosus*, *A. sclerotiorum*, *A. steynii*, and *A. westerdijkiae*. Some other species are variable producers of this toxin. Agricultural products damaged by *A. ochraceus*, *A. steynii*, and *A. westerdijkiae* are of greatest concern for OTA contamination (Visagie et al. 2014). The geographic distribution of some of these species remains unclear. OTA contamination of cocoa, coffee, and grapes and wine results from A. *carbonarius* and some strains of *A. niger* and this is an important problem mainly in warm grape-growing areas. The other species in section *Nigri* that produces OTA is *A. sclerotioniger* (isolates from coffee; Varga et al. 2011).

Ochratoxin in coffee and cocoa can be minimized by good agricultural practices during harvest, drying, and processing followed by sorting out damaged beans (Copetti et al. 2013; Taniwaki 2006). Preventing OTA contamination in grapes and wine is somewhat more difficult. The occurrence of the relevant species of black aspergilli is affected by several factors including weather, agronomic practice, and fungicide use (Battilani et al. 2003; Hocking et al. 2007). Grape damage from insects (Cozzi et al. 2006) and fungal diseases (e.g., mildew, Botrytis bunch rot) and rain damage predispose the grape to colonization by A. *carbonarius* and *A. niger*. Harvesting grapes with minimal damage, rapid processing, and good sanitation practices reduce the risk of OTA (Hocking et al. 2007; Visconti et al. 2008).

A few percent of surface-disinfected wheat and barley kernels collected at harvest in the UK, Denmark, and Sweden were contaminated by *P. aurantiogriseum* and *P. verrucosum* and this was similar in studies done in western Canada over many years. Infestation of some kernels by the ochratoxin-producing fungus *P. verrucosum* occurs from anthesis and surface contamination is common at harvest. The absolute level of pre-harvest infestation varies according to site and season (Miller 1995; Elmholt 2003).

## *The Toxin*

Ochratoxin is a potent nephrotoxin in swine and causes kidney cancer in male Fisher 344 rats (Benford et al. 2001). Pigs are affected at low exposures in terms of kidney damage, but typically there are no overt signs or biochemical/hematological changes. At higher concentrations (>2 μg/g), decreased weight grains occur (Prelusky et al. 1994; Stoev et al. 2002). Poultry are similarly affected with reduced growth rate and egg production at low ochratoxin concentrations>2 μg/g. Higher dietary ochratoxin concentrations are often fatal. Cattle are resistant to the ochratoxin concentrations typical of naturally contaminated grain (Prelusky et al. 1994). Ochratoxin is often found with other toxins such as citrinin, penicillic acid, and the naphthoquinone mycotoxins from *Penicillium aurantiogriseum* (Krogh 1991; Vrabcheva et al. 2000). Citrinin mimics the effects of ochratoxin, although it is less potent (Krogh 1991). The naphthoquinones xanthomegnin and viomellein from *P. aurantiogriseum* are nephrotoxic (Carlton et al. 1976). Interactions have been reported between ochratoxin and citrinin and ochratoxin and penicillic acid on renal toxicity in swine (Prelusky et al. 1994; Stoev et al. 2001).

Human exposure to ochratoxin primarily occurs from whole-grain breads. Some exposure comes from the consumption of animal products, especially pork and pig-blood-based products (Kuiper-Goodman and Scott 1989; Kuiper-Goodman et al. 2010). Ochratoxin has been long suspected as the cause of urinary tract cancers and kidney damage in areas of chronic past chronic exposure in parts of Eastern Europe (Anonymous 1977; Kuiper-Goodman and Scott 1989; IARC 1993). After investigations of Balkan Endemic Nephropathy (BEN) began in the late 1960s, various etiologies were suggested including viral disease. With the recognition that ochratoxin could produce serious renal pathology in swine, the ochratoxin hypothesis was strongly advanced by researchers (e.g., Krogh and Elling 1976; Pfohl-Leszkowicz et al. 2002). Various mechanisms have been advanced to explain the carcinogenicity including genotoxicity and the putative linkage to BEN (Haighton et al. 2012; Pfohl-Leszkowicz and Manderville 2011).

However, after 50 years, there is little evidence for OTA as causing human disease (Bui-Klimke and Wu 2014a). Historically, OTA has been a common exposure in parts of Eastern Europe where BEN occurs; exposure to aristolochic acid also may be common in some years. This is a potent renal toxicant and IARC class 1 human carcinogen. The literature suggested that *Aristolochia* was not an important weedy species in the region. However, that is the case (Dimitrova 2009; Markovic et al. 2005; Mehmeti et al. 2009). Seed set occurs just prior to harvest; thus, exposure to aristolochic acid can occur in the region where BEN has been seen. Aristolochic exposure results in tumors with a characteristic genetic signature which are seen in at least some BEN patients (Grollman et al. 2007; Savin et al. 2014). Bui-Klimke and Wu (2014b) concluded that on a weight of evidence basis, aristolochic acid is the probable cause of BEN (see also Wu and Wang 2013).

The last IARC evaluation of ochratoxin determined it to be a possible human carcinogen (IARC 1993). In 1995, the JECFA established PTWI of 100 ng/kg bw per week (Benford et al. 2001). In reaching this conclusion, "the Committee noted the large safety factor applied to the NOEL for nephrotoxicity in deriving the PTWI, which corresponds to a factor of 1500 applied to the NOEL for carcinogenicity in male rats, the most sensitive species and sex for this end-point." The last JECFA evaluation retained the PMTI which was supported using a study demonstrating the LOEAL in swine based on nephrotoxicity (Stoev et al. 2002). The JECFA panel remarked that "Although an association between the intake of ochratoxin A and nephropathy in humans has been postulated, causality has not been established." The current PTWI is 100 ng/kg bw (Benford et al. 2001).

## Other Mycotoxins that Can Occur in Food or Feed

### *T-2 Toxin*

T-2 toxin was isolated from "strain T-2" of *F. sporotrichioides* mis-identified as *F. tricinctum* isolated from corn associated with cow mortalities (Ueno 1983; Marasas et al. 1984). *F. sporotrichioides* also produces HT-2 toxin and typically both are present in affected commodities along with various other compounds (Greenhalgh et al. 1988). T-2 is also produced by *F. acuminatum* (Miller 1994). T-2 toxin has been the subject of considerable toxicological study (IARC 1993) because it is easy to isolate and purify. During World War II, there were large-scale poisonings of the rural population in the former Soviet Union caused by the consumption of overwintered grains left in the field over winter (estimates range to 1,000,000 victims). The disease was called Alimentary Toxic Aleukia (Joffe 1971). Samples of extracts made at the time were shown to contain T-2 and related toxins (Mirocha and Pathre 1973). In cool temperate and generally wet areas, T-2 toxin /HT-2 toxin can be a problem (Bertuzzi et al. 2014; Miller 1995; Pettersson et al. 2011). The toxicology of these trichothecenes toxin is similar in character to that described above for deoxynivalenol and is reviewed extensively by EFSA (2011b).

### *Patulin*

Patulin is primarily found in apple and grape juices where it occurs from the growth of *Penicillium expansum* on rotted fruit (Menniti et al. 2010). Aside from patulin, this fungus produces citrinin, chaetoglobosins, communesins, roquefortine C, and expansolides A and B (Andersen et al. 2004). Patulin is easily controlled by removing rotted fruit (Menniti et al 2010) and is further reduced by processing (Welke et al. 2009). The sparse toxicological data on this compound have been reviewed by Puel et al. (2010) and Brandon et al. (2012). The JEFCA PMTI is 400 ng/kg BW day which is based on growth retardation in mice (JECFA 1995).

## *Ergot Alkaloids*

Ergot alkaloids seldom appear in meaningful concentrations in food samples in the North America or Europe because the presence of *Claviceps* sclerotia in grains is a grading factor. Ergotism in humans has only rarely been reported in modern times in France, India, and parts of Africa (Beardall and Miller 1994; Belser-Ehrlich et al. 2013). Ergot sclerotia are efficiently removed during milling. Thus, the majority of products made from wheat or rye contain traces of ergot alkaloids in affected regions (Malysheva et al. 2014; Scott 2009; Scott et al 1992). Infections by ergot alkaloid-producing fungi remain common in the sense that plants on the edges of rye, barley, and wheat fields are often infected. There is some sense that ergot is getting more common again in parts of Europe (Krska and Crews 2008) and western North America (Menzies and Turkington 2014). Grains downgraded to animal feed can contain ergot; thus, this remains an animal health issue (Belser-Ehrlich et al. 2013). Cattle are more sensitive than mice, sheep, or swine and sclerotial contents of ca. 0.1 % appear to be tolerable. In cattle, the symptoms include lameness, and in other domestic animals, reduced weight gain can be expected (Prelusky et al. 1994).

## Future Prospects

Balbus et al. (2013) reviewed the implications of climate change in relation to management of human health risks of chemicals in the environment. They identified 8 contaminants as high risk of getting more serious, due to climate change, one of which was mycotoxins. Importantly, the global population has converged on a few sources of dietary starch (Khoury et al. 2014) most of which are rather susceptible to the toxins discussed here. Various authors have provided opinions on the potential relationship between climate change and mycotoxins (Magan et al. 2011; Miraglia et al. 2009; Paterson & Lima 2011; Wu et al. 2011) and food security (Marroquín-Cardona et al. 2014).

There are some examples of climate variability affecting the distribution of mycotoxins. Climate variability has played a role in the presence of mycotoxins in corn grown in Ontario. In the period 1972–1981, the prevalence and concentrations of zearalenone in corn were quite high (Andrews et al. 1981; Scott 1997). Sutton et al. (1980) found that zearalenone was associated with rainfall in August, but only moderately or weakly with rainfall for July, September, and October, and occurred during the latter part of the crop year. During 1970–1980, the weather was cooler. However, from 1980 to the present mean daily temperatures measured in several sites in Southwestern Ontario have increased (Environment Canada data for London airport; Hussell 2003, his figure 2). The biosynthesis of zearalenone has a requirement for high oxygen tension. As noted, zearalenone is typically accumulated in corn in the late summer when the crop is drying. Oxygen tensions are higher than in living plants and later in the season it is cooler than in July. Oxygen solubility in water is increased in colder versus warmer water. In contrast, deoxynivalenol is produced under conditions of low oxygen tension. Deoxynivalenol is seen in corn kernels concurrent with development of the infection when the plant is living and little zearalenone is being produced (Miller 2001). This warmer period has seen a possible change in the presence of fumonisin in corn. Analysis by Hooker and Schaafsma (2005) of 856 randomly selected Ontario corn fields between the years of 1993–2000 showed a total incidence of 23 % for fields contaminated with 1.0 FB1 ppm or greater. There was a large year-to-year variation, with the lowest incidence in 1993–1996. The highest incidence, 56 %, was observed in 1999. Mean concentrations varied between 1.0 ppm in 1994 and 2.3 ppm observed in 1995. The highest maximum concentration was seen in 1995 (7.0 ppm; Hooker and Schaafsma 2005). In 2012, summer mean temperatures were higher than the normal values in the Ontario corn-growing areas by as much as 4.3 °C (Environment Canada). In a sample of 84 corn samples taken at harvest 75 % were positive and six were >1 ranging to 4.2 μg/g (De Lamay Rios & Schaafsma unpublished data).

A more recent example has been the dramatic rise of aflatoxin in corn grown in parts of Europe. Prior to the early 2000s, aflatoxin was virtually unknown in European grown maize. In the last decade, this has changed such that in 2013, aflatoxin contamination became a substantial problem in some EU countries (Perrone et al. 2014).

## References

Adetuniji MC, Atanda OO, Ezekiel CN, Dipeolu AO, Uzochukwu SVA, Oyedepo J, Chilaka CA (2014) Distribution of mycotoxins and risk assessment of maize consumers in five agro-ecological zones of Nigeria. Eur Food Res Technol. doi:10.1007/s00217-014-2221-0

Amarasinghe CC, Tittlemier SA, Fernando WGD (2015) Nivalenol-producing *Fusarium cerealis* associated with Fusarium head blight in winter wheat in Manitoba. Can Plant Pathol 64:988–995

Amuzie CJ, Pestka JJ (2010) Suppression of insulin-like growth factor acid-labile subunit expression—a novel mechanism for deoxynivalenol-induced growth retardation. Toxicol Sci 113:412–421

Andersen B, Smedsgaard J, Frisvad JC (2004) *Penicillium expansum*: consistent production of patulin, chaetoglobosins, and other secondary metabolites in culture and their natural occurrence in fruit products. J Agric Food Chem 52:2421–2428

Andrews RI, Thompson BK, Trenholm HL (1981) A national survey of mycotoxins in Canada. J Am Oil Chem Soc 58:989–991

Angle JS, Dunn KA, Wagne GH (1982) Effect of cultural practices on the soil population of *Aspergillus flavus* and *Aspergillus parasiticus*. Soil Sci Soc Am J 46:301–304

Anonymous (1977) Balkan nephropathy. Lancet 309(8013):683–684

Armbrecht BH (1978) Aflatoxicosis in swine. In: Wyllie TD, Morehouse LG (eds) Mycotoxic fungi, mycotoxins, mycotoxicoses, vol 2. Marcel Dekker, New York, pp 227–235

Backhouse D (2014) Global distribution of *Fusarium graminearum*, *F. asiaticum* and *F. boothii* from wheat in relation to climate. Eur J Plant Pathol 139:161–173

Balbus JM, Boxall AB, Fenske RA, McKone TE, Zeise L (2013) Implications of global climate change for the assessment and management of human health risks of chemicals in the natural environment. Environ Toxicol Chem 32:62–78

Battilani P, Giorni P, Pietri A (2003) Epidemiology and toxin-producing fungi and ochratoxin A. Eur J Plant Pathol 109:723–730

Beardall JM, Miller JD (1994) Diseases in humans with mycotoxins as possible causes. In: Miller JD, Trenholm HL (eds) Mycotoxins in grain: compounds other than aflatoxin. Eagan Press, St. Paul, MN, pp 487–540

Belser-Ehrlich S, Harper A, Hussey J, Hallock R (2013) Human and cattle ergotism since 1900: symptoms, outbreaks, and regulations. Toxicol Ind Health 29:307–316

Benford D, Boyle C, Dekant W, Fuchs E, Gaylor DW, Hard G, McGregory DB, Pitt JI, Plestina R, Shephard G, Solfrizzo M, Verger PJP, Walker R (2001) Ochratoxin A safety evaluation of certain mycotoxins in food. WHO food additives series 47, FAO food and nutrition paper, vol 74. WHO Geneva, Switzerland, pp 281–415

Bennett JW, Klich M (2003) Mycotoxins. Clin Microbiol Rev 16:497–516

Bertuzzi T, Leggieri MC, Battilani P, Pietri A (2014) Co-occurrence of type A and B trichothecenes and zearalenone in wheat grown in northern Italy over the years 2009–2011. Food Addit Contam B 7:273–281

Bhat RV, Beedu SR, Ramakrishna Y, Munshi KL (1989) Outbreak of trichothecene mycotoxicosis associated with consumption of mould-damaged wheat in Kashmir Valley, India. Lancet 7:35–37

Bisby GR, James N, Timonin M (1933) Fungi isolated from Manitoba soil by the plate method. Can J Res 8:253–275

Bondy GS, Pestka JJ (2000) Immunomodulation by fungal toxins. J Toxicol Environ Health Part B 3:109–143

Bonnet MS, Roux J, Mounien L, Dallaporta M, Troadec JD (2012) Advances in deoxynivalenol toxicity mechanisms: the brain as a target. Toxins 4:1120–1138

Bouhet S, Oswald IP (2007) The intestine as a possible target for fumonisin toxicity. Mol Nutr Food Res 51:925–931

Brandon EFA, Baars AJ, Biesebeek JT, Oomen AG, Bakker MI, De Heer C (2012) Risk assessment of patulin intake from apple containing products by young children. World Mycotoxin J 5:391–403

Bui-Klimke TR, Wu F (2014a) Ochratoxin A and human health risk: a review of the evidence. Crit Rev Food Sci Nutr. doi:10.1080/10408398.2012.724480

Bui-Klimke T, Wu F (2014b) Evaluating weight of evidence in the mystery of Balkan endemic nephropathy. Risk Anal 34:1688–1705

Bulder AS, Arcella D, Bolger M, Carrington C, Kpodo K, Resnik S, Riley RT, Wolterink G, Wu F (2012) Fumonisins. In: Safety evaluation of certain food additives and contaminants. Seventy-fourth meeting of the Joint FAO/WHO Expert Committee on Food Additives (JECFA). WHO Food Additives Series No. 65. World Health Organization (WHO), Geneva, Switzerland, pp 325–527

Bullerman LB, Bianchini A (2007) Stability of mycotoxins during food processing. Int J Food Microbiol 119:140–146

Burdock GA, Flamm WG (2000) Review article: safety assessment of the mycotoxin cyclopiazonic acid. Int J Toxicol 19:195–218

Cabañes FJ, Bragulat MR, Castellá G (2010) Ochratoxin A producing species in the genus *Penicillium*. Toxins 2:1111–1120

Caldwell RW, Tuite J, Stob M, Baldwin R (1970) Zearalenone production by *Fusarium* species. Appl Microbiol 20:31–34

Carlton WW, Stack ME, Eppley RM (1976) Hepatic alterations produced in mice by xanthomegnin and viomellein, metabolites of *Penicillium viridicatum*. Toxicol Appl Pharmacol 38:455–459

Ceigler A, Bennett JD (1980) Mycotoxins and mycotoxicoses. Bioscience 30:512–515

Christensen M, Tuthill DE (1985) *Aspergillus*: an overview. In: Samson RA, Pitt JI (eds) Advances in *Penicillium* and *Aspergillus*. Plenum Press, New York, pp 195–209

Clear RM, Tucker JR, Gaba D, Patrick SK, Lee SJ, Demeke T, Gräfenhan T (2013) Deoxynivalenol levels and chemotype frequency in barley cultivars inoculated with two chemotypes of *Fusarium graminearum*. Can J Plant Pathol 35:37–45

Copetti MV, Iamanaka BT, Taniwaki MH (2013) Fungi and mycotoxin occurrence in cocoa. Chocolate in health and nutrition. Humana Press, London, pp 61–71

Cozzi G, Pascale M, Perrone G, Visconti A, Logrieco A (2006) Effect of *Lobesia botrana* damages on black aspergilli rot and ochratoxin A content in grapes. Int J Food Microbiol 111:S88–S92

Daamen RA, Langerak CJ, Stol W (1991) Surveys of cereal diseases and pests in the Netherlands. 3. *Monographella nivalis* and *Fusarium* spp. in winter wheat fields and seed lots. Neth J Plant Pathol 97:105–114

De La Campa R, Miller JD, Hendricks K (2004) Fumonisin in tortillas produced under small-scale facilities and the effect of traditional variations on masa production on this toxin. J Agric Food Chem 52:4432–4437

De La Campa R, Hooker DC, Miller JD, Schaafsma AW (2005) Modelling effects of environment, insect damage and Bt genotypes on fumonisin accumulation in maize in Agrentina and the Philippines. Mycopathologia 159:571–577

De Waal EJ (2002) Safety assessment of cyclopiazonic acid. Int J Toxicol 21:425–427

Del Ponte EM, Spolti P, Ward TJ, Gomes LB, Nicolli CP, Kuhnem PR, Tessmann DJ (2015) Regional and field-specific factors affect the composition of *Fusarium* head blight pathogens in subtropical no-till wheat agroecosystem of Brazil. Phytopathology 105:246–254

Deng F, Tao FB, Liu DY, Xu YY, Hao JH, Sun Y, Su PY (2012) Effects of growth environments and two environmental endocrine disruptors on children with idiopathic precocious puberty. Eur J Endocrinol 166:803–809

Dimitrova T (2009) Biological efficacy of herbicides for weed control in non-cropped areas. Pesticidi i fitomedicina 24:95–102

EFSA (2011a) Scientific opinion on the risks for public health related to the presence of zearalenone in food. EFSA J 9:2197 [124 pp]

EFSA (2011b) Scientific opinion on the risks for animal and public health related to the presence of T-2 and HT-2 toxin in food and feed. EFSA J 9:2481 [187 pp]

EFSA (2013) Scientific opinion on risks for animal and public health related to the presence of nivalenol in food and feed. EFSA J 11(6):3262 [119 pp]

Elmholt S (2003) Ecology of the ochratoxin A producing *Penicillium verrucosum*: occurrence in field soil and grain with special attention to farming system and on-farm drying practices. Biol Agric Hortic 20:311–337

Falkay G, Melis K, Foldesi I, Szuts P, Mesterhazy A (1993) Affinity of *Fusarium* toxins on human myometrial estradiol receptors. In: Gorog, S (ed) Proceedings of the 5th symposium on the analysis of steroids, Szombathely, Hungary, pp 25–32

Fink-Gremmels J, Malekinejad H (2007) Clinical effects and biochemical mechanisms associated with exposure to the mycoestrogen zearalenone. Anim Feed Sci Technol 137:326–341

Foroud NA, McCormick SP, MacMillan T, Badea A, Kendra DF, Ellis BE, Eudes F (2012) Greenhouse studies reveal increased aggressiveness of emergent Canadian *Fusarium graminearum* chemotypes in Wheat. Plant Dis 96:1271–1279

FSCJ (2010) Risk assessment report deoxynivalenol and nivalenol (mycotoxin). Food Safety Commission of Japan. Risk assessment report—veterinary medicines FS/872/2010

Gale LR, Ward TJ, Balmas V, Kistler HC (2007) Population subdivision of *Fusarium graminearum* sensu stricto in the upper midwestern United States. Phytopathology 97:1434–1439

Gale LR, Harrison SA, Ward TJ, O'Donnell K, Milus EA, Gale SW, Kistler HC (2011) Nivalenol-type populations of *Fusarium graminearum* and *F. asiaticum* are prevalent on wheat in southern Louisiana. Phytopathology 101:124–134

Gelderblom WC, Jaskiewicz K, Marasas WF, Thiel PG, Horak RM, Vleggaar R, Kriek NP (1988) Fumonisins—novel mycotoxins with cancer-promoting activity produced by *Fusarium moniliforme*. Appl Environ Microbiol 54:1806–1811

Gelineau-van Waes J, Starr L, Maddox JR, Aleman F, Voss KA, Wilberding J, Riley RT (2005) Maternal fumonisin exposure and risk for neural tube defects: disruption of sphingolipid metabolism and folate transport in an in vivo mouse model. Birth Defects Res (A) 73:487–497

Gelineau-van Waes J, Voss KA, Stevens VL, Speer MC, Riley RT (2009) Maternal fumonisin exposure as a risk factor for neural tube defects. In: Taylor S (ed) Advances Food Nutrition Res 56:145–181

Gelineau-van Waes J, Rainey MA, Maddox JR, Voss KA, Sachs AJ, Gardner NM, Wilberding JD, Riley RT (2012) Increased sphingoid base-1-phosphates and failure of neural tube closure after exposure to fumonisin or FTY720. Birth Defects Res A Clin Mol Teratol 94:790–803

Gilbert J, Abramson D, McCallum B, Clear R (2002) Comparison of Canadian *Fusarium graminearum* isolates for aggressiveness, vegetative compatibility, and production of ergosterol and mycotoxins. Mycopathologia 153:209–215

Gilbert J, Clear RM, Ward TJ, Gaba D, Tekauz A, Turkington TK, Woods SM, Nowicki T, O'Donnell K (2010) Relative aggressiveness and production of 3-or 15-acetyl deoxynivalenol and deoxynivalenol by *Fusarium graminearum* in spring wheat. Can J Plant Pathol 32:146–152

Gilbert J, Brûlé-Babel A, Guerrieri AT, Clear RM, Patrick S, Slusarenko K, Wolfe C (2014) Ratio of 3-ADON and 15-ADON isolates of *Fusarium graminearum* recovered from wheat kernels in Manitoba from 2008 to 2012. Can J Plant Pathol 36:54–63

Gong YY, Cardwell K, Hounsa A, Egal S, Turner PC, Hall AJ, Wild CP (2002) Dietary aflatoxin exposure and impaired growth in young children from Benin and Togo: cross sectional study. BMJ 325(7354):20–21

Gong Y, Hounsa A, Egal S, Turner PC, Sutcliffe AE, Hall AJ, Cardwell K, Wild CP (2004) Postweaning exposure to aflatoxin results in impaired child growth: a longitudinal study in Benin, West Africa. Environ Health Perspect 112:1334–1338

Greenhalgh R, Levandier D, Adams W, Miller JD, Blackwell BA, McAlees AJ, Taylor A (1985) Production and characterization of deoxynivalenol and other secondary metabolites of *Fusarium culmorum* (CMI 14764, HLX 1503). J Agric Food Chem 34:98–102

Greenhalgh R, Blackwell BA, Paré JRJ, Miller JD, Levandier D, Meier RM Taylor A and ApSimon JW (1986) Isolation and characterization by mass spectrometry and NMR spectroscopy of secondary metabolites of some *Fusarium* species. In: Proceedings of the 6th IUPAC international symposium on mycotoxins and phycotoxins. Steyn PS (ed). Elsevier Press, Amsterdam, pp 137–152

Greenhalgh R, Blackwell BA, Savard ME, Miller JD, Taylor A (1988) Secondary metabolites produced by *Fusarium sporotrichioides* DAOM 165006. J Agric Food Chem 36:216–219

Greenhalgh R, Fielder DA, Morrison LA, Charland J-P, Blackwell BA, Miller JD, Savard ME and ApSimon JW (1989) Apotrichothecenes. Minor metabolites of the *Fusarium* species. In: Proceedings of the 7th international symposium on mycotoxins and phycotoxins. Tokyo, Natori S, Hashimoto T, Ueno Y (eds). Elesevier, Amsterdam, pp 223–232

Grollman AP, Shibutani S, Moriya M, Miller F, Wu L, Moll U, Jelaković B (2007) Aristolochic acid and the etiology of endemic (Balkan) nephropathy. Proc Natl Acad Sci 104:12129–12134

Guo B, Chen ZY, Lee RD, Scully BT (2008) Drought stress and preharvest aflatoxin contamination in agricultural commodity: genetics, genomics and proteomics. J Integr Plant Biol 50:1281–1291

Haighton LA, Lynch BS, Magnuson BA, Nestmann ER (2012) A reassessment of risk associated with dietary intake of ochratoxin A based on a lifetime exposure model. Crit Rev Toxicol 42:147–168

Hammond BG, Campbell KW, Pilcher CD, Degooyer TA, Robinson AE, McMillen BL, Spangler SM, Riordan SG, Rice LG, Richard JL (2004) Lower fumonisin mycotoxin levels in the grain of Bt corn grown in the United States in 2000–2002. J Agric Food Chem 52:1390–1397

Haschek WM, Gumprecht LA, Smith G, Tumbleson ME, Constable PD (2001) Fumonisin toxicosis in swine: an overview of porcine pulmonary edema and current perspectives. Environ Health Perspect 109(S2):251–257

Heperkan D, Moretti A, Dikmen CD, Logrieco AF (2012) Toxigenic fungi and mycotoxin associated with figs in the Mediterranean area. Phytopathol Mediterr 51:119–130

Hocking AD, Leong SLL, Kazi BA, Emmett RW, Scott ES (2007) Fungi and mycotoxins in vineyards and grape products. Int J Food Microbiol 119:84–88

Hooker DC, Schaafsma AW (2005) Agronomic and environmental impacts on concentrations of deoxynivalenol and fumonisin B1 in corn across Ontario. Can J Plant Pathol 27:347–356

Horn BW (2003) Ecology and population biology of aflatoxigenic fungi in soil. Toxin Rev 22:351–379

Horn BW (2007) Biodiversity of *Aspergillus* section Flavi in the United States: a review. Food Addit Contam 24:1088–1101

Hussell DJT (2003) Climate change, spring temperatures, and timing of breeding of Tree Swallows (*Tachycineta bicolor*) in southern Ontario. Auk 120:607–618

IARC (1993) Some naturally occurring substances: food items and constituents, heterocyclic aromatic amines and mycotoxins. International Agency for Research on Cancer Monograph 56, Lyon, France

IARC (2002) Traditional herbal medicines, some mycotoxins, naphthalene and styrene. International Agency for Research on Cancer Monograph 82, Lyon, France

IARC (2012) Aflatoxins. International Agency for Research on Cancer Monograph 100F–23, Lyon, France

IPCS (2000) Fumonisin B1. Environmental health criteria 219, International Programme on Chemical Safety. WHO, Geneva

Iverson F, Armstrong C, Nera E, Truelove J, Fernie S, Scott P, Stapley R, Hayward S, Gunner S (1995) Chronic feeding study of deoxynivalenol in B6C3F1 male and female mice. Teratog Carcinog Mutagen 15:283–306

JECFA (1995) 44th report of the Joint FAO/WHO expert committee on food additives. Tech. Report Series, 859WHO, Geneva

JECFA (2001) Evaluation of certain food additives and contaminants in food. 56th report of the Joint Expert Committe on Food Additives and Contaminants. World Health Organization

JECFA (2011) Evaluation of certain food additives and contaminants in food: 74th report of the Joint FAO/WHO expert committee on food additives. World Health Organization

Joffe AZ (1971) Alimentary toxic aleukia. In: Kadis S, Ciegler A, Ajl SJ (eds) Microbial toxins. Academic, New York, pp 139–189

Kabak B (2009) The fate of mycotoxins during thermal food processing. J Sci Food Agric 89:549–554

Kasitu GC, ApSimon JW, Blackwell BA, Fielder DA, Greenhalgh R, Miller JD (1992) Isolation and characterization of culmorin derivatives produced by *Fusarium culmorum* CMI 14764. Can J Chem 70:1308–1316

Keller L, Bidochka MJ (1998) Habitat and temporal differences among soil microfungal assemblages in Ontario. Can J Bot 76:1798–1805

Kelly AC, Clear RM, O'Donnell K, McCormick S, Turkington TK, Tekauz A, Ward TJ (2015) Diversity of Fusarium head blight populations and trichothecene toxin types reveals regional differences in pathogen composition and temporal dynamics. Fungal Genet Biol 82:22–31

Keyl AC (1978) Aflatoxicosis in cattle. In: Wyllie TD, Morehouse LG (eds) Mycotoxic fungi, mycotoxins, mycotoxicoses, vol 2. Marcel Dekker, New York, pp 9–27

Khoury CK, Bjorkman AD, Dempewolf H, Ramirez-Villegas J, Guarino L, Jarvis A, Struik PC (2014) Increasing homogeneity in global food supplies and the implications for food security. Proc Natl Acad Sci USA 111:4001–4006

Kimanya ME, De Meulenaer B, Roberfroid D, Lachat C, Kolsteren P (2010) Fumonisin exposure through maize in complementary foods is inversely associated with linear growth of infants in Tanzania. Mol Nutr Food Res 54:1659–1667

King ED, Bassi AB Jr, Ross DC, Druebbisch B (2011) An industry perspective on the use of "atoxigenic" strains of *A. flavus* as biological control agents and the significance of cyclopiazonic acid. Toxin Rev 30:33–41

Klich MA (2002) Biogeography of *Aspergillus* species in soil and litter. Mycologia 94:21–27

Knudsen PB, Mogensen JM, Larsen TO, Nielsen KF (2010) Occurrence of fumonisins B2 and B4 in retail raisins. J Agric Food Chem 59:772–776

Krogh P (1991) Porcine nephropathy associated with ochratoxin A. In: Smith JE, Henderson RS (eds) Mycotoxins and animal foods. CRC Press, Boca Raton, FL, pp 627–646

Krogh P, Elling F (1976) Fungal toxins and endemic (Balkan) nephropathy. Lancet 308(7975):40

Krska R, Crews C (2008) Significance, chemistry and determination of ergot alkaloids: a review. Food Addit Contam 25:722–731

Kuhnem PR, Spolti P, Del Ponte E, Cummings JA, Bergstrom GC (2015a) Trichothecene genotype composition of *Fusarium graminearum* not differentiated among isolates from maize stubble, maize ears, wheat spikes, and the atmosphere in New York. Phytopathology 105:695–699

Kuhnem PR, Del Ponte EM, Dong Y, Bergstrom GC (2015b) *Fusarium graminearum* isolates from wheat and maize in New York show similar range of aggressiveness and toxigenicity in cross-species pathogenicity tests. Phytopathology 105:441–448

Kuiper-Goodman T, Scott PM (1989) Risk assessment of the mycotoxin ochratoxin A. Biomed Environ Sci 2:179–248

Kuiper-Goodman T, Hilts C, Billiard SM, Kiparissis Y, Richard IDK, Hayward S (2010) Health risk assessment of ochratoxin A for all age-sex strata in a market economy. Food Addit Contam 27:212–240

Liang JM, Xayamongkhon H, Broz K, Dong Y, McCormick SP, Abramova S, Kistler HC (2014) Temporal dynamics and population genetic structure of *Fusarium graminearum* in the upper Midwestern United States. Fungal Genet Biol 73:83–92

Liu Y, Wu F (2010) Global burden of aflatoxin-induced hepatocellular carcinoma: a risk assessment. Environ Health Perspect 118:818–824

Liu Y, Chang CCH, Marsh GM, Wu F (2012) Population attributable risk of aflatoxin-related liver cancer: systematic review and meta-analysis. Eur J Cancer 48:2125–2136

Logrieco A, Bottalico A, Altomare C (1988) Chemotaxonomic observations on zearalenone and trichothecene production by *Gibberella zeae* from cereals in southern Italy. Mycologia 80:892–895

Logrieco A, Ferracane R, Visconti A, Ritieni A (2010) Natural occurrence of fumonisin B2 in red wine from Italy. Food Addit Contam 27:1136–1141

Magan N, Medina A, Aldred D (2011) Possible climate-change effects on mycotoxin contamination of food crops pre-and postharvest. Plant Pathol 60:150–163

Malysheva SV, Larionova DA, Diana Di Mavungu J, De Saeger S (2014) Pattern and distribution of ergot alkaloids in cereals and cereal products from European countries. World Mycotoxin J 7:217–230

Manabe M, Tsuruta O (1978) Geographical distribution of aflatoxin-producing fungi inhabiting in Southeast Asia. Jpn Agric Res Q 12:224–227

Marasas WF (2001) Discovery and occurrence of the fumonisins: a historical perspective. Environ Health Perspect 109(S2):239–243

Marasas WFO, Nelson PE, Toussoun TA (1984) Toxigenic *Fusarium* species. Pennsylvania State University Press, University Park, PA

Marasas WFO, Riley RT, Hendricks KA, Stevens VL, Sadler TW, Gelineau-van Waes J, Missmer SA, Cabrera Valverde J, Torres OL, Gelderblom W, Allegood J, Martínez de Figueroa AC, Maddox J, Miller JD, Starr L, Sullards MC, Roman Trigo AV, Voss KA, Wang E, Merrill AH Jr (2004) Fumonisins disrupt sphingolipid metabolism, folate transport and development of neural crest cells in embryo culture and in vivo: a risk factor for human neural tube defects among populations consuming fumonisin-contaminated maize? J Nutr 134:711–716

Markovic M, Protic N, Protic R, Jankovic S (2005) New possibilities of weed control in wheat. Rom Agric Res 22:41–46

Marroquín-Cardona AG, Johnson NM, Phillips TD, Hayes AW (2014) Mycotoxins in a changing global environment–a review. Food Chem Toxicol 69:220–230

Massart F, Saggese G (2010) Oestrogenic mycotoxin exposures and precocious pubertal development. Int J Androl 33:369–376

Massart F, Meucci V, Saggese G, Soldani G (2008) High growth rate of girls with precocious puberty exposed to estrogenic mycotoxins. J Pediatr 152:690–694

Mehmeti A, Demaj A, Waldhardt R (2009) Plant species richness and composition in the arable land of Kosovo. Landscape Online 11:1–29

Menniti AM, Neri F, Gregori R, Maccaferri M (2010) Some factors influencing patulin production by *Penicillium expansum* in pome fruits. J Sci Food Agric 90:2183–2187

Menzies JG, Turkington TK (2014) An overview of the ergot (*Claviceps purpurea*) issue in western Canada: challenges and solutions. Can J Plant Path. doi:10.1080/07060661.2014.986527

Metzler M, Pfeiffer E, Hildebrand AA (2010) Zearalenone and its metabolites as endocrine disrupting chemicals. World Mycotoxin J 3:385–401

Miller JD (1994) Epidemiology of *Fusarium graminearum* diseases of wheat and corn. In: Miller JD, Trenholm HL (eds) Mycotoxins in grain: compounds other than aflatoxin. Eagan Press, St. Paul, MN, pp 19–36

Miller JD (1995) Fungi and mycotoxins in grain: implications for stored product research. J Stored Prod Res 31:1–6

Miller JD (2001) Factors that affect the occurrence of fumonisin. Environ Health Perspect 109(s. 2):321–324

Miller JD (2008) Mycotoxins in small grains and maize: old problems, new challenges. Food Addit Contam 25:219–230

Miller JD, Taylor A, Greenhalgh R (1983a) Production of deoxynivalenol and related compounds in liquid culture by *Fusarium graminearum*. Can J Microbiol 29(1171):1178

Miller JD, Young JC, Trenholm HL (1983b) *Fusarium* toxins in field corn. I. Parameters associated with fungal growth and production of deoxynivalenol and other mycotoxins. Can J Bot 61:3080–3087

Miller JD, Greenhalgh R, Wang YZ, Lu M (1991) Trichothecene mycotoxin chemotypes of three *Fusarium* species. Mycologia 83:121–130

Miller JD, Schaafsma AW, Bhatnagar D, Bondy G, Carbone I, Harris LJ, Harrison G, Munkvold GP, Oswald IP, Pestka JJ, Sharpe L, Sumarah MW, Tittlemier SA, Zhou T (2014) Mycotoxins that affect the North American Agri-Food sector: state of the art and directions for the future. World Mycotoxin J 7:63–82

Miraglia M, Marvin HJP, Kleter GA, Battilani P, Brera C, Coni C, Cubadda F (2009) Climate change and food safety: an emerging issue with special focus on Europe. Food Chem Toxicol 47:1009–1021

Mirocha CJ, Pathre S (1973) Identification of the toxic principle in a sample of poaefusarin. Appl Microbiol 26:719–724

Mirocha CJ, Abbas HK, Windels CE, Xie W (1989) Variation in deoxynivalenol, 15-acetyldeoxynivalenol, 3-acetyldeoxynivalenol, and zearalenone production by *Fusarium graminearum* isolates. Appl Environ Microbiol 55:1315–1316

Mishra PK, Tewari JP, Turkington KT, Clear RM (2009) Genetic evidence for a recent geographic expansion of 15-acetyldeoxynivalenol chemotypes of *Fusarium graminearum* in Canada. Can J Plant Pathol 31:468–474

Mogensen JM, Larsen TO, Nielsen KF (2010) Widespread occurrence of the mycotoxin fumonisin B2 in wine. J Agric Food Chem 58:4853–4857

Moore GG, Elliott JL, Singh R, Horn BW, Dorner JW, Stone EA, Chulze SN, Barros GG, Naik MK, Wright GC, Hell K, Carbone I (2013) Sexuality generates diversity in the aflatoxin gene cluster: evidence on a global scale. PLoS Pathog 9(8), e1003574

Moretti A, Ferracane L, Somma S, Ricci V, Mulè G, Susca A, Ritieni A, Logrieco AF (2010) Identification, mycotoxin risk and pathogenicity of *Fusarium* species associated with fig endosepsis in Apulia, Italy. Food Addit Contamt A 27:718–728

Morooka N, Uratsuji N, Yoshizawa T, Yamamoto H (1972) Studies on the toxic substances in barley infected with *Fusarium* species. J Food Hyg Soc Jpn 13:368–375

Mutiga SK, Were V, Hoffmann V, Harvey JW, Milgroom MG, Nelson RJ (2014) Extent and drivers of mycotoxin contamination: inferences from a survey of Kenyan maize mills. Phytopathology 104:1221–1231

Nielsen KF, Sumarah MW, Frisvad JC, Miller JD (2006) Production of metabolites by species in the *Penicillium roqueforti* complex. J Agric Food Chem 54:3756–3763

O'Donnell K, Kistler HC, Tacke BK, Casper HH (2000) Gene genealogies reveal global phylogeographic structure and reproductive isolation among lineages of *Fusarium graminearum*, the fungus causing wheat scab. Proc National Academy of Sciences 97:7905–7910.
O'Donnell K, Ward TJ, Geiser DM, Kistler HC, Aoki T (2004) Genealogical concordance between the mating type locus and seven other nuclear genes supports formal recognition of nine phylogenetically distinct species within the *Fusarium graminearum* clade. Fungal Genet Biol 41:600–623
Olarte RA, Horn BW, Dorner JW, Monacell JT, Singh R, Stone EA, Carbone I (2012) Effect of sexual recombination on population diversity in aflatoxin production by *Aspergillus flavus* and evidence for cryptic heterokaryosis. Mol Ecol 21:1453–1476
Olarte RA, Worthington CJ, Horn BW, Moore GG, Singh R, Monacell JW, Dorner JT, Stone EA, Xie D, Carbone I (2015) Enhanced diversity and aflatoxigenicity in interspecific hybrids of *Aspergillus flavus* and *Aspergillus parasiticus*. Mol Ecol 24(8):1889–1909
Panthi A, Hallen-Adams H, Wegulo SN, Hernandez Nopsa J, Baenziger PS (2014) Chemotype and aggressiveness of isolates of *Fusarium graminearum* causing head blight of wheat in Nebraska. Can J Plant Pathol 36:447–455
Park DL (2004) Effect of processing on aflatoxin. Adv Exp Med Biol 504:173–179
Parsons MW, Munkvold GP (2010) Associations of planting date, drought stress, and insects with *Fusarium* ear rot and fumonisin B1 contamination in California maize. Food Addit Contam 27:591–607
Parsons MW, Munkvold GP (2012) Effects of planting date and environmental factors on fusarium ear rot symptoms and fumonisin B1 accumulation in maize grown in six North American locations. Plant Pathol 61:1130–1142
Paterson RRM, Lima N (2011) Further mycotoxin effects from climate change. Food Res Int 44:2555–2566
Payne GA, Widstrom NW (1992) Aflatoxin in maize. Crit Rev Plant Sci 10:423–440
Perrone G, Gallo A, Logrieco AF (2014) Biodiversity of *Aspergillus* section Flavi in Europe in relation to the management of aflatoxin risk. Front Microbiol 5:377. doi:10.3389/fmicb.2014.00377
Perši N, Pleadin J, Kovačević D, Scortichini G, Milone S (2014) Ochratoxin A in raw materials and cooked meat products made from OTA-treated pigs. Meat Sci 96:203–210
Pestka JJ (2010) Deoxynivalenol: mechanisms of action, human exposure, and toxicological relevance. Arch Toxicol 84:663–679
Pestka JJ, Lin WS, Miller ER (1987) Emetic activity of the trichothecene 15-acetyldeoxynivalenol in swine. Food Chem Toxicol 25:855–858
Pettersson H, Brown C, Hauk J, Hoth S, Meyer J, Wessels D (2011) Survey of T-2 and HT-2 toxins by LC–MS/MS in oats and oat products from European oat mills in 2005–2009. Food Addit Contam B 4:110–115
Pfohl-Leszkowicz A, Manderville RA (2011) An update on direct genotoxicity as a molecular mechanism of ochratoxin A carcinogenicity. Chem Res Toxicol 25:252–262
Pfohl-Leszkowicz A, Petkova-Bocharova T, Chernozemsky IN, Castegnaro M (2002) Balkan endemic nephropathy and associated urinary tract tumours: a review on aetiological causes and the potential role of mycotoxins. Food Addit Contam 19:282–302
Piñeiro MS, Scott PM, Kanhere SR (1995) Mycotoxin producing potential of *Fusarium graminearum* isolates from Uruguayan barley. Mycopathologia 132:167–172
Pinton P, Oswald IP (2014) Effect of deoxynivalenol and other Type B trichothecenes on the intestine: a review. Toxins 6:1615–1643
Pitt JI, Wild CP, Baan RA, Gelderblom WCA, Miller JD, Riley RT, Wu F (2012). Improving public health through mycotoxin control. International Agency for Research on Cancer Scientific Publications Series, No. 158. Lyon, France, 151p
Prelusky DB (1994) Residues in food products of animal origin. In: Miller JD, Trenholm HL (eds) Mycotoxins in grain: compounds other than aflatoxin. Eagan Press, St. Paul, MN, pp 405–420
Prelusky DB, Rotter BA, Rotter RG (1994) Toxicology of mycotoxins. In: Miller JD, Trenholm HL (eds) Mycotoxins in grain: compounds other than aflatoxin. Eagan Press, St. Paul, MN, pp 359–404

Puel O, Galtier P, Oswald IP (2010) Biosynthesis and toxicological effects of patulin. Toxins 2:613–631

Rasmussen RR, Storm IMLD, Rasmussen PH, Smedsgaard J, Nielsen KF (2010) Multi-mycotoxin analysis of maize silage by LC-MS/MS. Anal Bioanal Chem 397:765–776

Rheeder JP, Marasas WF, Vismer HF (2002) Production of fumonisin analogs by *Fusarium* species. Appl Environ Microbiol 68:2101–2105

Rundberget T, Skaar I, Flåøyen A (2004) The presence of *Penicillium* and *Penicillium* mycotoxins in food wastes. Int J Food Microbiol 90:181–188

Russell L, Cox DF, Larsen G, Bodwell K, Nelson CE (1991) Incidence of molds and mycotoxins in commercial animal feed mills in 7 midwestern states, 1988–1989. J Anim Sci 69:5–12

Savin M, Džamić Z, Baralić M, Radojević-Škodrić S, Marinković J, Bumbaširević V (2014) Prognostic value of p53, c-ErbB2 and tunel data in upper urothelial carcinoma associated with Balkan nephropathy. Arch Biol Sci 66:641–649

Schaafsma AW, Hooker DC (2007) Climatic models to predict occurrence of *Fusarium* toxins in wheat and maize. Int J Food Microbiol 119:116–125

Schiefer HB (1990) Mycotoxins. The Wiley encyclopedia of food science & technology. Wiley, New York, 1862–1869

Schiff PL Jr (2006) Ergot and its alkaloids. Am J Pharm Educ 70(98):1–10

Schmale DG, Wood-Jones AK, Cowger C, Bergstrom GC, Arellano C (2011) Trichothecene genotypes of *Gibberella zeae* from winter wheat fields in the eastern USA. Plant Pathol 60:909–917

Scott PM (1997) Multi-year monitoring of Canadian grains and grain-based food for trichothecenes and zearalenone. Food Addit Contam 14:333–339

Scott PM (2009) Ergot alkaloids: extent of human and animal exposure. World Mycotoxin J 2:141–149

Scott PM, Lombaert GA, Pellaers P, Bacler S, Lappi J (1992) Ergot alkaloids in grain foods sold in Canada. J AOAC Int 75:773–779

Shirima CP, Kimanya ME, Routledge MN, Srey C, Kinabo JL, Humpf HU, Wild CP, Tu YK, Gong YY (2014) A prospective study of growth and biomarkers of exposure to aflatoxin and fumonisin during early childhood in Tanzania. Environ Health Perspect. doi:10.1289/ehp.1408097

Smart MG, Wicklow DT, Caldwell RW (1990) Pathogenesis in *Aspergillus* ear rot of maize: light microscopy of fungal spread from wounds. Phytopathology 80:1287–1294

Somma S, Petruzzella AL, Logrieco A (2012) Diversity of black *Aspergilli* and mycotoxin risks in grape, wine and dried vine fruits. Phytopathol Mediterr 51:131–147

Somma S, Petruzzella AL, Logrieco AF, Meca G, Cacciola OS, Moretti A (2014) Phylogenetic analyses of *Fusarium graminearum* strains from cereals in Italy, and characterisation of their molecular and chemical chemotypes. Crop Pasture Sci 65:52–60

Sonjak S, Ličen M, Frisvad JC, Gunde-Cimerman N (2011) Salting of dry-cured meat–a potential cause of contamination with the ochratoxin A-producing species *Penicillium nordicum*. Food Microbiol 28:1111–1116

Spolti P, Del Ponte EM, Cummings JA, Dong Y, Bergstrom GC (2014) Fitness attributes of *Fusarium graminearum* isolates from wheat in New York possessing a 3-ADON or 15-ADON trichothecene genotype. Phytopathology 104:513–519

Stoev SD, Vitanov S, Anguelov G, Petkova-Bocharova T, Creppy EE (2001) Experimental mycotoxic nephropathy in pigs provoked by a diet containing ochratoxin A and penicillic acid. Vet Res Comm 25:205–223

Stoev SD, Paskalev M, MacDonald S, Mantle PG (2002) Experimental one year ochratoxin A toxicosis in pigs. Exp Toxicol Pathol 53:481–487

Stroup A (1990) A company of scientists: botany, patronage, and community at the seventeenth-century Parisian Royal Academy of Sciences. University of California Press, Berkeley

Suarez L, Felkner M, Brender JD, Canfield M, Zhu H, Hendricks KA (2012) Neural tube defects on the Texas-Mexico border: what we've learned in the 20 years since the Brownsville cluster. Birth Defects Res 94:882–892

Suga H, Kitajima M, Nagumo R, Tsukiboshi T, Uegaki R, Nakajima T, Hyakumachi M (2014) A single nucleotide polymorphism in the translation elongation factor 1α gene correlates with the ability to produce fumonisin in Japanese *Fusarium fujikuroi*. Fungal Biol 118:402–412

Sugita-Konishi Y, Nakajima T (2010) Nivalenol: the mycology, occurrence, toxicology, analysis and regulation. In: Mycotoxins in food, feed and bioweapons. Springer, Berlin, pp 253–273

Sumarah MW, Miller JD, Blackwell BA (2005) Isolation and metabolite production by *Penicillium roqueforti*, *P. paneum* and *P. crustosum* isolated in Canada. Mycopathologia 159:571–577

Sutton JC (1982) Epidemiology of wheat head blight and maize ear rot caused by *Fusarium graminearum*. Can J Plant Pathol 4:195–209

Sutton JC, Baliko W, Funnell HS (1980) Relation of weather variables to incidence of zearalenone in corn in Southern Ontario. Can J Plant Sci 60:149–155

Takemura H, Shim JY, Sayama K, Tsubura A, Zhu BT, Shimoi K (2007) Characterization of the estrogenic activities of zearalenone and zeranol in vivo and in vitro. J Steroid Biochem Mol Biol 103:170–177

Tamburic-Ilincic L, Gaba D, Nowicki T, Schaafsma A, Fernando D (2008) Chemotypes of *Fusarium graminearum* isolates and accumulation of deoxynivalenol (DON), 15-ADON and 3-ADON in naturally infected and inoculated winter wheat in Ontario, Canada (Cereal Research Comm 36, SB:623–624)

Tamburic-Ilincic L, Muckle A, Schaafsma A (2015) Mycotoxins accumulation and *Fusarium graminearum* chemotype diversity in winter wheat grown in southwestern Ontario. Can J Plant Sci 95:931–938

Tanaka T, Hasegawa A, Yamamoto S, Sugiura Y, Ueno Y (1988) A case report on a minor contamination of nivalenol in cereals harvested in Canada. Mycopathologia 101:157–160

Tangni EK, Pussemier L, Van Hove F (2013) Mycotoxin contaminating maize and grass silages for dairy cattle feeding: current state and challenges. J Anim Sci Adv 3:492–511

Taniwaki MH (2006) An update on ochratoxigenic fungi and ochratoxin A in coffee. In Advances in food mycology. Springer, New York, pp 189–202

Tittlemier SA, Gaba D, Chan JM (2013) Monitoring of *Fusarium* trichothecenes in Canadian cereal grain shipments from 2010 to 2012. J Agric Food Chem 31:7412–7418

Torres O, Matute J, Gelineau-van Waes J, Maddox JR, Gregory SG, Ashley-Koch AE, Showker JL, Voss KA, Riley RT (2014) Human health implications from co-exposure to aflatoxins and fumonisins in maize-based foods in Latin America: Guatemala as a case study. World Mycotoxin J 2014:1–17

Uegaki R, Tohno M, Yamamura K, Tsukiboshi T (2014) Changes in the concentration of fumonisins in forage rice during the growing period, differences among cultivars and sites, and identification of the causal fungus. J Agric Food Chem 62:3356–3362

Ueno Y (1983) Trichothecenes: chemical, biological and toxicological aspects. Developments in food science vol 4. Elsevier, Amsterdam

Van der Fels-Klerx HJ, De Rijk TC, Booij CJH, Goedhart PW, Boers EAM, Zhao C, Van der Lee TAJ (2012) Occurrence of Fusarium Head Blight species and *Fusarium* mycotoxins in winter wheat in the Netherlands in 2009. Food Addit Contam A 29:1716–1726

Van Dongen PW, de Groot AN (1995) History of ergot alkaloids from ergotism to ergometrine. Eur J Obstet Gynecol Reprod Biol 60:109–116

Van Rensburg SJ, Altenkirk B (1974) *Claviceps purpurea*- ergotism. In: Purchase IFH (ed) Mycotoxins. Elsevier, New York, pp 69–96

Varga J, Frisvad JC, Samson RA (2009) A reappraisal of fungi producing aflatoxins. World Mycotoxin J 2:263–277

Varga J, Frisvad JC, Kocsubé S, Brankovics B, Tóth B, Szigeti G, Samson RA (2011) New and revisited species in *Aspergillus* section *Nigri*. Stud Mycol 69:1–17

Varga E, Wiesenberger G, Hametner C, Ward TJ, Dong Y, Schöfbeck D, McCormick S, Broz K, Stückler R, Schuhmacher R, Krska R, Kistler HC, Berthiller F, Adam G (2015) New tricks of an old enemy: isolates of *Fusarium graminearum* produce a type A trichothecene mycotoxin. Environ Microbiol 17(8):2588–2600. doi:10.1111/1462-2920.12718

Vesonder R, Ceigler A, Jensen AH (1973) Isolation of the emetic principle from *Fusarium graminearum*-infected corn. Appl Microbiol 26:1008–1010

Visagie CM, Varga J, Houbraken J, Meijer M, Kocsubé S, Yilmaz N, Samson RA (2014) Ochratoxin production and taxonomy of the yellow aspergilli *Aspergillus* section Circumdati. Stud Mycol 78:1–61

Visconti A, Perrone G, Cozzi G, Solfrizzo M (2008) Managing ochratoxin A risk in the grape-wine food chain. Food Addit Contam 25:193–202

Vogelgsang S, Widmer F, Jenny E, Enkerli J (2009) Characterization of novel *Fusarium graminearum* microsatellite markers in different *Fusarium* species from various countries. Eur J Plant Pathol 123:477–482

von der Ohe C, Gauthier V, Tamburic-Ilincic L, Brule-Babel A, Fernando WGD, Clear R, Ward TJ, Miedaner T (2010) A comparison of aggressiveness and deoxynivalenol production between Canadian *Fusarium graminearum* isolates with 3- acetyl and 15-acetyldeoxynivalenol chemotypes in field-grown spring wheat. Eur J Plant Pathol 127:407–417

Voss K, Riley R (2013) Fumonisin toxicity and mechanism of action: overview and current perspectives. Food Safety 1(1):49–69. doi:10.14252/foodsafetyfscj.2013006

Voss KA, Smith GW, Haschek WM (2007) Fumonisins: toxicokinetics, mechanism of action and toxicity. Anim Feed Sci Technol 137:299–325

Voss KA, Riley RT, Gelineau-van Waes J (2014) Fumonisin B1 induced neural tube defects were not increased in LM/Bc mice fed folate-deficient diet. Mol Nutr Food Res 58:1190–1198

Vrabcheva T, Usleber E, Dietrich R, Märtlbauer E (2000) Co-occurrence of ochratoxin A and citrinin in cereals from Bulgarian villages with a history of Balkan endemic nephropathy. J Agric Food Chem 48:2483–2488

Ward TJ, Clear RM, Rooney AP, O'Donnell K, Gaba D, Patrick S, Starkey DE, Gilbert J, Geiser DM, Nowicki TW (2008) An adaptive evolutionary shift in *Fusarium* head blight pathogen populations is driving the rapid spread of more toxigenic *Fusarium graminearum* in North America. Fungal Genet Biol 45:473–484

Welke JE, Hoeltz M, Dottori HA, Noll IB (2009) Effect of processing stages of apple juice concentrate on patulin levels. Food Control 20:48–52

Wilson TM, Ross PF, Rice LG, Osweiler GD, Nelson HA, Owens DL, Pickrell JW (1990) Fumonisin B1 levels associated with an epizootic of equine leukoencephalomalacia. J Vet Diagn Invest 2:213–216

Wogan GN (1966) Chemical nature and biological effects of the aflatoxins. Bacteriol Rev 30:460–469

Wu F, Wang T (2013) Risk assessment of upper tract urothelial carcinoma related to aristolochic acid. Cancer Epidemiol Biomarkers Prev 22:812–820

Wu F, Bhatnagar D, Bui-Klimke T, Carbone I, Hellmich R, Munkvold G, Takle E (2011) Climate change impacts on mycotoxin risks in US maize. World Mycotoxin J 4:79–93

Wu W, Bates MA, Bursian SJ, Flannery B, Zhou HR, Link JE, Zhang H, Pestka JJ (2013) Peptide YY3-36 and 5-hydroxytryptamine mediate emesis induction by trichothecene deoxynivalenol (vomitoxin). Toxicol Sci 133:186–195

Wyatt RD (1991) Poultry. In: Smith JE, Henderson RS (eds) Mycotoxins and animal foods. CRC Press, Boca Raton, FL, pp 553–606

Xu XM, Parry DW, Nicholson P, Thomsett MA, Simpson D, Edwards SG, Cooke BM, Doohan FM, Brennan JM, Moretti A, Tocco G, Mule G, Hornok L, Giczey G, Tatnell J (2005) Predominance and association of pathogenic fungi causing *Fusarium* ear blight in wheat in four European countries. Eur J Plant Pathol 112:143–154

Yoshizawa T, Jin Y (1995) Natural occurrence of acetylated derivatives of deoxynivalenol and nivalenol in wheat and barley in Japan. Food Addit Contam 2:689–694

Zhang H, Van der Lee T, Waalwijk C, Chen W, Xu J, Xu J, Feng J (2012) Population analysis of the *Fusarium graminearum* species complex from wheat in China show a shift to more aggressive isolates. PLoS One 7(2), e31722

# Chapter 20
# Inhalation Exposure and Toxic Effects of Mycotoxins

**Harriet M. Ammann**

## Introduction

The respiratory system differs from ingestion as a toxic exposure route because of its intimate, direct relationship with the blood in the general circulation through its gas exchange region. The nose and sinuses, with their baffle-like structures, are very good at trapping large particles to eliminate them via ciliated transport of mucus to be swallowed. However, because of nasal and sinus structure, particles, including mold spores and bacteria, can become trapped and colonize their mucous membranes. Additionally, the nose has axonal nerve endings that are directly connected with their cell bodies in the brain, along which particles can travel, while most nongaseous contaminants are excluded from the brain by the blood–brain barrier. Potency of toxic inhalants is greater than those ingested due to the structure and function of the respiratory system.

Toxic symptoms in man and farm animals have been known to be caused by molds for a long time before the identification of mycotoxins in 1960 in England when over 100,000 turkeys died as a result of "Turkey X disease" which was later found to be caused by the mycotoxin aflatoxin (UNJR 1970). Intense research on mycotoxin effects on crops and food animals and on human health from consumption of mycotoxin-containing foods has resulted due to the estimated mean annual cost of mycotoxin-induced losses of $932 million in food losses, $466 million in mitigation loss, and $6 million in livestock loss (pre-2003 dollars) (CAST 2003). While human disease impacts have also been investigated, most of the research on human health has focused on ingestion of contaminated food. Early reports from the Soviet Union had tracked a noninfectious disease of horses caused by feeding moldy straw, and a number of studies detailed local and systemic toxic effects in

H.M. Ammann, Ph.D. (✉)
Department of Environmental and Occupational Health Sciences, School of Public Health and Community Medicine, University of Washington, Seattle, WA, USA
e-mail: h.ammann@comcast.net



D.-W. Li (ed.), *Biology of Microfungi*, Fungal Biology,
DOI 10.1007/978-3-319-29137-6_20

humans also exposed to aerosols of toxic strains of *Stachybotrys chartarum* (reported as *S. atra*) (Forgacs 1972). Controlled animal experiments using the inhalation route of exposure are more difficult to do and generally more expensive than ingestion studies, and exposure through mycotoxin-contaminated animal feed and human food was thought to be a more significant source of damage exerted by mycotoxins. For these reasons relatively few inhalation experiments on animals were done in the 1960s and 1970s.

Ueno (1984) found that low-level T-2 toxin (140 ppb) inhalation exposure for 30 min killed mice within several days. In the 1980s and 1990s, the US Army Medical Research Institute of Infectious Diseases (USAMRIID) supported research on the inhalation and skin exposure for certain mycotoxins, especially the simple trichothecene T-2 toxin produced by certain *Fusarium* species, because of interest in the potential use of such a toxin as a battlefield weapon that could incapacitate enemy troops through aerosol dispersion. A seminal paper (Croft et al. 1986) drew attention to the potential exposure to mycotoxins from indoor mold growth, specifically that of *Stachybotrys atra* (=*S. chartarum*), with subsequent health effects. The energy crisis of the 1970s brought about a change in building construction, a tightening of buildings without sufficient consideration for ventilation needs, and the use of new, more moisture susceptible building materials, with the result that dampness and mold growth were recognized as an increasing problem as health effects in damp buildings were reported (IOM 2004; WHO 2009; NIOSH 2012). Considerable research on molds and mycotoxins in the built environment has been done (Nielsen and Frisvad 2011) and has recently been reviewed by Miller and McMullin (2014).

The work sponsored by USAMRIID used high concentrations of T-2 toxin in inhalation studies on mice, rats, guinea pigs, and swine to determine acute effects, including lethality, to determine utility of mycotoxins as a warfare agent. In addition to determining acute lethality and target tissues, these studies determined interesting differences between exposure routes.

For instance, inhalation of pure T-2 toxin in rats was 20 times more potently lethal at 10-min high-level exposures than other routes of exposure, including dermal application, ingestion, and systemic injection (Cresia et al. 1990). Because the swine respiratory system is more like humans than those of rodents, swine were also tested. Inhalation exposure of swine to T-2 toxin was as potent as intravenous injection for a number of the end points examined (Pang et al 1987, 1988), indicating direct access of the toxin from the lung to the circulation.

The primary function of the respiratory system is the efficient exchange of oxygen and carbon dioxide across the membranes of the alveoli and the capillaries which surround them. The blood carries oxygen (and other absorbed substances) to tissues throughout the body and carbon dioxide from those tissues back to the lung to be eliminated. However, the air is not composed of only nitrogen, oxygen, and a number of noble gases, but contains many other gases and particles produced by the natural environment and by human endeavors. The respiratory system takes in whatever is in the air, and the tissues ideal for necessary gas exchange present little barrier to small molecules, including mycotoxins, and, unlike the digestive system, allows direct entry into the circulatory system and the blood and thus to systemic target organs. Ingested materials are absorbed primarily through the small intestine into the enterohepatic circulation which routes blood to the liver where many poisons are detoxified, before blood is returned to the systemic circulation.

The nature of inhaled contaminants, especially the size and aerodynamic diameter of particles, determines where in the respiratory system particles land and what happens when they interact with the tissues there. Mycotoxins are not considered to be volatile and are associated with spores and other particles, which serve as a carrier into the respiratory system.

Mycotoxins were found in and on spores from molds, and because spores from many fungal genera are dispersed through aerosolization, it was assumed that measurement of spores in air was a means of determining exposure to mycotoxins. In itself, measurement of spores in air in buildings is fraught with difficulty since air moves through natural and mechanical ventilation, and its content of spores varies with both building and human activity. Those molds that disperse their spores through the air do so periodically through "blooms," and some of them show temporal patterns. Thus, concentrations of spores captured during quiescence can be very low, and those captured during blooms can be many orders of magnitude higher.

Mycotoxins are located not only in and on spores however but in and on mycelia, on fragments of both spores and mycelia, and on dust settled on surfaces on which molds grow (Palmgren and Lee 1986; IOM 2004). Mycotoxins are exotoxins, secreted onto the surfaces on which molds grow. These toxins are secondary metabolites, not needed for maintaining life of the fungal cells, yet costing the organism considerable metabolic energy (Bennett 1983). Secretion of exotoxins by molds onto the surfaces where they (and some bacteria) grow is thought to be a means by which bacteria and molds compete for resources that they need for survival, and this competition is one of the primary stimuli for toxin production in mixed cultures of microbes (Nielsen 2003). Temperature, water activity ($a_w$), and nature of the substrate are other major influences (Moss 1991; Nielsen et al 2004; Frazer et al 2012). While inhibiting the growth of competing microbes provides advantage to producers, other organisms such as vertebrates, including humans, can be harmed when exposure inhibits life processes, such as protein, RNA, or DNA synthesis, among others, that microbes share with other life forms. To distinguish secondary metabolites of molds that primarily affect other microbes (antibiotics) or plants (phytotoxins), mycotoxins are defined as low molecular weight secondary metabolites produced by molds that can harm animals, including humans, at low levels of exposure (Bennett and Klich 2003; Miller and McMullin 2014).

## Inhalable Particles and Mycotoxins

Because mold spores and mycelia fragment into smaller particles as spores germinate and mycelia grow and because small particles can dislodge from substrate, particles and dust with adsorbed toxins can be aerosolized through building and human activity; attempts to determine inhalation exposure need to account for all these potential exposure agents (Palmgren and Lee 1986). Mycotoxins are generally not known to be volatile, although some water-soluble toxins can form droplet aerosols (Pestka et al. 2008). Jarvis et al. (1998) noted that satratoxins H and G produced by highly toxic strains of *S. chartarum* are exported to the surface of spores where they become water soluble, probably because they are imbedded in surface polysaccharides.

Spores of molds can be relatively large compared to fragments of spores and mycelia. For instance, the aerodynamic diameter of *Aspergillus versicolor* spores is 2.5 μm, those of *Cladosporium cladosporioides* 1.8 μm, and that of *Penicillium melinii* 3.0 μm (Górny et al. 2002). *S. chartarum* spores are about 5 μm in aerodynamic diameter (Sorenson et al. 1987). Spore and mycelial fragments 0.3 μm (the limit of detection of the optical particle counter used) and smaller outnumbered spores by as many as 320 times in an experiment that studied the release of particles from moldy ceiling tiles from vibrations normally occurring in buildings, under various airflow conditions (Cho et al. 2005,2007; AIHA 2008). For *S. chartarum*, the number of fragments was 540 times that of spores.

Seo et al. (2007) developed a new field-compatible collection system for small fungal particles that could divide particles into three size fractions: spores (greater than 2.5 μm), fragment/spore mixtures (1.0–2.5 μm), and submicron fragments (less than 1.0 μm). This system was used in measuring fungal fragments in moldy houses during a field study in homes in New Orleans and Southern Ohio (Reponen et al. 2007) which found that the fungal fragment to spore ratio in actual moldy homes was much greater than had been determined under laboratory conditions by Cho et al. (2005). The spore to fragment ratio would be $10^3$ for the 0.3-μm fragment size and $10^6$ for the 0.03-μm fragment size, very much higher than the estimates from the earlier laboratory experiments. The authors hypothesize that naturally occurring air currents in homes and vibrations from activity and other disturbances account for the difference. Because sampling times ranged from 2 to 3 h, aerosolized spores could also have settled out, increasing the particle to spore ratio. Small particles, including mold fragments and dust from surfaces on which mold grows, are more important from a mycotoxin exposure perspective than mold spores, because small particles have a much larger aggregate surface area per unit mass than larger particles such as spores do. Small particles may convey much larger exposure to mycotoxins (and allergens) than spores and larger fragments do (AIHA 2008). Small particles are more buoyant and remain suspended in air for longer periods of time than large particles do and penetrate deeply into the lung, where natural defenses are fewer (IOM 2004; AIHA 2008).

Mycotoxins in settled dust and particulate air samples from damp indoor environments have been measured in a number of studies. Sterigmatocystin has been measured on building materials and dust (Andersson et al. 1997), in carpet dust (Englehart et al. 2002), and in moldy building (Bloom et al. 2007, 2009a, b). Toumi et al. (2000) found that of 79 bulk samples from moldy buildings, 43 % contained one or more mycotoxins and 15 % contained trichothecenes. The most prevalent toxin was sterigmatocystin, which was detected in 19 samples. The HITEA study

(Health Effects of Indoor Pollutants: Integrating microbial, toxicological, and epidemiological approaches) in Spain, the Netherlands, and Finland (Täubel et al. 2011) found a large number of fungal and bacterial secondary metabolites, including mycotoxins in dust samples from water-damaged schools.

Brasel et al. (2005a, b) found airborne particles smaller than conidia containing macrocyclic trichothecene mycotoxins from particle samples from damp indoor environments. Dustborne and airborne fungal particles were found to differ with differing housing characteristics, and mold, fungal particles, and endotoxin were found in homes investigated after hurricanes Katrina and Rita (Chew et al. 2003, 2006; Rao et al 2007). Spores, fragments, and dust particles can be re-entrained into the breathing space of occupants of damp buildings and be inhaled. Toivola and Nevalainen (2004) assessed personal exposures to microbes and particles and compared them to stationary particle monitors to obtain a better understanding of personal versus surrogate exposure. Particles from combustion (black smoke) correlated with personal monitors at home and at work. However, the stationary time-weighted microenvironmental model underestimated personal exposure to particle mass, viable fungi, total fungi, and total bacterial concentrations. Personal monitors gave better estimations of exposure to biological particles than the stationary monitors.

The size of particles determines their aerodynamic behavior and their deposition in the respiratory system. Cho et al. (2005) affirmed this for deposition of fungal fragments in the lung. Particles are deposited into the respiratory system depending on their size (studied as mean mass aerodynamic diameter or MMAD), the anatomy of the respiratory system, and air flow patterns that differ for nose, naso-oral or oral breathing, and respiratory rate and depth. Deposition is not homogenous, and larger particles (2–5 μm) tend to concentrate via impaction and deposition in the nose and upper airways of the lung (Balásházy et al. 2003).

The lower respiratory system branches out from the trachea below the glottis into two large bronchi and then into multiple generations of bronchiolar tubes (the number of which depends on the species of animal), ending in closed air sacs, the alveoli. All of the conducting airways are lined with ciliated epithelium and secretory cells that produce mucus, enzymes, and defensive molecules. Cilia beat in the direction of the glottis, moving a particle-trapping mucous blanket upward toward the esophagus where it is swallowed. The respiratory bronchioles nearest the alveoli lack cilia (St. George et al. 1993), but have outpouching alveoli in humans. As described above, the alveoli are a very fragile barrier between the air and the blood and are a portal to the bloodstream, which can carry contaminants to systemic target organs.

The MMAD determines whether particles deposit on the respiratory epithelium through impaction, gravitation, or diffusion. Large particles (˃5 μm) will mostly deposit in the nasopharyngeal region. About a 16 % fraction of 0.01-μm-size particles will also deposit in this region and about 40 % of 0.05-μm-size particles as well. The majority of submicron particle deposition occurs in the pulmonary region of the respiratory tract, that is, in the respiratory bronchi and alveoli (Bond 1993).

Impaction occurs as larger particles are pulled into the nose and down into the lung. Because of their inertia, particles encounter surfaces of the nasal turbinates, the bronchi, and bronchial junctions in the tracheobronchial area of the lung. Smaller particles deposit through the effect of gravity, settling on junctions where bronchioles branch in the lower airways (Kleinstreuer et al. 2008). Foci of impaction and sedimentation can represent hot spots of exposure to mycotoxins carried by particles, since they lodge there in larger numbers and may remain for longer periods of exposure (Balásházy et al. 1999, 2003; Miller and Ammann 2005). Such hot spots in the tracheobronchial regions can deliver large particle doses to small numbers of epithelial cells. Balásházy et al. (1999) calculated that as particle diameters increase from 0.01 to 10 μm, a small area of 0.1 by 0.1 mm can receive 50 to more than a hundred times the particle deposition than the average cell of the airway. Noting that neoplastic lesions of smokers are seen predominately at the bifurcations of the central airways, Churg (2000) has noted that particles landing at these junctions were slow to clear and were taken up by mucosal tissues in animals. They noted that in human autopsy sections of lungs, particles were concentrated in an almost 10:1 ratio at the branching of bronchioles compared to uptake in the straight, tubular portion of the airways.

Very fine particles (less than a micron in MMAD) enter the entire respiratory system, but especially the alveoli, through diffusion, behaving as a gas throughout the respiratory space until they eventually touch surfaces of the respiratory tract. In the alveoli they can be pinocytosed by alveolar cells or even absorbed directly through the two thin layers of alveoli and capillary squamous epithelium and enter the bloodstream of the general systemic circulation and thus reach specific target organs such as neural, immune or cardiac, and other tissues outside the lung (Oberdörster et al. 2002, 2004; Peters et al. 2006).

Only a fraction of particles inhaled deposit on respiratory surfaces; some fraction is exhaled again. Of the amount deposited, some will be trapped in the mucous of the mucociliary escalator and be transported to the oropharynx and swallowed into the digestive tract. Many mycotoxins have been shown to inhibit the function of ciliated cells or cause apoptosis of these, pulmonary macrophages, and mucus-producing cells, thus delaying or preventing clearance and prolonging exposure to the tissues where they land (Sorenson et al. 1986; Jakab et al. 1994; Amitani et al. 1995; Pestka et al. 2008).

Rate and efficiency of clearance mechanisms must also be considered. The mucociliary clearance mechanism works quite efficiently to trap larger particles in the nose and upper and tracheobronchial regions of the respiratory tract down to the respiratory bronchioles. Alveolar macrophages can capture some particles in the alveoli. The respiratory bronchioles in humans are not ciliated, and alveolar macrophages are generally not found there, so that area is especially vulnerable to particle deposition, longer-term residence time, and damage from toxic effect (Lippman et al. 1980; St. George et al. 1993; Oberdörster et al. 1994). Rate of ciliary beat and mucus movement, as well as uptake by alveolar macrophages, can vary depending on both physiological and toxicological influences.

In the alveolar area, alveolar macrophages can take up particles and remove them either to the interstitium of the lung or to the lymphatic system. Some macrophages may also crawl up to the mucociliary escalator and ride it to the oropharynx, where they are swallowed. Rates of particle uptake by macrophages, as well as movement and disposal, are influenced by the toxic potency of the particles taken up by these cells. Many mycotoxins have been shown to be cytotoxic, specifically to macrophages and other cells of the lung (Sorenson et al. 1986; Sorenson 1993, 1999; Yang et al. 2000). Toxic influences on the mucociliary escalator and on macrophages affect clearance. Longer residence time of particles on respiratory surfaces can allow for greater dose rate through leaching of adsorbed toxins off the particles into the interstitium, lymph or blood circulation, or increase direct damage to cells on which particles land, and into which they may be taken up. The effect on such cells for many mycotoxins, especially the trichothecenes, is apoptosis or premature cell death.

High particle concentrations of any nature, including those that are biologically derived, can have an effect on both mucociliary and macrophage particle clearance. High loads can inhibit the function of both or even damage the cells involved. Very high fine particle accumulation can lead to retention of particles in the interstitial tissues of the lung (Oberdörster et al. 1994).

Respiration rate and mode of breathing, that is, nasal, oronasal, or mouth breathing, can also affect not only the amount of particle deposit but also its location within the respiratory system. For instance, extrapolation of animal data on respiration to human must take into account such differences since particles inhaled by a two-legged human, with upright stance, will result in particles depositing in different areas of the lung than that which occurs with a four-legged animal. Rodents generally are obligate nose breathers, while humans can breathe through their nose at rest, their nose and mouth, or their mouth predominantly, depending on activity level and whether they are talking, singing, or crying or whether they are suffering some impairment due to allergy or disease (Pope 2000; American Academy of Pediatrics 2004; Salvi 2007).

For human children mouth breathing is more common than in adults. Human children (as well as the young in animal experiments) are more exposed than adults because they breathe more air per unit body weight than adults do. Children therefore are both more exposed to air contaminants than adults and more susceptible to toxins. Their respiratory system, immune system, and blood–brain barrier are not completely formed until months to years after birth, and defenses against inhaled toxicants are less effective than in adults (Pope 2000; American Academy of Pediatrics 2004; Salvi 2007).

Determination of dose to the respiratory surface from particles themselves is calculated by subtracting clearance from deposition. Attempts to determine dose from biological particles such as mold spores, hyphal fragments, or dust particles must take factors described above into account, no matter what health point is being investigated. However, residence time, solubility of toxic substances on and in spores, spore and hyphal fragments, and dust become an important consideration in determining the effects of mycotoxins that can be introduced to the respiratory system through inhalation of particulate matter.

## Nasal Exposure

The nose is the portal to the respiratory system. Its mucous membrane-lined baffles moisten and warm the air, and its mucus traps particles that can be cleared from the respiratory to the digestive system, through ciliated cells moving the particle-containing mucus downward to the oropharynx, where it can be swallowed. However, particles in inhaled air also move upward in the nasal passages and can enter the paranasal sinuses. Mold spores have been shown to colonize the mucous membranes of the sinuses, but may or may not elicit an immune response and play a role in the development of chronic fungal rhinosinusitis (Ponikau et al. 1999). However, in following 210 patients with chronic rhinosinusitis, Ponikau et al. (1999) found that 202 of the 210 patients tested positive for culturable fungi. Allergic fungal sinusitis was diagnosed in 94 of 101 consecutive surgical cases with chronic rhinosinusitis, based on histology and culture results. Polzehl et al. (2005) used both culture methods and polymerase chain reaction (PCR) in examining nasal lavage samples of patients with chronic rhinosinusitis and used the two methods to detect fungi in 50 % of patients examined that had been diagnosed with chronic rhinosinusitis.

Chronic rhinosinusitis can also involve bacterial colonization, or a mixture of microorganisms, and is characterized by prolonged inflammatory mucosal response (Harvey and Lund 2007). Chronic rhinosinusitis is often resistant to treatment through antibiotics, and surgical intervention, after which the condition frequently returns. Evidence exists for the formation of a biofilm on sinus mucous membranes as a part of this condition (Harvey and Lund 2007; Foreman et al. 2009, 2011, 2012; Suh et al. 2010; Boase et al. 2013). A biofilm is described as complex surface-associated populations of opportunistic or pathogenic microorganisms that are embedded in an extracellular matrix, whose members are different from free-living microorganisms (Harvey and Lund 2007; Hall-Stoodley and Stoodley 2009). Bacteria such as *Staphylococcus aureus*, *Haemophilus influenzae*, and *Pseudomonas aeruginosa* have been found in biofilms in patients suffering from this illness, as have a number of fungi such as yeasts of *Candida* species, *Aspergillus fumigatus*, *A. versicolor*, *Chaetomium*, and *Trichoderma*, among others (Foreman et al. 2009; Muszkieta et al. 2013).

Infections that involve biofilms are important clinically because bacteria and fungi that are part of a biofilm are recalcitrant against both antibiotic treatment and host defenses (Healey et al. 2008; Hall-Stoodley and Stoodley 2009). Biofilms protect bacteria against through a number of mechanisms, while the same bacteria as free-living organisms are susceptible (Parsek and Singh 2003). Similar mechanisms for fungal biofilm resistance are described by Ramage et al. (2012) for a number of nosocomial infections. Successful treatment of chronic rhinosinusitis with certain antifungal agents amphotericin B (Anyanwu et al. 2004; Ponikau et al. 2005) and itraconazole (Ponikau et al. 2006; Seiberling and Wormald 2009; Denning et al. 2009) has emphasized fungal role in this yet poorly understood disease.

Biofilms that contain mixed cultures of fungal and/or bacterial species may likewise be protected and may represent a means for increased mycotoxin production as a result of competition among such organisms within a biofilm. Bruns et al. (2010), using genomic techniques, showed that primary metabolism was reduced in mature biofilms containing *Aspergillus fumigatus,* but biosynthesis of secondary metabolites, especially gliotoxin, was upregulated. It is generally accepted that genes of fungi and bacteria that are silent in laboratory culture can be upregulated to produce secondary metabolites in mixed cultures (Schroeckh et al. 2009; Nützmann et al. 2011) as well as in biofilms (Zavahir and Seneviratne 2007; Bruns et al. 2010; Morales and Hogan 2010). With the sequencing of a number of fungal genomes, it has become obvious that genes coding for many secondary metabolites are only expressed when the organisms coexist in mixed microbial cultures, provided that water activity and temperature are favorable (Nielsen 2003; Schroeckh et al. 2009; Nützmann et al. 2011). The term "genome mining" is being used in efforts to awaken silent gene clusters through microbial cross-talk in the search for new antibiotics (Brakhage and Schroeckh 2010; Kück et al 2014; Seneviratne et al 2008).

Mycotoxins have been measured in patients with chronic rhinosinusitis (Lieberman et al. 2011). The role for particular secondary metabolites including particular mycotoxins in the formation of biofilms as part of chronic fungal sinusitis is being studied (Wargo and Hogan 2006; Zavahir and Seneviratne 2007; Shank and Kolter 2009). It is known that many mycotoxins have a modulating effect on immune function (Bondy and Pestka 2000; WHO 2009), which may in itself play a role in infection and the formation of biofilms in sinuses and in the nasal mucosa.

## Uptake from the Nose into the Brain

The nose not only is the portal to the respiratory system but also serves as a gateway to the nervous system. The olfactory nerve (cranial nerve I, the sense of smell) sensory receptors are located in the olfactory epithelium, and its afferent fibers lead to the olfactory bulb of the brain through the sieve-like bony cribriform plate of the ethmoid bone of the skull, located above the nose. The trigeminal nerve (cranial nerve V or common chemical sense that responds to pungency) also has fibers that innervate the nasal epithelium and dendrites that terminate in the olfactory bulb and elsewhere in the brain (Thorne et al. 2004; Brand 2006; Silver and Finger 2009).

It has long been known that a number of metals, small peptides, small particles, and viruses and a number of drugs can enter the brain directly via nasal neural transport (Shipley 1985; Tallkvist et al. 2002; Oberdörster et al. 2004; Lewis et al. 2005). These pathways are currently being explored to find means of treating central nervous system (CNS) disorders that are recalcitrant to treatment systemically because they are barred from the brain by the blood–brain barrier (Hanson and Frey 2007; Scranton et al. 2011). Axonal transport of particles, as described by Fechter et al. (2002) and Oberdörster et al. (2004), is especially relevant to mycotoxin exposure of the brain since currently inhalation exposure to mycotoxins is estimated to be primarily from small particles (Rand and Miller 2011).

Tracing of radiolabeled particles administered intranasally demonstrates that such small particles not only travel to the olfactory bulb of the brain but are distributed to the brain in general (Oberdörster et al. 2004; Scranton et al. 2011). The rostral migratory stream (RMS) is a structural entity that connects the olfactory bulb to periventricular regions of the brain. It is a pathway in mammals (though not well described in humans) several millimeters in length that connects the subventricular zone (SVZ) with the olfactory bulb, which has been largely studied in its transport of neuroblasts from the SVZ to the olfactory bulb (Scranton et al. 2011). Scranton et al. (2011) used low molecular weight fluorescent probes and radiolabeled ligands administered intranasally in mice to show that fluorescence and radioactive signal was found throughout the brain, but not in peripheral tissues like the lungs or the blood. Surgical interruption of the RMS prevented uptake into the brain.

Islam et al. (2006) instilled the macrocyclic trichothecene satratoxin G (SG) produced by certain strains of *S. chartarum* into the noses of mice and found that SG specifically induced apoptosis in olfactory sensory neurons in the olfactory epithelium in a dose–response manner. They found a lowest-effect level 24 h after exposure to be 25-μg/kg body weight and a no-effect level to be 5-μ/kg body weight, with severity increasing with dose. Time course of a single instillation of 500 μ/kg showed that maximum damage to the olfactory epithelium occurred at 3 days; exposure to lower doses (100 μg/kg) for 5 consecutive days resulted in the same degree of atrophy and apoptosis, indicating that the damage was cumulative. SG also induced neutrophilic rhinitis from the sensory epithelium, an indication of inflammation, and inflammatory cytokines were found in the olfactory bulb and within the brain. Satratoxins are not commercially available and are difficult to purify from fungal culture, so Islam et al. (2007) used roridin A (RA), a structurally related trichothecene produced by the fungus *Myrothecium* that is commercially available, and found similar neurotoxic effects in their mouse model. They also determined the kinetics of RA transport. RA-induced apoptosis of olfactory sensory neurons was observed 24 h after instillation and was maximal after 72 h. Atrophy of the olfactory nerve layer of the olfactory bulb in the brain occurred concurrently. Concomitant exposure to endotoxin (lipopolysaccharide found in cell walls of Gram-negative bacteria) potentiated the effect both in sensory neurons and the olfactory bulb. Certain Gram-negative bacteria grow in damp indoor spaces along with a number of mold species, including *S. chartarum*, so co-exposure to bacterial and fungal fragments is common. Microbial communities that include competing fungal and bacterial organisms also show upregulation and production of mycotoxins and bacterial exotoxins and are more likely to be sources of exposure via inhalation.

Kinetics of SG tissue distribution following intranasal exposure in mice was determined by Amuzie et al. (2010). SG was rapidly taken up from nasal tissues and distributed to tissues involved in respiratory, immune, and neuronal function, then cleared. The nasal turbinates retained a significant amount of the toxin, which may account for the toxin's ability to cause sensory neuron death. The toxin was rapidly cleared from plasma and low levels remained in blood. This can be explained by efficient distribution to various organs such as the kidney, lung, spleen, thymus, heart, olfactory bulb, and brain. As the authors point out, selective targeting of neurons in the nose and olfactory bulb of the brain is intriguing because olfactory function loss often occurs in the early stages of degenerative illnesses such as Parkinson's and Alzheimer's disease.

Rodents and humans differ in their breathing habit and have differences in the structure of their noses. Their nasal turbinates are complex and branching, and they have large amounts of olfactory epithelium which covers about 50 % of their intranasal surface. Humans and other primates such as rhesus monkeys have simple turbinate structures (Harkema et al. 2006), and their olfactory epithelium is only a small part of the nasal cavity, about 15 % of the total intranasal surface.

Carey et al. (2012) developed a young adult rhesus monkey model, whose nasal structures and airways more closely resemble that of humans, to study acute and repeated SG exposure from intranasal instillation. They found that low-dose (5 μg) exposure to SG caused neutrophilic rhinitis and apoptosis of olfactory sensory neurons in monkeys, similar to the injury described above in mice at comparable exposure concentrations. Four repeated daily doses of SG at the low concentration caused more damage than a single high dose of 20 μg, indicating cumulative damage. These experiments demonstrated that injury to the olfactory sensory epithelium and brain by SG occurs in primates whose nasal anatomy is similar to that of humans so that correlations to human injury can be made.

It has been argued by some authors that the number of *Stachybotrys* spores required to cause adverse human health effects in damp indoor spaces would be very large to be comparable to these pure toxin exposures (Chapman et al. 2003; Hossain et al. 2004; Kelman et al. 2004; Hardin et al. 2009). However, since SG is not only found in and on spores but in nonviable fungi, fungal fragments, and dust from surfaces on which *Stachybotrys* is growing, spores are not the only exposure agents to be considered. Concentrations of fungal fragments of *S. chartarum*, aerosolized through a fungal spore source strength tester (FSSST) and collected with an electrical low-pressure impactor (ELPI) that measured size distribution, were 514 times greater than that of spores (Cho et al. 2005). The particles were aerosolized from agar plates within the FSSST and thus were limited to fragments of *S. chartarum* spores and mycelia and did not include fine dust particles containing the exotoxins that could be aerosolized from moldy surfaces in the built environment. Bloom et al. (2007) demonstrated that mycotoxins, such as the macrocyclic trichothecenes SG and satratoxin H (SH), sterigmatocystin produced by *A. versicolor*, and citrinin, gliotoxin, and patulin produced by *Aspergillus* and *Penicillium* species, could be isolated from dust above floor level from moldy buildings. Gottschalk et al. (2008) used LC-MS/MS to measure airborne SG (0.25 $ng/m^3$) and SH (0.43 $ng/m^3$) in a water-damaged building. Satratoxin-equivalent concentrations ranging from 2 to 330 $ng/m^3$ have been estimated to occur in some water-damaged rooms within a home (Yike et al. 1999; Vesper et al. 2000; Carey et al. 2012).

About 70–90 % of inhalation exposure to molds is thought to be from fungal fragments measured as β-D-glucan, a structural molecule from fungal cell walls (Salares et al. 2009). Small particle air pollutants have been shown to translocate to the brain (Oberdörster et al. 2004) through the olfactory pathway and have been linked to cognitive deficits and brain abnormalities in children and dogs (Calderon-Garcidueñas et al. 2008). Epidemiological studies have linked traffic-related fine particles to cognitive decline in elderly men (Power et al. 2011) and women (Weuve et al. 2012), with mechanisms of oxidative stress and inflammation similar to those modeled in human neurological cells by Karunasena et al. (2010) for satratoxin H, at toxin levels found in wet and damp indoor spaces. The effect modeled in human cells, together with the neurological damage, demonstrated in mice and rhesus monkeys that satratoxins specifically target neural tissue in the nose and brain, and the neurological deficits reported in some occupants of moldy built environments (Johanning et al. 1996; Gordon et al. 2004) provide support that the nose–brain connection is an important route of inhalation exposure for mycotoxins.

## Uptake by the Lung

Mycotoxin effect on the respiratory system in general, and the lung in particular, has been studied intensively since an outbreak of pulmonary hemorrhage in 34 infants, 10 of whom died, in 1994–1998 in Cleveland, Ohio, in the USA (Dearborn et al. 1999; AIHA 2008). Because of the epidemiologic link of this outbreak to *S. chartarum* and subsequent case reports linking this mold with bleeding lungs in infants and children, much of this research has focused on the effect from *Stachybotrys* toxins (Nikulin et al. 1996, 1997; Jarvis et al. 1998; Rao et al. 2000; Gregory et al. 2004; Dearborn et al. 2002; Mader et al. 2007; Pestka et al. 2008).

Direct damage to various cells of the lung, as well as the effect of compromise of mechanical and immunological lung defenses, has been explored for macrocyclic trichothecenes from *S. chartarum.* Other toxins, for example, gliotoxin produced by *A. fumigatus* and some other molds, have been implicated as a virulence factor in aspergillosis, and its role in damage to ciliated respiratory cells, with decrease in particle clearance of the lung, has been studied (Amitani et al. 1995; Lewis et al. 2005; Amitani and Kawanami 2009). Aflatoxin $B_1$, produced by *A. flavus* and *A. parasiticus*, also has ciliostatic effects and impairs phagocytosis of particles (including viruses and bacteria) by alveolar macrophages and suppresses antibody response (Jakab et al. 1994).

Absorption of mycotoxins within the respiratory system is largely a factor of particle distribution since mycotoxins are not considered to be volatile (although some are semi-volatile). As described above, mycotoxins are in and on spores, on fragments of spores and mycelia. Because they are exotoxins secreted by fungi onto the surfaces on which molds grow, they are also found on dust associated with the moldy surfaces (Englehart et al. 2002; Nielsen 2004; Bloom et al. 2009a, b). Particles therefore act as carriers for the toxins. Some mycotoxins, such as the satratoxins, are water soluble and can form liquid aerosols and can be leached off particles into cells (Pestka et al. 2008). Whether and to what degree absorption of mycotoxin-associated particles occurs depends on a number of factors, of which the most important are the morphology and cellular makeup of the tissues where particles land, the physical and chemical nature of the toxins, and the ability of cells to take up toxins (Dahl and Gerde 1994; Miller 1999).

Recent epidemiological meta-analyses and reviews have linked exposure to damp indoor environments with worsening and development of new asthma and other respiratory health effects (Fisk et al. 2007; Mendell et al. 2011; Quansah et al. 2011). Respiratory exposure to mold and other microbes was implicated in these disease outcomes, although the role of specific mycotoxins had not been studied.

Mycotoxins and their effects associated to the lung and cells of the lung in culture of human and animal cells of the lung, and in animals, have recently been reviewed (Rand and Miller 2011; Miller and McMullin 2014). Evidence from such studies, especially those by Miller et al. (2010), indicates that low molecular weight compounds, including mycotoxins, produced by molds that grow in damp indoor environments are strongly pro-inflammatory and modulate genes in the lungs of mice for inflammatory cytokines and mucus production analogous to those in humans related to nonallergic, that is, toxin-induced, respiratory health effects.

There is ample evidence from animal studies that mycotoxins have systemic effects aside from damage to the route of entry. Early inhalation studies (Cresia et al. 1987, 1990; Pang et al. 1986, 1987, 1988) indicate that systemic effects occur more directly from this exposure route. Cresia et al. (1990) found that inhalation exposure of rats to the simple trichothecene T-2 toxin produced significant lesions to immune tissues throughout the body of these animals, while the lung remained intact. Myocardial and pancreatic lesions were demonstrated in swine (Pang et al. 1986, 1987) after intravenous administration of sublethal doses of T-2 toxin. Inhalation was equivalent to intravenous administration in later experiments in swine (Pang et al. 1988). Okai et al. (2008) found that inhalation of *S. chartarum* caused pulmonary arterial hypertension in mice.

Di Paolo et al. (1993) describe a case of acute renal failure from ochratoxin exposure in a woman who inhaled ochratoxin-containing wheat dust while working for 8 h in a granary that had been shut for several months. Guinea pigs and rabbits were experimentally exposed to the same dust and were shown to have kidney and liver lesions. The patient recovered with treatment after several months. Concentrations of toxins in the wheat dust were not determined.

Experimental exposures of humans to mycotoxins are not permissible, as mycotoxins are defined as causing harm to animals and humans at low doses, and some mycotoxins that have been isolated from damp indoor spaces are known to be potently toxic at low exposures. While there are some putative biomarkers for human exposure, such as DNA adducts for some genotoxic mycotoxins, it may be possible to determine exposure by measuring DNA adducts in bodily fluids, but not differentiate ingestion from inhalation or dermal exposure.

## Complexity of Damp Indoor Spaces

A review of the complexity of the damp indoor environment by Thrasher and Crawley (2009) discusses nine types of biocontaminants: indicator molds, Gram-negative and Gram-positive bacteria, microbial particulates, mycotoxins, volatile microbial compounds (both of microbial and nonmicrobial origin), proteins, galactomannans, 1→3-β-D glucans, and lipopolysaccharides (LPS), endotoxins, and toxic or health effects that have been associated with each category of contaminant, when such information was available. They could also have included allergens, which may exacerbate effects, or through inflammation, allow more exposure to the contaminants they discuss. People breathing air indoors are also exposed to particulates from combustion and animal detritus, VOCs from building materials, cleaning agents, paints, and odorants, among other air contaminants; exposure to many, or all of these, may have overlapping symptoms with those from biocontaminant exposure. The complexity of microbial growth in damp indoor environments also makes it difficult to identify specific agents related to disease. When health effects in damp structures are discussed, occupants' symptoms are often described as "nonspecific," when in fact the symptoms are specific to individual contaminants, but there is overlap of symptoms from exposure to multiple contaminants. There are many air contaminants that invoke "irritation of mucus membranes," for instance, and this may be related to a number of disease entities. Their combined exposure can exacerbate symptoms or worsen specific diseases, but the symptoms and disease are still specific to each contaminant exposure. What is "nonspecific" is that the description of exposure is incomplete.

Microbial communities in damp buildings are part of a larger ecosystem where populations of bacteria, fungi, and protozoans change over time in response to changes in resources and ambient conditions. For example, increases in moisture will allow molds that require a high water activity ($a_w$) to predominate, while xerophilic molds do not grow well. The reverse could prevent molds such as *S. chartarum* from growing. Specific nutrient availability will select for microbes in the building just as they do in laboratory media. Some molds and bacteria working to survive in a particular ecologic niche can produce secondary metabolites such as toxins to compete for resources. Water activity ($a_w$) and temperature changes allow different microbes to thrive (Nielsen et al. 2004).

The difficulty in establishing what the exposure agents from microbial growth in buildings is illustrated by Tuomi et al. (2000). They analyzed bulk samples from building materials from moldy interior surfaces in buildings with moisture damage for mycotoxins in Finland. They found sterigmatocystin, satratoxins G and H, diacetoxyscirpenol (DAS), deoxynivalenol (DON), and verrucarol in varying amounts in many samples. They also cultivated the samples on 2 % malt extract agar at 25 °C for 7 days before identification of fungal species. They found *Aspergillus versicolor*, a producer of sterigmatocystin, but also found this mycotoxin in samples that did not reveal this fungus through culture. Some sterigmatocystin-containing samples did not yield *Aspergillus* species, yet yielded *Penicillium* species; the majority of the samples containing *Penicillium* species yielded neither sterigmatocystin nor citrinin, the latter of which was found in a few of the *Penicillium* spp. samples. Satratoxins were found only in samples from which *S. chartarum* was grown, with one exception. Tuomi et al. (2000) point out that it is not surprising to find toxins without finding the known producer fungal species or to find the fungal species without finding the toxin, because of the dynamic nature of microbial communities in building ecosystems. The authors note that toxigenic species have different growth requirements and that the lack of correlation of species isolated with mycotoxins known to be produced by various mold species may also be due to the researcher's procedure of using a single growth medium and temperature for all their samples.

Instead of recognizing that a damp building is a complex ecosystem that changes with water, temperature, and microbial interactions, some authors have misinterpreted the paper by Tuomi et al. (2000) as stating that fungi growing on building materials do not always produce mycotoxins. This became a common view that has now been realized to be erroneous (Miller and McMullin 2014). The presence of other microbes competing for vital resources is one of the strong influences for toxin production in both microfungi and bacteria capable of producing exotoxins (Nielsen 2003; Wargo and Hogan 2006; Schroeckh et al. 2009; Bruns et al. 2010; Nützmann et al. 2011; Frey-Klett et al. 2011).

There is evidence that the presence of mixed microbial communities sends signals that activate genes coding for metabolites that are silent when isolated bacteria or fungi are grown in culture. Various techniques that take advantage of microbial interactions to awaken silent genes to produce secondary metabolites that might be useful in finding antibiotics are currently being exploited (Bergmann et al. 2010; Brakhage and Schroeckh 2010; Frey-Klett et al. 2011; Kück et al. 2014).

The complex spectrum of secondary metabolites measured in damp indoor spaces was recently elucidated as part of the HITEA study of schools in Finland, Spain, and The Netherlands. Using a multi-analyte, HPLC tandem-mass spectroscopy methodology to analyze metabolites in damp school environments, Vishwanath et al. (2009) were able to determine 159 fungal and 27 bacterial metabolites, many of which had never before been identified in samples from actual damp environments. In a subsequent study, the same methodology was applied to 69 samples from severely moisture-damaged homes, and these were found to have at least 1 of 186 targeted metabolites. Thirty-three bacterial metabolites were found co-occurring with mycotoxins. The bacterial compounds monactin, nonactin, staurosporine, and valinomycin were found in moist structures, and chloramphenicol was found in settled house dust. These highly bioactive compounds are produced by *Streptomyces* spp., bacteria which are considered to be indicators of dampness in buildings (Täubel et al. 2011).

Kirjavainen et al. (2015) used LC-MS/MS methodology to analyze dust samples from the living rooms of 93 homes of 1-year-old children in Finland, of which 15 had moisture damage, for 330 secondary metabolites. The purpose of this study was to characterize the presence of microbial secondary metabolites and to determine whether there was an association with molds and moisture damage and with the development of asthma. The study found that secondary metabolites were ubiquitously present in low concentrations in home floor dust, even in the absence of moisture damage. Forty-two different metabolites were found. The total load of metabolites in living room floor dust tended to be moderately increased if there was moisture damage in the living room or mold odor anywhere indoors, but this increase did not seem to have an adverse effect, but may rather have had a protective association with asthma in children. Of the children resident in the homes, 8 children had active asthma at age 6, and 15 had lifetime doctor diagnosed asthma. The macrocyclic trichothecene metabolites of *S. chartarum*, such as the satratoxins and verrucarol, which have been detected in dust samples from severely water-damaged homes and implicated in adverse health effects elsewhere (Bloom et al. 2007, 2009a, b; Peitzsch et al. 2012) were not detected, but the macrocyclic strains of *S. chartarum* are rarely found in the Nordic climate. Eleven of the samples in this study did contain stachybotrylactam, a toxic metabolite of *S. chartarum*, but this metabolite was not associated with moisture damage or asthma. The apparent protective finding of these low concentrations of secondary fungal and bacterial metabolites is of interest in view of the hygiene hypothesis, but the number of asthmatic children in the study is small, and further investigation must be done.

Some toxicological studies have attempted to determine the adverse effects of exposure to more than one mycotoxin, or combinations of bacterial toxins and mycotoxins, but none have been able to mirror the complexity of exposure described above (Speijers and Speijers 2004). Studies using the mouse macrophage cell line RAW 264.7 to investigate effects from *S. chartarum* and *Streptomyces californicus* that were grown in coculture showed synergistically increased markers of inflammation, cytotoxicity, and immuno- and genotoxic effects (Huttunen et al. 2004; Penttinen et al. 2005; Murtoniemi et al. 2005; Markkanen et al. 2009). Tissue culture evaluation of toxic interactions can determine additivity, antagonism, and synergism in such a model system. Relatively few mycotoxins have been investigated for additive or synergistic effects or antagonistic inhalation effects on animals, and critical effects have generally not been determined.

## Risk Assessment

Risk assessment is generally used to determine allowable levels of exposure to toxicants for public health protection or regulatory efforts (Ammann 2012). Calls for risk assessment of airborne mycotoxins have been heard for many years. The American Conference of Governmental Industrial Hygienists (ACGIH) Bioaerosols Committee issued a statement in 1999 that described why the organization considered a threshold limit value (TLV) for worker protection to be inappropriate for the complex mixture of bioaerosols found in the built environment as the result of dampness (ACGIH 1999; Ammann 1999). In 2004, the Institute of Medicine of the National Academies of Science, in its report "Damp Indoor Spaces and Health", made this recommendation: "Animal studies should be initiated to evaluate the effects of long-term (chronic) exposures to mycotoxins via inhalation. Such studies should establish dose–response, lowest-observed-adverse-effect levels, and no-observed-adverse effect levels for identified toxicologic endpoints in order to generate information for risk assessment that is not available from studies of acute, high-level exposures." No scientifically credible risk assessment for inhalation of any mycotoxins associated with damp indoor environments has been possible, because these kinds of studies are not currently available. Assessment, especially of the more potent mycotoxins from molds that thrive in damp indoor environments, may still be useful for immediate hazard assessment in building investigations and contribute to public health actions.

Guidance for risk assessment has been put forward by a number of public health and regulatory agencies. The risk assessment paradigm developed by the USEPA for allowable inhalation exposure levels of humans of various susceptibilities to single non-cancer-causing air toxicants (reference concentrations or RfCs) is used for regulatory purposes in the USA under the Clean Air Act and is generally accepted as a useful means of limiting toxic exposures. Carcinogenic air toxics are analyzed differently through probabilistic models to determine risk with 1 in a million chance of cancer considered a threshold for allowable emissions.

According to USEPA guidance, four specific steps are involved in RfC development:

1. Hazard evaluation identifying a critical effect and a critical study from an extensive review of the human and animal literature. Various tissues have different susceptibilities, so it is important to elucidate effects of toxins in whole animals in order to determine a critical effect, that is, the effect that occurs in the most sensitive animal and the most sensitive animal tissue, at the lowest level of exposure. The assumption in determining critical effect is that protecting against the most sensitive toxic end point will also protect against damage occurring in less sensitive tissue.

   Because of route of entry and physiological effects, including barriers to toxin access and the ability of different tissues to detoxify or to bind or eliminate toxins, such considerations must also be explored. Route of entry, i.e., oral, or inhalation methodologies differ because potency to system tissues varies by route of entry. Ability to repair damage also differs in tissues, so, for instance, lesions to the nervous system may have an overall more profound effect on the organism than damage to the lung. Developmental effects also differ profoundly depending on the stage of development when toxic impact occurs.
2. Dose–response assessment to determine no-observed-adverse-effect levels (NOAELs) and lowest-observed-adverse-effect levels (LOAELs), usually from chronic animal inhalation studies if no human data are available. Conversion of animal to human data through dosimetry adjustments produces a human equivalent concentration (HEC).
3. Exposure assessment consisting of exposure measurements.
4. Risk characterization results in an RfC after the HEC has been divided by appropriate uncertainty factors. The RfC is also ranked as having levels of confidence in the data available and the confidence in the RfC itself (USEPA 1994; Ammann 2012). The RfC, as defined by USEPA, "is an estimate (with uncertainty spanning perhaps an order of magnitude) of a continuous inhalation exposure to the human population (including sensitive subgroups) that is likely to be without appreciable risk of deleterious non-cancer health effects during a lifetime

(USEPA 1994)." In regulating air toxics, the RfC value is divided into modeled or measured concentrations in air produced by an industry to determine risk to populations near facilities. Development of RfCs for individual mycotoxins may help in determining relation of those toxins to health, but exposure in damp buildings is not to individual toxins.

New understanding of exposure through fine particles, assessment technology, genomics and proteomics tools, and knowledge of microbial interactions in the production of toxins by both fungi and bacteria indoors shows that exposure is even more complicated than previously thought. To date, attempts at risk assessment for microbial exposure indoors have failed to account for the complexity and variability of organisms and their products, which are currently being revealed. Long-term inhalation studies have not been performed, and no critical effects from inhalation of mycotoxins have been determined.

Screening tools suggested to be used in risk assessments for low levels of industrial emissions (Drew and Frangos 2007), such as the concentration of no toxicologic concern (CoNTC), are not applicable for mycotoxins, as proposed by Hardin et al. (2009). Unlike industrial emissions, toxin concentrations are not reliably measurable in air, nor are they predictable, and the large number of secondary metabolites and their toxicological and physiological interactions are unknown. The decision tree as developed by ILSI (International Life Sciences Institute) Europe (2005) for additives in food, and suggested as a screening tool for air toxics by Drew and Frangos (2007), also forbids the use of the process for genotoxic compounds and aflatoxin-like compounds and recommends compound-specific toxicity data be used. Sterigmatocystin is an aflatoxin-like carcinogenic compound frequently isolated from damp indoor spaces where its primary producer, *Aspergillus versicolor*, is considered an indicator of dampness (Toumi et al. 2000; Englehart et al. 2002).

Hardin et al. (2009) includes sterigmatocystin, aflatoxins $B_1$ and $B_2$, as well as other genotoxic and carcinogenic mycotoxins, such as ochratoxin A and citrinin that require toxin-specific analysis under the ILSI decision tree. Hardin et al. (2009) quotes ACOEM (2002) which "estimated that $10^{10}$ spores/m$^3$ (of *S. chartarum*) would be required to achieve a 1 mg satratoxin/m$^3$ which was the no-effect concentration of T-2 toxin in 10-min rat inhalation exposures (Cresia et al. 1987, 1990)." In fact, the Cresia et al. 1990 study cited in the ACOEM and in Hardin et al. (2009) determined an inhalation $LC_{50}$ (lethal concentration$_{50}$) of 0.02 mg/L (20 mg/m$^3$) of air for T-2 toxin, not satratoxins G or H, which are more potent macrocyclic trichothecenes than the simple trichothecene T-2 toxin. A no-effect concentration cannot be determined from an $LC_{50}$ experiment. The number 1 mg/m$^3$ for T-2 toxin described as a no-effect concentration for a 10-min exposure to T-2 in Hardin et al. (2009) is in fact the concentration at which no rats died within 24 h after exposure. No lethality after 24 h is not equivalent to "no effect" unless effect is defined as death.

The authors of the ACOEM position paper also cited papers by Nikulin et al. (1996) regarding satratoxin G and H concentrations in *S. chartarum* (reported as

*S. atra*) spores to "provide perspective relative to T-2 toxin, 1.0 mg satratoxin would require $10^{10}$ (ten trillion) *S. chartarum* spores/m$^3$." The comparison was between pure T-2 toxin concentrations for "no mortality in rats" and concentration of *S. chartarum* spores containing satratoxins G and H calculated from Nikulin et al.'s (1996) concentrations of toxins of *S. chartarum* (cited as *S. atra*) spores. There was no consideration of the relative toxicity of T-2 toxin and the satratoxins in the determination of the number of *S. chartarum* spores that would be equivalent to the T-2 toxin "not dead" concentration. In citing Rao et al. (2000a, b), the ACOEM authors state "A range of doses was administered in rat studies and multiple, sensitive indices were monitored, demonstrating a graded response, with $3 \times 10^6$ spores/kg being a clear no-effect level." Actually, Rao et al. (2000a, b) do not state that a "clear no-effect level" exists in either paper. Changes in numbers of neutrophils (polymorphonucleocytes), macrophages, albumin, and lactic acid dehydrogenase concentrations (all signs of inflammation) and increases in hemoglobin (a sign of bleeding) in bronchoalveolar lavage (BAL) fluid of rats exposed to "toxic" *S. chartarum* spores bear out the conclusion of Rao et al. (2000a) that "our data indicate that direct pulmonary exposure to *S. chartarum* spores can cause severe inflammatory effects in the lungs."

Acute studies of satratoxins in animals have primarily focused on damage to the lung. In contrast, the Cresia et al. (1990) $LC_{50}$ study found that T-2 toxin caused no respiratory lesions in the exposed rats, but necrosis of cells in immune tissues was observed, especially in the spleen and thymus gland. T-2 toxin and the satratoxins seem to have different target tissues, as well as significantly different potencies. No long-term inhalation studies at low enough exposures that could determine a NOAEL for satratoxins or other mycotoxins, suitable for risk assessment, have been performed to date. Kelman et al. (2004) also describes the concentration of T-2 toxin at which no rats died in the Cresia et al. (1990) $LC_{50}$ study as "the 10-min no-observed-effect level (NOEL)" in using it to substitute T-2 toxicity data for that of trichoverrols A and B. They repeat this description in Table 20 of the paper. The paper states: "Because we were unable to identify any toxicity studies done on mammals with these mycotoxins (trichoverrols A and B), we chose to compare them to T-2 toxin, a trichothecene mycotoxin produced by *Fusarium* species and purified for use as a biological warfare agent" (Kelman et al. 2004). The same paper states, "A maximal airborne mold spore concentration (*N*) of 200,000 spores/m$^3$ was assumed based on our experience collecting air samples in indoor environments with abundant visible surface mold." No published data are cited to support this number. The complexity of exposure to mycotoxins, especially from fragments and dust with adsorbed toxins, was not considered.

The descriptions of risk from the ACOEM position paper (Kelman et al. 2004, and Hardin et al. 2009) lack scientific credibility. Credible assessments of risk from inhalation exposure to individual mycotoxins from appropriate studies are lacking. Risk from the complex and variable mixture of mold and bacterial spores, fragments, and products is even more problematic to assess.

In the absence of the ability to evaluate risk from complex exposures of microbial toxins and the ability to determine what "safe" levels of exposure to such toxins are, what can be done to protect the public from exposure to airborne mycotoxins indoors? The ACGIH in 2008, in its book *Recognition, Evaluation, and Control of Indoor Mold*, described the processes that cause buildings to be wet (with resultant microbial growth) and those building and maintenance practices that keep buildings dry and clean. For the present, prevention of exposure is the best option for avoiding health effects from mycotoxins via inhalation in damp indoor spaces, but also for attendant allergens and other contaminants found there.

Exposure to damp spaces is a significant public health issue, for which upper and lower respiratory tract illness, in particular to the development and exacerbation of asthma, has been related causally in epidemiologic studies (Fisk et al. 2007; Mendell et al. 2011; Quansah et al. 2011). The relationship, however, is to dampness and mold, and the specific causal agent or agents are still to be identified. Only about 50 % of asthma is related to allergy worldwide (Douwes et al. 2002). Work performed by workers at the National Institute for Occupational Safety and Health (NIOSH) has shown that healthy nonallergic workers, who move into damp and moldy non-industrial workplaces such as offices, can develop asthma, which implies that the causative agents are toxic singly, or in combination, and not allergens (Park et al. 2008; Cox-Ganser et al. 2009). Other systemic effects from inhaled mycotoxins have not been as well investigated in humans, but have been reported in numerous clinical papers.

Reduction in microbial numbers and fragments has resulted from targeted renovation (Huttunen et al. 2008). The emphasis on prevention is supported by several well-designed and controlled intervention studies, in which sources of dampness and mold were remediated, moldy and damaged materials removed, and cleaning of visible mold accomplished, with a dramatic reduction in symptoms and worsening of asthma (Kercsmar et al. 2006; Krieger et al. 2010).

## Summary

Inhalation exposure of mycotoxins differs from ingestion and dermal exposure because inhalation allows direct access of gases and small particles to the bloodstream for distribution to all of the systemic circulation without passage of blood through the liver, the major detoxifying organ of the body. It also allows direct access to the brain by passage of particles and toxins along the olfactory and trigeminal cranial nerves, which are not protected by the blood–brain barrier.

While exposure to aerosolized spores from molds growing indoors has in the past been thought to be the way in which occupants of damp buildings were exposed to mycotoxins, it is now known that the primary agents of exposure are small particles that are fragments of spores and mycelia and re-entrained dust from surfaces onto which the fungi secreted their exotoxins. The aggregate surface area of a given mass of small particles is much greater than for it is an equal mass the size of spores, and this large surface area allows the particles to carry greater amounts of toxins to the lower respiratory tract, as well as to the nasal sensory epithelium, where the olfactory and trigeminal nerves can carry them into the brain.

The morphology and physiology of the respiratory system prevent homogeneous distribution of gases and particles, which allows accumulation of toxin-carrying particles at "hot spots" where damage and long-time residence can be greater. Very small toxin-containing particles act like a gas and are distributed through diffusion to the deep parts of the lung, the respiratory bronchioles, and alveoli, from where they can directly enter the bloodstream.

It is clear from inhalation studies in animals that mycotoxins cause systemic effects in target organs outside the lung. Few studies have attempted to trace such effects in humans.

Secondary metabolites, including mycotoxins, are more numerous and their mixtures more complex in damp buildings than was previously realized. Microbial interactions within human tissues, and within the built environment, appear to be a prime stimulus for both toxigenic microfungi and toxigenic bacteria to produce toxins in ecologically competitive growth. The additive, synergistic, or antagonistic effects of co-exposure to these metabolites are not known, except for a few highly toxic mycotoxins whose interactions and effects on the health of agricultural animals have been studied.

Prevention of exposure through building and maintenance practices that keep buildings dry and clean is currently the best action to prevent health effects in building occupants.

## References

American Council of Governmental Industrial Hygienists (ACGIH) (1989) Guidelines for the assessment of bioaerosols in the indoor environment. Cincinnati, OH

American College of Occupational and Environmental Medicine (ACOEM) (2002) Adverse human health effects associated with molds in the indoor environment (position paper prepared by Hardin BD, Kelman BJ, Saxon A under the auspices of the ACOEM Council on Scientific Affairs)

American Industrial Hygiene Association (AIHA) (2008) Recognition, evaluation, and control of indoor mold. In: Prezant, B, Weekes DM, Miller JD (eds). AIHA, Fairfax, VA

American Academy of Pediatrics (2004) Ambient air pollution: health hazards to children. Pediatrics 114:1699–1707

Amitani R, Taylor G, Elezis EN, Llewellyn-Jones C, Mitchell J, Kuze F, Cole PJ, Wilson R (1995) Purification and characterization of factors produced by *Aspergillus fumigatus* which affect human ciliated respiratory epithelium. Infect Immun 63(9):3266–3271

Amitani R, Kawanami R (2009) Interaction of *Aspergillus* with human respiratory mucosa: a study with organ culture model. Med Mycol 47(1):S127–S131

Ammann HM (1999) ACGIH TLV statement on bioaerosols. In: Johanning E (ed) Bioaerosols, fungi, and mycotoxins: health effects, assessment, prevention and control/Mount Sinai school of medicine. Eastern New York Occupational and Environmental Health Center, Albany, NY/ New York, NY

Ammann HM (2012) Risk and hazard assessment of molds growing indoors. In: Johanning E, Morey PR, Auger PL (eds) Bioaerosols, fungi, bacteria, mycotoxins in indoor and outdoor environments and human health. Fungal Research Group Foundation Inc., Albany, NY

Amuzie CJ, Islam Z, Kim JK, Seo J-H, Pestka JJ (2010) Kinetics of satratoxin G tissue distribution and excretion following intranasal exposure in the mouse. Toxicol Sci 116(2):433–440

Andersson MA, Nikulin M, Köljalg U, Andersson MC, Rainey F, Reijula K, Hintikka E-L, Salkinoja-Salonen M (1997) Bacteria, molds, toxins in water-damaged building materials. Appl Environ Microbiol 63(2):387–393

Anyanwu EC, Campbell AW, Ehin JE (2004) Mycotoxins and antifungal drug interactions: implications in the treatment of illness due to indoor chronic toxigenic mold exposures. Sci World J 4:167–177

Balásházy I, Hofmann W, Heistracher T (1999) Computation of local enhancement factors for the quantification of particle deposition patterns in airway bifurcations. J Aerosol Sci 30:185–203

Balásházy I, Hofmann W, Heistracher T (2003) Local particle deposition patterns may play a key role in the development of lung cancer. J Appl Physiol 94:1719–1725

Bennett JW (1983) Differentiation and secondary metabolism in mycelial fungi. In: Bennett JW, Ciegler A (eds) Secondary metabolism and differentiation in fungi. Chapter 1, vol 5, Mycology series. Marcel Dekker Inc, New York, NY

Bennett JW, Klich M (2003) Mycotoxins. Clin Microbiol Rev 16(3):497–516

Bergmann S, Funk AN, Scherlach K, Schroeckh V, Shelest E, Horn U et al (2010) Activation of silent fungal polyketide biosynthesis pathway through regulatory cross talk with a cryptic nonribosomal peptide synthetase gene cluster. Appl Environ Microbiol 76(24):8143–8149

Bloom E, Bal K, Nyman E, Must A, Larsson L (2007) Mass spectrometry-based strategy for direct detection and quantification of some mycotoxins produced by *Stachybotrys* and *Aspergillus* spp. in indoor environments. Appl Environ Microbiol 73(13):4211–4217

Bloom E, Grimsley LF, Pehrson C, Lewis J, Larsson (2009a) Molds and mycotoxins in dust from water-damaged homes in New Orleans after Hurricane Katrina. Indoor Air 19:153–158

Bloom E, Nyman E, Must A, Pherson C, Larsson L (2009b) Molds and mycotoxins in indoor environments- a survey of water-damaged buildings. J Occup Environ Hyg 6(11):671–678

Boase S, Foreman A, Cleland E, Tan L, Melton-Kreft R, Pant H et al (2013) The microbiome of chronic rhinosinusitis: culture, molecular diagnostics and biofilm detection. BMC Infect Dis 13:210–219

Bond JA (1993) Metabolism of xenobiotics by the respiratory tract. In: Gardner DE, Crapo JD, McClellan RO (eds) Toxicology of the lung, 2nd edn. Raven, New York, pp 187–215

Bondy GS, Pestka JJ (2000) Immunomodulation by fungal toxins. J Toxicol Environ Health Part B 3:109–143

Brakhage AA, Schroeckh V (2010) Fungal secondary metabolites- strategies to activate silent gene clusters. Fung Genet Biol 48(1):15–22

Brand G (2006) Olfactory/trigeminal interactions in nasal chemoreception. Neurosci Behav Rev 30:908–917

Brasel TL, Douglas DR, Wilson SC, Strauss DC (2005a) Detection of airborne *Stachybotrys chartarum* macrocyclic trichothecene mycotoxins on particulates smaller than conidia. Appl Environ Microbiol 71:114–122

Brasel TL, Martin JM, Carriker CG, Wilson SC, Straus DC (2005b) Detection of airborne *Stachybotrys chartarum* macrocyclic trichothecene mycotoxins in the indoor environment. Appl Environ Microbiol 71(11):7376–7388

Bruns S, Seidler M, Albrecht D, Salvenmosser S, Remme N, Hertweck C et al (2010) Functional genomic profiling of *Aspergillus fumigatus* biofilm reveals enhanced production of the mycotoxin gliotoxin. Proteomics 10(17):3097–3107

Calderon-Garcidueñas L, Mora-Tiscareño A, Ontiveros E, Gómez-Garza G, Barragán-Mejia G, Broadway J et al (2008) Air pollution, cognitive deficits and brain abnormalities: a pilot study with children and dogs. Brain Cogn 68(2):117–127

Carey SA, Plopper CG, Hyde DM, Islam Z, Pestka J, Harkema JR (2012) Satratoxin-G from the black mold *Stachybotrys chartarum* induces rhinitis and apoptosis of olfactory sensory neurons in the nasal airways of Rhesus monkeys. Toxicol Pathol 40:887–898

Council for Agricultural Science and Technology (CAST) (2003) Mycotoxins: risks in plant, animal and human systems. Task Force Report 139. Ames, IA

Chapman JA, Terr AI, Jacobs RL, Charlesworth EN, Bardana EJ Jr (2003) Toxic mold: phantom risk vs science. Ann Allergy Asthma Immunol 91:222–232

Chew GL, Rogers C, Burge HA, Muilenberg ML, Gold DR (2003) Dustborne and airborne fungal particles represent a different spectrum of fungi with differing relations to home characteristics. Allergy 58:13–20

Chew LG, Wilson J, Rabito FA, Grimsley F, Iqbal S, Reponen T, Muilenberg ML, Thorne PS, Dearborn DG, Morley RL (2006) Mold and endotoxin levels in the aftermath of Hurricane Katrina: a pilot project of homes in New Orleans undergoing renovation. Environ Health Perspect 114(12):1883–1889

Cho S-H, Seo S-C, Schmechel D, Grinshpun SA, Reponen T (2005) Aerodynamic characteristics and respiratory deposition of fungal fragments. Atmos Environ 39:5454–5465

Cho S-C, Grinshpun S, Iossifova Y, Schmechel D, Rao CY, Reponen T (2007) A new field-compatible methodology for the collection and analysis of fungal fragments. Aerosol Sci Tech 41:794–803

Churg A (2000) Particle uptake by epithelial cells. Lung Biol Health Dis 143:401–426

Cox-Ganser JM, Rao CY, Park J-H, Schumpert JC, Kreiss K (2009) Asthma and respiratory symptoms in hospital workers related to dampness and biological contaminants. Indoor Air 19(4):280–290

Cresia DA, Thurman JD, Jones LJ III, Nealley MI, York CG, Wannemacher RW Jr, Bunner DL (1987) Acute in halation toxicity of T-2 mycotoxin in mice. Fundam Appl Toxicol 8:230–235

Cresia DA, Thurman JD, Wannemacher RW Jr, Bunner DL (1990) Acute inhalation toxicity of T-2 mycotoxin in the rat and guinea pig. Fundam Appl Toxicol 14:54–59

Croft WA, Jarvis BB, Yatawara (1986) Airborne outbreak of trichothecene mycotoxicosis. Atmos Environ 20:549–552

Dahl AR, Gerde P (1994) Uptake and metabolism of toxicants in the respiratory tract. Environ Health Perspect 102(suppl 11):67–70

Dearborn DG, Yike I, Sorenson WG, Miller MJ, Etzel RA (1999) Overview of investigations into pulmonary hemorrhage among infants. Environ Health Perspect 107(Supp 3):495–499

Dearborn DG, Smith PG, Dahms BB, Allan TM, Sorenson WG, Montaña E, Etzel RA (2002) Clinical profile of 30 infants with acute pulmonary hemorrhage in Cleveland. Pediatrics 110:627–637

Denning DW, O'Driscoll BR, Powell G, Chew F, Atherton GT, Vyas A, Miles J, Morris J, Niven RM (2009) Randomized controlled trial of oral antifungal treatment for severe asthma with fungal sensitization the fungal asthma sensitization trial (FAST). Am J Respir Crit Care Med 179(1):11–18

Di Paolo N, Guarnieri A, Loi A, Sacchi G, Mangiarotti AM, Di Paolo M (1993) Acute renal failure from inhalation of mycotoxins. Nephron 64(4):621–625

Douwes J, Gibson P, Pekkanen P, Pearce N (2002) Non-eosinophilic asthma: importance and possible mechanisms. Thorax 57:643–648

Drew R, Frangos J (2007) The concentration of no toxicological concern (CoNTC): a risk assessment tool for air toxics. J Toxicol Environ Health 70:1584–1593

Englehart S, Loock A, Scutlarck D, Saguski H, Lommel A, Fächer H, Exner M (2002) Occurrence of toxigenic *Aspergillus versicolor* isolates and sterigmatocystin in carpet dust from damp indoor environments. Appl Environ Microbiol 68(8):3886–3890

Fechter LD, Johnson DL, Lynch RA (2002) The relationship of particle size to the olfactory nerve uptake of a non-soluble form of manganese into brain. Neurotoxicology 23(2):177–183

Fisk WJ, Lei-Gomez Q, Mendell MJ (2007) Meta-analysis of the associations of respiratory health effects with dampness and mold in homes. Indoor Air 17(4):284–296

Foreman A, Psaltis AJ, Tan LW, Wormald P-J (2009) Characterization of bacterial and fungal biofilms in chronic rhinosinusitis. Am J Rhinol Allergy 23(6):556–561

Foreman A, Jervis-Bardy J, Wormald P-J (2011) Do biofilms initiation and recalcitrance of chronic rhinosinusitis? Laryngoscope 121:1085–1091

Foreman A, Boase S, Psaltis A, Wormald P-J (2012) Role of bacterial and fungal biofilms in chronic rhinosinusitis. Curr Allergy Asthma Rep 12(2):127–135

Forgacs J (1972) Stachybotryotoxicosis. Microbiol Toxins 8:95–128

Frazer S, Pestka JJ, Kim J-K, Medina A, Alred D, Magan N (2012) Impact of environmental factors on growth and satratoxin G production by strains of *Stachybotrys chartarum*. World Mycotoxin J 5(1):47–53

Frey-Klett P, Burlinson P, Deveau A, Barret M, Tarkaa M, Sarniguet A (2011) Bacterial-fungal interactions: hyphens between agricultural, clinical, environmental, and food microbiologists. Microbiol Mol Biol Rev 75(4):583–609

Gordon WA, Cantor JB, Johanning E, Charatz HJ, Ashman TA, Breeze JL et al (2004) Cognitive impairment associated with toxigenic fungal exposure replication and extension of previous findings. Appl Neuropsychol 11(2):65–74

Górny RL, Reponen T, Willecke K, Schmechel D, Robine E, Boissier M, Grinshpun SA (2002) Fungal fragments as indoor air biocontaminants. Appl Environ Microbiol 68(7):3522–3531

Gottschalk C, Bauer J, Meyer K (2008) Detection of satratoxin G and H in indoor air from a water-damaged building. Mycopathologia 166(2):103–107

Gregory L, Pestka JJ, Dearborn DG, Rand TG (2004) Localization of satratoxin-G in *Stachybotrys chartarum* spores and spore-impacted mouse lung using immunocytochemistry. Toxicol Pathol 32(1):26–34

Hall-Stoodley L, Stoodley P (2009) Evolving concepts in biofilm infections. Cell Microbiol 11(7):1034–1043

Hanson LR, Frey WH II (2007) Intranasal delivery bypasses the blood–brain barrier to target therapeutic agents to the central nervous system and treat neurodegenerative disease. BMC Neurosci 9(suppl 3):55–59

Hardin BD, Robbins CA, Fallah P, Kelman BJ (2009) The concentration of no toxicologic concern (CoNTC) and airborne mycotoxins. J Toxicol Environ Health Part A 72:585–598

Harkema JR, Carey S, Wagner JG (2006) The nose revisited; a brief review of the comparative structure, function, and toxicologic pathology of the nasal epithelium. Toxicol Pathol 32:252–269

Harvey RJ, Lund VJ (2007) Biofilms and chronic rhinosinusitis: systematic review of evidence, current concepts and directions for research. Rhinology 45:3–13

Healey DY, Leid JG, Sanderson AR, Hunsaker DH (2008) Biofilms with fungi in chronic rhinosinusitis. Otolaryngol Head Neck Surg 138(5):641–647

Hossain MA, Ahmed MS, Ghannoum MA (2004) Attributes of *Stachybotrys chartarum* and its association with human disease. J Allergy Clin Immunol 113:200–209

Huttunen K, Pelkonen J, Nielsen KF, Nuutinen U, Jussila J, Hirvonen M-R (2004) Synergistic interaction in simultaneous exposure to *Streptomyces californicus* and *Stachybotrys chartarum*. Environ Health Perspect 112:659–665

Huttunen K, Rintala H, Hirvonen M-R, Vepsäläinen A, Hyvärinen MT, Toivola M, Nevalainen A (2008) Indoor air particles and bioaerosols before and after renovation of moisture-damaged buildings: the effect on biological activity and microflora. Environ Res 107:291–298

International Life Sciences Institute, Europe (ILSI Europe) (2005) Threshold of toxicological concern (TTC) a tool for assessing substances of unknown toxicity present at low levels in the diet, by Susan Barlow. ILSI Europe Brussels, Belgium

Institute of Medicine (IOM) National Academies of Science (2004) Damp indoor spaces and health. National Academy Press, Washington, DC

Islam Z, Harkema JR, Pestka JJ (2006) Satratoxin G from the black mold *Stachybotrys chartarum* evokes olfactory sensory neuron loss and inflammation in the murine nose and brain. Environ Health Perspect 114:1099–1107

Islam Z, Amuzie CJ, Harkema JR, Pestka JJ (2007) Neurotoxicity and inflammation of the nasal airways of mice exposed to the macrocyclic trichothecene mycotoxin roridin A: kinetics and potentiation by bacterial lipopolysaccharide co-exposure. Toxicol Sci 98(2):526–541

Jakab GJ, Hmieleski RR, Zarba A, Hemenway DR, Groopman JD (1994) Respiratory aflatoxicosis: suppression of pulmonary and systemic host defenses in rats and mice. Toxicol Appl Pharmacol 125:198–205

Jarvis BB, Sorenson WG, Hintikka E-L, Nikulin M, Zhou Y, Wang S, Hinckley S, Etzel R, Dearborn D (1998) Study of toxin production by isolates of *Stachybotrys chartarum* and

*Memnoniella echinata* isolated during a study of pulmonary hemosiderosis in infants. Appl Environ Microbiol 64:3620–3625

Johanning E, Biagini R, Hull D, Morey P, Jarvis BB, Landsbergis P (1996) Health and immunology study following exposure to toxigenic fungi (*Stachybotrys chartarum*) in a water-damaged office environment. Int Arch Occup Environ Health 68(4):207–218

Karunasena E, Larrañaga MD, Simoni JS, Douglas DR, Straus DC (2010) Building-associated neurological damage modeled in human cells: a mechanism of neurotoxic effects by exposure to mycotoxins in the indoor environment. Mycopathologia 170(6):377–390

Kelman BJ, Robbins CA, Swenson LJ, Hardin BD (2004) Risk from inhaled mycotoxins in indoor office and residential environments. Int J Toxicol 23:3–10

Kercsmar CM, Dearborn DG, Schluchter M, Xue L, Kirchner HC, Sobolewski J, Greenberg SJ, Vesper SJ, Allan T (2006) Reduction in asthma morbidity in children as a result of home remediation aimed at moisture sources. Environ Health Perspect 114:1574–1580

Kirjavainen PK, Täubel M, Karvonen AM, Sulyok M, Krska R, Hyvärinen A, Pekkanen J (2015) Microbial secondary metabolites in homes in association with moisture damage and asthma. doi:10.1111/ina.12213

Kleinstreuer C, Zhang Z, Li Z (2008) Modeling airflow and particle transport/deposition in pulmonary airways. Respir Physiol Neurobiol 163:128–138

Krieger J, Jacobs DE, Ashley PJ, Baeder A, Chew GL, Dearborn DG, Hynes HP, Miller JD, Morley R, Rabito F, Zeldin DC (2010) Housing interventions and control of asthma-related indoor biologic agents: a review of the evidence. J Public Health Manag Pract 16:11–20

Kück U, Bloemendal S, Teichert I (2014) Putting fungi to work: harvesting a cornucopia of drugs, toxins and antibiotics. PLoS Pathog 10(3):e1003950. doi:10.1371/journal.ppat.1003950

Lewis J, Bench G, Myers O, Staines W, Divine KK, Barrington W, Karlsson J (2005) Trigeminal uptake and clearance of inhaled manganese chloride in rats and mice. Neurotoxicology 26(1):113–123

Lieberman SM, Jacobs JB, Lebowitz RA, Fitzgerald MB, Crawford J, Feigenbaum BA (2011) Measurement of mycotoxins in patients with chronic rhinosinusitis. Otolaryngol Head Neck Surg 145(2):327–329

Lippman M, Yeates DB, Albert RE (1980) Deposition, retention, and clearance of inhaled particles. Br J Ind Med 37:337–362

Mader DR, Yike I, Distler AM, Dearborn DG (2007) Acute pulmonary hemorrhage during anesthesia in two cats exposed to toxic black mold (*Stachybotrys chartarum*). J Am Vet Med Assoc 231:731–735

Markkanen (Penttinen) P, Pelkonen J, Tapanianen M, Mäki-Paakkanen, Jalava PI, Hirvonen M-R (2009) Co-cultivated damp building related microbes *Streptomyces californicus* and *Stachybotrys chartarum* induce immunotoxic and genotoxic responses via oxidative stress. Inhal Toxicol 21(10):857–867

Mendell MJ, Mirer AG, Cheung K, Tong M, Douwes J (2011) Respiratory and allergic health effects of dampness, mold, and dampness-related agents: a review of the epidemiological evidence. Environ Health Perspect 119:748–756

Miller FJ (1999) Dosimetry of particles in laboratory animals and humans. In: Gardner DE, Crapo JD, McClellan RO (eds) Toxicology of the lung, 3rd edn. Taylor & Francis, Philadelphia, pp 513–555

Miller JD, Sun M, Gilyan A, Roy J, Rand TG (2010) Inflammation-associated gene transcription and expression in mouse lungs induced by low molecular weight compounds from fungi from the built environment. Chem Biol Interact 183:113–124

Miller JD, McMullin DR (2014) Fungal secondary metabolites as harmful indoor air contaminants: 10 years on. Appl Microbiol Biotechnol 98:9953–9966. doi:10.1007/s00253-014-6178-5

Miller RV, Ammann HM (2005) Risk assessment of fungal bioaerosols in indoor environments: current paradigms and evolving concepts. In: Johanning E (ed) Bioaerosols, fungi, bacteria, mycotoxins and human health: pathophysiology, clinical effects, exposure assessment, prevention and control in indoor environments and work. Fungal Research Group Foundation, Albany, NY, pp 474–481

Morales DK, Hogan DA (2010) *Candida albicans* interactions with bacteria in the context of human health and disease. PLoS Pathog 6(4):1–4

Moss MO (1991) The environmental factors controlling mycotoxin formation. In: Smith JE, Henderson RS (eds) Mycotoxins and animal food. CRC Press, Boca Raton, FL, pp 37–56

Murtoniemi T, Penttinen P, Nevalainen A, Hirvonen M-R (2005) Effects of microbial co-cultivation on inflammation and cytotoxic potential of spores. Inhal Toxicol 17(2):681–693

Muszkieta L, Beauvais A, Pähtz V, Gibbons JG, Anton Leberre V, Beau R, Shibuya K et al (2013) Investigation of *Aspergillus fumigatus* biofilm formation by various "omics" approaches. Front Microbiol 4(13):1–16. doi:10.3389/fmicb.2013.00013

Nielsen KF (2003) Mycotoxin production by indoor molds. Fungal Genet Biol 39:103–117

Nielsen KF, Holm G, Uttrup LP, Nielsen PA (2004) Mould growth on building materials under low water activities. Influence of humidity and temperature on fungal growth and secondary metabolism. Internat Biodegrad 54:325–336

Nielsen KF, Frisvad JC (2011) Mycotoxins on building materials. In: Adan OC, Samson RA (eds) Fundamentals of mold growth in indoor environments and strategies for healthy living. Springer Science and Business media, Wageningen, pp 245–266

Nikulin M, Reijula K, Jarvis BB, Hintikka E-L (1996) Experimental lung mycotoxicosis in mice induced by *Stachybotrys atra*. Int J Exp Pathol 77:213–218

Nikulin M, Reijula K, Jarvis BB, Veijalainen P, Hintikka E-L (1997) Effects of intranasal exposure to spores of *Stachybotrys atra* in mice. Fundam Appl Toxicol 35:182–188

National Institute for Occupational Safety and Health (NIOSH) (2012) Preventing occupational respiratory disease from exposures caused by dampness in office buildings, schools, and other nonindustrial buildings. NIOSH publication # 2013–102. Centers for disease Control and Prevention, National Institute for Occupational Safety and Health

Nützmann H-W, Reyes-Dominguez Y, Scherlach K, Schroeckh V, Horn F, Gazek A, Schümann J, Hertweck C, Strauss J, Brakhage AA (2011) Bacteria-induced natural product formation in the fungus *Aspergillus nidulans* requires Saga/Aga-mediated histone acetylation. Proc Natl Acad Sci USA 108(34):14282–14287

Oberdörster G, Ferin J, Lehnert BE (1994) Correlation between particle size, in vivo particle persistence, and lung injury. Environ Health Perspect 102(Supp 5):173–179

Oberdörster G, Sharp Z, Aturdorei AE, Elder A, Gelein R, Lunts A, Kreyling W, Cox C (2002) Extrapulmonary translocation of ultrafine carbon particles following whole body inhalation exposure of rats. J Toxicol Environ Health Part A 65:1531–1543

Oberdörster G, Sharp Z, Aturdorei AE, Gelein R, Kreyling W, Cox C (2004) Translocation of inhaled ultrafine particles to the brain. Inhal Toxicol 16(6–7):437–445

Okai E, Kamei K, Watanabe A, Nagayoshi M, Tada Y, Nagaoka T et al (2008) Inhalation of *Stachybotrys chartarum* causes pulmonary arterial hypertension in mice. Int J Exp Pathol 89:201–208

Palmgren MS, Lee LS (1986) Separation of mycotoxin-containing sources in grain dust and determination of their mycotoxin potential. Environ Health Perspect 66:105–108

Pang VF, Adams JH, Beasley VR, Buck WB, Haschek WM (1986) Myocardial and pancreatic lesions induced by T-2 toxin, a trichothecene mycotoxin, in swine. Vet Pathol 23:310–319

Pang VF, Lambert RJ, Beasley VR, Buck WB, Haschek WM (1987) Experimental T-2 toxicosis in swine following inhalation exposure: effects on pulmonary and systemic immunity, and morphologic changes. Toxicol Pathol 15(3):308–319

Pang VF, Lambert RJ, Felsburg PJ, Beasley VR, Buck WB, Haschek WM (1988) Experimental T-2 toxicosis in swine following inhalation exposure: clinical signs and effects on hematology, serum biochemistry, and immune response. Fundam Appl Toxicol 11(1):100–1007

Park J-H, Cox-Ganser JM, Kreiss K, White SK, Rao CY (2008) Hydrophilic fungi and ergosterol associated with respiratory illness in a water-damaged building. Environ Health Perspect 116(1):45–50

Parsek M, Singh PK (2003) Bacterial biofilms: an emerging link to disease pathogenesis. Annu Rev Microbiol 57:677–701

Peitzsch M, Sulyok M, Täubel M, Vishwanath V, Krop E, Borràs-Santos A, Hyvärinen A, Nevalainen A, Krska R, Larsson L (2012) Microbial secondary metabolites in school buildings inspected for moisture damage in Finland, The Netherlands and Spain. J Environ Monit 14:2044–2053

Penttinen P, Huttunen K, Pelkonen J, Hirvonen M-R (2005) The proportions of *Streptomyces californicus* and *Stachybotrys chartarum* in simultaneous exposure affect inflammatory responses in mouse RAW 264.7 macrophages. Inhal Toxicol 17:79–85

Peters A, Veronesi B, Calderón-Garcidueñas L, Gehr P, Chen LC, Geiser M, Reed W, Rothen-Rutishauser B, Schürch SH (2006) Translocation and potential neurological effects of fine and ultrafine particles a critical update. Part Fibre Toxicol 3:13–26

Pestka JJ, Yike I, Dearborn DG, Ward MDW, Harkema JR (2008) *Stachybotrys chartarum*, trichothecene mycotoxins and damp building-related illness: new insights into a public health enigma. Toxicol Sci 104(1):4–26

Polzehl D, Weschta M, Podbielski A, Riechelman H, Rimel D (2005) Fungus culture and PCR in nasal lavage samples of patients with chronic rhinosinusitis. J Med Mycol 54:31–37

Ponikau JU, Sherris DA, Kern EB, Homburger HA, Frigas E, Gaffey TA, Roberts GD (1999) The diagnosis and incidence of allergic fungal sinusitis. Mayo Clin Proc 74:877–884

Ponikau JU, Sherris DA, Weaver A, Kita H (2005) Treatment of chronic rhinosinusitis with intranasal amphotericin: a randomized, placebo-controlled, double-blind trial. J Allergy Clin Immunol 115(2):125–131

Ponikau JU, Sherris DA, Kephart GM, Adolphus C, Kita H (2006) The role of ubiquitous airborne fungi in chronic rhinosinusitis. Clin Rev Allergy Immunol 30(3):187–194

Pope CA III (2000) Epidemiology of fine particulate air pollution and human health: biologic mechanisms and who's at risk. Environ Health Perspect 108(Supp 4):713–723

Power MC, Weisskopf MC, Alexeeff SE, Coull BA, Spiro A III, Schwartz J (2011) Traffic-related air pollution and cognitive function in a cohort of older men. Environ Health Perspect 119(5):682–687

Quansah R, Jaakkola MS, Hugg TT, Heikkinen SAM, Jaakkola JJK (2012) Residential dampness and molds and the risk of developing asthma: a systematic review and meta-analysis. PLOS ONE 7(11)e47526. doi:10.1371/journal.pone.0047526

Ramage G, Rajendran R, Sherry L, Williams C (2012) Fungal biofilm resistance. Int J Microbiol 2012:528521. doi:10.1155/2012/528521

Rand TG, Miller JD (2011) Analysis for toxins and inflammatory compounds. In: Flannigan B, Samson RA, Miller JD (eds) Microorganisms in homes and indoor work environments: diversity, health impacts, investigation and control, 2nd edn. CRC Press/Taylor & Francis Group, Boca Raton, FL, pp 291–306

Rao CY, Burge HA, Brain JD (2000a) The time course of responses to intratracheally instilled toxic *Stachybotrys chartarum* spores in rats. Mycopathologia 149:27–34

Rao CY, Brain JD, Burge HA (2000b) Reduction of pulmonary toxicity of Stachybotrys chartarum spores by methanol extraction of mycotoxins. Applied Environ Microbiol 66(7):2817–2821

Rao CY, Riggs MA, Chew GL, Muilenberg ML, Thorne PS, Van Sickle D, Dunn KH, Brown C (2007) Characteristics of airborne molds, endotoxins, and glucans in homes in New Orleans after Hurricanes Katrina and Rita. Appl Environ Microbiol 73(5):1630–1634

Reponen T, Seo S-H, Grimsley F, Lee T, Crawford C, Grinchpun SA (2007) Fungal fragments in moldy houses: a field study in homes in New Orleans and southern Ohio. Atmos Environ 41(37):8140–8149

Salares VR, Hinde CA, Miller JD (2009) Analysis of settled dust in homes and fungal glucan in air particulate collected during HEPA vacuuming. Indoor Built Environ 18:485–491

Salvi S (2007) Health effects of ambient air pollution in children. Paediatr Respir Rev 8(4):275–280

Schroeckh V, Scherlach K, Nützmann H-W, Shelest E, Schmidt-Heck W, Schuemann J et al (2009) Intimate bacterial-fungal interaction triggers biosynthesis of archetypal polyketides in *Aspergillus nidulans*. Proc Natl Acad Sci USA 106(34):14558–14563

Scranton RA, Fletcher L, Sprague S, Jimenez DF, Digicaylioglu M (2011) The rostral migratory stream plays a key role in intranasal delivery of drugs into the CNS. PLoS One 6(4), e18711. doi:10.1371/journal.pone.0018711

Seneviratne G, Zavahir JS, Bandara WMMS, Weerasekara MLMAW (2008) Fungal-bacterial biofilms: their development for novel biotechnological applications. World J Microbiol Biotechnol 24:739–743

Seo S-C, Grinshpun SA, Iossifova Y, Schmechel D, Rao CY, Reponen T (2007) A new field-compatible methodology for the collection and analysis of fungal fragments. Aerosol Sci Technol 41(8):794–803

Shank EA, Kolter R (2009) New developments in microbial interspecies signaling. Curr Opin Microbiol 12(2):205–214

Shipley MT (1985) Transport of molecules from the nose to the brain: transneuronal anterograde and retrograde labeling in the rat olfactory system by wheat germ agglutinin-horseradish peroxidase applied to the nasal epithelium. Brain Res Bull 15(2):129–152

Seiberling K, Wormald P-J (2009) The role of itraconazole in recalcitrant fungal sinusitis. Am J Rhinol Allergy 23(3):303–306

Silver WL, Finger TE (2009) The anatomical and electrophysiological basis of peripheral and nasal trigeminal chemoreception. International Symposium on Olfaction and Taste. Ann N Y Acad Sci 1170:202–205

Sorenson WG, Gerberick GF, Lewis DM, Castranova V (1986) Toxicity of mycotoxins for the rat pulmonary macrophage in vitro. Environ Health Perspect 66:45–55

Sorenson, Frazer DG, Jarvis BB, Simpson J, Robinson VA (1987) Trichothecene mycotoxins in aerosolized conidia of *Stachybotrys atra*. Appl Environ Microbiol 53(6):1370–1375

Sorenson WG (1993) Mycotoxins toxic metabolites of fungi. In: Scranton RA, Fletcher L, Sprague S, Jimenez DF, Digicaylioglu M (eds) Fungal infections and the immune response. Plenum Press, New York, NY, pp 469–491

Sorenson WG (1999) Fungal spores: hazardous to health? Environ Health Perspect 107(Supp 3):469–472

Speijers GJA, Speijers MHM (2004) Combined toxic effects of mycotoxins. Toxicol Lett 153:91–98

St. George JA, Harkema JR, Hyde DM, Plopper CG (1993) Cell populations and structure/function relationships of cells in the airways. In: Gardner DE, Crapo JD, McClellan RO (eds) Toxicology of the lung, 2nd edn. Raven Press, New York, pp 81–110

Suh JD, Ramakrishnan V, Palmer JN (2010) Biofilms. Otolaryngol Clin N Am 4:521–530

Tallkvist J, Persson E, Henrikson J, Tjälve (2002) Cadmium-metallothionein interactions in the olfactory pathways of rats and pikes. Toxicol Sci 67:108–113

Täubel M, Sulyok M, Vishwanath V, Bloom E, Turunen M, Järvi K, Kauhanen E, Kriska R, Hyvärinen A, Larsson L, Nevalainen A (2011) Co-occurrence of toxic bacterial and fungal secondary metabolites in moisture-damaged indoor environments. Indoor Air 21(5):368–375

Thorne RG, Pronk GJ, Padmanabhan V, Frey WH II (2004) Delivery of insulin-like growth factor-1 to the brain and spinal cord along olfactory and trigeminal pathways following intranasal administration. Neroscience 127(2):481–496

Thrasher JD, Crawley S (2009) The biocontaminants and complexity of damp indoor spaces: more than what meets the eyes. Toxicol Ind Health 25(9–10):583–615

Toivola M, Nevalainen A (2004) Personal exposures to particles and microbes in relation to micro-environmental concentrations. Indoor Air 14(5):351–359

Toumi T, Reijula K, Johnsson T, Hemminki K, Hintikka E-L, Lindroos O, Kalso S, Koukila-Kähkölä, Mussalo-Rauhamaa H, Haahtela T (2000) Mycotoxins in crude building materials from water-damaged buildings. Appl Environ Microbiol 66(5):1899–1904

Ueno Y (1984) Toxicological features of T-2 toxin and related trichothecenes. Fundam Appl Toxicol 4(S):124

U.S.-Japan Cooperative Program in Natural Resources (UNJR) (1970) Proceedings of the first U.S.-Japan conference on toxic micro-organisms, 7–10 October 1968, Honolulu, HI. In: Herzberg M (ed) UNJR Joint Panels on Toxic Micro-organisms and the U.S. Department of the Interior

USEPA (1994) Methods for derivation of inhalation reference concentrations and application of inhalation dosimetry. EPA/600/8-90/066F. Office of Research and Development, Washington, DC

Vesper S, Dearborn DG, Yike I, Allan T, Sobolewski J, Hinckley SF, Jarvis BB, Haugland RA (2000) Evaluation of *Stachybotrys chartarum* in the house of an infant with pulmonary hemorrhage: quantitative assessment before, during and after remediation. J Urban Health 77:68–85

Vishwanath V, Sulyok M, Labuda R, Bicker W, Krska R (2009) Simultaneous determination of 186 fungal and bacterial metabolites in indoor matrices li liquid chromatography/tandem mass spectroscopy. Anal Bioanal Chem 395:1355–1372

World Health Organization (WHO) Europe (2009) WHO guidelines for indoor air quality: dampness and mould. In: Heseltine E, Rosen J (eds). WHO, Copenhagen

Wargo MJ, Hogan DA (2006) Fungal-bacterial interactions: a mixed bag of mingling microbes. Curr Opin Microbiol 9:359–364

Weuve J, Puett RC, Schwartz J, Yanosky JD, Laden F, Grodstein F (2012) Exposure to particulate air pollution and cognitive decline in older women. Arch Intern Med 172(3):219–227

Yike I, Allan T, Sorenson WG, Dearborn DG (1999) Highly sensitive protein translation assay for trichothecene toxicity in airborne particulates: comparison with cytotoxicity assays. Appl Environ Microbiol 65(1):88–94

Yang G-H, Jarvis BB, Chung Y-J, Pestka JJ (2000) Apoptosis induction by the satratoxins and other trichothecene mycotoxins: relationship to ERK, p38 MAPK, and SAPK/JNK activation. Toxicol Appl Pharmacol 164:149–160

Zavahir JS, Seneviratne G (2007) Potential of developed microbial biofilms in generating bioactive compounds. Res J Microbiol 2(4):397–401

# Chapter 21
# Fungi in Fermentation and Biotransformation Systems

**Carla C.C.R. de Carvalho**

## Introduction

Several microfungi species have been used successfully throughout human history to produce foods and beverages, such as bread, cheese, soy sauce, tea, beer, wine, and sake. Fungal metabolism and metabolites play fundamental roles in the manufacture of, e.g., ethanol, citric acid, and pharmaceutical drugs, as well as in the production of biocontrol agents, enzymes, and pigments. It has also been suggested that fungal enzymes are the most efficient lignocellulose degraders, allowing the conversion of biowaste and agriculture crop residues into, e.g., biochemicals and bioenergy (Lange 2010). Conventional production of bioethanol, the most common biofuel in use, from lignocellulosic material may apply fungal cellulases for biomass hydrolysis and yeast fermentation of the resulting glucose. Since the monosaccharides resulting from the hydrolysis of cellulosic materials cause feedback inhibition of the hydrolases used, it has been proposed to simultaneously perform hydrolysis and fermentation, thus preventing accumulation of glucose and disaccharides. Few individual microorganisms able to carry out simultaneous saccharification and fermentation have been reported, including the thermotolerant yeast strain *Kluyveromyces marxianus* CECT 10875 (Ballesteros et al. 2004). In simultaneous saccharification and cofermentation, the use of an organism able to ferment both glucose and pentoses released from biomass will result in an increased ethanol yield. In 1989, the cellulase hyperproducing strain *Fusarium oxysporum* F3 was reported as being able to ferment glucose, xylose, cellobiose, and cellulose directly to ethanol, reaching a maximum ethanol concentration of 14.5 g/L from 50 g/L of cellulose (53.2 % of the theoretical yield) in 6 days (Christakopoulos et al. 1989).

C.C.C.R. de Carvalho, Ph.D. (✉)
iBB-Institute for Bioengineering and Biosciences, Department of Bioengineering, Instituto Superior Técnico, Universidade de Lisboa, Av. Rovisco Pais, Lisbon 1049-001, Portugal
e-mail: ccarvalho@ist.utl.pt



D.-W. Li (ed.), *Biology of Microfungi*, Fungal Biology,
DOI 10.1007/978-3-319-29137-6_21

Nevertheless, the very low number of microorganisms able to carry out the entire process makes simultaneous saccharification and fermentation/cofermentation processes using enzymes for the saccharification step, or those using mixed cultures for the degradation of hexoses and pentoses, more common. A consolidated bioprocessing comprising (1) the production of saccharolytic enzymes, (2) the hydrolysis of the polysaccharides in the pretreated biomass, and (3) the fermentation of both hexose and pentose sugars in a single bioreactor, will most likely require synthetic biology techniques as no single microorganism with the desired features has been found. Among the candidates to be developed are *Saccharomyces cerevisiae* and *Kluyveromyces marxianus* (Karimi et al. 2006; Millati et al. 2008; Karimi and Zamani 2013). *Mucor indicus* is also able to ferment lignocellulosic hydrolysates, both hexoses and pentoses, while tolerating inhibiting compounds resulting in high yields of ethanol (Karimi et al. 2006; Millati et al. 2008). This species is an example of a microfungus with several applications from fish feed to wastewater treatment, being able to produce commercially interesting products from chitosan to polyunsaturated fatty acids (Karimi and Zamani 2013).

In wastewater treatment, fungal biomass has been used to reduce organic matter (at low pH where bacterial growth is inhibited), to sequester and adsorb suspended solids, to degrade recalcitrant pollutants, and to perform denitrification (More et al. 2010; Ryan et al. 2005; Mannan et al. 2007). However, the production of bioactive compounds is probably the most important feature of fungi. Terrestrial, marine, and plant endophytic fungi are a source of unique metabolites with, e.g., antimicrobial, insecticidal, and antitumour activities (Debbab et al. 2010; Wang et al. 2014).

To reach high titres of secondary metabolites during fermentation, a combination of conditions, including medium composition, substrate(s) type and concentration, aeration, stirring conditions, and addition of inducers, have to be found, implemented, and maintained. Process performance in filamentous fungi is clearly affected by strains and inocula, morphology, and rheology (Posch and Herwig 2014; Posch et al. 2013). In stirred tank reactors, microfungi are particularly sensible to shear force, but vigorous agitation is usually required to surpass the high viscosity and oxygen demands observed during submerged growth (Kelly et al. 2006; Ranjan 2008). Shear forces are probably responsible for differences in morphology observed during cell growth between bioreactor scales, making it necessary to use methodologies, such as microscopy and flow cytometry, soft sensors, and mathematical modelling to predict broth rheology and fungal morphology during scale-up of bioprocesses (Posch et al. 2013).

## Fermentation

In nature, most fungal genera evolved on moist, solid substrates. They should, therefore, be more apt to produce secondary metabolites and enzymes under these conditions than in liquid media. Solid-state fermentation (SSF), by simulating natural microbiological processes and by using the ability of fungal cells to adhere to solid particles, should result in higher productivities. The main SSF advantages when compared to submerged processes are the reduced bacterial contamination level, due to the low water activity used, the simple bioreactors necessary, and the low installation and running costs (Cannel and Moo-Young 1980; Viniegra-Gonzalez et al. 2003). It is particularly convenient when the solid phase is simultaneously the substrate, such as cellulose, lignin, or agricultural residues (Thomas et al. 2013; Pérez-Guerra et al. 2003). On the other hand, submerged fermentations allow better process control since parameters, such as temperature, pH, aeration, and mixture, are monitored and controlled in the bioreactor, allowing also a better scale-up (Papavizas et al. 1984; Soetaert and Vandamme 2010).

## *Solid-State Fermentation*

After centuries of being used for the production of, e.g., bread and cheese, SSF systems were considered "low-technology" processes by the end of the twentieth century (Pandey et al. 2000). However, during this century, SSF has received renewed attention since it allows stable productions in smaller fermenters, with less energy requirements, and causes less impact on the environment than liquid fermentations (Pérez-Guerra et al. 2003). SSF is defined as a culture where microbes thrive on the surface and at the interior of a solid matrix, in the absence of free water (Viniegra-Gonzàlez 1997; Lonsane et al. 1985). The system comprises solid, liquid, and gaseous phases, but the low amount of water in the system reduces the risk of bacterial contamination.

Depending on the nature of the substrate, two SSF systems may be considered: SSF on natural solid substrates and SSF on impregnated inert supports (Barrios-González 2012). Although the former should be a cheaper process, the latter may be an advantage for (1) the production of high-added-value products, as medium costs are usually only a fraction of the overall production costs (Barrios-González and Mejýa 2008), and (2) to study the system since the absorbed medium composition may be designed and monitored during fermentation (Barrios-González 2012).

Since the majority of fungal SSF are batch processes, the morphology and physiology of fungal cells will change with time. Studies aimed at understanding the biological aspects should be conducted, but most of the research already carried out aimed at the production of commercially interesting products from cheap substrates and at the optimisation of the bioprocess (Barrios-González 2012). Biomass has been estimated by, e.g., respirometry, infrared spectroscopy, and image analysis, whilst mass, water, and heat balances have been determined to accurately simulate and characterise SSF processes (Papavizas et al. 1984; Barrios-González and Mejýa 2008). Bioprocess engineering, biochemistry, molecular biology, and microbiology have to be successfully combined to help the understanding of these complex fermentations and to overcome the critical parameters affecting them (Table 21.1).

To understand the effect of the parameters and to optimise the design and operation of SSF, mathematical models have been developed. In these models, kinetics

**Table 21.1** Critical parameters influencing fungal morphology and productivity in solid-state fermentations

| Parameter | Considerations | Example |
|---|---|---|
| Strain selection | Natural production ability of the strain; possibility to improve it | Vu et al. (2010) |
| Solid substrate | Usually a natural agricultural or agro-industrial by-product/residue, or an inert synthetic material | Mitchell et al. (2004) |
| Aeration | Oxygen is crucial for fungal growth, production rates, and yields | Rahardjo et al. (2006), Coradin et al. (2011) |
| Water activity | Depends on the retention capacity of the substrate; influences fungal morphology | Casciatori et al. (2014), Mariano et al. (1995) |
| Physical parameters | Temperature, pH, and other environmental conditions influence fungal performance | Rodrigues et al. (2009) |
| Medium composition | Concentration and type of, e.g., phosphorous and nitrogen sources and minerals added should be optimised | Senthilkumar et al. (2005), Liu et al. (2007) |

may be described by empirical equations or by complex equation systems addressing micro-scale effects that affect bioreactor performance. Among the effects to include in a mathematical model are temperature, pH, aeration, water activity, and the nature of the solid substrate (e.g. chemical composition, mechanical properties, particle size) (Thomas et al. 2013). In SSF, mathematical models have to overcome the heterogeneity of fungal growth, substrate, and medium, which may result in heat and mass transfer limitations, inaccurate and non-reproducible measurements of process variables, and difficulties during scaling-up and bioreactor control. Since in SSF there is no convective transport of substrates and products, the necessary gradients to supply nutrients and remove the products may affect cellular growth and production and require models coupling reaction and diffusion phenomena (Rahardjo et al. 2006). Coradin and co-workers were able to develop a three-dimensional model, simulating the growth of the aerial hyphae in random directions, to study phenomena occurring at micro-scale and to help the preparation of substrate and fermenter operation (Coradin et al. 2011). The growth of aerial hyphae in the space between particles increases the pressure drop through aerated beds and may decrease oxygen transfer when agglomerates are formed.

When studying the impact of moisture content and packing technique on particle and bulk densities and porosities of beds packed with sugar cane bagasse, wheat bran, orange pulp and peel, and mixtures of them, the authors found that the macroscopic properties of the bed mixtures could be estimated as a weighted average of the properties of each individual material (Casciatori et al. 2014). Besides, the mycelial growth of *Myceliophthora thermophila* I-1D3b and *Trichoderma reesei* QM9414 did not affect the bed properties, but the cellulase yield of the latter was strongly affected by the ratio sugar cane bagasse/wheat bran and by the moisture content.

According to a mathematical model relating the growth of *Aspergillus niger*, in a packed bed with amberlite, with variables, such as sugar, water content, oxygen and carbon dioxide concentrations, and bulk temperature, the rate-limiting step for mycelial growth is sugar depletion (Mariano et al. 1995). The model was effective in predicting biological parameters difficult to determine experimentally, but failed when product formation occurred.

To improve xylanase production by *A. fischeri* in SSF with wheat bran, Senthilkumar et al. used response surface methodology and central composite rotary design (Senthilkumar et al. 2005). The analysis, involving the effects of four variables (sodium nitrite, potassium dihydrogen phosphate, magnesium sulphate, and yeast extract), allowed the optimisation of culture media for maximum xylanase production and minimum protease activity.

SSF has been found to be particularly useful for enzyme production, especially of ligninases, xylanases, and pectinases, to increase the value of agricultural by-products. The low-technology necessary allows its application in farms and could provide animal feeding in developing countries (Graminha et al. 2008).

## *Submerged Fermentation*

During submerged fermentation (SF) of filamentous fungi, the morphology plays a critical role (Gibbs et al. 2000). Several parameters influence the growth of fungi as free mycelia or pellets, such as shear stress, which should be high in stirred tanks (Kelly et al. 2006; Ranjan 2008; Xu et al. 2006). High viscosities are often observed, resulting in, e.g., decreased oxygen transfer, formation of nutrient gradients, high power requirements, and reduced productivity. Several methodologies have been proposed to prevent increased viscosity, including pulsed addition of the limiting carbon source (Bhargava et al. 2003), which results until the point of conidial formation (Bhargava et al. 2005). Since filamentous fungi require simultaneously low stirring speeds and high amounts of oxygen, while viscosity increases as a result of cellular growth, optimisation of fermentation conditions should be assessed for each strain and bioreactor (Coradin et al. 2011; Cho et al. 2006). If very high air flow rates and low stirring speeds are combined, air dispersion is reduced and impeller "flooding" (where air flows mainly up the stirrer shaft) results in poor mixing and reduced oxygen transfer rates (Michelin et al. 2011; Doran 1995).

Albaek et al. developed a mathematical model to simulate a 550 L pilot-scale fed-batch fermentation of recombinant *A. oryzae* at different stirring speeds and aeration rates, comprising the reaction equation stoichiometry, a mass transfer correlation, and viscosity prediction (Albaek et al. 2011). Two impeller types were tested in the 20 fermentations used to develop the model: a Rushton disc turbine and a hydrofoil Hayward Tyler B2. Contrarily to what was expected, analysis of the energy dissipation/circulation function (EDCF) showed that the apparent viscosity was lower when the B2 hydrofoil was used, whilst the lower number of blades and associated vortices would result in higher shear forces. Heo and co-workers also showed that the impeller type affects mycelial morphology: cells of *A. oryzae* grown in a submerged culture with propeller agitation grew in the form of a pellet whilst those in a bioreactor with turbine agitation grew as freely dispersed hyphae and in a clumped form (Heo et al. 2004). The former morphology allowed the highest protein production levels for both intracellular heterologous protein (β-glucuronidase) and the extracellularly homologous protein (α-amylase). Further improvement could be achieved by supplying the carbon source through pulsed feeding.

When the effects of increased agitation power on enzyme expression of *A. oryzae* were studied in an 80 $m^3$ fermenter operated in fed batch, it was found that increased power improved bulk mixing but resulted in lower recombinant enzyme productivity (Li et al. 2002). Biomass assays and image analysis showed that slower growth, altered morphology, or increased hyphal fragmentation were not responsible for the reduced productivity observed at higher power inputs. Since the impeller power was increased by 50 %, by increasing 10 % the impeller diameter while operating at lower stirring speeds, EDCF values dropped linearly from 40 to 15 kW $m^{-3}$ $s^{-1}$ during "high power" batches. In control fermentations the EDCF values were nearly constant at 25 kW $m^{-3}$ $s^{-1}$. The results thus showed that oxygen transfer was less efficient when higher agitation power was applied because of the lower impeller speed, making it important to study how power is applied to the media during scale-up of viscous fungal fermentations.

Low-shear environments may be provided by running fermentations in airlift bioreactors. Airlifts do not have mechanical stirrers, decreasing the risk of contamination, energy demand, and costs (Michelin et al. 2011, 2013). Enzyme productivities and production rates are, in general, higher in airlifts than in stirred tanks: lipase productivity by *Geotrichum candidum* was ca. 60 % higher (Burkert et al. 2005), xylanase production was higher, and both xylanase and β-xylosidase were produced at a faster rate by *Aspergillus terricola* (Michelin et al. 2011). However, higher xylanase levels were produced by *A. niger* in a stirred tank although when the enzymatic production was compared for the two types of bioreactors, at the same $k_La$ values, both xylanase and β-xylosidase productions were higher in the airlift (Michelin et al. 2013). During exopolysaccharide (EPS) production by *Paecilomyces tenuipes* C240, it was found that the specific production rate was significantly higher in the airlift reactor, but the final concentration of the mycelial biomass was much lower, whilst the carbohydrate composition of EPS produced in each reactor was quite different (Xu et al. 2006). On the other hand, the specific productivities and yield coefficients of both biomass and EPS of a submerged mycelial culture of the mushroom *Tremella fuciformis* were higher when the cells were grown in a 5 L airlift as compared to a 5 L stirred tank reactor (Cho et al. 2006). The results obtained in the different studies show that there is a link between mycelial morphology, which is affected by several parameters, such as shear stress and available oxygen concentration, and metabolic activities in filamentous fungi.

**Table 21.2** Examples of bioreactors used for fungal fermentations

| Bioreactor type | Remarks | Reference |
|---|---|---|
| Airlifts | Lower power requirements for the same $k_La$ than mechanically stirred tanks and lower shear stress | Cho et al. (2006), etc group (2013) |
| Bag reactors | Cell aggregation may be prevented; allow the growth of shear stress sensitive fungi | Jonczyk et al. (2013) |
| Laterally aerated moving bed | Can perform batch and continuous SSF | Wong et al. (2011) |
| Membrane reactors | Use the natural ability of fungal cells to adhere to surfaces | Hevekerl et al. (2014), Linde et al. (2014) |
| Microtiter plares | Allows to simplify and expedite process optimisation | Hevekerl et al. (2014), Linde et al. (2014) |
| Packed bed | High substrate/volume ratio allowed; forced aeration; porosity affects directly fluid dynamics and heat and mass transfer | Casciatori et al. (2014) |
| Rotary drum | Allows mixing with control of rotation speed | Diaz et al. (2009) |
| Stirred tanks | High biomass concentration possible but may affect rheology, mixing, and oxygen transfer; good control of parameters (temperature, pH, etc.) | Gibbs et al. (2000), Gabelle et al. (2012) |
| Tray bioreactors | Popular bioreactor for industrial SSF; scale-up achieved by increasing number of trays but difficult to predict cell behaviour; limited heat and mass transfer | Bhargav et al. (2008), Byndoor et al. (1997) |

## Bioreactor Design

As mentioned previously, the performance of fungi in bioreactors depends greatly on the rheological properties of the broth, which are also influenced by the concentration of biomass and cellular morphology. The relations between rheology and morphology should be studied and taken into consideration during the design of the bioreactor. Several types of reactors have been proposed for SSF and SF (Table 21.2), but the resulting productivities and yields obtained in each type vary considerably depending on fungal strains and products.

Tray bioreactors are possibly the simplest type of reactor for SSF, requiring low costs and operation skills, but the amount of substrate is limited and the number of trays necessary for scale-up requires space (Thomas et al. 2013). Heat and mass transfer in tray bioreactors may also be limited by intraparticle oxygen diffusion, growth rate of the fungus, and absence of heat exchange processes (Bhargav et al. 2008). Nevertheless, in tray bioreactors, the cells are not under mechanical stress; support agglomeration may be prevented by keeping the layer of substrate thin, and the low aeration may prevent fungal overgrowth (Rosales et al. 2007). For these reasons, laccase production by *Trametes hirsuta* is higher in this type of bioreactor than when the fungus was grown in Erlenmeyer flasks or on a fixed-bed tubular bioreactor (Rosales et al. 2007). In fact, tray reactors have been particularly successful for the production of industrial enzymes, such as phytases, pectinases, and cellulases (Thomas et al. 2013). To improve pectinase production from lemon peel pomace by *A. niger*, Ruiz et al. developed a column-tray bioreactor containing eight perforated base trays inside a vertical cylinder (Ruiz et al. 2012). The system used a compressor for forced aeration and a water jacket for temperature control.

To allow the mixing of the solid substrates in SSF, several groups have suggested rotating drum bioreactors. However, this reactor type seems to provide better results when static or low agitation conditions are implemented. The maximum hydrolytic enzyme activities from a mixture of 1:1 (w/w) of grape pomace and orange peels using *Aspergillus awamori* were attained in static conditions or an agitation as low as 1 min/day (Diaz et al. 2009). The production of cellulases and hemicellulases by the thermophilic microfungus *Thermoascus aurantiacus* was carried out in a drum bioreactor rotating at 10 rpm for 1 min every 3 h (Kalogeris et al. 2003).

In packed-bed reactors, the solid humidified substrate is retained on a perforated base through which air is forced to pass. To the glass or plastic body of the reactor, a water jacket may be added to control the fermentation temperature. However, in these reactors, non-uniform growth may be observed; the product may be difficult to recover and heat transfer and scale-up problems may occur (Couto and Sanromán 2006). When comparing coal biosolubilisation in a stirred tank, in a fluidised bed, and in a fixed-bed bioreactor, Oboirien and co-workers found that in the packed-bed bioreactor the biosolubilisation was minimal probably due to clogging of the bed particles by fungi which leads to unpredictable internal mass transport and due to the large size of the coal particles used (Oboirien et al. 2013). In this case, the stirred tank allowed the best fungal performance whilst the fluidised bed reactor demanded the highest aeration rates since air is also responsible for mixing. In fluidised beds, continuous agitation with forced air prevents adhesion and aggregation of the substrate particles, helping mass and heat transfer, but cellular damage and increased heat due to shear forces may decrease the expected product yield (Couto and Sanromán 2006).

Although stirred tank reactors may cause high shear stress to filamentous fungi, these reactors often overcome limitations caused by insufficient mixing and mass and heat transfer observed in other types of reactors. For example, high lactic acid concentration (85.7 g/L) and yield (86 %) were attained from waste potato starch in a mechanically stirred bioreactor using acid-adapted *Rhizopus arrhizus* (Zhang et al. 2008). Since mixing is the most important feature of stirred tanks, the type of impellers should be studied. During the fermentation of digested solka floc using a recombinant strain of *Zymomonas mobilis*, the bioreactor was tested with a Rushton turbine and a marine impeller, with and without wall baffles (Um and Hanley 2008). At 120 rpm, the enzymatic saccharification for glucose production attained much higher concentration of glucose in reactors with Rushton turbines than those with marine impellers. Furthermore, the presence of baffles in the reactor walls improved the fungal fermentation when the system was mixed by Rushton impellers but not when marine impellers were used. Since Rushton turbines performed poorly in a 75 L fermenter developed for the conversion of lignocellulosic agricultural materials by *Neurospora sitophila*, they were replaced by suitable axial flow impellers

(Chisti and Moo-Young 1994). Although no significant increase in fungal growth rate was observed between the two impellers, cellulose utilisation was ca. 86 % higher with the axial flow impeller as no agitation-associated mechanical damage was observed in the mycelia.

One of the advantages of using submerged cultures of filamentous fungi for the production of commercially interesting enzymes is the possibility to use gas–liquid mass transfer predictions to scale-up the system (Gabelle et al. 2012). A power law model to simulate the rheology of the model dependent on the biomass concentration could be determined during the growth of *Trichoderma reesei* (Gabelle et al. 2012). A model including the effects of dissolved oxygen, viscosity, temperature, pH, and dissolved carbon dioxide to describe penicillin fermentation by *Penicillium chrysogenum* could be developed to improve control and optimisation of biomass and penicillin production (Goldrick et al. 2015).

To prevent cell damaging in shear stress sensitive basidiomycetes such as *Flammulina velutipes*, a disposable bag bioreactor developed for cultivation of mammalian cells may be used (Jonczyk et al. 2013). Contrarily to what was observed for stirred tank reactors, dispersed pellets were observed and *F. velutipes* reached higher biomass concentrations and twofold higher peptidolytic activities.

During cultivation of filamentous fungi, the initial low viscosity Newtonian medium usually becomes a highly viscous non-Newtonian fluid and airlift bioreactors could be an efficient alternative to mechanically stirred reactors. In airlift reactors, the parameters affecting product formation are gas hold-up, liquid circulation velocity, and mixing (Kang et al. 2001). When studying an airlift bioreactor with internal recirculation loop for the production of a biopolymer with *Sclerotium glucanicum* NRRL 3006, Kang et al. noticed significant differences between experimental "real" fungal culture medium and simulated systems (Kang et al. 2001). While the latter are homogeneous, simple systems, real fungal cultures are highly heterogeneous, presenting several morphologies, and cells attached to substrate particles. Recirculation in the airlift may also be affected by biomass concentration: a fast growth rate of *Geotrichum candidum* observed at high $k_La$ values blocked biomass recirculation (Burkert et al. 2005). Nevertheless, the productivity in the airlift reactor was 60 % higher than that attained with a stirred fermenter.

What the published works demonstrate is that due to the heterogeneity of fungal strains, and the numerous types of morphological responses to the reaction conditions, several studies are required to select the best reactor design for each system as no simple prediction of cellular behaviour should be possible.

## Biotransformation Systems

Biotransformations using fungal cells allow the production of compounds that may be classified as "natural" products by the European and American food legislations (Krings and Berger 1998), reaching much higher market value than chemically synthesised compounds. One of the best examples is the production of vanillin

(3-methoxy-4-hydroxybenzaldehyde), the flavour of vanilla. Natural vanillin, extracted from the seed pods of the vanilla orchid, represents ca. 0.25 % of the global market (approximately 40–50 tons per year) and costs ca. US$ 4,000 per kg, whilst around 16,000 tons per year of synthetic vanillin are produced and sold at US$10–20 per kg (etc group 2013; Hansen et al. 2009). Solvay, the world's leading vanillin producer, now produces Rhovanil® Natural by bioconversion of ferulic acid, a natural organic compound found in rice bran (Solvay SA 2013). The Swiss biotech company Evolva developed a yeast-based fermentation route to produce both vanillin and other vanilla flavour components (evolva 2015).

A *de novo* pathway enabling one-cell microbial generation of vanillin from glucose was recently developed in the fission yeast *Schizosaccharomyces pombe* and also in baker's yeast, *Saccharomyces cerevisiae* (Hansen et al. 2009). The engineered pathways developed involved the incorporation of 3-dehydroshikimate dehydratase from *Podospora pauciseta*, of an aromatic carboxylic acid reductase from a bacterium of the *Nocardia* genus, and of an O-methyltransferase from *Homo sapiens*. The productivities reached 65 and 45 mg/L after introduction of 3 and 4 heterologous genes, respectively.

Several fungi, including species of *Aspergillus*, *Fusarium*, *Polyporus*, and *Rhodotorula* have been found able to convert ferulic acid, which results from lignin degradation of common agricultural residues like cereal brans and sugarbeet pulp (Rosazza et al. 1995). The fungus *Sporotrichum thermophile* produces vanillic acid during ferulic acid degradation via the propenoic chain degradation (Topakas et al. 2003). When the fungus was grown on glucose for 3 days, it was able to convert 80 mmol/L of ferulic acid with a molar yield of 35 %, reaching 4.8 g/L of vanillic acid.

Around one-third of the percentage of papers published in the first decade of this century on the production and/or biotransformation of terpenes were based on studies with fungi (de Carvalho and da Fonseca 2006). Terpenes are widely distributed in nature and some such as limonene and α-pinene are inexpensively available in large quantities, whilst their oxygenated derivatives terpenoids have been used as fragrances and flavours for centuries. Fungal biotransformations may provide terpenoids for the food, cosmetic, and pharmaceutical industries (Table 21.3) with high value to the increasingly demanding consumers looking for natural and healthy products. Furthermore, the compounds may be produced with high stereo-specificity and selectivity, under mild conditions (de Carvalho and da Fonseca 2006; de Carvalho 2009).

Fungal cells have also been a good source of interesting enzymes. The GRAS (generally regarded as safe) fungus *Aspergillus niger* was found as a good source of the enzyme tannase, produced by SSF from cheap tea by-products (Ni et al. 2015). After treating a tea infusion with tannase, the tea presented significant increases in clarity, 2,2′-azino-bis (3-ethylbenzothiazoline-6-sulphonate) and 2,2-diphenyl-1-picryl-hydrazyl inhibition activities, and hydroxyl radical scavenging capacity.

Chitosanases are able to catalyse the endohydrolysis of β-1,4-glycosidic bonds of partially acetylated chitosan to release chitosan oligosaccharides and are unable to hydrolyse chitin (Thadathil and Velappan 2014). The bioconversion of chitinous wastes from, e.g., the seafood processing industry to bioactive compounds seems promising. Among the fungi reported as chitosanase producers are strains belonging

**Table 21.3** Examples of biotransformations of terpenes carried out by fungi

| Fungus | Biotransformation | Reference |
|---|---|---|
| *Aspergillus niger* | Hydroxylation reactions of acyclic monoterpene alcohols | Madyastha and Murthy (1988) |
| *Aspergillus niger* | (+)-limonene to terpineols and carveols | Garcia-Carnelli et al. (2014) |
| *Botrytis cinerea* | α-santonin and sclareol | Farooq and Tahara (2000) |
| *Cyathus africanus* | Insecticidal sesquiterpenes cadina-4,10(15)-dien-3-one and aromadendr-1(10)-en-9-one | McCook et al. (2012) |
| *Mortierella minutissima* | (+)-limonene to perillyl alcohol and perillyl aldehyde | Trytek et al. (2009) |
| *Mucor plumbeus* | Conversion of sesquiterpenoid (−)-maaliol to (+)-7,8-didehydro-9β-hydroxymaalioxide and (−)-7,8-didehydro-1β-hydroxymaalioxide | Wang et al. (2009) |
| *Penicillium* and *Fusarium* strains | Biotransformations of (*RS*)-linalool, (*S*)-citronellal, and sabinene | Rueda et al. (2013) |
| *Penicillium digitatum* | D-limonene to α-terpineol | Lindmark-Henriksson et al. (2004), Ni et al. (2015) |
| *Penicillium digitatum* | Conversion of geraniol, nerol, "citrol", citral to 6-methyl-5-hepten-2-one | Demyttenaere and De Kimpe (2001) |
| *Pichia* and *Rhizopus* strains | Limonene, α- and β-pinene | Bier et al. (2011) |
| *Picea abies* | β-pinene to *trans*-pinocarveol and minor products | Lindmark-Henriksson et al. (2004) |

to the genera *Aspergillus, Gongronella*, and *Trichoderma* (Thadathil and Velappan 2014).

White-rot fungi have also been used to produce laccases, since they are the only known organisms able to fully mineralise all components of lignin, although yeasts have been used as suitable hosts for heterologous laccase production (Couto and Toca-Herrera 2007). Laccases have been mainly applied in the decolouration of dyes in the textile, dye, or printing industries and in the delignification and bleaching of paper pulps, with other applications including organic synthesis, the manufacture of biodevices, or the detoxification of pollutants (Couto and Toca-Herrera 2007; Cañas and Camarero 2010).

## Bioremediation

The enormous number of enzymes and substrates metabolised by fungi also make these microorganisms attractive for the bioremediation of recalcitrant compounds (Potin et al. 2004; Leyval et al. 1997; Gray 1998; Zhdanova et al. 2000).

As mentioned, white-rod fungi possess ligninolytic enzymes with broad substrate specificity, being able to degrade pesticides, polychlorinated biphenyls, polycyclic aromatic hydrocarbons, synthetic dyes, synthetic polymers, and wood preservatives (Pointing 2001). Strains of *Phanerochaete chrysosporium*, *Pleurotus ostreatus*, *Phellinus weirii*, and *Polyporus versicolor* have been shown able to mineralise DDT, whilst *P. chrysosporium*, *P. ostreatus*, and *Trametes versicolor* are capable of degrading polychlorinated biphenyls and polycyclic aromatic hydrocarbons.

Other fungi have been used for the bioconversion of soluble/insoluble (organic) substances in, e.g., activated domestic sludge: *Penicillium corylophilum* and *Aspergillus niger* increased filterability/dewaterability while decreasing considerably COD and turbidity (Mannan et al. 2005). Using response surface methodology, COD removal by *P. corylophilum* could be improved to 98.5 % by using an incubation temperature of 32.5 °C, the agitation at 105 rpm, and pH at 5.5 (Mannan et al. 2007).

Several fungal strains belonging to the genera *Cephalosporium*, *Paecilomyces*, *Penicillium*, *Aspergillus*, *Trichoderma*, and *Mucor* have been used in gas-phase biofilters for the treatment of waste gases (Liu et al. 2013; Kennes and Veiga 2004). They are able to remove malodorous sulphur compounds, which are the main cause of odour pollution, and may be used together with bacteria in biotrickling filters (Liu et al. 2013).

Fungal cells are also able to bioremediate contaminated soil with, e.g., polycyclic aromatic hydrocarbons (Potin et al. 2004), heavy metals (Leyval et al. 1997), and even radioactive material (Gray 1998). In fact, expensive fungal growth was observed on the walls of the inner part of the Shelter of the damaged fourth Unit of the Chernobyl Nuclear Power Plant (Zhdanova et al. 2000). Of the fungi isolated, about 80 % were melanin-containing and pigmented micromycetes.

## Final Remarks

Fungi are promising producers of commercially interesting compounds. These may be secondary metabolites or the result of enzymatic bioconversion. However, to achieve the best yields, environmental conditions must be adjusted and bioreactors should be chosen to allow the best productivities and activities. Fungal behaviour is difficult to model and predict. Nevertheless, several research groups are trying to provide further understanding on these complex microorganisms.

**Acknowledgements** The author acknowledges Fundação para a Ciência e a Tecnologia, I.P. (FCT), Portugal, for the exploratory project and financial support under program "Investigador FCT 2013" (IF/01203/2013/CP1163/CT0002).

## References


Albaek MO, Gernaey KV, Hansen MS, Stocks SM (2011) Modeling enzyme production with *Aspergillus oryzae* in pilot scale vessels with different agitation, aeration, and agitator types. Biotechnol Bioeng 108(8):1828–1840

Ballesteros M, Oliva JM, Negro MJ, Manzanares P, Ballesteros I (2004) Ethanol from lignocellulosic materials by a simultaneous saccharification and fermentation process (SFS) with *Kluyveromyces marxianus* CECT 10875. Process Biochem 39(12):1843–1848

Barrios-González J (2012) Solid-state fermentation: physiology of solid medium, its molecular basis and applications. Process Biochem 47(2):175–185

Barrios-González J, Mejýa A (2008) Production of antibiotics and other commercially valuable secondary metabolites. In: Pandey A, Soccol C, Larroche C (eds) Current developments in solid-state fermentation. Springer, New York, pp 302–336

Bhargav S, Panda BP, Ali M, Javed S (2008) Solid-state fermentation: an overview. Chem Biochem Eng Q 22(1):49–70

Bhargava S, Nandakumar MP, Roy A, Wenger KS, Marten MR (2003) Pulsed feeding during fed-batch fungal fermentation leads to reduced viscosity without detrimentally affecting protein expression. Biotechnol Bioeng 81(3):341–347

Bhargava S, Wenger KS, Rane K, Rising V, Marten MR (2005) Effect of cycle time on fungal morphology, broth rheology, and recombinant enzyme productivity during pulsed addition of limiting carbon source. Biotechnol Bioeng 89(5):524–529

Bier MCJ, Poletto S, Soccol VT, Soccol CR, Medeiros ABP (2011) Isolation and screening of microorganisms with potential for biotransformation of terpenic substrates. Braz Arch Biol Technol 54(5):1019–1026

Burkert JFD, Maldonado RR, Maugeri F, Rodrigues MI (2005) Comparison of lipase production by *Geotrichum candidum* in stirring and airlift fermenters. J Chem Technol Biotechnol 80(1):61–67

Byndoor MG, Karanth NG, Rao GV (1997) Efficient and versatile design of a tray type solid state fermentation bioreactor. In: Roussos S, Lonsane BK, Raimbault M, Viniegra-Gonzalez G (eds) Advances in solid state fermentation. Springer, Berlin, pp 113–119

Cañas AI, Camarero S (2010) Laccases and their natural mediators: biotechnological tools for sustainable eco-friendly processes. Biotechnol Adv 28(6):694–705

Cannel E, Moo-Young M (1980) Solid-state fermentation systems. Process Biochem 15:24–28

Casciatori FP, Laurentino CL, Taboga SR, Casciatori PA, Thomeo JC (2014) Structural properties of beds packed with agro-industrial solid by-products applicable for solid-state fermentation: experimental data and effects on process performance. Chem Eng J 255:214–224

Chisti Y, Moo-Young M (1994) Slurry bioreactor design for shear-sensitive mycoprotein production. In: Galindo E, Ramírez OT (eds) Advances in bioprocess engineering. Kluwer, Dordrecht, pp 25–28

Cho EJ, Oh JY, Chang HY, Yun JW (2006) Production of exopolysaccharides by submerged mycelial culture of a mushroom *Tremella fuciformis*. J Biotechnol 127(1):129–140

Christakopoulos P, Macris BJ, Kekos D (1989) Direct fermentation of cellulose to ethanol by *Fusarium oxysporum*. Enzyme Microb Technol 11(4):236–239

Coradin JH, Braun A, Viccini G, de Lima Luz LF Jr, Krieger N, Mitchell DA (2011) A three-dimensional discrete lattice-based system for modeling the growth of aerial hyphae of filamentous fungi on solid surfaces: a tool for investigating micro-scale phenomena in solid-state fermentation. Biochem Eng J 54(3):164–171

Couto SR, Sanromán MÁ (2006) Application of solid-state fermentation to food industry—a review. J Food Eng 76(3):291–302

Couto SR, Toca-Herrera JL (2007) Laccase production at reactor scale by filamentous fungi. Biotechnol Adv 25(6):558–569

de Carvalho CCCR (2009) Biotransformation of terpenes by fungi. In: Rai M (ed) Advances in fungal biotechnology. I.K. International Publishing House Pvt. Ltd., New Delhi, pp 253–267 de Carvalho CCCR, da Fonseca MMR (2006) Biotransformation of terpenes. Biotechnol Adv 24(2):134–142

Debbab A, Aly AH, Lin WH, Proksch P (2010) Bioactive compounds from marine bacteria and fungi. J Microbial Biotechnol 3(5):544–563

Demyttenaere J, De Kimpe N (2001) Biotransformation of terpenes by fungi: study of the pathways involved. J Mol Catal B: Enzym 11(4–6):265–270

Diaz AB, de Ory I, Caro I, Blandino A (2009) Solid state fermentation in a rotating drum bioreactor for the production of hydrolytic enzymes. In: Pierucci S (ed) Icheap-9: 9th international conference on chemical and process engineering, Pts 1–3, vol 17. Chemical Engineering Transactions, pp 1041–46

Doran PM (1995) Bioprocess engineering principles. Academic, London

etc group (2013) Vanilla and synthetic biology—a case study. etc group. http://www.etcgroup.org/sites/www.etcgroup.org/files/Vanilla_SynBio_case_study_Oct2013.pdf. Accessed 10 Jan 2015

evolva. http://www.evolva.com/products/vanilla#sthash.YHCPrTwx.dpuf. Accessed 10 Jan 2015

Farooq A, Tahara S (2000) Biotransformation of two cytotoxic terpenes, alpha-santonin and sclareol by *Botrytis cinerea*. Zeitschrift Fur Naturforschung C-a J Biosci 55(9–10):713–717

Gabelle JC, Jourdier E, Licht RB, Ben Chaabane F, Henaut I, Morchain J, Augier F (2012) Impact of rheology on the mass transfer coefficient during the growth phase of *Trichoderma reesei* in stirred bioreactors. Chem Eng Sci 75:408–417

Garcia-Carnelli C, Rodriguez P, Heinzen H, Menendez F (2014) Influence of culture conditions on the biotransformation of (+)-limonene by *Aspergillus niger*. Zeitschrift Fur Naturforschung Section C-a J Biosci 69(1–2):61–67

Gibbs PA, Seviour RJ, Schmid F (2000) Growth of filamentous fungi in submerged culture: problems and possible solutions. Crit Rev Biotechnol 20(1):17–48

Goldrick S, Ştefan A, Lovett D, Montague G, Lennox B (2015) The development of an industrial-scale fed-batch fermentation simulation. J Biotechnol 193:70–82

Graminha EBN, Gonçalves AZL, Pirota RDPB, Balsalobre MAA, Da Silva R, Gomes E (2008) Enzyme production by solid-state fermentation: application to animal nutrition. Anim Feed Sci Technol 144(1–2):1–22

Gray SN (1998) Fungi as potential bioremediation agents in soil contaminated with heavy or radioactive metals. Biochem Soc Trans 26(4):666–670

Hansen EH, Møller BL, Kock GR, Bünner CM, Kristensen C, Jensen OR, Okkels FT, Olsen CE, Motawia MS, Hansen J (2009) *De novo* biosynthesis of vanillin in fission yeast (*Schizosaccharomyces pombe*) and baker's yeast (*Saccharomyces cerevisiae*). Appl Environ Microbiol 75(9):2765–2774

Heo J-H, Ananin V, Park J-S, Lee C-R, Moon J-O, Kwon O, Kang H-A, Kim C, Rhee S (2004) Impeller types and feeding modes influence the morphology and protein expression in the submerged culture of *Aspergillus oryzae*. Biotechnol Bioprocess Eng 9(3):184–190

Hevekerl A, Kuenz A, Vorlop K-D (2014) Filamentous fungi in microtiter plates—an easy way to optimize itaconic acid production with *Aspergillus terreus*. Appl Microbiol Biotechnol 98(16):6983–6989

Jonczyk P, Takenberg M, Hartwig S, Beutel S, Berger RG, Scheper T (2013) Cultivation of shear stress sensitive microorganisms in disposable bag reactor systems. J Biotechnol 167(4):370–376

Kalogeris E, Iniotaki F, Topakas E, Christakopoulos P, Kekos D, Macris BJ (2003) Performance of an intermittent agitation rotating drum type bioreactor for solid-state fermentation of wheat straw. Bioresour Technol 86(3):207–213

Kang X, Wang H, Wang Y, Harvey LM, McNeil B (2001) Hydrodynamic characteristics and mixing behaviour of *Sclerotium glucanicum* culture fluids in an airlift reactor with an internal loop used for scleroglucan production. J Ind Microbiol Biotechnol 27(4):208–214

Karimi K, Zamani A (2013) *Mucor indicus*: biology and industrial application perspectives: a review. Biotechnol Adv 31(4):466–481

Karimi K, Emtiazi G, Taherzadeh MJ (2006) Production of ethanol and mycelial biomass from rice straw hemicellulose hydrolyzate by *Mucor indicus*. Process Biochem 41(3):653–658

Kelly S, Grimm LH, Bendig C, Hempel DC, Krull R (2006) Effects of fluid dynamic induced shear stress on fungal growth and morphology. Process Biochem 41(10):2113–2117

Kennes C, Veiga MC (2004) Fungal biocatalysts in the biofiltration of VOC-polluted air. J Biotechnol 113(1–3):305–319

Krings U, Berger RG (1998) Biotechnological production of flavours and fragrances. Appl Microbiol Biotechnol 49(1):1–8

Lange L (2010) The importance of fungi for a more sustainable future on our planet. Fungal Biol Rev 24(3–4):90–92

Leyval C, Turnau K, Haselwandter K (1997) Effect of heavy metal pollution on mycorrhizal colonization and function: physiological, ecological and applied aspects. Mycorrhiza 7(3):139–153

Li ZJ, Shukla V, Wenger KS, Fordyce AP, Pedersen AG, Marten MR (2002) Effects of increased impeller power in a production-scale *Aspergillus oryzae* fermentation. Biotechnol Prog 18(3):437–444

Linde T, Hansen NB, Lübeck M, Lübeck PS (2014) Fermentation in 24-well plates is an efficient screening platform for filamentous fungi. Lett Appl Microbiol 59(2):224–230

Lindmark-Henriksson M, Isaksson D, Vanek T, Valterova I, Hogberg HE, Sjodin K (2004) Transformation of terpenes using a *Picea abies* suspension culture. J Biotechnol 107(2):173–184

Liu CQ, Chen QH, Tang B, Ruan H, He GQ (2007) Response surface methodology for optimizing the fermentation medium of alpha-galactosidase in solid-state fermentation. Lett Appl Microbiol 45(2):206–212

Liu C, Liu J, Li J, He H, Peng S, Li C, Chen Y (2013) Removal of H2S by co-immobilized bacteria and fungi biocatalysts in a bio-trickling filter. Process Saf Environ Prot 91(1–2):145–152

Lonsane BK, Ghildyal NP, Budiatman S, Ramakrishna SV (1985) Engineering aspects of solid state fermentation. Enzyme Microb Technol 7(6):258–265

Madyastha KM, Murthy N (1988) Regiospecific hydroxylation of acyclic monoterpene alcohols by *Aspergillus niger*. Tetrahedron Lett 29(5):579–580

Mannan S, Fakhru'l-Razi A, Alam MZ (2005) Use of fungi to improve bioconversion of activated sludge. Water Res 39(13):2935–2943

Mannan S, Fakhru'l-Razi A, Alam MZ (2007) Optimization of process parameters for the bioconversion of activated sludge by *Penicillium corylophilum*, using response surface methodology. J Environ Sci 19(1):23–28

Mariano G-R, Richard A, Claude BJ, Sergio R (1995) A mathematical model for solid state fermentation of mycelial fungi on inert support. Chem Eng J Bioch Eng 60(1–3):189–198

McCook KP, Chen ARM, Reynolds WF, Reese PB (2012) The potential of *Cyathus africanus* for transformation of terpene substrates. Phytochemistry 82:61–66

Michelin M, Polizeli MLTM, da Silva DP, Ruzene DS, Vicente AA, Jorge JA, Terenzi HF, Teixeira JA (2011) Production of xylanolytic enzymes by *Aspergillus terricola* in stirred tank and airlift tower loop bioreactors. J Ind Microbiol Biotechnol 38(12):1979–1984

Michelin M, de Oliveira Mota AM, Teixeira de Moraes Polizeli ML, da Silva DP, Vicente AA, Teixeira JA (2013) Influence of volumetric oxygen transfer coefficient (k(L)a) on xylanases batch production by *Aspergillus niger* van Tieghem in stirred tank and internal-loop airlift bioreactors. Biochem Eng J 80:19–26

Millati R, Karimi K, Edebo L, Niklasson C, Taherzadeh MJ (2008) Ethanol production from xylose and wood hydrolyzate by *Mucor indicus* at different aeration rates. BioResources 3(4):1020–1029

Mitchell DA, von Meien OF, Krieger N, Dalsenter FDH (2004) A review of recent developments in modeling of microbial growth kinetics and intraparticle phenomena in solid-state fermentation. Biochem Eng J 17(1):15–26

More TT, Yan S, Tyagi RD, Surampalli RY (2010) Potential use of filamentous fungi for wastewater sludge treatment. Bioresour Technol 101(20):7691–7700

Ni H, Chen F, Jiang ZD, Cai MY, Yang YF, Xiao AF, Cai HN (2015) Biotransformation of tea catechins using *Aspergillus niger* tannase prepared by solid state fermentation on tea byproduct. LWT—Food Sci Technol 60(2):1206–1213

Oboirien BO, Ojumu TV, Obayopo SO (2013) Fungi solubilisation of low rank coal: performances of stirred tank, fluidised bed and packed bed reactors. Fuel Process Technol 106:295–302

Pandey A, Soccol CR, Mitchell D (2000) New developments in solid state fermentation: I-bioprocesses and products. Process Biochem 35(10):1153–1169

Papavizas GC, Dunn MT, Lewis JA, Beagle-Ristaino J (1984) Liquid fermentation technology for experimental production of biocontrol fungi. Phytopathology 74(10):1171–1175

Pérez-Guerra N, Torrado-Agrasar A, López-Macias C, Pastrana L (2003) Main characteristics and applications of solid substrate fermentation. Electron J Environ Agric Food Chem 2(3):343–350

Pointing SB (2001) Feasibility of bioremediation by white-rot fungi. Appl Microbiol Biotechnol 57(1–2):20–33

Posch AE, Herwig C (2014) Physiological description of multivariate interdependencies between process parameters, morphology and physiology during fed-batch penicillin production. Biotechnol Prog 30(3):689–699

Posch AE, Herwig C, Spadiut O (2013) Science-based bioprocess design for filamentous fungi. Trends Biotechnol 31(1):37–44

Potin O, Rafin C, Veignie E (2004) Bioremediation of an aged polycyclic aromatic hydrocarbons (PAHs)-contaminated soil by filamentous fungi isolated from the soil. Int Biodeter Biodegr 54(1):45–52

Rahardjo YSP, Tramper J, Rinzema A (2006) Modeling conversion and transport phenomena in solid-state fermentation: a review and perspectives. Biotechnol Adv 24(2):161–179

Ranjan AP (2008) Variation in mycelial growth and morphology of *Aspergillus niger* with changes in shear stress. Res J Biotechnol 3(1):22–26

Rodrigues C, de Souza Vandenberghe L, Teodoro J, Pandey A, Soccol C (2009) Improvement on citric acid production in solid-state fermentation by *Aspergillus niger* LPB BC mutant using citric pulp. Appl Biochem Biotechnol 158(1):72–87

Rosales E, Rodríguez Couto S, Sanromán MA (2007) Increased laccase production by *Trametes hirsuta* grown on ground orange peelings. Enzyme Microb Technol 40(5):1286–1290

Rosazza JPN, Huang Z, Dostal L, Volm T, Rousseau B (1995) Review: biocatalytic transformations of ferulic acid—an abundant aromatic natural product. J Ind Microbiol 15(6):457–471

Rueda MGM, Guerrini A, Giovannini PP, Medici A, Grandini A, Sacchetti G, Pedrini P (2013) Biotransformations of terpenes by fungi from Amazonian citrus plants. Chem Biodivers 10(10):1909–1919

Ruiz HA, Rodríguez-Jasso RM, Rodríguez R, Contreras-Esquivel JC, Aguilar CN (2012) Pectinase production from lemon peel pomace as support and carbon source in solid-state fermentation column-tray bioreactor. Biochem Eng J 65:90–95

Ryan DR, Leukes WD, Burton SG (2005) Fungal bioremediation of phenolic wastewaters in an airlift reactor. Biotechnol Prog 21(4):1068–1074

Senthilkumar SR, Ashokkumar B, Chandra Raj K, Gunasekaran P (2005) Optimization of medium composition for alkali-stable xylanase production by *Aspergillus fischeri* Fxn 1 in solid-state fermentation using central composite rotary design. Bioresour Technol 96(12):1380–1386

Soetaert W, Vandamme EJ (2010) Industrial biotechnology: sustainable growth and economic success. Wiley-VCH, Weinheim

Solvay SA (2013) Annual Report 2013. Solvay SA. http://www.solvay.cn/en/binaries/2013-annual-report-EN-164627.pdf. Accessed 10 Jan 2015

Thadathil N, Velappan SP (2014) Recent developments in chitosanase research and its biotechnological applications: a review. Food Chem 150:392–399

Thomas L, Larroche C, Pandey A (2013) Current developments in solid-state fermentation. Biochem Eng J 81:146–161

Topakas E, Kalogeris E, Kekos D, Macris BJ, Christakopoulos P (2003) Bioconversion of ferulic acid into vanillic acid by the thermophilic fungus *Sporotrichum thermophile*. LWT—Food Sci Technol 36(6):561–565

Trytek M, Fiedurek J, Skowronek M (2009) Biotransformation of (*R*)-(+)-limonene by the psychrotrophic fungus *Mortierella minutissima* in H2O2-oxygenated culture. Food Technol Biotechnol 47(2):131–136

Um BH, Hanley TR (2008) High-solid enzymatic hydrolysis and fermentation of solka floc into ethanol. J Microbiol Biotechnol 18(7):1257–1265

Viniegra-Gonzàlez G (1997) Solid state fermentation: definition, characteristics, limitations and monitoring. In: Roussos S, Lonsane BK, Raimbault M, Viniegra-Gonzalez G (eds) Advances in solid state fermentation. Springer, Berlin, pp 5–22

Viniegra-Gonzalez G, Favela-Torres E, Aguilar CN, Romero-Gomez SD, Diaz-Godinez G, Augur C (2003) Advantages of fungal enzyme production in solid state over liquid fermentation systems. Biochem Eng J 13(2–3):157–167

Vu VH, Pham TA, Kim K (2010) Improvement of fungal strain by repeated and sequential mutagenesis and optimization of solid state fermentation for the hyper-production of raw-starch-digesting enzyme. J Microbiol Biotechnol 20(4):718–726

Wang Y, Tan T-K, Tan GK, Connolly JD, Harrison LJ (2009) Microbial transformation of the sesquiterpenoid (−)-maaliol by *Mucor plumbeus*. Tetrahedron 65(29–30):5680–5683

Wang KW, Wang SW, Wu B, Wei JG (2014) Bioactive natural compounds from the mangrove endophytic fungi. Mini-Rev Med Chem 14(4):370–391

Wong Y, Saw H, Janaun J, Krishnaiah K, Prabhakar A (2011) Solid-state fermentation of palm kernel cake with *Aspergillus flavus* in laterally aerated moving bed bioreactor. Appl Biochem Biotechnol 164(2):170–182

Xu CP, Kim SW, Hwang HJ, Yun JW (2006) Production of exopolysaccharides by submerged culture of an enthomopathogenic fungus, *Paecilomyces tenuipes* C240 in stirred-tank and airlift reactors. Bioresour Technol 97(5):770–777

Zhang ZY, Jin B, Kelly JM (2008) Production of L(+)-lactic acid using acid-adapted precultures of *Rhizopus arrhizus* in a stirred tank reactor. Appl Biochem Biotechnol 149(3):265–276

Zhdanova NN, Zakharchenko VA, Vember VV, Nakonechnaya LT (2000) Fungi from Chernobyl: mycobiota of the inner regions of the containment structures of the damaged nuclear reactor. Mycol Res 104(12):1421–1426

# Chapter 22
# Microfungi in Biofuel and Bioenergy Research

**Richa Raghuwanshi, Shalini Singh, Mohd. Aamir, Amrita Saxena, Vijai Kumar Gupta, and R.S. Upadhyay**

## Introduction

The exigency for crude oil and the existing ebb of the petroleum reserve of the planet have necessitated the use of alternative sources of energy for fuel production. This can be inferred from the statistical data stated by the United Nations (UN), according to which the world primary energy consumption will increase twofold (24 billion tons coal equivalent per year) to the existing energy consumption calculated by World Energy Council (WEC) in 2009 due to the population increase to about 10 billion by 2050 (Dashtban et al. 2009). Among the existing renewable resources along with the close homology to the natural resource, biofuels have gained greater attention these days. Biofuels refer to the type of fuel whose energy is derived from biological carbon fixation. On this basis, this class includes fuels that are derived from biomass conversion or solid biomass, liquid fuels, and also various biogases.

On the basis of the mode of their production and depending on the starting material for their production, up until now, four generations of biofuels have been under consideration. First-generation biofuel refers to the biofuel derived from plant materials like cereals, sugar crops, and oil seeds. These types include biodiesel, green diesel, bioalcohols, bioethers, and biogas (Sims et al. 2010). Owing to the

R. Raghuwanshi, Ph.D.
Department of Botany, Mahila Mahavidyalaya, Banaras Hindu University, Varanasi 221005, India

S. Singh, Ph.D. • M. Aamir, Ph.D. • A. Saxena, Ph.D. • R.S. Upadhyay (✉)
Department of Botany, Banaras Hindu University, Varanasi 221005, India
e-mail: upadhyay_bhu@yahoo.co.uk

V.K. Gupta, Ph.D.
Molecular Glyco-biotechnology Group, Discipline of Biochemistry, School of Natural Sciences, National University of Ireland Galway, Galway, Ireland
e-mail: vijaifzd@gmail.com; vijai.gupta@nuigalway.ie



D.-W. Li (ed.), *Biology of Microfungi*, Fungal Biology,
DOI 10.1007/978-3-319-29137-6_22

decreasing availability of the plant materials, focus was shifted to the waste products for biofuel production that constituted the second-generation biofuels comprising mainly of cellulosic ethanol and biogas from municipal wastes, etc. The next step toward the development of biofuel owing to the high-cost conversion of lignocellulosic materials and decreased profit was the employment of microbes for the production of biofuel starting the third-generation biofuels. Numerous classes of bacteria and fungi were employed for successful conversion of raw materials into economic biofuel for commercial use. Lignin-degrading fungi, cellulose-decomposing bacteria, lipid-containing fungi, etc., were considered for effective and low-cost production of biofuels. The present concern has been shifted to the use of genetically developed photosynthetic cells for the conversion in biofuels, which brings the fourth generation.

Biofuels produced using lignocellulosic biomass provide several benefits to society such as (1) being renewable and sustainable, (2) indirectly helping a greenhouse gas (GHG) that is responsible for global warming for fixation, (3) facilitating local economy development, (4) reducing air pollution from burning of biomass, (5) bringing energy security for countries dependent on imported oil, and (6) creating jobs for engineers, fermentation specialists, process engineers, and scientists. Fungal biomass biofuel production represents a pivotal approach to face high energy prices and contribute to a net reduction of total greenhouse gas emissions. These microorganisms have the capability to accumulate significant quantities of storage triacylglycerols (TAGs), an attractive source of oil suitable for biodiesel production (Economou et al. 2011). The major reasons for the popularity and success in employing oleaginous fungi for biofuel production may be summarized in the following points:

- High growth rate
- Extensive enzymatic system for efficient lipid production and accumulation
- Ability to utilize cheap waste materials as substrates
- Growth and cultivation independent of seasonal variations
- Easy alteration in lipid production by manipulating nutrient conditions
- Probability of using mutation of specific enzymes at specific steps to increase lipid yield
- Act as potential hosts for cloning foreign genes related with lipid and PUFA production.

## Enzymes Involved in Biofuel Production

Though lignocellulosic biomass is a potential low-cost source of mixed sugars for fermentation to produce fuel ethanol, but it is a naturally recalcitrant material composed of cell walls having intricate network of celluloses, hemicelluloses (including xyloglucans, arabinoxylans, and glucomannans), pectins (like homogalacturonans, rhamnogalacturonan, and xylogalacturonans), lignins, and proteoglycans that creates technical barriers for the cost-effective transformation of biomass to fermentable sugars. Moderate yields and the resulting complex composition of sugars and inhibitory compounds lead to high processing costs. Cell walls in lignocellulosic biomass can be converted to mixed-sugar solutions with lignin-rich solid residues by sequential use of a range of thermochemical pretreatments and enzymatic saccharification. A wide variety of extracellular enzymes obtained from fungal biomass are implicated in lignocellulosic biomass decomposition, a potential low-cost source of mixed sugars for fermentation to fuel ethanol (Table 22.1).

## *Cellulases*

Cellulases are an O-glycoside hydrolase group of enzymes used to hydrolyze the cellulosic plant biomass to simple sugars that can be transformed (fermented) by microbes for biofuel, primarily ethanol generation. Cellulases hydrolyze the β-1,4 glucan linkage in cellulose and produce glucose, cellobiose, and cello-oligosaccharides. The cellulase enzymatic system consists of three enzymes acting synergistically to hydrolyze the crystalline cellulosic biomass into the small sugars Cellobiohydrolase (CBH), Endo β-1,4 glucanases (EG), and β- glucosidase (BGL). Endoglucanase hydrolyzes the middle of a low crystalline cellulose and starts cellulose breakdown by randomly attacking the interior of the amorphous regions of the cellulose fiber, thus releasing oligosaccharides occupying different degrees of polymerization and exposing its reducing and non-reducing ends (Sanchez et al. 2004). Cellobiohydrolases act processively on reducing and non-reducing chain ends to release mainly cellobiose. β-Glucosidases (BGLs) hydrolyze the β-1,4 glycosidic bond of cellobiose and cellodextrins to release glucose units. β-glucosidase is competitively inhibited by glucose (Dashtban et al. 2009). Enzymatic hydrolysis by cellulases depends on the physical properties of cellulose molecules such as crystallinity, degree of polymerization (DP), and accessible surface area, as these are the major factors responsible for controlling hydrolysis rate due to their effect on enzyme binding and accessibility of substrates to the cellulase enzymes. Crystallinity is a key factor affecting the hydrolysis of cellulose as the glycosidic bonds in crystalline regions are difficult to hydrolyze compared to those in the amorphous regions (Ahola et al. 2008). Several other novel enzymes assist in cellulose degradation by acting in synergy with the exo- and endoglucanases (Leggio et al. 2012) including copper-requiring polysaccharide monooxygenases. Some proteins have elastin-like properties including Swollenin and other cellulase-enhancing proteins, which contribute to its hydrolysis by increasing access of the cellulase enzymes to the cellulose chain ends (Nakatani et al. 2013).

The majority of cellulases have a characteristic two-domain structure with a catalytic domain and Cellulose Binding Domain (CBD). One of the major things that restricts or directs the degradation of native cellulose and thus regulates the catalytic activity of cellulolytic enzymes, is their different levels of orientation and different degrees of cooperation between individual units. Depolymerization of both crystalline and amorphous cellulose molecules to fermentable sugars is due to synergistic action of endoglucanases, exoglucanases, and β-glucosidases (Dyk and

**Table 22.1** Enzymes, characteristic properties, and their fungal sources involved in lignocellulosic biodegradation

| Enzymes | Principal types | Characteristic features and catalytic function | Fungal sources |
|---|---|---|---|
| Cellulolytic enzymes | Endoglucanases | Hydrolyzes accessible intramolecular β-1,4-glycosidic bonds of cellulose chains randomly to produce new chain ends, increases the accessible surface areas, and prevents the formation of inhibitory products | *Aspergillus niger, Penicillium, Fusarium, Coriolus, Volvariella, Rhizopus oryzae, Talaromyces emersonii, Chaetomium, Trichoderma, Phanerochaete, Schizophyllum, T. harzianum, Schizophyllum commune, Mucor, Pycnoporus, Bjerkandera, Trichoderma reesei, Phanerochaete chrysosporium, Sporotrichum pulverulentum, Penicillium pinophilum, Aureobasidium pullulans* (yeast) |
| | Exoglucanases | Cleaves cellulose chains at the reducing and non-reducing to produce cellobiose or glucose | |
| | β-glucosidases | Hydrolyze cellobiose and cellodextrins to glucose to prevent cellobiose inhibition | |
| | Cellobiose phosphorylase | Dephosphorylates the phosphorylated moiety and thus favors the hydrolysis by other cellulolytic enzymes | **Anaerobic Fungi**<br>*Orpinomyces, N. patriciarum, Caecomyces, Neocallimastix, Orpinomyces*, and *Piromyces* |
| Hemicelluloytic *Enzymes* | Endoxylanases }<br>Exoxylanases<br>α- xylosidase | Generates oligosaccharides from the cleavage of xylan oligosaccharides, producing xyloses. All these act synergistically to depolymerize hemicellulosic sugars to their corresponding fermentable monomers to produce biofuels and other valuable co-products | *Polyporus sulfurous, Aspergillus niger, Pycnoporus sanguineus, Neocallimastix frontalis, Trichoderma reesei, B. adusta, Anaeromyces mucronatus, P. chrysosporium* |
| | α-glucuronidase | Hydrolyzes the **α** 1,2 linkages between D-glucuronic acid and xylose | |
| | β-mannosidases}<br>β -glycosidase | Hydrolyzes the endomannase-generated oligomers **β**-1,4 bonds | |
| | α-L-arabinofuranosidase<br>Acetyl xylan esterase | Hydrolyzes the terminal arabinose residues from side chain of Xylan De-acetylates O-acetyl group from acetylated Xylan | |
| | Ferulic acid esterase | Cleaves ester bond between arabinose side chains and ferulic acids | |

| | | | |
|---|---|---|---|
| Lignin-degrading or lignin-modifying enzymes (LMEs) | Lignin peroxidases | Oxidizes non-phenolic part of lignin and along with Laccases and MnPs causes degradation and delignification of woody component | *Phanerochaete chrysosporium, Fomitopsis palustris, Orpinomyces* (anaerobic), *Phlebia, Physisporinus rivulosus, Dichomitus squalens, Piptoporus betulinus, Laetiporus portentosus, Trametes versicolor, Fusarium solani, Ceriporiopsis, Gloeophyllum trabeum, Pycnoporus coccineus, Pycnoporus sanguineus, Cyathus, Coniophora puteana, Magnaporthe grisea, Myrothecium verrucaria, Neurospora crassa* |
| | Manganese peroxidases | Extracellular glycoproteins; catalyzes the peroxide-dependent oxidation of Mn(II) (as the reducing substrate) to Mn(II) and increases accessible surface for lignin degradation along with LiP | |
| | Laccases | Blue multicopper oxidoreductases; uses molecular $O_2$ to oxidize organic compounds with the coupling of electron reduction of dioxygen to water and the oxidation of a vast variety of substrates like phenols, lignin, amines, arylamines. | |
| | Versatile peroxidases | Dual oxidative ability and oxidizes various substrates of other lignin peroxidases | |

Pletschke 2012). All the commercially available glycoside hydrolases are principally isolated from fungi and it was estimated that in 2012, cellulases accounted for 20 % of the total enzyme market which was approximately 6 billion dollars (Mathew et al. 2008).

## Cellulase System of Fungi

Filamentous fungi are a potent source of cellulases and secrete non-complexed cellulases (not bounded to cell wall) into their extracellular environment. A wide variety of fungi, such as *Aspergillus* spp., *Trichoderma reesei*, *Penicillium pinophilum* (Korotkova et al. 2009), *Trichoderma viride* (Gusakov et al. 2007), *Phanerochaete chrysosporium* (Tsukada et al. 2006), *Fomitopsis pinicola* (Joo et al. 2010), *Fomitopsis palustris* (Yoon et al. 2008), *Talaromyces emersonii* (Murray et al. 1984), *Neocallimastix*, *Orpinomyces*, and *Piromyces* (Steenbakkers et al. 2002) have been reported with extensive ability of secreting extracellular cellualse enzymes. Filamentous fungi are the major producer of cellulase and hemicellulase enzymes, but the enzymatic component and their corresponding activity may differ in different species and are also affected by various parameters like product inhibition. *Hypocrea jecorina* (*Trichoderma reesei*) is the most extensively studied cellulolytic fungus that secretes an array of cellulases (Kubicek 2013) and is the main industrial source of cellulases. *Aspergillus* produces the enzymatic component of cellulase and has strong hydrolytic activity, but it mainly produces BGLs compared to *T. reesei*, as *T. reesei* is subject to product inhibition. When there is need for biomass saccharification with *T. reesei,* it is often supplemented with *Aspergillus* BGLs. A complex array of cellulases, hemicellulases, and ligninases are produced by *Phanerochaete chrysosporium* (Broda et al. 1996). High titre of cellulase production is also reported from *Penicillium* species like *P. brasilianum* (Jorgensen et al. 2007). For the efficient hydrolysis of lignocellulosic biomass, the major factor that determines the lignocellulose saccharification is the catalytic efficiency of an individual enzyme and its percentage composition in cocktail when a multienzyme catalyst is used for conversion. The ideal cellulase complex must be highly active in intended biomass feedstock, must be able to completely hydrolyze the feedstock, withstand mild acidic pH, and finally must be cost-effective.

## Approaches to Increase Cellulase Production

There are a wide variety of fungal species capable of cellulase production, but the enzymatic yield and level of individual cellulase components are not satisfactory for complete lignocellulosic biomass saccharification, so there is a need to enhance the production of cellulase enzyme. One of the keys for developing cellulase is to construct them either by assembly of enzymes to form cocktail or to engineer cellulase producers to express the desired combination of cellulases. For instance, an enzymatic cocktail has been produced by mixing *T. reesei* cellulases with other enzymes like xylanases, pectinases, and BGLs and this cocktail is tested for lignocellulosic biomass saccharification from different feedstocks. Solid-state fermentation (SSF) is one of the most important cost-effective technologies for biodegradation of lignocellulosic biomass employing cellulolytic microbes. Other approaches to enhance cellulase production include increasing the copy number of the BGL gene and the amount of BGL in cellulase enzymes secreted by *T. reesei* (Fowler and Brown 1992) or altering the cellulose mixture profile of *T. reesei* by introducing glucose-tolerant BGL gene into the fungus (White et al. 2000). Site-directed mutagenesis, expression cassettes, and antisense technology are some recent techniques used frequently to modulate the fungal biosystem for enhanced cellulase production (Gincy and Sasikumar 2007). Potent cellulase genes from different fungi can be isolated, cloned, and expressed in their fungal hosts. Enhanced production of cellulases can also be achieved by the use of promoters that are insensitive to glucose repression (Watanabe and Tokuda 2001).

## Hemicellulases

Hemicelluloses are highly branched and mostly noncrystalline heteropolysaccharides. Sugar units that are generally found in hemicellulose are pentoses (D-xylose, L-arabinose), hexoses (D-galactose, L-galactose, D-mannose, L-rhamnose, L-fucose), and uronic acids (D-glucuronic acid) (Glazer and Nikaido 2007). Hemicellulose generally comprises 15–35 % of plant biomass. Xylan and glucomannan are amongst the most relevant hemicellulose units. Xylan is a major structural heteropolysaccharide in plant cells composed of homopolymeric backbone chains of 1,4 linked Β-D-Xylopyranose units (Koukiekolo et al. 2005). An enzyme or a group of enzymes that hydrolyzes the Xylan and glucomannan units to simple sugars are called hemicellulases. Xylan is found at the interface between lignin and cellulose and is believed to be accountable for fiber cohesion and overall plant cell-wall stability (Collins et al. 2005). More complex structures of hemicellulose require synergistic action of various types of hemicellulases for efficient degradation. Xylanases like cellulases can also be divided into endo-acting xylanases (E.C.3.2.1.8 available in GH5,8,10,11,43), exo-acting xylanases (E.C.3.2.1.156) found in GH8 acting from the reducing end, and often complemented with xylosidases (E.C.3.1.2.37), for example, GHI 3,39,52,54,116,120 acting from non-reducing ends. Use of other hemicellulosic degrading enzymes is classified under glycoside hydrolase and carbohydrate esterase (CE) families for degrading cellulose and hemicellulose moieties. For instance, mannanases (E.C.3.2.1.78) catalyze different mannan-containing hemicelluloses and classified under the category (GH 26,113), Mannosidases (E.C.3.2.1.25) (GHI2,5), galactosidases (E.C.3.2.1.23) (GH2,3,35,42), and others including Arabinofuranosidases (E.C.3.2.1.55) (GH3,43,51,54,62) together with other enzymes degrade hemicelluloses to monomeric sugars (Table 22.2, Fig. 22.1). Fungal xylanases are generally active at mesophilic temperature (40–60 °C) with a

**Table 22.2** Glyoside hydrolase and carbohydrate esterase family for cellulosic degradation

| Enzyme | Enzyme families |
|---|---|
| Endoxylanase | GH-5,8,10, 11,43 |
| Beta-xylosidase | GH-3,39,43,52,54 |
| Alpha-L-arabinofuranosidase | GH-3,43,41,54,62 |
| Alpha glucuronidase | GH-4,67 |
| Alplia-galactosidase | GH-4,36 |
| Acetyl xylan esterase | CE 1,2,34,5.6,7 |
| Feruloyl esterase | CE1 |

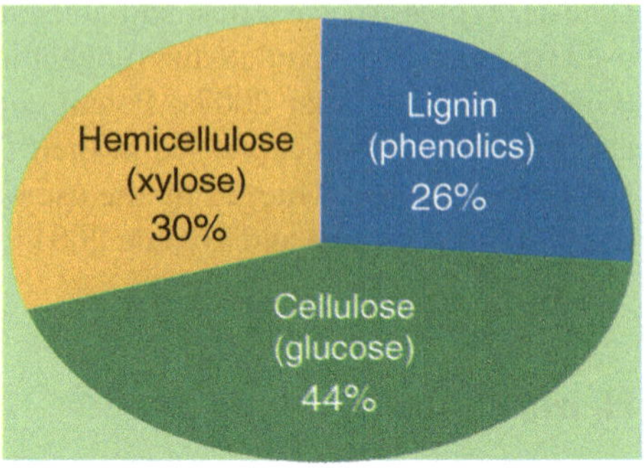


**Fig. 22.1** Composition of lignocellulosic biomass

slightly acidic pH; however, xylanases have also been reported to be active in extreme environments. Psychrophilic fungi, such as *Penicillium* sp., *Alternaria alternata*, and *Phoma* sp., have been isolated from the Antarctic environment.

Endo-1, 4-β-xylanase (1, 4-β-D-xylan xylanohydrolase; EC 3.2.1.8) hydrolyzes the glycosidic bonds in the xylan backbone to reduce its degree of polymerization and thus increasing accessibility of cellulose to enzymatic hydrolysis releasing xylo-oligomers which can further produce other small sugars (Polizeli et al. 2005). Exo-1, 4-β-D-xylosidase (1,4-β-D-xylan xylohydrolase; EC 3.2.1.37) hydrolyzes xylobiose and short-chain xylooligosaccharides generated by the action of endoxylanases, releasing D-xylose residues from the non-reducing end. β-xylosidases hydrolyze only xylobiose and their affinity for xylooligosaccharides decreases with the increasing degree of polymerization. Xylose-utilizing species including *Candida* sp., *Geotrichum* sp., *Sporopachydermis* sp., *Trichosporon* sp., *Pichia* sp., and *Sugiyamaella* sp. have been isolated from buffalo feces (Wanlapa et al. 2013) or from soil (Zhang et al. 2014). Other enzymes involved in hemicellulosic degradation includes α-L-arabinofuranosidases (EC 3.2.1.55) hydrolyzing the terminal arabinose residues from the side chains of xylan and other arabinose containing polysaccharides (Saha 2003). α-D-glucuronidases (EC 3.2.1.139) hydrolyze the α-1,2 linkages between the 4-*O*-methylglucuronic/D-glucuronic acid and xylose residues in glucuronoxylan. The hydrolysis of the stable α-(1, 2)-linkage is the bottleneck in the enzymatic hydrolysis of xylan. Acetyl xylan esterase (EC 3.1.1.6) removes the *O*-acetyl groups from acetylated xylan. This enzyme plays an important role in the hydrolysis of xylan, since the acetyl groups can interfere in the action of enzymes that cleave the xylan backbone, and so their removal facilitates the action of xylanases (Polizeli et al. 2005).

## Lignin-Modifying Enzymes (LMEs)

Lignin is the second most abundant constituent of the plant cell wall, where it protects cellulose against hydrolytic attack by saprobic and pathogenic microbes. Lignin degradation plays a major role in carbon recycling in the ecosystem as well as converting plant biomass for second-generation biofuel (ethanol) production. The fungal lignin degradation process is oxidative and nonspecific, which decreases methoxy, phenoxy, and aliphatic content of lignin, cleaves aromatic rings, and forms new carbonyl groups by the action of enzymes, namely, laccase, lignin peroxidase, and manganese peroxidase, etc. These changes in the lignin molecule result in depolymerization and carbon dioxide production (Timothy et al. 2011).

White rot fungi are reported to be efficient in secreting LMEs with some examples like *Auricularia polytricha*, *Flammulina velutipes*, *Ganoderma lucidum*, *G. applanatum*, *G. australe*, *G. capense*, *G. carnosum*, *G. fornicatum*, *G. gibbasum*, *G. resinaceum*, *G. stipitatum*, *G. trabeum*, *Irpex lacteus*, *Phanerochaete chrysosporium*, *Pleurotus sajor-caju*, *Pleurotus ostreatus*, *Pleurotus dryinus*, *Pleurotus tuberregium*, *Lentinula edodes*, *Trametes hirsuta*, and *Trametes versicolor* (syn. *Coriolus versicolor*) (Kannan et al. 1990; Fang et al. 1997; Elissetche et al. 2006; Arboleda et al. 2008; Elisashvili et al. 2008, 2009; Dinis et al. 2009; Erden et al. 2009; Asgher et al. 2012; Pinto et al. 2012; Manavalan et al. 2013; Salvachua et al. 2013; Shevchenko et al. 2013).

## Lignin Peroxidases (LiPs)

Lignin peroxidases are heme containing glycoproteins also known as Heme peroxidases that catalyze the $H_2O_2$-dependent oxidative depolymerization of a vast variety of non-phenolic lignin compounds (Wong 2009). LiPs oxidize the substrates in multistep electron transfers and have high reduction potential in comparison to other heme peroxidases and also do not require any chemical mediator for its reaction. Among fungal genera mostly White Rot Fungi are the major producer of Lignin peroxidases or ligninases. *Phanerochaete chrysosporium* (Pointing et al. 2005) and *Trametes versicolor* (syn. *Coriolus versicolor*) are major fungal lignin degraders. *P. chrysosporium* is a potent source of extracellular ligninases with some peroxide generating enzyme glyoxal oxidase (GLOX) along with other ligninases.

## Manganese Peroxidases (MnPs)

MnPs are extracellular glycoproteins and also classified with Heme Peroxidases, secreted in multiple isoforms, and contain one molecule of Heme as Protoporphyrin (IX). MnP catalyzes the peroxide-dependent oxidation of Mn (II) (as the reducing substrate) to Mn (III) released from the enzyme surface by forming complexes with oxalate or other metal chelators. This chelated Mn (III) complex then acts as a reactive low-molecular-weight, diffusible redox mediator of phenolic substrates including simple phenols, amines, dyes, and phenolic lignin substructures. Fungal genera that secrete MnPs include *P. chrysosporium*, *Panus tigrinus* (Lisov et al. 2003), *Lenzites betulina* (also reported as *Lenzites betulinus*) (Hoshino et al. 2002), *Phanerochaete flavido-alba* (de la Rubia et al. 2002), *Agaricus bisporus* (Lankinen et al. 2001), and *Bjerkandera* sp. (Palma et al. 2000).

## Versatile Peroxidases (VPs)

Versatile Peroxidases are glycoproteins with hybrid properties capable of oxidizing substrates of other peroxidases like LiPs and MnPs. VPs form an attractive ligninolytic enzyme group due to their dual oxidative ability to oxidize Mn (II) and other phenolic and non-phenolic aromatic compounds. VPs can oxidize substrates of both high and low redox potentials. The major fungal genera secreting VPs include *Pleurotus eryngii, P. ostreatus* (Cohen et al. 2001), *Bjerkandera adusta* (Wang et al. 2003), *Bjerkandera fumosa*, and *Pleurotus pulmonarius*; basidiomycetous fungal genera secreting VPs include *P. tigrinus* (Lisov et al. 2007). VPs have a wide variety of biotechnological applications due to their catalytic versatility in catalyzing those reactions where other peroxidases fail to catalyze the same.

## Phenol Oxidases (Laccases)

Laccases are glycosylated blue multicopper oxidoreductases using molecular oxygen to oxidize various aromatic and non-aromatic compounds by coupling the electron reduction of dioxygen into water with the simultaneous oxidation of a vast variety of substrates, such as phenols, arylamines, anilines, thiols, and lignins. Laccases are produced and secreted by a wide range of fungal genera including white rot fungi *Lentinus tigrinus* (Farroni et al. 2007), *T. versicolor* (Necochea et al. 2005), and *Cyathus bulleri* (Mishra and Bisaria 2006). Several brown rot fungi are also potent producers of laccases including *Coniophora puteana.* Many ascomycetous genera are also reported as potent producer of laccase *Melanocarpus albomyces* (Hakulinen et al. 2006), *Chaetomium thermophile* (Ishigami and Yamada 1986), *Magnaporthe grisea* (Iyer and Chattoo 2003), *Myrothecium verrucaria* 24G-4 (Sulistyaningdyah et al. 2004), and *Neurospora crassa* (Schilling et al. 1992).

## Potential Strains of Oleaginous Microfungi Employed in Biofuel Production

An emerging potential alternative for biodiesel production is represented by microbial lipids single-cell oils (SCOs), which oleaginous microorganisms can accumulate up to 70 % or more of their biomass of their dry weight. Some of the oleaginous species show the ability to metabolize pentoses, demonstrating the potential to produce triacylglycerol (TAG) from lignocellulosic biomass. Microbial TAGs may be a prospective alternative feedstock and proven to be a pivotal for a sustainable biodiesel industry. Nowadays, the possibility to track lipid production in real time can be achieved by Fourier transform infrared spectroscopy, which is essential for a viable and successful development of an industrial production process (Ami et al. 2014). Important oleaginous fungi employed for biodiesel production are as follows:

## *Rhodosporidium toruloides*

The red yeast *Rhodosporidium toruloides* is an oleaginous mesophilic species of class Ustilaginomycetes which can accumulate lipids to above 70 % of its dry cell weight from a wide variety of carbon sources and can transform carbohydrates from lignocellulosic hydrolysate into long-chain fatty acids that contribute to biodiesel production. This species is able to carry out diverse biochemical reactions such as biodegradation of epoxides, biphenyls, and oxiranes, biosynthesis of carotenoids, and other types of biotransformations (Yu et al. 2010). Yang et al. (2014) studied the two-stage process of *Rhodosporidium toruloides* Y4 for the conversion of crude glycerol into lipid. This yeast strain is capable of accumulating lipids up to 76 % of cell dry weight and provides an attractive route to integrate biodiesel production with microbial lipid technology for better resource utilization efficiency and economical viability (Fig. 22.1).

Novel biochemical approaches remain to be developed to improve microbial lipid technology (Yang et al. 2014). Sulfate limitation technology was found to be effective to promote accumulating substantial amounts of intracellular lipid in *Rhodosporidium toruloides* Y4. The sulfate-limitation approach to control lipid biosynthesis should be valuable to explore nitrogen-rich raw materials as the feedstock for lipid production (Wu et al. 2010). Gas chromatography analysis revealed that lipids from *R. toruloides* Y4 contained mainly long-chain fatty acids, such as oleic acid, palmitic acid, stearic acid, and linoleic acid, and act as a potential alternative oil resource for biodiesel production (Li et al. 2007). On the other hand, Kumar et al. (Kumar et al. 2014) reported the 20.05-Mb draft genome of the red yeast *Rhodosporidium toruloides* MTCC 457, predicted to encode 5993 proteins, 4 rRNAs, and 125 tRNAs and found to be valuable for molecular genetic analysis and manipulation of lipid accumulation. Fourier Transform Infrared Spectroscopy (FTIR) can be proposed as a powerful tool for the development of a viable biodiesel production in oleaginous yeasts, such as *Cryptococcus curvatus* and *Rhodosporidium toruloides*, and of non-oleaginous yeast *Saccharomyces cerevisiae*.

## *Yarrowia lipolytica*

*Yarrowia lipolytica* is tropical marine yeast and can undergo extensive modifications for converting a wide range of hydrophobic and hydrophilic biomass to fatty acid-based products and enhance its potential in the sustainable production of biodiesel, functional dietary lipid compounds, and other value-added oleochemical compounds. *Y. lipolytica* could be used as a potential feedstock for biodiesel production when grown on glucose and inexpensive wastes (Katre et al. 2012). All of these features make *Y. lipolytica* suitable in industrial applications of lipid-modifying enzymes and highlight its potential for use in sustainable production of second-generation biofuels. Simple and renewable substrates, such as molasses, *N*-acetylglucosamine, sewage sludge, palm oil mill, olive oil mill wastewater, whey, municipal wastewater, industrial fats, etc., have been demonstrated to increase the economy of the bioconversion process by *Y. lipolytica* (Liang and Jiang 2013). It is also known to convert glycerol to SCO and hence can play dual roles in the upstream and downstream of the biodiesel industry. The nitrogen-limited cultures of *Y. lipolytica* demonstrated increased production of organic acids with enhanced accumulation of lipids (Papanikolaou et al. 2002). Use of genetic engineering under a nitrogen-limiting condition has proved efficient in augmenting the biomass and cellulosic lipid yields following the co-fermentation strategies (Sestric et al. 2014). Two-stage fed-batch system for conversion of volatile FAs to the SCO has been proved useful to obtain high biomass using *Y. lipolytica* (Cherry and Fidantsef 2003). Abghari and Chen (2014) demonstrated the use of *Y. lipolytica* to produce nanoproducts and lipid-based bioproducts that provide an effective tool for bridging between biotechnology and nanotechnology through development of novel biocatalysts and nano-oils (Fig. 22.2)

## *Cryptococcus* Species

The oleaginous yeast *Cryptococcus curvatus* (Diddens & Lodder) Golubev is a promising feedstock for biofuel production. These species are used to hydrolyze biomass that has been pretreated using dilute acid and produce lipids that may be used as feedstock for producing biofuels (Yu et al. 2010). In *Cryptococcus curvatus* volatile fatty acids (VFA) from dark fermentation hydrogen production were tested as carbon sources for the culture of oleaginous yeast, which is a promising feedstock for biofuel production. On the other hand, *Cryptococcus laurentii* 11 biomass when supplemented with sugarcane molasses indicated a predominant (86 %) presence of neutral lipids with high content of 16- and 18-carbon-chain saturated and monosaturated fatty acids and thus can be considered suitable for the production of biodiesel (Castanha et al. 2014). High cell density fed-batch cultivation on low-cost

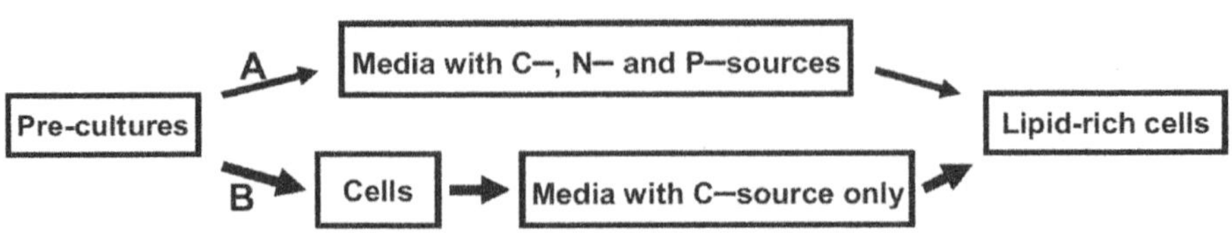


**Fig. 22.2** Microbial lipid production based on conventional (path A) and two-stage (path B) process (Yang et al. 2014)

substrate, viz. crude glycerol, indicated a high oleic acid content followed by palmitic acid, stearic acid, and linoleic acid, and the oil was transesterified to biodiesel (Thiru et al. 2011). Lipid from *C. curvatus* was found to be a quality-sufficient source of oil as a transportation fuel in terms of cetane, iodine values, and oxidation stability (Ryu et al. 2013). Dairy wastes were also used as substrates and achieved production of high concentrations of sophorolipids using a two-stage cultivation process for the yeast *Cryptococcus curvatus* ATCC 20509 (Johny 2013). *C. curvatus* lignocellulosic materials, such as corn fiber and sweet sorghum, are preprocessed and can be used to produce liquid biofuels (Liang et al. 2014). *C. terricola* used for fuel production through consolidated bioprocessing produced high proportions of C16:0 and C18 fatty acids when grown on starch and proved to be a promising alternative source for biodiesel production (Tanimura et al. 2014).

## *Cunninghamella echinulata and Mortierella isabellina*

These strains are considered potential producers of single-cell oil (SCO) containing $\gamma$-linolenic acid, a polyunsaturated fatty acid. *Mortierella isabellina* (current name: *Umbelopsis isabellina* (Oudem.) W. Gams) can be considered as a promising producer of SCO that can be subsequently converted into second-generation biodiesel (Chatzifragko et al. 2011). The growth of *Cunninghamella echinulata* on various nitrogen containing raw materials (corn gluten, corn steep, whey, yeast extract, and tomato waste hydrolysate) yielded various quantities of $\gamma$-linolenic acid (GLA)-rich cellular lipids. Growth on tomato waste hydrolysate yielded 17.6 g/L of biomass containing 39.6 % oil and 800 mg/L GLA. On xylose containing media *M. isabellina* accumulated 65.5 % and *C. echinulata* 57.7 % of lipid and produced 6.7gL$^{-1}$ of single-cell oil and 1119 mgL$^{-1}$ of $\gamma$-linolenic acid. On the side, *M. isabellina* produced GLA-rich SCO from pear pomace, an agro-industrial waste accumulating in large amounts in several Mediterranean countries (Fakas et al. 2009). *C. echinulata* when grown on tomato waste hydrolysate medium rapidly, took up glucose and produced large amounts of lipids that contain GLA-rich triacylglycerols (TAG) and hence may be of commercial interest. Actually, *C. echinulata* has been successfully used by several researchers for GLA production (Papanikolaou et al. 2008), but cultivation of this mold on media containing organic nitrogen has been rarely reported. Certik and Shimizu (1999) compared the effects of various organic and inorganic nitrogen sources on lipogenesis in a strain of *C. echinulata* and found that the use of organic nitrogen increased lipid accumulation.

## *Aspergillus spp. and Penicillium spp.*

*Penicillium* spp. and *Aspergillus* spp. strains have been revealed as appropriate lipase producers, but the studies dealing with the production of microbial lipids by *Penicillium* spp. and *Aspergillus* spp. are insufficient. Studies dealing with the production of fat by *Aspergillus nidulans*, *Aspergillus flavus*, *Aspergillus fumigatus*, *Aspergillus sydowii*, or *Aspergillus niger* have been reported (Hui et al. 2010). Waste cooking olive oil was an adequate substrate for the growth of *Aspergillus* sp. and *Penicillium expansum* producing remarkable quantities of lipid-rich biomass (Papanikolaou et al. 2011). Studies were done on the use of potato processing wastewater for microbial lipid production by *Aspergillus oryzae* with the purpose of recycling potato processing wastewater for biodiesel production (Muniraja et al. 2011). In *Aspergillus oryzae* whole-cell biocatalyst, coexpression of *Fusarium heterosporum* lipase (FHL) and mono- and di-acylglycerol lipase B (mdlB) has been developed to improve biodiesel production. For enzymatic biodiesel production from plant oil hydrolysates, an *Aspergillus oryzae* whole-cell biocatalyst expresses the *Candida antarctica* lipase B (r-CALB) with high esterification activity which resulted in the biodiesel production (Adachi et al. 2013). In *Aspergillus oryzae*, highly efficient biodiesel production was achieved by using the lipase activity of the whole-cell biocatalyst. In *Aspergillus awamori,* Taguchi orthogonal array (OA) is useful for optimizing the number of functional factors at a time which were involved in the lipid production and the pH individually showed significant influence on the lipid synthesis (Prakasham et al. 2007). *Aspergillus awamori* enhanced the lipid production by 31 % and the lipids with fatty acid ester linkages and free fatty acids can produce fatty acid methyl esters (FAME) after transesterification that can be used as biodiesel. Also, the mutant strain of *A. niger* NMG12/4 showed maximum lipase activity of 15.5 U $cm^{-3}$ at 96 h, which is prospective for the development of industrial biotechnology for production of extracellular lipase (Toscano et al. 2015). In *Penicillium decumbens,* production of cellulases effectively degrades the lignocellulose for the second-generation biofuel production (Liu et al. 2013).

## *Rhodotorula glutinis*

*Rhodotorula glutinis* is an oleaginous yeast that produces copious quantities of lipids in the form of triacylglycerols (TAG) and can be used to make biodiesel via a transesterification process. The ester bonds in the TAG are broken leaving behind two products, i.e., fatty acid methyl esters and glycerol, which provide an inexpensive carbon source (Easterling et al. 2009). In *R. glutinis*, fatty acids are activated in an ATP-dependent manner and an enzyme acyl–acyl carrier protein (ACP) plays a role in activating fatty acids for triacylglycerol biosynthesis. There is plenty of evidence to suggest that this organism has the potential to be a source of fatty acids for the production of biodiesel. *R. glutinis* have the ability to grow and accumulate lipids when grown on glycerol. Over 70 % of biodiesel production costs are due to the expense of feedstocks such as soybean and rapeseed oil. Also, the lipid production potential of the red yeast *Rhodotorula glutinis* grown on non-detoxified hydrolysates from wheat straw and miscanthus as carbon sources showed the lipid composition of C 16:0, C 18:1, and C 18:2 indicating a good level of suitability for biodiesel production (Mast et al. 2014).

### *Trichoderma reesei*

The enzyme producer *T. reesei* stands out among industrially applied microorganisms because it can degrade cellulose at the rates sufficient for industrial use and a wide range of mutants have been developed for *T. reesei. T. reesei* serves today as a model organism for the regulation and biochemistry of (hemi) cellulose degradation (Kubicek et al. 2009). The microfungus *Trichoderma reesei* is a well-known producer of lignocellulolytic enzymes that are used for depolymerization of plant lignocellulosic biomass (Martinez et al. 2008). *T. reesei* is already the main industrial source for cellulases and hemicellulases, but despite the fact that industrial strains produce more than 100 g/l of cellulases (Cherry and Fidantsef 2003), efforts are needed to reduce costs and maximize yield and efficiency of the produced enzyme mixtures. Furthermore, *T. reesei* has the ability to utilize all the lignocellulose sugars for producing ethanol (Xu et al. 2009). Huang et al. (2014) demonstrated direct ethanol production from lignocellulosic sugars and sugarcane bagasse by a recombinant *Trichoderma reesei* strain HJ48. In *T. reesei*, cost-effective lignocellulolytic enzyme is produced when supplemented on a cane molasses medium (He et al. 2014). On the other side, Javanovic et al. (Jovanović et al. 2014) focused on the potential of producing erythritol in *T. reesei* from lignocellulosic biomass. As such a strong producer of cellulases and hemicellulases revealed for *T. reesei*, 10 cellulolytic and 16 xylanolytic enzyme-encoding genes, it is likely that *T. reesei* is able to grow on cheap biowaste material like wheat straw as the sole carbon source (Dashtban et al. 2013). The high lipid containing biomass in *T. reesei* can be used to extract oil and the contents can be termed as bio-oil (or biodiesel or myco-diesel after transesterification). The resulting bio-oil production from wastewater treatment by *T. reesei* reactors was found to be 74.1 mg/L, whereas biomass containing bio-oil contents (%w/w) was 9.82 % in 96 h. (Bhanja et al. 2014). This study suggests that wastewater can be used as a potential feedstock for bio-oil production with the use of oleaginous fungal strains and which could be a possible route of waste to energy.

## Novel Technologies to Meet the Challenges of Biofuel Research

### *Metabolic Engineering*

Metabolic engineering is a novel approach that has developed microbial cell factories for converting renewable carbon sources into biofuels, widely useful for industrial application (Jang et al. 2012). Major use of metabolic engineering in biofuel synthesis is to engineer obese microbes for overproduction of fat and oils, improving tolerance of microbes to biofuels (ethanol) and lastly high-throughput screens for various extracellular compounds. Improving tolerance of microbes to biofuel generated and enhancing robust growth and production under adverse industrial conditions are still one of the major challenges for metabolic engineering, as adverse and hostile conditions may elicit multigenic responses coordinated at the transcriptomics and proteomics level. Microbial metabolism is quite complex and the biosynthetic pathway to produce biofuels requires a complex array of multiple enzymatic reactions. Current molecular biology techniques can effectively alter enzyme levels to increase the flux toward biofuel synthesis. Metabolic engineering allows fine-tuning and modulation of both expression level and target activity of proteins/enzymes with the help of some traditional techniques that enable engineering and *de novo* synthesis of promoters, ribosome binding sites, and entire coding regions or by regulating the choice of plasmids and their copy numbers, promoter engineering, codon optimization, synthetic scaffolds, directed evolution or modification of key enzymes, and knockout/knockdown of competitive pathways (Nowroozi et al. 2014). Some new tools that enable genome engineering and system-wide identification of genes that confer target traits are very useful. These emerging techniques include multiscale analysis of library enrichment (SCALEs) that provide quantification of the effects of expression of specific genes, trackable multiplex engineering (TRMR) (Warner et al. 2010), coexisting/coexpressing genomic libraries (CoGeL) (Nicolaou et al. 2011), genome-scale analysis of library sorting (GALibSo) (Stadlmayr et al. 2010), multiplex automated genome engineering (MAGE) (Wang and Chen 2009), conjugative assembly genome engineering (CAGE) (Isaacs et al. 2011), and global transcription machinery engineering (gTME) (Alper et al. 2006). New genetic techniques, such as RNA Interference, CRISPRs, or TALENs, offer new capabilities to edit microbial metabolisms (Sun and Zhao 2013).

Ethanol fermentation by yeast is the most developed biofuel process, but low combustion energy and high purification costs prevent the wide use of ethanol as an economical fuel. Therefore, researchers have engineered microbes to produce new fuels. Advanced biofuel examples include higher alcohols via the keto-acid and the Ehrlich pathway, terpene-based fuels (e.g., isopentenol) from the mevalonate pathway, and fatty acid ethyl esters and alkanes from fatty acid biosynthesis pathways. Despite the development of these diverse biofuel producers, it is still challenging to commercialize biofuel processes due to the poor microbial productivity in large bioreactors and the low profit margins of biofuels (Lamonica 2014).

## *Global Transcription Machinery Engineering (gTME)*

Global transcription machinery engineering (gTME) is an approach that alters the essential proteins regulating the transcriptomes and introduces transcriptional-level modifications that are transferable between strains. Mutagenesis of the basal transcription factors leads into global reprogramming of transcription of genes and results into new phenotypes which can be isolated by various screening methods. Since gTME is independent of preliminary information like enzymes and their kinetic parameters, biochemical routes, and pathways, multigenic or polygenic traits such as tolerance toward different types of stresses can also be addressed. The genetic change that causes the optimal global reprogramming can then be transferred to other strains (Alper and Stephanopoulos 2007). In contrast to other evolutionary engineering approaches where random mutations accumulate in the entire genome, the gTME approach allows for genotype–phenotype correlations traceable to a single mutant protein. Due to the lack of regulation of the transcription factor, the gTME strategy is able to change the metabolic strength and direction. gTME has been shown as an efficient solution to improve substrate utilization, product tolerance, and production in yeast (Çakar et al. 2012). gTME enables the creation and isolation of polygenic mutants under several conditions, thus facilitating phenotypes that would be difficult to obtain with conventional gene modification techniques. gTME approach has been successful in improving the tolerance of yeasts to high concentration of sugar and ethanol (Tyo et al. 2007).

## *Enzyme Engineering*

One of the major limitations of manufacturing biofuels from plant biomass is the presence of highly refractive lignocellulosic components causing major technical hurdles in the biomass saccharification and hence production of biofuels. Cellulases and hemicellulases used in biomass saccharification have optimum temperature and pH range with low hydrolytic efficiency. Biomass saccharification at high temperatures decreases their activity and catalytic efficiency. For complete and efficient bioconversion of lignocellulosic biomass to biofuels, there is need to develop an ideal cellulase system that must be highly active on desired biomass feedstock, completely hydrolyze biomass, must be active at mildly acidic pH, withstand process stress, and most importantly to be cost-effective. Another approach has been to prepare "Enzymatic cocktail." Enzyme cocktails have been developed by mixing *T. reesei* cellulases with other enzymes like pectinases, xylanases, and BGLs, and these cocktails were employed to hydrolyze various feedstocks (Berlin et al. 2007).

Biomass degradation enzymes from thermophilic fungi demonstrate higher hydrolytic capacity despite the fact that extracellular enzyme titres are typically lower than more conventionally used species (Berka et al. 2011). Several approaches have been pursued, including directed evolution using error-prone Polymerase Chain Reaction-based mutagenesis of cellulase genes, adaptive evolution using natural selection to specific environmental conditions, or rational protein design to improve the enzymatic activity of cellulases or to expand the physiological conditions at which the enzymes are active. "Rational Design" approach of engineering cellulases involves site-directed mutagenesis of conserved residues (identified by sequence comparison between homologues from different species). Site-directed mutagenesis of non-active site residues to amide carboxylate pairs enhanced catalytic activity and increased pH susceptibility. Similarly in Directed evolution approach catalytic efficiency of cellulases is improved by random mutations along the length of given DNA sequences. "DNA shuffling" approach selects DNA sequences that are randomly fragmented using DNaseI and then recombined by annealing and extension of DNA strands using self-primed PCR, followed by selection of clones with desired properties from a library of recombined fragments (Patten et al. 1997).

## System-Level Approaches and Biofuel Production

Computational tools like genomics, transcriptomics, proteomics, glycomics and lignomics, and fluxomics can be helpful in understanding core system biology. The system-level approach to understanding complex metabolic and regulatory networks enables the design of more efficient microorganisms for the production of valuable molecules, including biofuels. Sequencing industrialized strains and then comparing it with their progenitors may give vital information in designing new microbial systems that increase bioconversion efficiency and lower biofuel cost. Significant progress in the development of functional genomic tools in recent years has powered the elucidation of complex phenotypes and the engineering of new ones for the development of promising industrial strains (Zhang et al. 2006). Transcriptomics focuses on measuring expression of RNA under specified conditions and determines the full and coordinated set of molecular response which helps to elucidate the regulatory network and evaluate the models of cellular response. The use of transcriptional profiling tools, such as DNA microarrays, allows the study of gene expression by identifying differentially expressed genes under different experimental conditions or resulting from certain genetic perturbations. Currently, microarray technology is widely employed for analyzing the gene expression profiles from heterogeneous system critical to biomass conversion. A proteomic study reveals the identification and quantification of protein complexes present in both plant and microbial systems. Proteomics can be used to explore a microbe's protein-expression profile under various environmental conditions as the basis for identifying protein function and understanding the complex network of processes facilitated by multiprotein molecular machines (Delaunois et al. 2014).

Metabolomics studies target fuel production, including methods to isolate, extract, and analyze labile metabolites such as those involved in cellular energy metabolism (e.g., ATP, GTP, NADP, and NADPH). Metabolomics makes use of high-throughput analyses like Nuclear Magnetic Resonance (NMR) and Gas or Liquid Chromatography coupled to Mass Spectrometry (GC-MS and LC-MS) for the quantification of intra- and extracellular small-molecular-weight metabolites. Metabolic studies also serve as a screening tool to investigate the unknown effect of a specific genetic modification on the phenotype of the organism, and as a means to describe metabolism kinetics by performing pulse experiments (Buchholz et al. 2002). Glycomics (profiling materials related to structural polysaccharides) and lignomics (profiling materials related to lignin) have essential capabilities and information which will help us understand native and modified pathways for the synthesis of cell-wall polymers by tracking precursor consumption, generating and utilizing intermediate structures, and exploring their connection to plant cell-wall chemical composition and physical-chemical structure and also elucidate substrate modification by tracking structural changes and concentration fluxes in saccharification products that may be linked to harsh pretreatments. Fluxomics studies analyze the fluxes of diverse substrates through complex networks of metabolic pathways. Quantifying metabolic fluxes in microorganisms allows identification of rate-limiting steps in a biosynthetic pathway that could be improved by genetic manipulation or by alterations in cultivation conditions (Tang et al. 2009).

## Genomics and Metagenomics

Next generation DNA sequencing approach has explored the widespread possibilities of degradation of lignocellulolytic biomass by comprehensive analysis of their genomes, transcriptomes, proteomes, and interactomes. This technique is nowadays used to determine the whole-genome sequence related to lignocellulosic biomass bioconversion. For example, the genome of white rot fungus *P. chrysosporium* and brown rot fungus *Postia placenta* has been sequenced to explore the presence of a wide array of genes related to biomass degradation and conversion. Genome sequencing tools have revealed that *T. reesei* (syn. *Hypocrea jecorina*) is a powerful degrader of agricultural crop residues encoding fewer plant cell-wall polysaccharide degrading cellulases and hemicellulases than any other sequenced fungus (*P. chrysosporium* and brown rot fungus *Postia placenta*) (Martinez et al. 2009).

Metagenomics provides a culture-independent genome analysis of entire microbial communities of a particular environmental niche. Metagenomic studies for lignocellulosic biomass degradation can explore the potential and novel bioagents for effective hemicellulose and lignin conversion to improve biomass utilization for cost-effective biofuel production from communal or unculturable populations and also play a key role in sequencing new genomes harboring cellulolytic components acquired by lateral gene transfer (Fig. 22.3). The outcomes of these functional and comparative studies will include a repertoire of new enzymes and proteins available for engineering approaches (e.g., designer cellulosomes or free cellulase systems). Biomass conversion to sugars and biofuels requires optimizing microbial breakdown of structural sugars and fermentation of complex sugar mixtures. A number of microbial communities have evolved over millions of years to maximize and coordinate these capabilities, with several of the better-studied ones associated with ruminants and the hindgut of termites, and hence a reasonable number of model fermentative communities that can degrade lignocelluloses are critical targets for metagenomic analysis (Mhuatong et al. 2015; DeAngelis et al. 2010).

## Biotechnology in Fungal Biofuel Production

One of the major barriers in the efficient use of biomass-derived sugars is the lack of microbial biocatalysts that can grow and function optimally in highly stressed and harsh environments created by both biomass hydrolysis and cellular metabolism.

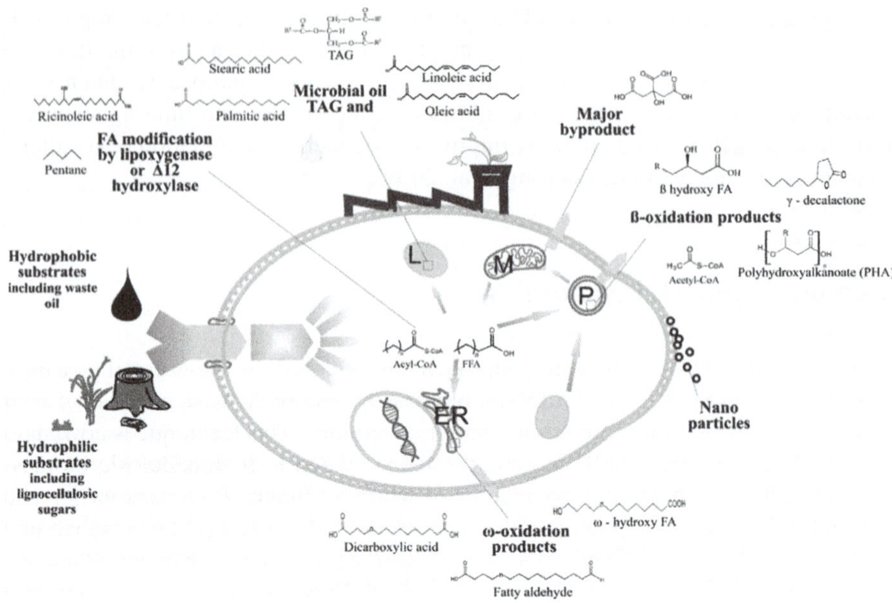


**Fig. 22.3** Schematic representation of the fatty acid-derived bioproducts production through engineered lipid pathway using *Y. lipolytica* cell factory (Abghari and Chen 2014)

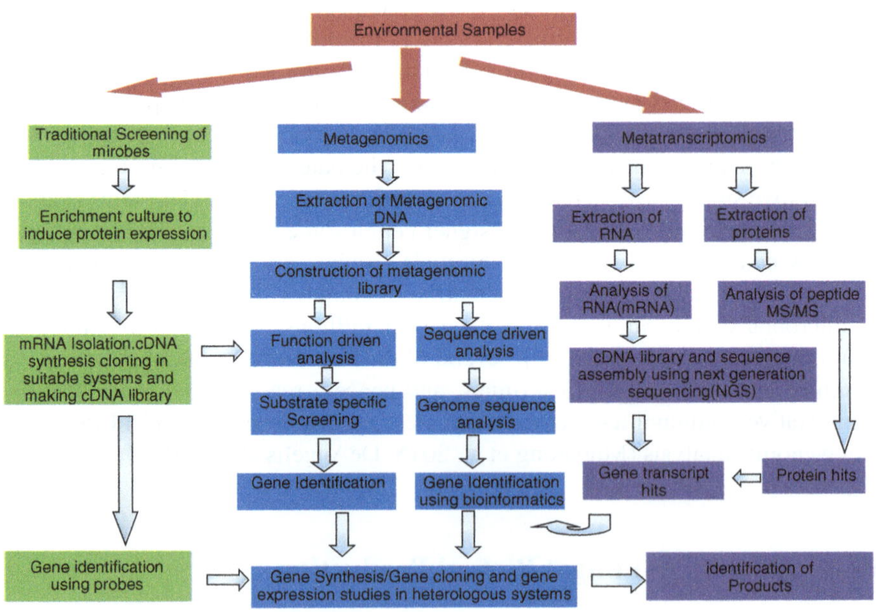


**Fig. 22.4** Metagenomic approaches for discovery of novel enzymes (modified from Sebastian et al. 2013)

Genetic engineering plays a key role in the transformation of microbes into the desired cell factories with high efficiency of bioenergy generation. Genetic engineering has improved the microbial biocatalysts to enhance the extent of biomass saccharification. A wide variety of examples are available for gene transfer technologies for the expression of genes from different microbes to increase the catalytic properties of Glycosyl hydrolases. One of the most current usages of gene transfer technology is the development of ideal microbes for consolidated bioprocessing (CBP). CBP has been shown to offer large cost benefits relative to other process configurations in both near-term and futuristic contexts (Lynd et al. 2005). Selection of suitable and efficient cellulolytic enzymes is a key factor in the process of engineering non-cellulolytic organisms with high product yields. CBP where the lignocellulosic biomass utilization and product formation properties are sequestered in one single microorganism is widely considered an ultimate low-cost configuration for cellulose hydrolysis and fermentation. One major disadvantage using mycelial fungi for ethanol production is their slow bioconversion rate compared to yeasts and high tolerance to bioethanol and currently we do not have any natural microbe available for CBP at a desired efficiency for industrial bioethanol production. The main two strategies can be employed to develop CBP organisms: (1) engineering natural cellulolytic microorganisms to improve product-related properties, i.e., so-called native host strategy, which engineers organisms with their native ability to use cellulose or pentose sugars to improve product-related properties, and (2) engineering non-cellulolytic organisms by the use of gene transfer technology to develop recombinant microbes. Ethanol yields in *A. niger* can be increased by expression of a pyruvate decarboxylase gene from *Zymomonas mobilis*. However, the major focus for producing CBP microbe is the heterologous expression of the cellulase gene in natural ethanologens and the most commonly used host for heterologous expression of cellulase genes for CBP is the baker's yeast, *S. cerevisiae* (Nakatani et al. 2013). All three classes of cellulase genes have been expressed in *S. pastorianus* (Fitzpatrick et al. 2014). Similarly, the key point of CBP for biodiesel production is the engineering of a microorganism that can efficiently depolymerize biomass polysaccharides to fermentable sugars and efficiently convert this mixed-sugar hydrolysate into FAEEs (Lin et al. 2013). Another approach of genetic engineering for enhancing cellulase activity is to increase the gene copy number. Episomal plasmids have been extensively used to express cellulase genes. Genetic engineering also offers to develop chimeric cellulosomes for efficient degradation of biomass substrate either by incorporating bacterial or fungal cellulases.

Biomass conversion to bioenergy and biofuels requires an array of enzymatic system. The catalytic efficiency and efficacy of enzymes decrease with reaction time. The nanotechnology approach in biofuel production mainly focuses on the development and application of reusable nanocatalysts in bioconversions of biomass to biofuels. Designing nanocatalysts for biomass degradation and biofuel generation is one of the most important fast-growing and unexplored research areas of Nanotechnology. The approach utilizes the assessment of the activity of designed hydrolytic nanocatalyst by adsorbing them or ligating them through chemical means on carbon nanotubes (CNT). Optimization of such chemical ligation will be very important because it would offer the advantage of performing repeated uses of the same CNT–enzyme conjugate after their isolation from processed biomasses (Wabeke et al. 2014; Neto and De Andrade 2013).

## Challenges in Fungal Biofuel Production

Microbial biofuels are a good alternative to petroleum-based fuels. They offer several benefits to society and the environment. Countries in the world have set their own targets to replace petroleum fuel by biofuels. Today, the contribution of enzyme costs to the economics of lignocellulosic biofuel production continues to be a much-debated topic. Some authors argue that the cost of enzymes is a major barrier for biofuel production (Brijwani et al. 2010), while others assume that it is not, as it will decrease with technological innovation or other advances (Aden and Foust 2009). The cost of enzymes for biofuel applications seriously hampers robust techno-economic analysis of biofuel production processes. The contribution can be lowered by shifting to lower cost feedstocks, reducing the fermentation times, and reducing the complexity of the process to drive down capital costs. Producing as many co-products as possible in a biorefinery will help to reduce the cost of biofuel production. It is important that a biorefinery should be established in an appropriate location that has good water resources, access to feedstocks, and energy that is needed to process the feedstock.

## Future Prospects

Keeping in view the fast depleting conventional resources of energy from the planet along with the hazardous impact of their use on the environment, the abilities of microbes to produce biofuels appear as an eco-friendly alternative. Conversion of plant biomass using microbial systems provides an excellent alternative for production of biofuel on a commercial basis. Recommended aims of future efforts to upgrade the fungal biofuel production include appropriate genetic and metabolic manipulation of the strains, applying protein engineering strategies, and achieving high productivity and product recovery at a larger scale. An effort to increase biofuel production has led scientists to discover genes in microfungi that improve their tolerance to ethanol, allowing them to produce more ethanol from the same amount of nutrients. Genetic engineering may be used to manipulate the lipid-associated metabolic pathway. Further exploration on regulatory and transport mechanisms, transcriptional machinery, and signal transduction pathways involved in lipid accumulation and degradation will pave the way to better understanding and utilization of this platform.

## Conclusion

Microfungi provide a promising platform to produce a wide range of lipid-based bioproducts. Microfungal lipids would be the best solution to produce biodiesel, to overcome the conflict between food and fuel. They could compete with conventional fuels only by improving the technology and increasing both the biomass and the lipid yields. Consequently, the development of oleaginous microorganism strains offering high lipid content for biodiesel production is of critical importance for an effective industrial application.

**Acknowledgment** The authors wish to thank the Department of Botany, Banaras Hindu University (BHU), Varanasi-221005, India, and DST-INSPIRE fellowship scheme for the financial assistance.

## References

Abghari A, Chen S (2014) *Yarrowia lipolytica* as an oleaginous cell factory platform for production of fatty acid-based biofuel and bioproducts. Front Energy Res 2:21–32

Adachi D, Hama S, Nakashima K, Bogaki T, Ogino C, Kondo A (2013) Production of biodiesel from plant oil hydrolysates using an *Aspergillus oryzae* whole-cell biocatalyst highly expressing *Candida antarctica* lipase B. Bioresour Technol 135:410–416

Aden A, Foust T (2009) Technoeconomic analysis of the dilute sulfuric acid and enzymatic hydrolysis process for the conversion of corn stover to ethanol. Cellulose 16:535–545

Ahola S, Turon X, Osterberg M, Laine J, Rojas O (2008) Enzymatic hydrolysis of native cellulose nanofibrils and other cellulose model films: effect of surface structure. Langmuir 24:11592–11599

Alper H, Stephanopoulos G (2007) Global transcription machinery engineering: a new approach for improving cellular phenotype. Metab Eng 9(3):258–267

Alper H, Moxley J, Nevoigt E, Fink GR, Stephanopoulos G (2006) Engineering yeast transcription machinery for improved ethanol tolerance and production. Science 314:1565–1568

Ami D, Posteri R, Mereghetti P, Porro D, Doglia SM, Branduardi P (2014) Fourier transform infrared spectroscopy as a method to study lipid accumulation in oleaginous yeasts. Biotechnol Biofuels 7:12

Arboleda C, Mejia AI, Franco-Molano AE, Jimenez GA, Penninckx MJ (2008) Autochthonous white rot fungi from the tropical forest of Colombia for dye decolourisation and ligninolytic enzymes production. Sydowia 60:165–180

Asgher M, Iqbal HMN, Irshad M (2012) Characterization of purified and xerogel immobilized novel lignin peroxidase produced from *Trametes versicolor* IBL-04 using solid state medium of corncobs. BMC Biotechnol 12:46

Berka RM, Grigoriev IV, Otillar R, Salamov A, Grimwood J, Darmond C, Moisan M-C, Henrissat B, Coutinho PM, Lambard V, Natvig DO, Lindquist E, Schmutz J, Lucas S, Harris P, Powlowski J, Bellemare A, Taylor D, Butler G, de Vries RP, Allijin IE, van den Brink J, Ushinsky S, Storm R, Powell AJ, Paulsen IT, Elbourne LDH, Baker SE, Magnuson J, La Boissiere S, Clutterbuck AJ, Martinez D, de Leon AL, Rey MW, Tsang A (2011) Comparative genomic analysis of the thermophilic biomass-degrading fungi *Myceliophthora thermophila* and *Thielavia terrestris*. Nat Biotechnol 29:922–927

Berlin A, Maximenko V, Gilkes N, Saddler J (2007) Optimization of enzyme complexes for lignocellulose hydrolysis. Biotechnol Bioeng 97:287–296

Bhanja A, Minde G, Magdum S, Kalyanraman V (2014) Comparative studies of oleaginous fungal strains (*Mucor circinelloides* and *Trichoderma reesei*) for effective wastewater treatment and bio-oil production. Biotechnol Res Int 2014:1–7

Brijwani K, Oberoi HS, Vadlani PV (2010) Production of a cellulolytic enzyme system in mixed-culture solid-state fermentation of soybean hulls supplemented with wheat bran. Process Biochem 45:120–128

Broda P, Birch PRJ, Brooks PR, Sims PFG (1996) Lignocellulose degradation by *Phanerochaete chrysosporium*: gene families and gene expression for a complex process. Mol Microbiol 19:923–932

Buchholz A, Hurlebaus J, Wandrey C, Takors R (2002) Metabolomics: quantification of intracellular metabolite dynamics. Biomol Eng 19(1):5–15

Çakar ZP, Turanlı-Yıldız B, Alkım C, Yılmaz UI (2012) Evolutionary engineering of *Saccharomyces cerevisiae* for improved industrially important properties. FEMS Yeast Res 12:171–182

Castanha RF, Mariano AP, de Morais LAS, Scramin S, Monteiro RTR (2014) Optimization of lipids production by *Cryptococcus laurentii* 11 using cheese whey with molasses. Braz J Microbiol 45(2):379–387

Certik M, Shimizu S (1999) Biosynthesis and regulation of microbial polyunsaturated fattyacid production. J Biosci Bioeng 87:1–14

Chatzifragko A, Makri A, Belha A, Bellou S, Mavrou M, Mastoridou M, Mystrioti P, Onjaro G, Aggelis G, Papanikolaou S (2011) Bio technological conversions of biodiesel derived waste glycerol by yeast and fungal species. Energy 36(2):1097–1108

Cherry JR, Fidantsef AL (2003) Directed evolution of industrial enzymes: an update. Curr Opin Biotechnol 14:438–443

Cohen J, Saito K, Entekhabi D (2001) The role of the Siberian high in Northern Hemisphere climate variability. Geophys Res Lett 28:299–302

Collins T, Gerday C, Feller G (2005) Xylanases, xylanase families and extremophilic xylanases. FEMS Microbiol Rev 29:3–23

Dashtban M, Schraft H, Qin W (2009) Fungal bioconversion of lignocellulosic residues; opportunities and perspectives. Int J Biol Sci 5:578–595

Dashtban M, Kepka G, Seiboth B, Qin W (2013) Xylitol production by genetically engineered *Trichoderma reesei* strains using barley straw as feedstock. Appl Biochem Biotechnol 4(2):554–569

de la Rubia T, Linares A, Perez J, Munoz-Dorado J, Romera J, Martinez J (2002) Characterization of manganese-dependent peroxidase isoenzymes from the ligninolytic fungus *Phanerochaete flavido-alba*. Res Microbiol 153:547–554

DeAngelis KM, Gladden JM, Allgaier M, D'haeseleer P, Fortney JL, Reddy A, Hugenholtz P, Singer SW, Gheynst JSV, Silver WL, Simmons BA, Hazen TC (2010) Strategies for enhancing the effectiveness of metagenomic-based enzyme discovery in lignocellulolytic microbial communities. Bioenerg Res 3:146–161. doi:10.1007/s12155-010-9089-z

Delaunois B, Jeandet P, Clement C, Baillieul F, Dorey S, Cordelier S (2014) Uncovering plant-pathogen crosstalk through apoplastic proteomic studies. Front Plant Sci 5(249):1–18

Dinis MJ, Bezerra RM, Nunes F, Dias AA, Guedes CV, Ferreira LM, Cone JW, Marques GS, Barros AR, Rodrigues MA (2009) Modification of wheat straw lignin by solid state fermentation with white-rot fungi. Bioresour Technol 100(20):4829–4835

Dyk JS, Pletschke BI (2012) A review of lignocellulose bioconversion using enzymatic hydrolysis and synergistic cooperation between enzymes: factors affecting enzymes, conversion and synergy. Biotechnol Adv 30:1458–1480

Easterling ER, French WT, Hernandez R, Licha M (2009) The effect of glycerol as a sole and secondary substrate on the growth and fatty acid composition of *Rhodotorula glutinis*. Bioresour Technol 100(1):356–361

Economou CN, Aggelis G, Pavlou S, Vayenas DV (2011) Modelling of single-cell oil production under nitrogen limited and substrate inhibition conditions. Biotechnol Bioeng 108:1049–1055

Elisashvili V, Penninckx M, Kachlishvili E, Tsiklauri N, Metreveli E, Kharziani T, Kvesitadze G (2008) *Lentinus edodes* and *Pleurotus* species lignocellulolytic enzymes activity in submerged and solid-state fermentation of lignocellulosic wastes of different composition. Bioresour Technol 99(3):457–462

Elisashvili V, Kachlishvili E, Tsiklauri N, Metreveli E, Khardziani T, Agathos SN (2009) Lignocellulose-degrading enzyme production by white-rot Basidiomycetes isolated from the forests of Georgia. World J Microbiol Biotechnol 25(2):331–339

Elissetche JP, Ferraz A, Freer J, Mendonca R, Rodriguez J (2006) Thiobarbituric acid reactive substances, $Fe^{3+}$ reduction and enzymatic activities in cultures of *Ganoderma australe* growing on *Drimys winter* wood. FEMS Microbiol Lett 260(1):112–118

Erden E, Ucar MC, Gezer T, Pazarlioglu NK (2009) Screening for ligninolytic enzymes from autochthonous fungi and applications for decolorization of Remazole Marine Blue. Braz J Microbiol 40(2):346–353

Fakas S, Papanikolaou S, Batsos A, Galiotou-Panayotou M, Mallouchos A, Aggelis G (2009) Evaluating renewable carbon sources as substrates for single cell oil production by *Cunninghamella echinulata* and *Mortierella isabellina*. Biomass Bioenergy 33(4):573–580

Fang J, Qu Y, Gao P (1997) Wide distribution of cellobiose oxidizing enzymes in wood-rot fungus indicates a physiological importance in lignocellulosics degradation. Biotechnol Tech 11(3):195–198

Farroni T, Enrica M, Silvia R, Johanson HM (2007) Perception of facial expressions in new borns. Eur J Dev Psychol 4(1):2–13

Fitzpatrick J, Kricka W, James T, Bond U (2014) Expression of three *Trichoderma reesei* cellulase genes in *Saccharomyces pastorianus* for the development of a two-step process of hydrolysis and fermentation of cellulose. J Appl Microbol 43:351–358

Fowler T, Brown RD Jr (1992) The bgl1 gene encoding extracellular -glucosidase from *Trichoderma reesei* is required for rapid induction of the cellulase complex. Mol Microbiol 6:3225–3235

Gincy MM, Sasikumar JM (2007) Antioxidant activity of *Pseudarthria viscid*. Indian J Pharm Sci 64(4):581–582

Glazer A, Nikaido H (2007) Microbial biotechnology: fundamentals of applied microbiology, 2nd edn. W.H. Freeman and Company, New York

Gusakov AV, Salanovich TN, Anton AF, Ustinov BB, Ukunev ON, Burlingame R, Baze M, Sinitsyn (2007) Designing highly efficient cellulase mixtures for enzymatic hydrolysis of cellulases. Biotechnol Bioeng 97(5):1028–1032

Hakulinen N, Kruus K, Koivula A, Rouvinen J (2006) A crystallographic and spectroscopic study on the effect of X-ray radiation on the crystal structure of *Melanocarpus albomyces* laccase. Biochem Biophys Res Commun 350:929–934

He J, Wu A, Chen D, Yu B, Mao X, Zheng P, Yu J, Tian G (2014) Cost-effective lignocellulolytic enzyme production by *Trichoderma reesei* on a cane molasses medium. Biotechnol Biofuels 7:43

Hoshino F, Kajino T, Sugiyama H, Asami O, Takahashi H (2002) Thermally stable and hydrogen peroxide tolerant manganese peroxidase (MnP) from *Lenzites betulinus*. FEBS Lett 530:249–252

Huang J, Chen D, Wei Y, Wang Q, Li Z, Chen Y, Huang R (2014) Direct ethanol production from lignocellulosic sugars and sugarcane bagasse by a recombinant *Trichoderma reesei* strain HJ48. Sci World J 2014:1–8

Hui L, Wan C, Hai-tao D, Xue-jiao C, Qi-fa Z, Yu-hua Z (2010) Direct microbial conversion of wheat straw into lipid by a cellulolytic fungus of *Aspergillus oryzae* A-4 in solid-state fermentation. Bioresour Technol 101:7556–7562

Isaacs FJ, Carr PA, Wang HH, Lajoie MJ, Sterling B, Kraal L, Tolonen AC, Gianoulis TA, Goodman DB, Reppas NB, Emig CJ, Bang D, Hwang SJ, Jewett MC, Jacobson JM, Church GM (2011) Precise manipulation of chromosomes in vivo enables genome-wide codon replacement. Science 333(6):348–353

Ishigami T, Yamada Y (1986) Purification and properties of polyphenol oxidase from *Chaetomium thermophile*, a thermophilic fungus. J Gen Appl Microbiol 2:293–301

Iyer G, Chattoo BB (2003) Purification and characterization of laccase from the rice blast fungus, *Magnaporthe grisea*. FEMS Microbiol Lett 227:121–126

Jang YS, Kim B, Shin JH, Choi YJ, Choi S, Song CW, Lee J, Park HG, Lee SY (2012) Bio-based production of C2–C6 platform chemicals. Biotechnol Bioeng 109(10):2437–2459
Johny JM (2013) Inhibitory effect of biosurfactant purified from probiotic yeast against biofilm producers. J Environ Sci Toxicol Food Technol (IOSR-JESTFT) 6(1):5155
Joo AR, Jeya M, Lee KM, Moon HJ, Kim YS, Lee JK (2010) Production and characterization of beta-1-4-glucosidase from a strain of *Penicillium pinophilum*. Process Biochem 45:851–858
Jorgensen H, Kristensen JB, Felby C (2007) Enzymatic conversion of lignocellulose into fermentable sugars: challenges and opportunities. Biofuels Bioprod Biorefin 1:119–134
Jovanović B, Mach RL, Mach-Aigner AR (2014) Erythritol production on wheat straw using *Trichoderma reesei*. AMB Express 4:34
Kannan K, Oblisami G, Loganathan BG (1990) Enzymology of ligno-cellulose degradation by *Pleurotus sajor-caju* during growth on paper-mill sludge. Biol Waste 33(1):1–8
Katre G, Joshi C, Khot M, Zinjarde S, Ravikumar A (2012) Evaluation of single cell oil (SCO) from a tropical marine yeast *Yarrowia lipolytica* NCIM 3589 as a potential feedstock for biodiesel. AMB Express 2(1):36–42
Korotkova OG, Semenova MV, Morozova VV, Zorov IN, Sokolova LM, Bubnova TM, Okunev ON, Sinitsyn AP (2009) Isolation and properties of fungal beta-glucosidases. Biochem (Moscow) 74:569–577
Koukiekolo R, Cho HY, Kosugi A, Inui M, Yukawa H, Doi RH (2005) Degradation of corn fiber by *Clostridium cellulovorans* cellulases and hemicellulases and contribution of scaffolding protein CbpA. Appl Environ Microbiol 71:3504–3511
Kubicek CP (2013) Systems biological approaches towards understanding cellulase production by *Trichoderma reesei*. J Biotechnol 163(2):133–142
Kubicek CP, Mikus M, Schuster A, Schmoll M, Seiboth B (2009) Metabolic engineering strategies for the improvement of cellulase production by *Hypocrea jecorina*. Biotechnol Biofuels 2:19–24
Kumar S, Kushwaha H, Bachhawat AK, Raghava GPS, Ganesan K (2014) Genome sequence of the oleaginous red yeast *Rhodosporidium toruloides* MTCC 457. Eukaryot Cell 11(8):1083–1094
Lamonica M (2014) Why the promise of cheap fuel from super bugs fell short? MIT Technology Review, Cambridge, MA
Lankinen VP, Bonnen AM, Anton LH, Wood DA, Kalkkinen N, Hatakka A, Thurston CF (2001) Characteristics and N-terminal amino acid sequence of manganese peroxidase from solid substrate cultures of *Agaricus bisporus*. Appl Microbiol Biotechnol 55:170–176
Leggio LL, Welner D, De Maria L (2012) A structural overview of GH61 proteins—fungal cellulose degrading polysaccharide monooxygenases. Comput Struct Biotechnol J 2:1–8
Li Y, Zhao ZK, Bai F (2007) High density cultivation of oleaginous yeast *Rhodosporidium toruloides* Y4 in fedbatch culture. Enzyme Microb Technol 41:312–317
Liang MH, Jiang JG (2013) Advancing oleaginous microorganisms to produce lipid via metabolic engineering technology. Prog Lipid Res 52:395–408
Liang Y, Jarosz K, Wardlow AT, Zhang J, Cui Y (2014) Lipid Production by *Cryptococcus curvatus* on hydrolysates derived from corn fiber and sweet sorghum bagasse following dilute acid pretreatment. Appl Biochem Biotechnol 173(8):20862098
Lin H, Wang Q, Sang Q, Zhan J, Zhao Y (2013) Genetic engineering of microorganisms for biodiesel production. Bioengineered 4(5):292–304
Lisov AV, Leontievsky AA, Golovleva LA (2003) Hybrid Mn-peroxidase from the ligninolytic fungus *Panus tigrinus*: isolation, substrate specificity, and catalytic cycle. Biochem (Moscow) 68:1027–1035
Lisov AV, Leontievsky AA, Golovleva LA (2007) Hybrid Mn-peroxidases from basidiomycetes. Appl Biochem Microbiol 43:536–543
Liu G, Zhang L, Wei X, Zou G, Qin Y, Ma L, Li J, Zheng H, Wang S, Wang C, Xun L, Zhao GP, Zhou Z, Qu Y (2013) Genomic and secretomic analyses reveal unique features of the lignocellulolytic enzyme system of *Penicillium decumbens*. PLoS One 8(2):55185

Lynd LR, Zyl WHV, McBride JE, Laser M (2005) Consolidated bioprocessing of cellulosic biomass: an update. Curr Opin Biotechnol 16:577–583

Manavalan T, Manavalan A, Thangavelu KP, Heese K (2013) Characterization of optimized production, purification and application of laccase from *Ganoderma lucidum*. Biochem Eng J 70:106–114

Martinez D, Berka RM, Henrissat B, Saloheimo M, Arvas M, Baker SE, Chapman J, Chertkov O, Coutinho PM (2008) Genome sequencing and analysis of the biomass-degrading fungus *Trichoderma reesei* (syn. *Hypocrea jecorina*). Nat Biotechnol 26:553–560

Martinez D, Challacombe J, Hibbett D, Schmoll M, Kubicek CP, Ferreira P, Ruiz-Duenas FJ, Martinez AT, Kersten P et al (2009) Genome, transcriptome, and secretome analysis of wood decay fungus *Postia placenta* supports unique mechanisms of lignocellulose conversion. Proc Natl Acad Sci USA 106:1954–1959

Mast B, Zohrens N, Schmids F, Hernandez R, French WT, Merkt N (2014) Lipid production for microbial biodiesel by the oleagenious yeast *Rhodotorula glutinis* using hydrolysates of wheat straw and miscanthus as carbon sources. Waste Biomass Valoriz 5:955–962

Mathew GM, Sukumaran RK, Singhania RR, Pandey A (2008) Progress in research on fungal cellulases for lignocellulose degradation. J Sci Ind Res 67:898–907

Mhuatong W, Charoensawan V, Kanokratana P, Tangphatsornruang S, Champreda V (2015) Comparative analysis of sugarcane bagasse metagenome reveals unique and conserved biomass-degrading enzymes among lignocellulolytic microbial communities. Biotechnol Biofuels 8:16

Mishra S, Bisaria VS (2006) Production and characterization of laccase from *Cyathus bulleri* and its use in decolourization of recalcitrant textile dyes. Appl Microbiol Biotechnol 71:646–653

Muniraja IK, Xiaob L, Huc Z, Zhana X, Shid J (2011) Microbial lipid production from potato processing wastewater using oleaginous filamentous fungi *Aspergillus oryzae*. Water Res 110(5):1138–1150

Murray WD, Sowden LC, Colvin JR (1984) *Bacteroides cellulosolvens* sp. nov., a cellulolytic species from sewage sludge. Int J Syst Bacteriol 34:185–187

Nakatani Y, Yamada R, Ogino C, Kondo A (2013) Synergetic effect of yeast cell-surface expression of cellulase and expansin-like protein on direct ethanol production from cellulose. Microb Cell Fact 12:66

Necochea R, Valderrama B, Diaz-Sandoval S, Folch-Mallol JL, Vazquez-Duhalt R, Iturriaga G (2005) Phylogenetic and biochemical characterisation of a recombinant laccase from *Trametes versicolor*. FEMS Microbiol Lett 244:235–241

Neto SA, De Andrade AR (2013) New energy sources: the enzymatic biofuel cell. J Braz Chem Soc 24(12):1891–1912

Nicolaou SA, Gaida SM, Papoutsakis ET (2011) Coexisting/coexpressing genomic libraries (CoGeL) identify interactions among distantly located genetic loci for developing complex microbial phenotypes. Nucleic Acids Res 39(22):152

Nowroozi F, Baidoo EK, Ermakov S, Redding-Johanson A, Batth T, Petzold C, Keasling J (2014) Metabolic pathway optimization using ribosome binding site variants and combinatorial gene assembly. Appl Microbiol Biotechnol 98:1567–1581

Palma C, Martinez AT, Lema JM, Martinez MJ (2000) Different fungal manganese-oxidizing peroxidases: a comparison between *Bjerkandera* sp. and *Phanerochaete chrysosporium*. J Biotechnol 77:235–245

Papanikolaou S, Chevalot I, Komaitis M, Marc I, Aggelis G (2002) Single cell oil production by *Yarrowia lipolytica* growing on an industrial derivative of animal fat in batch cultures. Appl Microbiol Biotechnol 58:308–312

Papanikolaou S, Fakas S, Fick M, Chevalot I, Galiotou-Panayotou M, Komaitis M, Marcand I, Aggelis G (2008) Biotechnological valorisation of raw glycerol discharged after bio-diesel (fatty acid methyl-esters) manufacturing process: production of 1,3- propanediol, citric acid and single cell oil. Biomass Bioenergy 32:60–71

Papanikolaou S, Dimou A, Fakas S, Diamantopoulou P, Philippoussis A, Galiotou-Panayotou M (2011) Biotechnological conversion of waste cooking olive oil into lipid-rich biomass using *Aspergillus* and *Penicillium* strains. J Appl Microbiol 110(5):1138–50

Patten P, Howard R, Stemmer W (1997) Applications of DNA shuttling to pharmaceuticals and vaccines. Curr Opin Biotechnol 8:724–733

Pinto PA, Dias AA, Fraga I, Marques G, Rodrigues MA, Colaco J, Sampaio A, Bezerra RM (2012) Influence of ligninolytic enzymes on straw saccharification during fungal pretreatment. Bioresour Technol 111:261–267

Pointing SB, Pelling AL, Smith GJ, Hyde KD, Reddy CA (2005) Screening of basidiomycetes and xylariaceous fungi for lignin peroxidase and laccase gene-specific sequences. Mycol Res 109:115–124

Polizeli ML, Rizzatti AC, Monti R, Terenzi HF, Jorge JA, Amorim DS (2005) Xylanases from fungi: properties and industrial applications. Appl Microbiol Biotechnol 67(5):577–591

Prakasham RS, Rao CS, Rao RS, Sarma PN (2007) Enhancement of acid amylase production by an isolated *Aspergillus awamori*. J Appl Microbiol 102(1):204–211

Ryu BG, Kim J, Kim K, Han JI, Yang YE (2013) High-cell-density cultivation of oleaginous yeast *Cryptococcus curvatus* for biodiesel production using organic waste from the brewery industry. Bioresour Technol 135:357–364

Saha BC (2003) Hemicellulose bioconversion. J Ind Microbiol Biotechnol 30:279–291

Salvachua D, Martinez AT, Tien M, Lopez-Lucendo MF, Garcia F, de Los RV, Martinez MJ, Prieto A (2013) Differential proteomic analysis of the secretome of *Irpex lacteus* and other white-rot fungi during wheat straw pretreatment. Biotechnol Biofuels 6(1):115

Sanchez G, Pilcher L, Roslander C, Modig T, Galbe M, Liden G (2004) Dilute acid hydrolysis for fermentation of the Bolivian straw material *Paja brara*. Bioresour Technol 93(3):249–256

Schilling B, Linden RM, Kupper U, Lerch K (1992) Expression of *Neurospora crassa* laccase under the control of the copper-inducible metallothionein-promoter. Curr Genet 22(3):197–203

Sebastian R, Kim JY, Kim TH, Lee KT (2013) Metagenomics: a promising approach to assess enzymes biocatalyst for biofuel production. Asian J Biotechnol 5:33–50

Sestric R, Munch G, Cicek N, Sparling R, Levin DB (2014) Growth and neutral lipid synthesis by *Yarrowia lipolytica* on various carbon substrates under nutrient-sufficient and nutrient-limited conditions. Bioresour Technol 164:41–46

Shevchenko EA, Bessolitsyna EA, Darmov IV (2013) Identification of genes encoding ligninolytic enzymes in naturally occurring basidiomycete isolates. Prikl Biokhim Mikrobiol 49(3):285–291

Sims REH, Mabee W, Saddler JN, Taylor M (2010) An overview of second generation biofuel technologies. Bioresour Technol 101:1570–1580

Stadlmayr G, Benakovitsch K, Gasser B, Mattanovich D, Sauer M (2010) Genome-scale analysis of library sorting (GALibSo): isolation of secretion enhancing factors for recombinant protein production in *Pichia pastoris*. Biotechnol Bioeng 105(3):543–555

Steenbakkers PJ, Freelove A, Van Cranenbroek B, Sweegers BM, Harhangi HR, Vogels GD (2002) The major component of the cellulosomes of anaerobic fungi from the genus *Piromyces* is a family 48 glycoside hydrolase. DNA Seq 13:313–320

Sulistyaningdyah WT, Ogawa J, Tanaka H, Maeda C, Shimizu S (2004) Characterization of alkaliphilic laccase activity in the culture supernatant of *Myrothecium verrucaria* 24G-4 in comparison with bilirubin oxidase. FEMS Microbiol Lett 230:209–214

Sun N, Zhao H (2013) Transcription activator-like effector nucleases (TALENs): a highly efficient and versatile tool for genome editing. Biotechnol Bioeng 110:1811–1821

Tang Y, Martin HG, Myers S, Rodriguez S, Baidoo EEK, Keasling JD (2009) Advances in analysis of microbial metabolic fluxes via $^{13}C$ isotopic labeling. Mass Spectrom Rev 28:362–375

Tanimura A, Takashima M, Sugita T, Endoh R, Kikukawa M, Yamaguchi S, Sakuradani E, Ogawa J, Ohkuma M, Shima J (2014) *Cryptococcus terricola* is a promising oleaginous yeast for biodiesel production from starch through consolidated bioprocessing. Sci Rep 4:4776

Thiru M, Sankh S, Rangaswamy V (2011) Process for biodiesel production from *Cryptococcus curvatus*. Bioresource Technol 102(22):10436–10440

Timothy DHB, Mark A, Elizabeth MH, Raman R (2011) Pathways for degradation of lignin in bacteria and fungi. J Nat Product Rep 28:1883–1896

Toscano L, Gochev V, Montero G, Stoytcheva M (2015) Enhanced production of extracellular lipase by novel mutant strain of *Aspergillus niger*. Biotechnol Biotechnol Equip 25(1): 2243–2247

Tsukada T, Igarashi K, Yoshid M, Samejima M (2006) Molecular cloning and characterization of two intracellular beta-glucosidases belonging to glycoside hydrolase family 1 from the basidiomycete *Phanerochaete chrysosporium*. Appl Microbiol Biotechnol 73:807–814

Tyo KE, Alper HS, Stephanopoulos GN (2007) Expanding the metabolic engineering toolbox: more options to engineer cells. Trends Biotechnol 25(3):132–137

Wabeke JT, Al-Zubaidi H, Clara P, Adams CP, Ariyadasa LAW, Nick ST, Bolandi A, Ofoli RY, Obare SO (2014) Synthesis of nanoparticles for biomass conversion processes. ACS Symp Ser 12(1186):219–246. doi:10.1021/bk-2014-1186.ch012

Wang Z, Chen S (2009) Potential of biofilm-based biofuel production. Appl Microbiol Biotechnol 83:1–18

Wang Y, Vazquez-Duhalt R, Pickard MA (2003) Manganese-lignin peroxidase hybrid from *Bjerkandera adusta* oxidizes polycyclic aromatic hydrocarbons more actively in the absence of manganese. Can J Microbiol 49:675–682

Wanlapa L, Ancharida A, Motofumi S, Moriya O, Somboon T (2013) Diversity and fermentation products of xylose-utilizing yeasts isolated from buffalo feces in Thailand. Microbes Environ 28:354–360

Warner JR, Reeder PJ, Karimpour-Fard A, Woodruff LBA, Gill RT (2010) Rapid profiling of a microbial genome using mixtures of barcoded oligonucleotides. Nat Biotechnol 28(8): 856–862

Watanabe H, Tokuda G (2001) Animal cellulases. Cell Mol Life Sci 58:1167–1178

White A, Cannell MGR, Friend AD (2000) The high-latitude terrestrial carbon sink: a model analysis. Global Change Biol 6:227–245

Wong DW (2009) Structure and action mechanism of ligninolytic enzymes. Appl Biochem Biotechnol 157:174–209

Wu S, Zhao X, Shen H, Wang Q, Zhao ZK (2010) Microbial lipid production by *Rhodosporidium toruloides* under sulfate-limited conditions. Bioresour Technol 102(2):1803–1807

Xu Q, Singh A, Himmel ME (2009) Perspectives and new directions for the production of bioethanol using consolidated bioprocessing of lignocellulose. Curr Opin Biotechnol 20:364–371

Yang X, Jin G, Gong Z, Shen H, Bai F, Zhao ZK (2014) Recycling biodiesel-derived glycerol by the oleaginous yeast *Rhodosporidium toruloides* Y4 through the two-stage lipid production process. Biochem Eng J 91:86–91

Yoon JJ, Cha CJ, Kim YS, Kim W (2008) Degradation of cellulose by the major endoglucanase produced from the brown-rot fungus *Fomitopsis pinicola*. Biotechnol Lett 30:1373–1378

Yu X, Wen Z, Lin Y, Tu S, Wang Z, Yan J (2010) Intensification of biodiesel synthesis using metal foam reactors. Fuel 89:3450–3456

Zhang S, Ye BC, Chu J, Zhuang YP, Guo MJ (2006) From multi-scale methodology to systems biology: to integrate strain improvement and fermentation optimization. J Chem Technol Biotechnol 81(5):734–745

Zhang G, Lin Y, He P, Li L, Wang Q, Ma Y (2014) Characterization of the sugar alcohol-producing yeast *Pichia anomala*. J Ind Microbiol Biotechnol 41:41–48

# Chapter 23
# Interactions of Microfungi and Plant-Parasitic Nematodes

**James LaMondia and Patricia Timper**

## Fungal- and Plant-Feeding Nematodes

Fungal-feeding and plant-parasitic nematodes use a specialized mouthpart, the needle-like stylet, along with modifications of the pharynx to pierce either fungal hyphal cells or plant cells and ingest the cell contents (Hussey et al. 2002). Many different stylet-bearing nematodes have been shown to feed on fungi. In most cases, ecologists consider that nematodes with stylets which have not been shown to feed on plants should be classified as fungal feeders. There is considerable overlap, as some fungal feeders are also known plant parasites and others are associated with and suspected to feed on plants. The genera *Aphelenchus*, *Aphelenchoides*, *Ditylenchus*, and *Tylenchus* are among the most common plant- and fungal-feeding nematodes in this category (Freckman and Caswell 1985). Fungivorous nematodes commonly exist in soil in association with small populations of many different fungal species, so it is not surprising that most fungal-feeding nematodes can feed on and complete a life cycle by feeding on a wide variety of fungi (Ruess and Dighton 1996; Ruess et al. 2000; Townshend 1964) with varying degrees of success. It has been demonstrated that not all fungi result in the same level of nematode population increase and reproductive success, so it may not be surprising that nematodes may prefer to feed on multiple fungi, especially as the suitability of individual fungi as a food source may change over time depending on whether (or when) compounds

J. LaMondia, Ph.D. (✉)
The Connecticut Agricultural Experiment Station Valley Laboratory,
153 Cook Hill Road, Windsor, CT 06095, USA
e-mail: James.Lamondia@ct.gov

P. Timper, Ph.D.
USDA ARS, Tifton, GA 31793, USA
e-mail: Patricia.Timper@ars.usda.gov

 
D.-W. Li (ed.), *Biology of Microfungi*, Fungal Biology,
DOI 10.1007/978-3-319-29137-6_23

toxic to nematodes may be produced (Mankau 1969; Ciancio 1995; Hasna et al. 2007). *Ditylenchus* or *Aphelenchus* nematodes had different preferences of fungi for feeding and development (Pillai and Taylor 1967). For example, *Fusarium solani* and *Alternaria solani* were excellent hosts for both nematodes, but *Phytophthora cactorum* only supported *Ditylenchus* whereas *Rhizoctonia solani* resulted in large populations of *Aphelenchus*. Hasna et al. (2007) studied the attraction of *Aphelenchus avenae* and *Aphelenchoides* sp. nematodes to different fungal species as food sources and subsequently measured population increases on those fungi. They determined that both nematode species were differentially attracted to all seven fungi that they tested; however, nematode fecundity was not necessarily correlated to nematode attraction to the fungus as a food source. Ikonen (2001) observed similar results with six different fungi, not all fungi supported equal reproduction, or even resulted in nematode reproduction. In fact, nematodes were also attracted to fungi which did not serve as a feeding host and did not result in nematode reproduction (Townshend 1964).

Most nematode species in the genus *Aphelenchoides* are fungivores (perhaps more than >100 different species). While most nematodes do not feed on pure cultures of fungi in nature, *Aphelenchoides* (and also *Ditylenchus* species) can feed directly on commercial mushroom beds and consume hyphal cell contents, resulting in the destruction of mycelium and the loss of the mushroom crop (Hesling 1966). A few species of foliar bud and leaf nematodes such as *Aphelenchoides fragariae*, *A. besseyi*, and *A. ritzemabosi* are well known as important plant pathogens and result in serious plant diseases. Some fungal-feeding nematodes can indirectly reduce plant growth and vigor by feeding on and damaging the mycorrhizal fungi responsible for increased nutrient uptake by plants (Ingham 1988; Riffle 1967; 1971; Ruess et al. 2000; Sutherland and Fortin 1968). *Aphelenchoides composticola, Aphelenchus avenae,* and *Ditylenchus myceliophagus* fed on fungi in soil, including mycorrhizal fungi, and damaged mycelia, but the effects observed on plant growth were small (Giannakis and Sanders 1989). *Aphelenchoides bicaudatus* was able to parasitize endomycorrhizal fungi and restrict fungal growth in a manner similar to ectomycorrhizal fungi (Shafer et al. 1981). The arbuscular mycorrhizal fungi *Gigaspora margarita* and *Glomus coronatum* supported population increases of *Aphelenchus avenae* in pots that did not occur when these fungi were not present on plants. The nematodes reduced subsequent root colonization by the fungi and *G. coronatum* spores were visibly damaged by nematode feeding, while *Gi. margarita* was not (Bakhtiar et al. 2001). Ingham (1988) documented that plant-parasitic nematodes and mycorrhizal fungi are usually mutually inhibitory and nematodes feeding on fungi typically reduced fungal populations while high levels of mycelium associated with roots reduced nematode feeding and infection. In some cases, better root growth resulting from mycorrhizae increased nematode populations; in other instances, nematode feeding and associated wounding increased the incidence of root infection by mycorrhizal fungi.

There are few morphological differences to separate *Aphelenchoides* species, so Rybarczyk-Mydłowska et al. (2012) utilized small subunit rDNA sequences to distinguish species. Two plant-parasitic nematodes, *Aphelenchoides besseyi* and

*A. ritzemabosi,* grouped together in a cluster phylogenetically isolated from *A. fragariae, A. subtenuis*, and other fungal-feeding (or fungal- and plant-feeding) species. Even so, *A. besseyi,* which causes leaf and bud crimp disease on strawberry, was able to complete its life cycle on a pure culture of *Alternaria citri* (Wang et al. 1993). *Aphelenchoides besseyi* can survive in the environment in the absence of strawberry in a number of vegetable crops, in ornamental and weed species, or in strawberry residue, straw, or even in soil alone. When viable roots or root hairs are not present, these obligate parasitic nematodes are assumed to feed on the fungi living on incorporated organic matter in soil.

Fungal-feeding nematodes in soil are often assumed not to be parasites or to be only minor plant parasites or pathogens. Sutherland (1967) did not observe *Aphelenchus avenae* feeding or invasion of roots of seven species of pine and spruce seedlings, concluding that the nematodes survived by feeding on the fungi associated with the plant rhizospheres. Other researchers have observed that *A. avenae* may be associated with necrotic roots, as the nematode was observed moving in and laying eggs in the cortex of roots, and was a "facultative plant parasitic with dominant necrobiotic tendencies" (Steiner 1936). Christie and Arndt (1936) observed both *Aphelenchus* and *Aphelenchoides* spp. feeding and reproducing on fungi in culture and on plant cells in the early stages of necrosis as well as adjacent healthy plant cells. They concluded that these nematodes were more than just saprobes. Additional researchers have shown that *A. avenae* could feed and reproduce on sterile plant tissue cultures in the total absence of fungi (Barker and Darling 1965), and in addition to feeding on fungi, could feed on root hairs, and enter and feed on healthy root tissues (Chin and Estey 1966).

Karim et al. (2009) identified more than 5000 expressed sequence tags representing more than 2200 genes from *A. avenae*. These ESTs included genes encoding cell wall-degrading enzymes that were similar to those previously found in other plant-parasitic nematodes, supporting the observed dual role of this nematode as both a fungal feeder and a plant parasite. They postulated that plant parasitism by nematodes evolved from fungal feeding and that *A. avenae* may be an important animal for future studies of the evolution of plant parasitism in nematodes.

To further support this hypothesis, the *A. avenae*-expressed sequence tags were more similar to the pinewood nematode *Bursaphelenchus xylophilus* than to the potato cyst nematode *Globodera pallida* or the root-knot nematode *Meloidogyne arenaria. Bursaphelenchus xylophilus* is also known to be a mycophagous nematode that feeds on fungi such as the blue stain *Ceratocystis* fungus symbiotic with bark beetles. This nematode likely also has evolved from a fungal feeder to be an important plant parasite. The pinewood nematode was originally described as *Aphelenchoides xylophilus*, a fungal-feeding nematode associated with bark beetles. The beetles transmitted both the nematode and blue stain fungi (Steiner and Buhrer 1934) to dead or dying pines. Pinewood nematodes were later shown to be a very serious pathogen of pines in Japan (Mamiya 1972), feeding on living plant tissue to cause tree death (Kiyohara and Suzuki 1978). There are a number of *Bursaphelenchus* species, most of which feed on fungi in dead and dying trees. *Bursaphelenchus fungivorous* can reproduce on at least 54 different fungal species

(Townshend 1964). Apparently, the pinewood nematode *B. xylophilus* evolved from a fungal-feeding ancestor to feed on plant tissue, but in the absence of living plant cells it too can have a completely mycophagous life cycle. For example, if it is transmitted to dead or dying pines by beetle oviposition it can complete its life cycle without living plant tissues (Graham 1967). The *B. xylophilus* nematode can utilize wood-inhabiting fungi such as *Ceratocystis* blue stain fungi to increase populations in dead trees, and the presence of the fungus also increased spread of the nematode by beetle vectors (Maehara and Futai 1997). The presence of additional wood-inhabiting fungi such as *Trichoderma virens* (J.H. Mill. et al.) Arx (reported as *Gliocladium virens*) enhanced the survival of *Bursaphelenchus xylophilus* in drying pine tissues and also served as a food source for the nematode, increasing populations and nematode recovery from pine seedlings (McGawley et al. 1985).

Alternatively, transmission of pinewood nematodes to living trees during maturation feeding would require plant feeding to complete the life cycle. Cellulases produced by *B. xylophilus* associated with plant feeding were likely acquired from fungi through horizontal gene transfer (Jones et al. 2005). Researchers used expressed sequence tags to study the proteins nematodes use to parasitize plants and demonstrated that genes encoding for cell wall-degrading enzymes appear in plant-parasitic nematodes but not in other nematodes or in other animals studied. A number of genes have been identified; some are similar to bacterial sequences and others are similar to known fungal sequences. Both are likely due to independent horizontal gene transfer. In *B. xylophilus*, a cellulase similar to fungi and associated with the ability to parasitize living plants was likely the result of horizontal gene transfer and, like with the *Aphelenchoides* example cited above, associated with the evolution of plant parasitism (Kikuchi et al. 2004). In addition, a glucanase present in a number of *Bursaphelenchus* species may be the result of horizontal gene transfer from bacteria. As β-1,3-glucanases are constituents of fungal cell walls, these proteins are likely associated with the fungal-feeding lifestyle of the genus (Kikuchi et al. 2005).

The genus *Ditylenchus* is another example of a stylet-bearing nematode containing both plant-parasitic and fungal-feeding species. *Ditylenchus dipsaci* and *D. destructor* (stem and bulb nematodes) are economically important pathogens of many plants (Faulkner and Darling 1961). *Ditylenchus dipsaci* was able to reproduce on *Musicillium theobromae* (Turconi) Zare & W. Gams (reported as *Verticillium theobromae*) and *Cladosporium* spp. (Viglierchio 1971). *Ditylenchus destructor* was also shown to feed and reproduce on at least 64 species of fungi representing 40 genera, 8 orders, and all classes (Anderson 1964; Faulkner and Darling 1961). Undoubtedly, the ability to feed on many different fungi in plants, decomposing plant tissues, or soil plays an important role in the survival of many plant-parasitic nematodes over time when living host plants are not present (Hooper and Southey 1978). This is certainly an under-investigated area of research and may be a mechanism responsible for survival of plant-parasitic nematodes despite management with non-host rotation crops.

## Nematodes as Biocontrols of Plant Pathogenic Fungi

Because fungal-feeding nematodes can be attracted to and actively feed on plant pathogenic fungi, they may have some abilities to reduce pathogen biomass and act as biological controls of plant disease. Rhoades and Linford (1959) were likely the first to demonstrate that *Aphelenchus avenae* could control Pythium root rot in corn. A number of mycophagous nematodes were shown to feed and reproduce on root rotting and vascular wilt fungi (Nickle and McIntosh 1968). *Aphelenchus avenae* was shown to control other root rot fungi such as *Rhizoctonia solani* and *Fusarium solani* in pot experiments with alfalfa seedlings (Barnes et al. 1981) and *R. solani, F. solani*, and *F. oxysporum* in beans or peas (Hong and Estey 1985; Klink and Barker 1968), and in growth chambers at a range of soil moisture conditions when added to soils at realistic population levels observed under field conditions (Lootsma and Scholte 1997). *Aphelenchus avenae* could control *R. solani* on beans although different isolates of the fungus differed in ability to support nematode populations (Barker 1964). The level of *R. solani* control was not correlated with ability of the isolate to serve as a nematode host. Potato stem canker caused by *Rhizoctonia solani* inoculated to soil as sclerotia was significantly suppressed by *A. avenae* feeding and nematode populations increased over time. Klink and Barker (1968) determined that early feeding by *A. avenae* could inhibit the formation of sclerotia by *R. solani.* Both *A. avenae* and *Aphelenchoides* spp. suppressed *Rhizoctonia solani* and reduced damping off of cauliflower seedlings. In a series of experiments, Lagerlof et al. (2011) concluded that *A. avenae* resulted in better disease control than the *Aphelenchoides* spp. tested and that there was no interaction with the disease-suppressive activity of various composts.

*Ditylenchus* spp. also fed on and inhibited the development of Fusarium wilt of carnation (Schindler and Stewart 1956). Jun and Kim (2004) investigated combinations of different *Trichoderma* species with *Aphelenchus avenae* for control of *Pythium* and damping off of radish seedlings. They found that *T. harzianum* plus nematodes resulted in the best disease control, while other *Trichoderma* species either had no impact or resulted in reduced disease control. Their conclusion was that the *T. harzianum* was a good food source that increased nematode populations which also fed on *Pythium*, where the other *Trichoderma* species were less favorable or even antagonistic to the nematodes.

*Cryphonectria parasitica*, the fungal pathogen which causes chestnut blight, can be partially controlled using hypovirulent strains of the fungus as a biological control (Hogan and Griffin 2008). These hypovirulent strains are infected with a virus that reduces virulence and results in nonlethal superficial cankers on the trees (Elliston 1985). Both virulent and hypovirulent strains of *C. parasitica* are good hosts of the fungivorous nematode *Aphelenchoides hylurgi* (Griffin et al. 2009, 2012), and experiments demonstrated that the nematodes were able to spread propagules of the hypovirulent strain. The direct effects of nematode feeding on the hyphae on canker development were not determined, but perhaps more importantly, the researchers postulated that *Aphelenchoides hylurgi* may be an effective means of spreading hypovirulence to new cankers, increasing the incidence and efficacy of biological control under natural conditions.

## Inhibition of Fungal Biocontrol

Alternatively, fungal-feeding nematodes may feed on the biocontrol fungi which themselves parasitize plant pathogenic fungi. Bae and Knudsen (2001) collected an *Aphelenchoides* sp. from field soils that fed on the biocontrol fungus *Trichoderma harzianum*. In culture, the nematode fed on *T. harzianum* hyphae, increased in number, and reduced radial growth of the fungus. In both untreated and heat-treated soils, the *Aphelenchoides* sp. increased populations over time, reduced biological control activity against *Sclerotinia sclerotiorum*, and reduced recovery of *T. harzianum* from soil, indicating that natural soil populations of fungivorous nematodes may reduce the biocontrol activity of other soil fungi against plant pathogens.

Nematodes may also interact with endophytic fungi. Endophytic nonpathogenic *Fusarium oxysporum* has been shown to negatively affect *Meloidogyne incognita*, *M. javanica*, *Pratylenchus goodeyi*, and *Radopholus similis* in banana, tomato, and rice (Dababat and Sikora 2007b; Mwaura et al. 2010; Sikora 2008). Endophytes may produce metabolites toxic to nematodes or systemically alter root exudates to reduce attraction or induce repellency (Vu 2005). Endophytes may also be mutualistic with nematodes. While not a plant parasite, the fungal-feeding nematode *Paraphelenchus acontioides* appears to interact with a fungal endophyte, *Fusarium* cf. *torulosum*, of brome grass *Bromus tectorum*, in a manner advantageous to both the nematode and the fungus (Baynes et al. 2012). The presence of the nematode increased the incidence of infection of the endophyte in the plant by some as yet unknown mechanism, and the nematode preferentially grazed on this endophyte.

## Nematode Interactions with Plant Pathogenic Fungi in Disease Complexes

In a Presidential paper presented at the 1930 meeting of the American Phytopathological Society, H. S. Fawcett stated the oft-cited quote "Nature does not work with pure cultures alone but most frequently with associations." He also stated that "… a number of diseases may require an association of organisms for their occurrence and cannot be produced by infection of one organism alone" (Fawcett 1931). Since that time, many diseases of complex etiology have been discovered. While none of the examples of "synergism" that were cited in his 1930 address included nematodes, nematodes certainly interact with fungi in complex diseases. In the process of feeding on plants, plant-parasitic nematodes necessarily cause wounds and can also result in numerous changes in host physiology. The damage to and effects on plants can differ substantially based on how different nematodes feed. Ectoparasites, nematodes which feed on plants while remaining for the most part outside the root, typically feed on plant cells by insertion of the stylet, a sclerotized mouthpart, into cells. Repeated insertion into the same or different cells causes microinjection wounds which occur in conjunction with salivation and pharyngeal gland secretions while cutting into or through cell walls, and the stylet is used to ingest cell contents (Wyss 2002). Endoparasitic nematodes enter the roots and can be migratory, moving into plant tissues while feeding on cell contents in a manner similar to ectoparasites, or sedentary. Sedentary endoparasites enter roots and settle into one location for feeding and development. These nematodes can secrete effector proteins to actively suppress plant defense responses and can change cellular identity and metabolism to induce the development of specialized feeding cells which concentrate nutrients for developing nematodes (Ali et al. 2015; Mitchum et al. 2013).

Nematode-induced root wounding, cell senescence and death, local or systemic host physiological changes, and nematode-induced suppression of plant defense responses may each act to increase the incidence or severity of diseases that are primarily caused by fungal pathogens. Nematode–fungal interactions in complex disease can be additive in effect or synergistic when disease is significantly greater than the sum of disease caused by each pathogen individually. Many examples of complex diseases caused by the interaction of more than one pathogen and including nematodes have been proposed. Sikora and Carter (1987) proposed limiting the definition of complex diseases involving plant-parasitic nematodes to those based on field observations of disease where both nematodes and additional pathogens can be accounted for, environmental factors and population densities of both nematodes and fungal propagules that are realistic and in a range based on actual observed data, timing of inoculations that are realistic and determined by environmental conditions and phenology of the crop/pathogen(s) interaction, and where experiments were conducted with appropriate factorial/multifactorial designs used to produce statistical evidence of a synergistic complex interaction.

What is likely the very first example of a nematode–fungal disease complex was described by Atkinson in 1892 and involved an increased severity of Fusarium wilt of cotton caused by *Fusarium oxysporum* f. sp. *vasinfectum* when plants were also infected with root-knot nematodes (*Meloidogyne* spp.). Subsequently, most cases of synergistic interactions between nematodes and fungi involve the sedentary endoparasitic root-knot and cyst nematodes increasing disease caused by *Fusarium* or *Verticillium* wilt fungi. *Meloidogyne* spp. have been shown to interact with Fusarium wilt in a number of crops, including alfalfa (Griffin 1986), bananas (Jonathan and Rajendran 1998), beans (France and Abawi 1994), chickpeas (Uma Maheswari et al. 1997), coffee (Bertrand et al. 2000), cotton (Atkinson 1892; De Vay et al. 1997; Abd-El-Alim et al. 1999), lentils (De et al. 2001), peas (Siddiqui and Mahmood 1999), soybeans (Xing and Westphal 2013), tobacco (LaMondia 1992; LaMondia and Taylor 1987), tomatoes (Abawi and Barker 1984; Suleman et al. 1997), and watermelon (Sumner and Johnson 1973). Cyst nematodes can act in a similar manner to increase wilt diseases. The potato cyst nematodes *Globodera rostochiensis* (Evans 1987) and *G. pallida* (Hide et al. 1984; Storey and Evans 1987) interact with *Verticillium* spp. in potato plants to increase disease severity, and Fusarium wilt of tobacco is most damaging in the presence of the tobacco cyst nematode *G. tabacum* (LaMondia and Taylor 1987).

In addition to wilt diseases, a number of complex nematode–fungal root rot diseases have been described. *Meloidogyne incognita* has been associated with black root rot of cotton caused by *Thielaviopsis basicola* (Walker et al. 2000). Neither pathogen causes significant disease individually; plant mortality only occurs when both pathogens are present. Management of the nematode component alone can reduce the severity of the fungal root rot disease (Fichtner et al. 2005). *Meloidogyne enterolobii* (Gomes et al. 2013) and *M. mayaguensis* (Gomes et al. 2011) interact with *Fusarium solani* to result in a decline of guava due to a complex root rot syndrome. *Meloidogyne javanica* can interact with *Rhizoctonia solani* to increase root rot and pod disease in peanut (Abdel-Momen and Starr 1998), and *M. incognita* increases the severity of root rot by *R. solani* in grapevines (Walker 1997). Early infection with the potato cyst nematode *Globodera pallida* increased root rot in potato by *R. solani* (Bhattarai et al. 2010), and the sugar beet cyst nematode *Heterodera schachtii* increased infection and disease severity caused by *R. solani* in beets (Polychronopoulos et al. 1969). The soybean cyst nematode *Heterodera glycines* interacts with *Fusarium virguliforme* (=*F. solani* f. sp. *glycines*) to result in the sudden death syndrome of soybean, significantly increasing root and crown rot, chlorosis, defoliation, and pod abortion symptom development when both pathogens are present (McLean and Lawrence 1993; Xing and Westphal 2013).

Nematodes may influence the development of fungal diseases in several ways. Root wounding occurs as infective juveniles enter and migrate through the root cortex to suitable feeding locations behind the root tip. Nematodes migrating through root tissue may also act to disperse fungal inoculum (Neher 2010). Inagaki and Powell (1969) suggested that wounding alone was the primary mechanism for the migratory endoparasite *Pratylenchus brachyurus*-incited increases in infection by *Phytophthora parasitica* in tobacco. They investigated artificial wounding and timing of the interaction and concluded that nematode infection 1 week earlier but not 2–3 weeks increased infection by *P. parasitica* was consistent with mechanical damage due to invasion. However, additional effects of artificial wounding and nematodes could also occur. A stronger argument for wounding as the primary reason for increasing fungal infection was presented by observation of *Rhizoctonia solani* growing in those areas of the root damaged by cyst nematode (*Heterodera schachtii*) invasion in sugar beet roots (Polychronopoulos et al. 1969). Likewise, *Verticillium* infected via *Globodera pallida* juvenile invasion routes in potato (Storey and Evans 1987). Infection by the fungus was most likely if nematode wounds were fresh and infection was reduced over time after roots responded to damage. In addition to nematode movement through tissues, root wounding may also be caused by physical damage to the cortex as the developing female inside the root expands greatly in size as she develops to maturity and produces eggs. Cortical cells can be damaged as they are compressed, split, or separated from adjacent cells. *R. solani* was shown to infect areas of the root damaged by the development of the northern root-knot nematode *M. hapla* inside root galls (Fagbenle and Inskeep 1987). *Heterodera daverti* interacts synergistically with *Fusarium oxysporum* or *F. avenaceum* to decrease subterranean clover plant dry weight, but only if the fungal pathogens were inoculated 1–2 weeks prior to the nematode (Nordmeyer and Sikora 1983a). Rather than the nematode increasing infection by the fungal pathogen, the fungi may make roots more attractive and likely to be infected by the nematode. In fact, in vitro research (Nordmeyer and Sikora 1983b) showed that nematodes moved in greater numbers toward culture filtrate treated roots and diffusates from infected roots than control roots. Fungal factors such as cell wall-degrading enzymes may increase chemical orientation factors or make roots more easily penetrated.

Migratory endoparasitic nematodes such as lesion nematodes, *Pratylenchus* spp., have been widely studied as important factors increasing potato early dying disease in combination with *Verticillium dahlia* or *V. albo-atrum* (Martin et al. 1982; Wheeler et al. 1992). In this disease complex, nematode infection acts to increase the impact of relatively low populations of *Verticillium* spp. to result in significantly increased early plant senescence and tuber yield losses that would not otherwise occur in the absence of the nematode. Root wounding is often cited as the likely mechanism of this interaction as *Pratylenchus* juveniles and adults infect, feed, and migrate intercellularly through the root cortex. However, certain species of *Pratylenchus*, including *P. penetrans* and *P. scribneri*, can increase potato early dying, whereas *P. crenatus*, which causes similar wounds due to cortical invasion, does not (Riedel et al. 1985). In fact, further research has shown that geographically isolated distinguishable populations of *P. neglectus* can differ in ability to interact with *Verticillium* spp. to incite early dying (Hafez et al. 1999). The fungal pathogen can also differ in ability to interact with a nematode partner in a disease complex. *Pratylenchus penetrans* synergistically increased Verticillium wilt of peppermint and spearmint by *V. dahlia* VCG 2B but not with VCG 4A (Johnson and Santo 2001). Faulkner et al. (1970) utilized split-root systems in peppermint to demonstrate that *Pratylenchus brachyurus* could systemically increase the incidence and severity of Verticillium wilt on different root systems of the same plant. Clearly, migratory endoparasites can do more than just wound roots to impact subsequent disease by *Verticillium*.

*Pratylenchus penetrans* has been associated with significantly increased root rot in perennial strawberry beds caused by *Rhizoctonia fragariae*, often leading to removal of fields from production. Lesion nematode feeding and movement in strawberry roots directly results in cell damage and death. The indirect effects of lesion nematode infection are discoloration of the endodermis and early polyderm formation, followed by localized areas of secondary growth and cortical cell senescence or death. Senescing root tissue or dying cells resulting from the direct and indirect effects of *P. penetrans* were more susceptible to *R. fragariae*, leading to increased infection and cortical root rot (LaMondia 2004). Unlike the *Pratylenchus*-*Verticillium* system (Faulkner et al. 1970), inoculation of lesion nematodes and *R. fragariae* on separate root systems of the same plant did not result in increased disease, indicating that local rather than systemic effects of *P. penetrans* were most important in the interaction of these pathogens in strawberry black root rot (LaMondia 2003).

The stem and bulb nematode *Ditylenchus dipsaci* can feed on different plant tissues than the nematodes discussed to this point, but *Ditylenchus* spp. can also interact with fungi to cause or increase plant disease. Fusarium basal rot of onion was shown to be significantly greater when both *D. dipsaci* and *Fusarium* spp. were present (Trifonova and Koleva 2002). Synergistic interactions were observed when nematodes were inoculated concomitantly or when nematodes followed inoculation with the fungus. Crown and root rot of sugar beet caused by *Rhizoctonia solani* was also synergistically increased by coinfection with *D. dipsaci* (Hillnhutter et al. 2011). The authors concluded that the fungus gained access to the plant through wounds caused by the nematode based on fungal infection at those sites. Foliage blight of *Phlox subulata* resulting from dual infection of *D. dipsaci* and *Botrytis cinerea* resulted in severe dieback and the failure of *Botrytis* gray mold control with appropriate fungicides (LaMondia unpublished). This lack of disease control was assumed to be due to fungicide resistance in the *Botrytis* pathogen, but was determined instead to be due to ongoing damage by the stem nematode that resulted in secondary infection of dead and dying leaves and stems by *Botrytis cinerea*. Fusarium wilt of alfalfa by *F. oxysporum* f. sp. *medicaginis* was increased by infection with *D. dipsaci* in both field and greenhouse studies (Griffin 1992). Vrain (1987) determined that *D. dipsaci* and not *Pratylenchus penetrans* interacted with *V. albo-atrum* to increase the severity of wilt symptoms and affect forage yields.

Root-knot and cyst nematode-induced giant cells and syncytia increase metabolic activity and effectively concentrate nutrients as a food source for the stationary developing nematodes (Jones 1981; Jones and Northcote 1972). Many researchers have reported that fungi such as *F. oxysporum*, *F. solani*, *Pythium* spp., and *R. solani* colonize nematode-infected tissues in a manner not observed in the absence of nematode infection (Abdel-Momen and Starr 1998; McLean and Lawrence 1993; Meléndez and Powell 1970; Negrón and Acosta 1989). In addition to physical cell damage, leakage of cell contents from feeding cells may also influence infection. Golden and Van Gundy (1972) determined that *R. solani* and *T. basicola* responded to *M. incognita*-induced galls and increased leakage of electrolytes and organic compounds by growing more vigorously and actively colonizing nematode-infected roots, especially as females expanded and matured. In fact, perhaps because damaged tissues are such good sources of nutrition, it has been reported that root-knot nematode infection can result in root disease by fungi usually considered to be nonpathogenic or weakly pathogenic, such as *Curvularia*, *Penicillium*, and *Trichoderma* (Combettes 1983).

## Fungal Antagonism of Nematodes

Antagonistic interactions between microfungi and plant-parasitic nematodes are as numerous as they are varied. While some facultative plant-parasitic nematodes, such as *Aphelenchoides fragariae* and *Bursaphelenchus xylophilus*, feed on fungi as well as plants, the majority of fungal–nematode interactions are detrimental to the nematode. Fungi kill or disrupt the life cycle of nematodes by parasitism, by production of toxins or extracellular enzymes, and by inducing resistance in plants. Many reviews have been written about the interaction between microfungi and plant-parasitic nematodes (Liu et al. 2009; Stirling 2014; Davies and Spiegel 2011; Dong and Zhang 2006; Morton et al. 2004; Lopez-Llorca et al. 2008). Hence, this section is not intended to be a comprehensive literature review, but to provide the reader with a detailed overview of the ecological, cellular, and molecular interactions between fungi and their host nematodes.

## Parasites of Motile Nematode Stages

Except for the eggs, most nematode life stages are actively moving in their environment, which presents a challenge for relatively slow-growing fungal parasites. Fungi have adapted to parasitize mobile stages of nematodes by employing trapping structures or adhesive conidia to immobilize or make firm contact with nematodes that touch them. These fungi have a broad host range, parasitizing microbivorous nematodes as well as plant- and animal-parasitic nematodes.

### *Fungi Producing Traps*

Fungi that produce trapping structures are often referred to as predatory or carnivorous because they trap, wound, or paralyze nematodes before consuming them in a manner reminiscent of carnivorous plants. Most trapping fungi are true pathogens that infect a living host; however, some kill the host prior to infecting. Trapping fungi are found primarily in the Orbiliaceae (Ascomycota) and in several families in the Basidiomycota (Table 23.1). The taxonomy of the trapping fungi in the Orbiliaceae has undergone numerous changes in the last 20 years. Previously, the genera were classified based on morphology of their conidia (Drechsler 1937; Rubner 1996); however, current classification of these fungi is based on morphology of the trapping structures and molecular similarity (Ahren et al. 1998; Scholler et al. 1999; Li et al. 2005). Scholler et al. (1999) accepted 82 species of trapping fungi in the Orbiliaceae. *Dactylellina* spp. produce stalked adhesive knobs with some species also producing nonconstricting rings or adhesive branches. *Arthrobotrys* spp. produce adhesive networks, while *Dreschlerella* spp. produce constricting rings that dramatically clamp around the nematode body. Traps are typically produced on hyphae; however, under conditions of low nutrition or intense competition, traps are produced directly from germinating conidia (Persmark and Nordbring-Hertz 1997). Additionally, the traps of *Dactylellina* and *Dreschlerella* are continuously produced regardless of the environment, whereas the traps of *Arthrobotrys* are produced in response to a combination of low nutrition and the presence of nematodes (Nordbring-Hertz 1977; Scholler and Rubner 1994).

**Table 23.1** Genera of nematode-trapping fungi

| Genus | Order (family) | Trap type | Citation |
|---|---|---|---|
| *Dactylellina* | Helotiales (Orbiliaceae) | Stalked adhesive knobs; nonconstricting rings | Li et al. (2005) |
| *Arthrobotrys* | Helotiales (Orbiliaceae) | Adhesive networks | Li et al. (2005) |
| *Dreschlerella* | Helotiales (Orbiliaceae) | Constricting rings | Li et al. (2005) |
| *Stylopage* | Zoopagales (Zoopagaceae) | Adhesive hyphae | Barron (1977) |
| *Resupinatus* | Agaricales (Tricholamataceae) | Hourglass-shaped adhesive cells | Thorn and Barron (1984) |
| *Pleurotus* | Agaricales (Pleurotaceae) | Paralytic droplets | Barron and Thorn (1987) |
| *Hohenbuehelia* | Agaricales (Pleurotaceae) | Hourglass-shaped adhesive cells | Thorn and Barron (1984) |
| *Coprinus* | Agaricales (Agaricaceae) | Spiny balls | Luo et al. (2007) |
| *Stropharia* | Agaricales (Strophariaceae) | Spiny acanthocytes | Luo et al. (2006) |
| *Hyphoderma* | Polyporales (Meruliaceae) | Adhesive stephanocysts | Tzean and Liou (1993) |

Recent evidence suggests that nematode pheromones known as ascarosides are involved in triggering trap production in *Arthrobotrys* (Hsueh et al. 2013). Some genera in the Basidiomycota and Zygomycotous fungi also produce adhesive hyphae or knobs to secure nematodes prior to infection, though these trapping structures are morphologically distinct from the Orbiliaceae (Table 23.1). Basidiomycetes also form unique structures which kill or immobilize nematodes prior to consuming them. *Pleurotus ostreatus* produces droplets of toxin on hyphal stalks which instantly paralyze nematodes that come in contact with the toxin (Barron and Thorn 1987). Recently, spiny acanthocytes of *Stropharia* spp. and spiny balls of *Coprinus comatus* have been shown to mortally wound nematodes with their sharp projections before rapidly colonizing their prey (Luo et al. 2004, 2006, 2007; Zouhar et al. 2013). Additional descriptions and images of trapping structures can be found in (Barron 1977; Stirling 1991; Nordbring-Hertz et al. 2001; Tunlid and Ahren 2011).

The mechanism of trapping and infection has primarily been studied in *Arthrobotrys oligospora* and related Orbiliaceae (see Tunlid and Ahren 2011 for review). Early studies indicated that adhesion of *A. oligospora* to the nematode cuticle involved lectins on the traps binding with carbohydrates on the nematode surface (Nordbring-Hertz and Mattiasson 1979; Premachandran and Pramer 1984). An N-acetylgalactosamine-binding lectin (AOL) was subsequently isolated from *A. oligospora*; however, AOL-deletion mutants showed no loss of nematode parasitism (Balogh et al. 2003). Moreover, none of the seven genes encoding for lectins that were identified in *A. oligospora* by Yang et al. (2011) were upregulated during trap formation. The mechanism of adhesion is still unresolved. Following capture, the fungus penetrates the nematode cuticle by means of mechanical force and extracellular proteases (Veenhuis et al. 1985). The nematode is paralyzed during the initial penetration of the cuticle by the fungus; evidence indicates that a serine protease is involved in the paralysis (Ahman et al. 2002). At the point of entry, an infection bulb is formed from which trophic hyphae develop. Nematode death and colonization of the cadaver is rapid (48–60 h) and vegetative hyphae soon emerge from the degraded cadaver (Dijksterhuis et al. 1994).

Despite the broad host range of trapping fungi, some nematodes are resistant to capture. Gaspard and Mankau (1987) evaluated 16 species of nematodes and found that nematodes in the class Secernentea were trapped by *Dactylellina ellipsospora* (*Monacrosporium ellipsosporum*), whereas nematodes in Adenophorea were not trapped. Even within Secernentea, nematodes can exhibit resistance to trapping fungi. Cyst nematodes (*Globodera* and *Heterodera* spp.), for example, are able to evade capture by fungi with adhesive traps (Wimble and Young 1983; Den Belder and Jansen 1994; Jaffee 1998), but not those with constricting rings (Jaffee 1998). Nematodes that are resistant to trapping fungi may possess surface components that mask receptors for fungal attachment on the nematode cuticle. Mendoza de Gives et al. (1999) demonstrated that three mutants of *Caenorhabditis elegans*, which had altered surface coats, were more efficiently captured than the wild type by *Arthrobotrys* (*Duddingtonia*) *flagrans*. The authors hypothesized that the mutants had lost a component of their surface coat or cuticle that in the wild type provides a barrier to recognition by the fungus.

Plant-parasitic nematodes, for the most part, are found in and around living plants, while several of the nematode-trapping basidiomycetes (*Hohenbuehelia*, *Hyphoderma*, *Pleurotus*, and *Resupinatus*) are found in decaying wood, compost, and barnyard soil (Thorn and Barron 1984, 1986; Tzean and Liou 1993). Although these fungi can trap and consume plant-parasitic nematodes, they typically do not co-occur in the same habitat. The exception is *Hohenbuehelia* (syn. *Nematoctonus*), which has been isolated in low frequencies from plant rhizospheres (Persmark and Jansson 1997). It is unclear how commonly the spiny traps of *Coprinus* or *Stropharia* are found in association with plant-parasitic nematodes because, until recently, they had not been recognized as nematode-killing structures. Luo et al. (2006) demonstrated that *Stropharia rugosoannulata* readily colonized woodland soil where it produced abundant acanthocytes. The trapping fungi in Orbiliaceae and *Stylopage* are also found in decaying wood, dung, and humus (Duddington 1951); however, they have been regularly isolated from agricultural soils including the rhizosphere and endosphere of plant roots where they could potentially interact with plant-parasitic nematodes. The abundance of these fungi is generally greater in the rhizosphere than in the bulk soil (Persmark and Jansson 1997; Peterson and Katznelson 1965; Gaspard and Mankau 1986). This is not surprising given that the rhizosphere is more densely populated with host nematodes (Persmark and Jansson 1997), both plant parasites and microbivorous, than the bulk soil and some of the trapping fungi may obtain nutrition from the root exudates. Interestingly, *A. oligospora* was found to penetrate barley and tomato roots and colonize the cortex but not the vascular tissue (Bordallo et al. 2002). The fungus demonstrated directed growth towards roots and penetrated the epidermis via appressoria. Two other trapping fungi, *A. musiformis* and *A. psychrophila*, did not show directed growth towards roots.

Addition of organic matter to soil often increases the abundance of trapping fungi (Cooke 1962a, b; Wang et al. 2003a; Jaffee 2002, 2004). The organic matter can serve as a source of nutrition for the fungi but also increases numbers of microbivorous nematodes which serve as hosts (Linford et al. 1938). Moreover, in agricultural soils, abundance of trapping fungi has been found to be greatest in the top 10 cm where organic matter content is highest (Persmark et al. 1996; Stirling et al. 2011). Abundant trapping fungi, however, does not always lead to suppression of nematode populations. These fungi show varying degrees of dependence on nematodes for nutrition, both within and between genera. Species of *Arthrobotrys* tend to be competitive saprobes and consume nematodes as an alternative food source during periods of intense competition (Jansson and Nordbring-Hertz 1979; Cooke 1963). Their abundance may increase in response to a carbon source without producing traps (Jaffee 2002, 2004). Conversely, species of *Dactylellina* and *Dreschlerella* tend to be weak saprobes and are more dependent on nematodes for nutrition. These predominantly parasitic species of trapping fungi increase in response to nematode density, whereas more saprobic species do not (Jaffee et al. 1993; Persmark et al. 1996).

Although trapping fungi in the Orbiliaceae are frequently encountered in agricultural soils and in close association with plant roots, they have not proven to be reliable in suppressing populations of plant-parasitic nematodes (Stirling 1991). Many of the early attempts to control these nematodes with trapping fungi focused on species of *Arthrobotrys*, which are not principally parasites. Galper et al. (1995) found that in soil, *Dactylellina* spp. and *Dreschlerella* spp. were more effective in trapping the root-knot nematode, *Meloidogyne javanica*, than were *Arthrobotrys* spp. In other studies, *Dactylellina cionopaga* and *Dreschlerella dactyloides*, applied in alginate pellets, reduced root penetration and galling by root-knot nematodes (Stirling and Smith 1998; Jaffee and Muldoon 1997). However, Persson and Jansson (1999) were unable to demonstrate suppression of root-knot nematodes by *D. dactyloides* or *Dactylellina* spp. formulated in alginate pellets even though several of the fungi, particularly *D. ellipsospora*, abundantly colonized the tomato rhizosphere. Attempts to determine whether native trapping fungi are suppressing populations of plant-parasitic nematodes have also yielded inconsistent results. Amending soil with sunn hemp residue increased the abundance of parasitic trapping fungi, but improved suppression of the reniform nematode, *Rotylenchulus reniformis*, was not observed (Wang et al. 2003b). Yet, in soils cultivated with sugarcane, both the diversity and the abundance of trapping fungi were correlated to suppression of *Radopholus similis* (Stirling et al. 2011). The latter study utilized a bioassay to measure nematode suppression which reduced the variability associated with determining suppression based on resident nematode populations.

## *Fungi Producing Adhesive Conidia*

Fungi that attach to nematodes via adhesive conidia are often referred to as "endoparasites"; however, this term is misleading because the trapping fungi are also endoparasites of nematodes. Compared to the trapping fungi, there are only a few nematophagous fungi producing adhesive conidia. Two species in the Entomophthorales and Kickxellomycotina, *Meristacrum asterospermum* and *Zygnemomyces echinulatus*, and one species in the Agaricales, *Nematoctonus leissporus* (Pleurotaceae), produce adhesive conidia (Barron 1977; Saikawa et al. 1997); however, very little is known about their ecology or impact on nematode populations. Most of the research has focused on a handful of fungi in the Hypocreales that produce adhesive conidia, *Drechmeria coniospora, Drechmeria* (*Verticillium*) *balanoides,* and *Hirsutella* spp.; both genera are in the Ophiocordycipitaceae, which contains other parasites of invertebrates (Quandt et al. 2014). The new genus and species *Esteya vermicola*, described by Liou et al. (1999) from infected pinewood nematode (*Bursaphelenchus xylophilus*), also produces adhesive lunate conidia. This fungus is a member of the Ophiostomataceae, which are typically found growing saprophytically in living trees in association with bark beetles.

*Drechmeria coniospora* is an obligate pathogen of nematodes that is readily isolated from different soils (Glockling and Holbrook 2003). The fungus produces clusters of club-shaped conidia on a conidiophore. These conidia are not immediately infective and must undergo a maturation process to produce an adhesive knob on the narrow end of the conidium (Van den Boogert et al. 1992). Knob formation is not influenced by the presence of nematodes and occurs only after the conidia are released from the conidiophore and dispersed in the environment; aggregated conidia are inhibited from maturation. Conidia of *D. coniospora* primarily adhere to the head region of adults and juveniles, and additionally to the tails of male nematodes; however, in some nematode species, the conidia adhere over the entire cuticle (Jansson et al. 1985). The mechanism of adhesion involves proteins in the adhesive material binding to proteins excreted from nematode sensory organs and does not involve a lectin–carbohydrate interaction as previously hypothesized (Jansson 1993). Conidial attachment to a particular nematode species does not always lead to infection; the recognition signal for infection by *D. coniospora* appears to be more specific than the recognition signal for adhesion (Jansson et al. 1985, 1987). On a susceptible host, the conidium will form an appressorium that presses firmly against the nematode cuticle, possibly with the aid of an adhesive, before forming a penetration tube into the pseudocoel (Dijksterhuis et al. 1990). The nematode pseudocoel is then colonized by trophic hyphae with a characteristic wavy appearance. Nematode death occurs within 48 h of initial infection. Interestingly, the fungus does not fully colonize the nematode cadaver before the outgrowth of conidiophores and sporulation (Dijksterhuis et al. 1991). Approximately, 5000–10,000 conidia are produced per nematode. Because *D. coniospora* infects only a few species of plant-parasitic nematodes, it has not been frequently evaluated as a biological control organism.

*Drechmeria balanoides* has acorn-shaped conidia and is an obligate pathogen of nematodes. The adhesive material on the conidia is formed shortly after their release from the phialide (Sjollema et al. 1993). Infection is initiated by formation of an appressorium followed by a penetration tube, but unlike *D. coniospora*, an infection bulb forms after penetration. Conidiophores emerge from the cadaver about 60 h after initial infection. Atkinson and Dürschner-Pelz (1995) determined that cadavers of *Globodera rostochiensis* and *Ditylenchus dipsaci* yielded 12,000–16,000 conidia per cadaver (Atkinson and Dürschner-Pelz 1995). Little is known of the host range of *D. balanoides*. Durschnerpelz and Atkinson (1988) indicated that this fungus had a broader host range among plant-parasitic nematodes than did *D. coniospora*. However, few studies have evaluated the biological control potential of this fungus except against the stem nematode *D. dipsaci*. The fungus was found infecting *D. dipsaci* within the foliage of white clover; moreover, inoculation of white clover seed with *D. coniospora* reduced nematode numbers in the foliage compared to plants without fungal inoculation (Hay and Regnault 1995; Hay and Bateson 1997).

Three species of *Hirsutella* are obligate parasites of nematodes: *H. rhossiliensis* and *H. minnesotensis* have broad host ranges including numerous plant-parasitic nematodes, while *H. vermicola* is primarily a pathogen of bacterivorous nematodes and will not be discussed further (Xiang et al. 2007). *Hirsutella rhossiliensis* has been found worldwide parasitizing a variety of nematode species, while *H. minnesotensis* has been found predominantly parasitizing the soybean cyst nematode (*Heterodera glycines*) in China and the USA (Liu and Chen 2000; Xiang et al. 2007; Costa et al. 2012). The conidia of these two species are produced singly on phialides and are unable to attach to nematodes once they are dislodged from the phialides (McInnis and Jaffee 1989). Attachment of conidia and infection can occur anywhere on the nematode cuticle. Little is known of the mechanism of infection. Two serine proteases have been isolated from *H. rhossiliensis* and presumably are involved in the infection process (Wang et al. 2007, 2009). Following infection, the trophic hyphae colonize the body of the nematode killing it in a few days. When the cadaver is completely colonized by the fungus, hyphae containing sparsely spaced phialides and conidia radiate out from the cadaver. Conidia production by *H. rhossiliensis* is determined by the size of the nematode; 112 and 700 conidia/cadaver were produced from parasitized sugar beet cyst nematode (*Heterodera schachtii*) and ring nematode (*Mesocriconema xenoplax*), respectively (Jaffee and Zehr 1983; Jaffee et al. 1990). Although both *H. rhossiliensis* and *H. minnesotensis* can parasitize a broad range of nematodes, there is some host specialization among isolates of these fungi (Tedford et al. 1994; Liu and Chen 2001; Xiang et al. 2007). It is noteworthy that cyst nematodes in the genus *Heterodera* tend to be more susceptible to these two *Hirsutella* spp. than root-knot nematodes (*Meloidogyne* spp.), the reverse of that observed for trapping fungi (Xiang et al. 2007).

*Hirsutella rhossiliensis* and *H. minnesotensis* have been the subject of more studies to determine their potential for biological control of plant-parasitic nematodes than any other nematophagous fungi producing adhesive conidia. Environmental factors can have large effects on rates of nematode parasitism; soil texture, moisture, and temperature influence both fungal growth and nematode movement. Acquisition of *H. rhossiliensis* conidia by *H. schachtii* was greater in silty clay and loamy sand than in coarse sand (Jaffee et al. 1990). Similarly, parasitism of *H. glycines* by *H. minnesotensis* declined with increasing sand content (Xiang et al. 2010). The large pore sizes in the sand both limit the movement of *H. schachtii,* and thus contact with the adhesive conidia, and also allow the nematode to move through the pore without touching conidia. Rates of nematode parasitism by *Hirsutella* spp. are reduced at both high and low soil water contents. Nematodes require water films to move in order to contact conidia; however, when soil pores are filled with water, nematode movement is limited and the fungus does not produce conidia (Tedford et al. 1992; Xiang et al. 2010). Warmer soil temperatures should increase fungal growth and nematode activity leading to greater acquisition of conidia. However, Xiang et al. (2010) found that persistence of *H. minnesotensis* was greatest at cooler soil temperatures (5–10 °C). Rates of parasitism by both *H. rhossiliensis* and *H. minnesotensis* are greater in acidic soil than in neutral and basic soils (Jaffee and Zasoski 2001; Liu and Chen 2009). Jaffee and Zasoski (2001) provided convincing evidence that pH has an indirect effect on *H. rhossiliensis* by influencing antagonists of the fungus; pH had no influence on nematode parasitism in soil that had been heated to kill most organisms. Nevertheless, other research indicates that pH may also have a direct effect on these fungi (Jaffee and Zehr 1983; Liu and Chen 2009). Soil fauna and flora can substantially reduce the activity of *H. rhossiliensis*. Sporulation of the fungus from alginate pellets was inhibited by both micro- and macro-organisms (Jaffee 2000; 1999; Jaffee et al. 1997b). Enchytraeids, in particular, frequently contributed to the decline of pelletized *H. rhossiliensis* in soil (Jaffee et al. 1997a). The fungus appears to be more sensitive to biotic inhibition when it is formulated in alginate pellets than when it is growing from infected nematodes (Jaffee 2000).

*Hirsutella* spp. are frequently found in agricultural fields, sometimes parasitizing large numbers of cyst (*Heterodera* spp.) and ring (*Mesocriconema* spp.) nematodes (Muller 1985; Jaffee et al. 1988, 1989; Liu and Chen 2000). Moreover, soils containing high densities of *H. rhossiliensis* collected from the field suppressed penetration of cabbage roots by *H. schachtii* and egg production by *H. glycines* (Jaffee and Muldoon 1989; Chen 2007). Despite the promising results in naturally infested soil, applications of *H. rhossiliensis* as pelletized hyphae to microplot or field soil failed to suppress populations of plant-parasitic nematodes (Tedford et al. 1993; Jaffee et al. 1996) due, in part, to biotic inhibition of the formulated fungus.

Unlike the other fungi producing adhesive conidia, *Esteya vermicola* appears to be a facultative parasite of nematodes (Wang et al. 2011b). The fungus has been found on three continents (Asia, Europe, and South America) and isolated from wood, pinewood nematodes, and nematodes in pine forest soil (Wang et al. 2014). *Esteya vermicola* produces two types of conidia: one type is lunate with a central structure that resembles an endospore and a second type is cylindrical to baciloid; only the lunate conidia are adhesive to nematodes (Liou et al. 1999). The adhesive material is formed on the concave side of the conidia. Similar to *Hirsutella* spp., the conidia are unable to attach to nematodes if they are dislodged from the conidiophore (Wang et al. 2008). Following attachment to the nematode cuticle, *E. vermicola* forms an infection peg which then expands into a bulb. The fungus grows endoparastically before emerging from the cadaver to produce only lunate conidia. The cylindrical conidia are formed on nutrient-rich media and do not infect nematodes (Liou et al. 1999). Although the host range of *E. vermicola* has not been fully characterized, it was shown to be a virulent pathogen of both the pine wilt nematode and microbivorous nematodes (Wang et al. 2008). Of the eight plant-parasitic nematodes tested by Wang et al. (2014), the fungus infected *B. xylophilus*, *B. mucronatus*, *Aphelenchoides besseyi*, and *Ditylenchus destructor*, but not *Meloidogyne incognita*, *Heterodera avenae*, or *Pratylenchus penetrans*. However, the rate of infection by *E. vermicola* was greater for *B. xylophilus* than for any of the other nematode hosts suggesting some host specialization. The fungus shows promise as a biological control agent of the pinewood nematode. Injecting logs with a conidial suspension of *E. vermicola* reduced numbers of pine wilt nematode by 49–79 % after 2 months. Moreover, spraying pine seedlings with a suspension of the fungus 1 month prior to inoculating with pinewood nematode reduced wilting and increased tree survival (Wang et al. 2011a).

### *Attraction of Nematodes to Nematophagous Fungi*

In addition to traps and adhesive conidia, some nematophagous fungi also utilize another tactic to increase encounters with their hosts: they produce substances that attract nematodes to them. The attraction intensity is related to the dependency of the fungus on nematodes for nutrition (Jansson and Nordbring-Hertz 1979, 1980; Jansson 1982a). Facultative saprobes such as *Arthrobotrys* spp. are generally less attractive than obligate parasites such as *Drechmeria* spp.; *Dactylellina* spp. show an intermediate level of attractiveness. The mycelia of nematophagous fungi attract nematodes, but the presence of either traps or adhesive conidia increases the level of attraction (Jansson 1982a, b). In more recent studies, attraction of pinewood nematodes was greatest for *E. vermicola*, intermediate for *Dreschlerella* (= *Dactylaria*) *brochopaga*, and least for *Botrytis cinerea* (Wang et al. 2010a). Interestingly, *E. vermicola* is able to mimic the scent of pine trees by emitting volatile compounds (two monoterpenes and a terpenoid) which are among the compounds that attract pinewood nematodes to their host trees (Lin et al. 2013).

## Parasites of Sedentary Nematode Stages

Root-knot nematodes (*Meloidogyne* spp.) and cyst nematodes (*Heterodera* and *Globodera* spp.) are among the most economically important plant-parasitic nematodes worldwide. After penetrating the roots of a host plant and establishing a feeding site, these nematodes become sedentary. Both root-knot and cyst nematodes lay their eggs in a cluster surrounded by a gelatinous matrix; some eggs are also retained within the body of female cyst nematodes. These tight clusters of eggs are particularly vulnerable to fungal parasitism. The gelatinous matrix provides the eggs some protection from microorganisms (Orion et al. 2001); however, specialized pathogens are able to overcome this protection. There is also circumstantial evidence that some bacteria found within egg masses may also protect the eggs from nematophagous fungi (Kok et al. 2001).

Fungi that are specialized for parasitizing sedentary stages of nematodes (eggs, sedentary juveniles, and females) are from different lineages within the Hypocreales including *Pochonia* spp. (Clavicipitaceae), *Purpureocillium lilacinum* (syn. *Paecilomyces lilacinus*; Ophiocordycipitaceae), and *Trichoderma* spp. (Hypocreaceae). *Brachyphoris* (syn. *Dactylella*) *oviparasitica* (Orbiliaceae), which is related to trapping fungi but does not form traps, also parasitizes sedentary stages. All of these fungi are facultative parasites of sedentary nematode stages; their ability to grow saprobically depends on the isolate, with some isolates exhibiting greater levels of saprobic growth, often at the expense of parasitic activity (Siddiqui et al. 2009). The fungi listed above are commonly found in agricultural soils, often associated with root-knot and cyst nematodes (Stirling 2014).

There are several species of *Pochonia* known to parasitize nematodes (Zare et al. 2001); however, most of the research has focused on *P. chlamydosporia*. Members of the genus produce conidia on verticillate phialides as well as stalked, multicellular chlamydospores (dictyochlamydospores). The two subspecies of *P. chlamydosporia* are differentiated by whether conidia are produced in heads (var. *chlamydosporia*) or in chains (var. *catenulate*), and both subspecies are parasites of nematodes.

In the absence of nematodes, *P. chlamydosporia* is able to grow saprobically in the rhizosphere of plants. Presumably, the fungus obtains nutrition from root exudates and is often found in greater abundance in the rhizosphere of plants than in bulk soil (De Leij et al. 1993; Mauchline et al. 2002). Colonization of the rhizosphere by *P. chlamydosporia* varies among plant species (Manzanilla-Lopez et al. 2011; Bourne et al. 1996). For example, Bourne et al. (1996) observed greater saprobic colonization on brassicas (kale and cabbage) than on solanaceous plants (tomato and eggplant). The presence of nematodes within the root system also increases the abundance of *P. chlamydosporia*, particularly after egg deposition begins; this increase may be due to parasitic growth or additional nutrient leakage from galls (Bourne et al. 1996). The fungus has also been observed to grow endophytically within the roots of both monocots and dicots (Bordallo et al. 2002; Manzanilla-Lopez et al. 2011; Escudero and Lopez-Llorca 2012).

Upon contact with nematode eggs, *P. chlamydosporia* colonizes the surface of the egg forming numerous appressoria and penetration pegs (Escudero and Lopez-Llorca 2012; Segers et al. 1996). An alkaline serine protease designated VCP1 is involved in infection of nematode eggs by *P. chlamydospora* (Morton et al. 2003a). Polymorphisms within the gene encoding this enzyme are related to host preference for either cyst or root-knot nematodes among fungal isolates. Other extracellular enzymes (chitinases, lipases, and esterases) are also produced by *P. chlamydosporia*, but their role in pathogenicity is not known (Esteves et al. 2009). Although it is possible for isolates from root-knot nematode to infect cyst nematodes and vice versa, the infection efficiency is greater when fungal isolates are obtained from the same nematode genus (Siddiqui et al. 2009). In addition to differences in the VCP1 gene, isolates of *P. chlamydosporia* from root-knot nematodes can also be distinguished from those isolated from cyst nematodes with DNA fingerprinting (Morton et al. 2003b).

*Pochonia chlamydosporia* has been a primary contributor in soils suppressive to the cereal cyst nematode (*Heterodera avenae*) in England and Europe (Kerry et al. 1982; Kerry and Crump 1998), as well as soils suppressive to the southern root-knot nematode (*Meloidogyne incognita*) in California (Bent et al. 2008). The fungus has also been evaluated in numerous greenhouse and field studies for suppression of plant-parasitic nematodes (see reviews by Siddiqui and Mahmood 1996; Timper 2011). The effectiveness of *P. chlamydosporia* in suppressing nematode populations depends on several factors in addition to the fungal isolate. The fungus is more effective in suppressing nematode populations on moderate to poor host plants for the nematode than on good host plants because of lower reproductive rates of the nematode and, in the case of root-knot nematodes, smaller galls (Bourne and Kerry 1999). Bourne et al. (1996) observed greater parasitism of *M. incognita* eggs on potato, which produced smaller galls with more exposed egg masses, than on tomato, which produced large galls with more egg masses embedded in the gall tissue. Similarly, *P. chlamydosporia* is less effective in reducing nematode populations in soils with heavy nematode infestations (De Leij et al. 1992). The presence of the fungus in the rhizosphere early in the growing season is also critical. Suppression of the cyst nematode *H. schachtii* by *P. chlamydosporia* was correlated with the number of pre-gravid females infected rather than later infections of eggs (Kerry 1988). Atkins et al. (2003) took advantage of the ability of *P. chlamydosporia* to colonize roots in the absence of nematodes by applying the fungus to non-host crops of *M. incognita* grown in rotation with tomato. The non-host crops reduced populations of the nematode while supporting growth of the fungus leading to greater suppression of *M. incognita* by *P. chlamydosporia* in the tomato crop.

*Purpureocillium lilacinum* has been isolated worldwide from soil, decaying vegetation, insects, nematodes, and mammals (Luangsa-Ard et al. 2011) and is commonly associated with plant-parasitic nematodes in agricultural fields (Kilama et al. 2007; Gaspard et al. 1990; Stirling and West 1991; Bernard et al. 1996). Given the diverse substrates on which *P. lilacinum* is found, it is not surprising that there is considerable genetic variability among isolates (Tigano-Milani et al. 1995b). Isolates of the fungus also vary in their pathogenicity to nematode eggs; however, there is no relationship between genetic clustering and pathogenicity (Tigano-Milani et al. 1995a; Gunasekera et al. 2000). Unlike *P. chlamydosporia* or *B. oviparasitica*, there is no indication of host specificity among isolates of *P. lilacinum* for particular nematode genera. Encounters with nematode eggs and other sedentary stages appear to be random, as no directed growth of *P. lilacinum* toward eggs has been observed (Holland et al. 1999). When hyphae of the fungus grow over nematode eggs, the hyphae become appressed to the egg surface and appressoria form at the site of penetration. Although *P. lilacinum* is able to penetrate the nematode cuticle and infect all sedentary stages, appressoria were only observed forming on eggs (Holland et al. 1999; Khan et al. 2006). The fungus is believed to utilize both mechanical pressure and enzymes to penetrate the nematode cuticle and egg shell (Holland et al. 1999).

There is good evidence that some of the chitinases and proteases produced by *P. lilacinum* are involved in pathogenicity. A serine protease from the Samson strain of the fungus degraded the vitelline layer of *M. hapla* eggs and killed the embryos (Bonants et al. 1995). Overexpression of the serine protease gene increased parasitism of *M. incognita* eggs by 20 % (Wang et al. 2010b). Khan et al. (2004) observed similar effects of a serine protease from strain 251 on nematode eggs, but also observed greater structural damage to the eggs when both the protease and chitinases from the fungus were combined than when the enzymes were applied separately. Moreover, among isolates of *P. lilacinum* and *P. chlamydosporia*, efficacy against *M. incognita* was strongly correlated with in vitro protease and chitinase production (Wei et al. 2009). In some cases, *P. lilacinum* may kill nematodes with toxins prior to infecting them. Both acetic acid and leucinostatins have been implicated as the primary toxic metabolite in culture filtrates of *P. lilacinum* (Djian et al. 1991; Park et al. 2004); however, it is not clear whether these toxins are involved in nematode mortality in the rhizosphere. Isolates of *P. lilacinum* varied in leucinostatin production, but those isolates producing the toxin were only weakly pathogenic to nematodes eggs (Park et al. 2004). Isolates may vary in their mode of action with some being primarily pathogenic, while others are primarily toxic to nematodes. Similarly, different modes of action may be required for consuming different hosts. For example, strain 251 readily infected eggs of *Meloidogyne* spp. but not eggs of *R. similis*. Nevertheless, the eggs of *R. similis* appeared abnormal and other stages of the nematode were immobilized when in contact with *P. lilacinum* hyphae; these stages were eventually colonized by the fungus (Khan et al. 2006).

There have been conflicting reports about whether *P. lilacinum* colonizes the rhizosphere or endosphere of plant roots. Some of these discrepancies may be due to differences among isolates of the fungus; however, inconsistencies in root colonization have also been observed for a single isolate. In two studies, rhizosphere colonization of several host plants was not observed for *P. lilacinum* strain 251 and persistence of the strain in soil was not increased in the presence of different crop plants compared to fallow soil (Rumbos and Kiewnick 2006; Kiewnick and Gullino 2009). In a third study, however, abundance of strain 251 was greater in the rhizosphere of sugar beet and rape than in non-rhizosphere soil (Manzanilla-Lopez et al. 2011). Rhizosphere colonization by strain 251 may depend on crop species or cultivar or environmental factors such as soil type. For example, Manzanilla-Lopez et al (2011) reported that strain 251 preferentially colonized the rhizosphere of rape and sugar beet, but not the rhizosphere of potato or wheat. Cabanillas et al. (1988) observed endophytic colonization of excised tomato roots by a Peruvian strain of *P. lilacinum*. In a more extensive study with 10 crop species, *P. lilacinum* strain 251 was found endophytically at low levels in some, but not all crop species, and at high levels in barley (Rumbos and Kiewnick 2006). These results were in contrast to

Holland et al. (2003) who did not observe endophytic growth of strain 251 in eight crop species, including tomato and barley.

Although *P. lilacinum* is commonly found parasitizing nematode eggs in agricultural soils, it has never been associated with soils that are suppressive to nematodes. Abundance of *P. lilacinum* strain 251 in soil was not influenced by the presence of host nematodes, suggesting that the fungus is not strongly dependent on nematodes as a food source (Rumbos et al. 2008). The importance of soil organic matter as an alternative food source of the fungus is unclear. Increasing the organic matter in soil enhanced persistence of the *P. lilacinum*, while increasing the sand content diminished persistence (Rumbos et al. 2008). However, persistence may have also been influenced by leaching of conidia, which would have been reduced by the organic matter and increased by the sand content. Application of organic matter in the form of cow manure had no influence on either the percentage suppression of nematodes by *P. lilacinum* or re-isolation of the fungus from *M. incognita* females and egg masses (Siddiqui and Futai 2009).

*Purpureocillium lilacinum* is commercially available in several countries for control of plant-parasitic nematodes and numerous studies have evaluated its efficacy under greenhouse and field conditions (see Stirling 1991; Siddiqui and Mahmood 1996; Stirling 2014 for reviews). In general, the fungus shows greater efficacy in suppressing nematode populations when it is applied 1–2 weeks before planting to allow longer exposure of nematode eggs to infection (Anastasiadis et al. 2008; Mendoza et al. 2007). As mentioned earlier, *P. lilacinum* has low persistence in soil; therefore, multiple applications early in the season have shown greater reduction in nematode numbers than single applications (Cabanillas and Barker 1989; Mendoza et al. 2007; Anastasiadis et al. 2008; Udo et al. 2013).

*Brachyphoris oviparasitica* and the sterile fungus designated Arkansas Fungus (ARF) are closely related based on phylogenic analysis of rRNA genes (Yang et al. 2012) and will be discussed together as related *Brachyphoris* species. Both *B. oviparasitica* and ARF parasitize eggs of cyst and root-knot nematodes (Kim and Riggs 1991; Stirling et al. 1979). Nematodes that are embedded in the root are protected from infection; however, juveniles and females of cyst nematodes break through the epidermis of the root as they develop and are readily parasitized by these fungi (Kim and Riggs 1991; Timper et al. 1999; Becker et al. 2013; Stirling et al. 1979). Additionally, ARF parasitizes eggs and sedentary females of the reniform nematode, *R. reniformis* (Wang et al. 2004b). Though not well studied, there appears to be some host specificity among isolates of ARF. For example, isolates from the soybean cyst nematode (*H. glycines*) infected the eggs of several other species of *Heterodera* and *M. incognita*, but did not infect eggs of the reniform nematode or the tobacco cyst nematode (*Globodera tabacum*) (Kim and Riggs 1991; Wang et al. 2004b). The mechanism of infection by *B. oviparasitica* and ARF is unknown. When infecting juveniles and females of cyst nematodes, both fungi produce dense mats of mycelium on the nematode cuticle (Timper et al. 1999; Becker et al. 2013). Penetration holes are formed under these mats in a manner suggestive of enzymatic activity (Kim et al. 1992).

The fungus ARF is able to colonize and decompose organic matter in bulk soil (Wang et al. 2004a). After application of mycelium to soil without nematodes, ARF formed abundant mycelial mats primarily in the bulk soil with only a few mats attached to roots (Timper et al. 1999). Similarly, *B. oviparasitica* did not show tropism towards roots (Becker et al. 2013), suggesting that, unlike *P. chlamydosporia*, these fungi do not preferentially colonize the rhizosphere. Stirling et al. (1979), however, observed greater abundance of *B. oviparasitica* in rhizosphere soil from peach than in non-rhizosphere soil, possibly due to the presence of parasitized egg masses on the root. More research is needed to determine whether *B. oviparasitica* and ARF colonize roots in the absence of nematodes.

*Brachyphoris oviparasitica* was first identified in peach soils that were suppressive to root-knot nematodes in California (Stirling and Mankau 1978). The fungus was later shown to be the primary organism responsible for the natural suppression of root-knot nematodes in peach and the suppression of the sugar beet cyst nematode (*H. schachtii*) in an experimental field site in California (Stirling et al. 1979; Westphal and Becker 2001; Yin et al. 2003). The fungus ARF has been associated with natural suppression of the soybean cyst nematode (*H. glycines*) in Arkansas (Kim and Riggs 1991). Additionally, ARF has been isolated from *H. glycines* from several locations in the Mid-South region of the USA. (Kim et al. 1998). Although *B. oviparasitica* and ARF have shown promise as biological control agents of cyst and reniform nematodes in greenhouse and microplot experiments, no studies have evaluated these fungi under field conditions (Olatinwo et al. 2006a, b, c; Wang et al. 2004b; Timper and Riggs 1998). As with *P. chlamydosporia*, the host plant contributes to the level of nematode suppression by *B. oviparasitica*. A greater percentage of *M. incognita* eggs were parasitized by *B. oviparasitica* on peach than on tomato (96 % vs. 57 %), presumably because the fungus was more efficient at colonizing the smaller egg masses on peach than the larger egg masses on tomato (Stirling et al. 1979).

The genus *Trichoderma* is well known for containing widely distributed free-living soil fungi, but species and strains in this genus are also associated with other habitats, as plant symbionts, and as fungal parasites. The *Trichoderma* anamorph has been linked with the teleomorph *Hypocrea*. The taxonomy of these fungi is very difficult due to limited and variable morphological characters and is becoming increasingly reliant on molecular analyses for species identification (Samuels 2006).

*Trichoderma* species are widely adapted to interact with plants and plant pathogens over a wide range of environments to increase plant growth and reduce plant disease and have many attributes which allow them to be successful in different environments. These fungi can grow saprobically, endophytically, or in the rhizosphere and phyllosphere transition zones; they produce large numbers of propagules and antibiotic compounds that result in competitive advantages against many other microorganisms, and they can be successful mycoparasites (Howell et al. 2000; Howell 2006; Woo et al. 2006). As a result, *Trichoderma* strains have been developed as successful biological controls of fungal plant pathogens (Harman 2006).

*Trichoderma* spp. antagonize nematode by multiple mechanisms. Many of the attributes which allow *Trichoderma* to be successful biological control agents of fungi may also result in control of plant-parasitic nematodes (Sharon et al. 2011). Certain *Trichoderma* spp. produce proteases and chitinases which may play a role in suppression of nematode populations (Sharon et al. 2011). *Trichoderma* strains that can grow in the gelatinous matrix of egg masses previously thought to protect eggs from microbes may directly parasitize root-knot nematode eggs (Sharon et al. 2001). Howell (2002) reported that *Trichoderma* can metabolically break down root exudates as they diffuse into the rhizosphere or spermosphere before they can stimulate the germination of pathogen propagules such as oospores, effectively masking plant roots and preventing infection. This same process may be utilized to reduce egg hatch and reduce or delay root infection by nematodes which hatch in response to host root exudates. In addition, *T. atroviride* produces nematicidal compounds that reduce hatch of immature eggs, but increase hatch of mature eggs (Sharon et al. 2007).

## Antagonistic Plant Symbionts

Several fungi which form symbiotic associations with plants resulting in improved plant growth also suppress populations of plant-parasitic nematodes; they are often endophytic within plant roots. Unlike the fungi discussed previously, these fungi are not specialized for parasitizing nematodes, but instead antagonize nematodes by producing toxins, altering root exudates, or inducing a resistance response in plants.

### *Toxins and Extracellular Enzymes*

While numerous fungi produce metabolites in liquid media that are toxic to nematodes, it is not clear if these metabolites are produced in sufficient quantities in roots to affect nematode viability or behavior. Therefore, only studies demonstrating activity of metabolites against nematodes at concentrations found in root tissue or root exudates will be discussed below.

*Neotyphodium* spp. are asexual forms of *Epichloë* (Clavicipitaceae) which are mutualistically associated with grasses in the family Poaceae. *Neotyphodium* spp. are exclusively endophytic and seed transmitted. After the seed germinates, the mycelium grows intercellularly in the stem and leaf sheath, but does not grow into the leaf blade or roots. The most well studied of these grass/endophyte associations has been tall fescue (*Schedonorus arundinaceus*) and its endophyte *N. coenophialum*, and perennial ryegrass (*Lolium perenne*) and its endophyte *N. lolii*. These endophytes confer resistance to many abiotic and biotic stresses, including herbivory from mammals, insects, and nematodes. *Neotyphodium* spp. produce four groups of alkaloids which are involved in either toxicity or feeding deterrence to vertebrates and invertebrates that feed on grasses containing them (Schardl et al. 2004). The endophyte in tall fescue is able to suppress populations of plant-parasitic nematode, notably *Pratylenchus scribneri* and *Meloidogyne marylandi*, even though the fungus is not present in the roots (West et al. 1988; Kimmons et al. 1990; Timper et al. 2005). Although the mechanism by which the fungus confers nematode resistance to the plant is unknown, it is believed that one or more of the biologically active alkaloids are involved in the suppression. Root extracts of endophyte-infected tall fescue paralyzed *P. scribneri* and reduced egg hatch and viability of *M. incognita* J2 (Bacetty et al. 2009; Meyer et al. 2013). Root exudates of endophytic tall fescue also reduced egg hatch and J2 viability, suggesting that compounds both within the root and exuded from the root are involved in nematode suppression. Except for peramine, most of the other alkaloids (ergovaline, lolitrem B, and the lolines) are found in low concentrations in roots (Siegel and Bush 1996). Knockout mutants of *Neotyphodium* sp. lacking ergot alkaloids, including ergovaline, still conferred resistance to *P. scribneri* in annual ryegrass, indicating that the ergot alkaloids are not essential for nematode suppression (Panaccione et al. 2006). Lolitrem B is not produced in the tall fescue/*N. coenophialum* association and is, therefore, not a likely source of resistance to nematodes. Of the bioactive alkaloids produced by *Neotyphodium* spp., the lolines have the greatest potential to affect nematodes. These alkaloids are found in both root tissue and exudates of endophyte infected tall fescue at concentrations that are toxic to *P. scribneri* (Bush et al. 1993; Bacetty et al. 2009).

## Alteration of Root Exudates

*Fusarium* spp. are cosmopolitan, ecologically diverse fungi in the family Nectriaceae (Hypocreales). Although *Fusarium* spp. are important plant pathogens, many species and strains are nonpathogenic to plants. *Fusarium* spp. are commonly found growing endophytically within the roots of a number of plant species with *F. oxysporum* being the most common species (Macia-Vicente et al. 2008). Hallmann and Sikora (2011) have reviewed research evaluating the efficacy and mode of action of *Fusarium* endophytes for suppression of plant-parasitic nematodes. One of the mechanisms of nematode suppression appears to be altered root exudates. Fewer *M. incognita* and *Radopholus similis* were attracted to and infected roots containing *F. oxysporum* strain 162 than control plants (Dababat and Sikora 2007b; Hallmann and Sikora 2011). Moreover, in the absence of plants, root exudates from tomato containing *F. oxysporum* attracted fewer *M. incognita* than exudates from control plants. It is not clear whether the exudates contain a toxin which repels the nematodes or contains components that are less attractive to the nematodes. The effect of *F. oxysporum* in reducing attraction and penetration of roots may be dependent on the strain of the fungus or the plant species. In banana, there was no difference in attraction or penetration by *R. similis* between roots without endophytes and roots containing one of three different strains of *F. oxysporum* (Athman et al. 2006, 2007).

Arbuscular mycorrhizal fungi (AMF) form obligate mutualistic relationships with plants. These fungi live within the cortical tissue of roots and enhance plant growth through nutrient uptake and suppression of pathogens. There are several genera of AMF, all in the order Glomerales (Morton and Benny 1990), including *Gigaspora*, *Scutellospora* (Gigasporaceae), *Glomus*, *Sclerocystis* (Glomaceae), *Acaulospora*, and *Entrophospora* (Acaulosporaceae). Numerous studies have investigated the interaction between AMF and plant-parasitic nematodes. Recent reviews of the literature indicate that in most studies, plants colonized by AMF have lower nematode populations than plants without AMF (Hol and Cook 2005; Hallmann and Sikora 2011; Veresoglou and Rillig 2012). Whether or not AMF confer resistance to plant-parasitic nematodes depends on the genus of AMF, AMF-plant specificity, genus of nematode, and the timing of AMF inoculation. The mechanism by which AMF influences nematode populations is also likely dependent on the genus of AMF and plant. In both tomato and banana, AMF colonization of roots appears to alter the root exudates leading to fewer nematodes penetrating AMF compared to non-AMF roots (Vos et al. 2012a, b). On agar plates, root exudates from AMF plants repelled *R. similis*, whereas exudates from non-AMF plants were either neutral or attracted the nematode.

*Piriformospora indica* (Basidiomycota: Sebacinaceae) is a plant growth-promoting fungus that endophytically colonizes roots of a wide range of plants. This fungus has been shown to increase plant tolerance to stress, induce disease resistance, and increase plant yields (Waller et al. 2005). In two greenhouse studies, *P. indica* reduced populations of *H. schachtii* in *Arabidopsis* and *H. glycines* in soybean (Daneshkhah et al. 2013; Bajaj et al. 2015). One mechanism of suppression appears to be altered root exudates; exudates of colonized soybean roots were less attractive to infective juveniles of *H. glycines* than exudates from control roots (Daneshkhah et al. 2013).

## Induced Resistance

In the last 20 years, it has become increasingly clear that many nonpathogenic microorganisms living in the rhizosphere or endosphere of plant roots trigger a heightened immune response in the plant (Pieterse et al. 2014). This phenomenon is referred to as induced systemic resistance (ISR). In ISR, the microbes prime the plant defense system leading to a faster or stronger resistance response to pathogen infection. Using a split-root system to physically separate the inducing organism from plant-parasitic nematodes, several studies have demonstrated that AMF-colonized plants elicit a systemic resistance response in grapevine, tomato, and banana against nematodes (Elsen et al. 2008; Hao et al. 2012; Vos et al. 2012c). In both tomato and grapevine, expression of defense-related genes was greater in AMF plants challenged with nematodes than in AMF alone or nematode alone plants, suggesting that AMF prime the plant defenses for subsequent nematode infection (Li et al. 2006; Hao et al. 2012; Vos et al. 2013).

Other endophytic antagonists of nematodes are also capable of inducing systemic resistance in plants. Numerous studies with *Trichoderma* spp. have demonstrated

ISR responses against a wide range of plant pathogens (Harman 2006; Woo et al. 2006). The ability to colonize plant tissue as an endophyte may be key to induction of plant defense responses (Howell 2006). *Trichoderma harzianum* (strain T10) was shown to induce systemic resistance to the root-knot nematode *M. javanica* in a split-root tomato system, alone or in combination with salicylic acid or jasmonic acid (Selim et al. 2014). Similarly, endophytic, nonpathogenic strains of *F. oxysporum* also induce resistance to plant-parasitic nematodes in banana and in tomato (Vu et al. 2006; Dababat and Sikora 2007a; Martinuz et al. 2012). Paparu et al. (2013) further showed that the *F. oxysporum* primed banana for greater expression of defense-related genes when infected by *R. similis*. Although not demonstrated with plant-parasitic nematodes, Pedrotti et al. (2013) used a split-root hydroponic system to show that root infection by *P. indica* triggers a priming of defense responses in *Arabidopsis*. It is likely that plants colonized by *P. indica* exhibit ISR responses against nematodes given the pervasive association between endophytes and ISR responses.

Interestingly, trapping fungi and parasites of sedentary stages of nematodes, particularly endophytic strains of these fungi, are also capable of inducing systemic resistance to nematodes. Colonization of roots by both endophytic and rhizospheric strains of *A. oligospora* reduced nematode numbers and increased defense-related enzymes in tomato compared to plants inoculated only with *M. incognita* or without nematode and fungus (Singh et al., 2013). The endophytic strain was superior to the rhizospheric strain both in terms of suppressing nematode numbers and enhancing the immune response of the plant. An endophytic strain of *P. chlamydosporia* caused a moderate induction of genes involved in ISR in barley (*Hordeum vulgare*); however, the plants were not challenged with plant-parasitic nematodes to conclusively demonstrate priming for resistance to nematodes (Larriba et al. 2015).

## References

Abawi GS, Barker KR (1984) Effects of cultivar, soil temperature and population levels of *Meloidogyne incognita* on root necrosis and *Fusarium* wilt of tomatoes. Phytopathology 74:433–438

Abd-El-Alim FF, Barker KR, Ibrahim IK, Darwish AK, Michail SH (1999) Interactions of *Fusarium oxysporum* f.sp. *vasinfectum* and *Meloidogyne incognita* on selected cotton genotypes. Pak J Nematol 17:51–60

Abdel-Momen SM, Starr JL (1998) *Meloidogyne javanica–Rhizoctonia solani* disease complex of peanut. Fundam Appl Nematol 21:611–616

Ahman J, Johansson T, Olsson M, Punt PJ, van den Hondel C, Tunlid A (2002) Improving the pathogenicity of a nematode-trapping fungus by genetic engineering of a subtilisin with nematotoxic activity. Appl Environ Microbiol 68:3408–3415

Ahren D, Ursing BM, Tunlid A (1998) Phylogeny of nematode-trapping fungi based on 18S rDNA sequences. FEMS Microbiol Lett 158:179–184

Ali S, Magne M, Chen S, Côté O, Gerič Stare B, Obradovic N, Jamshaid L, Wang X, Bélair G, Moffett P (2015) Analysis of putative apoplastic effectors from the nematode, *Globodera rostochiensis*, and identification of an expansin-like protein that can induce and suppress host defenses. PLoS One 10(1):e011504. doi:10.1371/journal.pone.0115042

Anastasiadis IA, Giannakou IO, Prophetou-Athanasiadou DA, Gowen SR (2008) The combined effect of the application of a biocontrol agent *Paecilomyces lilacinus*, with various practices for the control of root-knot nematodes. Crop Prot 27:352–361
Anderson, RV (1964) Feeding of *Ditylenchus destructor*. Phytopathol. 54:1121–1126.,
Athman SY, Dubois T, Coyne D, Gold CS, Labuschagne N, Viljoen A (2006) Effect of endophytic *Fusarium oxysporum* on host preference of *Radopholus similis* to tissue culture banana plants. J Nematol 38:455–460
Athman SY, Dubois T, Coyne D, Gold CS, Labuschagne N, Viljoen A (2007) Effect of endophytic *Fusarium oxysporum* on root penetration and reproduction of *Radopholus similis* in tissue culture-derived banana (*Musa* spp.) plants. Nematology 9:599–607
Atkins SD, Hidalgo-Diaz L, Kalisz H, Mauchline TH, Hirsch PR, Kerry BR (2003) Development of a new management strategy for the control of root-knot nematodes (*Meloidogyne* spp.) in organic vegetable production. Pest Manag Sci 59:183–189
Atkinson GF (1892) Some diseases of cotton. Alabama Polytechnical Institute of Agriculture—Exp Station Bull 41:61–65
Atkinson HJ, Dürschner-Pelz U (1995) Spore transmission and epidemiology of *Verticillium balanoides*, an endozoic fungal parasite of nematodes in soil. J Invertebr Pathol 65:237–242
Bacetty AA, Snook ME, Glenn AE, Noe JP, Hill N, Culbreath A, Timper P, Nagabhyru P, Bacon CW (2009) Toxicity of endophyte-infected tall fescue alkaloids and grass metabolites on *Pratylenchus scribneri*. Phytopathology 99:1336–1345
Bae Y-S, Knudsen GR (2001) Influence of a fungus-feeding nematode on growth and biocontrol efficacy of *Trichoderma harzianum*. Phytopathology 91:301–306
Bajaj R, Hu W, Huang Y, Chen S, Prasad R, Varma A, Bushley KE (2015) The beneficial root endophyte *Piriformospora indica* reduces egg density of the soybean cyst nematode. Biol Control 90:193–199
Bakhtiar Y, Miller D, Cavagnaro T, Smith S (2001) Interactions between two arbuscular mycorrhizal fungi and fungivorous nematodes and control of the nematode with fenamifos. Appl Soil Ecol 17:107–117
Balogh J, Tunlid A, Rosen S (2003) Deletion of a lectin gene does not affect the phenotype of the nematode-trapping fungus *Arthrobotrys oligospora*. Fungal Genet Biol 39:128–135
Barker KR (1964) On the disease reduction and reproduction of the nematode *Aphelenchus avenae* on isolates of *Rhizoctonia solani*. Plant Dis Rep 48:428–432
Barker KR, Darling HM (1965) Reproduction of *Aphelenchus avenae* on plant tissue in culture. Nematologica 11:162–166
Barnes CL, Russel C, Foster CD, McNew RW (1981) *Aphelenchus avenae*, a potential biological control agent for root rot fungi. Plant Dis 65:423–424
Barron GL (1977) The nematode-destroying fungi. Topics in mycobiology, No. 1. Canadian Biological Publications Ltd., Ontario
Barron GL, Thorn RG (1987) Destruction of nematodes by species of *Pleurotus*. Can J Bot 65:774–778
Baynes MA, Russell DM, Newcombe G, Carta LK, Rossman AY, Ismaiel A (2012) A mutualistic interaction between a fungivorous nematode and a fungus within the endophytic community of *Bromus tectorum*. Fungal Ecol 5:610–623
Becker JS, Borneman J, Becker JO (2013) *Dactylella oviparasitica* parasitism of the sugar beet cyst nematode observed in trixenic culture plates. Biol Control 64:51–56
Bent E, Loffredo A, McKenry MV, Becker JO, Borneman J (2008) Detection and investigation of soil biological activity against *Meloidogyne incognita*. J Nematol 40:109–118
Bernard EC, Self LH, Tyler DD (1996) Fungal parasitism of soybean cyst nematode, *Heterodera glycines* (Nemata: Heteroderidae), in differing cropping-tillage regimes. Appl Soil Ecol 5:57–70
Bertrand B, Nunez C, Sarah JL (2000) Disease complex in coffee involving *Meloidogyne arabicida* and *Fusarium oxysporum*. Plant Pathol 49:383–388
Bhattarai S, Haydock PPJ, Back MA, Hare MC, Lankford WT (2010) Interactions between field populations of the potato cyst nematode *Globodera pallida* and *Rhizoctonia solani* diseases of potatoes under controlled environment and glasshouse conditions. Nematology 12:783–790

Bonants PJM, Fitters PFL, Thijs H, Denbelder E, Waalwijk C, Henfling J (1995) A basic serine-protease from *Paecilomyces lilacinus* with biological activity against *Meloidogyne hapla* eggs. Microbiology 141:775–784

Bordallo JJ, Lopez-Llorca LV, Jansson HB, Salinas J, Persmark L, Asensio L (2002) Colonization of plant roots by egg-parasitic and nematode-trapping fungi. New Phytol 154:491–499

Bourne JM, Kerry BR (1999) Effect of the host plant on the efficacy of *Verticillium chlamydosporium* as a biological control agent of root-knot nematodes at different nematode densities and fungal application rates. Soil Biol Biochem 31:75–84

Bourne JM, Kerry BR, deLeij F (1996) The importance of the host plant on the interaction between root-knot nematodes (*Meloidogyne* spp.) and the nematophagous fungus, *Verticillium chlamydosporium* Goddard. Biocontrol Sci Technol 6:539–548

Bush LP, Fannin FF, Siegel MR, Dahlman DL, Burton HR (1993) Chemistry, occurrence and biological effects of saturated pyrrolizidine alkaloids associated with endophyte-grass interactions. Agr Ecosyst Environ 44:81–102

Cabanillas E, Barker KR, Daykin ME (1988) Histology of the interactions of *Paecilomyces lilacinus* with *Meloidogyne incognita* on tomato. J Nematol 20:362–365

Chen S (2007) Suppression of *Heterodera glycines* in soils from fields with long-term soybean monoculture. Biocontrol Sci Technol 17:125–134

Chin DA, Estey RH (1966) Studies on the pathogenicity of *Aphelenchus avenae* Bastian, 1865. Phytoprotection 47:66–72

Christie JR, Arndt CH (1936) Feeding habits of the nematodes *Aphelenchoides parietinus* and *Aphelenchus avenae*. Phytopathology 26:698–701

Ciancio A (1995) Observations on the nematicidal properties of some mycotoxins. Fundam Appl Nematol 18:451–454

Combettes S (1983) Utilization des relations nematodes mycophages—*Trichoderma* dans la lutte contre les champignons phytopathogenes. In: Le antagonisms microbiens, Calloque dl'INRA, Paris, 8:129–210

Cooke RC (1962a) Behaviour of nematode-trapping fungi during decomposition of organic matter in the soil. Trans Br Mycol Soc 45:314–320

Cooke RC (1962b) The ecology of nematode-trapping fungi in the soil. Ann Appl Biol 50:507–513

Cooke RC (1963) Ecological characteristics of nematode-trapping Hyphomycetes: preliminary studies. Ann Appl Biol 52:431–437

Costa SR, Kerry BR, Bardgett RD, Davies KG (2012) Interactions between nematodes and their microbial enemies in coastal sand dunes. Oecologia 170:1053–1066

Dababat A, Sikora RA (2007) Induced resistance by the mutualistic endophyte, *Fusarium oxysporum* strain 162, toward *Meloidogyne incognita* on tomato. Biocontrol Sci Technol 17:969–975

Dabatat AEA, Sikora RA (2007) Influence of the mutualistic endophyte *Fusarium oxysporum* on *Meloidogyne incognita* attraction and invasion. Nematology 9:771–776

Daneshkhah R, Cabello S, Rozanska E, Sobczak M, Grundler FMW, Wieczorek K, Hofmann J (2013) *Piriformospora indica* antagonizes cyst nematode infection and development in *Arabidopsis* roots. J Exp Bot 64:3763–3774

Davies K, Spiegel S (2011) Biological control of plant-parasitic nematodes: building a coherence between microbial ecology and molecular mechanisms. Springer, London

De Leij FAAM, Kerry BR, Dennehy JA (1992) The effect of fungal application rate and nematode density on the effectiveness of *Verticillium chlamydosporium* as a biological control agent for *Meloidogyne incognita*. Nematologica 38:112–122

De Leij FAAM, Kerry BR, Dennehy JA (1993) *Verticillium chlamydosporium* as a biological control agent for *Meloidogyne incognita* and *M. hapla* in pot and micro-plot tests. Nematologica 39:115–126

De Vay JE, Gutierrez AP, Pullman GS, Wakeman RJ, Garber RH, Jeffers DP, Smith SN, Goodell PB, Roberts PA (1997) Inoculum densities of *Fusarium oxysporum* f. sp. *vasinfectum* and *Meloidogyne incognita* in relation to the development of *Fusarium* wilt and the phenology of cotton plants (*Gossypium hirsutum*). Phytopathology 87:341–346

De RK, Ali SS, Dwivedi RP (2001) Effect of interaction between *Fusarium oxysporum* f. sp. *lentis* and *Meloidogyne javanica* on lentil. Indian J Pulses Res 14:71–73

Den Belder E, Jansen E (1994) Capture of plant-parasitic nematodes by an adhesive hyphae forming isolate of *Arthrobotrys oligospora* and some other nematode-trapping fungi. Nematologica 40:423–437

Dijksterhuis J, Veenhuis M, Harder W (1990) Ultrastructural study of adhesion and initial stages of infection of nematodes by conidia of *Drechmeria coniospora*. Mycol Res 94:1–8

Dijksterhuis J, Harder W, Wyss U, Veenhuis M (1991) Colonization and digestion of nematodes by the endoparasitic nematophagous fungus *Drechmeria coniospora*. Mycol Res 95:873–878

Dijksterhuis J, Veenhuis M, Harder W, Nordbringhertz B (1994) Nematophagous fungi—physiological aspects and structure-function relationships. Adv Microb Physiol 36:111–143

Djian C, Pijarowski L, Ponchet M, Arpin N, Favre-Bonvin J (1991) Acetic acid: a selective nematicidal metabolite from culture filtrates of *Paecilomyces lilacinus* (Thom) Samson and *Trichoderma longibrachiatum* Rifai. Nematologica 37:101–112

Dong LQ, Zhang KQ (2006) Microbial control of plant-parasitic nematodes: a five-party interaction. Plant Soil 288:31–45

Drechsler C (1937) Some hyphomycetes that prey on free-living terricolous nematodes. Mycologia 29:447–552

Duddington CL (1951) The ecology of predacious fungi. Trans Br Mycol Soc 34:322–331

Durschnerpelz VV, Atkinson HJ (1988) Recognition of *Ditylenchus* and other nematodes by spores of the endoparasitic fungus *Verticillium balanoides*. J Invertebr Pathol 51:97–106

Elliston J (1985) Characteristics of dsRNA-free and dsRNA-containing v strains of *Endothia parasitica* in relation to hypovirulence. Phytopathology 74:151–158

Elsen A, Gervacio D, Swennen R, De Waele D (2008) AMF-induced biocontrol against plant parasitic nematodes in *Musa* sp.: a systemic effect. Mycorrhiza 18:251–256

Escudero N, Lopez-Llorca LV (2012) Effects on plant growth and root-knot nematode infection of an endophytic GFP transformant of the nematophagous fungus *Pochonia chlamydosporia*. Symbiosis 57:33–42

Esteves I, Peteira B, Atkins SD, Magan N, Kerry B (2009) Production of extracellular enzymes by different isolates of *Pochonia chlamydosporia*. Mycol Res 113:867–876

Evans K (1987) The interactions of potato cyst nematodes and *Verticillium dahliae* on early and main crop potato cultivars. Ann Appl Biol 110:329–39

Fagbenle HH, Inskeep PM (1987) Scanning electron microscopy of concomitant infections of *Meloidogyne hapla* and *Rhizoctonia solani* on peanut. Phytopathology 77:1714

Faulkner LR, Darling HM (1961) Pathological histology, hosts and culture of potato rot nematode. Phytopathology 51:778–786

Faulkner LR, Bolander WJ, Skotland CB (1970) Interaction of *Verticillium dahliae* and *Pratylenchus minyus* in Verticillium wilt of peppermint: influence of the nematode as determined by a double root technique. Phytopathology 60:100–103

Fawcett HS (1931) The importance of investigations on the effects of known mixtures of organisms. Phytopathology 21:545–550

Fichtner SM, Isakeit T, Wheeler TA, Kaufman HW, Gannaway JR (2005) Evaluation of several approaches to manage *Meloidogyne incognita* and cotton seedling disease complexes in the high plains of Texas. J Nematol 37:66–70

France RA, Abawi GS (1994) Interaction between *Meloidogyne incognita* and *Fusarium oxysporum* f. sp. *phaseoli* on selected bean genotypes. J Nematol 26:467–474

Freckman DW, Caswell EP (1985) The ecology of nematodes in ecosystems. Annu Rev Phytopathol 23:275–296

Galper S, Eden LM, Stirling GR, Smith LJ (1995) Simple screening methods for assessing the predacious activity of nematode-trapping fungi. Nematologica 41:130–140

Gaspard JT, Mankau R (1986) Nematophagous fungi associated with *Tylenchulus semipenetrans* and the citrus rhizosphere. Nematologica 32:359–363

Gaspard JT, Mankau R (1987) Density-dependence and host-specificity of the nematode-trapping fungus *Manacrosporium ellipsosporum*. Rev Nematol 10:242–246

Gaspard JT, Jaffee BA, Ferris H (1990) Association of *Verticillium chlamydosporium* and *Paecilomyces lilacinus* with root-knot nematode infested soil. J Nematol 22:207–213

Giannakis N, Sanders FE (1989) Interactions between mycophagous nematodes, mycorrhizal and other soil fungi. Agr Ecosyst Environ 29:163–167

Glockling SL, Holbrook GP (2003) Endoparasites of soil nematodes and rotifers 1: the common and the rare. Mycologist 17:150–154

Golden JK, Van Gundy SD (1972) Influence of *Meloidogyne incognita* on root rot development by *Rhizoctonia solani* and *Thielaviopsis basicola* in tomato. J Nematol 4:225

Gomes VM, Souza RM, Mussi-Dias V, da Silveira SF, Dlinski C (2011) Guava decline: a complex disease involving *Meloidogyne mayaguensis* and *Fusarium solani*. J Phytopathol 159:45–50

Gomes VM, Souza RM, da Silveira SF, Almeida AM (2013) Guava decline: effect of root exudates from *Meloidogyne enterolobii*-parasitized plants on *Fusarium solani* in vitro and on growth and development of guava seedlings under controlled conditions. Eur J Plant Pathol 137:393–401

Graham K (1967) Fungus insect mutualism in trees and timber. Annu Rev Entomol 12:105–126.

Griffin GD (1992) Stem nematode-Fusarium wilt complex in alfalfa as related to irrigation management at harvest time. J Nematol 24:315–320

Griffin GD (1986) The importance of nematode resistance on the interaction of *Meloidogyne hapla* and *Fusarium oxysporum* on alfalfa. Phytopathology 76:843

Griffin GJ, Eisenback JD, Yancey MM, Templeton J (2009) *Aphelenchoides hylurgi* as a carrier of white, hypovirulent *Cryphonectria parasitica* and its possible role in hypovirulence spread on blight-controlled American chestnut trees. J Nematol 41:267–273

Griffin GJ, Eisenback JD, Oldham K (2012) Widespread distribution of fungivorus *Aphelenchoides* spp. in blight cankers on American chestnut trees. J Nematol 44:316–320

Gunasekera TS, Holland RJ, Gillings MR, Briscoe DA, Neethling DC, Williams KL, Nevalainen KMH (2000) Phenotypic and genetic characterization of *Paecilomyces lilacinus* strains with biocontrol activity against root-knot nematodes. Can J Microbiol 46:775–783

Hafez SL, Al-Rehiayani S, Thornton M, Sundararaj P (1999) Differentiation of two geographically isolated populations of *Pratylenchus neglectus* based on their parasitism of potato and interaction with *Verticillium dahliae*. Nematropica 29:25–36

Hallmann J, Sikora RA (2011) Endophytic fungi. In: Davies K, Spiegel Y (eds) Biological control of plant-parasitic nematodes: building coherence between microbial ecology and molecular mechanisms. Springer, London, pp 227–258

Hao ZP, Fayolle L, van Tuinen D, Chatagnier O, Li XL, Gianinazzi S, Gianinazzi-Pearson V (2012) Local and systemic mycorrhiza-induced protection against the ectoparasitic nematode *Xiphinema index* involves priming of defence gene responses in grapevine. J Exp Bot 63:3657–3672

Harman GE (2006) Overview of mechanisms and uses of *Trichoderma* spp. Phytopathology 96:190–194

Hasna MK, Insunza V, Lagerlof L, Ramert B (2007) Food attraction and population growth of fungivorous nematodes and different fungi. Ann Appl Biol 151:175–182

Hay FS, Bateson L (1997) Effect of the nematophagous fungi *Hirsutella rhossiliensis* and *Verticillium balanoides* on stem nematode (*Ditylenchus dipsaci*) in white clover. Australas Plant Pathol 26:142–147

Hay FS, Regnault L (1995) Introduction of the endoparasitic nematophagous fungus *Verticillium balanoides* into white clover (*Trifolium repens*) by its host, stem nematode (*Ditylenchus dipsaci*). Australas Plant Pathol 24:249–251

Hesling JJ (1966) Preliminary experiments on the control of mycophagous eelworms in mushroom beds, with a note on their swarming. Plant Pathol 15:163–167

Hide GA, Corbett DCM, Evans K (1984) Effects of soil treatments and cultivars on early dying disease of potatoes caused by *Globodera rostochiensis* and *Verticillium dahliae*. Ann Appl Biol 104:277–289

Hillnhutter C, Albersmeier A, Berdugo CA, Sikora RA (2011) Synergistic damage by interactions between *Ditylenchus dipsaci* and *Rhizoctonia solani* (AG 2-2IIIB) on sugar beet. J Plant Dis Prot 118:127–133

Hogan EP, Griffin GJ (2008) Importance of *Cryphonectria parasitica* stromata production and intermediate-pigmented isolates in spread of *Cryphonectria* hypovirus 1 on grafted American chestnut trees. For Pathol 38:302–313

Hol WHG, Cook R (2005) An overview of arbuscular mycorrhizal fungi-nematode interactions. Basic Appl Ecol 6:489–503

Holland RJ, Williams KL, Khan A (1999) Infection of *Meloidogyne javanica* by *Paecilomyces lilacinus*. Nematology 1:131–139

Holland RJ, Williams KL, Nevalainen KMH (2003) *Paecilomyces lilacinus* strain Bioact251 is not a plant endophyte. Australas Plant Pathol 32:473–478

Hong CP, Estey RH (1985) Control of the damping-off disease of pea by *Aphelenchus avenae*. Indian J Nematol 15:1–4

Hooper DJ, Southey JF (1978) *Ditylenchus*, *Anguina* and related genera. In: Southey JF (ed) Plant nematology, vol 1, Ministry of agriculture, fisheries and food reference Book GD. Her Majesty's Stationary Office, London, pp 78–97

Howell CR (2002) Cotton seedling preemergence damping-off incited by *Rhizopus oryzae* and *Pythium* spp. and its biological control with *Trichoderma* spp. Phytopathology 92:177–180

Howell CR (2006) Understanding the mechanisms employed by *Trichoderma virens* to effect biological control of cotton diseases. Phytopathology 96:178–180

Howell CR, Hanson LE, Stipanovic RD, Puckhaber LS (2000) Induction of terpenoid synthesis in cotton roots and control of *Rhizoctonia solani* by seed treatment with *Trichoderma virens*. Phytopathology 90:248–252

Hsueh Y-P, Mahanti P, Schroeder FC, Sternberg PW (2013) Nematode-trapping fungi eavesdrop on nematode pheromones. Curr Biol 23:83–86

Hussey RS, Davis EL, Baum TJ (2002) Secrets in secretions: genes that control nematode parasitism of plants. Brazil J Plant Physiol 14:123–135

Ikonen EK (2001) Population growth of two aphelenchid nematodes with six different fungi as a food source. Nematology 3:9–15

Inagaki H, Powell NT (1969) Influence of the root-lesion nematode on black shank symptom development in flue-cured tobacco. Phytopathology 59:1350–1355

Ingham RE (1988) Interactions between nematodes and vesicular arbuscular mycorrhizae. Agr Ecosyst Environ 24:169–182

Jaffee BA (1998) Susceptibility of a cyst and a root-knot nematode to three nematode-trapping fungi. Fundam Appl Nematol 21:695–703

Jaffee BA (1999) Enchytraeids and nematophagous fungi in tomato fields and vineyards. Phytopathology 89:398–406

Jaffee BA (2000) Augmentation of soil with the nematophagous fungi *Hirsutella rhossiliensis* and *Arthrobotrys haptotyla*. Phytopathology 90:498–504

Jaffee BA (2002) Soil cages for studying how organic amendments affect nematode-trapping fungi. Appl Soil Ecol 21:1–9

Jaffee BA (2004) Do organic amendments enhance the nematode-trapping fungi *Dactylellina haptotyla* and *Arthrobotrys oligospora*? J Nematol 36:267–275

Jaffee BA, Muldoon AE (1989) Suppression of cyst nematode by natural infestation of nematophagous fungus. J Nematol 21:505–510

Jaffee BA, Muldoon AE (1997) Suppression of the root-knot nematode *Meloidogyne javanica* by alginate pellets containing the nematophagous fungi *Hirsutella rhossiliensis*, *Monacrosporium cionopagum* and *M. ellipsosporum*. Biocontrol Sci Technol 7:203–217

Jaffee BA, Zasoski RJ (2001) Soil pH and the activity of a pelletized nematophagous fungus. Phytopathology 91:324–330

Jaffee BA, Zehr EI (1983) Sporulation of the fungus *Hirsutella rhossiliensis* from the nematode *Criconemella xenoplax*. Plant Dis 67:1265–1267

Jaffee BA, Gaspard JT, Ferris H, Muldoon AE (1988) Quantification of parasitism of the soil-borne nematode *Criconemella xenoplax* by the nematophagous fungus *Hirsutella rhossiliensis*. Soil Biol Biochem 20:631–636

Jaffee BA, Gaspard JT, Ferris H (1989) Density-dependent parasitism of the soil-borne nematode *Criconemella xenoplax* by the nematophagous fungus *Hirsutella rhossiliensis*. Microb Ecol 17:193–200

Jaffee BA, Muldoon AE, Phillips R, Mangel M (1990) Rates of spore transmission, mortality, and production for the nematophagous fungus *Hirsutella rhossiliensis*. Phytopathology 80: 1083–1088

Jaffee BA, Tedford EC, Muldoon AE (1993) Tests for density-dependent parasitism of nematodes by nematode-trapping and endoparasitic fungi. Biol Control 3:329–336

Jaffee BA, Muldoon AE, Westerdahl BB (1996) Failure of a mycelial formulation of the nematophagous fungus *Hirsutella rhossiliensis* to suppress the nematode *Heterodera schachtii*. Biol Control 6:340–346

Jaffee BA, Muldoon AE, Didden WAM (1997a) Enchytraeids and nematophagous fungi in soil microcosms. Biol Fertil Soils 25:382–388

Jaffee BA, Santos PF, Muldoon AE (1997b) Suppression of nematophagous fungi by enchytraeid worms: a field exclosure experiment. Oecologia 112:412–423

Jansson HB (1982a) Attraction of nematodes to endoparasitic nematophagous fungi. Trans Br Mycol Soc 79:25–29

Jansson HB (1982b) Predacity by nematophagous fungi and its relation to the attraction of nematodes. Microb Ecol 8:233–240

Jansson H-B (1993) Adhesion to nematodes of conidia from the nematophagous fungus *Drechmeria coniospora*. J Gen Microbiol 139:1899–1906

Jansson HB, Nordbring-Hertz B (1979) Attraction of nematodes to living mycelium of nematophagous fungi. J Gen Microbiol 112:89–93

Jansson HB, Nordbring-Hertz B (1980) Interactions between nematophagous fungi and plant-parasitic nematodes: attraction, induction of trap formation and capture. Nematologica 26:383–389

Jansson HB, Jeyaprakash A, Zuckerman BM (1985) Differential adhesion and infection of nematodes by the endoparasitic fungus *Meria coniospora* (Deuteromycetes). Appl Environ Microbiol 49:552–555

Jansson H-B, Dackman C, Zuckerman BM (1987) Adhesion and infection of plant parasitic nematodes by the fungus *Drechmeria coniospora*. Nematologica 33:480–487

Johnson DA, Santo GS (2001) Development of wilt in mint in response to infection by two pathotypes of *Verticillium dahlia* and co-infection by *Pratylenchus penetrans*. Plant Dis 85: 1189–1192

Jonathan EI, Rajendran G (1998) Interaction of *Meloidogyne incognita* and *Fusarium oxysporum* f.sp. *cubense* on banana. Nematol Mediterr 26:9–12

Jones MGK (1981) Host cell responses to endoparasitic attack: structure and function of giant cells and syncytia. Ann Appl Biol 97:353–372

Jones MGK, Northcote DH (1972) Nematode-induced syncytium—a multinucleate transfer cell. J Cell Sci 10:789–809

Jones JT, Furlanetto C, Kikuchi T (2005) Horizontal gene transfer from bacteria and fungi as a driving force in the evolution of plant parasitism in nematodes. Nematology 7:641–646

Jun OK, Kim YH (2004) *Aphelenchus avenae* and antagonistic fungi as biological control agents of *Pythium* spp. Plant Pathol J 20:271–276

Karim NJ, Jones T, Okada H, Kikuchi T (2009) Analysis of expressed sequence tags and identification of genes encoding cell-wall-degrading enzymes from the fungivorous nematode *Aphelenchus avenae*. BMC Genomics 10:525. doi:10.1186/1471-2164-10-525

Kerry BR (1988) Two microorganisms for the biological control of plant parasitic nematodes. Proc Brit Crop Prot Conf Pests Dis, London

Kerry BR, Crump DH (1998) The dynamics of the decline of the cereal cyst nematode, *Heterodera avenae*, in four soils under intensive cereal production. Fundam Appl Nematol 21:617–625

Kerry BR, Crump DH, Mullen LA (1982) Studies of the cereal cyst-nematode, *Heterodera avenae* under continuous cereals, 1975–1978. II. Fungal parasitism of nematode females and eggs. Ann Appl Biol 100:489–499

Khan A, Williams KL, Nevalainen HKM (2004) Effects of *Paecilomyces lilacinus* protease and chitinase on the eggshell structures and hatching of *Meloidogyne javanica* juveniles. Biol Control 31:346–352

Khan A, Williams KL, Nevalainen HKM (2006) Infection of plant-parasitic nematodes by *Paecilomyces lilacinus* and *Monacrosporium lysipagum*. Biocontrol 51:659–678

Kiewnick S, Gullino ML (2009) Importance of multitrophic interactions for successful biocontrol of plant parasitic nematodes with *Paecilomyces lilacinus* strain 251. In: Gisi U, Chet I (eds) Recent developments in management of plant diseases, vol 1. Springer, London, pp 81–92

Kikuchi T, Jones JT, Aikawa T, Kosaka H, Ogura N (2004) A family of glycosyl hydrolase family 45 cellulases from the pine wood nematode *Bursaphelenchus xylophilus*. FEBS Lett 572: 201–205

Kikuchi T, Shibuya H, Jones JT (2005) Molecular and biochemical characterization of an endo-β-1,3-glucanase from the pine wood nematode *Bursaphelenchus xylophilus* acquired by horizontal gene transfer from bacteria. Biochem J 389:117–125

Kilama P, Dubois T, Coyne D, Nicre B, Gold CS, Adipala E (2007) Antagonism of *Paecilomyces* spp. isolated from banana (*Musa* spp.) roots and rhizosphere against *Radopholus similis*. Nematropica 37:215–225

Kim DG, Riggs RD (1991) Characteristics and efficacy of a sterile hyphomycete (ARF18), a new biocontrol agent for *Heterodera glycines* and other nematodes. J Nematol 23:275–282

Kim DG, Riggs RD, Kim KS (1992) Ultrastructure of *Heterodera glycines* parasitized by Arkansas Fungus 18. Phytopathology 82:429–433

Kim DG, Riggs RD, Correll JC (1998) Isolation, characterization and distribution of a biocontrol fungus from cysts of *Heterodera glycines*. Phytopathology 88:465–471

Kimmons CA, Gwinn KD, Bernard EC (1990) Nematode reproduction on endophyte-infected and endophyte-free tall fescue. Plant Dis 74:757–761

Kiyohara T, Suzuki K (1978) Nematode population growth and disease development in the pine wilt disease. Eur J For Pathol 8:285–292

Klink JW, Barker KR (1968) Effect of *Aphelenchus avenae* on the survival and pathogenic activity of root-rotting fungi. Phytopathology 58:228–232

Kok CJ, Papert A, Hok-A-Hin CH (2001) Microflora of *Meloidogyne* egg masses: species composition, population density and effect on the biocontrol agent *Verticillium chlamydosporium* (Goddard). Nematology 3:729–734

Lagerlof J, Insunza V, Lundegardh B, Ramert B (2011) Interaction between a fungal plant disease, fungivorous nematodes and compost suppressiveness. Acta Agr Scand B S P 61:372–377

LaMondia JA (1992) Predisposition of broadleaf tobacco to Fusarium wilt by early season infection with tobacco cyst nematodes. J Nematol 24:425–431

LaMondia JA (2003) Interaction of *Pratylenchus penetrans* and *Rhizoctonia fragariae* in strawberry black root rot. J Nematol 35:17–22

LaMondia JA (2004) Strawberry black root rot. Adv Strawberry Res 23:1–10

LaMondia JA, Taylor GS (1987) Influence of the tobacco cyst nematode (*Globodera tabacum*) on Fusarium wilt of Connecticut broadleaf tobacco. Plant Dis 71:1129–1132

Larriba E, Jaime M, Nislow C, Martin-Nieto J, Lopez-Llorca LV (2015) Endophytic colonization of barley (*Hordeum vulgare*) roots by the nematophagous fungus *Pochonia chlamydosporia* reveals plant growth promotion and a general defense and stress transcriptomic response. J Plant Res 128:665–678

Li Y, Hyde KD, Jeewon R, Cai L, Vijaykrishna D, Zhang KQ (2005) Phylogenetics and evolution of nematode-trapping fungi (Orbiliales) estimated from nuclear and protein coding genes. Mycologia 97:1034–1046

Li HY, Yang GD, Shu HR, Yang YT, Ye BX, Nishida I, Zheng CC (2006) Colonization by the arbuscular mycorrhizal fungus *Glomus versiforme* induces a defense response against the root-knot nematode *Meloidogyne incognita* in the grapevine (*Vitis amurensis* Rupr.), which includes transcriptional activation of the class III chitinase gene VCH3. Plant Cell Physiol 47:154–163

Lin F, Ye JL, Wang HG, Zhang AJ, Zhao BG (2013) Host deception: Predaceous fungus, *Esteya vermicola*, entices pine wood nematode by mimicking the scent of pine tree for nutrient. PLoS One 8, e71676

Linford MB, Yap F, Oliveira JM (1938) Reduction of soil populations of the root-knot nematode during decomposition of organic matter. Soil Sci 45:127–140

Liou JY, Shih JY, Tzean SS (1999) *Esteya*, a new nematophagous genus from Taiwan, attacking the pinewood nematode (*Bursaphelenchus xylophilus*). Mycol Res 103:242–248

Liu XZ, Chen SY (2000) Parasitism of *Heterodera glycines* by *Hirsutella* spp. in Minnesota soybean fields. Biol Control 19:161–166

Liu XZ, Chen SY (2001) Screening isolates of *Hirsutella* species for biocontrol of *Heterodera glycines*. Biocontrol Sci Technol 11:151–160

Liu SF, Chen SY (2009) Effectiveness of *Hirsutella minnesotensis* and *H. rhossiliensis* in control of the soybean cyst nematode in four soils with various pH, texture, and organic matter. Biocontrol Sci Technol 19:595–612

Liu X, Xiang M, Che Y (2009) The living strategy of nematophagous fungi. Mycoscience 50:20–25

Lootsma M, Scholte K (1997) Effect of soil moisture content on the suppression of Rhizoctonia stem canker on potato by the nematode *Aphelenchus avenae* and the springtail *Folsomia fimetaria*. Plant Pathol 46:209–215

Lopez-Llorca LV, Macia-Vicente JG, Jansson HB (2008) Mode of action and interactions of nematophagous fungi. In: Ciancio A, Mukerji KG (eds) Integrated management and biocontrol of vegetable and grain crops nematodes. Springer, London, pp 51–76

Luangsa-Ard J, Houbraken J, van Doorn T, Hong SB, Borman AM, Hywel-Jones NL, Samson RA (2011) *Purpureocillium*, a new genus for the medically important *Paecilomyces lilacinus*. FEMS Microbiol Lett 321:141–149

Luo H, Mo MH, Huang XW, Li X, Zhang KQ (2004) *Coprinus comatus*: a basidiomycete fungus forms novel spiny structures and infects nematode. Mycologia 96:1218–1224

Luo H, Li X, Li GH, Pan YB, Zhang KQ (2006) Acanthocytes of *Stropharia rugosoannulata* function as a nematode-attacking device. Appl Environ Microbiol 72:2982–2987

Luo H, Liu YJ, Fang L, Li X, Tang NH, Zhang KQ (2007) *Coprinus comatus* damages nematode cuticles mechanically with spiny balls and produces potent toxins to immobilize nematodes. Appl Environ Microbiol 73:3916–3923

Macia-Vicente JG, Jansson HB, Abdullah SK, Descals E, Salinas J, Lopez-Llorca LV (2008) Fungal root endophytes from natural vegetation in Mediterranean environments with special reference to *Fusarium* spp. FEMS Microbiol Ecol 64:90–105

Maehara N, Futai K (1997) Effect of fungal interactions on the numbers of the pinewood nematode, *Bursaphelenchus xylophilus* (Nematoda: Aphelenchoididae), carried by the Japanese pine sawyer, *Monochamus alternatus* (Coleoptera: Cerambycidae). Fundam Appl Nematol 20:611–617

Mamiya Y (1972) Pine wood nematode, *Bursaphelenchus lignicolus* Mamiya and Kiyohara, as a causal agent of pine wilting disease. Rev Plant Prot Res 5:46–60

Mankau R (1969) Toxicity of culture filtrates of *Aspergillus niger* to the mycophagous nematode, *Aphelenchus avenae*. Phytopathology 59:13

Manzanilla-Lopez RH, Esteves I, Powers SJ, Kerry BR (2011) Effects of crop plants on abundance of *Pochonia chlamydosporia* and other fungal parasites of root-knot and potato cyst nematodes. Ann Appl Biol 159:118–129

Martin MJ, Riedel RM, Rowe RC (1982) *Verticillium dahliae* and *Pratylenchus penetrans*: interactions in the early dying complex of potato in Ohio. Phytopathology 72:640–644

Martinuz A, Schouten A, Sikora RA (2012) Systemically induced resistance and microbial competitive exclusion: implications on biological control. Phytopathology 102:260–266

Mauchline TH, Kerry BR, Hirsch PR (2002) Quantification in soil and the rhizosphere of the nematophagous fungus *Verticillium chlamydosporium* by competitive PCR and comparison with selective plating. Appl Environ Microbiol 68:1846–1853

McGawley EC, Winchell KL, Jones JP, Birchfield W, Bergen GT (1985) Population development and influence of *Bursaphelenchus xylophilus* on *Gliocladium virens*. J Nematol 17:69–76

McInnis TM, Jaffee BA (1989) An assay for *Hirsutella rhossiliensis* spores and the importance of phialides for nematode inoculation. J Nematol 21:229–234

McLean KS, Lawrence GW (1993) Interrelationship of *Heterodera glycines* and *Fusarium solani* in sudden-death syndrome of soybean. J Nematol 25:434–439
Meléndez PL, Powell NT (1970) Histological studies of the *Pythium*–root knot complex in tobacco. Phytopathology 60:1303
Mendoza De Gives P, Davies KG, Clark SJ, Behnke JM (1999) Predatory behaviour of trapping fungi against srf mutants of *Caenorhabditis elegans* and different plant and animal parasitic nematodes. Parasitology 119:95–104
Mendoza AR, Sikora RA, Kiewnick S (2007) Influence of *Paecilomyces lilacinus* strain 251 on the biological control of the burrowing nematode *Radopholus similis* in banana. Nematropica 37:203–213
Meyer SLF, Nyczepir AP, Rupprecht SM, Mitchell AD, Martin PAW, Brush CW, Chitwood DJ, Vinyard BT (2013) Tall fescue 'Jesup (Max-Q)': *Meloidogyne incognita* development in roots and nematotoxicity. Agron J 105:755–763
Mitchum MG, Hussey RS, Baum TJ, Wang X, Elling AA, Wubben M, Davis EL (2013) Nematode effector proteins: an emerging paradigm of parasitism. New Phytol 199:879–94
Morton JB, Benny GL (1990) Revised classification of arbuscular mycorrhizal fungi (Zygomycetes): a new order, Glomales, two new suborders, Glomineae and Gigasporineae, and two new families, Acaulosporaceae and Gigasporaceae, and with emendation of Glomaceae. Mycotaxon 37:471–491
Morton CO, Hirsch PR, Peberdy JP, Kerry BR (2003a) Cloning of and genetic variation in protease VCP1 from the nematophagous fungus *Pochonia chlamydosporia*. Mycol Res 107:38–46
Morton CO, Mauchline TH, Kerry BR, Hirsch PR (2003b) PCR-based DNA fingerprinting indicates host-related genetic variation in the nematophagous fungus *Pochonia chlamydosporia*. Mycol Res 107:198–205
Morton CO, Hirsch PR, Kerry BR (2004) Infection of plant-parasitic nematodes by nematophagous fungi—a review of the application of molecular biology to understand infection processes and to improve biological control. Nematology 6:161–170
Muller J (1985) The influence of two pesticides on fungal parasites of *Heterodera schachtii*. Les Colloques de L'INRA 31:225–231
Mwaura P, Dubois T, Losenge T, Coyne D, Kahangi E (2010) Effect of endophytic *Fusarium oxysporum* on paralysis and mortality of *Pratylenchus goodeyi*. Afr J Biotechnol 9:1130–1134
Negrón JA, Acosta N (1989) The *Fusarium oxysporum* f.sp. *coffeae*–*Meloidogyne incognita* complex in bourbon coffee. Nematropica 19:161–168
Neher DA (2010) Ecology of plant and free-living nematodes in natural and agricultural soil. Annu Rev Phytopathol 48:371–394
Nickle WR, McIntosh P (1968) Studies on the feeding and reproduction of seven mycophagous nematodes on *Rhizoctonia, Fusarium,* and *Verticillium*. Nematologica 14:11–12
Nordbring-Hertz B (1977) Nematode-induced morphogenesis in the predaceous fungus *Arthrobotrys oligospora*. Nematologica 23:443–451
Nordbring-Hertz B, Mattiasson B (1979) Action of a nematode-trapping fungus shows lectin-mediated host-microorganism interaction. Nature 281:477–479
Nordbring-Hertz B, Jansson H-B, Tunlid A (2001) Nematophagous fungi. In: Encyclopedia of life sciences. Wiley, Chichester
Nordmeyer D, Sikora RA (1983a) Studies on the interaction between *Heterodera daverti, Fusarium avenaceum*, and *Fusarium oxysporum* on *Trifolium subterraneum*. Rev Nematol 6:193–198
Nordmeyer D, Sikora RA (1983b) Effect of culture filtrate from *Fusarium avenaceum* on the penetration of *Heterodera daverti* into roots of *Trifolium subterraneum*. Nematologica 29:88–94
Olatinwo R, Becker JO, Borneman J (2006a) Suppression of *Heterodera schachtii* populations by *Dactylella oviparasitica* in four soils. J Nematol 38:345–348
Olatinwo R, Borneman J, Becker JO (2006b) Induction of beet-cyst nematode suppressiveness by the fungi *Dactylella oviparasitica* and *Fusarium oxysporum* in field microplots. Phytopathology 96:855–859

Olatinwo R, Yin B, Becker JO, Borneman J (2006c) Suppression of the plant-parasitic nematode *Heterodera schachtii* by the fungus *Dactylella oviparasitica*. Phytopathology 96:111–114

Orion D, Kritzman G, Meyer SLF, Erbe EF, Chitwood DJ (2001) A role of the gelatinous matrix in the resistance of root-knot nematode (*Meloidogyne* spp.) eggs to microorganisms. J Nematol 33:203–207

Panaccione DG, Kotcon JB, Schardl CL, Johnson RD, Morton JB (2006) Ergot alkaloids are not essential for endophytic fungus-associated population suppression of the lesion nematode, *Pratylenchus scribneri*, on perennial ryegrass. Nematology 8:583–590

Paparu P, Dubois T, Coyne D, Viljoen A (2013) Differential gene expression in East African highland bananas (*Musa* spp.): interactions between non-pathogenic *Fusarium oxysporum* V5w2 and *Radopholus similis*. Physiol Mol Plant Pathol 82:56–63

Park JO, Hargreaves JR, McConville EJ, Stirling GR, Ghisalberti EL, Sivasithamparam K (2004) Production of leucinostatins and nematicidal activity of Australian isolates of *Paecilomyces lilacinus* (Thom) Samson. Lett Appl Microbiol 38:271–276

Pedrotti L, Mueller MJ, Waller F (2013) *Piriformospora indica* root colonization triggers local and systemic root responses and inhibits secondary colonization of distal roots. PLoS One 8, e69352

Persmark L, Jansson H-B (1997) Nematophagous fungi in the rhizosphere of agricultural crops. FEMS Microbiol Ecol 22:303–312

Persmark L, Nordbring-Hertz B (1997) Conidial trap formation of nematode-trapping fungi in soil and soil extracts. FEMS Microbiol Ecol 22:313–323

Persmark L, Banck A, Jansson H-B (1996) Population dynamics of nematophagous fungi and nematodes in an arable soil: vertical and seasonal fluctuations. Soil Biol Biochem 28:1005–1014

Persson C, Jansson HB (1999) Rhizosphere colonization and control of *Meloidogyne* spp. by nematode-trapping fungi. J Nematol 31:164–171

Peterson EA, Katznelson H (1965) Studies on the relationship between nematodes and other soil microorganisms. IV. Incidence of nematode-trapping fungi in the vicinity of plant roots. Can J Microbiol 11:491–495

Pieterse CMJ, Zamioudis C, Berendsen RL, Weller DM, Van Wees SCM, Bakker P (2014) Induced systemic resistance by beneficial microbes. Annu Rev Phytopathol 52:347–375

Pillai JK, Taylor DP (1967) Influence of fungi on host preference, host suitability, and morphometrics of five mycophagous nematodes. Nematologica 13:529–540

Polychronopoulos AG, Houston BR, Lownsbery BF (1969) Penetration and development of *Rhizoctonia solani* in sugar beet seedlings infected with *Heterodera schachtii*. Phytopathology 59:482–485

Premachandran D, Pramer D (1984) Role of N-acetylgalactosamine-specific protein in trapping of nematodes by *Arthrobotrys oligospora*. Appl Environ Microbiol 47:1358–1359

Quandt CA, Kepler RM, Gams W, Araujo JPM, Ban S, Evans HC, Hughes D, Humber R, Hywel-Jones N, Li Z, Luangsa-Ard JJ, Rehner SA, Sanjuan T, Sato H, Shrestha B, Sung G-H, Yao Y-J, Zare R, Spatafora JW (2014) Phylogenetic-based nomenclatural proposals for Ophiocordycipitaceae (Hypocreales) with new combinations in *Tolypocladium*. IMA Fungus 5(1):121–134

Rhoades HL, Linford MB (1959) Control of Pythium root rot by the nematode *Aphelenchus avenae*. Plant Dis Rep 43:323–328

Riedel RM, Rowe RC, Martin MJ (1985) Differential interactions of *Pratylenchus crenatus*, *P. penetrans*, and *P. scribneri* with *Verticillium dahliae* in potato early dying disease. Phytopathology 75:419–422

Riffle J (1967) Effect of an *Aphelenchoides* species on the growth of a mycorrhizal and a pseudo-mycorrhizal fungus. Phytopathology 57:541–544

Riffle J (1971) Effect of nematodes on root-inhabiting fungi. In: Hacskaylo E (ed) Mycorrhizae: proceedings of the first North American Conference on Mycorrhizae, USDA Forest Service Miscellaneous Publication 1189, pp 97–113

Rubner A (1996) Revision of predacious hyphomycetes in the *Dactylella-Monacrosporium* complex. Stud Mycol 39:1–134

Ruess L, Dighton J (1996) Cultural studies on soil nematodes and their fungal hosts. Nematologica 42:330–346

Ruess L, Zpata EJG, Dighton J (2000) Food preferences of a fungal-feeding *Aphelenchoides* species. Nematology 2:223–230

Rumbos CI, Kiewnick S (2006) Effect of plant species on persistence of *Paecilomyces lilacinus* strain 251 in soil and on root colonization by the fungus. Plant Soil 283:25–31

Rumbos C, Mendoza A, Sikora R, Kiewnick S (2008) Persistence of the nematophagous fungus *Paecilomyces lilacinus* strain 251 in soil under controlled conditions. Biocontrol Sci Technol 18:1041–1050

Rybarczyk-Mydłowska K, Mooyman P, van Megen H, van den Elsen S, Vervoort M, Veenhuizen P, van Doorn J, Dees R, Karssen G, Bakker J, Helder J (2012) Small subunit ribosomal DNA-based phylogenetic analysis of foliar nematodes (*Aphelenchoides* spp.) and their quantitative detection in complex DNA backgrounds. Phytopathology 102:1153–1160

Saikawa M, Oguchi M, Ruiz RFC (1997) Electron microscopy of two nematode-destroying fungi, *Meristacrum asterospermum* and *Zygnemomyces echinulatus* (Meristacraceae, Entomophthorales). Can J Bot 75:762–768

Samuels GJ (2006) *Trichoderma*: systematics, the sexual state, and ecology. Phytopathology 96:195–206

Schardl CL, Leuchtmann A, Spiering MJ (2004) Symbioses of grasses with seedborne fungal endophytes. Annu Rev Plant Biol 55:315–340

Schindler AF, Stewart RN (1956) Fusarium wilt of carnations retarded by fungus-eating nematodes, *Ditylenchus* spp. Phytopathology 46:469

Scholler M, Rubner A (1994) Predacious activity of the nematode-destroying fungus *Arthrobotrys oligospora* in dependence of the medium composition. Microbiol Res 149:145–149

Scholler M, Hagedorn G, Rubner A (1999) A reevaluation of predatory orbiliaceous fungi. II. A new generic concept. Sydowia 51:89–113

Segers R, Butt TM, Kerry BR, Beckett A, Peberdy JF (1996) The role of the proteinase VCP1 produced by the nematophagous *Verticillium chlamydosporium* in the infection process of nematode eggs. Mycol Res 100:421–428

Selim ME, Mahdy ME, Sorial ME, Dababat AA, Sikora RA (2014) Biological and chemical dependent systemic resistance and their significance for the control of root-knot nematodes. Nematology 16:917–927

Shafer SR, Rhodes LH, Riedel M (1981) In-vitro parasitism of endomycorrhizal fungi of ericaceous plants by the mycophagous nematode *Aphelenchoides bicaudatus*. Mycologia 73:141–149

Sharon E, Bar-Eyal M, Chet I, Herrera-Estrella A, Kleifeld O, Spiegel Y (2001) Biological control of the root-knot nematode *Meloidogyne javanica* by *Trichoderma harzianum*. Phytopathology 91:687–693

Sharon E, Chet I, Viterbo A, Bar-Eyal M, Nagan H, Samuels GJ, Spiegel Y (2007) Parasitism of *Trichoderma* on *Meloidogyne javanica* and role of the gelatinous matrix. Eur J Plant Pathol 118:247–258

Sharon E, Chet I, Spiegel Y (2011) *Trichoderma* as a biological control agent. In: Davies K, Spiegel Y (eds) Biological control of plant-parasitic nematodes: building coherence between microbial ecology and molecular mechanisms. Springer, London, pp 183–201

Siddiqui ZA, Futai K (2009) Biocontrol of *Meloidogyne incognita* on tomato using antagonistic fungi, plant-growth-promoting rhizobacteria and cattle manure. Pest Manag Sci 65:943–948

Siddiqui ZA, Mahmood I (1996) Biological control of plant parasitic nematodes by fungi: a review. Biores Technol 58:229–239

Siddiqui ZA, Mahmood I (1999) Effects of *Meloidogyne incognita*, *Fusarium oxysporum* f. sp. *pisi*, *Rhizobium* sp. and different soil types on growth, chlorophyll, and carotenoid pigments of pea. Isr J Plant Sci 47:251–256

Siddiqui IA, Atkins SD, Kerry BR (2009) Relationship between saprotrophic growth in soil of different biotypes of *Pochonia chlamydosporia* and the infection of nematode eggs. Ann Appl Biol 155:131–141

Siegel MR, Bush LP (1996) Defensive chemicals in grass-fungal endophyte associations. In: Romeo JT, Saunders JA, Barbosa P (eds) Phytochemical diversity and redundancy in ecological interactions. Plenum Press, New York, pp 81–119

Sikora RA (2008) Mutualistic endophytic fungi and in-planta suppressiveness to plant parasitic nematodes. Biol Control 46:15–23

Sikora RA, Carter WW (1987) Nematode interactions with fungal and bacterial plant pathogens—fact or fantasy. In: Veech JA, Dickson DW (eds) Vistas on nematology. Society of Nematology, Hyattsville, pp 307–312.

Singh UB, Sahu A, Sahu N, Singh RK, Renu S, Singh DP, Manna MC, Sarma BK, Singh HB Singh KP (2013) *Arthrobotrys oligospora*-mediated biological control of diseases of tomato (*Lycopersicon esculentum* Mill.) caused by *Meloidogyne incognita* and *Rhizoctonia solani*. J Appl Microbiol 114:196–208.

Sjollema KA, Dijksterhuis J, Veenhuis M, Harder W (1993) An electron microscopical study of the infection of the nematode *Panagrellus redivivus* by the endoparasitic fungus *Verticillium balanoides*. Mycol Res 97:479–484

Steiner G (1936) The status of nematode *Aphelenchus avenae* Bastian, 1865 as plant parasite. Phytopathology 26:294–295

Steiner G, Buhrer EM (1934) *Aphelenchoides xylophilus*, n. sp., a nematode associated with blue-stain and other fungi in timber. J Agr Res 48:949–951

Stirling GR (1991) Biological control of plant parasitic nematodes: progress, problems and prospects. CABI, Wallingford

Stirling GR (2014) Biological control of plant-parasitic nematodes: soil ecosystem management in sustainable agriculture, 2nd edn. CABI, Wallingford

Stirling GR, Mankau R (1978) *Dactylella oviparasitica*, a new fungal parasite of *Meloidogyne* eggs. Mycologia 70:774–783

Stirling GR, Smith LJ (1998) Field tests of formulated products containing either *Verticillium chlamydosporium* or *Arthrobotrys dactyloides* for biological control of root-knot nematodes. Biol Control 11:231–239

Stirling GR, West LM (1991) Fungal parasites of root-knot nematode eggs from tropical and subtropical regions of Australia. Australas Plant Pathol 20:149–154

Stirling GR, McKenry MV, Mankau R (1979) Biological control of root-knot nematodes (*Meloidogyne* spp.) on peach. Phytopathology 69:806–809

Stirling GR, Rames E, Stirling AM, Hamill S (2011) Factors associated with the suppressiveness of sugarcane soils to plant-parasitic nematodes. J Nematol 43:135–148

Storey GW, Evans K (1987) Interactions between *Globodera pallida* juveniles, *Verticillium dahliae* and three potato cultivars, with descriptions of associated histopathologies. Plant Pathol 36:192–200

Suleman P, Sardanelli S, Krusberg LR, Straney DC (1997) Variability among *Fusarium oxysporum* f. sp. *lycopersici* isolates in their ability to interact with *Meloidogyne incognita* race 1. Kuwait J Sci Eng 24:299–307

Sumner DR, Johnson AW (1973) Effect of root-knot nematodes on Fusarium wilt of watermelon. Phytopathology 63:857–861

Sutherland JR (1967) Failure of the nematode *Aphelenchus avenae* to parasitize conifer seedling roots. Plant Dis Rep 51:367–369

Sutherland JR, Fortin JA (1968) Effect of the nematode *Aphelenchus avenae* on some ectotrophic, mycorrhizal fungi and on a red pine mycorrhizal relationship. Phytopathology 58:519–523

Tedford EC, Jaffee BA, Muldoon AE (1992) Effect of soil moisture and texture on transmission of the nematophagous fungus *Hirsutella rhossiliensis* to cyst and root-knot nematodes. Phytopathology 82:1002–1007

Tedford EC, Jaffee BA, Muldoon AE, Anderson CE, Westerdahl BB (1993) Parasitism of *Heterodera schachtii* and *Meloidogyne javanica* by *Hirsutella rhossiliensis* in microplots over two growing seasons. J Nematol 25:427–433

Tedford EC, Jaffee BA, Muldoon AE (1994) Variability among isolates of the nematophagous fungus *Hirsutella*. Mycol Res 98:1127–1136

Thorn RG, Barron GL (1984) Carnivorous mushrooms. Science 224:76–78

Thorn RG, Barron GL (1986) *Nematoctonus* and the tribe Resupinateae in Ontario, Canada. Mycotaxon 25:321–453

Tigano-Milani MS, Carneiro RG, Faria MR, Frazao HS, McCoy CW (1995a) Isozyme characterization and pathogenicity of *Paecilomyces fumosoroseus* and *P. lilacinus* to *Diabrotica speciosa* (Coleoptera: Chrysomelidae) and *Meloidogyne javanica* (Nematoda: Tylenchidae). Biol Control 5:378–382

Tigano-Milani MS, Samson RA, Martins I, Sobral BWS (1995b) DNA markers for differentiating isolates of *Paecilomyces lilacinus*. Microbiology 141:239–245

Timper P (2011) Utilization of biological control for managing plant-parasitic nematodes. In: Davies K, Spiegel S (eds) Biological control of plant-parasitic nematodes: building coherence between microbial ecology and molecular mechanisms. Springer, London, pp 259–289

Timper P, Riggs RD (1998) Variation in efficacy of isolates of the fungus ARF against the soybean cyst nematode *Heterodera glycines*. J Nematol 30:461–467

Timper P, Riggs RD, Crippen DL (1999) Parasitism of sedentary stages of *Heterodera glycines* by isolates of a sterile nematophagous fungus. Phytopathology 89:1193–1199

Timper P, Gates RN, Bouton JH (2005) Response of *Pratylenchus* spp. in tall fescue infected with different strains of the fungal endophyte *Neotyphodium coenophialum*. Nematology 7:105–110

Townshend JL (1964) Fungus hosts of *Aphelenchus avenae* Bastian, 1865 and *Bursaphelenchus fungivorous* Franklin & Hooper, 1962 and their attractiveness to these nematode species. Can J Microbiol 10:727–737

Trifonova Z, Koleva K (2002) Interaction of *Ditylenchus dipsaci* (Kuhn) Filipjev and *Fusarium* spp. on onion (*Allium cepa* L.). Acta Zool Bulgar 54:67–72

Tunlid A, Ahren D (2011) Molecular mechanisms of the interaction between nematode-trapping fungi and nematodes: lessons from genomics. In: Davies K, Spiegel S (eds) Biological control of plant-parasitic nematodes: building coherence between microbial ecology and molecular mechanisms. Springer, London, pp 145–169

Tzean SS, Liou JY (1993) Nematophagous resupinate basidiomycetous fungi. Phytopathology 83:1015–1020

Udo IA, Uguru MI, Ogbuji RO (2013) Comparative efficacy of arbuscular mycorrhizal fungi in combination with bioformulated *Paecilomyces lilacinus* against *Meloidogyne incognita* on tomato in two Ultisols of South-eastern Nigeria. Biocontrol Sci Technol 23:1083–1097

Uma Maheswari T, Sharma SB, Reddy DDR, Haware MP (1997) Interaction of *Fusarium oxysporum* f. sp. *ciceri* and *Meloidogyne javanica* on *Cicer arietinum*. J Nematol 29:117–126

Van den Boogert PHJF, Dijksterhuis J, Velvis H, Veenhuis M (1992) Adhesive knob formation by conidia of the nematophagous fungus *Drechmeria coniospora*. Anton Leeuw 61:221–229

Veenhuis M, Nordbring HB, Harder W (1985) An electron microscopic analysis of capture and initial stages of penetration of nematodes by *Arthrobotrys oligospora*. Anton Leeuw 51: 385–398

Veresoglou SD, Rillig MC (2012) Suppression of fungal and nematode plant pathogens through arbuscular mycorrhizal fungi. Biol Lett 8:214–217

Viglierchio DR (1971) Race genesis in *D. dipsaci*. Nematologica 17:386–387

Vos C, Claerhout S, Mkandawire R, Panis B, De Waele D, Elsen A (2012a) Arbuscular mycorrhizal fungi reduce root-knot nematode penetration through altered root exudation of their host. Plant Soil 354:335–345

Vos C, Van den Broucke D, Lombi FM, De Waele D, Elsen A (2012b) Mycorrhiza-induced resistance in banana acts on nematode host location and penetration. Soil Biol Biochem 47:60–66

Vos CM, Tesfahun AN, Panis B, De Waele D, Elsen A (2012c) Arbuscular mycorrhizal fungi induce systemic resistance in tomato against the sedentary nematode *Meloidogyne incognita* and the migratory nematode *Pratylenchus penetrans*. Appl Soil Ecol 61:1–6

Vos C, Schouteden N, van Tuinen D, Chatagnier O, Elsen A, De Waele D, Panis B, Gianinazzi-Pearson V (2013) Mycorrhiza-induced resistance against the root-knot nematode *Meloidogyne incognita* involves priming of defense gene responses in tomato. Soil Biol Biochem 60:45–54

Vrain TC (1987) Effect of *Ditylenchus dipsaci* and *Pratylenchus penetrans* on Verticillium wilt of alfalfa. J Nematol 19:379–383

Vu TT (2005) Modes of action of non-pathogenic *Fusarium oxysporum* endophytes for bio-enhancement of banana toward *Radopholus similis*. Dissertation, University of Bonn. Bonn, Germany

Vu T, Hauschild R, Sikora RA (2006) *Fusarium oxysporum* endophytes induced systemic resistance against *Radopholus similis* on banana. Nematology 8:847–852

Walker GE (1997) Effects of *Meloidogyne* spp. and *Rhizoctonia solani* on the growth of grapevine rootings. J Nematol 29:190–198

Walker NR, Kirkpatrick TL, Rothrock CS (2000) Influence of *Meloidogyne incognita* and *Thielaviopsis basicola* populations on early-season disease development and cotton growth. Plant Dis 84:449–453.

Waller F, Achatz B, Baltruschat H, Fodor J, Becker K, Fischer M, Heier T, Huckelhoven R, Neumann C, von Wettstein D, Franken P, Kogel KH (2005) The endophytic fungus *Piriformosa indica* reprograms barley to salt-stress tolerance, disease resistance, and higher yield. Proc Natl Acad Sci USA 102:13386–13391.

Wang KM, Tsay TT, Lin YY (1993) The occurrence of *Aphelenchoides besseyi* on strawberry and its ecology in Taiwan. Plant Prot Bull (Taichung) 35:14–29

Wang KH, McSorley R, Gallaher RN (2003a) Effect of *Crotalaria juncea* amendment on nematode communities in soil with different agricultural histories. J Nematol 35:294–301

Wang KH, Sipes BS, Schmitt DP (2003b) Enhancement of *Rotylenchulus reniformis* suppressiveness by *Crotalaria juncea* amendment in pineapple soils. Agr Ecosyst Environ 94:197–203

Wang KN, Riggs RD, Crippen D (2004a) Decomposition of plant debris by the nematophagous fungus ARF. J Nematol 36:263–266

Wang KN, Riggs RD, Crippen D (2004b) Suppression of *Rotylenchulus reniformis* on cotton by the nematophagous fungus ARF. J Nematol 36:186–191

Wang B, Wu W, Liu X (2007) Purification and characterization of a neutral serine protease with nematicidal activity from *Hirsutella rhossiliensis*. Mycopathologia 163:169–176

Wang CY, Fang ZM, Sun BS, Gu LJ, Zhang KQ, Sung CK (2008) High infectivity of an endoparasitic fungus strain, *Esteya vermicola*, against nematodes. J Microbiol 46:380–389

Wang B, Liu X, Wu W, Liu X, Li S (2009) Purification, characterization and gene cloning of an alkaline serine protease from a highly virulent strain of the nematode-endoparasitic fungus *Hirsutella rhossiliensis*. Microbiol Res 164:665–673

Wang CY, Wang Z, Fang ZM, Zhang DL, Gu LJ, Liu L, Sung CK (2010a) Attraction of pinewood nematode to endoparasitic nematophagous fungus *Esteya vermicola*. Curr Microbiol 60:387–392

Wang J, Wang J, Liu F, Pan C (2010b) Enhancing the virulence of *Paecilomyces lilacinus* against *Meloidogyne incognita* eggs by overexpression of a serine protease. Biotechnol Lett 32:1159–1166

Wang CY, Fang ZM, Wang Z, Zhang DL, Gu LJ, Lee MR, Liu L, Sung CK (2011a) Biological control of the pinewood nematode *Bursaphelenchus xylophilus* by application of the endoparasitic fungus *Esteya vermicola*. Biocontrol 56:91–100

Wang Z, Wang CY, Yang ZH, Fang ZM, Moon YJ, Sun BS, Lee MR, Sung CK (2011b) Viability and pathogenicity of *Esteya vermicola* in pine trees. Biocontrol Sci Technol 21:387–393

Wang X, Wang T, Wang J, Guan T, Li H (2014) Morphological, molecular and biological characterization of *Esteya vermicola*, a nematophagous fungus isolated from intercepted wood packing materials exported from Brazil. Mycoscience 55:367–377

Wei B-Q, Xue Q-Y, Wei L-H, Niu D-D, Liu H-X, Chen L-F, Guo J-H (2009) A novel screening strategy to identify biocontrol fungi using protease production or chitinase activity against *Meloidogyne* root-knot nematodes. Biocontrol Sci Technol 19:859–870

West CP, Izekor E, Oosterhuis DM, Robbins RT (1988) The effect of *Acremonium coenophialum* on the growth and nematode infestation of tall fescue. Plant Soil 112:3–6

Westphal A, Becker JO (2001) Components of soil suppressiveness against *Heterodera schachtii.* Soil Biol Biochem 33:9–16

Wheeler TA, Madden LV, Rowe RC, Riedel RM (1992) Modelling of yield loss in potato early dying caused by *Pratylenchus penetrans* and *Verticillium dahliae.* J Nematol 24:99–102

Wimble DB, Young TWK (1983) Capture of nematodes by adhesive knobs in *Dactylella lyspiga.* Microbios 36:33–39

Woo SL, Scala F, Ruocco M, Lorito M (2006) The molecular biology of the interactions between *Trichoderma* spp., phytopathogenic fungi, and plants. Phytopathology 96:181–185

Wyss U (2002) Feeding behaviour of plant-parasitic nematodes. In: Lee D (ed) Biology of nematodes. Taylor & Francis, London, pp 233–259

Xiang MC, Yang XH, Wang ZX, Liu XZ, Chen SY, Xiao QM (2007) Variability of morphology, parasitism, and nucleotide sequences among isolates and species of nematophagous *Hirsutella.* Biol Control 41:110–119

Xiang M, Pa X, Liu X, Zhang L (2010) Effect of environment on the abundance and activity of the nematophagous fungus *Hirsutella minnesotensis* in soil. FEMS Microbiol Ecol 71:413–417

Xing L, Westphal A (2013) Synergism in the interaction of *Fusarium virguliforme* with *Heterodera glycines* in sudden death syndrome of soybean. J Plant Dis Prot 120:209–217

Yang JK, Wang L, Ji XL, Feng Y, Li XM, Zou CG, Xu JP, Ren Y, Mi QL, Wu JL, Liu SQ, Liu Y, Huang XW, Wang HY, Niu XM, Li J, Liang LM, Luo YL, Ji KF, Zhou W, Yu ZF, Li GH, Liu YJ, Li L, Qiao M, Feng L, Zhang KQ (2011) Genomic and proteomic analyses of the fungus *Arthrobotrys oligospora* provide insights into nematode-trap formation. PLoS Pathog 7, e1002179

Yang JI, Benecke S, Jeske DR, Rocha FS, Becker JS, Timper P, Becker JO, Borneman J (2012) Population dynamics of *Dactylella oviparasitica* and *Heterodera schachtii*: toward a decision model for sugar beet planting. J Nematol 44:237–244

Yin B, Valinsky L, Gao XB, Becker JO, Borneman J (2003) Identification of fungal rDNA associated with soil suppressiveness against *Heterodera schachtii* using oligonucleotide fingerprinting. Phytopathology 93:1006–1013

Zare R, Gams W, Evans HC (2001) A revision of *Verticillium* section Prostrata. V. The genus *Pochonia*, with notes on *Rotiferophthora*. Nova Hedwig 73:51–86

Zouhar M, Douda O, Novakova J, Doudova E, Mazakova J, Wenzlova J, Rysanek P, Renco M (2013) First report about the trapping activity of *Stropharia rugosoannulata* acanthocytes for northern root knot nematode. Helminthologia 50:127–131

# Chapter 24
# Pathogenic Microfungi Associated with *Spartina* in Salt Marshes

**Wade H. Elmer**

## Introduction

Diebacks and/or disease outbreaks caused by fungal pathogens in natural ecosystems, such as salt marshes, are not well studied, in part, due to the rarity of these events, and their perceived economic insignificance when compared to agricultural systems. However, it may be surprising to many that salt marshes are one of the most valuable ecosystems and outrank tropical rain forests and coral reefs for productivity. Carbon sequestration can approach more 3 kg dry matter/m$^2$/year which outperforms coral reefs by 40 % (Bertness 2007). Furthermore, their value in absorbing excess nitrogen and phosphorus, detoxifying pollutants, and providing habitat for marine animals elevates the salt marsh ecosystem as the most valuable natural ecosystem on the planet in terms of productivity.

What is also surprising is the low number of dieback events or disease outbreaks that have been documented in salt marshes. This may be, in part, due to confusion in recognizing a disease outbreak versus the usual death and recovery observed during accretion and subsidence that is common to a salt marsh. It is interesting that salt marshes are exclusively dominated by only a few foundation plant species. In temperate climates, species in the genera *Spartina* and *Juncus* compose the majority of acreage, whereas in tropical regions, salt marshes yield to the species of mangroves (*Rhizophora* spp.). Most of the low marshes in northern and southern temperate climes are dominated by species of *Spartina*. In the northwestern Atlantic down through the Gulf, and spotted locales in Southwestern Atlantic marshes, *Spartina alterniflora* Loisel dominates the low marsh habitat. Northeastern Pacific marshes are dominated by *S. foliosa* whereas tidal marshes on the west and east coast of the

W.H. Elmer, Ph.D. (✉)
Department of Plant Pathology and Ecology, The Connecticut Agricultural Experiment Station, P. O. Box 1106, New Haven, CT 06504, USA
e-mail: Wade.Elmer@ct.gov



D.-W. Li (ed.), *Biology of Microfungi*, Fungal Biology,
DOI 10.1007/978-3-319-29137-6_24

South American continent are colonized by *S. densiflora*. In the northeastern Atlantic marshes in Europe, S. *maritima* dominates the low marsh.

However, over the past century, *S. alterniflora* has invaded many foreign marshes and hybridized with native species to generate fertile hybrids (Ainouche et al. 2009; Sloop et al. 2009). This has resulted in a wide genetic assortment of genotypes that confer both selective advantages (resistance) and disadvantages (susceptibilities) to fungi and stressors. For example, the first recorded account of a dieback was observed on a *Spartina* hybrid in the 1950s along the marshes in the Hampshire basin in England (Goodman 1959; Goodman et al. 1959; Goodman and Williams 1961). A hybridized *Spartina* species called *S. anglica*, an allopolyploid hybrid derived from the chromosome doubling of *S.* x *townsendii* [a cross between *S. alterniflora* and *S. maritima* (Ayers and Strong 2001)], had dominated southern England marshes. Over 200 ha of marshes declined and died. Since the attempt by Goodman et al. (1959) to transmit the dieback from unhealthy plants to healthy plants failed, they concluded a parasite was not involved. This conclusion was further supported by the observation that, of over 20 fungal species isolated from dead rhizomes onto nonselective media, none was demonstrated to be pathogenic. Although many potential pathogens like *Fusarium* spp. were not reported, these species might easily have been missed in their study, since dead, not declining, plants were sampled, and since selective agars were not used.

However, even when a fungal pathogen is identified, the actual contributions of the disease to the dieback of the marsh are often indirect or not clear. Examples below will depict different cases. Sometimes a pathogen is clearly present, yet ecological costs are marginal whereas in other examples, a combination of factors, such as genetic shifts in the host species, herbivores, and/or other plant stressors like drought, rising sea levels, excessive salinity, nitrogen, and phosphorus, may interact with a fungal pathogen and lead to a major dieback. Deciphering the actual contributions of the fungal pathogen becomes a daunting task in light of climate change stressors. The following chapter highlights four examples where microfungi were associated with dieback/disease events in salt marshes. When possible the life cycles, genetic relationships, and ecological implications are presented.

## *Fusarium palustre* (W. H. Elmer & R. E. Marra) and Other *Fusarium* spp.

### *Distribution*

*Fusarium palustre* is an endophytic fungus that is closely associated with *S. alterniflora* in Northern hemispheres (Elmer and LaMondia 2011; Elmer and Marra 2011; Elmer et al. 2013) (Fig. 24.1). It was recently described and found exclusively in salt marshes extending from the Gulf of Mexico to New England. It has also been isolated from the invasive *S. alterniflora* plantings that have invaded salt marshes in

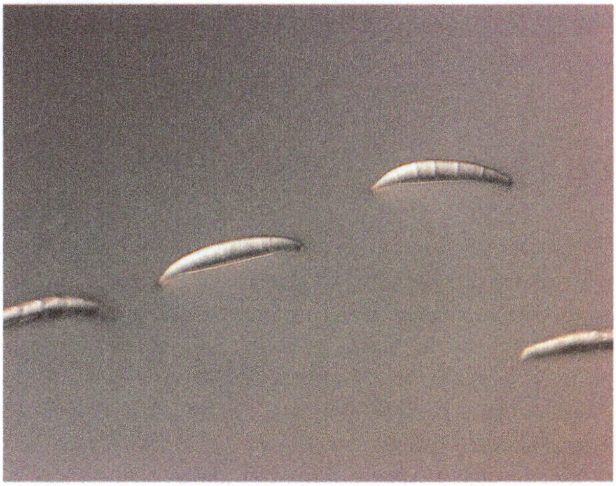

**Fig. 24.1** Macroconidia of *Fusarium palustre*

Brittany, France (unpublished data). A survey of salt marshes in Argentina did not find *F. palustre* on *S. alterniflora* or *S. densiflora*, indicating the fungus may have geographical limits (unpublished data).

On Dongtong wetland on Chongming Island in Shanghai, China, the invasive *S. alterniflora* was deliberately introduced and spread into areas where other native species began to die back. One such species was the native common reed grass, *Phragmites australis*. It was revealed that *F. palustre* had spread from *S. alterniflora* onto *Phr. australis* and was associated with a major dieback of that plant (Li et al. 2014). The dieback was only observed in marshes where *S. alterniflora* had invaded and was never isolated in pure stands of *Phr. australis.* Aside from *S. alterniflora, Phr. australis* is the only other host reported for *F. palustre*.

Along with *F. palustre* are many species of *Fusarium* that are found colonizing *S. alterniflora* that appear to be avirulent or only slightly pathogenic (Elmer and LaMondia 2011; Elmer and Marra 2011; Elmer et al. 2013). These species fall into the *F. incarnatum-equiseti* species complex (O'Donnell et al. 2009). There have been other reports where *Fusarium* spp. associated with salt marsh plants were reported in the literature. The first report was a brief description of a *Fusarium* sp., called *F. spartinae* Ellis & Everh., observed on leaves of *Spartina stricta* Roth (syn *S. maritima*) (Ellis and Everhart 1902). However, the species description, made in vivo, was far too generic to be considered synonymous with *F. palustre* and no isolates were saved. Other surveys have mentioned *Fusarium* sp. (Gessner and Goos 1973; Gessner and Kohlmeyer 1976), but the isolates were not identified and are no longer available. Therefore, it is not certain if these previous reports had been on *F. palustre.*

There was another interesting association between a *Fusarium* species and the ergot fungus (*Claviceps purpurea*) that was found colonizing *S. anglica* in England (Preece et al. 1994). *F. heterosporum* Nees & T. Nees was found associated with the ergots, but no pathology between the *F. heterosporum* and *C. purpurea* or *S. anglica* was documented (Preece et al. 1994; Raybould et al. 1998).

## Description

When DNA sequence queries of the translation elongation factor (*tef*1) of *F. palustre* isolates did not closely match any known species of *Fusarium*, a phylogenetic analysis was performed (Elmer and Marra 2011). Combined partial sequences of three genes, b-tubulin, calmodulin, and tef1, were aligned for 20 *F. palustre* isolates along with other *Fusarium* species. Strong bootstrap support provided evidence that the *F. palustre* isolates along the Atlantic coast were closely related and represented a new species (Elmer and Marra 2011). Its closest known relative based on these sequences was *F. sporotrichioides* and *F. langsethiae*.

The fungus produces macro- and meso-conidia in monophialides in vitro, but the fungus has not been observed sporulating on the host. This may be due to the frequent washing and removal of spores from tidal action. It is not clear what role spores function in infection or dispersal. The fungus has been isolated from marsh water, but the propagule was not identified [unpublished data]. Chlamydospores are produced intercalary in mycelium and these propagules may function more in dispersal on pieces of tissue than the other conidia. Chlamydospores and thick-walled hyphae may also function during periods of stress as survival propagules. During low tides when drought conditions prevail, saline conditions can be excessive on and around *S. alterniflora* plants. *F. palustre* is more saline tolerant than most *Fusarium* spp. Elmer and LaMondia (2014) found that hyphae of *F. palustre* were uninhibited on NaCl-amended agar at levels of 0.27 M NaCl (equivalent to marsh water) whereas genetically similar terrestrial species, *F. sporotrichioides*, showed immediate inhibition at 0.14 M NaCl. *F. palustre* has also been found to produce T-2 toxins (personal communication with Dr. Susan McCormick, ARS, Peoria, IL), but the role of these toxins on pathogenicity or survival is not clear.

## Pathology and Life Cycle

In the Western Atlantic, *F. palustre* has been implicated in Sudden Vegetation Dieback (SVD) (Fig. 24.2). Of over 10 species of *Fusarium* found colonizing *Spartina* only *F. palustre* isolates were capable of inciting stem rot and plant stunting (Elmer and LaMondia 2011; Elmer and Marra 2011; Elmer et al. 2013). However, in other studies, isolates of *Fusarium* in the *Gibberella fujikuroi* species complex (O'Donnell et al. 1998) could also incite disease, but these isolates may be more restricted to southern locales (Elmer et al. 2013).

SVD affects primarily *S. alterniflora* and is characterized as a rapid decline that begins with thinning and/or browning of the aboveground foliage of *S. alterniflora*

**Fig. 24.2** Sudden vegetation dieback of *Spartina alterniflora* along intertidal creek bank in Branford, CT

followed by death of the rhizomes (Alber et al. 2008; Elmer et al. 2013; Smith and Carullo 2007). Once plants die, barren areas of remnant peat remain indefinitely. The defining signature of SVD is death of the rhizomes and a very slow recovery that can take from 1 to more than 10 years. It was originally called Brown marsh (McKee et al. 2004). Surveys have reported up to 10 *Fusarium* species have been isolated from *S. alterniflora* in SVD sites, but 3 out of 4 colonies isolated from S. *alterniflora* give rise to *F. palustre* (Elmer et al. 2013). The fungus was consistently recovered from plants in SVD sites, but can still be found in low densities in marshes where no SVD occurs. The fungus has been isolated from roots, crowns, stems, and seeds, but the incidence is usually greatest in the basal stem sections (Elmer and LaMondia 2011). Isolation from marsh soil is relatively rare, so *F. palustre* may not persist as a typical soilborne pathogen.

Inoculation of healthy plants with *F. palustre* rarely results in death. Stem lesions do result from stab inoculations and generally stunting is observed following conidial drenches of roots (Elmer 2014; Elmer and Marra 2011; Elmer et al. 2013) (Fig. 24.3). Infected plants usually have less vigor than healthy plants and symptoms are always greater when plants are stressed by drought or poor nutrition [(Elmer 2014; Elmer and LaMondia 2011); unpublished data]. Therefore, *F. palustre* may operate more as a component of a multilayered ecosystem under multiple stressors. Recent studies have shown that fungal infection can render plants more vulnerable to

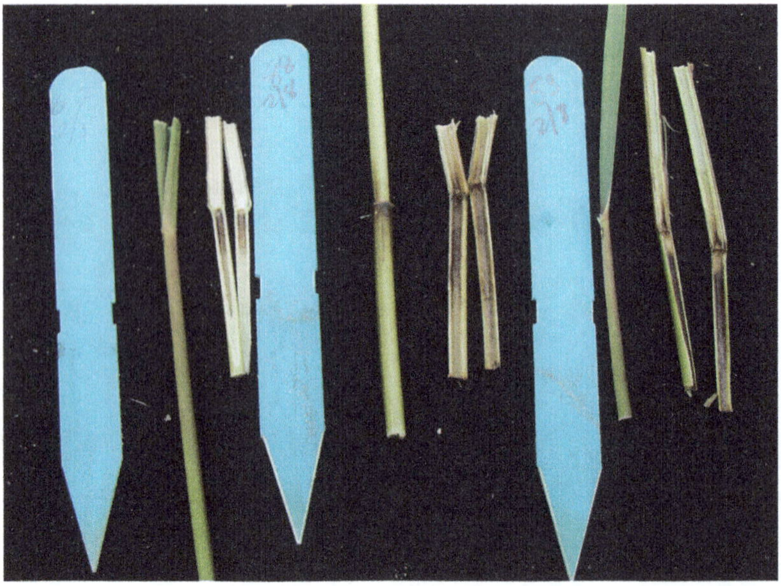

**Fig. 24.3** Stem lesions on *Spartina alterniflora* following pathogenicity tests with different isolates of *Fusarium palustre*

herbivory by the purple marsh crab (*Sesarma reticulatum*) (Elmer 2014). Intense grazing by the purple marsh crab was strongly correlated with SVD sites (Altieri et al. 2012; Holdredge et al. 2009) and controlled studies found that the purple marsh crab grazed more on disease-stressed *S. alterniflora* plants than on healthy plants (Elmer 2014). One possible mechanism that could explain the increased attraction is chemotaxis where stressed plants may emit volatiles that attract crabs. No such attractants have been identified. However, *S. alterniflora* is unique in that it contains dimethylsulfoniopropionate (DMSP), a naturally occurring putative osmolyte, which is oxidized to dimethylsulfoxide (DMSO) during periods of stress (Husband et al. 2012). Studies in Georgia found that the DMSO:DMSP ratio was a sensitive indicator of presymptomatic stress in *S. alterniflora* and consistently greater in leaves and stems of plants in dieback sites (McFarlin and Alber 2013). In one preliminary trial, we have found the healthy *S. alterniflora* transplants sprayed with DMSO at 2.5 μmoles/ml and set in mecocosms with purple marsh crabs were grazed significantly more in the first 24 h than untreated control plants (unpublished data, $P > 0.001$). The role of volatile compounds as a chemoattractant in *S. alterniflora* is still not clear.

**Fig. 24.4** Ergot of *Spartina alterniflora* caused by *Claviceps purpurea* var. *spartinae* (Courtesy of Court Stevenson and Lorie Staver, University of Maryland)

## *Claviceps purpurea var. spartinae* R.A. Duncan & J.F. White

### *Distribution*

Ergot, caused by the ascomycetous fungal pathogen, *Claviceps purpurea,* was first discovered along the Gulf of Mexico in 1895 on *S. alterniflora* (Tracy and Earle 1895) (Fig. 24.4). *C. purpurea* appeared to be a resident fungus in most US and European marshes causing disease on *S. alterniflora*, *S. foliosa*, and Spartina hybrids, but incidence can be very low to over 96 % depending on host and environmental conditions (Eleuterious 1970; Eleuterius and Meyers 1974; Ellis and Everhart 1902; Fisher et al. 2005a). Eleuterius and Meyer (1974) reported that *Distichlis spicata, Spartina patens,* and *Spartina cynosuroides* could also serve as hosts of *C. purpurea.*

## Description

Duncan et al. (2002) proposed listing the pathogen on *Spartina* as a separate variety of *C. purpurea* and named it *C. purpurea* var. *spartinae*. They found that sequences of the ITS regions placed the *Spartina* pathogen into the clade of *C. purpurea*, but distinguished it based on morphological differences in the sclerotia and unique alkaloid profiles. It is interesting that the sclerotia from isolates of *C. purpurea* var. *spartinae* were able to float in saline water whereas sclerotia from isolates from other hosts sank. This adaption to the marsh environment provides a selective advantage and aids in dispersal. Pažoutová et al. (2002) further classified the *C. purpurea* population colonizing *S. anglica* in Britain and *S. alterniflora* in the USA. In addition to floating sclerotia, they noted unusually long cylindrical conidia. Molecular assays utilizing RAPDs, AFLPs, and sequences from rDNA compared isolates from other hosts and concluded that the Spartina isolates were a genetically distinct, homogeneous population of *C. purpurea.* The same morphological and genetic markers were found also in *S. alterniflora* isolates from *Spartina* from the USA. All *Spartina* isolates belonged to a fungal chemotype that produces the alkaloids ergocristine and ergocryptine (Pažoutová et al. 2002). Given the similarity between the Spartina isolates, it was speculated that a common origin was likely and that the British stands of *S. anglica* were likely colonized by isolates introduced from America on *S. alterniflora*.

## Pathology and Life Cycle

Infection in the UK was greatest on *S. alterniflora* florets when the plant had recently recolonized manmade beaches and sites where only barren peat persisted (Preece et al. 1994). They also noted greater incidence on plants closer to the water than more inland, but offered the observation that this may be due more to conditions that affect flowering than environmental conditions that affect ergot infection and development. However, these observations led the authors to suggest that plant stress might increase infection and to be mindful of how marsh disturbances like canals and other ecological modifications might respond to ergot infection. In general, outbreaks are considered relatively rare (Fisher et al. 2005b). However, the ergot disease of *Spartina* can rapidly spread when cool, wet conditions prevail and when a more susceptible homogeneous germplasm dominates the marsh.

Not much information is available on the actual infection cycle on *Spartina*, but it likely follows the same patterns known for most terrestrial plants (Tudzynski and Scheffer 2004). Gray et al. (1990) studied ergot on *S. anglica* in England and stated that windblown ascospores derived from flask-shaped perithecia on overwintering sclerotia land on grass florets at anthesis in the spring to germinate. Once the cuticle has been invaded, the hyphae colonize the ovarian tissue, grow down toward the base of the ovary, and colonize the vascular tissue. The pathogen develops a

**Fig. 24.5** Lesions of Spartina rust caused by *Puccinia sparganioidis* in initial stages (Courtesy of Carrie Knott, University of Kentucky)

mycelial stroma, called a sphacelium, in the ovary and produces masses of conidia that are exuded into a sugar-rich fluid derived from phloem sap. These conidia provide the summer inoculum that initiates the spread of the disease during the growing season. The ergot fungus is homothallic and the perithecia are produced on the sclerotia which drop from the plant as the plant senesces. Sclerotia float with the tides into rack lines and barren mud flats to overwinter (Gray et al. 1990). It is unclear how long the sclerotia would remain viable in the saline water.

In England and on the west and east coast of the USA, widespread ergot epidemics have been recorded in salt marshes (Fisher et al. 2005b; Gray et al. 1990; Pažoutová et al. 2002; Raybould et al. 1998; Van Dyke and Amerson 1976). From 1983 to 1995, a detailed survey was conducted in England on the incidence of ergot on the hybridized *S. anglica* (Raybould et al. 1998). Raybold et al. (1998) reported that the disease caused no overall differences between the number of seed set on infected compared to uninfected inflorescences. However, when the rate of infection was considered, heavily infected inflorescences had less seed set whereas incidences less than 10 % produced more seed per inflorescence. Each year infection by ergot was relatively uniform on *S. anglica* so no wide diversity in susceptibility was thought to exist in the host population (Raybould et al. 1998).

Conversely, in the San Francisco Estuary, outbreaks of ergot regularly occurred on the native *S. foliosa*, but the hybrids that formed between the introduced *S. alterniflora* and the native *S. foliosa* were more resistant to ergot and sustained lower levels of infection (Fisher et al. 2005a, b). Since these hybrids were more robust and competitively superior to *S. foliosa*, they have spread extensively throughout these marshes. In marshes where the native and hybrid coexist, the higher rates of

**Fig. 24.6** Lesions of Spartina rust caused by *Puccinia sparganioidis* in later stages (Courtesy of Carrie Knott, University of Kentucky)

infection on the native *S. foliosa* reduce plant fecundity more than on the hybrid, which in turn speed the displacement of the native *S. foliosa* in these ecosystems.

## *Puccinia sparganioidis* Ellis & Barthol

### *Distribution*

*Puccinia sparganioidis* (often reported as *Puccinia sparganioidis* Ellis & Tracy) is a macrocyclic rust disease of *Spartina* spp. that shares three of five spore stages on its alternate host, ash (*Fraxinus* spp.) (Figs. 24.5 and 24.6). It was first observed by W. G. Farlow in 1883 in Massachusetts and labeled as *Uromyces spartinae* Farl. (Arthur 1902). His report coincided with a major outbreak of the fungus on white ash (*Fraxinus americanus*) (Anonymous 1916; Arthur 1902). At present, the known distribution of *Pu. sparganioidis* includes the USA and Canada east of the Rocky Mountains, Mexico, and Brazil (Anonymous 1916; Arthur 1902; Gray et al. 1990). The disease has previously been reported on *S. alterniflora* in Connecticut, Delaware, Florida, Louisiana, Maine, Maryland, Massachusetts, Mississippi, North Carolina, New Hampshire, Rhode Island, Vermont, and Virginia (Davelos et al. 1996). There are no reports of it occurring in Asia, Europe, or South America on *Spartina* spp. or *Fraxinus* spp.

## *Description, Pathology, and Life cycle*

As an obligate parasite, *Pu. sparganioidis*, is only found in association with its hosts (*Fraxinus* spp. and *Spartina* spp.). It is a heteroecious macrocyclic rust possessing five spore stages in succession, two of which must occur on a *Spartina* spp. and the other three on ash (*Fraxinus* spp.) (Arthur 1902) In the spring, overwintering teliospores germinate on *Spartina* residues and the basidiospores are released where they infect the current-year tissues of ash, causing spermogonia. Aecia develop in these lesions on ash and release aeciospores that must infect a species of *Spartina*. Once infection has occurred, uredinia develop in early summer releasing urediniospores that repeatedly infect *Spartina* spp. causing numerous orange, long, hypophyllous lesions. The uredinial lesions become erumpent giving rise to new colonies of urediniospores. Uredinia eventually develop into brownish-black telia in the fall. However, Kaur et al. (2010) did not observe telia on *S. alterniflora* in Louisiana. Basidiospores then infect the ash the following spring if weather conditions are favorable for infection.

The disease has been noted in the salt marsh on *S. alterniflora, S. cynosuroides*, and *S. patens*. There are few reports stating it developed on *Distichlis spicata*. The fungus is also reported throughout the Midwest where it completes its life cycle on prairie cord grass (*Spartina pectintata*). On ash, it is reported on several (*Fraxinus*) species including white, green, and occasionally, black ash. Although outbreaks are relatively common, there is no evidence, thus far, that Spartina rust is limiting or has major ecological costs to *Spartina*. Van Dyke and Amerson (1976) found more rust infection of *S. alterniflora* on plants grown with higher soil water salinity and suggested surface salts may be inhibitory to aeciospores and urediniospores.

However, the damage caused to *Fraxinus* can be aesthetically limiting causing disfigurement and premature defoliation. Heavy infections over several years could weaken the tree. It is likely that the same scenarios could result in weakening *Spartina* spp. in the marsh if infections were heavy and prolonged over many seasons.

Studies to determine whether clonal selection in the host could occur and result in more virulent pathotypes found no pattern between clones of *S. pectinata* (Davelos et al. 1996). Phylogenetic analyses based on 5.8S rDNA, ITS found that *Pu. sparganioidis* belonged to a highly supported clade (Group 1) within the family Pucciniaceae. Its closest relative was *Puccinia physalidis* (Dixon et al. 2010).

**Fig. 24.7** The salt marsh periwinkle snail *Littoraria irrorata* grazing on stems of *Spartina alterniflora* colonized by *Phaeosphaeria spartinicola*

## *Phaeosphaeria spartinicola* Leuchtm

### *Distribution and Description*

*Phaeosphaeria spartinicola* is an ascomycetous endophyte of *Spartina* spp. and presumably found wherever *Spartina* is grown. It was first described in 1991 ((Leuchtmann and Newell 1991) Kohlmeyer and Volkmann-Kihlmeter 2002). Reports of *Pha. spartinicola* and many of its synonyms are listed as species of *Leptosphaeria* and *Pleospora* and have been reported along the Pacific and Atlantic Coast of the USA (Gessner and Goos 1973; Gessner and Kohlmeyer 1976; Jones et al. 2002), England (Goodman 1959), and South America. Reports by Gessner and Goos (1973) and Gessner and Kohlmeyer 1976) found that 75–100 % of the aboveground biomass was colonized by *Leptosphaeria* spp. and *Pleospora* spp., which are likely synonyms of *Pha. spartinicola*. Lyons (2007) found *Pha. spartinicola* was ubiquitous on *S. alterniflora* in all Georgia marshes sampled. In addition, different species of *Mycosphaerella* are often found as well on *Spartina,* but the identity and taxonomic status of these species are not as clear (Kaur et al. 2010; Li et al. 2014). Although Koch's Postulates have never been satisfied with *Pha. spartinicola* on healthy plants, the ubiquity of this species in salt marshes raises several important questions regarding their ecological role.

*Life cycle and ecology.* Most discussions on *Pha. spartinicola* centered on its primary role as a saprobe and its function as a secondary decomposer of *S. alterniflora* and other marsh grass species (McKee et al. 2004; Newell 1996, 2001a; Newell and Barlocher 1993). Inclusion in this chapter (a chapter that is devoted to pathogens) is still warranted due to the important role this species plays in the dieback events that occur in the southern USA marshes where a facultative mutualism exists between herbivorous periwinkle snail (*Littoraria irrorata*) (Fig. 24.7). *Pha. spartinicola* provides much of the dietary sustenance for the snail (Newell 2001b; Raybould et al. 1998). During periods of drought, grazing by the snail was associated with major dieback of S. *alterniflora* in southern marshes (Silliman et al 2005; Silliman and Newell 2003). Snails wound *S. alterniflora* leaves with their radula and then deposit fungal-infested fecal matter into freshly grazed wounds. The snail then returns to the plant after the fungus has sporulated in the wounds and selectively consumes the fungal mycelium and spores. The fungus appears to colonize the wound only as a necrotroph. As a result, the impact of this facultative mutualism between snails and *Pha. spartinicola* led to major destruction of certain salt marsh communities.

Other associations have been documented between *Pha. spartinicola* and herbivores on *S. densiflora* in Argentina (Daleo et al. 2009). Crab grazing facilitated colonization of necrotic tissues by *Pha. spartinicola* which in turn reduced productivity by interpreting the photosynthates production by more than 50 % (Daleo et al. 2009). Although a true disease condition has not been verified by Koch's postulates, the role of *Pha. spartinicola* in marsh ecology may extend beyond its primary role in decomposition.

## References


Ainouche ML, Fortune PM, Salmon A, Parisod C, Grandbastien MA, Fukunaga K, Ricou M, Misset MT (2009) Hybridization, polyploidy and invasion: lessons from *Spartina* (Poaceae). Biol Invasions 11:1159–1173

Alber M, Swenson EM, Adamowicz SC, Mendelssohn IA (2008) Salt marsh dieback: an overview of recent events in the US. Estuar Coast Shelf Sci 80:1–11. doi:10.1016/j.ecss.2008.08.009

Altieri AH, Bertness MD, Coverdale TC, Herrmann NC, Angelini C (2012) A trophic cascade triggers collapse of a salt-marsh ecosystem with intensive recreational fishing. Ecology 93:1402–1410

Anonymous (1916) Diseases of white ash. The CT. Agric Exp Sta Bull 43:412–414

Arthur JC (1902) The Uredineae occurring upon *Phragmites*, *Spartina*, and *Arundinaria* in American. Bot Gaz 34:1–20

Ayers DR, Strong DR (2001) Origin and genetic diversity of *Spartina anglica* (Poaceae) using nuclear DNA markers. Am J Bot 88:1863–1867

Bertness MD (2007) *Atlantic Shorelines*. Princeton University Press, Princeton, NJ, 431 pp

Daleo P, Silliman B, Alberti J, Escapa M, Canepuccia A, Peña N, Iribarne O (2009) Grazer facilitation of fungal infection and the control of plant growth in south-western Atlantic salt marshes. J Ecol 97:781–787. doi:10.1111/j.1365-2745.2009.01508.x

Davelos AL, Alexander HM, Slade NA (1996) Ecological genetic interactions between a clonal hosts plant (*Spartina pectinata*) and associated rust fungi *Puccinia seymouriana* and *Puccinia sparganioides*. Oecologia 105:205–213

Dixon LJ, Canterbury LAM, Aime MC, Glynn NC, Comstock JC (2010) Phylogentic relationships of sugarcane rust fungi. Mycol Prog 9:459–468. doi:10.1007/s11557-009-0649-6

Duncan RA Jr, Sullivan R, Alderman SC, Spatafora JW, White JF Jr (2002) *Claviceps purpurea* var. *spartinae* var. nov.: an ergot adapted to the aquatic environment. Mycotaxon 81:11–25

Eleuterius LN (1970) Observations on *Claviceps purpurea* on *Spartina alterniflora*. Gulf Res Rep 3:105–109

Eleuterius LN, Meyers SP (1974) *Claviceps purpurea* on *Spartina* in coastal marshes. Mycologia 66:979–986

Ellis JB, Everhart BM (1902) New species of fungi from various locales. J Mycol 8:10–19

Elmer WH (2014) A tripartite interaction between *Spartina alterniflora, Fusarium palustre* and the purple marsh crab (*Sesarma reticulatum*) contributes to sudden vegetation dieback of salt marshes in New England. Phytopathology 104:1070–1077, http://dx.doi.org/10.1094/PHYTO-08-13-0219-R

Elmer WH, LaMondia JA (2014) Comparison of saline tolerance of genetically similar species of *Fusarium* and *Meloidogyne* recovered from marine and terrestrial habitats. Estuar Coast Shelf Sci 149:320–324. doi:10.1016/j.ecss.2014.09.013

Elmer WH, Marra RE (2011) New species of *Fusarium* associated with dieback of *Spartina alterniflora* in Atlantic salt marshes. Mycologia 103:806–819

Elmer WH, LaMondia JA, Caruso F (2011) Association between *Fusarium* spp. on *Spartina alterniflora* and dieback sites in Connecticut and Massachusetts. Estuar Coasts 35:436–444. doi:10.1007/s12237-011-9448-9

Elmer WH, Useman S, Schneider RW, Marra RE, LaMondia JA, del Mar Jimenez-Gasco M, Mendelssohn IA, Caruso FL (2013) Sudden vegetation dieback in Atlantic and Gulf coast salt marshes. Plant Dis 97:436–444, 10.1094/PDIS-09-12-0871-FE

Fisher AJ, DiTomaso JM, Gordon TR (2005a) Sub-specific groups of *Claviceps purpurea* associated with grass species in Willapa Bay, Washington, and the prospects for biological control of invasive *Spartina alterniflora*. Biol Control 34:170–179

Fisher AJ, Gordon TR, DiTomaso JM (2005b) Geographic distribution and diversity in *Claviceps purpurea* (Fr.) Tul from salt marsh habitats and characterization of Pacific Coast populations. Mycol Res 109:439–446

Gessner RV, Goos RD (1973) Fungi from *Spartina alterniflora* in Rhode Island. Mycologia 65:1296–1301

Gessner RV, Kohlmeyer J (1976) Geographical distribution and taxonomy of fungi from salt marsh *Spartina*. Can J Bot 54:2023–2037

Goodman PJ (1959) The possible role of pathogenic fungi in dieback of *Spartina townsendii*. Trans Br Mycol Soc 42:409–415

Goodman PJ, Williams WT (1961) Investigations into 'die-back' in *Spartina townsendii* Agg. J Ecol 49:391–398

Goodman PJ, Braybrooks EM, Lambert JM (1959) Investigations into dieback in *Spartina townsendii* Groves 1. The present status of *Spartina townsendii* in Britain. J Ecol 47:651–677

Gray AJ, Drury M, Raybould AF (1990) *Spartina* and the ergot fungus *Claviceps purpurea*—a singular contest? In: Burdon JJ, Leather SR (eds) Pests, pathogens and plant communities. Blackwell Scientific Publications, Oxford, pp 63–79

Holdredge C, Bertness MD, Altieri AH (2009) Role of crab herbivory in die-off of New England salt marshes. Conserv Biol 23:672–679

Husband JD, Kiene RP, Sherman TD (2012) Oxidation of dimethylsulfoniopropionate (DMSP) in response to oxidative stress in *Spartina alterniflora* and protection of a non-DMSP-producing grass by exogenous DMSP+acrylate. Environ Exp Bot 79:44–48, 10.1016/j.envexpbot.2012.01.006

Jones GEB, Hyde KD, Pang K-L, Suetrong S (2012) Phylogeny of the Dothideomycetes and other classes of marine Ascomycota. In: Jones GEB, Pang K-L (eds) Marine fungi and fungal-like organisms. Walter de Grulyer GmbH & Co, Boston, MA, pp 17–34

Kaur R, Knott C, Aime MC (2010) First report of rust disease caused by *Puccinia sparganioidis* on *Spartina alterniflora* in Louisiana. Plant Dis 94:636

Kohlmeyer J, Volkmann-Kihlmeter B (2002) Fungi on *Juncus* and *Spartina*: new marine species of *Anthostomella*, with a list of marine fungi known on *Spartina*. Mycol Res 106:365–374. doi:10.1017/S0953756201005469

Leuchtmann A, Newell SY (1991) *Phaeosphaeria spartinicola*, a new species on *Spartina*. Mycotaxon 41:1–7

Li H, Zhang X, Zheng R, Li X, Elmer WH, Wolfe LM, Li B (2014) Indirect effects of non-native *Spartina alterniflora* and its fungal pathogen (*Fusarium palustre*) on native salt marsh plants in China. J Ecol 102:1112–1119. doi:10.1111/1365-2745.12285

Lyons JI (2007) Molecular description of ascomycete fungal communities on *Spartina* spp. in the U.S. Ph.D Thesis, University of Georgia Athens, GA

McFarlin CR, Alber M (2013) Foliar DMSO:DMSP ratio and metal content as indicators of stress in *Spartina alterniflora*. Mar Ecol Prog Ser 474:1–13

McKee KL, Mendelssohn IA, Materne MD (2004) Acute salt marsh dieback in the Mississippi River deltaic plain: a drought-induced phenomenon? Glob Ecol Biogeogr 13:65–79. doi:10.1111/j.1466-882X.2004.00075.x

Newell SY (1996) Established and potential impacts of eukaryotic mycelial decomposers in marine/terrestrial ecotones. J Exp Mar Biol Ecol 200:187–206

Newell SY (2001a) Multiyear patterns of fungal biomass dynamics and productivity within naturally decaying smooth cordgrass shoots. Limnol Oceanogr 46:573–583

Newell SY (2001b) Spore-expulsion rates and extents of blade occupation by ascomycetes of the smooth-cordgrass standing-decay system. Bot Mar 44:277–285

Newell SY, Barlocher F (1993) Removal of fungal and total organic matter from decaying cordgrass leaves by shredder snails. J Exp Mar Biol Ecol 171:39–49

O'Donnell K, Cigelnik E, Nirenberg HI (1998) Molecular systematics and phylogeography of the *Gibberella fujikuroi* species complex. Mycologia 90:465–493. doi:10.2307/3761407

O'Donnell K, Sutton DA, Rinaldi MG, Gueidan C, Crous PW, Geiser DM (2009) Novel multilocus sequence typing scheme reveals high genetic diversity of human pathogenic members of the *Fusarium incarnatuirm-F. equseti* and *F. chlamydosporum* species complexes within the United States. J Clin Microbiol 47:3851–3861. doi:10.1128/JCM.01616-09

Pažoutová S, Raybould AF, Honzátko A, Kolínská R (2002) Specialised population of *Claviceps purpurea* from salt marsh *Spartina* species. Mycol Res 106:210–214

Preece TF, Pettit TR, Biggs DT (1994) *Fusarium heterosporum* growing on ergots (*Claviceps purpurea*) in spikelets of common cord-grass (*Spartina anglica*) in the Isle of Wight. Mycologist 8:9–11. doi:10.1016/S0269-915X(09)80673-3

Raybould AF, Gray AJ, Clarke RT (1998) The long-term epidemic of *Claviceps purpurea* on *Spartina anglica* in Poole Harbour: pattern of infection, effects on seed production and the role of *Fusarium heterosporum*. New Phytol 138:497–505. doi:10.1046/j.1469-8137.1998.00123.x

Silliman BR, Newell SY (2003) Fungal farming in a snail. Proc Natl Acad Sci USA 100:15643–15648

Silliman BR, van de Koppel J, Bertness MD, Stanton LE, Mendelssohn IA (2005) Drought, snails, and large-scale die-off of southern U.S. salt marshes. Science 310:1803–1806

Sloop CM, Ayres DR, Strong DR (2009) The rapid evolution of self-fertility in *Spartina* hybrids (*Spartina alterniflora*×*foliosa*) invading San Francisco Bay, CA. Biol Invasions 11:1131–1144. doi:10.1007/s10530-008-9385-0

Smith JP, Carullo M (2007) Survey of potential marsh dieback sites in coastal Massachusetts. Massachusetts Bays National Estuary Program and Massachusetts Office of Coastal Zone Management, 25 pp. (http://www.mass.gov/czm/docs/pdf/marsh_dieback.pdf)

Tracy SM, Earle FS (1895) Mississippi fungi. Agric Mech Coll Exp Sta Bull 34:80–122

Tudzynski P, Scheffer J (2004) *Claviceps purpurea*: molecular aspects of a unique pathogenic lifestyle. Mol Plant Pathol 5:377–388
Van Dyke CG, Amerson HV (1976) Interactions of *Puccinia sparganioides* with smooth cordgrass (*Spartina alterniflora*). Plant Dis Rep 60:670–674

# Index




D.-W. Li (ed.), *Biology of Microfungi*, Fungal Biology,
DOI 10.1007/978-3-319-29137-6

**B**

**D**

**G**

## H



## I

**J**

Lightning Source UK Ltd.
Milton Keynes UK
UKOW06n2149050416

271588UK00001B/42/P

9 783319 291352